B.S. Patil's Building and Engineering Contracts, 7th Edition

B.S. Patil's Building and Engineering Contracts, 7th Edition

B. S. Patil
S. P. Woolhouse

CRC Press
Taylor & Francis Group
Boca Raton London New York

CRC Press is an imprint of the
Taylor & Francis Group, an **Informa** business

CRC Press
Taylor & Francis Group
52 Vanderbilt Avenue,
New York, NY 10017

CRC Press is an imprint of Taylor & Francis Group, an Informa business

No claim to original U.S. Government works

Printed on acid-free paper

International Standard Book Number-13: 978-0-367-13336-8 (Hardback)

International Standard Book Number-13: 978-0-367-13331-3 (Paperback)

Library of Congress Cataloging-in-Publication Data

LoC Data here

Visit the Taylor & Francis website at
http://www.taylorandfrancis.com

and the CRC Press website at
http://www.crcpress.com

Dedicated to Mrs. Suman Patil, who published all the previous editions and kept the book in demand.

Contents

Preface

In the seventh edition reference is made to the new standard forms of construction contracts introduced in India and the second edition of FIDIC Conditions of Contract 2017. References to the earlier forms and editions are retained as these may be in use for some more years. The case law has been updated in this edition to include important cases reported up to 2018. Some illustrations in the earlier editions have been pruned or deleted. A number of foreign judgments have been included to make this edition useful to international readers.

A couple of notable features of this edition that need special mention include:

1. The "no damages for delay clause", the American position, has been declared by the Supreme Court of India as invalid under the Indian Contract Act, given that the Act contains express provisions to govern the situation. The English position on the "no damages" clause is more or less similar to the law declared by the Supreme Court of India. There is no use denying the fact that regardless of its proper purpose, the typical structure of this kind of a clause provides many opportunities for abuse or injustice. Besides, in the opinion of the authors, such clauses should be regarded as commercially unacceptable.

2. The award of quantum of damages for breach of contract, particularly for loss of overheads and expected profit, on the basis of well-known formulae like Hudson, Emden, etc., has been upheld by the Supreme Court of India. There is a much controversy about *consequential damages, foreseeable damages* and award of *loss of profit or profitability*. The popular formulae, too, have limitations and drawbacks, particularly when the delay is not proportionate to the cost of balance work as pointed out in this title. A non-controversial method based on scope of initial damages contemplated at the time of signing of the agreement has been suggested. The Patil Form suggested in this title has been formulated on the basis of the universally accepted principle of ascertaining damages, which has also been accepted by the Supreme Court of India as being applicable under the Indian Contract Act. The suggested Patil Form is flexible to accommodate differing situations. Use of the said form is usually accompanied by proof of loss, which helps to eliminate unwarranted controversies pertaining to the award of damages.

The authors' companion volume on The Arbitration and Conciliation Act 1996 is under revision and awaiting the latest amendments to be passed by the Legislature to be included in its seventh edition.

We express sincere thanks to Ms Aafreen Ayub, Associate Editor(Co-Publication), Taylor & Francis Group for persuading the previous publisher and the author to submit the book proposal to CRC Press and to Dr. Gagandeep Singh, Editorial Manager, CRC Press for his cooperation and patience throughout the project. Our thanks are also due to Ms. Mouli Sharma and Ms. Divya Sharma for their guidance in the preparation and submission of manuscript, Ms. Rachel Carter, the copy-editor for her painstaking efforts and patience, and above all to Ms. Julie Willis, Managing Director and Senior Production Editor, at Swales & Willis for her support and guidance in the production of this edition. Of course, no project of this size could have been completed without the loving support of our family members: Suman, Matthew, Emily and Rosalyn.

Pune (India), May 2019
The authors

Reasonable care has been taken to eliminate errors and omissions, however, if inadvertently some remain, the authors shall be obliged to anyone who brings them to their notice so the same may be corrected in the next edition.

About the Authors

Mr. B. S. Patil, B.E.(Civil), LL.B., F.I.E.

Mr Patil has had a brilliant career. After working for nearly 20 years as a civil engineer, he became a pioneer legal practitioner in India, specializing in construction and arbitration law. This seventh edition benefits from his in-depth study of the subject and rich experience as a consultant, conciliator, arbitrator and advocate. Apart from his other books, *Civil Engineering Contracts and Estimates* and *The Law of Arbitration and Conciliation*, he has contributed several papers and articles to engineering journals. He is a frequent speaker at conferences on the subject of law applicable to engineering and construction contracts. He was a member of the Faculty of Engineering, Marathwada University, Aurangabad, and at the College of Engineering, Pune (India). He is a Fellow of the Institution of Engineers, India, member of the Maharashtra Bar Council, accredited mediator of the Centre for Effective Dispute Resolution, London, and a life member of the Indian Council of Arbitration.

Dr. S. P. Woolhouse, B.S.L., LL.B., M.Phil.(Cantab), Ph.D.

Sarita is dual qualified in India and England and Wales. Sarita has a number of years of experience in practice, specializing in Indian construction and arbitration disputes followed by over 15 years in England, specializing in cross-border dispute resolution across various sectors including construction, energy, telecoms, etc. She has acted for both private and public sector entities in negotiation, mediation, litigation and arbitration. She has advised on draft contracts and treaties. The current edition benefits from her experience as a counsel, consultant, and an arbitrator. She has contributed numerous papers to various journals and conferences in Europe and India. The second edition of *B.S. Patil's Law of Arbitration*, revised by her, was very successful and she is working with Mr Patil on the current edition of the title. She is a Senior Lecturer in Law and is based in Cambridge, UK.

Table of Cases

M

R

T

Introduction

0.1 The Nature of the Law of Contract

The law of contract differs from the other branches of law inasmuch as it does not lay down a number of rights or duties which the law will enforce, but consists of a number of limiting principles subject to which the parties may create rights and duties for themselves which the law will uphold. This branch of law is said to be a child of commerce. It grew with the development of commerce and industry in England. The principles of the English law of contract were given effect to, with minor modifications or changes thought necessary, in the Indian Contract Act, 1872 (hereinafter called the Act).

A contract involves a minimum of two parties although more than two may be involved. The basic principle of law is that when one person (promisor) gives a promise or makes an offer and creates a reasonable expectation of performance in the mind of the other person (promisee) to whom it was made, law enforces the promise. This objective theory of contract contradicts the subjective theory. According to the proponent of the subjective theory, the essence of a contract was said to be "the meeting of the wills of the parties; agreement was the outcome of free and consenting minds".[1] The law of contract, on the other hand, recognizes a promise and will aid its fulfilment or at least enforce a payment of compensation on the breach of a promise, even if the parties were manifestly not ad idem. For example, law recognizes that a binding contract comes into existence when an offer is accepted by post, as soon as the letter of acceptance is posted, even if the letter has not yet reached the addressee. However, a letter revoking the offer has no effect until it has been brought to the notice of the promisee. Thus, it is quite probable that an offer might get accepted when the letter of its revocation is being posted. The contract is valid though no consensus exists in such a case. English and Indian law, in fact, adopt neither a wholly subjective nor a wholly objective approach to the question of contractual obligations. To alleviate the difficulty of combining the elements of the subjective and the objective theories in a definition of contract, the Indian Contract Act ("the Act") simply defines a contract as "an agreement enforceable by law". The Act lays down essential requirements of a valid contract and then the various defences that may be raised to show that no contract was concluded. This approach has an advantage in view of the fact that not all defences render a contract void or illegal, and therefore, unenforceable by either party. In some cases, the contract may not be so completely without effect. It may be voidable, that is to say enforceable at the option of one or more parties but not at the option of the other or others.

0.2 Freedom of Contract

Law of contracts is based on the notion that the parties have, by an agreement, created rights and obligations, which are purely personal in nature, that is, are only enforceable by action against the party in default. It is also underpinned by the political philosophy of the eighteenth century, namely, the concept of human liberty. Every man, it was said, should be free to pursue his own interest in his own way. This resulted in freedom of contract, that is to say freedom to contract on

---

[1] *Anson's Law of Contract*, 24th Ed., p. 5.

whatever terms might seem advantageous to the individual. This party autonomy is the foundation of the principles of contract law.

Freedom to enter into a contract became the cornerstone of nineteenth-century laissez-faire economics. This concept essentially presupposes that a binding contract is a result of individual negotiations. This classical concept of freedom of contract suffers from certain grave weaknesses. It takes no account of social and economic pressures, which in many circumstances might virtually force a person to enter into a contract. Thus, the idea that a contract is based on mutually negotiated terms is only true in a very restricted sense.

0.3 Contract of Adhesion

In a great many commercial and other everyday transactions, the common man rarely has a say in finalizing the terms of the agreement he enters into. Many examples can be cited. For example, when he applies for a gas connection, power connection, mobile phone contract, or for that matter steps onto a bus or train or an airplane on a journey, he accepts the conditions imposed by the operator. The very nature and volume of transactions demand that such contracts be standardized with their terms and conditions drawn in advance. The role of the party accepting those terms and conditions is rendered to be that of a consumer required to sign on the dotted line. Such contracts, which force one of the two parties to the contract merely to adhere to the printed terms, are called contracts of adhesion or standard form contracts.

0.4 Exclusion or Exception Clauses Vis-à-Vis Sanctity of Contract

While none can find fault with the idea and concept of having standard form contracts in trade and commerce, with a view to saving time and money and avoiding possible disputes arising out of the terms being misunderstood, such forms invariably suffer from a serious drawback. The forms are drafted at the instance of one party and the drafters are tempted to insert in the terms some conditions which exempt that party from legal liabilities in certain eventualities. Such provisions are termed exemption clauses or exclusion clauses. These provisions pose a challenge to the judiciary. The judiciary, the world over, generally attempts to hold a contract between parties as a sacrosanct document and tries to give effect to the provisions thereof with the view to giving effect to the parties' intention, no matter how harsh the terms may be to one party or the other. This principle of dispensation of justice had its origin in the basic belief that an agreement is a bilaterally negotiated document with all the parties having agreed to the terms of their own free will.

Standard form contracts falsified this basic belief with the result that if effect is given to exception or exclusion clauses, the law to that effect is rendered impotent and injustice will be caused to the innocent party signing the contract on the dotted lines. The judiciary accepted this challenge and found ways and means to avert patent injustice being dispensed to a party seeking its intervention. How the judiciary worldwide dealt with this challenge has been the subject matter of exhaustive discussion in relevant chapters.

0.5 Classification of Contracts

0.5.1 Formal and Informal Contracts

Where the law requires a contract to be under seal it is called a formal contract. All others are informal contracts. Contracts under seal, such as deeds and bonds, are instruments that are sealed by the party bound thereby, and delivered to the one to whom the liability is incurred. All deeds are documents under seal but all documents under seal are not deeds, for example, an award of an arbitrator, a company's Memorandum of Association, a share certificate, etc. are not deeds.

0.5.2 Express and Implied Contracts

Contracts can be classified on the basis of the way in which the consent of the parties is manifested. Contracts are said to be express when the parties state their terms in words. They are said to be implied when their terms are not so stated, as, for example, when a person is permitted to board a bus or enter into a restaurant; from the conduct of the parties the law implies a promise by the person to pay the bus fare or the cost of eatables and services received in the restaurant, as the case may be. An express contract to last for a fixed period, when allowed by the parties to continue beyond the said period, as though the contract still continued to be binding, may result in an implied contract for the extended term. In such a case, the court may infer that the parties, by their conduct, agreed to renew the express contract for a new term.

The term implied contract is used in yet another sense to denote a situation in which the law imposes an obligation on one party to pay money to the other, though there is no semblance of an agreement between them, either express or implied. For example, "A, tradesman, leaves goods at B's house by mistake, B treats the goods as his own. He is bound to pay A for them".[2] It needs to be clarified that a contract may be express or implied; the law does not distinguish between the two when enforcing the contract. Thus, the difference between the two, if any, is of little importance, except that some contractual disputes (and the related expense and aggravation of litigation) can be avoided if the express terms of the contract are well drafted, rendering it unnecessary to rely upon implied contracts.

0.5.3 Unilateral and Bilateral Contracts

A unilateral contract is one in which only one party is bound and does not mean a contract made by one party alone. For, as already mentioned, there must be at least two parties to make a valid contract. A bilateral contract, on the other hand, is one in which both the parties are bound. Examples of unilateral contracts include an offer of a reward for the return of lost property. Bilateral contracts include contracts consisting of reciprocal promises and are by far the most common.

0.5.4 Building and Engineering Contracts

To build is to construct, erect or make by assembling separate parts or materials an edifice for any use, such as a dwelling house, a factory, a dam, a bridge, a road, etc. For the purpose of this book the meaning assigned to the word "engineering" is limited to the art of designing, building or erecting civil engineering works. The scope of civil engineering itself is, however, very wide and includes a large variety of works such as residential and public buildings, roads, bridges, railways, dams, docks, airports, power stations, water supply and drainage schemes, tunnels, sea defences, land reclamation projects, etc. Roughly 40 to 50% of India's Five Year Plan expenditure falls within the purview of construction activity. Those who carry out this colossal work of nation building are engineers, architects and contractors.

0.6 Parties to Building and Engineering Contracts: Employer/Owner, Engineer/ Architect and Contractor

No matter whether the work is public or private, the scheme must, in the first instance, be set in motion by some individual or some body or government department having the necessary capital

---

[2] Illustration under Section 70 of the Indian Contract Act.

to pay for it. The term "Employer/Owner" used hereafter in this book refers to such an individual, firm, company or public body which initiates and finances the work. The word commonly used in this connection is "employer" because the person who initiates and finances the project really employs all the necessary agencies or organizations for the execution of the project.

Individuals or firms of consulting architects or engineers or civil engineers in the employment of private and public bodies carry out the design and supervision of civil engineering works. The term "engineer" (which includes architect) used in this book refers to the individuals or firms of consulting engineers employed by the owner in order to advise him, negotiate for him, supervise the work, and to protect his interest in general. Under most standard form contracts, the engineer also plays the role of deciding disputes between the employer and the contractor in the first instance.

The term "contractor" is used to designate independent businessmen or companies who enter into a suitable agreement with the owner to shoulder the responsibility of converting works from the blueprints to reality by organization and execution of the contract work.

0.7 Public Private Partnership (PPP) Contracts

With the advancement of science and technology, increase in trade and commerce and industrialization, the demand for developmental projects has been on the increase. If public authorities were to depend on public funds for executing all such projects, the execution of projects would get delayed for lack of funds. The concept of the BOT (Build, Operate and Transfer) contract, therefore, developed in the Western economies and is taking hold in the newly emerging economies that historically carried out infrastructure works through public agencies. Private entrepreneurs finance certain works likely to generate revenue from their end users. They also maintain and operate the facility for a mutually agreed period and thereafter the project is transferred to the public authority. Considering the nature of a PPP contract, one would expect reduced scope for disputes in the PPP contracts. However, the type of cases handled by the authors show that there is a delay in execution of the projects involving damages in terms of the agreement or total termination of the contract on account of non-satisfaction of conditions precedent agreed to by the parties, such as non-availability of land for implementing the project. Despite the agreements incorporating adequate provisions to meet each and every eventuality, liability is disputed, resulting in arbitration and litigation.

0.8 Development of a Project

It is obvious that no scheme of an appreciable size can be planned in a day. Most schemes take months, or sometimes years, before they reach the construction stage. During this period of development, the project goes through the following stages:

1. If the owner is an individual or a body without its own engineering staff, the owner employs an engineer or architect to investigate and plan the project for him.

2. The engineer carries out preliminary investigations, prepares an approximate estimate of the cost, and determines the financial feasibility of the project.

3. If the engineer recommends the project as technically and financially feasible, the owner may give his consent to go ahead with it. In the case of public works such an approval is called an "Administrative approval".

4. The engineer then undertakes the work of detailed investigations and preparation of drawings, and all other papers containing the data needed to enable the contractors to bid on the job. All these papers are referred to as the "tender or contract documents".

5. In case the owner is a government department, the detailed plans and estimates have to be checked and approved by a competent authority. This approval is called "technical sanction".

6. When all is set for the construction of the project, contractors are invited to bid for the job. The contractors, after making all their business arrangements, submit their offers, which are known as "tenders". The engineer scrutinizes the proposals and recommends a particular tender for acceptance by the owner who then enters into an agreement (the "contract") with the contractor.

7. The contractor sets his organization into motion to carry out the project in the most efficient and economic way.

0.9 Standard Forms

With globalization, the increase in the volume of trade and commerce, and advancements in technology, the use of standard form contracts has become very common. The construction industry, second largest in the world next to agriculture, is no exception to this practice. There are a number of standard contract forms in use in the world and in India too. Almost every major government department, such as the Central Public Works Department (CPWD), the Military Engineering Services (MES), the Railways, etc. have evolved special forms for their use. Besides, each State Public Works Department (PWD), and Irrigation and Power Departments use their own forms. Public Sector Undertakings, municipalities and other government and semi-government organizations, such as the State Electricity Boards, etc., use forms similar to one or the other standard forms. For private works, the standard form suggested by the Indian Institute of Architects is generally adopted.

Historically, standard form contracts were drafted for the employer and contained many one-sided conditions protecting the employer's interest alone. Over the years it was realized that such one-sided contracts increased the cost of work inasmuch as the risks were to be borne by the contractor. There have been unsuccessful attempts to rationalize and draft one standard form applicable to all construction contracts in India. The standard form contracts propagated by the International Federation of Consulting Engineers (FIDIC) have become popular and are used in several countries, including India. FIDIC conditions are analysed and commented upon in this book along with the other standard form contracts frequently used. Similarly, the latest forms developed in India, such as those of NITI Aayog and the Government of India Ministry of Statistics and Programme Implementation Infrastructure and Project Monitoring Division, have also been discussed. References to old forms used by various departments of the Government of India and the state governments contained in the earlier editions of this book have also been retained.

0.10 Need for Unified Standard Form and Change in Types of Contract

Due to the vast scope for massive multi-crore rupee projects in India, a number of companies and firms have grown up and operate nationally and internationally. It is always desirable that the engineers and executives of such companies have a thorough knowledge of the conditions incorporated in the Standard Printed Agreements. There seems to be no specific reason as to why one uniform standard form cannot be evolved and adopted in India, when for international contracts over fifty different countries agree to adopt one uniform standard form and conditions such as FIDIC.

There are, it seems, mainly two causes for the non-adoption of one form for all contracts in the country as a whole. The first could be normal human nature to resist change. The second could be the unwillingness to give up seemingly unlimited powers and exemptions from liability that benefit the engineer and the owner under the age-old one-sided conditions of contract. This continues in complete disregard of many judicial pronouncements and arbitration awards which negate or minimize the effect of such one-sided powers or exemption clauses.

There is yet another perspective from which the very types of agreements in use deserve to be looked at. These agreements incorporate the seeds of disputes and disagreements right at the time of signing by the parties. The most commonly used type of contracts include:

1. Lump sum, with or without Bill of Quantities.
2. Item rate.
3. Percentage rate.

By its very nature, each of the above types requires the contractor and the owner to agree upon either the price or the rates payable in future based on the probable expenditure of executing the works. All experience shows that the consideration so based rarely proves to be realistic. There are many reasons for this. However, the main ones include:

1. Inaccurate estimate of quantities, changes required to be made in the designs, drawings and at times in specifications, mainly due to inadequate investigations at the preparatory stage.
2. Unprecedented trend of rise in prices of materials, machinery and equipment, petrol, oil lubricants and labour wages, making it virtually impossible for anyone to make a correct guess of the probable cost of construction, coupled with the fact that there has been no satisfactory solution to work out a reasonable and realistic escalation formula to take care of the probable rise in cost. A study conducted by the lead author showed that the escalation formulae in vogue yielded results approximately 50% of the actual price rise.
3. Deliberate unbalancing of tenders by intending tenderers in cut-throat competition somehow to grab a contract and then to rely on the first or the second factor above to make good the probable loss or walk away with a reasonable profit due to successful gamble of unbalancing the rates.

The real solution to the innumerable problems faced by both the owner and the contracting companies the world over would lie in developing and adopting the cost plus type contract form. With determined efforts, it is possible to evolve a suitable cost plus form for public works also.

An attempt is made in this book to incorporate the problem situations that arise in implementing the various forms of contracts in use, illustrate the outcomes with decided court cases so that readers can avoid the legal pitfalls that entrapped other members of his profession/business.

***

1

The Tender

1.0 Introduction

The first step that leads to the formation of a building or engineering contract is an invitation to tender. The owner or employer of the project sends it to known agencies by post in the case of private works and, generally, by public advertisement if the employer happens to be a public body. The advertisement is called a tender notice. The party, to whom the notice is addressed, sends its response in the form of an offer to the employer, which response/offer is called a "tender". Thus, a tender is an offer made by one party to another. In engineering practice, the word tender usually means an offer by a contractor to undertake a work in return for a sum of money. The law applicable to such types of offers in India is the Indian Contract Act, 1872 (the Act). The relevant provisions of the Act that are referred to adjudicate disputes that may arise between the parties to a contract, during the formation of a binding contract between them, are considered in this chapter and Chapters 2 and 3. The provisions are illustrated with a number of decided court cases.

1.1 Provisions of the Act

The following provisions of the Indian Contract Act are considered in the paragraphs below as follows.

Section of the Act	Paragraph No.
2(a)	1.2
2(b)	1.16
3	1.16.1
4	1.16.1; 1.18.1
5	1.19
6	1.13.2 to 1.13.7
7	1.16.7
8	1.16.3
9	1.16.2

Provisions of the Standard Form of Contracts in use, such as Central Public Works Department (CPWD), Fédération Internationale Des Ingénieurs-Conseils (FIDIC), etc., are also referred to at appropriate places.[1] The Consulting Engineers Association of India is a member of FIDIC which is the international federation of consulting engineers set up in 1913. FIDIC allows one national association per country to join as a member.

---

[1] Unless expressly stated, all references are to the FIDIC Red Book.

1.2 Tender – Proposal – Defined

In the Act, the word "Proposal" which is synonymous in English usage with "offer" is used. Section 2(a) of the Act defines a proposal as follows:

> When one person signifies to another his willingness to do or abstain from doing anything, with a view to obtaining the assent of that other to such act or abstinence, he is said to make a proposal.

This definition clearly indicates that the word proposal is confined to an offer to be bound by promise. Thus, for an offer to be legally operative, requisite intent to contract must be present. In other words, the offer is an expression of willingness to contract made with the intention (actual or apparent) that it shall become binding on the person making it as soon as the person to whom it is addressed accepts it. It is for this reason that a quotation of prices, which is a statement of intention but not requiring acceptance, is not an offer but an invitation to offer. Similarly, a supplier of materials who displays samples of materials with marked prices in his window is not bound to sell the materials at the rates marked, or at any other rates. It is the buyer who has to make an offer to purchase. The same is true of many common forms of advertisements.

1.2.1 Requisites of a Valid Tender

Some of the requisites of a valid tender, as laid down by the Supreme Court of India, are as follows.[2] A tender must be made:

1. Unconditionally
2. At a proper place
3. At a proper time
4. In a proper form
5. To a proper person
6. Confirming the terms of obligation
7. By a person who is able and willing to perform his obligations
8. Offering a reasonable opportunity for inspection, and
9. For full amount.

For example, in response to an advertisement by a Development Authority to allot plots on a first come, first served basis on payment of full consideration, the petitioner, in the case, had submitted an application along with payment of full consideration; it was held to amount to an offer. However, there is no concluded contract between the parties till the said offer is accepted by the Development Authority.[3]

An assurance given by an official not authorized to make agreement on behalf of a company cannot be termed as an offer much less a contract, but merely a sincere advice.[4]

---

[2] Tata Cellular v. Union of India; 1994 (6) SCC 51: AIR 1996 SC 11; R.S. Daikho v. State of Manipur, (Manipur) (Imphal Bench) 2018 (1) BC 166.

[3] Adikanda Biswal v. B.D.A. Bhubaneswar, AIR 2006 Ori. 36.

[4] I.B.P. Co. Ltd. v. Ramashish Prasad Singh, AIR 2006 Pat. 91.

1.3 Invitation to Tender

An invitation to tender, whether by private negotiations or by public advertisement, is not an offer. It is an attempt to ascertain if an offer can be received from interested builder(s) to execute the work within the estimated limit of time and finance.[5] It must be noted that a tender creates no legal right so long as it is not accepted in the recognized manner. There is no obligation on the part of the owner of a proposed project to accept the lowest or any tender.[6] It is thus clear that a reservation of right to reject the lowest or all tenders in the invitation to tender is an exercise in caution.

EXAMPLES FROM CASE LAW

1.   A tender was submitted for supply of sleepers. The Electricity Board called a supplier for negotiations with 2% amount as security deposit. The supplier paid the deposit. The negotiations led to an agreement for a supply of 6,000 wooden sleepers. A purchase order was duly sent by the Electricity Board. However, the supplier refused to accept the purchase order on the ground that it contained terms that were entirely different from those agreed. The Electricity Board contended that there was no substantial variation in the terms. The agreed terms contained an arbitration clause. The supplier ignored the arbitration clause and filed a suit at his place of business for recovery of the security deposit. The trial court held that there was no concluded contract and therefore the arbitration clause could not be invoked. On appeal, the High Court confirmed the judgment and decree of the trial court by taking notice of two provisions of tender notice and acceptance.[7] First, the acceptance order gave an option to the supplier to accept it within seven days, and if no communication was received within seven days it would be deemed to have been accepted. Second, the contract included a provision that in case the terms and conditions as specified in the conditions differed from those in the purchase order, the latter shall prevail. The supplier was held entitled to refund of their security deposit.

2.   "A" had made an offer of Rs. 6,000/- for purchase of a property to its owner. He wrote to the agent of the owner asking whether his offer was accepted and saying that he was prepared to accept any higher price found reasonable. To this letter, the agent replied: "In reply to your letter – I received yesterday a cable (from the owner of the property) regarding your offer of Rs. 6000/- which reads as follows: 'Won't accept less than Rs. Ten Thousand'." A agreed to pay Rs. 10,000/- as what he termed as a counter-offer made by the owner. In a suit brought by A for specific performance, it was observed by the Supreme Court of India:[8] "the real question is whether defendant No. 1 made a counter-offer in his cable – or he was merely inviting offers ...", and held that on the entire facts of the case there was no concluded contract.

3.   P received a letter dated 1st February, from the DFO, stating: "Kindly inform whether you are ready to pay further Rs. 17,000/- for the contract – which (contract) is under dispute at present. This contract can be given to you on this compromise only." On 5th February, P wrote in reply: "I am ready to pay Rs. 17,000/- provided my claim to have the refund Rs. 17,000/- already paid ... or any other relief consequential to the judgment of that case remains unaffected. I reserve my right to claim the said or like amount. Subject to these

[5] Anil Kumar Srivastava v. State of U.P., AIR 2004 SC 4299.
[6] N.P. Singh v. Forest Officer, AIR 1962 Manipur 47 (50, 51).
[7] Rajasthan State Electricity Board v. M/s Dayal Wood Works, AIR 1998 A.P. 381.
[8] Mac Pherson v. Appanna, AIR 1951 SC 184 (185, 186).

conditions I shall pay Rs. 17,000/- as required in your referred letter." On these facts, the Supreme Court of India held:[9] "It is extremely doubtful if the letter dated February 1, is an offer. It seems to be an invitation to make offer. Be that as it may, even if it is treated as an offer there was no unconditional acceptance by the letter dated February 5."

1.3.1 Revocation of Tender Notification

Generally, the owner may at any moment revoke the invitation to tender without making himself responsible for any expenses incurred by persons in connection with the preparation of the tenders. The cost of tendering is an expense undertaken by the contractor, and should he fail to do so the owner will not be liable to him for the expenses involved. In a sale of property by open auction, the highest bidder has no right to seek a confirmation of sale.

EXAMPLES FROM CASE LAW

1. A government authority decided to sell certain surplus land by auction. However, the sale was not confirmed when the authority deemed the bid of the highest bidder to be too low. Mere participation in an auction and the deposit of earnest money did not confer any vested right in the petitioner to seek a confirmation of sale. It was held that it was for the vendor, i.e. the government authority, to grant an approval of sale, keeping in view its interest in the property, the sale price.[10]

2. An auction sale notice clearly stated that the accepting authority of the bid reserved the right to accept or reject the highest or any bid without adducing any reason thereof. Pursuant to the said notice, the plaintiff paid the necessary deposit and participated in the bid. His bid was the highest. He subsequently deposited 25% of the bid amount as required under the auction sale notice. Thereafter, the plaintiff approached the authority to accept the balance money and deliver the materials. The plaintiff was informed that the auction had been cancelled. The Orissa High Court held:[11]

> The auctioneer's request for a bid is not an offer which can be accepted by the highest bidder. It is the bid that constitutes an offer. Therefore, unless that bid is accepted and acceptance is communicated to the bidder, there is no binding contract between the parties and, therefore, no title to the goods accrues in favour of the highest bidder and he had deposited the earnest money in accordance with the conditions laid down in the sale notice.

The Orissa High Court decision in Example (2) above explains the correct legal position, if an auction sale is to be cancelled for *bona fide* reasons. However, if there is an attempt to use legal principle to justify arbitrary conduct, a court can interfere as is clear from what follows.

---

[9] Badri Prasad v. State of M. P., AIR 1970 SC 706 (712).

[10] Laxmi Narain v. State of Haryana, (P. & H.) (DB) 2009(1) R.C.R.(Civil) 556: 2008(4) PLR 529; reliance was placed upon Laxmikant v. Satyawan, AIR 1996 Supreme Court 2052.

[11] Executive Engineer, Sundargarh v. Mohan Prasad Sahu, AIR 1990 Ori. 26. Also see M/s. Alok Enterprises v. Rajasthan Financial Corporation, AIR 2003 Raj. 199.

The Allahabad High Court held:[12] In an auction by a public body, cancelling the result of auction when the highest bidder has been made to act upon his obligations is unfair and illegal. More so, when a unilateral order was secretly made after the close of a public auction. Prior to an attempt to cancel the result of the auction, the plaintiff respondent, the highest bidder, was entitled to an explanation of why his highest bid was not being acted upon. And this should have been done immediately. Reliance was placed upon observations of the Supreme Court in Ram and Shyam Co. v. State of Haryana,[13] a matter relating to a public auction made by the state as follows: "At one stage, it was observed that the Government is not free like an ordinary individual, in selecting recipient for its largesse and it cannot choose to deal with any person it pleases in its absolute and unfettered discretion. The law is now well settled that the Government need not deal with anyone, but if it does so, it must do so fairly and without discretion and without unfair procedure."

In a British case, the tendering was subject to an express code forbidding the tenders from being reduced arbitrarily, and yet, tenderers were given a chance to lower their tenders after the claimant had been the original lowest tenderer. When others were given a chance to lower their tenders, the claimant did not win the bid. He sued to recover damages for the cost of having made the abortive tender.[14] This is an area where the doctrine of legitimate expectation may be of assistance. This doctrine is discussed in Chapter 3.

1.3.2 Requirements to Be Fulfilled before Inviting Tenders

The person inviting tenders needs to exercise care that all necessary requirements prior to the invitations are fulfilled so that the tenderers are not put to unnecessary trouble and expense, if lack of preparedness requires revoking the tender invitation. One may be advised to follow the guidelines contained in Section 17 of the CPWD Manual, which *inter alia* lays down the prerequisites including:

1. Sanction to be obtained to a detailed estimate showing the quantities, rates and amount of various items of work and also the specifications to be adopted in respect of each item.

2. In the case of urgent works, with no time to prepare a detailed estimate, rough quantities and rates for the main items covering the major part of cost should be worked out to facilitate comparison.

3. Splitting of work for the purpose of inviting tenders is to be discouraged and, where necessary, prior approval of the competent authority must be obtained.

4. Before tenders are invited the following tender documents should be ready:

   a) Notice inviting tenders (NIT) in appropriate form.

   b) Standard Form of tender to be used along with General and Special Conditions, if any.

   c) The Schedule of Quantities of Works.

   d) A complete set of drawings referred to in the Schedule of Quantities.

   e) A complete set of Specifications of the work to be done.

   f) Essential architectural drawings (such as, Site Layout Plan, Foundation Plan) and adequate structural drawings for commencing the works.

The CPWD Manual further stipulates that no tender should be invited unless (i) stipulated materials are available or likely to be received before the work commences, and (ii) all tender documents are ready.

---

[12] Zila Parishad, Muzaffarnagar v. Udai Veer Singh, (Allahabad) (DB) AIR 1989 (Allahabad) 64 (72).

[13] Ram and Shyam Co. v. State of Haryana, AIR 1985 Supreme Court 1147.

[14] J. & A. Developments Ltd. v. Edina Manufacturing Ltd., [2006] N.I. 85 (QBD (NI)).

It is suggested that a similar stipulation to the effect that no tender should be invited unless the site of work is fully available or likely to be made available before the work commences will go a long way in avoiding delays, cost escalations and disputes. In this respect, the CPWD form No. 6 incorporates Clause 2A which provides the following two alternatives:

a) The site for the work is available, or
b) The site for the work shall be made available in parts as specified ...

Prior to handing over blank tender forms to the intending tenderers only one of the above two alternatives is required to be kept and the second scored out. It is often seen in practice that this care is not taken. Since the second alternative requires parts to be specified, if nothing is specified underneath, the tenderers, in such cases, are entitled to a presumption that the full site for the work is available or will be made available prior to commencement of the work.

A welcome change is found in the standard agreement for Public Private Partnership (PPP) contracts introduced by the National Highway Authority of India (NHAI) providing *inter alia* that the date of commencement would only be decided after the major part, say 90%, of the site is made available to the concessioner. A further provision is made for payment of damages at the agreed rate for delay beyond 90 days in handing over the balance of the site.

1.4 Types of Tender Documents

The standard forms in which tenders are invited include the following ten types:

1.4.1 Lump Sum Tender

In this form, the intending tenderers are required to quote a lump sum figure for completing the works in accordance with the given drawings, designs, specifications, etc. It is essential, from the very nature of this form, that it should be used rarely and only in cases where full scope of the work is not only known beforehand but is also not likely to undergo any major change during execution of the work. Any change, when made, will give rise to a dispute about the cost implications.

To improve upon its inherent drawback, this form is sometimes used with an agreed Bill of Quantities (BOQ) and rates (Lump Sum with BOQ) so that any changes or modifications, if made, would be valued in accordance with the said BOQ and rates. This form (Lump Sum with BOQ) is used by FIDIC Red Book which expressly provides for re-measurements and a method of costing the variations in quantities.

1.4.2 Item Rate Tender

In an item rate tender form, the intending tenderers are supplied detailed BOQ showing various items of work to be executed, the approximate quantity under each item, and the unit in which the work done will be measured. The tenderers are invited to quote rates against each item both in words and figures, and work out the cost at the quoted rates for the quantity mentioned under each item.

This is by far the most widely used form of tender for Engineering Procurement Construction (EPC) contracts for major works. There is, however, an inherent drawback in this form inasmuch as it allows the intending tenderers to unbalance their tenders. The tenders are unbalanced for various reasons including the following major ones:

(a) "Front-loading" is one way of unbalancing the tender wherein the tenderer quotes high rates for those items of works that will be completed at an early stage and low rates for items that will be tackled at the end of the project. This is done with a view to generating excess cash-flow to start with.

(b) If the quantities shown in the BOQs are expected to vary, the tenderer will quote high rates for underestimated quantities and low rates for overestimated quantities with the aim to generate excessive profit, if the gamble succeeds. In such a case, the lowest tender is likely to be higher than the highest tender received.[15]

Attempts to overcome this drawback can be made by taking an additional security deposit for performance of the items for which quoted rates are abnormally low so that the contractor is not tempted to abandon work on such items and, in case he does so, the employer is protected against the increased cost. To eliminate the above drawback in particular cases where quantities are not correctly estimated, this form can be used with some modification, and in that new form it is called Percentage Rate Tender.

1.4.3 Percentage Rate Tender

This form is similar to an item rate tender with the modification that the BOQ supplied is filled with rate per unit and cost of each item including total cost. The tenderers are asked to quote on the basis of percentage rate over and above or below the rates mentioned in the BOQ up to two decimal places only. In other words, a tenderer is not given freedom to change rates of individual items. This avoids submission of a potentially unbalanced tender.

1.4.4 Supply of Materials

The Public Works Departments (PWD) of state governments have a special form for use where only the purchase of materials is involved. The material suppliers are required to quote rates for supply of the required materials.

1.4.5 Piecework

For small works in which it is necessary to start the work in anticipation of formal acceptance or for running contracts such as laying of pipes for water supply or sewerage, etc. this form is adopted. The piecework is cancelled as soon as a regular contract is signed. In this form, the agreement is in respect of payment at stipulated or agreed rates without specific quantity or time being agreed upon. When the quantity and time limit are mentioned, it contains a provision entitling the engineer to put an end to the agreement at his option at any time.

Work order is yet another form in vogue for very small works. A separate form is used in PWD for inviting tenders for works such as demolition of a building and the removal of debris from the site, etc.

The above main forms of contracts, such as lump sum, item rate, etc., require the parties to agree upon the price to be paid in future at the time of signing of the contract. This was suitable in the past for works likely to be completed within one year or so and where the prices were expected to change only after annual budgets. In the present era, when nobody can predict with a reasonable degree of accuracy the possible price rises, and escalation formulae are not adequate to cover the actual price rises, disputes are increasing and so also the resultant arbitration or litigation proceedings. Under these circumstances, the cost plus form of contract recommends itself highly, with proper safeguards like defining the "cost".

---

[15] For illustrations see the author's book: *Civil Engineering Contracts*, University Press, 4th Ed., pp. 44–45.

1.4.6 "Cost Plus" Type of Contract

This form, of comparatively recent origin, provides for the basis of payments to be made to a contractor for work done on actual "prime cost" plus a fee to cover his overheads and profit. The cost is to cover cost of materials transported to the site or ex-godown, labour, cost, machinery, equipment cost on "day work". In no case should the supervision charges, overheads and profit be considered as cost. These are to be covered by the fee to be paid. The fee may be a fixed lump sum or variable, being expressed as a percentage of the prime cost. Depending upon the mode of payment of fee this form is divisible into the following categories.

1. **Cost plus fixed fee:** Under this form, the owner agrees to pay to the contractor the prime cost on actual basis plus a fixed sum to cover the contractor's overheads and profit.

2. **Cost plus percentage:** In this form, the contractor is entitled to charge a percentage of the prime cost to be actually worked out to cover his overheads and profit. This has the inherent drawback that the contractor does not have any incentive for economizing in the cost of construction.

3. **Cost plus fluctuating fee:** To eliminate the drawback of the cost plus percentage form of the contract, the fee is agreed to be paid on a sliding scale, the higher the cost the lower the percentage and vice versa. The best form incorporating this concept is the target cost type.

4. **Target cost:** In this type, a pre-estimated cost is fixed as a target cost. Over and above the prime cost, the contractor is paid a percentage of the prime cost. It is further agreed that in case there is a saving, a part of the saving will be paid to the contractor over and above his fees. If there is an excess expenditure on prime cost over and above the target amount, an agreed percentage of the excess will be deducted from the fee otherwise due and payable. This form thus introduces an element of incentive to the contractor to keep the prime cost down to the minimum.

One reason as to why the cost plus forms are not becoming popular even in the private works is the anxiety of the owner to know his liability to pay for the work in advance and not to keep it uncertain. However, experience shows that works are seldom completed at the cost either tendered or estimated.

This form is also ideally suitable for works that cannot be assessed on an ordinary pricing basis, for example, to bridge a gap in a canal embankment or a riverbank without there being time available to assess the quantum of work and to invite tenders, etc. It is also ideal for building special monumental or palatial structures without any budgetary restraint but allowing changes to be made during construction.

1.4.7 "Package Deal" or Industrialized Building Contracts

In all the six forms discussed above, it is necessary for the owner to employ professional staff to investigate the project, including surveying, preparation of designs, drawings, BOQs, and specifications, etc. before tenders are invited. In the "package deal" or "industrialized building" or "turn key" contracts, the necessity on the part of the owner to employ professional staff is eliminated. By broad outline specifications, the owner's requirements are made known to the intending tenderers and it is for the tenderers to produce designs, drawings and specifications and quote the price for completion of the work. Although BOQs are not necessary in this type, occasionally they are included to cover possible variations. This type of contract generally includes a stage-wise payment schedule. The right to approve the designs, sometimes reserved by the owner, is of no particular benefit to the owner either in terms of the cost or the time. Examples of works where such types of contracts are extensively used in India are overhead service reservoirs, bridges, apartments, etc. Other examples include PPP contracts on DBFOT (Design, Build, Finance, Operate and Transfer) basis.

In this form of contract, there is an implied warranty of suitability of the work for the required purpose, absolute and free from fault, available to the owner for a substantial period of time. This warranty, of course, will not cover normal wear and tear or annual maintenance of the structure, or of parts, which might reasonably be expected to have a limited life.

A great drawback of this form is observed when working conditions, in particular, subsoil conditions, are found to be totally different from those contemplated, requiring new design or methods or superior quality work. Invariably, in such cases, the contractors put forth claims for compensation resulting in disputes. To avoid this situation, if likely to be met with, a "mixed package deal" contract can be adopted. In this form, works likely to vary, including foundations, are kept under the design control of the owner in the usual way as for the other types of contracts, possibly with BOQ, but the superstructure is provided under a package deal or industrialized building design of the contractor.

It has been noticed that except for PPP contract forms, no standard form has been evolved for this kind of contract for public works and use is made of the standard form for lump sum or other type without scoring out the irrelevant parts or contradictory terms, giving rise to avoidable disputes and differences.

1.4.8 Standing Offer

A standing offer is an offer which is submitted in response to an invitation to tender, quoting unit prices for the supply, as required from time to time and on order, of goods or materials and/or labour, over a fixed period of time. It should be noted that an acceptance of a standing offer in general terms will not create a binding contract. The person to whom such a tender is submitted does not incur any liability merely by accepting it. The parties submitting the tender can also withdraw before a definite order is placed. In short, an acceptance in general terms, followed by placing of an order creates a contract. Also, each separate order and its acceptance constitute a different and distinct contract.[16]

Once an order is placed, the contract is binding on both the parties and it is incumbent on the tenderer to offer and the promisee to accept the supplies in pursuance of that order. A stipulation reserving the right to cancel the contract created by a formal order is void.

EXAMPLE FROM CASE LAW

A contractor submitted a tender for the supply of 14,000 imperial maunds[17] of cane jaggery during February and March 1948, which was duly accepted by the Southern Mahratta Railway. One of the terms was that the administration reserved the right to cancel the contract at any stage during the tenure of the contract without calling up the outstandings on the unexpired portion of the contract. On 8th March, the contractor was informed by the Railway Administration that the remaining jaggery outstanding on that date against the order of 16th February 1948 be treated as cancelled and the contract closed. Aggrieved by the letter, the contractor filed a suit for damages for breach of contract. The trial court dismissed the suit; but the Madras High Court held that a clause reserving the right to cancel a contract was void. The Supreme Court upheld this contention and held:[18]

> Condition referred to a right in the appellant to cancel the agreement for such supply of jaggery for which no formal order has been placed. Once the order was placed that was a binding contract making it incumbent upon the respondent to supply and on the appellant to accept the jaggery supplied in pursuance of the orders.

---

[16] Chaturbhuj Vithaldas v. Moreshwar Parshram, AIR 1954 SC 236.

[17] A maund was equal to 82.28 lb or 37.32 kg.

[18] Union of India v. M. Thathiah, AIR 1966 SC 1724.

Another noteworthy feature of this form is that the person who has accepted a standing offer does not bind himself to any maximum or minimum supply of goods or materials ordered. However, during the time contracted for, he cannot purchase the thing contracted for in the standing offer elsewhere. Such an act may amount to a breach of contract.

It must, however, be noted that a tender of this type and its general acceptance may not always amount to a "standing offer". Much will depend upon the language of the original invitation to tender. For example, where a corporation advertises for 1,000 tonnes of cement to be supplied during the period of 1st April to 31st March, and accepts a tender, the result is a concluded contract. On the other hand, if the above advertisement were to invite tenders for the supply of cement not exceeding 1,000 tonnes altogether, deliveries to be made if and when demanded, the tender and its acceptance will merely amount to a standing offer, which could mature into a contract as and when orders would be placed.

1.4.9 Labour Contract

The special feature of this type of contract is that the owner agrees to supply all the requisite materials to the contractor and the latter agrees to supply all the labour and workmanship necessary to complete the work according to the drawings and specifications. This form of contract is suitable for those cases where the owner is in a position to buy large quantities of materials at favourable rates and also deliver them to the site of work most economically; for example, spreading and compacting metal and laying and fixing sleepers on a railway track.

1.4.10 Build Operate and Transfer (BOT) Contract

In this type of PPP contract, the only major difference from the other types of contract is in respect of the mode of payment of cost of the work to the contractor.

Nature of Agreement

The contractor undertakes to design, finance, construct, operate and maintain the works for a concession period in consideration of the exercise and/or enjoyment of the rights, powers, benefits privileges, authorities and entitlements including the amount receivable from the collection of charges levied on the beneficiaries who use the work and in some cases annuity payment each year. For example, in the case of a road or bridge, the contractor constructs the structure and is entitled to collect tolls from the road or bridge users for the concession period.

The cost of construction of the project is ascertained in the manner for a lump sum or item rate contract. The tenderer then works out the cost of financing the project and its recovery from the collection of tolls or similar method till the cost of construction as well as maintenance, periodic renewals together with the cost of finance and expected profit is fully recovered. Each year's income will reduce the cost till the cost becomes zero or negative. The number of years required to recover the total cost is generally called a "concession period". The concession period will depend upon the cost of construction worked out by each tenderer and also the rate of interest considered for financing the project and expected profit. The tenderer who submits the "cash-flow projections" seeking the minimum period of concession is generally awarded the contract.

Sometimes, the authority fixes the concession period and seeks offers on the basis of grant in the form of "annuity" expected from the employer. The tenderer expecting the least grant is generally awarded the contract. In addition, income tax holidays for predetermined years to be selected by the tenderers may be offered as incentives to reduce the grant amount. The parties agree for release of the grant in the phased manner as required by the contractor.

Contract Documents in BOT Form

In addition to the usual contract documents, cash-flow projections form an integral part of the contract; these set out the details of how the contractor intends to generate finance and use it till the recovery of his capital, investment cost and profit during the concession period. The main agreement contains several schedules describing the project site, project facilities, site delivery schedule, design requirements, construction requirements, operation and maintenance requirement, cash-flow projections, annuity payment schedule, performance security, State Support Agreement, Substitution Agreement and Hand Back Requirements, etc.

State Support Agreement: The implementation of a BOT agreement requires extensive and continued support and grant of certain rights and authorities by the government. For this purpose the State Support Agreement is signed and included in the contract documents.

Substitution Agreement: The contractor generally obtains the large amount of finance required for the project from Financial Institutions/banks who are called "lenders". The lender proposes a person called the "selectee" who is approved by the employer and who is empowered to substitute the contractor for the balance period, in the event of the contractor's failure to cure a default leading to the suspension of the concession in the contractor's favour. An agreement signed by the employer, the contractor and the financial institutions to this effect also forms the part of contract agreement.

Hand Back Requirements list the minimum requirements and item-wise inventory of all assets to be handed to the employer by the contractor at the end of the concession period.

Advantages

The main advantage is that the state or public authority is not required to finance the project immediately, and the project is made to generate the funds and be self-financing to the extent required. The developmental works which would otherwise be delayed can be undertaken forthwith.

Disadvantages

The main drawback of this form of contract is that it is not suitable for each and every type of work. In particular, works that are not likely to generate capital for self-financing are not suitable for this form. Examples include rural water supply projects, village roads, city roads, irrigation projects, etc.

Suitability

This form of contract is ideal for highways and expressways, bridges, tunnels, electricity power generation and supply, etc. that save the cost of operation and/or time of travel for which users will pay reasonable charges raising adequate cash-flows to make the project self-financing. Care should be taken to invite tenders publicly, after defining the requisite parameters. Failure to do so may result in an illegal contract.

EXAMPLE FROM CASE LAW

Construction of an underground shopping complex in a park of historical importance, against the Municipal Act and Master Plan for a city, was awarded on BOT basis without inviting tenders and on totally one-sided conditions favouring the builder (e.g. the period of concession was not defined, the builder was allowed to lease shops on his own terms and conditions, etc.). The contract was held illegal and the construction work done ordered to be demolished.[19]

---

[19] M.I. Builders Pvt. Ltd. v. Radhey Shyam Sahu, AIR 1999 SC 2468.

1.5 Modes of Inviting Tenders

There are three possible modes of inviting tenders. These are briefly discussed below.

1.5.1 Negotiated Tenders

The owner may decide to call for discussions and entrust the execution of the work to a previously known contractor or firm or company in whom he has developed confidence. The rates and or cost are mutually agreed and an agreement is entered into. This mode may also be suitable when the proposed work is of a special type requiring expertise in the form of specialist workmen and/or machinery and there is only one firm or company capable of executing the work. This mode is not suitable for public works and when in an extreme emergency it is required to be adopted, the procedure prescribes "If due to very great urgency competitive rates cannot otherwise be fixed, quotations may be obtained". Powers of the PWD engineers are restricted to certain fixed sums per annum.

However, the rule is not applicable to special agreements such as the setting up of industrial units in a state. The Bombay High Court refused to interfere with the decision in a case wherein a Power Purchase Agreement was entered into between the Maharashtra State Electricity Board (MSEB) and a power company by negotiations.[20] In the course of the decision, reference was made to few Supreme Court decisions. In Kasturilal v. State of Jammu and Kashmir, the Supreme Court of India – a three-judge bench – had laid down the law as follows:[21]

> [B]ut in a case like this where the State is allocating resources such as water, power, raw materials etc. for the purpose of encouraging setting up of industries within the State, we do not think that the State is bound to advertise and tell the people that it wants a particular industry to be set up within the State and invite those interested to come up with proposals for the purpose. The State may choose to do so, if it thinks fit and in a given situation, it may even turn out to be advantageous for the State to do so, but if any private party comes before the State and offers to set up an industry, the State would not be committing any breach of any constitutional or legal obligation if it negotiates with such party and agrees to provide resources and other facilities for the purpose of setting up of industry.

In another case, the Supreme Court in categorical terms observed:[22] "We may also point out that when the State Government is granting license for putting up a new industry, it is not at all necessary that it should advertise and invite offers for putting up such industry."

1.5.2 Limited Competition

Instead of inviting tenders openly, an employer may invite tenders from a few selected contractors by inviting them to submit the tenders in limited competition. This method has the advantage that the owner may not be required to incur expenses and spend time in public invitation when his intention is to award the work to one of the few known reputed contractors. Care must be exercised in such a case to keep the names of contractors competing in the invitation secret. This is with an obvious view to preventing them from getting together and deciding to raise the bids and share the extra money so raised.

---

[20] Ramdas Shrinivas Nayak v. Union of India, AIR 1995 Bom. 235. Also see: Centre of Indian Trade Unions v. Union of India, AIR 1997 Bom. 79.
[21] Kasturilal v. State of Jammu and Kashmir, AIR 1980 SC 1992.
[22] State of M.P. v. Nandalal Jaiswal, AIR 1987 SC 251.

This method also recommends itself when there are only few recognized firms or companies capable of executing the work. In the case of public works this mode may be adopted under the following circumstances:

(i) The work is required to be executed with great speed that not all contractors are in a position to generate.

(ii) Where the work is of a special nature requiring specialized equipment, which is not likely to be available with all contractors.

(iii) Where the work is of a secret nature and public announcement is not desirable.

The CPWD Manual providing for the above situations came for consideration before the Delhi High Court, which upheld the right to shortlist contractors for restricted competition but quashed the action to invite tenders from the shortlisted contractors on previous occasion for specific works. It was held that public notice has to be issued inviting prequalification applications generally or in respect of the work in question.[23]

EXAMPLE FROM CASE LAW

After inviting public tenders for supply of masts to the department of telecommunication, a counter-offer was given to 17 parties, which was repeated. Since no worthwhile progress was made in the first half of the year, it was decided to call for limited tenders from parties excluding the petitioners. The reason given for the exclusion was that although the petitioners had their design approved, a prototype was required to be submitted to the respondent, which the petitioner had not done when tenders were invited. It was observed that getting a prototype approved is in the nature of sample approval before bulk supplies are made. It can be done only when an order is placed and specifications are known. In the absence of a procedure prescribed for giving the type of approval as to how and when it is done and how much time it takes as well as absence of administrative instructions under which it was required to be given, it was held that the condition was arbitrary and without any rational basis. The condition was struck down and the direction was given to issue tender documents to the petitioners.[24]

1.5.3 Open Competition

By far the most common mode of inviting tenders for major works is by inserting an advertisement in newspapers, periodicals, or trade journals, inviting all interested contractors to participate in the invitation. Depending upon the magnitude of the work involved this mode may be of either of the two forms as follows:

i) *Local Competition Bids (LCB):* restricting the participation of the contractors of the country in which the work is to be executed; or

ii) *International Competitive Bids* (ICB): allowing contracting firms or companies from all over the world to participate in the tender invitation.

---

[23] M/s Globe Construction Co. v. Govt. of NCT of Delhi, AIR 1999 Delhi 322.
[24] Victoria Engineering Works v. Union of India, AIR 1995 Delhi 253.

1.5.4 Mode of Publication of Tender Notice for Public Works

In the case of public works and other related transactions by the public body, tender notification must be published sufficiently in advance and given publicity in national newspapers having wide circulation in the cities and places where the major likely participants have offices. In the case of annual purchases etc., individual intimation to the parties who have given best service in the past is also highly recommended. The purpose is not only to get competitive offers but also to prevent arbitrariness or favouritism by excluding some of the competitors from the process. The activities of the government have a public element and, therefore, there should be fairness and equality. For this reason, judicial review is available to an affected party and the court can review the decision-making process itself. If the court finds that there is *mala fide* or arbitrary exercise of power, the court can quash such an action.[25]

In the case of public works, failure to publish a tender notice may render the action invalid.[26] However, the mode and manner of publicity may depend upon the nature and importance of the work.

EXAMPLES FROM CASE LAW

1. Publication of tender notice by displaying it on the notice board and publishing it in one newspaper was held to have complied with the yardstick of due publicity.[27]

2. A municipality allotted the work of construction and repairs of roads and drains worth Rs. 25,00,000/- without publication of notice in daily newspapers. It was urged by the municipality that by sending notices to several offices for display on their respective notice boards as well as publishing on the notice board of the municipality, all the prescribed procedures for allotment of the work were followed. The allotment of the work was challenged in a writ petition. The Patna High Court observed: "the act of municipal authorities, in inviting tenders for allotment of works must not only be fair but it also must appear to be fair." It was held that under the facts and circumstances of the case, the action of the authorities in allotment of work cannot be held to be fair. However, in view of the fact that there had been a huge investment and the construction work had already progressed, no relief was granted.[28]

3. The Allahabad High Court quashed a fishery lease granted without advertising it in well-known newspapers having wide circulation and thereafter holding public auction/public tender.[29]

4. A tender notice was published in two local dailies. The Allahabad High Court observed that advertisement in an unknown newspaper stands on the same footing as no advertisement at all since the purpose of the advertisement is that there should be wide publicity otherwise Art. 14 of the Constitution will be violated. On appeal, the Supreme Court set

---

[25] Netalbag v. State of West Bengal, (2000) 8 SCC 262: AIR 2000 SC 3313; Tata Cellular v. Union of India, 1994 (6) SCC 51: AIR 1996 SC 11; Arphi Electonics Pvt. Ltd. v. Union of India, AIR 1995 Delhi 388; Ram and Shyam Company v. State of Haryana AIR 1985 SC 1147: (1985) 3 SCC 267. Union of India v. Mithiborwla, AIR 1975 SC 266; Ramana Dayaram shetty v. The International Airport Authority of India, AIR 1979 SC 1628; also see the cases referred to in Chapter 3.

[26] Nex Tenders (I) Pvt. Ltd. v. Ministry of Commerce & Industries, 2009(1) CTLJ 309 (Delhi) (DB).

[27] Siddhartha Kalita v. State of Assam, AIR 2002 Gau. 27. Also see: Mahendra Kumar Mohanty v. Collector Khurda, 2008(1) CTLJ 442 (Orissa)(DB).

[28] Sudhir Prasad Singh v. State of Bihar, AIR 1994 Pat.72.

[29] Matsya Jivi Sahakari Samiti, Belepar v. Sub-Divisional Officer, AIR 2003 All. 27.

aside the said order and held that the High Court did not consider the material placed before it, namely that the two newspapers had circulations of 30,000 and 39,600, and had been in circulation for half a century and 11 years, respectively. Also, 12 persons had responded to the advertisement.[30]

This method has the drawback of requiring the issue of tenders to all those who may apply for the same and perhaps to deal with a totally unknown company which may prove to be incapable of executing the work for want of expertise, experience or adequate resources. This drawback can be eliminated by following the procedure of pre-qualifying the bidders.

1.6 Prequalification of Tenderers

For large and complex works or highly specialized jobs, prequalification of tenderers is carried out. In this procedure the invitation in the first instance is only for prequalification. The advertisement should give full particulars of the works to be carried out, the expected qualifications of those desirous of applying and also if joint ventures will be allowed and, if allowed, the particulars to be furnished by the applicants. The advertisement should be published in newspapers with wide circulation and in case of specialized agencies also by addressing individual communications.

Prequalification application forms are issued to those desirous of applying. The applicants furnish the information in the prescribed format along with certificates and supporting details. The employer scrutinizes the information received and applicants fulfilling the selection criteria are qualified and notified accordingly. The blank tender forms are then issued to prequalified bidders.

A government order was issued in 1989 stipulating that any work costing more than one crore of rupees (1 crore is equal to 10 million) be allotted only after prequalification. The Madras High Court upheld the government's decision not to allow the parties who had not participated in prequalification to tender for the work.[31]

Certain guiding principles emerge from the decided cases, which also highlight the likely pitfalls in the process. Some principles are considered below.[32]

1. **Need to have a rational nexus between demand made and conditions imposed:**

While deciding qualifications, there should be a rational nexus between demand made and conditions imposed or else it is likely to be held arbitrary.

EXAMPLES FROM CASE LAW

1. A tender notice for the supply of 4,000 metric tonnes (MT) of cement in a period of four months and another notice for supply of 1,000 and 4,000 MT in six months included a condition under which only manufacturers with installed capacity of 400,000 MT per

---

[30] Business Link v. A.S. Advertising Co. and ors., 2004(1) CTLJ 503 (SC).
[31] S. P. Natarajan v. Highways and Rural Works Dept., Chennai, AIR 1999 Mad. 458.
[32] Also see cases in Chapter 3.

annum or above, were eligible to quote. A manufacturer who had a plant with the capacity to produce 100,000 MT of cement of the requisite quality per year, filed a writ petition in the M. P. High Court. It was held (relying on AIR 1985 SC 1147 and AIR 1993 SC 1601) that the condition of eligibility suffers from arbitrariness "because there is absolutely no nexus between the demand ... and ... the condition of production capacity ... imposed in the tender". The condition was quashed and the petitioner's tender was directed to be considered.[33]

2. The minimum qualification requirement in a tender invitation for coal transportation included a condition that a tenderer "should have supplied one lakh tonnes of coal per month to a thermal power station by the rail-sea-rail route". It was urged that the said condition was highly unreasonable, arbitrary and violative of Art. 14 of the Constitution of India. The Karnataka High Court upheld the above contentions on the ground that imposing the condition of handling coal from coal fields to thermal power stations had become very narrow eliminating all but one tenderer. Also, a person with the necessary experience to handle coal by a rail route can be considered to be competent to move the coal by sea, and vice versa. The Karnataka High Court rejected the contention that the prescription of the eligibility criteria was a justifiable policy decision. The conditions were held to be arbitrary, illegal and unenforceable.[34]

3. A contractor registered under "A" category, deemed capable of executing works up to Rs. 60 lacs, was refused the blank tender forms for works costing less than 40 lacs. While holding the condition that a contractor must have executed single work of a particular magnitude as reasonable, the authorities were directed to take into consideration the fact of the contractor holding "A" Class registration and allow him to participate.[35]

2. An executive authority must be vigorously held to the standards by which it professes its action to be judged

The rule is that an executive authority must be vigorously held to the standards by which it professes its action to be judged, and it must scrupulously observe those standards on pain of invalidation of an act in violation of them. But, it cannot follow that the executive authority cannot at all deviate from or relax the standards. That would depend upon the nature of the act, the necessity for making the deviation or relaxation, and the effects that the deviation or relaxation is likely to cause.

A distinction must be made between those terms and conditions that are essential terms and conditions of eligibility, and those that are incidental or inconsequential in nature. If the relaxation is made in the essential conditions, such as conditions relating to turnover, fairness and equal treatment require that the process should be carried out afresh.[36]

---

[33] M/s Dhar Cement Ltd. v. State of M.P., AIR 1993 M.P. 251. Also see Association of UPS & PCSM v. Society of AMEE & R(Sameer), 2003(1) Arb. LR 399 (Delhi) (DB).

[34] M/s Esteco Coal Services Ltd. v. Karnataka Power Corp. Ltd., AIR 1997 Kant. 220.

[35] Patitpaban Mohapatra v. Bhubaneswar Development Authority, AIR 2001 Ori.59. Also see: M/s Ritz Construction Co. v. Union of India; AIR 2001 J. & K. 7; Dewan and Co. v. Municipal Corporation of Delhi, AIR 1995 Delhi 1.

[36] Konark Infrastructure Pvt. Ltd. v. Commissioner Ulhasnagar Municipal Corpn., AIR 2000 Bom. 389.

1.7 Requirements of Notice Inviting Tenders

In the case of public works, notices calling for tenders are generally in the standard form and serially numbered and entered into a register maintained for that purpose. The notice should be issued only after all tender documents are ready for issue. These documents invariably form part of the agreement to be signed subsequently and therefore must be prepared with great care and caution. When old documents are used to prepare draft of documents, care must be taken to see that the irrelevant part is deleted or else it is likely to create difficulties in the future.

1.7.1 Advance Publicity

Notice should be published sufficiently in advance to enable the intending tenderers to carefully work out their tenders. A time less than one month is seldom sufficient. However, in an exceptional circumstance, such as the prior contract coming to an end, seven days' time allowed may be sufficient.[37] However, cases like this are exceptions to the rule because the entire process of invitation and finalization of tenders exercised hurriedly come under suspicion.[38]

1.7.2 Contents of the Tender Notice

The contents of the notice should be so drafted as to attract the attention of the tenderers for whom it is intended and also to caution others who may not be interested or not qualified to tender. In addition, the tender notice should give sufficient information about the proposed work so as to avoid answering a number of queries from the intending tenderers. The guiding principle to remember can be formulated thus, that on reading the notice the prospective tenderer should not hastily and/or wrongly conclude that he may not be interested in submitting the tender on the one hand, and a person not interested might not be tempted to buy the tender form hoping the work may fall within his capacity or specialization.

In the case of public works and government contracts, the notice should incorporate the words "On behalf of the President of India" or "On behalf of the Governor of the State" as the case may be. Other information to be given may include:

1. Qualifications of tenderers.
2. Name of the work, its situation and brief description. Estimated cost may have to be mentioned if the description is not adequate to enable its reader to ascertain the volume or scope of the work.
3. Earnest money to accompany the tender, its mode of payment.
4. Security deposit that may have to be paid at the time of signing of the agreement and mode of the payment such as cash, FDR or Bank Guarantee.
5. Time limit for completion of the work.
6. From whom and at what cost, if any, blank tender forms may be obtained and up to which date.
7. To whom the completed tender form should be submitted and up to what time and date.
8. Mode of submission of tender, namely one cover, two covers or three covers.
9. Conditional tenders, alternative design offers, if acceptable or not.
10. Date, time and place of opening tenders and if the representatives of tenderers will be allowed to be present. Lack of full particulars in the tender notification is likely to vitiate the entire exercise.

---

[37] M/s Vinod & Associates v. State of M.P., AIR 2004 M.P. 65.

[38] M/s Alpha Engineer v. State of Rajasthan, AIR 2004 Raj. 23. Also see discussion in Chapter 3.

EXAMPLES FROM CASE LAW

1. The petitioners were the manufacturers of PVC insulated galvanized M.S. wires manufactured by them for supply to the department of telecommunications, the only buyer of the product in the country. On all previous occasions the department sold tender forms at a nominal price of Rs. 200/- and the earnest money payable was a fixed amount of Rs. 5,000/-. The department changed the cost of the earnest money for the supply of 2 mm-diameter wire, to Rs. 8,000/- for a 12-page bid document and the earnest money amount was fixed at Rs. 20 lacs irrespective of the value of the offer. A bidder could bid for the entire quantity or part thereof. No floor quantity for bidding was fixed, thereby treating the bidders for the entire quantity and for small part thereof on par. The notice further described the terms of dispatch as "F.O.R. destination" without disclosing the destination. The changed conditions under challenge were held to be highly illogical and grossly irrational and infringing the equality clause in more respects than one and were struck down.[39]

2. Tenders were invited for supply of milk, etc. to meet the requirements of about 300 students of a residential school. One of the conditions mentioned in the tender schedule for supply of milk was that "the rates quoted should not be less than the rates prevailing in the locality". In other words the condition of the tender schedule was that if the rate is less than the local market rate, such tender will not be taken into consideration. However, no specifications for the milk to be supplied, such as the fat content etc. or the quantity to be supplied were given in the tender schedule. No criterion was laid down for arriving at the market rates. On these facts, it was held by the A. P. High Court that the lapses vitiated the entire exercise of calling for tenders.[40]

3. A notice inviting tenders stipulated a condition that no contract would be awarded to any contractor unless he furnishes the Provident Fund Code Number allotted to him for his workmen. Under the instructions issued by the Government of India and followed by the Department of Provident Fund, no separate EPF Code Numbers were to be issued to contractors, any person employed by or through a contractor being also an employee of the principal employer. Allowing the petition, the respondent was directed not to insist upon the condition of separate EPF Code Number for issuing tenders.[41]

4. A clause in an advertisement for tenders stipulated that tenderers were required to deposit earnest money amounting to over Rs. 30 lacs only in the shape of 6-year National Savings Certificates (NSCs) duly pledged in favour of the employer. It was highlighted that NSC amount would remain blocked for six years, unsuccessful tenderers would also not be able to encash the amount earlier. There is a bar that NSCs are not to be issued to companies and corporations so that they would not be eligible to tender for the work. It was held the condition was irrational, illogical and had no nexus with the object sought to be achieved.[42]

5. A condition of deposit of earnest money by way of bank draft/call deposit only from certain specified nationalized banks and no other bank was held without any basis. However, the tenderer who, knowing full well such a requirement, submitted his bid enclosing therewith pay orders issued by some other bank and did not challenge the

---

[39] Flyfot (I) Ltd. v. Union of India, AIR 1996 Cal. 291.

[40] Shaikh Fareed Saheb v. Jt. Collector Nellore, AIR 1994 A.P.160.

[41] Rakesh Kumar v. National Hydro Electric Power Corporation, AIR 2002 H.P. 70.

[42] Bigla Kachhap v. State of Jharkhand, AIR 2002 Jhar. 81.

condition before submitting his tender papers was not granted any relief as, in the meantime, the work was awarded to other contractor.[43]

6. When proposals were invited from manufacturers/dealers, the contention that manufacturers alone satisfied the tender requirement was rejected on the ground that it would make the use of the expression "dealers" redundant.[44]

7. A tender notification stated that tender documents were to be issued from 13th June 1994 to 21st June 1994. 21st and 22nd June were declared public holiday and local holiday, respectively. The authority refused to issue tenders stating that the last date was preponed to 20th June 1994. The Madras High Court held that the tender schedules could have been issued on 23 June 1994, since the last two days turned out to be holidays. It was further held that when tender schedules were refused illegally to an intending participant, there could be no doubt that he was an aggrieved party.[45]

8. A tenderer was obstructed by certain persons when he visited the office of the authority to submit his tender. He was permitted to submit his tender, which had been duly purchased within the time allowed, after two intervening days which happened to be holidays. A petition was filed against the action of the authority allowing an extension of two days. The Calcutta High Court held:[46] the petitioner failed to establish that for violation of time schedule mentioned in the NIT, he suffered any actual prejudice. The work order issued in favour of the respondent was, therefore, not liable to be interfered with.

1.7.3 Forfeiture of Earnest Money Clause – If Illegal?

In the earlier editions of this title, it was suggested that a condition in the tender notification stipulating that all earnest money deposited by the tenderer would be forfeited if he withdraws or modifies his tender after submission and before acceptance is illegal for want of consideration.[47] It would be so where the submission and receipt of bids are not subject to the condition that in the event of a withdrawal of the bid the earnest money would stand forfeited. However, almost all tender invitations incorporate the condition of forfeiture of earnest money, if tender is withdrawn during its validity period.

The law seems to be well settled by a series of decisions of the Supreme Court of India to the effect that where the tender invitation stipulates keeping an offer open for a particular time and in case the offer is withdrawn before the said validity period, earnest money can be forfeited.[48] The upshot of the above decisions is that it is no longer possible for offerors to contend that the right to withdraw the bid in terms of Section 5 of the Contract Act, 1872 would entitle them to withdraw without suffering forfeiture of the earnest money, even in cases where the submission and receipt of bids is itself subject to the condition that in the event of a withdrawal of the bid the earnest money would stand forfeited.

---

[43] Rafique Ahmed Mazmuder v. Hindusthan Paper Corpn. Ltd,. AIR 2008 Gau. 85.

[44] M/s Doshi Ion Exchange & Chemical Inds. Ltd. v. Union of India, AIR 2001 Raj. 276.

[45] K.Soosalrathnam v. Divisional Engineer, AIR 1995 Mad. 90.

[46] Utpal Mitra v. Chief Executive Officer, AIR 2006 Cal. 74.

[47] M/s Krishnaveni Constructions v. Executive Engineer, Panchayat Raj, Darsi, AIR 1995 A.P. 362.

[48] National Highways Authority of India v. Ganga Enterprises and another, 2003 (4) R.C.R.(Civil) 575: (2003) 7 SCC 410; State of Maharashtra and others v. A.P. Paper Mills Ltd., (2006) 4 SCC 209; and State of Haryana and others v. Malik Traders, 2011(4) R.C.R.(Civil) 873: 2011(6) Recent Apex Judgment (R.A.J.) 95: (2011) 13 SCC 200.

The logic behind the proposition is that bid security is given by the offeror and taken by the offeree to ensure that the offer is not withdrawn during the bid validity period of 90 days and a contract comes into existence. Such conditions are included to ensure that only genuine parties make the bids. In the absence of such conditions, persons who do not have the capacity or have no intention of entering into the contract will make bids. The very purpose of such a condition in the offer/bid will be defeated, if forfeiture is not permitted when the offer is withdrawn in violation of the agreement.

It was observed and held by the Apex Court:[49]

> A person may have a right to withdraw his offer but if he has made his offer on a condition that some earnest money will be forfeited for not entering into contract or if some act is not performed, then even though he may have a right to withdraw his offer, he has no right to claim that the earnest/security be returned to him. Forfeiture of such earnest/security in no way affects any statutory right under the Indian Contract Act. Such earnest/security is given and taken to ensure that a contract comes into existence. ... It is settled law that a contract of guarantee is a complete and separate contract by itself. The law regarding enforcement of an "on demand bank guarantee" is very clear. If the enforcement is in terms of the guarantee, then court must not interfere with the enforcement of bank guarantee. The court can only interfere if the invocation is against terms of the guarantee or if there is any fraud. Courts cannot restrain invocation of an "on demand guarantee" in accordance with its terms by looking at terms of the underlying contract. The existence or non-existence of any underlying contract becomes irrelevant when the invocation is in terms of the bank guarantee. The bank guarantee stipulated that if the bid was withdrawn within 120 days or if the performance security was not given or if an agreement was not signed, the guarantee could be enforced. The bank guarantee was enforced because the bid was withdrawn within 120 days. Therefore, it could not be said that the invocation of the bank guarantee was against the terms of the bank guarantee.

1.7.4 Mistake in Tender Notice

A tender notice published on 23rd October 1992 mentioned the date of submission of tenders as 2nd November 1992 and the latter part stated that the blank form could be purchased on 1st December 1992. It was stated that the date 2nd November 1992 was a printing error and the date of submission was 2nd December 1992. The Allahabad High Court held: If due to a printing error in the notice, there is possibility of people being misled, it is incumbent on the authorities to correct the mistake by publishing a fresh notice or by publishing corrigendum so as to enable more persons to submit tenders.[50] In the said notice, each tenderer was required to submit a certificate of ownership of boats without specifying any authority from whom the certificate was to be obtained. It was held that statement of ownership made by the tenderer in form of affidavit was a valid certificate.

1.8 Preparation and Submission of Tender

The steps generally taken by the tenderer on reading the notice inviting tenders, if he is interested in submitting a tender, include:

---

[49] State of Haryana v. M/s. Malik Traders, (SC): 2011 AIR (SCW) 5094: 2011 AIR (SC) 3574: 2011 AIR SC (Civil) 2309; 2011(13) SCC 200: 2011(10) SCR 372: 2011(5) Cal. H.C.N. 45: 2011(9) Scale 1. For further discussion see paragraph 12.20 under Chapter 12 Forfeiture of earnest money.
[50] Gorakhnath Upadhyaya v. State of U.P., AIR 1994 All. 283.

1. Purchase of blank tender form.
2. Study of tender drawings (if they are not attached to the blank tender form, at the office where they are kept).
3. Visit to the site of the work to ascertain:

    (a) Nature of site, access to it, foundation conditions, if exposed or by taking exploratory holes or bores if so mentioned in the tender notice, extent of site available for site office, labour camp, material dumps, stores, storage of machinery, equipment or workshop shade and movement of men and machinery in general.

    (b) Nearest railway station, bus station, petrol depots, banks, etc.

    (c) Sources and prices of raw material required, suppliers with their addresses, transportation facilities and cost involved.

    (d) Availability of local labour, skilled and unskilled, prevailing wages, piece rate workers and their addresses.

    (e) Nearest sources of power, water, etc.

4. Rough verification of estimated quantities vis-à-vis drawings to ascertain if any item has been overestimated or underestimated or omitted from the BOQ. However, in the case of lump sum contracts, preparing a detailed estimate from the drawings is a must.
5. Analysis of rates or cost to ascertain the prime cost which includes: cost of materials, labour, machinery and equipment.
6. Deciding the percentage of the basic or the prime cost worked out in step (5) above to be added to it to cover overheads and profit expected.
7. Working out the lump sum or item rates or percentage rate to be tendered.
8. Checking of the rates so worked out for omissions, errors and units of measurement in particular.
9. Copies of documents required to be attached such as registration in a particular class, list of works executed, works in hand, list of machinery and equipment available/proposed to be acquired, etc. are generally kept ready and are cross-checked with the notice requirements of the tender notice.
10. Earnest money to be paid in the form of a bank guarantee – Demand Draft to be obtained.
11. The final sum, rates or percentages to be actually tendered are a closely guarded secret and the person to submit it is given eleventh hour instructions as to what sum, rates or percentage is to be quoted so that the secrecy is maintained till the end.
12. Care must be exercised while submitting the sum, rates or percentage so that no error creeps into the rates, or in totalling.
13. The envelope(s) are sealed and handed over before the appointed time on the last date (general practice followed by many contractors, although early submission is not ruled out) or dropped in the box kept for the purpose.

With the advent and advancement of internet technology, tenders are nowadays invited and submitted on line.[51]

---

[51] For online tendering procedure see the lead author's book *Civil Engineering Contracts and Estimates*, 4th Ed., pp. 40–42.

1.9 Irregularity in Submission of Tenders

In building and engineering contracts, offers are usually invited in prescribed forms. So long as the specified procedure is observed and essential requirements are complied with while submitting the tender, no legal problem tends to crop up. It is only when some irregularity is committed, whether knowingly or unknowingly, that validity of the tender has to be decided. This will be clear from the following illustrations.

1.9.1 Unsigned Tender May Be Binding

A written tender by a contractor undertaking to construct a building at a stated price, though not signed by the contractor or his authority, becomes a binding and enforceable contract on its acceptance.[52] In the case of government contracts, however, care must be taken to sign the tender in the manner specified. In one case the tender notification stated that all corrections of rates and items in the tender should be initialled by the contractor and every page of the rate schedule be signed by the contractor in full. The contractor had signed with initials and surname. His tender was rejected on that count. The writ petition was dismissed by holding that full signature means first name, middle name and the surname. The fact that on earlier occasions the tenders submitted by the petitioner with similar signature were accepted was held to be of no consequence.[53]

The author is in respectful disagreement with the views expressed in the above decision inasmuch as it held that the definition of "sign" with its grammatical variations and cognate expressions given in the General Clauses Act was not attracted. The provisions of Art. 299 of the Constitution require a formal contract to be signed which stage is reached only after a tender is accepted. The M. P. High Court has, in a case wherein the tenderer had not put his signature on each and every page, held that want of signature at certain places is no ground for non-acceptance of the tender. It was observed that a tenderer should expressly mention about non-acceptance of such terms and conditions in specific.[54]

1.9.2 Estimates as Offers Are Binding

In response to a letter inviting tenders, a firm of contractors wrote in reply, "Estimate – our estimate to carry out the sundry alterations to the above premises according to the drawings and specifications, amounts to the sum of 1230 pounds". The next day the owner wrote back accepting the offer "to execute ... For the sum of 1230 pounds ... the work in question." Subsequently the contractor wrote that they had made an error in their figures and under the circumstances they must withdraw their estimates. When they refused to do the work, the owner got it done by another contractor whose charge was higher, and sued the original contractors to recover the difference between the two charges as damages for breach of contract. The defendants contended that they had used the word "estimate" advisedly, and did not intend it as a tender to contract. In this contention they were supported by several other contractors who stated that the distinction between an "estimate" and a "tender" was recognized in the building trade, the one recognized as not binding and the other as binding. It was, however, held that the defendant's offer was a tender and its acceptance by the plaintiff converted it into a legal contract. The decision of this case was mainly based not upon the supposed custom of the trade but on the language of the letters, which had passed between the two parties.[55] This decision could be contrasted with a South Australian Supreme Court decision 2006 in which an estimate was held

---

[52] Webb and Sons v. Williams, (1907) NZLR.

[53] R. N. Ghosh v. State of Tripura, AIR 2000 Gauhati 114.

[54] Shivmangal v. Narayanprasad & Ors., AIR 2007 (NOC) 1878 (M.P.).

[55] Croshaw v. Pritchard and Renwick, (1899) 16 TLR 45.

to be just an estimate and not an offer that could, upon acceptance, lead to a binding agreement.[56] In this case, the word estimate was repeated several times in a fax but the price quoted was not found to be binding. If making a pure estimate, the intention should be made clear beyond doubt instead of relying on the word "estimate" alone.

1.10 Payment of Earnest Money

In order that there may be a valid tender, so that the tender or offer can be considered or accepted, a contractor is usually required to deposit the earnest money or security. Most tender invitations include this stipulation. Compliance with this condition will be necessary to make the tender valid.[57] If an opportunity is given to the tenderer to make good deficiency, as it is a pre-requisite condition for the entertainment of the offer itself, the demand for compliance of such a condition does not amount to the making of a counter-offer.[58]

EXAMPLES FROM CASE LAW

1. A co-operative society submitted the highest bid in one case but claimed an exemption to deposit earnest money, under *bona fide* belief that a co-operative society was exempt from depositing earnest money, but agreed to deposit in the event of intimation by the government. The bid was rejected without giving an intimation or an opportunity to make such deposit. The Orissa High Court observed that the offer of the petitioner being about Rs. 30 lacs more than the next bid, the interest of the public revenue should not be ignored on mere technical grounds and directed the parties to hold fresh negotiations including the petitioner.[59]

2. A tenderer submitted his tender without payment of earnest money deposit as per the condition of the tender invitation. He failed to make the deposit in spite of reminders. His tender was accepted. He declined to perform on the ground that there was an error in his tender, the rate quoted per metre length was in fact per foot and not per metre as unit. A suit was filed for recovery of the excess expenditure incurred over and above the tendered cost. It was held that the condition of payment of earnest money was an essential condition and it was not waived or dispensed with at any point of time and therefore there was no concluded contract.[60]

1.11 Offer to Be Accompanied by Certain Documents – When a Condition Precedent?

Instructions to tenderers in a case clearly mentioned that if the tender is not accompanied by their earnest money deposit, the tender shall not be considered, and with regard to the documents to be submitted along with the tender it was mentioned that the tender was liable to be rejected if the documents were not attached along with the tender form. It was held that acceptance of a higher

---

[56] Kyren Pty Ltd. v. Built Projects Pty Ltd., [2006] SASC 204.
[57] SPC Engineering Co. v. Union of India, AIR 1966 Cal. 259(263).
[58] UPSE Board v. Goel Electric Stores, AIR 1977 ALL 494.
[59] Welfare Thrift & Multipurpose Co-op. Society Ltd. v. State of Orissa, AIR Ori. 134.
[60] GSRTC Ahmedabad v. B. Arunchandra & Co., AIR 2001 Guj. 343.

bid for operating a cycle-park, subject to the production of requisite documents, was valid and not in breach of Art. 14 of the Constitution of India.[61]

1.12 Conditions Offered by Contractor in His Tender

A contractor may submit his tender accepting, without reservations, all the conditions put before him, in which case the tender is said to be unqualified. He may, however, desire to make the tender subject to certain conditions. In such a case, the owner should carefully consider the alternative conditions offered. On the other hand, a contractor should be careful to see that the conditions he has put forth are included in and form part of the contract documents. The following cases will serve as examples and a warning to those contractors who wish to qualify tenders by superimposing conditions.

EXAMPLES FROM CASE LAW

1. At the time when the tender was floated, there was no condition incorporated therein that Modified Value Added Tax (Modvat) credit would be claimed by the appellant – Security Press. When the terms and conditions were reduced in the supply order dated 31st May 1995, therein the condition of Modvat credit was incorporated and it was accepted by the claimant. The Supreme Court held: First, the contract had come into existence and the supply had been started on the basis of that supply order. Though the claimant had protested with regard to this clause, the appellant did not accede to the request for deleting that clause and the appellant had informed the claimant on 31st December 1995 that there was no change in the conditions of the supply order. Still, the claimant continued to supply the goods as per the order. Therefore, on the face of this condition, there was no going back from that. In case the claimant was not inclined to accept this clause he could have very well withdrawn from the contract. But he did not do so and continued with the contract. Therefore, on the basis of the clear terms of the contract, the claimant was bound by it and had to restore whatever the Modvat credit received by it to the appellant – Security Press.[62]

2. The plaintiff issued its NIT on 17th January 1969. The defendant made its offer by submission of the tender. In addition to the tender form, the defendant qualified the tender by setting out its additional conditions for supply of the material by its communications dated 17th January 1969 and 6th February 1969. It was insisting on supplying the ingots in its make "Zamak" as well as in the description as detailed in its communications. Its further letter dated 7th March 1969 also reiterated that it would effect supplies of the items in terms of its letter dated 17th January 1969 read with letter dated 6th February 1969. It was apparent that there was no meeting of minds with regard to the nature of the ingots that were to be supplied. The defendant's offer as contained in its tender was clearly a conditional offer and was subject to the conditions set down by it in its letters dated 17th January 1969, 6th February 1969 and 20th February 1969 which were never accepted by the plaintiff. It was held that no concluded contract came into

---

[61] Mukul Kumar v. Divisional Railway Manager, Northern Railway, AIR 1995 All. 72. Also see paragraph 3.15 in Chapter 3 for further discussion and case law.

[62] Security Printing & Minting Corp. of India Limited v. M/s. Gandhi Industrial Corporation, (SC): 2008(3) BCR 339: 2008 (1) R.C.R.(Civil) 16: 2007(6) Recent Apex Judgments (R.A.J.) 204: 2007(13) SCC 236: 2007(12) Scale 488: 2007(11) SCR 86: 2007 DNJ 965: 2007(4) R.A.J. 536: 2007(217) ELT 489: 2007(4) ArbiLR 65: 2008(1) AIR Bom. R 44: 2008(1) All WC 426: 2007 AIR (SCW) 6854.

existence. It was observed: where an offer is made in alternative terms, the acceptance must make clear to which set of terms the assent is directed.[63]

3. In a letter accompanying a tender, the tenderer wrote, "I have quoted my rates for each slab and add for every extra one and part thereof". The tender was for the transportation of iron and steel materials from various stockyards to the destination source. The authorities declined payment on multi-slab basis contending that the multi-slab rates had never been accepted and were not covered by the agreement. The tenderer had accepted payments till the end of the contract on a single-slab basis. The recommendations made by one of the officers to make payment on a multi-slab basis, it was urged, was not determinative of the terms and conditions of a written contract. The learned single judge accepted the defence. The division bench allowed the appeal against the said decision and dismissed cross-objections. The Supreme Court allowed the appeal and set aside the judgment of the division bench. It was held that the written agreement nowhere indicated that the rate to be paid to the plaintiff was on a multi-slab basis. The terms of the contract were also not susceptible of such a construction. The letter accompanying the tender contained the above quoted sentence in ink, signed by neither party. The Supreme Court rejected oral evidence of the defence witness that it was written after negotiations.[64]

4. A large contracting company sent in a signed tender on the appropriate form undertaking the construction of the proposed work at the quoted price and within the time limit specified. The tender was accompanied by a covering letter of the same date which read: "We have pleasure in enclosing our tender ..." and included a condition that the tender was subject to adequate supplies of material and labour being available. The tender was not immediately accepted. Negotiations followed and finally an agreement was drawn up. One of the clauses of the agreement named the documents which were deemed to form part of the agreement and be construed as such. The letter containing the condition was not included in the list. The work which was to be completed in eight months took 22 months for its completion. The main reason was the lack of skilled labour. The contractors claimed payment on a *quantum meruit* basis that was the actual cost of the work to them. To sustain this claim, the contractors alleged that their letter accompanying the tender formed part of the contract. However, it was held that the letter forwarding the tender was merely to propose a qualification to the expected contract which the offeree did not accept. It was part of the preliminary negotiations. The contractors failed in their case.[65]

5. In a similar case in India, a tender embodying certain terms was submitted and accepted. Both parties were agreed on all the matters contained in it and their conduct showed that both parties were agreeable to reduce the contract to writing. Before the agreement was signed, one party wanted to include a further condition in the contract. The Supreme Court of India held:

> We will assume that the request was made to the other contracting party. But without waiting for the assent of the other side, both sides accept and sign the contract as it existed before the fresh suggestion was made. It is an error in law to reduce from this that there was acceptance of the fresh proposal. On the contrary the legal conclusion is that the new suggestion was dropped and that the contractor was content to accept the contract as it was without this condition.[66]

---

[63] Binani Metals Limited v. Union of India (Delhi) (D.B.): 2004(114) DLT 637: 2004(8) AD (Delhi) 191: 2004(78) DRJ 264: 2005(1) R.A.J. 30.
[64] T.N. Electricity Board v. N. Raju Reddiar, AIR 1996 SC 2024.
[65] Davis Construction Ltd. v. Fareham Urban District Council, 1956 AC 696.
[66] Thawardas v. Union of India, AIR 1955 SC 468 (476).

1.13 Withdrawal of Tender

A tenderer is entitled to withdraw his offer at any time prior to its acceptance. An offer does not impose any obligation on the person making it until it is accepted by the other to whom it is made. Even a condition in the notice inviting tenders that a tenderer shall not withdraw his offer or keep it open for a certain period is a promise by the tenderer but without any consideration from the person to whom it is made and therefore not binding on the tenderer. On the contrary such a provision can operate only for the benefit of the tenderer and as a warning that an acceptance after the specified time will be too late. Such a promise, however, can be made binding for a distinct consideration, such as the owner offering a certain sum to the tenderer to keep his tender valid for acceptance for a certain period before definitely deciding whether to accept or finally reject the same. This view, expressed in the earlier editions, needs to be reviewed in the light of a change in the law brought about by the decisions of the Supreme Court of India discussed in paragraph 1.7.3 above.

1.13.1 Sections 4, 5 and 6 of the Indian Contract Act

The law relating to revocation of offers and acceptances is contained in Sections 4, 5 and 6 of the Indian Contract Act, the relevant parts of the said sections read as below:

---

5. *Revocation of proposals and acceptances*

A proposal may be revoked at any time before the communication of its acceptance is complete as against the proposer, but not afterwards.

An acceptance may be revoked at any time before the communication of the acceptance is complete as against the acceptor, but not afterwards.

ILLUSTRATIONS

(a) A proposes, by a letter sent by post, to sell his house to B.
(b) B accepts the proposal by a letter sent by post.
(c) A may revoke his proposal at any time before or at the moment when B posts his letter of acceptance, but not afterwards.

Once the proposal made by the proposer is accepted and a communication of acceptance to the proposer is out of control of the acceptor, a concluded contract between the parties comes into existence and thereafter both the parties stand precluded from revoking the same except in accordance with the terms and conditions of the contract.[67] When the agreement is in writing, in terms of a letter of intent containing an arbitration clause, a party to it is not entitled to say that because of a repudiation of contract, no contract exists.[68]

If a proof of posting of the letter of acceptance on the address given by the offeror is produced, the acceptance is complete even though the letter of acceptance was not delivered to the offeror.[69]

---

[67] Vipin Mehra v. Star India Pvt. Ltd., (Delhi): 2003(106) DLT 516: 2003(6) AD (Delhi) 109: 2003(70) DRJ 677: 2003(3) ArbiLR 178: 2003 (3) R.A.J. 435: 2003(2) ILR (Delhi) 209.

[68] Gulf (Oil) Corporation Ltd. v. Steel Authority, (Calcutta): 2006 (4) ICC 470: 2006 (3) Cal. H.C.N. 718.

[69] J.K. Enterprises, M/s. v. State of M.P. (M.P.): 1997 AIR (M.P.) 68: 1997(2) M.P.L.J. 31: 1997(1) ArbiLR 683: 1997 (3) Civil LJ 649.

4. *Communication when complete*

The communication of proposal is complete when it comes to the knowledge of the person to whom it is made.

The communication of acceptance is complete

... as against the proposer, when it is put in the course of transmission to him so as to be out of the power of the acceptor.

... as against the acceptor, when it comes to the knowledge of the proposer.

The communication of revocation is complete, as against the person who makes it, when it is put into a course of transmission to the person to whom it is made so as to be out of power of the person who makes it as against the person to whom it is made, when it comes to his knowledge.

ILLUSTRATION

(a) ...

(b) ...

(c) A revokes his proposal by telegram. The revocation is complete as against A when the telegram is dispatched. It is complete as against B when B receives it.

The provisions are self-explanatory.

1.13.2 Withdrawal or Revocation How Made

A proposal may stand revoked by many ways other than by express communication. Section 6 of the Indian Contract Act deals with this aspect. The said section reads:

6. *Revocation how made*:

A proposal is revoked – (1) By the communication of notice of revocation by the proposer to the other party;

(2) by the lapse of time prescribed in such proposal for its acceptance, or, if no time is so prescribed, by the lapse of a reasonable time, without communication of the acceptance;

(3) by the failure of the acceptor to fulfill a condition precedent to acceptance; or

(4) by the death or insanity of the proposer, if the fact of his death or insanity comes to the knowledge of the acceptor before acceptance.

These provisions are briefly discussed below:

The accepting authority's power of acceptance is terminated the moment communication of notice of revocation by the tenderer is complete as against the accepting authority. This communication, as stated in Section 3 of the Indian Contract Act, can be by any act or omission by which it is intended to be communicated.[70] Letter by post, hand delivery, fax message,[71] telex, telephone used to be the usual modes. In the case of revocation by telephone, the proof of conversation could pose a problem if the other party denies it or takes the stand that acceptance was already posted.

The other acts of revocation could include a suggestion of modification or changes to the originally submitted tender or alterations in the terms of the offer.

In respect of public works executed by government authorities, statutory corporations, etc., the present day practice of accepting tenders leaves much to be desired. The tenderers are asked to hold their tenders open for about three to six months. There are several instances when a decision is not taken even after such a long period and further extension of time for validity of tender is sought.

---

[70] For Section 3, see paragraph 1.16.1 below.

[71] Quadricon Pvt. Ltd. v. Bajarang Alloys Ltd., AIR 2008 Bom. 88.

During this period, a contractor is entitled, under ordinary law, to revoke his offer. In the meantime, if prices of materials, and petroleum products in particular, rise, having an effect of overall price rise, the tenderer may inform that the tender will be kept open subject to rates/cost being increased by a specified percentage. On receipt of such a communication decision is sometimes taken hurriedly to convey acceptance of the original tender within the validity period. Invariably a question is likely to arise whether the acceptance has converted the original offer into a binding contract. The answer to such a question will depend upon the wording of the notice of modification. To avoid controversy, a tenderer is well advised to inform in no uncertain words that his original tender stands modified by his communication in question so as not to give scope for further dispute, if his intention is indeed to revise his original offer. If he intends to be bound by his original offer provided acceptance is communicated within the initial validity period, he may make it clear. Even in such cases the parties cannot avoid problems and difficulties because the acceptance is given without the authority being in a position to allow the contractor to commence the work. The site might not be available, the drawings might not be ready; these are a few common hindrances.

Sometimes, the delay in acceptance takes place due to judicial intervention, as happened in one case. The Supreme Court of India observed: "Considering the tremendous importance of the project, it (THDC) has yet not been able even to finalise the tender. Three and a half years have rolled by and yet no progress has been made, thanks to the legal battles in between the two giants called Voith GMBH (respondent No.1) and Alstom (respondent No.2)".[72]

The acceptance letters are also generally standardized with an order to start the work. The provision of the Institution of Civil Engineers (ICE) or FIDIC standard form is generally incorporated, obliging the contractor to commence the work within the stipulated number of days of receiving the work order. If for any reason the work cannot be so commenced, the contractor may seek an extension of time plus compensation for the losses on several counts.

EXAMPLES FROM CASE LAW

1. A tender notification contained a condition that the quotations would be valid for a minimum of three months, which validity would be extended by the tenderer up to six months if so required. The tenders were opened on 14th October 1999. The price bid could not be opened, because of the stay by Civil Court. The injunction was finally vacated on 27th April 2000. Letter of intent was issued on 28th April 2000. The tenderer refused to execute the work. His earnest money was forfeited. Penalty of Rs. 2.81 lacs was sought to be recovered and the firm blacklisted. The learned single judge set aside the order of black-listing but refused to interfere with the balance order. Aggrieved, an appeal was filed against the said order. Allowing the appeal it was held that any party, while submitting a tender, is always conscious of the prevailing market conditions, pricing and the rates structure as on the date of submission of the tenders. With the passage of time, these elements and components tend to fluctuate. It is with a view to protecting the interest of these tenderers that the tender notification specifies a period of three months or such other outer limit of validity of tenders. In the case under consideration, the authority had not even sought an extension of time of validity of tender beyond three months. It was held that issuing a letter of intent after the validity period was over, though through no fault of either party, was clearly in violation of Art. 14 of the Constitution. The order was quashed and set aside.[73]

---

[72] THDC India Ltd. v. Voith Hydro GMBH Co., (SC) 2011 AIR (SCW) 2243: 2011(4) SCC 756: 2011(3) BC 6: 2011 AIR SC (Civil) 1022: 2011(5) R.C.R. (Civil) 189: 2011(3) R.A.J. 76: 2011(5) SCR 618: 2011(3) Scale 677: 2011(3) JT 592: 2011(2) SCC (Civil) 470.
[73] Essem Transport & Construction Pvt. Ltd. v. Central Coal fields Ltd., AIR 2003 Jhar. 13.

2. In response to invitation of tenders, the plaintiff submitted his offer on 8th August 1990 making it valid for a period of 90 days. The tenders were opened on 9th August 1990. Certain clarifications were sought and correspondence took place between the parties. On 20th November 1990, the plaintiff modified his offer by enhancing his rates. The parties further negotiated about the supply of part quantity at the old price. On 24th November 1990 the defendant accepted the offer at the old rates and sent a telex message to that end. The plaintiff replied that there was no question of acceptance of its offer, much less offer at the old rates. The defendant invoked bank guarantee. It was held that the original offer was revised prior to acceptance and therefore a binding concluded contract did not come into existence and as such the bank guarantee could not be encashed.[74]

3. In another case, a tender was submitted without complying with the condition of payment of earnest money. The tenderer was not exempted from payment of earnest money. Nevertheless, the tender was accepted. The tenderer took certain objections, including that there was a mistake in quoting the rate. The authority insisted on performance and, failing this, purchase of the contract goods from the open market. The tenderer failed and the authority filed a suit for compensation for the loss incurred in procuring the goods from the open market. It was held that the contract was not concluded when the tender was not accompanied with earnest money which was a precondition.[75]

4. Where acceptance by a letter is followed by revocation by telegram, if the telegram reaches first, acceptance stands revoked. A tender was submitted by S. Its acceptance by the government was communicated to S by a letter dispatched on 28th April. S revoked his tender by telegram on 13th May. It reached the government after the letter of acceptance was posted. He claimed the letter of 28th April was not received. The concerned officer gave a statement on oath that the letter was duly posted. It was held the contract was completed as soon as the letter of acceptance was posted and the contractor could not revoke.[76]

5. A tender submitted by the plaintiff firm was not signed by all the partners of the firm. The firm's annual requirement under registration had been assessed at 15,000 quintals (1,500 MT) of food grains. Tenders were submitted and opened on 29th June 1983. The plaintiff, by his letter dated 8th July 1983, informed the defendant that the offer was for 1,500 MT of stocks and not over and above the specific quantity. This letter was received by the defendant on 11th July 1983. The defendant sent an acceptance telegram on 22nd July 1983 for 6,200 MT. The plaintiff alleged that there was no valid offer and, further, that even if it was, it stood modified by the letter of 8 July and that the defendant's acceptance for quantity 6,200 MT could be a counter-offer not accepted by the plaintiff. The defendant threatened to sell 6,200 MT of paddy at the risk and cost of the plaintiff and to cancel his registration. The plaintiff successfully filed a suit in the Delhi High Court for a declaration that there was no legal, valid and binding contract.[77]

6. A landowner, during the course of an inquiry to fix compensation to be awarded for his land, stated to the Land Acquisition Officer concerned that he was willing to agree to the land being acquired, provided he was given compensation in lump sum. The landowner withdrew this concession before any award was made. Subsequently, he filed a writ petition questioning the validity of the land acquisition proceedings. It was held by the Supreme Court:[78]

---

[74] Oil and Natural Gas Commission v. Balaram Cements Ltd., AIR 2001 Guj. 287.

[75] G.S.R.T.C., Ahemedabad v. B. Arunchandra & Co., AIR 2001 Guj. 343.

[76] Sadhoo Lal Motilal v. State of M.P., AIR 1972 ALL 137.

[77] M/s Suraj Besan & Rice Mills v. Food Corpn. of India, AIR 1988 Del. 224 (227).

[78] Nutakki Sesharatanam v. Sub-Collector, Land Acquisition, Vijaywada, AIR 1992 SC 131.

The statement of the appellant amounted in law to no more than an offer in terms of the Contract Act. The said offer was never accepted by the Land Acquisition Officer to whom it was made. ... Till the offer was accepted there was no contract between the parties and the appellant was entitled to withdraw his offer. There was nothing inequitable or improper in withdrawing the offer, as the appellant was in no way bound to keep the offer open indefinitely.

7. D made an offer to P to lease 184 acres for the mining of iron and manganese ores subject to P paying Rs. 1,80,000/- at his option within three months from the date of the offer. D posted a letter revoking the offer, which reached P after P filed a suit against D claiming that D was entitled to remain in possession of the mining area. Before the revocation letter reached P, P had not accepted the offer either orally or by any letter sent to D. The High Court held that P accepted the offer by the plaint and this acceptance was communicated to D by serving a copy of the plaint by the letter of revocation of offer reached D. It was held by the Supreme Court: "It is not usual to accept a business offer by a plaint, nor is it usual to communicate an acceptance by serving a copy of the plaint through the medium of the court".[79] The appeal was dismissed on the ground that there was no concluded contract.

From Example (7) above, it is clear that for a revocation to be operative it must be communicated to and actually brought to the knowledge of the offeree before he dispatches the letter of acceptance or communicates the acceptance in any other way. For, as already mentioned, the effective date on which a party exercises the option of acceptance is to be ascertained from the date when the acceptance is put in transmission and the letter is posted.

1.13.3 Conditions in NIT – Tender Shall Not Be Withdrawn – When Effective?

If there is a statutory power to make rules for the conduct of the departmental business, the government may be justified in prescribing among the conditions of tenders for public works that a tender shall not be withdrawn before acceptance or refusal.[80] In the absence of a statutory rule, mere inclusion of a condition to that effect in a tender notice may not suffice.

EXAMPLES FROM CASE LAW

1. A condition in the tender notification stipulating that all earnest money deposited by the tenderer will be forfeited if he withdraws or modifies his tender after submission and before acceptance was held to be illegal for want of consideration.[81]

2. In a case decided by the Madhya Pradesh High Court,[82] the tender notice contained a clause whereby a tenderer was allowed to withdraw his tender before the commencement of the opening of the tenders on the condition that on opening the remaining tenders, there should be at least one valid tender complete in all respects available for consideration. The contention of the

---

[79] Viswesardas Gokuldas v. B.K.N. Singh, (1969) 1 SCC 547 (549) 58. Baroda Oil Cakes Traders v. Purushottam Narayandas Baghulia, AIR 1954 Bom. 491.
[80] Secretary of State v. Bhojkar Krishnaji, (1925) 49 Bom. 759.
[81] M/s Krishnaveni Constructions v. Executive Engineer, Panchayat Raj, Darsi, AIR 1995 A.P. 362.
[82] Rajendra Kumar Verma v. State of Madhya Pradesh, AIR 1972 Madh. Pra. 131.

government was that tender notices were issued under Section 12 of the M.P. Tendu Patta Adhiniyam of 1964 and consequently the terms thereof should be treated as law on the subject and enforceable as such. The High Court observed.

> We are unable to agree with this contention. No rules have been framed for the disposal of the Tendu leaves. Section 12 only authorizes the Government to dispose of Tendu leaves as it considered proper. The terms given in the tender notice are merely executive directions laid down for the purpose of receiving offers. From a perusal of tender notice, it is clear that a tender notice cannot have a status of law and could not be enforced as such.

1.13.4 Consequences of Revocation of Tenders

If a tender is validly accepted, a binding contract comes into existence. A tenderer whose tender is accepted cannot revoke his offer without facing the consequences of breach of contract. The question to be considered at this stage is about the validity of a stipulation in invitation to tenders about earnest money paid by the tenderer. Most tender notices stipulate that in the event of a successful tenderer backing out, his earnest money would be forfeited in full. In view of the concept of damages for breach, the authors are of the opinion that earnest money cannot automatically be forfeited without any proof of damage. The aggrieved party, however, would be entitled to a reasonable compensation.[83]

1.13.5 Effect of Acceptance beyond the Validity Period: Section 6(2) of the Act

Generally, notices inviting tenders clarify the time within which a tender is likely to be accepted. "A proposal is revoked by the lapse of time prescribed in such a proposal for its acceptance".[84] In other words, the proposals received may cease to exist after their validity period is over.

What happens if a proposal is accepted beyond the validity period needs to be considered. In such a case, it will be open to the tenderer to ignore acceptance, or inform it was too late and not be bound by it. If, on the other hand, the tenderer conducts himself in such a manner say by fulfilling the condition of the tendering security deposit, etc., it might amount to a waiver of the revocation brought about by the law or an implied consent to extend the validity of his tender.

In certain cases, if the notice inviting tenders is silent in this respect, it is advisable that the tenderer should prescribe the time limit within which he wants the acceptance to be communicated. If he fails to do so, it is not that he agrees to keep it open for an indefinite period. The provisions of the law are clear and read: "a proposal is revoked … if no time is prescribed, by the lapse of a reasonable time, without communication of acceptance".[85]

The difficulty is the answer to the question as to what is a reasonable time. This is almost always a question of fact to be determined in each case on its merits. However, broadly it can be suggested that the nature of work and the time limit within which it is to be completed may be particularly relevant. In a construction contract, for example, an offer to complete the work in eight months or less, before the rainy season, would not be likely to be held open for acceptance if the rainy season is likely to intervene. Other relevant factors may be the presence or absence of provision for escalation in cost/rates, size of the plant, labour force, and site organization that may have to be organized and deployed.

---

[83] See paragraphs 1.8 and 12.20 for further discussion.
[84] Section 6(2) of the Indian Contract Act.
[85] Ibid.

EXAMPLES FROM CASE LAW

1.   In a tender for cotton bales for export, the exporters applied for allotment of cotton bales against orders from foreign purchasers for shipments to be made between April to August 1989. The applications were duly accompanied by bank guarantees of the requisite sums. The tenders were opened on 3rd April 1989. However, tenderers were informed that the fate of their tenders would be communicated within a reasonable time. In spite of persistent follow up the decision was not received. On 13th April 1989 the petitioner tenderer was informed by the foreign buyer that if by 14th April 1989 confirmation was not received the contract would be cancelled. On the same day the petitioner tenderer brought this fact to the notice of the Authority by hand delivered letter. On 18th April 1989 the foreign purchaser cancelled the contract. On 19th April 1989, the petitioner tenderer revoked the bid and asked for the return of bank guarantees and relevant documents. A reminder was sent on 2nd May 1989. On 12th May 1989, the Authority issued letters of allotment without making any reference to earlier letters. In response, the petitioner raised the fact that the acceptance was subsequent to revocation of tender and requested the return of bank guarantees. The Authority, on the other hand, threatened to encash the said bank guarantees. That is why the petitioner filed a writ. It was held that after the tenderers withdrew the tenders, there was no offer left for acceptance that could give rise to a concluded contract. The bank guarantees could not be invoked. The action of the Authority was quashed and set aside with the direction to return the bank guarantees.[86]

2.   The Forest Manual provided that the Tendu trees should be pollarded and coppiced during the period 15th February to 15th March. A bid to carry out the work, submitted in Tendu Patta auction by the petitioner, was highest and he had paid earnest money on 5th February 1992. The tender was neither accepted nor rejected and when the period was about to expire, the petitioner sent a telegram requesting an immediate order of acceptance or else he would not be liable to accept the offer. On 9th March 1992 the petitioner wrote a letter to the effect that if he was not allowed to culture the leaves at his cost, his bid should be treated as withdrawn and earnest money refunded. The petitioner was informed that since the bid was not accepted the permission to culture could not be granted. In a letter dated 14th March 1992, the petitioner withdrew his offer and sought a refund of his earnest money with interest. It was only on 3rd April 1992, without taking cognizance of the earlier letters, that the respondent informed the petitioner that his tender was accepted and he should pay an additional security amount. The petitioner received this letter on 7th April 1992 and he did not reply to it. A subsequent letter, dated 23rd April 1992, informed the petitioner that if he was not prepared to carry out the work, it would be re-auctioned. The petitioner did not reply to this letter and, instead, filed a writ petition requesting a refund of earnest money, which was rejected on 5th February 1995 on the ground that he had an alternative remedy to file a civil suit. The division bench rejected an appeal. The respondent issued notice on 29th September 1995 seeking payment of the damages for breach of contract, failing which the amount would be recovered as land revenue under the Land Revenue Act. The petitioner filed a writ, which was allowed, and the respondent's notice was quashed and set aside. It was held that the respondent could act in accordance with the law by filing a civil suit provided it was within the period of limitation.[87]

3.   In an auction to collect tolls over a bridge, the petitioner was the second highest bidder. The authorities accepted the highest bid whereupon the petitioner applied on 20th August 1983 for refund of earnest money. On 5th September 1983, he submitted a reminder but without any

---

[86] M/s Sekhsaria Exports v. Union of India, AIR 2004 Bom. 35.
[87] Satya Narayan v. State of Rajasthan, AIR 2000 Raj. 302.

result. A writ petition was filed seeking direction to refund earnest money. In reply, the respondent contended that after the highest bidder declined to accept the contract, the petitioner's bid was accepted by letter dated 15th October 1983. The petitioner denied the receipt of any acceptance letter. It was held that the bid of the petitioner stood withdrawn by his letters demanding the refund of earnest money and the respondent was bound to refund the earnest money with 12% interest p.a.[88]

4. A tender was submitted on 7th May 1997 with earnest money. After tenders were opened the said tender was found to be the lowest. An error was pointed out by the tenderer which was sought to be corrected the next day. After the validity period of 60 days elapsed, the tenderer refused to extend the validity and sought refund of earnest money. About three months thereafter the tenderer was informed that his earnest money stood forfeited in accordance with Clause 4 of the terms of invitation. The said provision stipulated that the earnest money "shall be liable to forfeiture if the tenderer after submitting his tender resiles from or modifies his offer". The condition further provided for refund of earnest money to all unsuccessful bidders after the decision is taken to accept the tender of the successful bidder and he has paid a security deposit. It was held that the earnest money could not be forfeited because until the offer submitted was accepted there was no contract and, as such, no loss on account of breach of contract.[89] The second lowest tender was accepted after negotiations and resulted in no loss to the authorities. During the course of judgment, it was observed that forfeiture of the earnest money arose not out of contract but pre-contract understanding that the tenderer was allowed to submit the tender in consideration of his agreeing to forfeit his earnest money in case he modifies his offer before acceptance. It was also observed that should the action of the tenderer result in loss to the authority, earnest money could be forfeited.

5. An appeal filed before the Supreme Court against the decision of the Delhi High Court directing the appellant to refund a security deposit was heard by a three-judge bench. The brief facts were as follows: The appellant proposed to allot land to about 200 co-operative housing societies. When the proposal was first made the cost was fixed at Rs. 875/- to Rs. 950/- per sq. m. Interested societies were to deposit Rs. 5 lacs by way of earnest money and to apply for allotment. Once the interested societies accepted the offer, formal allotment was made by communication dated 25th January 1991. Before the possession of the land was delivered, the appellant, by communication dated 3rd November 1992, increased the price of the land to Rs. 1,650.65/- per sq. m. Some societies, aggrieved by the said enhancement, approached the High Court. The High Court upheld the enhancement. The Supreme Court disposed of the special leave petition against the said judgment by extending the time limit to pay the first instalment with interest up to 31st May 1993 from 31st April 1993 allowed by the High Court, with direction that no further extension of time would be granted. The respondent before the Supreme Court failed to pay the amount, the appellant forfeited 10% of the amount payable at Rs. 1,650.65/- per sq. m. by way of earnest money. The High Court directed to refund the entire earnest money without any deduction. The Supreme Court had, in the appeal, to decide the validity of the said direction. On the above facts, the court came to the conclusion that the respondent society had accepted the offer of revised price dated 3rd November 1992 and, as such, they were bound to pay the enhanced price. As they did not comply, the respondent made itself liable to forfeiture of earnest money. However, it was held that earnest money deposited was only Rs. 5 lacs and not 10% of the revised premium. The appellant was held entitled to refund of the amount in excess of Rs. 5

---

[88] Sardar Surjeet Singh v. State, AIR 1995 All. 146.
[89] Food Corporation of India v. Sujit Roy, AIR 2000 Gau. 61.

lacs and, if not paid within 4 weeks from the said date of judgment, with interest at 18% per annum from the date of judgment till the date of payment.[90]

6.   The Karnataka High Court, in a petition for winding up of a company, ordered the official liquidator to advertise for sale the property of the company. Accordingly, the highest bid was received and accepted but it was subject to acceptance by the High Court. The bidder had paid earnest money of Rs. 5 lacs and also 25% of the bid amount, the balance being payable after the High Court approved the bid acceptance. The High Court did not approve the bid and directed for re-advertisement. The company judge moved for refund of money deposited. The creditors resisted the application on the ground that the bidder was liable for damages due to adopting dilatory tactics in postponing the proceedings for confirmation of sale by the High Court. The company judge made an order directing the secured creditors to be made parties and to refund Rs. 50 lacs out of Rs. 59 lacs paid and Rs. 5 lacs earnest money. The said order was not appealed against. An application was filed before a division bench by the highest bidder seeking a refund of the entire amount. It was held that in the absence of confirmation of sale by the High Court, the only conclusion to be drawn was that the applicant is *prima facie* entitled to a refund of the entire amount. It was observed[91] that the right to claim liquidated damages would exclude the right to claim an unascertained sum of money, the respondent could not insist on retention of money in excess of Rs. 5 lacs earnest money. However, since the applicant had not challenged the order of the company judge, the refund of the entire amount could not be ordered.[92]

7.   The Andhra Pradesh High Court ordered a refund of earnest money with interest from the date the tender was withdrawn. The date of opening of the tenders was postponed because of orders of the High Court. A tenderer sought refund of earnest money before the tenders were opened on the postponed date. It was held that retention of earnest money after withdrawal of tender was unreasonable.[93]

1.13.6 Proposal Revoked by Failure to Fulfil Condition Precedent

Section 6(3) declares that a proposal is revoked by the failure of the acceptor to fulfil a condition precedent to acceptance. For example, if the tender is subject to a condition calling upon the acceptor to do one or the other thing requested before acceptance, failure to do so will result in revocation of the proposal.

1.13.7 Death or Insanity of Tenderer

The provision made in this respect in Section 6(4) of the Indian Contract Act is self-explanatory and reads: "A proposal is revoked by the death or insanity of the proposer, if the fact of his death or insanity comes to the knowledge of the acceptor before acceptance."

English law on the point is different, for in English law it is understood that "death of either party before acceptance causes an offer to lapse."[94] Further, supervening insanity of the offeror, in English law, does not operate as revocation.

---

[90] Delhi Development Authority v. Grihsthapana Co-op. Group Housing Socy. Ltd., AIR 1995 SC 1312.
[91] Relying on the Sir Chunilal V Mehta & Sons Ltd. v. Century Spinning and Manufacturing Co. Ltd., AIR 1962 SC 1314.
[92] Narendra Kumar Nakhat v. M/s Nandi Hasbi Textile Mills Ltd., AIR 1997 Kant. 185.
[93] Aditya Mass Communicated Pvt. Ltd. v. A.P.S.R.T.C., Hyderabad, AIR 1998 A.P. 125.
[94] In Re Hemp, Yarn and Cordage Company, Hindley's Case (1896) 2 Ch 121 C.A.

As regards death of the promisee, the provisions of the English law and the Indian law are identical. If a proposal is addressed to a man who dies without having accepted or refused it, his executors have no power to accept it.

A partnership firm or a company, either limited or unlimited, comes to an end by dissolution or by winding up. Thus, where a contracting firm is a partnership or a company, and the tender is submitted in the name of the firm or the company, the tender stands revoked if the firm or the company is dissolved and this is communicated to the owner before he has accepted the tender. The mere fact of death or insanity of tenderer by itself does not amount to revocation if it does not come to the knowledge of the acceptor before acceptance.

1.14 Consideration and Scrutiny of Tenders before Acceptance

Unless good and sufficient reasons are recorded, at least in the case of public works, all tenders of the estimated value of a specified sum or above, as a rule, are opened in the presence of authorized representatives of tenderers. Percentage and lump sum tenders are read out to the tenderers and in the case of item rate tenders, the total amount worked out by different tenderers is read out.

A comparative statement of all tenders received in response to the notice is prepared. Standard forms are available for that purpose in the case of Public Works Departments. The person authorized to open the tender should sign the statement. Care needs to be exercised in preparing and scrutinizing comparative statements to guard against arithmetical and other mistakes.

It is observed that most of the tenderers are in the habit of stipulating a number of conditions while submitting their tenders. These conditions are generally to be found in the letters forwarding their tenders. If any conditions are found in the body of the tender, the person concerned has to make a note of such a condition to remove any possible doubts that the same might have been introduced later. Similarly, for the same reasons, all corrections and over writings must also be initialled by the person opening the tenders.

It becomes essential to calculate the financial effect of such conditions with a view to "load" the tendered sum. Sometimes rebates are offered by the tenderers and may have to be taken note of. Rebates may be offered when the NITs prohibit stipulating of conditions by the tenderers. The tenderers draft the conditions so as not to make them appear mandatory but in the form of offer giving a special rebate. It is open to the accepting authority to reject or accept such conditions. It is permissible to accept conditions giving rebate for regular payment of running account bills and final bill.

The fact that finalized effect of conditions is taken into account to determine the lowest tender does not make it obligatory on the part of the accepting authority to accept all such conditions. In cases where all conditions are not acceptable, negotiations are held and the tenderer is asked to bring the conditions in an acceptable form. The CPWD Manual mentions that "in all cases the authority vested with the officer competent to accept a tender is restricted to carry out negotiations only with the lowest tenderer".

This directive, *prima facie*, looks designed to maintain the sanctity of open competition. If negotiations are to be held with the second and third or the lowest tenderers, the tenderers might not take care to be competitive in their offers in the first instance. However, there seems to be no bar to holding negotiations with all tenderers. This aspect is further discussed in Chapter 3.

All tenders should be properly scrutinized for omissions, errors, etc. If on checking it is found that if there is a difference in the rates tendered by the tenderer in words and figures or in the amount worked out by him, the CPWD Manual lays down the following rules:

(i) If the rates in figures and words do not match, the rates, which correspond to the amounts worked out by the contractor, shall be taken as correct. (If it is made clear in the NIT conditions that the rates quoted in words shall be taken as correct, the said condition has to be followed).

(ii) When the amount worked out does not correspond to the rate either in words or in figures, the rate quoted in words shall be taken as correct.

(iii) When the rate quoted in words and in figures is identical but the amount is not worked out correctly, the rate quoted shall be taken as correct and not the amount.

(iv) In the case of percentage rate tender, if the amount is erroneous the tendered percentage shall be taken as correct.

It is important that the comparative statement correctly incorporates the total as checked on individual tenders. In this connection, it may be noted that the CPWD Manual Rules are not to be read in a hyper technical and pedantic manner leading to absurdity.

EXAMPLES FROM CASE LAW

1. The rate quoted in an original bid against an item of structural steel was Rs. 2,700/- per MT. However, on "the photo-copy of bid", the rate quoted was Rs. 27,000/- per MT. As per the tender notification, the original tender and the copy of tender were to be in different covers and sealed. However, the tenderer sealed both the copies in one envelope. After the tenders were opened this tender was considered as non-responsive. However, in a petition filed in Delhi High Court, it was contended by the petitioner that the quoted rate in the original at Rs. 2,700/- per MT was correct and that the Tender Committee had accepted the said rate and recommended award of tender to the petitioner. However, the petitioner could not prove it. The petitioner desired that the amount column be corrected by multiplying the quantity by the allegedly correct rate of Rs. 2,700/- in which event the petitioner would be the lowest. The writ was rejected by holding that it cannot be said that the decision-making process was faulty or the decision was *mala fide*, arbitrary or was so perverse that no prudent man could take it.[95]

2. A tenderer submitted his item rate tender for construction of a telephone exchange building in which one of the items in the BOQ was supply and installation of syntax plastic water storage tanks for which the tenderer quoted his rate as Rs. 9,500/- both in words and in figures. The unit mentioned was "one Ltrs". In the amount column the sum worked out was Rs. 19,000/-. The quantity column mentioned two numbers of 2,000 litres. The concerned authority calculated the amount of two tanks at Rs. 3,80,00,000/- treating Rs. 9,500/- as rate per litre rather than rate per tank. The tender was treated as the highest and not considered. The lower tender was accepted and agreement executed. When the tenderer came to know the developments, he informed that his rate was per tank and not per litre. This was stamped as an "afterthought" and his tender was ignored. Aggrieved, the tenderer approached the Gujarat High Court in writ petition under Arts 14, 226, 229 of the Constitution. It was held:

> If the rules and guidelines are properly read and understood, it is difficult, rather impossible, to infer that there is a mandate to the officers concerned to interpret the tender documents in unreasonable and fanciful manner. On the contrary, the guidelines require intelligent scrutiny of tenders. ... the department has failed to take into consideration the substance of the rules and guidelines. ... Overall reading of the guidelines show that mechanical and too technical or pedantical approach is not permissible.[96]

[95] Ranjit Construction Co. Ltd. v. National Highways Authority of India, AIR 2004 Del. 64; 2004 (1) CTLJ 51 (Delhi) (DB).
[96] Vadi and Patwa v. Union of India, AIR 1993 Guj. 100.

While in respectful agreement with the above observations, it is pointed out that the unit mentioned "one litres" under unit column led to confusion, which was referred to and figured in the judgment. The quantity "two numbers of 2,000 litres" permitted one to take number as a unit. There was one tenderer who so understood the provision. The decision is based on these facts peculiar to the case.

1.15 Decision to Accept a Tender

The decision to accept a tender should be taken after thorough scrutiny of all tenders and after the comparative statement is prepared. In the case of private works where the terms of employment empower an architect/engineer to accept tender for and on behalf of the owner, he may do so. However, it is advisable that before communicating the acceptance, the owner is consulted or told of the probable choice so that should he have any preference it is known at the right time. In the case of public works, the rules of procedure empower certain officers to accept the tender and the lowest ranked officer who invited tenders should submit the results of the scrutiny with recommendations, if any, to the authority competent to accept the tender. In the case of municipalities or corporations the power is vested with the Standing Committee for acceptance of tenders and it is advisable that such sanction of the Standing Committee be obtained before accepting a tender.

A tender notice required that only a *bona fide* consumer of molasses could make an offer. An offer was submitted by a transport business in the name of another firm, a distillery, on the condition that the molasses would be purchased for the benefit of the distillery. The offer was accepted by the Cane Commissioner and not by the committee constituted for the purpose of acceptance. On these facts, it was held by the Supreme Court that there was no valid tender and valid acceptance and no concluded contract resulted.[97]

The practice of accepting the lowest tender needs to be reviewed in the light of experience gained in that respect. A study, if conducted, would probably confirm the opinion of the authors that generally the second lowest tender is very near to the reasonable and workable cost of actual execution of the work. The law and the rules can be suitably modified to this effect with a view to avoiding problems created by reckless tendering somehow to quote the lowest sum/rates. Of course rules do permit acceptance of a tender higher than the lowest, if proper reasons are available.

1.15.1 Reverse Auction

After sealed tenders are invited in the normal way, the successful bidder is decided by conducting a reverse auction. It is a process where the participants are required to bid down the price to be selected. Contrary to a normal auction, where the price bid is upward, in a reverse auction the price bid is downward. This process is generally conducted online, the venue for the auction being a website. The employer engages a service provider who provides auction engine, a software program developed for this purpose that encapsulates the entire auction environment, processing logic and information flows. The service provider provides access through user ID protected by password. The participating bidders have to give written consent accepting the general rules and procedure. The participant has to ensure that the password and user ID are not disclosed to an unauthorized person. Before the auction is conducted bidders will be told the start date, time and duration of reverse bidding. Start bid price is generally specified. The service provider generally retains the right to cancel or reschedule a reverse auction for reasons specified in the rules and regulations, such as inability of bidders to access the website due to infrastructure failure. The

---

[97] Kisan Sahakari Chini Mills Ltd. v. Vardan Linkers, 2008(1) CTLJ 347 (SC).

bidders have to participate from their own office using their own IP (Internet Protocol) addresses. If required, the help of the service provider's technicians can be called for during the auction process. The minimum and maximum bidding quantities are generally specified as a percentage of price bid. The bidder quoting the lowest price is generally selected for award of contract. The process appears to be a sophisticated and convenient way of conducting negotiations with all participant bidders. If the reverse auction fails for any reason, including no bidder participating to lower the price from that quoted, the sealed bid price covers are opened and the tenders processed in the normal way.

The system has many objectionable features. The advantages of inviting sealed bids, not allowing the bid price to be changed after opening the tenders, negotiating with only the lowest bidder, etc., are lost. The expectation that the owner gets the advantage of the lowest possible price may be illusory. A bidder badly in need of a contract to keep his idling resources engaged might quote a recklessly low price and would try to make good the loss by compromising on the quality or speed of the work. In the opinion of the authors, the system is not at all advisable to be adopted for construction of major projects, as it discourages healthy competition in tenders. It may successfully yield the desired result for procuring bought out items where the manufacturer's standard specifications can be rigorously adhered to and quality check ensured.

1.15.2 Acceptance When Only One Tender Is Received

When tenders are invited for public works and in response only one tender is received, can such a tender be considered and accepted or tenders should be re-invited? This question cannot be answered generally. The answer will depend upon the facts and circumstances of each case. If the tender notification had been given wide publicity in appropriate media allowing sufficient time for the intending tenderers to submit tenders, the authority may negotiate with the single tenderer and if the offer is found to be reasonable vis-à-vis the estimate prepared by the authority, one might not be able to find fault with such a decision. Similarly, there may be cases where for supply of equipment or execution of specialized work there might not be many firms likely to submit offers, in such cases also a single tender can be considered as valid and accepted after negotiations. However, if there is a possibility that many more persons are likely to be interested in the work and complain about inadequacy of publicity or on any other ground, the authorities are well advised to invite fresh offers.[98]

1.16 Acceptance of Tender

What Is Acceptance?

An acceptance is the final and unqualified expression of assent to the terms of an offer. Section 2 (b) of the Indian Contract Act defines acceptance as follows: "When the person to whom the proposal is made signifies his assent thereto, the proposal is said to be accepted". Sections 3, 7, 8 and 9 of the Indian Contract Act, which are relevant for the discussion are dealt with below.

1.16.1 Section 3 of the Act

> 3. *Communication, acceptance and revocation of proposals*
>   The communication of proposals, the acceptance of proposals and the revocation of proposals and acceptances, respectively, are deemed to be made by any act or omission of

---

[98] Also see paragraph 1.20 below.

the party proposing, accepting or revoking by which he intends to communicate such proposals, acceptance or revocation, or which has the effect of communicating it.

This section deals with communication of

(i) proposals

(ii) acceptance of proposals

(iii) revocation of proposals and

(iv) revocation of acceptance.

This section does not lay down any specific mode of acceptance but declares that such communications are deemed to be made by:

(a) any act, or

(b) omission of the party proposing, accepting or revoking:

(i) by which he intends to communicate such proposal, acceptance or revocation or

(ii) which has the effect of communicating it.

The former contemplates an intention to communicate and the latter the actual communication. However, the word "or" does not necessarily mean "and" but in appropriate cases can be considered to be "and". The law in this respect is stated by the Supreme Court of India in these words: "A contract being the result of an offer made by one party and acceptance of that very offer by the other, acceptance of the offer and intimation of the acceptance which the law regards as sufficient is necessary."[99]

In short, the acceptance has to be communicated.[100] If an offer is accepted using the means or modes of communiucation prescribed by the offeror, using those means or modes is sufficient for the purpose of communication of that acceptance as against the offeror (it may not be sufficient to convey acceptance of offer in other cases). The normal mode is to convey it by words, i.e. an express acceptance. If acceptance is conveyed by conduct, it is said to be tacit or implied.

A railway administrator issued a cheque toward settlement of claim with the condition that if the amount was not acceptable to the claimant/offeree, the cheque should be returned. The claimant, before encashment of the cheque, protested the condition. It was held the action of the claimant did not amount to acceptance of offer by conduct.[101] He had to protest by positive action before encashment.

EXAMPLES FROM CASE LAW

1. A partnership firm of Chartered Accountants wrote a letter suggesting the client for "Voluntary Disclosure Income Scheme" and stating therein that professional fees would be 1% of the total liabilities. The client accepted the advice telephonically and advised to proceed in the matter. It was held that there was a concluded contract between the parties when, after receipt of letter containing terms and conditions direction to proceed was given, the offer stood accepted by conduct.[102]

---

[99] Bhagwandas v. Giridharilal and Co., AIR 1966 SC 543 (547).

[100] Kalluram Keshvani v. State, AIR 1986 M.P. 204.

[101] Union of India v. M/s Jain Enterprises, AIR 2008 (NOC) 2266 (Gau.).

[102] S. Butail and Company v. H.P. State Forest Corpn., AIR 2002 H.P. 1.

2. An insurance premium received and held in a suspense account till the proposal was scrutinized does not amount to acceptance of the proposal and no concluded contract comes into existence. Death of the proposer before final acceptance does not entitle the beneficiary to any relief except to refund of the premium paid.[103]

3. If the offer and acceptance were completed earlier and the placing of the order was only a formality, the place of acceptance is the place of contract for the purpose of determining the court's jurisdiction.[104]

1.16.2 Implied Acceptance

Section 9: of the Act deals with implied acceptance and it reads as under:

> 9. *Promises, express and implied*
> In so far as the proposal or acceptance of any promise is made in words, the promise is said to be express. In so far as such proposal or acceptance is made otherwise than in words, the promise is said to be implied.

When acceptance is expressed by words either written or spoken, it is said to be express. Acceptance when made otherwise than in words is said to be implied. A common example of either proposal or acceptance being made otherwise than by words is by the conduct of a party, which conduct takes the place of written or spoken words, either in the proposal or its acceptance. In such a case an agreement can be inferred from conduct alone.

1.16.3 Conduct May Be Proof of Acceptance

It is a matter of common experience that the communication of intention may be effectually made in many ways other than by written or spoken or signalled words. The Indian Contract Act takes note of this fact. Section 8 provides for acceptance of a proposal by conduct as opposed to other modes of acceptance, such as verbal or written communication. It reads: "Performance of the conditions of a proposal or the acceptance of any consideration for a reciprocal promise which may be offered with a proposal is an acceptance of the proposal."

EXAMPLES FROM CASE LAW

1. The plaintiff required large quantities of cement for the construction of a plant. The defendant was a manufacturer of cement. The plaintiff negotiated with the defendant for supply of cement. Negotiations were done verbally. In a letter dated 24th October 1997 the defendant offered to supply 1,500 MT of cement of a particular quality at the rate of L 1,900/- per MT. The plaintiff issued a letter dated 24th October 1997 (Exhibit B) which the plaintiff claimed to be a counter-offer where the plaintiff wanted supply of 1,500 MT of the same quality of cement at the same price. It was specified that the price would remain firm during the pendency of the order. According to the plaintiff, it sent a cheque for Rs. 3,00,000/- along with the letter dated 24th October 1997 to the defendant. Such cheque was

[103] Life Insurance Corporation of India v. Brazinha D'souza, AIR 1995 Bom. 223.
[104] State of Orissa v. Pratibha Prakash Bhavan, AIR 1995 Ori.62.

encashed by the defendant. The plaintiff claimed that, with the encashment of such cheque, a concluded contract came into being on the terms contained in the letter of the plaintiff dated 24th October 1997. According to the plaintiff, the parties acted on the basis of the letter dated 24th October 1997 of the plaintiff. Held:

> In the instant case Exhibit "B" was a counter offer of the defendant. The plaintiff on receipt of such counter offer did not reject the same nor did he make any other offer. The defendant on the contrary accepted and acted on Exhibit "B". It could, therefore, be safely held that Exhibit "B" was the contract between the parties.[105]

2. Between A and B there was prior agreement relating to forward transactions that if the contract form is not returned unsigned with a letter it shall amount to acceptance of transactions noted therein. It was held that all the subsequent transactions must be read in the light of this agreement and, consequently, when "A" did not return the contract form, in the eye of the law he accepted the proposal or, in other words, entered into transactions noted in the contract form.[106]

3. An engineering company involved in the construction of a bank in Saudi Arabia needed cast steel nodes. Negotiations were held with a steel corporation and a letter of intent was sent informing the company's intention to enter into a subcontract with the corporation for supply and delivery of the steel castings. The letter stated:

> We understand that you are already in possession of a complete set of our node detail drawings and we request that you proceed immediately with the works pending the preparation and issuing to you of the official form of sub-contract.

The steel corporation commenced the work and within a week sought the sequence in which the deliveries were needed by the engineering company and continued to manufacture the nodes. No official form of subcontract was sent, and nor did the parties agree upon the price or delivery dates. Litigation resulted in which the steel corporation claimed the price of the nodes delivered under the contract and alternatively upon a *quantum meruit*. The engineering corporation admitted liability in part subject to a plea of set off substantially in excess of the cost claimed by way of a counter-claim on the ground that the steel corporation committed a breach of contract by delivering the nodes late and out of sequence. It was held that no contract had come into existence between the parties on the basis of the letter of intent and consequent performance by the other side, but the steel corporation was entitled to be paid upon a *quantum meruit*.[107]

1.16.4 When Acceptance Cannot Be Implied

In this connection the following guiding principles from the Allahabad High Court should be noted:[108]

(i) The seeking clarification of an offer cannot amount either to the acceptance of an offer or to the making of a counter-offer.

---

[105] S. Fertilisers Private Ltd. v. Associated Cement Companies Ltd., 2014 AIR CC 1376.
[106] Gaddarmal Hiralal v. M/s C. Agarwal & Co., AIR 1968, All 292 (294).
[107] British Steel Corporation v. Cleveland Bridge & Engineering Co. Ltd., (1984) 1 All ER 504.
[108] UPSE Board v. Goel Electric Stores, AIR 1977 ALL 494.

(ii) While seeking clarification of the offer and exercising the option to reject any modification of the conditions the offeree did so, his act does not amount to rejection of the original tender.

(iii) Offeree's giving opportunity to correct defects in the offer made does not amount to making a counter-offer, especially when the communication ended with the words "failing which your tender will not be considered".

(iv) Where only a bank guarantee was given and no money was tendered or deposited along with the tender, the non-refund of the deposit does not amount to implied acceptance of the tender.

(v) Where tenders had been invited on behalf of a corporate body any one tender could be accepted by a definite statement on that behalf.

1.16.5 When Acceptance Can Be Implied

The following examples will help clarify the above situation.

EXAMPLES FROM CASE LAW

1. A tenderer, having purchased tender documents for a project, was approached by another company for joint participation. The tenderer sent a draft agreement for signature by the second company containing an arbitration clause. The said second company returned the agreement duly signed but with material variations converting the joint liability of the parties to unilateral liability of the tendering company. The tenderer did not communicate with the other company but unilaterally submitted the tender and subsequently withdrew the same. The second company claimed damages and invoked the arbitration clause. The tenderer filed an application under Section 33 in Delhi High Court. The High Court held the agreement had come into force the moment it was signed and sent and the tenderer, though he did not sign it, acted on it, which amounted to indirect acceptance. A three-judge bench of the Supreme Court held that the material alterations in the contract had a substantial effect on whether one could infer a concluded contract in these circumstances.. The joint liability of the parties was made unilateral liability of the parties. A clause containing a material part of the terms for performance of the contract was deleted. Thereby there was no consensus *ad idem* on the material terms of the contract. It was further held that even if proposal and counter-proposal were assumed to have concluded into an agreement, it was a contingent contract and by operation of Section 32 of the Contract Act, the counter-proposal could not be enforced.[109]

2. Where a part of the offer was disputed at the negotiation stage and the original offeree communicated that fact to the offerer showing that he understood the offer in a particular sense, it was held by the Supreme Court that this communication will probably amount to a counter-offer, in which case it may be that mere silence of the original offerer will constitute his acceptance.[110]

---

[109] U.P. Rajkia Nirman Nigam Ltd. v. Indure Pvt. Ltd., AIR 1996 SC 1373. Section 32 of the Contract Act deals with contingent contracts and declares that the contingent contracts cannot be enforced unless and until the contingent event has happened.

[110] Ramji Dayawala & Sons Pvt. Ltd. v. Invest Import AIR 1981 SC 2085 (2091); See *Halsbury's Laws of England*, 4th Edition, Vol. 9, Para. 251).

1.16.6 Acceptance to Be in Reasonable Time

As already stated, an offer or tender, unless withdrawn, remains in force until it is accepted or lapses by effluxion of time, which in the absence of an express provision will be a reasonable time. In deciding what a reasonable time for this purpose is, regard will be given to the nature of the work and the period in which the contract is to be completed. A tender submitted in response to an invitation which includes a condition that the tender will remain open for acceptance for a period of 90 days, cannot be accepted after that time. Unless the tenderer extends the validity period of the tender, a later acceptance will be clearly ineffective. Where a tender could not be accepted within the specified time because of writ petitions to which the tenderer was not a party, a writ of *mandamus* directing refund of earnest money was issued.[111]

1.16.7 Acceptance Must Be Absolute

Section 7 of the Act reads:

> 7 *Acceptance must be absolute*
> In order to convert the proposal into a promise, the acceptance must: (i) be absolute and unqualified and (ii) be expressed in some usual and reasonable manner, unless the proposal prescribes the manner in which it is to be accepted. If the acceptance is not made in such manner, the proposer may within a reasonable time after the acceptance is communicated to him, insist that his proposal shall be accepted in the prescribed manner, and not otherwise, but if he fails to do so, he accepts the acceptance.

A proposal must be accepted or rejected in entirety. This is what Section 7 lays down. The acceptance must comply within the terms of the proposal. Where the tenderer agreed to reduce the rates in the original tender and the authority accepted part of his tender, the Patna High Court held it was not a valid acceptance under Section 4 read with Section 7 and earnest money could not be forfeited.[112]

It is now a well settled principle of law that the acceptance of an offer must be in terms of the offer and a conditional acceptance will amount to a counter-offer. In Mayawanti v. Kaushalya Devi, the Supreme Court of India held as follows: "It is settled law that if a contract is to be made, the intention of the offeree to accept the offer must be expressed without leaving room for doubt as to the fact of acceptance or to the coincidence of the terms of acceptance with those of the offer. The rule is that the acceptance must be absolute, and must correspond with the terms of the offer."[113]

The offer or proposal had to be accepted in its entirety with the condition or not at all and, if the offer was not accepted in its entirety, then it would be a deemed refusal and earnest money cannot be forfeited.[114]

If the acceptance is with certain conditions or reservations, it amounts to a counter-proposal, which in turn must be accepted by the first party to convert it into a binding contract.[115]

Section 8, acceptance by conduct, must be read with Section 7. In order that acceptance of the proposal be inferred in terms with Section 8, the acceptance of the consideration must be unconditional.[116]

---

[111] Shyam Biri Works Pvt. Ltd., M/s. v. U.P. Forest Corpn., (Allahabad) (DB): 1990 AIR (Allahabad) 205: 1990(3) UPLBEC 1643.

[112] Technocom v. Railway Board, AIR 2009 Pat.15.

[113] In Mayawanti v. Kaushalya Devi, (1990) 3 SCC 1.

[114] D. S. Constructions Ltd., M/s. v. Rites Ltd., AIR 2006 Del. 98.

[115] Vishwa Industrial Company Pvt. Ltd. v. Mahanadi Coalfields Ltd., AIR 2007 Ori. 71.

[116] See paragraph 1.16.3 above for Section 8.

EXAMPLES FROM CASE LAW

1. The appellant claimed the value of consignments not delivered by the railway. The railway admitted part of the claim and enclosed two cheques. It was clarified:

> In case the above offer is not acceptable to you, the Cheque should be returned forthwith to this office: failing which it will be deemed that you have accepted the offer in full and final satisfaction of your claim. The retention of this cheque and/or encashment thereof will automatically amount to acceptance in full and final satisfaction of your above claim without reason and you will be estopped from claiming any further relief on the subject.

On receipt of two letters along with the two cheques, the appellant wrote to the Railways two identical letters of 20th August 1993 stating that the claims were placed under PROTEST and could not be accepted and that the balance amount should be remitted within 15 days. The Supreme Court held:

> In the absence of any pleading or evidence to establish that the encashment of the cheques was subsequent to the protest letters by the appellant, it is not possible to hold that by encashing the cheques the appellant had not adopted the mode of acceptance prescribed in the letters of the Railways dated 7th April 1993. In the absence of such evidence it must be held that by encashing the cheques received from the Railways, the appellant accepted the offer by adopting the mode of acceptance prescribed in the offer of the Railways.[117]

The case was decided on the basis of evidence or lack of it, it is submitted.

2. A Development Authority advertisement invited applications for allotment of flats to be constructed by it. A brochure was issued giving the terms and conditions that would govern the applications. The brochure stated the estimated cost and further clearly mentioned that the final costing would be done later and the right to amend the cost was reserved. After the lots were drawn the allotees were informed the estimated cost. The allotment letter clearly stipulated "the final cost would be intimated to you on the basis of actual costing after the completion of the Scheme, which would be payable by you." The Scheme originally was based on four-storey buildings. Only two-storey buildings were built and, as a result, the entire cost of the flats was distributed among fewer occupiers resulting in an increase in the cost. The applicants accepted the enhancement by filing affidavits. After taking possession the applicants protested against the final cost. When recovery proceedings were commenced, a petition was filed before the High Court challenging enhancement in the final cost. The High Court stayed the recovery proceedings and subsequently by final order allowed the writ petition. Aggrieved, the Authority filed an appeal before the Supreme Court. The Supreme Court held on the basis of the terms and conditions to which the applications were subject, the contents of the allotment letter and acceptance of enhanced price by affidavits that the Authority was entitled to collect the enhanced price. However, the rate of interest was reduced from 18% to 9% p.a.[118]

3. A tender was submitted with a bank guarantee by way of earnest money. At the request of the Oil and Natural Gas Commission (ONGC), the tenderer extended the validity up to 15th September 1999 as opposed to 15th November 1999 sought by ONGC. On 6th September 1999,

---

[117] Bhagwati Prasad Pawan Kumar M/s v. Union of India, AIR 2006 SC 2331=2006AIR SCW 3101. See Example (7) below.
[118] Bareilly Development Authority v. Vrinda Gujarati, AIR 2004 SC 1749.

a further extension up to 31st October 1999 was sought to be confirmed on or before 9th September 1999. On 15th September 1999, the tenderer expressed inability to extend the period of validity of its tender. On the same day ONGC wrote a letter, alleged to be a letter of acceptance, which concluded by stating, "ONGC has considered your above proposal for award of contract and the Notification of Award is likely to be issued on or before 30th September 1999." It was held:

> It is well settled that the offer and acceptance must be based or founded on three components: Certainty, Commitment and Communication. ... if any one of the three components is lacking either in the offer or the acceptance, there cannot be a valid contract. One of the important components that is lacking in this case is certainty.

It was further held that the offer was not valid beyond 15th September 1999. The letter of acceptance cannot be treated as acceptance of offer and consequentially the contract never came into existence. No excuse arose to invoke the bank guarantee and the bank guarantee ceased to operate.[119]

4. The Union of India invited tenders for the construction of a bridge. A construction company submitted its tender for a sum of Rs. 48,73,800/- according to its own design accompanying the tender. Thereafter there were prolonged negotiations for about 17 months. Finally, a letter of acceptance was communicated to the firm reading:

> Your tender for the work ... has been accepted on behalf of the President of India for Rs. ... made up as under ... You are requested to submit your design for the super-structure as envisaged in the scope of your tender subject to the conditions enumerated below ... The award is subject to the following conditions.

The letter mentioned 13 conditions and ended as follows:

> You are required to attend this office to complete the formal agreement ... being prepared on the basis of the tender conditions read with the conditions embodied in this letter of acceptance, and in case of discrepancy between tender conditions and conditions of acceptance, the latter will prevail.

It was held by the Delhi High Court that, "this letter called letter of acceptance, is only a counter-offer and it cannot reasonably be construed to be acceptance of an offer."[120] Since this counter-offer was not absolutely and unequivocally accepted by the construction company, there was no concluded contract between the parties. It was observed:

> As a matter of law, when there is variance between the offer and acceptance even in respect of any material term, acceptance cannot be said to be absolute and unqualified and the same will not result in the formation of a legal contract.[121]

---

[119] Kilburn Engineering Ltd. v. Oil and Natural Gas Corpn. Ltd., AIR 2000 Bom. 405.

[120] Union of India v. U. S. Duggal & Co., AIR 1972 Delhi 110; Rawatsons Engineers (P) Ltd. v. Union of India & Ors., 2008(NOC) 2009 (PAT.): National Coop. Consumer's Federation of India Ltd. v. Union of India, (Delhi) 2008 (Sup2) ArbiLR 521: 2008(3) R.A.J. 69: 2008(51) R.C.R.(Civil) 504.

[121] Union of India v. U. S. Duggal & Co., AIR 1972 Delhi 110; Rawatsons Engineers (P) Ltd. v. Union of India & Ors., 2008(NOC) 2009 (PAT.): National Coop. Consumer's Federation of India Ltd. v. Union of India, (Delhi) 2008 (Sup 2) ArbiLR 521: 2008(3) R.A.J. 69: 2008(51) R.C.R.(Civil) 504.

5.   In one case, the letter of acceptance stated that the offer was accepted "subject to your depositing 10 per cent as security", but in the last paragraph it also stated that "the contract was concluded by the acceptance and a formal acceptance of tender will follow immediately on receipt of Treasury Receipt." It was held that,

> reading the letter as a whole it amounted to an absolute and unqualified acceptance of the offer or tender and was not intended to make a substantial variation in contract making the deposit of security as a condition precedent instead of a condition subsequent.[122]

6.   The AP Government directed a work order to be issued after the signing of a Memorandum of Understanding (MOU) and the finalization of terms and conditions of the components. No MOU was entered into, nor were terms and conditions finalized, and no work order was issued. It was held that the principle of promissory estoppel had no application and that the state could not be compelled to enter into contract with the tenderer.[123]

7.   The plaintiff sued the Union of India claiming costs of goods lost in transit by the Railway Administration. The Railway Administration sent the plaintiff a cheque stating that it was in full and final settlement of the plaintiff's claim. The plaintiff alleged that the Railway Administration was in the habit of sending cheques in this manner and made it clear that the cheque would be accepted only as a part payment. On these facts, it was held that the plaintiff's action of accepting the cheque for a smaller sum could not preclude the plaintiff from recovering the balance by filing a suit.[124] The principle is that a person making a proposal cannot impose upon a party to whom it is addressed an obligation to refuse it under the penalty of imputed assent, or to attach to his silence the legal result that he must be deemed to have accepted it.[125]

8.   Where, in a case of allotment of an industrial plot, a notice was given to make payment within seven days, it was held that the period of seven days for payment could not begin to run earlier than the date of receipt of communication by addressee. It is wrong to cancel allotment by calculating seven days from the date of dispatch of notice.[126]

1.17 Correspondence between Acceptance and Offer

Negotiations for a contract can be long and complicated; and in such cases it is a difficult question to answer when (if at all) an offer has been made and accepted. A court then has to look at the whole course of negotiations and determine at what point (if any) the parties reached the agreement. The parties may continue their negotiation after this point has been reached, but this fact will not affect the existence of contract between them unless it can be construed as an agreement to rescind the contract.[127] An agent who contracts in his own name may become personally bound by the contract.[128]

---

[122] Jawahar Lal Bhawan v. Union of India, All. L.J. 411; AIR 1962 SC 378(385).

[123] M/s Lotus Constructions v. Govt. of Andhra Pradesh, AIR 1997 A.P. 200.

[124] Union of India v. M/s Babulal, AIR 1968 Bom. 294.

[125] Haji Mahomed v. E. Spineer, (1900) ILR 24 Bom. 510.

[126] RIICO Ltd. v. Goyal Handicrafts Pvt. Ltd., (Rajasthan)(Jaipur Bench)(DB) 2014(1) W.L.C. 44: 2013(24) R.C.R. (Civil) 14.

[127] Perry v. Suffields Ltd., (1916) 2 Ch 187 (CA). Also see, Global Asset Capital Inc. v. Aabar Block Sarl, [2017] EWCA Civ. 37; [2017] 4 WLR 163.

[128] Davis v. Sweet, (1962) 2 WLR 525.

EXAMPLES FROM CASE LAW

1.   The Union of India invited tenders for the supply of stone ballast. The plaintiff submitted his tender on 11th June 1966 which was to remain open for acceptance for 90 days, i.e. until 9th October 1966. The respondent's Divisional Superintendent wrote to the plaintiff on 26th September 1966 to ask him to provide basic labour amenities, a qualified blaster and a qualified manager as required under the Mines Act. If the plaintiff was agreeable to this, he was called for negotiations. The plaintiff informed the respondents that he was not willing to agree to the new conditions introduced in the letter of 26th September 1966. In spite of this refusal by the plaintiff, the Divisional Superintendent wrote to him on 7th October 1966 that his tender had been accepted and again requested the plaintiff to appoint a quarry manager, a qualified blaster, and to provide the basic amenities. The plaintiff insisted that the new conditions were not acceptable to him and he asked for refund of his earnest money and security deposit. Before the Allahabad High Court, the respondents alleged that no new condition was introduced while accepting the tender and that the plaintiff was bound by his original tender which was accepted by the respondent. The plaintiff contended that the Mines Act did not apply to the case of the plaintiff. The High Court observed that the conditions of the original tender did not provide for the appointment of any qualified manager or blaster or other supervisory staff before starting the work in case the plaintiff was able to manage his affairs in such a manner that the provisions of the Mines Act would not apply to his case. It was, therefore, held that no concluded contract had taken place between the parties. The security deposit and earnest money of the plaintiff was liable to be refunded.[129]

2.   Where the invitation for tenders provided that the tenders unaccompanied by the security deposit were liable to be rejected summarily and yet the defendant's tender without the security deposit was accepted by the plaintiff, the Kerala High Court held that there was a concluded contract between the parties. The condition regarding security deposit was held to give the plaintiff an option to waive the security deposit.[130]

3.   The Union of India invited tenders for supply of aluminium conductor steel reinforced cables. The appellants submitted their tender subject to the condition that they would not furnish any security deposit. This offer was kept open for acceptance up to 15th August 1979. The appellants were informed that their tender was accepted on the terms and conditions specified in the schedule to the tender form, Clause 8 of which provided for giving of security deposit. They also asked the appellants to keep their offer open until 15th September 1979. The appellants sent the following telegram in reply thereto: "We accept your advance order ... dt 13th August 1979. Further in reference to your letter ... we extend the validity of our offer up to 15th September 1979 from 15th August 1979." The Union of India contended that this telegram constituted an absolute acceptance of the counter-offer. This contention did not find favour with the Supreme Court which held that if the appellants intended to accept the counter-offer unconditionally there was no reason for the appellants to extend the period for acceptance of their original offer, which clearly included the term that no security deposit would be furnished. It was held that there was no concluded contract between the parties.[131]

4.   An authority issued a public notice, inviting tenders from intending bidders for the award of the right to collect tolls on a bridge. The auction was postponed from time to time till it was

---

[129] Chhotey Lal v. Union of India, AIR 1987 All. 329.
[130] M/s Bismi Abdullah & Sons v. Regional Manager, F.C.I., Traivandrum, AIR 1987 Ker. 56.
[131] M/s Zodiac Electrical Pvt. Ltd. v. Union of India, AIR 1986 SC 1918.

finally held on 14th November 1990. In the first instance, seven persons had obtained tenders but only two filed tenders in time. At the auction held on 14th November 1990 only one tenderer participated and offered a bid of Rs. 75 lacs per year. The second tenderer aired a grievance that the postponement was done without due publicity and offered a bid of Rs. 80 lacs. Thereafter the authority approached the tenderer who had originally offered Rs. 75 lacs and he in due course made a final offer of Rs. 80.27 lacs, which was accepted. The third tenderer, who did not participate in the auction, came up before the High Court with the grievance that he was virtually prevented from participation due to a strike in government offices preventing him from fulfilling the precondition of paying a security deposit. A cash security of Rs. 7 lacs was offered in seeking permission to participate. This request was turned down. Thereafter, on the same day, an application was given by the third tenderer making a bid of Rs. 86 lacs per year. The High Court dismissed the writ petition. The Supreme Court ordered re-auction on the terms given in the judgment.[132]

1.17.1 Letter of Intent (LOI)

A letter of intent (LOI) at best can be said to be no more than the expression in writing of a party's present intention to enter into a contract at a future date. Save in exceptional circumstances it can have no binding effect. LOI is not intended to bind either party ultimately to enter into any contract.[133]

It is no doubt true that an LOI may be construed as a letter of acceptance if such intention is evident from its terms. It is not uncommon in contracts involving detailed procedure, in order to save time, to issue an LOI communicating the acceptance of the offer and asking the contractor to start the work with a stipulation that the detailed contract would be drawn up later. If such a letter is issued to the contractor, though it may be termed as an LOI, it may amount to acceptance of the offer, resulting in a concluded contract between the parties. But the question whether the LOI is merely an expression of an intention to place an order in future or whether it is a final acceptance of the offer thereby leading to a contract, is a matter that has to be decided with reference to the terms of the letter. Where parties to a transaction exchanged LOI, the terms of such letters may, of course, have negative contractual intention; but, on the other hand, where the language does not show negative contractual intention, it is open to the courts to hold the parties are bound by the document; and the courts will, in particular, be inclined to do so where the parties have acted on the document for a long period of time or have expended considerable sums of money in reliance on it.[134] Be that as it may, the answer to the question whether in a case any contract has come into existence must depend on a true construction of the relevant communications which have passed between the parties and the effect of their actions pursuant to those communications.

EXAMPLES FROM CASE LAW

1.   A tender invitation stipulated that acceptance of bid could be intimated to the successful bidder through an award, and the award would conclude the contract. It further stipulated that the contractor would be required to execute a formal agreement within the

---

[132] M/s Rajshita v. State of U.P., AIR 1992 SC 1600.

[133] Dresser Rand S. A. v. M/s. Bindal Agro Chem Ltd., AIR 2006 SC 871= 2006 AIR SCW 277.

[134] *Chitty on Contracts*, paragraph 2.115 in Volume 1, 28th edition.

time specified in the award. The managing director defined the award to mean the acceptance. It further stated that the effective date of the agreement would be the date of issue of an LOI. On these facts, the Delhi High Court held that a concluded contract was reached between the parties on issue of the LOI, though a formal contract was not signed.[135]

2. In a case, a letter of provisional acceptance was followed by another letter asking to submit a performance bank guarantee within 15 days. The tenderer failed to submit the bank guarantee. The acceptance letters were revoked. Aggrieved, the tenderer sought reference to arbitration on the ground that expenses were incurred but performance was prohibited. It was held that the deposit of the performance guarantee was a condition precedent before the final letter of acceptance and award of work to the bidder. Since the bidder failed to submit a performance guarantee, no concluded contract came into existence.[136]

3. If offer and acceptance were completed earlier and placing of order was only a formality, the agreement became complete by acceptance.[137]

R invited T to tender for the design and construction of a new factory building. The tender was finalized in May 1969. On 2nd June 1969, R informed T that R wished T to carry out the project so as to complete it in 1972. T asked for an early LOI to cover for the work T would then be undertaking. On 17th June 1969 R sent the letter to the effect: "As agreed ... on 2nd June 1969, it is (our) intention to award a contract to (you)." The letter ended thus: "All this subject to obtaining agreement on the land and leases ... And the site investigation being undertaken by ... The whole to be subject to agreement on an acceptable contract." On these facts, it was held that R was liable for the work carried out. It was observed "Since the parties discussed the matter, the question is whether their discussion resulted in some, and if so what, contractual obligation." It was held that T had offered to carry out the preparation work subject to "R" assuming liability and indicating that T would regard an LOI as an acceptance of T's offer. The LOI did not negate this presumption. R's liability was held to be based on the resultant ancillary contract covering the interim costs.[138]

4. Where two alternative tenders were submitted, one on a fixed price basis, and the other on a "cost plus" basis, and the letter of acceptance was issued without specifying which tender was accepted, it was held that no concluded contract resulted and the work done was to be paid for on a *quantum meruit* basis.[139]

1.18 Mode of Communication of Acceptance

The law requires an absolute and unqualified acceptance, within a reasonable time and to be expressed in the usual reasonable manner or in the manner prescribed by the proposal to be communicated to the offerer. For example, if the conditions under which a tender is submitted

---

[135] M/s Progressive Constructions Ltd. v. Bharat Hydro Power Corpn. Ltd., AIR 1996, Delhi 92.

[136] M/s. Ganesh Shanker Pandey & Co. v. Union of India, AIR 2004 All. 26.

[137] State of Orissa v. Pratibha Prakash Bhavan, AIR 1995 Ori. 62.

[138] Turriff Construction Ltd. v. Regalia Knitting Mills Ltd., Q. B. (1971) 9 BLR 20. Also see, AC Controls Ltd. v. BBC, [2002] EWHC 3132 (TCC); 89 Con LR 52.

[139] Peter Lind & Co. Ltd. v. Mersey Docks & Harbour Board Q.B., (1972) 2 Lloyd's Rep. 234.

provide for a written acceptance, a verbal acceptance is not sufficient.[140] In such a case, the proposer may, within a reasonable time after acceptance is communicated to him, insist that his proposal shall be accepted in the prescribed manner, and not otherwise, but if he fails to do so, he accepts the acceptance.

In this connection, it is necessary to clarify that a proposer may prescribe a form or time of acceptance but he cannot prescribe a form or time of refusal so as to conclude a contract if the other party does not refuse in some particular way or within a given time.[141] Mere silence or inaction may not operate as acceptance. In the absence of a duty to speak, silence cannot amount to a representation on which a plea of estoppel can be founded. However, circumstances can arise where acceptance could more legitimately be presumed from silence. Previous dealings may be one such circumstance.[142]

1.18.1 Communication When Complete

When the contracting parties are face to face, we can easily say the acceptance is communicated the moment he acceptor says "Yes" to the offer. But if parties happen to be at a distance from each other, communication through a messenger, by post or telephone or telegram or telex can be considered to be in "usual and reasonable manner". When communication is made by any one of those modes, the question arises as to when the communication is complete. Section 4 of the Indian Contract Act answers this question in the following words:

> 4. *Communication when complete*
> The communication of a proposal is complete when it comes to the knowledge of the person to whom it is made.

Illustration (a) reads:

> A proposes by letter, to sell a house to B at a certain price. The communication of the proposal is complete when B receives the letter.

The section further provides:

> The communication of acceptance is complete, as against the proposer, when it is put in a course of transmission to him so as to be out of the power of the acceptor; as against the acceptor, when it comes to the knowledge of the proposer.

The section gives the following illustration:

> B accepts A's proposal by a letter sent by post. The communication of the acceptance is complete as against A, when the letter is posted; as against B, when the letter is received by A.

Provisions of Section 4 – Territorial Jurisdiction of Court

Section 4 of the Act, 1872 provides that communication of an acceptance is complete as against a proposer when it is put in the course of transmission to him so as to be out of the power of the acceptor and as against the acceptor, when it comes to the knowledge of the proposer.[143] For deciding

---

[140] Fripp v. Upton, (1884) 3 NZLR 237.

[141] *Anson's Law of Contract*, 24th Ed., p. 42.

[142] See Example (2) in paragraph 1.16.3 of this chapter.

[143] OK Play India Limited v. Pradeep Tayal, (Delhi): 2015(219) DLT 513: 2015(3) BC 9: 2015(6) AD (Delhi) 22: 2015(19) R.C.R.(Civil) 315. Also see: Shree Balaji Mining Pvt. Ltd. v. Extec Screens, (Orissa) 2015(151) AIC 628: 2015(1) ILR (Cuttack) 1035: 2015(76) R.C.R.(Civil) 383: 2015(120) CutLT 969: 2015 (sup 2) *Ori. Law Rev.* 512. Also see Greentose Pvt. Ltd. v. Gujarat Narmada Valley Fertilizers Co. Ltd., (Gujarat) 2015(3) GujLH 580.

territorial jurisdiction of the court to entertain a suit for damages the Karnataka High Court observed that jurisdiction lay with the court within whose territory the cause of action arose (wholly or in part). It was further held that a contract is concluded when an offer is accepted. It is the acceptance of the offer that gives rise to a cause of action and not merely the making of an offer.

EXAMPLE FROM CASE LAW

A tender for construction of building was called for by the defendant Society from its office situated at Soundatti in Belgaum District. It is also seen that the tender for accepting the contract was submitted within the territorial jurisdiction of the Soundatti Court. The contract was also entered between the parties and the same was executed within the jurisdiction of Soundatti. In that view of the matter, the finding of the trial court was unsustainable to the effect that several correspondences had taken place between plaintiff and defendant, some of them originating from the Bangalore branch office of the plaintiff, and therefore the plaintiff had cause of action to initiate the suit within the territorial jurisdiction of Bangalore. A decree passed by the Civil Court at Bangalore was held without jurisdiction and set aside.[144]

For all practical purposes it can be said that the contract is concluded at the time when, and in the place where, the letter or telegram is posted.[145] However, a different view has been expressed by the Cuttack High Court under the facts of the case stated thus:

The cause of action in a suit for damage for the breach of contract arises where the contract is made, where the contract is to be performed and where the contract is breached ... it is seen in the case in hand that neither the contract can be said to have been made at Rourkela nor it can be said to have been agreed to be performed there nor the breach to have occurred. For mere communication of the acceptance of the offer by making correspondence from a place, it cannot be said that the contract was made at that place. Contract is made at a place when acceptance is received and part of the cause of action for suit for damage for breach arises at that place. As per the provision of Section 4 of the Contract Act, the communication of a proposal is complete when it comes to the knowledge of the person to whom it is made. So by dispatch of letter of acceptance of offer from Rourkela, the contract cannot be said to have been made there. Also remittance of money from that place would not suffice the purpose of bringing in the jurisdiction of said court to entertain the suit claiming damage for breach of contract saying that cause of action has arisen there since the payment by that cannot be said to have been made to the defendant at Rourkela which is not the place where the money being remitted was payable. In the instant case there being late delivery the breach is said to have been committed and damage is claimed on account of that. Therefore, the trial court has rightly accepted the prayer of the respondent that it has no jurisdiction to try the suit and has accordingly followed the right path of returning the plaint to the appellant for being presented in the proper court. In view of above, this court finds no such illegality or infirmity with the order impugned in this appeal so as to be interfered with.[146]

When the offerer and offeree are not at one place they have to exchange the offer and acceptance through other means. When by agreement, course of conduct, or usage of trade, acceptance by post or telegram is authorized, the bargain in acceptance is put into a course of transmission by the offeree by

---

[144] Malaprabha Co-Op. Spinning Mills Ltd. v. Buildmet Pvt. Ltd., (Karnataka)(DB): 2012(5) KantLJ 445: 2012 AIR (Karnataka) 165: 2013(122) AIC 383: 2013 ILR (Karnataka) 94: 2012(13) R.C.R.(Civil).

[145] Parbati Choudhary v. Bhagwan Das Gupta, 2002(3) Arb. LR 159 (Gauhati) (DB). Acceptance in the case of public contracts is more fully dealt with in Chapter 3.

[146] Shree Balaji Mining Pvt. Ltd. v. Extec Screens (Orissa), 2015 (76) RCR Civil 383.

posting a letter or dispatching a telegram or telex. In such an eventuality, generally the contract would be deemed to have been entered into at the place where the offer was received and acceptance was posted.[147]

1.18.2 Acceptance by Letter

This is by far the most common mode of communication of acceptance. Once the letter of acceptance is sent on the address given by the offeror, acceptance is completed even though it was not delivered to the offeror.[148]

1.18.3 Acceptance by Telegram/Telex

This is also a mode of communication of acceptance of proposal. Acceptance by telegram on the address given in the tender amounts to acceptance having been conveyed to the tenderer.[149] In case of communication of acceptance by telegram, place of contract is the place from where the telegram started its journey.[150]

In this connection it may be noted that a telegram by itself is not an authentic document. It is like an unsigned anonymous document. "Unless a telegram is confirmed by a subsequent signed application, representation or an affidavit, the contents of the telegram have no authenticity at all and the same cannot be taken into consideration for assessing the value of the other authentic documents on the record."[151] The practice of sending a letter confirming the dispatch and the contents of the telegram is necessary. It is submitted that the same care may have to be taken to confirm a telex message.

Where negotiations took place between the parties after the tender was submitted and the tenderer received a fax of acceptance which stated that "regular purchase order follows", it was held that the contract stood concluded. Conditions as to joint signatures of tenderer and its collaborators or furnishing of bank guarantee, were all held to be subsequent conditions. Even no intimation of acceptance by the collaborator of the tenderer was held immaterial as the tenderer represented on their behalf.[152]

Where a telex message of the acceptance of a bid was sent at place "D" and a payment under a bank guarantee was also made at place "D", it was held that the contract could be said to have been completed at place "D". A clause in the agreement providing for the venue of arbitration at place "H" could not take away jurisdiction of the court at place "D".[153]

1.18.4 Exchange of Emails

When a contract is concluded by exchange of emails, the absence of a formal/signed contract is of no consequence. Existence of an arbitration agreement, in the absence of a signed agreement between parties, can be inferred from various documents signed by the

---

[147] Progressive Constructions Ltd., M/s. v. Bharat Hydro Power Corpn. Ltd., (Delhi) 1996 AIR (Delhi) 92: 1995(59) DLT 290: 1995(3) AD (Delhi) 325: 1995(2) DL 189: 1995(2) ArbiLR 233: 1996(1) ILR (Delhi) 23.

[148] J. K. Enterprises, M/s. v. State of M.P. (M.P.), 1997 AIR (M.P.) 68: 1997(2) M.P.L.J. 31: 1997(1) ArbiLR 683: 1997(3) CivilLJ 649.

[149] M/s Indian Hardware and Forgings v. Haryana State Electricity Board, (P. & H.) 2003(2) R.C.R.(Civil) 689: 2003(2) PLR 731: 2003(2) CivCC 609: 2004(2) R.A.J. 462.

[150] O.N.G.C. v. Modern Construction and Co., Mansa (Gujarat) (DB): 1998 AIR (Gujarat) 46: 1997(3) GLR 1855: 1998(1) CivilLJ 742: 1998(3) CurCC 544.

[151] Dist. Magistrate v. G.J. Jothisankar, AIR 1993 SC 2633 (2635).

[152] D. Wren International Ltd. v. Engineers India Ltd., (Calcutta): 1996 AIR (Calcutta) 424: 1997(1) ArbiLR 241.

[153] Trivent Oil Field Service Ltd. v. Oil and Natural Gas Commission, (Delhi): 2006 AIR (Delhi) 331: 2006(128) DLT 541: 2006(3) AD (Delhi) 19: 2006(87) DRJ 482: 2006(1) ArbiLR 360: 2006(2) PLR 46: 2006(4) BC 429: 2006(1) R.A.J. 412.

parties in form of exchange of emails, letters, telex, telegrams and other means of telecommunications.[154]

1.18.5 Acceptance in Telephonic Conversation

In the case of telephonic conversation, in a sense the parties are in the presence of each other: each party is able to hear the voice of the other. There is instantaneous communication of speech intimating offer and acceptance, rejection or counter-offer. The intervention of an electrical impulse which results in the instantaneous communication of messages from a distance does not alter the nature of the conversation, making it analogous to that of an offer and acceptance through post or by telegraph. The question assumes importance as to the place where the agreement took place. By majority view, a three-judge bench of the Supreme Court in Bhagwandas v. Giridharilal & Co. held that the contract was complete when the acceptance was communicated by telephone conversation. It was further held the contract was made at the place where the acceptance was received and a part of cause of action for suit for damages for breach of contract arose at that place. Analogy of contracts by post and telegram was held to be inapplicable.[155] The view expressed by His Lordship Mr. Justice Hidayatullah in his minority judgment in the above case is noteworthy, it is submitted. His Lordship held:

> In my opinion, the language of S.4 of the Indian Contract Act covers the case of communication over the telephone. Our Act does not provide separately for post, telegraph, telephone or wireless. Some of these were unknown in 1872 and no attempt has been made to modify the law. It may be presumed that the language has been considered adequate to cover cases of these new inventions. ... If the rule suggested is accepted it would put a very powerful defence in the hands of the proposer if his denial that he heard the speech could take away the implications of our law that acceptance is complete as soon as it is put in course of transmission to the proposer.

1.19 Revocation of Acceptance

Just as a tenderer is at liberty to revoke his tender before its acceptance, the accepting authority is also entitled to revoke communication of acceptance. Section 5 of the Indian Contract Act provides as under:

> 5. An acceptance may be revoked at any time before the communication of the acceptance is complete as against the acceptor, but not afterwards.

The relevant part of the illustration below the section reproduced partly reads:

> B accepts the proposal by a letter sent by post. B may revoke his acceptance at any time before or at the moment when the letter communicating it reaches A but not afterwards.

Revocation of acceptance obviously contemplates a faster mode than the one adopted for communicating acceptance, such as by telegram, telex, fax message, telephone, etc. If the revocation of acceptance is communicated before acceptance reaches, the tenderer will have no

---

[154] Trimex International FZE Ltd. Dubai v. Vedanta Aluminium Ltd. India, (SC): 2010(1) ArbiLR 286: 2010(1) JT 474: 2010(3) SCC 1: 2010(1) SCC (Civil) 570:: 2010 (Sup) AIR (SC) 455: 2010 AIR (SCW) 909.

[155] Bhagwandas v. Giridharilal & Co., AIR 1966 SC 543 (549). Also see S. Oil Mills v. R. Oil Mills, AIR 1952 Mysore III Kamisetti Subbiah v. Katha Venkataswamy (1903) ILR 27 Mad. 355.

right under the transaction except that the principles of natural justice may demand an opportunity of hearing to the person affected.[156] If, however, the communication of acceptance reaches the tenderer before its revocation, the contract stands concluded entitling the tenderer to claim compensation for the loss, if any, under the concluded contract if the contract is cancelled.

The law as regards the mode of communication etc. is similar to communication of revocation of offer as is made clear by the language of Sections 3 to 6 of the Act.

1.20 Rejection of Tender

When considering the merits or validity of a decision of the acceptance or rejection of tenders, a distinction needs to be drawn between private and public works. It is true that ordinarily a private individual would be guided by economic considerations of self-gain in any action taken by him. However, it is always open to him under the law to act contrary to his self-interest or to oblige another in entering into a contract or dealing with his property.

The government, on the other hand, is subject to restraints inherent in its position in a democratic society. Every action taken by the government, public agencies or authorities must be in the public interest and it would be liable to be tested for its validity on the touchstones of reasonableness and public interest. If it fails to satisfy either test, the action would be unconstitutional and invalid. In public works contracts, ordinarily the lowest bidder is awarded the contract although it is not an invariable rule in all cases.[157]

In taking the decision to reject the lowest tender or all tenders or to accept a tender higher than the lowest, there are certain considerations which are common to either private or public works and those which are applicable to public works alone. The considerations in respect of public works are dealt with in Chapter 3. The common considerations are dealt with below.

1.20.1 Rejection of All Tenders

All tenders may have to be rejected in some circumstances, such as the following:

1. No tender is received conforming to the conditions of the tender notice.

2. The minimum number of tenders to assure adequacy of competition not received.

3. The lowest tender received is higher than the budget provision thereby requiring a reconsideration of the project plans and estimates with a view to reducing the scope and extent, if possible, or to select alternative specification which could be cheaper than those originally provided to the tenderers.

4. A change in the policy decision or technical or other requirements of the owner, necessitating the abandonment of the proposed work.

5. There is a justifiable suspicion that all the tenderers have formed a cartel and that rates/cost tendered is elevated as a result of such a cartel.

1.20.2 Cartel Formation – An Unfair Trade Practice?

All tenders may have to be rejected if there exists a well-grounded suspicion that the tenderers have formed a "cartel". Cartels are deemed to be anti-competitive in most jurisdictions. The

---

[156] Ponaram Hazrika v. State of Assam and others, AIR 1994 NOC 74 Gau.

[157] M/s. BECIL v. Arraycom India Ltd., (SC) 2010(1) SCC 139: 2010(1) R.C.R.(Civil) 50: 2009(6) Recent Apex Judgments (R.A.J.) 556: 2009(163) DLT 764: 2010(1) PLR 398: 2009 (Sup) AIR (SC) 2878: 2010(3) All WC 2328: 2009 AIR (SCW) 6532: 2009(13) Scale 322: 2009(4) R.A.J. 657: 2010(78) ALR 238: 2010(29) VST 555.

European Union Damages Directive[158] defines cartels as "agreements and/or concerted practices between two or more competing undertakings which coordinate their behavior in order to influence the competitive structure of the market to their own advantage, by engaging in different anti-competitive practices, such as, fixing prices, allocating customers, sharing markets". In fact, the construction industry received special attention from the European Commission which imposed significant fines in the light of the substantial size of the market, the effect of the illicit agreements on consumers and the repeated anti-competitive conduct of a few companies. For example, in the early 2000s, a plasterboard cartel was the subject of a €478 million fine.

Since in a price fixing conspiracy, the conduct is illegal per se, further inquiry on the issues of intent or the anti-competitive effect is not required. The mere existence of a price fixing agreement establishes the illegal purpose, since the aim and result of every price fixing agreement, if effective, is the elimination of one form of competition.[159]

Cartels are prohibited in India under the Competition Act, 2002 as amended by the Competition (Amendment) Act, 2007. Section 2(c) of the Indian Competition Act defines a cartel to include, "an association of producers, sellers, distributors, traders or service providers who, by agreement amongst themselves, limit, control or attempt to control the production, distribution, sale or price of, or, trade in goods or provision of services". Even before the enactment of this Act, the Supreme Court of India had observed[160] that the objective of a cartel amounts to an unfair trade practice which is not in the public interest. However, the determination whether such an agreement unreasonably restrains the trade depends on the nature of the agreement and on the surrounding circumstances that give rise to an inference that the parties intended to restrain trade and monopolies.

The Indian cement industry is second to China in terms of production, according to the website of the Cement Manufacturers' Association (CMA). In 2002, Builders' Association of India filed a complaint of anti-competitive conduct against CMA and 11 cement manufacturers. It was alleged that the cement prices increased not as a result of high demand but as a result of cartelization by the manufacturers.

If it is found that all or some tenderers have formed a cartel and rigged their bids, all such tenders could be justifiably rejected.

1.20.3 Conditional Acceptance/Rejection of Tender – Effect

Once the offeree rejects an offer, he cannot subsequently accept it. A conditional acceptance of an offer amounts to a counter-offer. A counter-offer by the offeree does not amount to an acceptance of the offer. It also amounts to a rejection of the original offer which, therefore, cannot subsequently be accepted.

EXAMPLES FROM CASE LAW

1. A offered to sell a farm to B for $1,000. B said he would give $950. A refused and B then said he would give $1,000 and when A declined to adhere to his original offer, B tried to obtain specific performance of the alleged contract. It was held that an offer to buy at $950 in response to an offer to sell for $1,000 was a refusal followed by counter-offer and that no contract had come into existence.[161]

---

[158] Directive 2014/104/EU, Art. 2(14).

[159] United States v. Society of Ind. Gasoline Marketers, 624 F. 2d 461, 465 (4th Cir. 1979) Cert. Den.

[160] Union of India v. Hindustan Development Corporation, AIR 1994 SC 988 (1013).

[161] Hyde v. Wrench, (1840) 3 Bear 334.

2. The tender document in a case contained a clause reading "Do you agree to sole arbitration by an Officer in the Ministry of Law to be appointed in the manner specified"? The tenderer had positively stated NO to this condition. An acceptance letter was issued adding a clause of arbitration which acceptance was acted upon by the tenderer. The Delhi High Court held that even though Section 7 of the Contracts Act requires an acceptance to be absolute and unconditional, if there is a conditional acceptance or a counter-offer and it is acted upon without protest, a concluded contract results.[162]

1.20.4 Rejecting the Lowest Tender – Grounds

The lowest tender should be rejected if, while submitting it, the following contingencies occur:

1. Tender is at variance with the conditions of invitation such as alterations in designs, or specifications or time schedule.
2. Any of the pages of the blank tender issued are removed or replaced.
3. The tender is not properly signed.
4. Any erasures made in the tender.
5. The tenderer has not signed all corrections and additions or pasted slips.

If there was no prequalification of tenderers the lowest tender may be rejected on the ground of:

(i) Improper offer such as "so many Rupees below the lowest offer".
(ii) Inadequate finance.
(iii) Lack of experience in a particular type of work.
(iv) Unsatisfactory reputation.
(v) Inadequate staff and equipment.

However, adequate care and caution needs to be taken while rejecting the lowest tender, particularly in respect of public works, as the action is open to be questioned in a court.

EXAMPLES FROM CASE LAW

1. A Public Sector Undertaking invited tenders for the transportation of food grains, etc. Seven tenders were received. The tender accepting committee decided to call the second lowest tenderer for negotiations. The second lowest tenderer gave in writing that he was not ready to reduce the tendered rates. His tender was recommended for acceptance. The acceptance was communicated telegraphically. The lowest tenderer filed a writ petition in the High Court. The acceptance was revoked and the second lowest tenderer was called for renegotiations and was persuaded to accept lower rates than those quoted by the lowest tenderer. The lowest tenderer was not called for negotiations. The allegation of poor financial position of the lowest tenderer was rejected as no standard for fiscal fitness was

---

[162] Union of India v. Peeco Hydraulic Pvt. Ltd., AIR 2002 Delhi 367. 2002(3) Arb. LR 678 (Delhi).

laid down. It was held that the action of accepting the second lowest tender was arbitrary. The authorities were directed to invite fresh tenders.[163]

2.   A tender notice for printing, binding and supply of yellow page telephone directories contained a stipulation that tenderers should substantiate their experience with documentary proof by furnishing credentials in the field of having compiled, printed and supplied telephone directories to the large telephone systems with the capacity of more than 50,000 lines. A company, which had no experience but attempted to depend upon the experience of its shareholders, offered the highest royalty; its offer was rejected. In a writ petition, the Delhi High Court held that the approach of the authorities in arriving at a conclusion that the petitioner did not satisfy the eligibility condition was not arbitrary.[164]

1.20.5   Rejection of a Tender – The Doctrine of Legitimate Expectation

The doctrine of legitimate expectation is a rule of law that recognizes the expectation of a person to be lawful or legal. This concept and case law are discussed at length in Chapter 3.

1.21   Cost of Tendering

The cost of tendering is generally borne by the contractor. It gets added to his overheads, which he ultimately recovers from the works awarded to him. As such, there is no implication that the costs incurred by a tenderer in preparing his tender will be reimbursed by the owner. However, the contractor may be entitled to a reasonable payment if (a) he performs additional services at the request of the owner and/or (b) he performs substantial preparatory work over and above the normal tender preparation and no contract is ever placed. The following examples will clarify the point.

EXAMPLES FROM CASE LAW

1.   W tendered for the reconstruction of war damaged premises belonging to D who led W to believe that it would receive the contract. At D's request, W calculated the requirements of timber and steel and prepared further schedules and estimates. D used the information in negotiations with the War Damage Commission. Eventually, W was informed that D intended to employ another contractor to do the work. In fact, D sold the premises. W claimed damages for breach of contract, and, alternatively, remuneration on a *quantum meruit*. It was held that although no binding contract had been concluded between W and D, a promise should be implied that D would pay a reasonable sum to W in respect of the services rendered.[165]

2.   A tender for designing and building a factory for K, to replace one which had burnt to the ground, was submitted by M. The submission of tender was followed by a meeting on 18th December 1986, when M was the only tenderer invited for discussions. K's Chairman made it clear that no contract to rebuild the factory would be entered into until he had

[163] Pawan Kumar v. Food Corporation of India, AIR 1999 P. & H. 76.
[164] New Horizons Limited v. Union of India, Air 1994 Delhi 126.
[165] William Lacey (Hounslow) Ltd. v. Davis, [1957] 1 WLR 932.

obtained the insurance money to pay for the new building. However, nothing was said to the effect that if the insurance money had not been forthcoming all the preparatory work done by M up to that point would be at M's risk. K was well aware that M would have to start preparatory work before the contract was signed. The contract was agreed to be signed on 16th January 1987. No contract was, however, signed because the proceeds of the insurance money from the old factory were insufficient to cover the costs. On these facts it was held that there was an express request made by K to M to carry out a small quantity of design work and there was an implied request to carry out preparatory works in general. Both the requests gave rise to payment of a reasonable sum.[166]

3.   Where a successful tenderer for the fabrication of steel work entered into complicated negotiations with the owner and received an LOI to commence work on the construction of some items pending contract, the tenderer was entitled to claim compensation on a *quantum meruit* basis for the manufacture of those items when the negotiations were unsuccessful in leading to a binding contract.[167]

1.22 Misrepresentation and Fraud

If an owner or his agent has, by fraudulent representations as to the facts, induced a tenderer to submit a tender that is disadvantageous to the tenderer, the tenderer may revoke his tender on discovering the fact. He can do so even if his tender has been accepted and a contract entered into. Section 19 of the Indian Contract Act explicitly declares that: "[w]hen consent to an agreement is caused by coercion, fraud or misrepresentation, the agreement is a contract, *voidable* at the option of the party whose consent was so caused." Where the contractors were misled by the owner's assistant engineer into thinking that the site conditions were ordinary ones, but they were found to be very unusual, it was held that the contractors were entitled to terminate the contract and recover the sum paid by them as deposit.[168] The law in this respect is fully considered in Chapter 2.

1.23 Bribery and Secret Commission

It may be said as a general rule that an owner can recover all profits and advantages made by an engineer or architect in his dealings with the work of his client beyond his ordinary fee, or a secret commission paid to an engineer or architect by a contractor in order unfairly to obtain the contract the latter has tendered for. Also, the receipt of such a bribe by an engineer or architect could justify his immediate dismissal without notice. Further, the owner may repudiate the contract negotiated by the said engineer/architect. As against the contractor, it could be presumed that the engineer/architect was influenced by the bribery. In addition, if the owner is a public body, the engineer/architect who has accepted the bribe and the contractor who has given it will both be liable for criminal proceedings. On the other hand, an employer who has paid money to his engineer/architect for an illegal purpose cannot recover the same.[169]

---

166 Marston Ltd. v. Kigrass Ltd., QB (1989) 15 Con. LR 116.
167 British Steel Corp. v Cleveland Bridge & Engineering Co. Ltd., [1984] 1 All E.R. 504.
168 Moss and Co, Ltd. v. Swansea Corporation, (1910) 94 JP 351.
169 Harry Parker Ltd. v. Mason, (1940) 2 KB 590 CA; also see, Patel v. Mirza, [2013] EWHC 1892 (Ch); [2013] Lloyd's Rep. FC 525.

1.24 Interference by High Court

The High Court will interfere with the acceptance or rejection of tenders by the government or other public authorities, when it is seen that a decision is influenced by extraneous considerations which ought not to have been taken into consideration. Even if the action was taken in good faith, it is bad in law and cannot be upheld. For further discussion, see Chapter 3.

1.25 Re-advertisement

When all tenders are rejected, re-advertisement becomes essential unless it is proposed to abandon or postpone the project. If the project is advertised for a second time on an "as is" basis, the tenders that will be received are likely to be influenced by the knowledge obtained from the former proposals with the result that the prices tendered may be higher than justified by the mere passage of time. It is advisable to make some changes in the major features of the work or its design so that the new tenders would be made on an independent basis.

2

The Contract

2.0 Introduction

A contract is an agreement. An agreement arises out of the acceptance of a proposal or offer. Thus, an agreement is a promise or a set of reciprocal promises. It necessarily involves a minimum of two persons, although a greater number may be involved. The person who makes a proposal is called the "promisor" and the person who accepts the proposal is called "promisee". Consequently, one person cannot enter into an agreement with himself even if discharging duties in different roles.

Although all contracts are agreements, the converse is not true. All agreements are not contracts. Section 2(h) of the Indian Contract Act defines a contract as an agreement enforceable by law. An agreement not enforceable by law is said to be void; whereas an agreement which is enforceable by law at the option of one or more of the parties thereto, but not at the option of other or others, is called a voidable contract. These terms are discussed in this chapter. Other terms, including contingent contracts, collateral contracts, quasi-contracts, implied contracts, contracts of indemnity and guarantee are dealt with below. Necessity of written contracts, contract documents and types of engineering contracts are other topics considered in this chapter.

2.1 Provisions of the Act

The following provisions of the Indian Contract Act are referred to and/or discussed in the paragraphs below as follows.

2.2 Essentials of a Valid Contract

Section 10 of the Indian Contract Act sets out the essential features of a valid contract. It provides as follows (see below and over page):

10. *What Agreements Are Contracts*
All agreements are contracts if they are made by the free consent of parties competent to contract, for a lawful consideration and with a lawful object, and are not hereby expressly declared to be void.

Nothing herein contained shall affect any law in force in India, and not hereby expressly repealed, by which any contract is required to be made in writing or in the presence of witnesses, or any law relating to the registration of documents.

A valid contract thus requires the following elements:

1. There must be an agreement or meeting of minds.
2. The agreement must be between parties competent to enter into a contract.

Section of the Act	Paragraph No.
2	2.0, 2.6.1, 2.8
10	2.2
12	2.4.2
13	2.5
14	2.5
15	2.5.1
16	2.5.2
17	2.5.3
18	2.5.4
19	2.8
19(A)	2.8
20	2.5.6, 2.8
21, 22	2.5.6
23–30, 35, 36, 56	2.6.1, 2.7, 2.8
31–366	2.9
51, 52	2.12
56	2.8
62, 63	2.14
64	2.8.6
65	2.8.7
66	2.8.5
68–70	2.12, 2.13
124, 125	2.14.2
238	2.5.4

3. The parties must give their free consent.

4. The agreement must be supported by consideration, which must also be lawful.

5. The subject matter of the agreement must be definite and lawful.

6. The agreement does not fall into any category, which the Contract Act has declared as void.

Before considering the differences between valid, voidable or void contracts, it may be noted that every contract need not be in writing. An oral agreement is binding unless a law requires the agreement to be in writing.[1] For example, contracts by the Government of India are required to be in writing and comply with a special procedure which has to be followed for making a government contract valid. Contracts for the sale of land are required to be in writing. However, once a contract is concluded orally or in writing, the mere fact that a formal contract has to be prepared and initialled by the parties would not affect either the acceptance of the contract so entered into or implementation thereof, even if the formal contract has never been initialled.[2] This is true even in the case of government contracts so long as the offer and its acceptance by the person competent to accept are in writing and other requirements are satisfied. This is discussed in detail in Chapter 3.

---

[1] Tarsem Singh v.Sukhminder Singh, AIR 1998 SC 1400.

[2] Trimex International FZE Ltd. Dubai v. Vedanta Aluminium Ltd. India, (SC): 2010(1) ArbiLR 286: 2010(1) JT 474: 2010(3) SCC 1: 2010(1) Scale 574: 2010(Sup) AIR (SC) 455: 2010(2) All WC 1170: 2010 AIR (SCW) 909: 2010(1) Apex Court Judgment (SC) 505.

2.2.1 Non-Statutory and Statutory Contracts

The Government of India contracts fall into two categories.

1. Some contracts executed in exercise of the executive powers of the Union or the State are governed by Art. 299 of the Constitution of India, and are not subject to estoppel or ratification.

2. Statutory contracts contain their terms and conditions as prescribed in the relevant Statute. The provisions of the relevant Statutes and Rules framed thereunder govern the rights of the parties.

Merely because a contract is entered into, in exercise of an enacting power conferred by a statute, that by itself cannot render the contract into a statutory contract. If a contract has to contain the terms and conditions prescribed in a statute, that contract becomes a statutory contract. A contract may contain certain other terms and conditions which may not be of a statutory character and which have been incorporated therein as a result of a mutual agreement between the parties. A statute may expressly or impliedly confer power on a statutory body to enter into contracts in order to enable it to discharge its functions. Every act of a statutory body need not necessarily involve an exercise of a statutory power. Such bodies, like private parties, have power to contract or deal with property. Such contracts fall in the realm of the private law.

EXAMPLES FROM CASE LAW

1. The Madhya Pradesh (MP) Government invited private companies to set up power plants to increase its power generating capacity. The MP Electricity Board (MPEB) invited offers from potential investors for prequalification including one from the petitioner. The MP Government issued a letter of intent. Similar letters of intent were issued for other projects to Independent Power Producers (IPPs). Thirteen IPPs entered into power purchase agreements (PPA) with MPEB. One of the terms in the PPAs pertained to opening of escrow accounts. While the terms and conditions were under discussion between the IPPs and Financial Institutes, the Government of India, in exercise of its power under the Electricity Act, issued a notification amending its earlier tariff notification. The MP Government invited all IPPs for a discussion to alter the terms for PPAs in view of the amendment made by the central government to the tariff notification. The petitioner contended that the terms of the PPA remained unchanged. MPEB decided to prioritize the projects with the least tariff criteria out of all projects, whether using liquid fuel, hydel (hydroelectric) or coal. A writ petition was filed challenging its action. The division bench of the High Court held that the PPAs were statutory contracts. This decision was appealed before the Supreme Court. The Supreme Court held:[3]

> If a contract incorporates certain terms and conditions in it which are statutory then the said contract to that extent is statutory. A contract may contain other terms and conditions which may not be of statutory character and which have been incorporated therein as a result of a mutual agreement between the parties. Therefore, the PPAs can be regarded as statutory only to the extent that they contain provisions regarding

---

[3] India Thermal Power Ltd. v. State of M.P., AIR 2000 SC 1005.

determination of tariffs and other statutory requirements of Section 43-A(2). Opening and maintaining of an escrow account ... are not the statutory requirements and, therefore, ... that obligation cannot be regarded as statutory.

The action of MPEB giving priorities to the projects on the least tariff criteria was held to be rational.

2. A contract with an electricity board for the construction of a dam is essentially a non-statutory contract. Disputes arising out of the terms of such contracts have to be settled by the ordinary principles of law of contract. Where the payment of escalation due to enhanced wages was involved, the High Court directed the board to make a payment to the contractor. The Supreme Court did not interfere with the direction of the High Court except as to the rate of interest payable given that at the late stage other remedies would have become foreclosed.[4]

The essential elements of a valid contract as listed under Section 10 are considered below in detail.

2.3 Proposal and Acceptance – Meeting of Minds

To constitute an agreement there must be, first, a proper proposal by one party, and second, its absolute and unqualified acceptance by the other party. In other words, there must be a meeting of the minds of the contracting parties with reference to the subject of the agreement. If a finder of a lost article is to return it to the owner unaware of the reward declared by the owner to the finder, he may not be entitled to the reward. The obvious reason being there had been no acceptance of the proposal of the reward. In building and engineering contracts, a contract comes into existence once there is an acceptance of an offer. It can be inferred from the parties' correspondence. A tender and its acceptance conclude a contract.[5] Where, however, the acceptance is not absolute or unqualified there is no concluded contract.[6]

EXAMPLE FROM CASE LAW

The state government of Jharkhand decided to construct 14 hospitals in rural areas. The petitioners and various other organizations submitted their Expressions of Interest. The petitioners were informed that they had been selected. However, subsequently, pursuant to a review meeting held with the Chief Minister of Jharkhand, a notice inviting tender (NIT) was published. A Memorandum of Understanding (MOU) was sent to the petitioners but not signed. It was held that forwarding of a draft MOU to the petitioners could not have resulted in a concluded contract. In government contracts, there is no concept of an oral agreement. Held: No enforceable legal right had accrued in favour of the petitioners to seek an order directing the respondents to execute the MOU.[7]

---

[4] Kerala State Electricity Board v. Kurien E. Kalathil, AIR 2000 SC 2573.
[5] Gujarat State Fertilizers Co. Ltd. v. H.J. Baker & Bros., AIR 1999 Guj. 209; Quadricon Pvt. Ltd. v. Bajarang Alloys Ltd., AIR 2008 Bom. 88.
[6] Rawatsons Engineers (P) Ltd. v. Union of India & Ors., 2008(NOC) 2009 (PAT.).
[7] Institute of Continuing Education, Dist. R. v. State of Jharkhand, (Jharkhand): 2015(1) JLJR 750: 2015(2) AIR Jhar R. 92: 2015(2) JBCJ 140: 2015(3) BC 361: 2015(2) J.C.R. 351: 2015(16) R.C.R.(Civil) 400.

Where an owner/employer under a construction contract made direct payments to a subcontractor with whom the main contractor had entered into contract, it was held that this did not establish a privity of contract between the employer and the subcontractor, and only for this reason alone the employer could not be saddled with any further liability to pay the amounts due to the subcontractor from the main contractor. It is important to note that the payments were made to the subcontractor only upon the certification of work done by the main contractor and there was no other document or correspondence establishing a direct contractual relationship between the employer and the subcontractor.[8]

All agreements of sale are bilateral contracts containing reciprocal promises. Even an oral agreement to sell is valid. As such, a written agreement signed by one of the parties, if it evidences such an oral agreement, will also be valid. An agreement of sale signed by the vendor alone and delivered to the purchaser, and accepted by the purchaser is a valid contract in India. There is, however, no practice of a purchaser alone signing an agreement of sale.[9] A division bench of the Calcutta High Court held in a case that simply because it is a unilateral agreement signed by purchaser alone, it cannot be said to be invalid if it was handed over to the owner who did not dispute it.[10] An agreement to agree does not result in a concluded contract.[11]

EXAMPLES FROM CASE LAW

1. An estate officer was asked to allot a plot to the petitioner. He submitted a report to the chief administrator that the plot in question was allotted to some other person. It was held the petitioner did not acquire any right because no regular allotment letter was issued. The amount paid by the petitioner was refunded as there was no concluded contract for the allotment.[12]

2. A successful bidder for three items was offered orders for only two items, which he refused. Thereafter, a third order was placed and it was clarified that each order is a subject matter of independent contract. The bidder refused to furnish a performance guarantee. It was held that there was no concluded contract.[13]

3. An agreement incorporated a provision for the renewal of agreement by mutual consent between the parties for a further period of five years. It was held by the Allahabad High Court that the said provision does not create any obligation or rights in either parties for the renewal of such an agreement. In other words, the parties had provided for a need of further meeting of minds before the contract stood renewed.[14]

2.4 Who Are Competent to Enter into a Contract?

Every person is competent to enter into a contract if he is not a minor or is not of unsound mind or is not disqualified from contracting by any law to which he is subject.[15]

---

[8] Essar Oil Ltd. v. Hindustan Shipyard Ltd., 2015 AIR (SCW) 4765: 2016(1) Recent Apex Judgments (R.A.J.) 51: 2015(6) JT 367: 2015 DNJ 732: 2015(4) ArbiLR 1: 2015(7) Scale 631: 2015(5) MLJ 634: 2015 All SCR 2357:: 2015(10) SCC 642: 2015(3) Apex Court Judgments (SC) 479: 2016(1) CivilLJ 843.

[9] Aloka Bose v. Parmatma Devi, AIR 2009 SC 1527.

[10] Kanchilal Paul v. Sasthi Charan Banerjee & Ors., AIR 2009 (NOC) 1055 (CAL.).

[11] Punit Beriwala v. Suva Sanyal, AIR 1998 Cal. 44.

[12] Smt. Saroj Sharma v. Haryana Urban Development Authority, AIR 2009 (NOC) 1807 (P. & H.).

[13] Uniflex Cables Ltd. v. MTNL & Anr., AIR 2009 (NOC) 1732 (DEL.).

[14] BDA Limited v. State of Uttar Pradesh, AIR 1995 All.277.

[15] Section 11 of the Indian Contract Act.

2.4.1 Minor's Agreement

A person domiciled in India is deemed to have attained majority when he reaches the age of 18 years and not before. Any agreement entered into by him during his minority is void under Indian law.[16]

2.4.2 Agreement by a Person of Unsound Mind

In India, a person of unsound mind is held to be absolutely incompetent to enter into a contract.[17] A person is said to be of unsound mind for the purpose of making a contract if, at the time when he makes it he is incapable of understanding it and of forming a rational judgement as to its effect upon his interests. A person, who is usually of unsound mind, but occasionally of sound mind, may make a contract when he is of sound mind.[18] Where a person, who is usually of sound mind, is alleged to have been of unsound mind at the time of entering into a contract, the burden of proving the fact lies on him who challenges the validity of the contract.[19]

2.4.3 Agreement by a Person of Unsound Mind – When Binding?

A person incapable of entering into a contract cannot, however, escape liability in respect of necessities supplied to him. Under Section 68 of the Indian Contract Act, the property of such a person is liable for such necessities. However, no personal liability is incurred by him. Section 70, discussed in paragraph 2.12, is also not applicable in such a case. Necessities must be things which the minor actually needs; it would, therefore, be difficult to prove that "necessities" include execution of building works. Sale of property of a person of unsound mind by his wife was held to be in violation of Section 75 of the Indian Lunacy Act, 1912. The contract was held to be void under Section 23 of the Indian Contract Act.[20]

2.4.4 Members of Parliament/Legislative Assembly

A Member of Parliament or of a legislative assembly is prohibited from being interested in contracts for the public service. If, however, he is a member of an incorporated company, the prohibition is not applicable to him.

EXAMPLE FROM CASE LAW

The election of a candidate to the legislative assembly of the Karnataka legislature was set aside by the Karnataka High Court[21] as he had, on the dates of filing the nomination paper, scrutiny and election, a subsisting contract with the appropriate government for execution of government contract works.

2.4.5 Contract of Corporations

A corporation is a legal person, an entity created by law although there is also an argument that a corporation is a creature of contract. Its contractual powers are limited by the terms of

---

[16] Mohiri Bibee v. Dhurmodas Ghose, (1903), 30 Cal. 539 LR 30 I1 114.
[17] Machaima v. Usman Beari (1907), 17. Mad. L. J. 78.
[18] Section 12 of the Indian Contract Act.
[19] Tilok Chand v. Mahanadu, 33 AL 458.
[20] Johri v. Mahila Draupati, AIR 1991 M. P. 340.
[21] Vijayakumar Khandre v. Prakash Khandre, AIR 2002 Kant. 145.

its incorporation. If a corporation enters into a contract which is beyond the scope of its limited powers, such contract is *ultra vires* and ordinarily unenforceable.

Persons desirous of entering into a contract with a corporation must satisfy themselves that the activities of the corporation are in compliance with the terms of its charter and that the individuals purporting to act on behalf of the corporation are duly authorized.

EXAMPLE FROM CASE LAW

A bidder participated in an auction and his bid was accepted whereupon he paid the earnest money as well. The acceptance of the bid contained signatures of the Chairman of the Board and the bidder. The Municipal rules contemplated that in case of acceptance of the bid, an agreement would be executed on stamp paper but no such agreement was executed. The rules in the case contained a general provision that every contract entered into by or on behalf of the Municipal Corporation had to be signed by its chairman and executive officer, and also had to be sealed with the common seal of the Municipality. Relying upon two decisions of the Supreme Court,[22] it was held that there was no concluded contract.[23]

2.4.6 Contract by One Partner of a Firm – When Binding?

A contract entered into by one partner of a firm would be binding on the other partners of the firm when validity of the contract was not denied by the other partners nor any objection was raised as to signing of the tender by the partner on behalf of the firm. It would be assumed that the other partners had ratified the contract. Accordingly, a reference made by the court under Section 20 of the Arbitration Act pursuant to an arbitration clause contained in such contract is not without jurisdiction.[24]

2.4.7 Contract by One Company, Bill and Supply to Another Company – Joint Responsibility?

Sometimes a consultancy firm is authorized to enter into a contract with a third party for or on behalf of the client. If any dispute arises out of the said contract, question arises as to against whom the proceedings are to be launched? It is advisable to join both the parties. Once it is successfully demonstrated that there was privity of contract among the parties, the defendants will be jointly liable.

EXAMPLE FROM CASE LAW

A paper mill, acting through its authorized representatives, corresponded with a machinery supplier. Thereafter the officials of the mill and an engineering firm jointly visited the machinery supplier's unit. The mill placed an order for supply of machinery which stipulated that the bill should be raised and machinery supplied to the engineering firm. A cheque issued by the

---

[22] Union of India v. M/s Uttam Singh Dugal and & Co., AIR 1972 Delhi 110 and State of Uttar Pradesh v. Kishori Lal Minocha, AIR 1980 SC 680.
[23] Municipal Board Khumber v. Ydu Bnath Singh, AIR 2004 Raj. 79.
[24] Sanganer Dal and Flour Mill v. F.C.I., AIR 1992 SC 481.

engineering firm bounced but the mill showed willingness to pay the advance amount. Bills and notices were sent to the engineering firm. When a suit was filed for the recovery of sale consideration, the mill contended that there was no privity of contract between the mill and the supplier. This argument was rejected and both the paper mill and the engineering firm were held liable for payment of sale consideration.[25]

2.4.8 Contract by Executor

An executor or a trustee should not put himself in such a position in which his personal interest and his duties under the will of a deceased come into conflict. The testamentary court must give effect to the will and not an agreement by and between the executor and a third party, which would be contrary to the wishes of the testator.[26]

2.5 Free Consent

Section 13 of the Indian Contract Act defines consent in these words:

> 13. *"Consent" defined*
> Two or more persons are said to consent when they agree upon the same thing in the same sense.

For validity of a contract, free consent is required.

Section 14 of the Indian Contract Act defines free consent thus:

> 14. *"Free Consent" defined*
> Consent is said to be free when it is not caused by
>
> (1) Coercion, as defined in Section 15, or
> (2) Undue Influence, as defined in Section 16, or
> (3) Fraud, as defined in Section 17, or
> (4) Misrepresentation, as defined in Section 18, or
> (5) Mistake, subject to the provisions of Sections 20, 21 and 22.
>
> Consent is said to be so caused when it would not have been given but for the existence of such coercion, undue influence, fraud, misrepresentation or mistake.

Section 14 considers consent as an act of reason, accompanied by deliberation, presupposing free use of moral and physical power by the person giving it. In the case of consent, the burden of proof rests on those who assert it. Once consent is said to exist, the court presumes it to be free unless there are circumstances justifying departure from the general rule. Where there are no such circumstances, the burden of proof that consent was not free rests on those who assert it.

---

[25] Tamilnadu Card Boards and Paper Mill Ltd. v. Sirpur Paper Mills Ltd., AIR 2003 A.P. 438.
[26] Chandrabhai K. Bhoir v. Krishna Arjun Bhoir, AIR 2009 SC 1645.

Coercion, undue influence, fraud, misrepresentation and mistake are all separate categories in law and they may overlap or may be combined, but they must be separately specifically pleaded with precision. Each of these is considered below in brief.

2.5.1 Coercion

Section 15 of the Act defines coercion in these words:

---

15. *"Coercion" defined*

Coercion" is the committing, or threatening to commit, any act forbidden by the Indian Penal Code (XLV of 1860) or the unlawful detaining, or threatening to detain, any property to the prejudice of any person whatever, with the intention of causing any person to enter into an agreement.

Explanation: It is immaterial whether the Indian Penal Code (XLV 1860) is or is not in force in the place where coercion is employed.

ILLUSTRATION

A, on board an English ship on the high seas, causes B to enter into an agreement by an act amounting to criminal intimidation under the Indian Penal Code (XLV 1860).

A afterwards sues B for breach of contract at Calcutta.

A has employed coercion, although his act yet is not an offence by the law of England, and although Section 506 of the Indian Penal Code was not in force at the time when, or place where, the act was done.

Where a property purchased for Rs. 3,12,370/- was sold for Rs. 1,12,750/- allegedly under coercion, force or undue influence, it was held the defence of the seller was probable, though no police complaint was lodged and specific performance was not granted.[27]

The plea of coercion in building and engineering contracts is resorted to by the parties, not so much at the time of making the agreement, as at the time of settlement of the final bill. Invariably, the owner refuses to make the payment of the final bill unless the contractor signs a "no dues" or "no claim" certificate. In other words, the owner threatens to detain money due and payable unless a "no dues" certificate is signed. Section 14 uses the phrase "any property". Property, according to *Black's Law Dictionary*, is the ownership of a thing, "the right of one or more persons to possess and use it to the exclusion of others."[28] It is common understanding now that property includes money, things in action (shares, copyrights, goodwill, opportunities and so on) and securities for money.

More often than not, courts refer to arbitration the question whether there was a full and final settlement in case where a "no dues" or any other version of a "full and final settlement" statement is signed. Such a dispute is then submitted to arbitration under the applicable contracts. In the absence of an arbitration clause, the court with jurisdiction will have to decide the question whether a consent to the final settlement was free or under coercion. There is no presumption in favour of the contractor merely because he pleads coercion. Such an allegation must be proved to the satisfaction of the arbitrator or the court. In one case the Supreme Court held:[29]

---

[27] Munusamy v. Nava Pillai, AIR 2008 (NOC) 1401 (MAD).
[28] Retrieved from www.thelawdictionary.org/property on 16th October 2018.
[29] State of Kerala v. M. A. Mathai, AIR 2007 SC 1537 = 2007 AIR SCW 2158.

In the instant case both the trial court and the High Court have without any basis come to hold that the supplemental agreement was due to coercion etc. For coming to such conclusion, material had to be placed, evidence had to be led. Mere assertion by the plaintiff without any material to support the said stand should not have been accepted by the trial court and the High Court.

2.5.2 Undue Influence

Section 16 of the Indian Contract Act defines undue influence in these words:

---

16. *"Undue influence" defined*

  (1) A contract is said to be induced by "undue influence" where the relations subsisting between the parties are such that one of the parties is in a position to dominate the will of the other and uses that position to obtain an unfair advantage over the other.

  (2) In particular and without prejudice to the generality of the foregoing principle, a person is deemed to be in a position to dominate the will of another:

    (a) Where he holds a real or apparent authority over the other, or where he stands in a fiduciary relation to the other; or

    (b) Where he makes a contract with a person whose mental capacity is temporarily or permanently affected by the reason of age, illness, or mental or bodily distress.

  (3) Where a person who is in a position to dominate the will of another, enters into contract with him, and the transaction appears, on the face of it or on the evidence adduced, to be unconscionable, the burden of proving that such contract was not induced by undue influence shall lie upon the person in a position to dominate the will of the other.

Nothing in this sub-section shall affect the provisions of Section 111 of the Indian Evidence Act, (1 of 1872).

ILLUSTRATIONS

  (a) A pays a sum of money in advance to his son B during his minority, upon B's coming of age A obtains, by misuse of parental influence, a bond from B for a greater amount than the sum due in respect of the advance. A employs undue influence.[30]

  (b) A, a man enfeebled by disease or age, is introduced, by B's influence over him as his medical attendant, to agree to pay B an unreasonable sum for his professional services. B employs undue influence.

  (c) A being in debt to B, the money-lender o.f his village, contracts a fresh loan on terms which appear to be unconscionable. It lies on B to prove that the contract was not induced by undue influence.

  (d) A applies to a banker for a loan at a time when there is stringency in the market. The banker declines to make the loan except at an unusually high rate of interest. A accepts the loan on these terms. This is a transaction in the ordinary course of business and the contract is not induced by undue influence.

---

[30] Niko Devi v. Kripa, AIR 1989 H.P. 5.1; Section 16 of the Indian Contract Act.

The onus is cast upon a person holding the position of confidence or trust to show that the transaction in question is a fair one, not brought forth by reason of the fiduciary relation and that he has not taken any advantage of his position.[31] For example, a minor female child having lost parents was living with her cousin who brought her up and also managed her movable and immovable properties. She was asked to execute a general power of attorney in his favour and later on she came to know that it was in fact a gift deed in respect of her property in favour of the defendant. The Himachal Pradesh High Court held that there could be no better example of transaction vitiated by undue influence.[32]

Building and engineering contracts are invariably in writing and when parties express their intention by certain words, they cannot deny the reasonable effect of those words. In case of disputes, the parties will be bound by the interpretation which a court of law may put upon the language of the instrument. It may, therefore, be difficult to set aside a building or engineering contract on the ground of want of consent. However, a contract can be set aside at the desire of the party whose consent was not free.

Undue Influence and Coercion – When May Not Vitiate the Agreement

It is essential, in cases where a party claims that an agreement was signed under undue influence or coercion, to take appropriate action to question the validity of the agreement soon after the factors causing undue influence or coercion cease to exist, or in other words sufficient time and opportunities are available to challenge the validity of the agreement. In the absence of any such action it will be difficult to hold that the agreement was vitiated by undue influence or coercion.[33]

In order to prove undue influence two things must be shown:

1. The relationship between the parties was such that one was in a position to dominate the will of the other, and
2. He has used that position to obtain an unfair advantage.[34]

2.5.3 Fraud

Section 17 of the Indian Contract Act defines fraud in these words:

---

17. *"Fraud" defined*

"Fraud" means and includes any of the following acts committed by a party to a contract, or with his connivance, or by his agent, with intent to deceive another party thereto or his agent, or to induce him to enter into the contract:

(1) The suggestion, as to a fact, of that which is not true by one who does not believe it to be true;

(2) The active concealment of a fact by one having knowledge or belief of the fact;

(3) A promise made without any intention of performing it;

(4) Any other act fitted to deceive;

(5) Any such act or omission as the law specially declares to be fraudulent.

---

[31] Saguna v. Vinod G. Nehemiah, 2008(1) CTLJ 223 (Madras) (DB).

[32] Niko Devi v. Kripa, AIR 1989 H. P. 51.

[33] P.C.K. Muthia Chettiar v. V.E.S. Shanmugham Chettiar, (SC) 1969 AIR (SC) 552: 1969(1) SCR 444: 1968(2) AIR (SCW) 351: 1969(2) SCJ 105: 1969(2) MLJ 28: 1969(2) An.WR 28.

[34] Habeeb Khan v. Valasula Devi, AIR 1997 A.P. 53; Also see as illustration: Marci Celine D'Souza v. Renie Fernadez, AIR 1998 Ker. 280.

Explanation: Mere silence as to facts likely to affect the willingness of a person to enter into a contract is not fraud, unless the circumstances of the case are such that regard being had to them, it is the duty of the person keeping silence to speak, or unless his silence is in itself, equivalent to speech.

ILLUSTRATIONS

(a) A sells, by auction, to B, a horse which A knows to be unsound. A says nothing to B about the horse's unsoundness. This is not fraud in A.

(b) B is A's daughter and has just come of age. Here, the relation between the parties would make it A's duty to tell B if the horse is unsound.

(c) B says to A – "If you do not deny it, I shall assume that the horse is sound." A says nothing. Here A's silence is equivalent to speech.

(d) A and B being traders, enter upon a contract. A has private information of a change in prices which would affect B's willingness to proceed with the contract. A is not bound to inform B.

Fraud and Constructive Fraud Distinguished

Fraud as defined in the Indian Contract Act means actual fraud. In equity too courts have developed the doctrine of "constructive fraud". In equity the term fraud embraces not only actual fraud but also certain other conduct which falls below the standard demanded by equity. This does not just mean a moral fraud as understood in its plain and ordinary meaning but includes "a breach of the sort of obligation which is enforced by a Court that from the beginning regarded itself as a Court of conscience."[35] Constructive fraud in English law is conduct which falls below the standards demanded by equity manifested by, for example, undue influence, abuse of confidence, unconscionable bargain and so on.[36]

The doctrine of constructive fraud was pressed into service by the Allahabad High Court in a case in which a decree was passed in a suit for recovery by a bank against a firm. An appeal filed by the bank against the decree held by the firm's predecessor firm in another matter was withdrawn on the assurance of the new firm that the new firm would agree to a one-off settlement under which it would not enforce the decree against the bank. However, as soon as the appeal was withdrawn by the bank, the defendant firm filed its execution application for the decree against the bank. In result, the order dismissing the appeals as withdrawn was recalled and the appeal restored to its original number and position.[37]

2.5.4 Misrepresentation

Section 18 of the Indian Contract Act defines misrepresentation thus:

18. *"Misrepresentation" defined*
    "Misrepresentation" means and includes:

---

[35] Nocton v. Lord Ashburton, [1914] AC 932.

[36] Hart v. O'Connor, [1985] A.C. 1000 at p. 1024.

[37] State Bank of India v. Firm Jamuna Prasad Jaiswal & Sons, AIR 2003 All.337. Also see Nedungadi Bank Ltd. v. M/s. Ezhimala Agri. Products, AIR 2004 Ker. 62.

(1) The positive assertion, in a manner not warranted by the information of the person making it, of that which is not true, though he believes it to be true.

(2) Any breach of duty which, without an intent to deceive, gains an advantage to the person committing it, or any one claiming under him, by misleading another to his prejudice or to the prejudice of any one claiming under him;

(3) Causing, however innocently, a party to an agreement to make a mistake as to the substance of the thing which is the subject of the agreement.

From the above definitions of fraud and misrepresentation, it is clear that in the case of the former an intent to deceive or intent to induce the other party to enter into a contract, is an essential element whereas in the latter, "intent to deceive" is not necessary.

Analysis of this portion of the section reveals the following ingredients:

1. There should be a suggestion as to a fact.

2. The fact suggested should not be true.

3. The suggestions should have been made by a person who does not believe it to be true; and

4. The suggestion should be found to have been made with intent either to deceive or to induce the other party to enter into the contract in question.

Level of Proof

It is well-settled law that even within the province of civil litigation, when an allegation of misrepresentation or fraud is made, the level of proof required is extremely high and is rated on par with a criminal trial.[38]

Application in Construction Contracts

In building and engineering contracts, an occasion to refer to Section 17 or 18 arises in respect of statements made in the documents contained in the NITs, particularly with reference to sub-surface explorations, foundations and/or availability of site and/or general working conditions. It is necessary to exercise due care while pleading fraud or misrepresentation. For example, if sub-surface exploratory bores were taken and difficult or unusual nature of strata met with or likely to be met with was not disclosed, it may amount to fraud; on the other hand if representations are made in respect of strata likely to be met with, without having made exploration and on mere belief of it being true, it could amount to misrepresentation. It is further to be noted that the owner may also be liable for misrepresentation, either fraudulent or innocent, made by his engineer or architect. Section 238 of the Indian Contract Act provides:

---

238. *Effect, on agreement, of misrepresentation of fraud, by agent*

Misrepresentations made, or frauds committed, by agents acting in the course of their business for their principals, have the same effect on agreement made by such agents as if such misrepresentations or frauds had been made, or frauds committed by the principals; but misrepresentations made, or frauds committed, by agents, in matters which do not fall within their authority, do not affect their principals.

---

[38] Savithramma v. H. Gurappa Reddy, AIR 1996 Kant. 99. As an illustration of undue influence having been proved in a case of sale deeds see: Kartick Prasad Gorai v. Neami Prasad Gorai, AIR 1998 Cal. 278.

ILLUSTRATIONS

(a) A, being B's agent for the sale of goods, induces C to buy them by a misrepresentation, which he was not authorized by B to make. The contract is voidable as between B and C at the option of C.

(b) A, the captain of B's ship, signs bills of lading without having received on board the goods mentioned therein. The bills of lading are void as between B and the pretended consignor.

In construction contracts, invariably, reliance is placed on one of the conditions of invitation of tenders which asks tenderers to make their own investigations and not to rely upon the information supplied. The question is, if it be so, why supply the information at all? Is it not with a hope that it will be relied upon and no independent investigations will be made? From the decided cases, it appears that a court or arbitrator is not likely to accept a defence based on such a clause, if it could be shown that the innocent party did rely on the false statement. The person who is found to have made a misrepresentation cannot say that the other person ought to have realized that it was false. It is possible that such a provision may give protection against honest mistakes but none against fraud by the owner or his agent.[39]

EXAMPLE FROM CASE LAW

The Gujarat Housing Board invited tenders for construction of tenements in lump sum form. In the suit filed it was urged that the Board, by making *mala fide* and misleading references in the details of the tenders to a schedule of rates, actively and fraudulently led the intending contractors including the plaintiffs into the belief that the defendant Board's estimated costs were worked out on the basis of the said schedule of rates. In defence, it was alleged that the tender notices and other documents relating to the tender contract costs were shown only for the purpose of fixing the security deposits and earnest money. The Gujarat High Court, observed:[40]

> We are not at all impressed by this ... It is undoubtedly true that the estimate of cost shown in the tender notices would be useful in fixing the amount of security deposit as well as earnest money to be taken from the concerned contractor but that does not exhaust the utility of the estimate of the cost because the said estimate does not cease to represent a very important fact relating to the contract, namely that the execution of the contract would involve a particular amount of cost. ... We find in this contention that though it is true that in lump sum contracts, it is not obligatory on the department to mention any amount of estimated cost in the tender notice and other documents relating to the tender, the fact remains that the department has preferred to make a reference to these estimates of cost. Therefore, even if the department was not bound to mention this fact, it has preferred to reveal this fact, and hence if this fact is found to be wrong or incorrect or misleading, the department has to take the consequences thereof ...
>
> In our opinion, the finding that the false representation was knowingly made by the party concerned is by itself sufficient to bring the case within the mischief of Section 17 of the Contract Act. It is, therefore, not necessary to make any further probe into the motive of the person making false representation. Even if the motive of the person indulging in deceit is very laudable, he cannot escape from the consequences of his action."

---

[39] S Pearson and Son Ltd v. Dublin Corporation, [1907] AC 351, p. 488.
[40] R. C. Thakkar v. Gujarat Housing Board, AIR 1973 Guj. 34.

This judgment, it is submitted, does not spell out a general principle of law that mere mention of the estimated cost in the tender notices will amount to a fraudulent representation if the estimated cost provided be incorrect. It would be a fraudulent misrepresentation in a case where, for example, the engineer does not show or supply drawings on demand, for inspection of the tenderers and makes a statement that his estimate is correct and may be relied on. For in that case tenderers would be deprived of a legitimate way to discover the truth with ordinary diligence.

Critical Study of the Above Case

The judgment in the example above, based on the peculiar facts of the case, nevertheless raises certain important questions from the standpoint of a practising engineer. Some of these questions are:

(i) What is "estimated" cost?

(ii) Does an engineer, in the ordinary course, guarantee its accuracy?

(iii) Why are estimates prepared?

(iv) Is a contractor supposed to be guided by the estimated cost in arriving at his tendered sum?

(v) Is it necessary to mention estimated cost in an NIT?

(vi) Does the mention of estimated cost amount to a representation of the cost of construction?

These questions are considered below:

1. What is "Estimated Cost"?

The Oxford English Dictionary meaning of the word "estimate" is: "[a]n approximate calculation or judgement of the value, number, quantity, or extent of something."[41] A contractor's estimate is a statement of the sum for which he will undertake any specified work. It would, therefore, not be wrong to define an engineer's estimate of cost of a project as his approximate judgement of the probable expenses likely to be involved in the execution of a project. It is well understood by all concerned in engineering practice that an estimated cost is only a close approximation of the actual cost of construction. How close it will be naturally depends upon the skill and judgement of the estimator. It is also a well-settled principle in law that in all such cases the skill required is of those *ordinarily skilled* in the business. Further, there is also no implied promise that miscalculation may not occur.

In estimating cost of a work so many variable factors come into the picture that one is not surprised to see a large variation between the estimated cost as fixed by the engineer on the one hand and among the sums tendered by the different tenderers on the other hand. Why is it so?

To answer this question, one must have knowledge of the different factors which constitute the cost of an item of work. The main factors that constitute the cost of an item of work as revealed by an analysis are:

(i) Cost of materials

(ii) Cost of labour

(iii) Cost of deployment of plant, machinery and equipment

(iv) Overhead expenses

(v) Profit.

---

[41] Retrieved from https://en.oxforddictionaries.com/definition/estimate on 21st October 2018.

All these factors are to a great extent influenced by many factors including the time limit fixed for the completion of work, size of the firm undertaking the work, its efficiency in organization and management of site operations, profit expected by it which, in turn, may depend upon contracts already in hand. Naturally the cost estimated by an engineer, without the knowledge of some of the above factors, cannot possibly be the actual cost of the work. This being the position, unless there is a contract to the contrary, the engineer does not in the ordinary course guarantee that the actual cost will not vary from his estimated cost (see paragraph 16.10.4).

2. Purpose of an Estimate?

An engineer's estimate is necessary to give the owner of the proposed project a reasonably accurate idea of the probable expenses, to help him decide whether work can be undertaken as proposed or needs to be curtailed or abandoned, depending upon the availability of funds and prospective direct and indirect benefits. For public works, estimates are required to obtain prior sanction for allocating the required amount of money. Thus, basically an engineer's estimate is prepared for the benefit of the owner. True, in an item rate contract, the detailed quantities worked out in the detailed estimate form part of the contract, but invariably such contracts include a stipulation to the effect that the quantities are approximate and likely to vary. A percentage rate contract is a step further and besides the quantities even estimated rates are part of the contract. The contractor undertakes to execute the work at estimated rates plus or minus the percentage figure by which these rates are to be increased or decreased.

Where tenders are proposed to be invited in a lump sum form, it is not at all necessary for the engineer to prepare a detailed estimate, as approximate methods of estimating costs are available to serve the purpose mentioned earlier. In public works, such estimates may be required to meet the requirements of the procedure specified by the government.

In view of the above points the authors are of the opinion that, irrespective of the form of a contract, a tenderer is supposed to arrive at his tendered sum on the basis of his independent judgement. It is he, and not the engineer, who is in a better position to estimate the probable expenses involved in the execution of a project. This is why there exist several different forms of contract or else the owner would ask the willingness of tenderers to undertake the work at the estimated cost and select one from among them known for reliability, reputation, etc. The discussion brings us to an important question as to whether the estimated cost should be mentioned in the tender notice.

3. Should Estimated Cost Be Mentioned in Notices Inviting Tenders?

In the judgment cited above, the Gujarat High Court held that the estimate of the cost shown in an NIT represents a very important fact relating to the contract, namely the cost of execution of work. The authors respectfully differ from this view. The estimated cost is mentioned in the NIT, in their opinion, for the following reason:

The estimated cost briefly indicates the magnitude of the work and enables the intending tenderer to decide whether the work is too small to interest him or perhaps too large for him to handle. If the estimated cost represents anything, the representation may be that the owner is most likely to award a contract to the one whose tendered amount would be nearly equal to or less than the estimated cost. If it were so as held, the estimated cost should not be mentioned in NITs. As an alternative to it a brief description of the work, which would enable its reader to imagine the magnitude of the work for the purpose mentioned above, should be included in the NIT. It is not necessary that the earnest money or security deposit should bear a definite percentage to the estimated cost. They can be so decided that at the most they may disclose a range within which the estimated cost might lie.

2.5.5 Mistake

Black's Law Dictionary defines a mistake as "[s]ome unintentional act, omission, or error arising from ignorance, surprise, imposition, or misplaced confidence."[42] Mistake may not be caused by neglect of a legal duty on the part of the person making the mistake. It consists in (i) an unconscious ignorance of a fact, past or present, material to the contract, or (ii) belief in the present existence of a thing material to the contract which does not exist, or in the past existence of a thing which has not existed.

A mistake of law, on the other hand, happens when a person, having full knowledge of the facts, comes to an erroneous conclusion as to their legal effect. It is a mistaken opinion or inference, arising from an imperfect or incorrect exercise of the judgement that the person forming it is in full possession of the facts.

Mistake can be mutual or unilateral. Mutual mistake is where the parties have a common intention, but it is induced by a common or mutual mistake. "Mutual", the word, as used in the expression mutual mistake of fact, expresses reciprocity and distinguishes it from a mistake which is a common mistake of both the parties. In the case of a common mistake, both the contracting parties make the same mistake.

Mistake of fact as a ground of defence need be neither "mutual" nor common in the strict sense because it may be wholly the mistake of one of the parties, the other being totally ignorant of the fact that the other party is labouring or acted under a mistake. Unilateral mistake is mistake by only one party to an agreement. The difference between a mutual mistake and a unilateral mistake is that in the former the judicial approach is objective whereas in the latter it is subjective. In the former, the question is whether the agreement must be taken to have been reached. In the latter, the innocent party is allowed to prove its effect upon his mind, in the hope of avoiding its consequences.

2.5.6 Effect of Mistake by Tenderer

The Indian Contract Act provides for the effect of a mistake on the validity of a contract in Sections 20, 21 and 22. These sections read as under:

---

20. *Agreement void where both parties are under mistake as to matter of fact*
    Where both the parties to an agreement are under a mistake as to a matter of fact essential to the agreement, the agreement is void.

Explanation: An erroneous opinion as to the value of the thing which forms the subject matter of the agreement is not to be deemed a mistake as to a matter of fact.

ILLUSTRATIONS

(a) A agrees to sell B a specific cargo of goods supposed to be on its way from England to Bombay. It turns out that, before the day of the bargain, the ship conveying the cargo had been cast away and the goods lost. Neither party was aware of these facts. The agreement is void.

(b) A agrees to buy from B a certain horse. It turns out that the horse was dead at the time of the bargain, though neither party was aware of the fact. The agreement is void.

(c) A, being entitled to an estate for the life of B, agrees to sell it to C. B was dead at the time of the agreement, but both parties were ignorant of the fact. The agreement is void.

---

[42] Retrieved from https://thelawdictionary.or/mistake/on 20th October 2018.

EXAMPLE FROM CASE LAW

An agreement to sell land was entered into with the price to be calculated. The rate was understood by each party in different units; one party treating it per bigha (0.4 acres) and the other on the basis of kanal (0.12 acres). Since the units relate to the area, the dispute pertained to the area under sale, the subject matter of the contract. It was held by the Supreme Court that the agreement was void from its inception though discovered to be void at a much later stage.[43]

21. *Effect of mistake as to law*
    A contract is not voidable because it was caused by a mistake as to any law in force in India; but a mistake as to a law not in force in India has the same effect as a mistake of fact.

ILLUSTRATION

A and B make a contract grounded on the erroneous belief that a particular debt is barred by the Indian Law of Limitation; the contract is not voidable.

22. *Contract caused by mistake of one party as to matter of fact*
    A contract is not voidable merely because it was caused by one of the parties to it being under a mistake as to a matter of fact.

It is desirable that a tenderer must bring a unilateral mistake committed by him in his tendered price to the notice of the accepting authority immediately on its being discovered. If the mistake is pointed out and a correction is sought to be made in the tender before the acceptance of the tender is communicated, the tender cannot be accepted in its original form. However, if the owner discovers an error in the tender he cannot accept the tender if the mistake is in respect of the terms of the tender itself and not merely as to the motive of underlying assumptions on which the tender is based.

EXAMPLES FROM CASE LAW

1.   While submitting his tender, a Canadian contractor omitted the entire first page, which included a price escalation clause. He hastened to make it known to the employer as soon as he discovered the mistake, long before the tender was accepted. In spite of this knowledge, the tender without the first page was accepted. It was held:

> To put it simply, this is a case where one party intended to make a contract on one set of terms and the other intended to make it upon another set of terms, with the result that there is lack of consensus, the parties were not ad idem ... .

Under such conditions, the mistake was a good defence to any action brought to enforce the alleged contract or to obtain damages for its breach. The tender document was held to be *prima facie* incomplete.[44]

---

[43] Tarsem Singh v. Sukhminder Singh, AIR 1998 SC 1400.
[44] McMaster University v. Wilchar Construction Ltd., Ontario High Court [1971 3 O.R. 801 (Ont. HC) affirmed 12 O. R, (2d) 512n (Ont. C.A.).

2. The Supreme Court of Canada dealt with a different set of facts in another case.[45] In that case, the plaintiff's tender and a deposit of $150,000 were submitted at 3.00 p.m. on the day set for the submission of tenders. When the tenders were opened and the plaintiff's tender was the lowest by quite a margin, the plaintiff's employee realized that there must have been a mistake in the calculation of the price. It should have been 27% larger than that quoted. The document that governed the submission of tenders "Information for Tenderers" was relied upon by the owner to argue that the tender could not be revoked and that if the contractor refused to execute a contract, they could forfeit the deposit. The Canadian Supreme Court had to consider the issue of the protection of the integrity of the tendering process. The difference from the previous case was that the tender was submitted as a deed (under the corporate seal of the plaintiff) thus avoiding the problem of consideration.

3. A tenderer computed his tender cost incorrectly. As a result, while he thought he was tendering for the construction of houses at £1,670 a pair, in fact, it was £1,613 per house because of the way in which the Bill of Quantities (BOQ) was made up. It was held that the parties were not labouring under a mutual mistake, the contract expressed what the defendants (employer) intended it to express and the contractor was bound by his mistake. The result was that the contract could neither be set aside nor rectified.[46]

The English case law on mistake was summarized by the Court of Appeal in 2002 as follows:

> [T]he following elements must be present if common mistake is to avoid a contract: (i) there must be a common assumption as to the existence of a state of affairs; (ii) there must be no warranty by either party that that state of affairs exists; (iii) the non-existence of the state of affairs must not be attributable to the fault of either party; (iv) the non-existence of the state of affairs must render contractual performance impossible; (v) the state of affairs may be the existence, or a vital attribute, of the consideration to be provided or circumstances which must subsist if performance of the contractual adventure is to be possible.[47]

2.6 Consideration

Although there may be an agreement between two competent parties with reference to a common purpose fulfilling other requisites thus far discussed, the agreement will not be legally binding unless it meets yet another test which the law has imposed. Section 25 of the Indian Contract Act declares that an agreement (subject to strictly limited exceptions) without consideration is void.

2.6.1 What Is Consideration?

Consideration is defined in Section 2(d) of the Indian Contract Act as below:

---

[45] The Queen v. Ron Engineering v. Construction (Eastern) Ltd., (1981) 35 N.R. 40.
[46] W. Higgins Ltd. v. Northampton Corporation Chancery Division, (1927) 1 Ch. 128.
[47] Great Peace Shipping Ltd. v. Tsavliris Salvage (International) Ltd., [2002] EWCA Civ 1407; [2003] QB 679.

2(d). *"Consideration"*

When, at the desire of the promisor, the promisee or any other person has done or abstained from doing, or does or abstains from doing, or promises to do or abstain from doing something, such act or abstinence or promise is called a consideration for the promise.

This definition can be analysed as below:

A consideration may be an act or abstinence (i.e. not doing a certain thing) or a promise.

The act or abstinence must have been done at the desire or request of the promisor by the promisee or any other person on his behalf.

It is well settled that consideration can be past or present or future.[48]

The following examples, reproduced from illustrations in Section 25 of the Act will help clarify the concept of consideration.

ILLUSTRATIONS

1 A promises, for no consideration, to give to B Rs. 1,000/-. This is a void agreement.

2 A finds B's purse and gives it to him. B promises to give A Rs. 50/-. This is a contract.

3 A supports B's infant son, B promises to pay A's expenses in so doing. This is a contract.

4 A agrees to sell a horse worth Rs. 1,000/- for Rs. 10/-. A's consent to the agreement was freely given. The agreement is a contract notwithstanding the inadequacy of the consideration.

5 A agrees to sell a horse worth Rs. 1,000/- for Rs. 10/-. A denies that his consent to the agreement was freely given. The inadequacy of the consideration is a fact which the court should take into account in considering whether or not A's consent was freely given.

Consideration even in the Indian context would mean a reasonable equivalent or other valuable benefit passed on by the promisor to the promisee or by the transferor to the transferee. Love and affection is also a consideration within the meaning of Sections 122 and 123 of the Transfer of Property Act.[49]

EXAMPLES FROM CASE LAW

1. In one case, a property purchased for Rs. 3,12,370/- was sold for Rs. 1,12,750/- allegedly under coercion, force, or undue influence. It was held the defence of the seller was probable, though no police complaint was lodged. Specific performance was not granted.[50]

2. The appellant before the Supreme Court of India had agreed to supply dal to the Army Purchase Organization and for that purpose invited tenders. The respondent supplied dal, which was accepted by the appellant and supplied to the Army. The Army rejected part of it. A supplementary agreement was entered into to upgrade and replace the rejected dal. Part of the rejected dal was replaced but 89.607 tonnes was not replaced. A suit was filed to recover the cost and damages with interest. It was urged that the supplementary agreement

---

[48] Central Bank of India v. Tarseema Compress Wood Mfg Co., AIR 1997 Bom. 225.

[49] Ranganayakamma v. K.S. Prakash, (D) by L.Rs., 2008(4) Recent Apex Judgments (R.A.J.) 244; 2008(8) JT 510: 2008(9) Scale 144: 2008(9) SCR 297; 2008(15) SCC 673: 2009 (sup) AIR (SC) 1218; 2008 AIR (SCW) 6476.

[50] Munusamy v. Nava Pillai, AIR 2008 (NOC) 1401 (MAD). Also see: M/s Q-Soft System & Solutions Pvt. Ltd. v. H. N. Giridhar, AIR 2008 (NOC) 1294 (Kar.).

was signed under duress and coercion. It was held by the Andhra Pradesh High Court that there was no duress or coercion but there was no consideration for the supplementary agreement. The cost of un-replaced dal was decreed but the compensation was denied. On appeal, the Supreme Court upheld the finding that the supplementary agreement had no consideration.[51] The plea that non-encashment of bank guarantee was a consideration was rejected because the agreement did not mention so, and also that after acceptance of dal the original agreement came to an end of which no breach was committed to enable encashment of bank guarantee.

It will be noticed from the examples given above that the essence of consideration is that the promisee takes on him some kind of burden, or detriment. The detriment may consist of something that has been done, or actually parting with something of value, or in undertaking a legal responsibility or in forgoing the exercise of a legal right. In the last case, where one forgoes a legal right, the other party may have no actual benefit, yet according to the law it is a consideration.

EXAMPLE FROM CASE LAW

An excellent illustration of the surrender of legal right as a consideration is furnished by an old American case.[52] In this case, an uncle promised his nephew $5,000 if the latter refrained from smoking until he was 21 years old. When the uncle did not pay him $5,000 as promised, he sued him for breach of the contract. The uncle put forth the defence of lack of consideration. It was held by the court that the nephew had given up a legal right and was entitled to recover. It will be noticed that the consideration was valid although the uncle received no benefit.

However, where a businessman paid £3,579,375 to a construction contractor in respect of five property developments, without requiring any detailed investigation of the figures and later found that a detailed investigation would have reduced the sum from £3,579,375 to £3,281,82, the Court of Appeal in England rejected the argument that there was a partial failure of consideration.[53]

In building and engineering contracts, a contractor's promise to execute the work according to drawings and specifications is binding because it is generally given in consideration of the owner's promise to pay for the cost of the work. A consideration, to be enforceable by law must be lawful, definite and certain. It is settled that a definite price is an essential element of a binding agreement. Although a definite price need not be stated in the contract, assertion thereof either expressly or impliedly is imperative.[54]

What happens where the consideration consists of performance of a pre-existing duty? The classic English law was that this is not a valid consideration for a promise.[55] A claimant undertook a subcontract to complete work on a block of flats but at an undervalued sum of £20,000. Inevitably, he came up against financial difficulties to complete the rest of the work, and threatened to stop work without further payment. The defendant – the main contractor – agreed to pay a further £10,300 to him to avoid triggering a penalty clause with the main contractor's other contracting party. It was held

---

[51] Food Corporation of India v. Surana Commercial Company, 2005 (1) CTLJ 107 (SC).
  Also see Vijaya Minerals Pvt. Ltd. v. Bikash Chandra Deb, AIR 1996 Cal. 67.
[52] Haner v. Sindways, 124, N.Y. 538 (1891).
[53] Graham Leslie v. Farrar Construction Limited, [2016] EWCA Civ. 1041.
[54] Delhi Development Authority, N.D. and Anr. v. Joint Action Committee, Allottee of SFS Flats and Ors., AIR 2008 SC 1343.
[55] Stilk v. Myrick, [1809] 170 ER 1168.

that the claimant was entitled to claim that extra payment (even if it was granted to complete the work originally undertaken). Here, the consideration for the bonus payment was the practical benefit of avoiding the penalty clause.[56] This is a difficult decision that applied a flexible approach to the consideration rule. Had the facts involved a contractor that had not under-bid in the original contract but simply held the main contractor to ransom, the court might have come to a different conclusion.

Every agreement of which the object (i.e. purpose or "design") or consideration is unlawful is void. The following illustrations selected from Section 23 of the Indian Contract Act will help clarify the matter.

1. A agrees to sell his house to B for Rs. 10,000. Here B's promise; to pay the sum of Rs. 10,000 is the consideration for A's promise to sell the house, and A's promise to sell the house is the consideration for B's promise to pay Rs. 10,000. These are lawful considerations.

2. A promises to B to obtain an employment in the public service, and B promises to pay Rs. 1,000 to A. The agreement is void, as the consideration for it is unlawful.

A few more examples will help further clarify the law.

EXAMPLES FROM CASE LAW

1. An agreement was entered into between the Societies and the agents under which the purport of one of the clauses was that the agent, who was paid a huge amount, should influence the government and procure Preliminary and Final Notifications under Sections 4 and 6 of the Land Acquisition Act. It was held that the agreement was opposed to public policy and the impugned Notifications being the product or fruits of such an agreement being injurious to public interest and detrimental to purity of administration could not be allowed to stand.[57]

2. In an agreement for a sale of land, the construction work near the sea was stayed because of the impact of Costal Zone Regulations. After the stay was vacated, the buyer was directed to pay the amount with penal interest. His plea that the penal interest claim was unreasonable was denied and the petition was dismissed on assurance of the authority that interest on interest would not be charged. It was held that Section 23 is attracted only when the consideration or object is unlawful. The difficulty faced by the buyer did not attract the vice of Section 23.[58]

3. The plaintiff bank gave a loan for the construction of a building of which lease deed was duly registered. It was subsequently revealed that the building could only be used for residential purpose under the Delhi Development Act. It was held that the agreement and the lease deed were void as per Section 23 and the defendant had to repay the loan amount with interest.[59]

4. A contract for hiring vessels for three years contained a provision enabling the hirer to terminate the contract after one year without assigning any reason. It was held by a division bench of the Bombay High Court that such a provision in the agreement is not unconscionable or opposed to public policy but covered by Section 14(1)(c) of Specific

---

[56] Williams v. Roffey Bros & Nicholls Ltd., [1990] 2 WLR 1153.

[57] Anjanappa v. Vyalikaval House Building Co-operative Society Limited, (SC): 2012(5) KantLJ 462: 2012(10) SCC 184: Recent Apex Judgments (R.A.J.) 456: 2012(2) Scale 504: 2013(1) Apex Court Judgments (SC) 687.

[58] T.H. & R. Pvt. Ltd., Ernakulam v. Greater Cochin Devpt. Authority, AIR 2001 Ker. 279.

[59] Bank of Rajasthan Ltd. v. Sh Palaram Gupta, AIR 2001 Delhi 58.

Relief Act. The said provision stipulates that "a contract which is in its nature determinable" cannot be specifically enforced. Interim injunction issued by the single judge under Section 9 of the Arbitration and Conciliation Act, 1996 was set aside and injunction vacated by the division bench.[60]

2.6.2 Distinction between Executed and Executory Consideration

The consideration may be either executed or executory. An executed consideration consists of an act or forbearance in return for a promise. It is the act which forms the consideration. No contract is formed unless the act is performed. The act stipulated for exhausts the consideration, so that any subsequent promise, without further consideration is merely a *nudum pactum*. In a contract with executed consideration the liability is outstanding on one side only. On the other hand, in an executory consideration, the liability is outstanding on both sides. It is in fact a promise for a promise. One promise is bought by the other. The contract is concluded as soon as the promises are exchanged. In mercantile contracts this is by far the most common variety.[61]

In a case, there was a clear stipulation in the policy of a Land Development Authority that the rate prevalent on the date of allotment would apply to a lease. An offer of allotment was made to the respondent who did not accept the same. The whole of the pre-paid deposit was refunded and accepted by the respondent without objection. Fresh allotment was made to the respondent three years after the earlier offer and it clearly specified the rate which was accepted by the respondent and the lease executed. A complaint was made against the charging of a higher rate than that in the earlier offer and was accepted by the Monopolies and Restrictive Trade Practices Commission. On appeal, it was challenged in the Supreme Court. It was held the complaint was not tenable since the earlier offer did not culminate in a concluded contract.[62]

2.7 Subject Matter

It is obvious that for a contract to be valid, its subject matter must be legal. The terms of a contract must expressly lay down the requirements with sufficient clarity and enough detail so as not to leave any doubt as to what is wanted. An agreement, the meaning of which is not certain, or capable of being made certain, is not enforceable by law and is void.

The price in a building or construction contract is of fundamental importance; it is so essential a term that there is no contract unless there is an agreement as to what the price is to be or at least there is an agreed method of ascertaining it, not dependent on the negotiations of the two parties themselves.

EXAMPLE FROM CASE LAW

In a suit for specific performance of contract, it was found that there was no concluded contract between the prospective purchasers of flats and the promoters of building since there was no consensus between the parties as regards the price payable for the transfer of the flats. It was held that in the absence of a concluded contract, specific performance could

---

[60] O.N.G.C. Ltd. v. Streamline Shipping Co. Pvt. Ltd., AIR 2002 Bom. 420.

[61] Union of India v. M/s Chaman Lal Loona, AIR 1957 SC 652 (655).

[62] New Okhla Industrial Development Authority v. Arvind Sonekar, AIR 2008 SC 1983. Also see Kohli Housing & Dev. P. Ltd. v. Convenience Enterprises Pvt. Ltd., 2009(1) CTLJ 379 (Delhi) (DB).

not be ordered. The builder was directed to refund the amount received by him from plaintiffs/purchasers and on receipt of the same the plaintiffs were to vacate the flats in their possession.[63]

Scope for Interpretation

Throughout the law of contract there is respect for the sanctity of the contract and the need to give effect to the reasonable expectation of an honest man. It is important that law ought to uphold rather than destroy apparent contracts. Solemn contracts entered into between parties are not to be readily declared invalid for uncertainty or vagueness. If the court is satisfied that there was ascertainable and determinative intention, it must give effect to that intention. A contract will become void only when its terms are vague, uncertain and cannot be made certain. Mere difficulty in interpretation is not synonymous with vagueness. Documents embodying business agreements should be construed fairly and broadly so as to give business efficacy. It is true that if any of the terms of the document is clearly uncertain and incapable of being made certain it may not be open to the parties to attempt to remove the vagueness of uncertainty by adducing other evidence.[64]

2.7.1 Subject Matter Must Be Lawful

Section 23 of the Indian Contract Act declares when an agreement is unlawful and therefore void. The section reads as below:

> 23. The consideration or object of an agreement is lawful unless it is forbidden by law, or is of such a nature that, if permitted, it would defeat the provisions of any law, or is fraudulent, or involves or implies injury to the person or property of another, or the Court regards it as immoral, or opposed to public policy.
>
> In each of these cases, the consideration or object of an agreement is said to be unlawful. Every agreement of which the object or consideration is unlawful is void.

The provisions are clear and self-explanatory. However, the phrases "forbidden by law" and "if permitted, it would defeat the provisions of any law" refer to defeating the intention which the legislature has expressed, or which is necessarily implied from the express terms of an Act. An agreement will not be void merely because it tends to defeat some purpose ascribed to the legislature by conjecture, from extraneous evidence, such as legislative debates or preliminary memoranda, not forming part of the enactment. What makes an agreement, which is otherwise legal, void is that its performance is impossible except by disobedience of law. Clearly, no question of illegality can arise unless the performance of the unlawful act was necessarily the effect of an agreement.[65] "Nor must it be forgotten that the rule by which contracts not expressly forbidden by statute or declared to be void are in proper cases nullified for disobedience to a statute is a rule of public policy only, and public policy understood in a wider sense may at times be better served by refusing to nullify a bargain save on serious and sufficient grounds." This was the observation of Lord Wright.[66] The following examples will make the point clear.

---

[63] M/s Mirahul Enterprises v. Vijaya Sirivastava, AIR 2003 Delhi 15.
[64] Kandamath Cine Enterprises (Pvt.) Ltd v. John Philipose, AIR 1990 Ker. 198 (201).
[65] Shri Lachoo Mal v. Shri Radhey Shyam, 1971(8) R.C.R.(Rent) 320: (1971) 1 SCC 619.
[66] Vita Food Products Incorporated v. Unus Company Ltd. (in liquidation), [1939] AC 277 at p. 293.

EXAMPLES FROM CASE LAW

1. Ministry of Defence issued instruction to the officers of the Defence Department to reject a tender where the rate quoted by the tenderer was more than 20% below reasonable rates. The Supreme Court held that the letter was not an Act of legislature declaring that any supply made at a rate below 20% of the reasonable rates was unlawful.[67]

2. At the time a person was promoted as junior engineer by way of a stopgap arrangement, he had given an undertaking that he would not claim promotion as of right or any benefit pertaining to the post. He was continued on the said post but without benefits of the said post. The Supreme Court held that the agreement that he would not claim higher salary on being promoted was opposed to public policy and thus unenforceable under Section 23 of the Contract Act. The person was held to be entitled to salary of promotional post and to be considered for regular promotion.[68]

Applicability of some of the principles embodied in Section 23 to the cases in building and engineering contracts are considered below in paragraphs 2.7.2 and 2.7.3.

2.7.2 An Agreement between Intending Tenderers that One Should Refrain from Tendering Is Not Unlawful

A tenderer, say, desperately wanting a contract, enters into an agreement with another tenderer whereby the other agrees not to submit a tender, or to withdraw his tender in consideration of an agreed sum. What is the effect of such an agreement? It is clear that the owner inviting tenders would be deprived of the benefits of a clean and fair competition among tenderers. It is also likely that he may have to incur additional expenses in executing the work. If the owner happens to be a public body or a government, the interest of the public at large is involved. Therefore, a question arises as to whether such contracts are lawful. This should be considered in the light of Section 23 of the Indian Contract Act. The consideration in such a contract is the sum of money payable by one party to another. The consideration from the other party is its promise to forgo the exercise of a legal right to submit the tender. Thus, *prima facie*, the mutual considerations appear legal and enforceable by law.

The section, however, does not stop at that. It includes the word "object". The word "object" is not used in the same sense as "consideration" but means "purpose" or "design". It is, therefore, necessary to examine whether the purpose of agreement is invalid on any of the grounds mentioned in the section. When there was no Competition Act in India,[69] the only ground on which such contracts were challenged was that the court regarded the object of such a contract as "immoral or opposed to public policy". In a number of decisions by the Indian Courts, it has been held that such agreements are valid and not opposed to public policy. The Kerala High Court, in a case decided in 1973, relied upon the earlier decisions of the Madras High Court and the Nagpur High Court. The short facts of the case are given in the example below.

---

[67] Union of India v. Col. L.S.N. Murthy, 2011(4) ArbiLR 288:: 2012(1) SCC 718; 2012(3) Recent Apex Judgments (R. A.J.) 297: 2011(13) SCR 295.

[68] Secretary cum Chief Engineer v. Hari Om Sharma, AIR 1998 SC 2909.

[69] Ramalingiah v. Subbarami Reddi, AIR 1951 Mad. 390.

EXAMPLE FROM CASE LAW

The parties were PWD contractors. The defendant had tendered for a contract along with the plaintiff and two others. The defendant's tender was at first accepted and he commenced work. But later, the work was re-tendered. On that occasion to dissuade the plaintiff from tendering, the defendant issued a cheque for Rs. 1,000/-. Two other contractors were also scared away like this. This was how the case developed at the trial. The plaintiff's suit initially was for a recovery of Rs. 1,000/- purported to have been borrowed by the defendant undertaking to repay within a month. Instead of parting with cash, the defendant had issued a cheque to the plaintiff. The bank dishonoured the cheque. At the trial the defendant put forth above mentioned facts and pleaded that the agreement was opposed to public policy hence unsustainable in law. These contentions were rejected and the suit was decreed. The appellate court confirmed the decree of the trial court. An appeal was filed in the High Court. Confirming the decree and dismissing the second appeal the Kerala High Court observed,[70] "[i]t is clear that the transaction is not opposed to public policy. After having benefited by the abstention of the plaintiff from tendering, the defendant now wants to wriggle out and befool the plaintiff."

It may, however, be noted that the owner has a right to reject all tenders. He may very well adopt this course of action if there exists a well-grounded suspicion of collusion between the tenderers.

2.7.3 An Agreement to Execute Work in the Land Not Acquired, If and When Invalid

Section 23 declares that the consideration or object of an agreement is unlawful if it involves injury to person or property of another. It is not uncommon in public works that an agreement is signed even before the land in question is acquired from its owners. The example below explains the applicability of Section 23 to the contracts so entered. The authors agree with the views expressed therein.

EXAMPLE FROM CASE LAW

The Railway Administration entered into a contract with the respondent-contractor for construction of a railway line. Due to various reasons, including the inability of the Railway Administration to make available the entire land after acquisition, the work was delayed. Therefore, a fresh agreement with revised rates was entered into. After several extensions of the time limit for completion, only a small portion of the work was done. The contract was terminated at the risk and cost of the contractor. When steps were taken to encash the bank guarantee, the contractor persuaded the bank not to respond. The respondent filed a suit to declare the agreement null and void and for an injunction against encashment of bank guarantee. It was contended that the agreement involved executing the work in the property belonging to a third person and was, therefore, unlawful. The trial court dismissed the application. The District Judge in appeal allowed it. In civil revision petition before the Kerala High Court, it was urged that the contract itself was null and void for the reason that it involved injury to the property of the third person and hence was unlawful under Section 23 of the Contract Act. Reliance was placed upon the decision in K. Abdulkhadar v. Plantation Corporation of Kerala Ltd.[71] It was held by the Kerala High Court that[72]

---

[70] M. Mohammed v. A. Narayana Rao, AIR 1973 Ker. 266.
[71] K.F. Abdulkhadar v. Plantation Corporation of Kerala Ltd., AIR 1989 Ker. 1.
[72] Union of India v. Philips Construction, AIR 1989 Ker. 152.

the agreement was only to do the work in the properties after they are acquired and handed over. The agreement itself contemplated delays in that process and by mutual consent it was provided that in case of delay in so doing time will be extended without the right to get compensation. Even though not within the period stipulated in the agreement the properties were acquired and made available. On account of the delay in acquiring and handing over the land, the contractor had not rescinded the contract and he did not claim any enhanced amount. That means the contract was subsisting even after the land was made available. Thereafter the contractor sought several extensions of the term of contract on account of his own inconvenience and obtained extensions also. He is not justified in contending that the contract is void under S. 23 of the Contract Act. That is evidently only a device to wriggle out of the contract and escape liability. S. 23 of the Contract Act has no application in this case for the reason that the work agreed to be done was only on the properties after they are acquired and made available. Such attitudes on the part of the contractors are uncharitable and not conducive to public interest. I find no merit in the contention.

The earlier decision was distinguished on the facts in these words:

The agreement in that case was for the construction of a Road for the Plantation Corporation through properties of strangers. Even before acquiring and handing over some portions of those properties the Corporation cancelled the agreement on the allegation of breach of agreement by the contractor and claimed damages. It was in that context that the decision held that it involves injury to the properties of other persons and hence void under S. 23.

2.8 Void and Voidable Contracts

Section 2(g) of the Indian Contract Act defines void agreement thus:

An agreement not enforceable by law is said to be void.

The word void means total absence of legal effect. It is a nullity. The Indian Contract Act incorporates the following provisions, which declare which agreements are void.

Section 20 where both parties are under mistake;

Section 23: Consideration and object unlawful;

Sections 24 and 25: Agreement without consideration;

Section 26: Agreement in restraint of marriage;

Section 27: Agreement in restraint of trade;

Section 28: Agreement in restraint of legal proceedings;

Section 29: Agreement void for uncertainty;

Section 30: Agreement by way of wager void;

Section 35: Agreement becomes void being dependent upon contingent event happening;

Section 36: Agreement contingent on impossible event;

Section 56: Agreement to do impossible act.

2.8.1 Voidable Contract

Section 2(i) of the Indian Contract Act defines a voidable contract in these words: "An agreement which is enforceable by law at the option of one or more of the parties thereto, but not at the option of the other or others, is a voidable contract."

A voidable contract is not a nullity. It is enforceable by law at the option of one of the parties to the contract but such option is not available to the other party.

Sections 14, 15, 16, 17, 18 and 19 of the Indian Contract Act deal with voidable contracts. These have been considered in paragraph 2.5 above. Section 19 of the Indian Contract Act describes the circumstances in which a contract is rendered voidable. The said section reads:

19. *Voidability of agreements without free consent –*

When consent to an agreement is caused by coercion, fraud or misrepresentation, the agreement is a contract voidable at the option of the party whose consent was so caused.

A party to a contract, whose consent caused by fraud or misrepresentation, may, if he thinks fit, insist that the contract shall be performed, and that he shall be put in the position in which he would have been if the representations made had been true.

Exception: If such consent was caused by misrepresentation or by silence, fraudulent within the meaning of section 17, the Contract nevertheless, is not voidable, if the party whose consent was so caused had the means of discovering the truth with ordinary diligence.

Explanation: A fraud or misrepresentation which did not cause the consent to contract of the party on whom such fraud was practised, or to whom such misrepresentation was made, does not render a contract void.

ILLUSTRATIONS

(a) A intending to deceive B, falsely represents that five hundred maunds of indigo are made annually at A's factory, and thereby induces B to buy the factory. The contract is voidable at the option of B.

(b) A, by a misrepresentation, leads B erroneously to believe that 500 maunds of indigo are made annually at A's factory. B examines the accounts of the factory, which show that only 400 maunds of indigo have been made. After this, B buys the factory. The contract is not voidable on account of A's misrepresentation.

(c) A fraudulently informs B that A's estate is free from encumbrance. B thereupon buys the estate. The estate is subject to a mortgage. B may either avoid the contract, or may insist on its being carried out and the mortgage-debt redeemed.

(d) B, having discovered a vein of ore on the estate of A, adopts means to conceal, and does conceal, the existence of the ore from A. Through A's ignorance B is enabled to buy the estate at an under-value. The contract is voidable at the option of A.

(e) A is entitled to succeed to an estate at the death of B. B dies. C, having received intelligence of B's death, prevents the intelligence reaching A, and thus induces A to sell him his interest in the estate. The sale is voidable at the option of A.

The explanation enacts that a fraud or misrepresentation does not render a contract voidable, if it does not cause the consent to a contract of the party to whom such misrepresentation was made. One cannot complain of having been misled by a representation, which did not lead him at all.

The exception lays down that the contract is not voidable if the party whose consent was caused by misrepresentation or by silence (fraudulent within the meaning of Section 17), had the means of discovering the truth with ordinary diligence. It is now well established that the exception does not apply to the cases wherein there is a positive case of an active fraudulent representation which has been responsible for bringing about the contract between the parties. This position has been established in various decisions, including the decisions given by Allahabad High Court in Niaz Ahmed Khan v. Parsottam Chandra, AIR 1931 Allahabad 154, by Calcutta High Court in John Minas Apcar v. Louis Caird Malchus, AIR 1939 Calcutta 473, and by Madras High Court in Kopparthi Venkataratnam v. Palleti Sivaramudu, AIR 1940 Madras 560.[73]

The right to rescind the contract for fraud, misrepresentation or coercion is lost if the person having the right to rescind affirms the contract. Until the contract is avoided, it is valid.[74]

In respect of building or engineering contracts, as already observed the misrepresentation, either fraudulent or innocent, is likely to be discovered after some progress is achieved in execution of the work. Besides, the misrepresentation may affect the execution of one or more items which may not justify the contractor to claim rescission of the contract as a whole. In such a case, it is submitted that Section 19 enables a contractor to carry out the work and claim damages. Care may have to be taken to give notice of the intention to claim damages soon after the misrepresentation or fraud is discovered so that it cannot be said that the right is waived and the contract affirmed.[75]

EXAMPLES FROM CASE LAW

1. Two years after the contractors completed the contract for a branch railway, they claimed rescission of the contract on the ground of innocent misrepresentation made by the engineer of the Railway Company as to the nature of strata through which the railway alignment passed. Fraud was not proved. It was held that the contractors by completing the contract with full knowledge of the facts had rendered *restitutio in integrum* (restoring the parties to their original positions) impossible. The claim of the contractor failed.[76]

2. The Maharashtra State Government approved a scheme of a landowner for development of land under which some land was reserved for public utility. The agreement provided that the landowner would transfer the public utility land to an improvement trust free of cost and without compensation. The validity of the agreement was questioned. Both the High Court and the Supreme Court of India held that the agreement was neither void nor illegal for want of consideration.[77]

The effect on validity of an agreement induced by undue influence is dealt with in Section 19 (A) of the Act. That section provides:

---

[73] Niaz Ahmed Khan v. Parsottam Chandra, AIR 1931 Allahabad 154 in John Minas Apcar v. Louis Caird Malchus, AIR 1939 Calcutta 473 and Madras High Court in Kopparthi Venkataratnam v. Palleti Sivaramudu, AIR 1940 Madras 560 referred to in M/s. R.C. Thakkar v. (The Bombay Housing Board by its successors) now The Gujarat Housing Board, AIR 1973 Gujarat 34.

[74] Dhurandar Prasad Singh v. Jai Prakash University, AIR 2001 SC 2552.

[75] See, for example, P.C.K. Muthia Chettiar v. V.E.S. Shanmugham Chettiar, (SC) 1969 AIR (SC) 552: 1969(1) SCR 444: 1968(2) AIR (SCW) 351: 1969(2) SCJ 105: 1969(2) MLJ 28: 1969(2) An.WR 28.

[76] Glasgow and South Western Rly. v. Boyd & Forrest, [1915] AC 526.

[77] Narayanrao Jagobaji Gowande Public Trust v. State of Maharashtra, (SC): 2016(1) R.C.R.(Civil) 1015: 2016(2) Recent Apex Judgments (R.A.J.) 51: 2016(2) Scale 149: 2016 AIR (SCW) 823: 2016(2) JT 159: 2016(1) Law Herald (SC) 700: 2016(2) AIR Bom.R 509: 2016 AIR (SC) 823: 2016(2) ALL MR 938: 2016(160) AIC 208: 2016(4) SCC 443: 2016(161) AIC 110: 2016(2) KCCR 158: 2016(1) PLJ 689: 2016(2) C.W.C. 142.

19(A). *Power to set aside contract induced by undue influence*

When consent to an agreement is caused by undue influence, the agreement is a contract voidable at the option of the party whose consent was so caused.

Any such contract may be set aside either absolutely or, if the party who was entitled to avoid it has received any benefit thereunder, upon such terms and conditions as to the court may deem just.

ILLUSTRATIONS

(a) A's son has forged B's name to a promissory note. B, under threat of prosecuting A's son, obtains a bond from A, for the amount of the forged note. If B sues on this bond, the court may set the bond aside.

(b) A, a money-lender, advances Rs. 100 to B, an agriculturist, and, by undue influence, induces B to execute a bond for Rs.200 with interest at 6 per cent per month. The court may set the bond aside, ordering B to repay the Rs. 100 with such interest as may seem just.

An agreement may be a valid contract when it was made but may become legally non-enforceable subsequently due to a change in governing circumstances. In such a case, the contract attracts the provision of Section 2(j).

2.8.2 Void Contract

Section 2(j) defines void contract thus: *"Void Contract:* A contract which ceases to be enforceable by law becomes void when it ceases to be enforceable."

An example of such a contract is a contract with an alien friend, which was valid when it was made, but ceases to be enforceable by law when, in wartime, the alien friend becomes alien enemy.[78]

2.8.3 Material Alterations in an Agreement Renders it Void

Material alterations in an agreement subsequent to its execution render it void. A material alteration is one (i) which varies the rights, liabilities or the legal position of the parties as ascertained by the deed in its original state; or (ii) otherwise varies the legal effects of the instruments originally expressed; or (iii) reduces to certainty some provision which was originally unascertained and as such void; or (iv) otherwise prejudices the party bound by the deed as originally executed. An alteration made in a deed, after its execution in some particular, which is not material, does not in any way affect the validity of the deed.[79] Neither party to a void contract can claim any relief in a court of law.[80]

---

[78] Mahanth Singh v. U. Ba Yi, AIR 1939 PC 110; also see B.O.I. Finance Ltd. v. The Custodian, AIR 1997 SC 1952.

[79] Halsbury's Law of England 287; Kalianna Gounder v. Palani Gounder, AIR 1970 SC 1942; Sardar v. Ram Khilauna, AIR 1997 All. 268.

[80] Khaja Moinuddin Khan v. S.P. Ranga Rao, AIR 2000 A.P. 344. R.G. Jogdand v. K.J. Naikwade, AIR 2002 Bom. 149. Parakkate Shankaran Keshavan v. T.A. Sukumaran, AIR 1997 Bom. 381. However, see: Tarsem Singh v. Sukhminder Singh, AIR 1998 SC 1400.

2.8.4 Difference between Void and Illegal Contract

The distinction between void and illegal contracts may be very thin but it exists. Under Section 23, contracts that are opposed to public policy or immoral are void. A void contract is one that has no legal effect. Illegality of a contract arises as a result of infraction, contravention or breach of any express or implied provisions of law. The "law" means a constitutionally valid enactment made by legislature or by subordinate legislation, i.e. rules, byelaws, regulations or usages and customs having the force of law. An illegal contract resembles a void contract in that it also has no legal effect as between the immediate parties. An illegal contract has this further effect that even transactions collateral to it become tainted with illegality and are, therefore, in certain circumstances not enforceable. A collateral contract to a void contract, on the other hand, may be enforceable.

EXAMPLE FROM CASE LAW

Petitioners had opened accounts in a post office monthly income scheme. Under the scheme there was a capping of the maximum balance over and above which no interest would be payable. The petitioners went on depositing amounts in excess of the said limit and also received interest thereon. Subsequently, the post office sought refund of the excess interest paid. The Orissa High Court held that Section 23 of the Indian Contract Act did not cover the case and therefore the direction for recovery was not sustainable.[81]

2.8.5 Mode of Communicating or Revoking Rescission of Voidable Contract

Section 66 of the Indian contract Act provides as under.[82]

The rescission of a voidable contract may be communicated or revoked in the same manner, and subject to the rules as apply to the communication or revocation of a proposal.[83]

2.8.6 Consequences of Rescission of Voidable Contract

Section 64 of the Indian Contract Act stipulates the consequences in simple words as under:

When a person at whose option a contract is voidable rescinds it, the other party thereto need not perform any promise therein contained in which he is promisor. The party rescinding a voidable contract shall, if he has received any benefit thereunder from another party to such contract, restore such benefit, so far as, may be, to the person from whom it was received.

2.8.7 Consequences of Void Agreement

Section 65 of the Indian Contract Act provides for obligations of the parties to void agreement or contracts in these words:

65. *Obligation of person who has received advantage under void agreement or contract that becomes void*
   When an agreement is discovered to be void or when a contract becomes void, any person who has received any advantage under such agreement or contract is bound to restore it, or to make compensation for it, to the person from whom he received it.

---

[81] Rajat Kumar Rath v. Government of India, AIR 2000 Ori.32.
[82] For full discussion see Ch. 1 paragraphs 1.13 and 1.13.2.
[83] See discussion under Section 4 of the Act and 1.13.1.

ILLUSTRATIONS

(a) A pays B Rs. 1,000 in consideration of B's promising to marry C, A's daughter. C is dead at the time of the promise. The agreement is void, but B must repay A Rs. 1,000.

(b) A contracts with B to deliver to him 250 maunds of rice before the 1st of May. A delivers 130 maunds only before that day, and none after. B retains the 130 maunds after the 1st of May. He is bound to pay A for them.

(c) A, a singer, contracts with B, the manager of a theatre, to sing at his theatre for two nights every week during the next two months, and B engages to pay her Rs. 100 for each night's performance. On the sixth night, A wilfully absents herself from the theatre, and B, in consequence, rescinds the contract. B must pay A for the five nights on which she had sung.

(d) A contracts to sing for B at a concert for Rs. 1,000, which is paid in advance. A is too ill to sing. A is not bound to make compensation to B for the loss of the profits which B would have made if A had been able to sing, but must refund to B the Rs. 1,000 paid in advance.

EXAMPLE FROM CASE LAW

A portion of land belonging to a company under liquidation was sold in an auction. The entire sale amount was deposited. However, it was discovered that the land was not only less than that mentioned in the notice but there was a title suit pending against it unknown to the purchaser and the official liquidator. It was held that the sale was liable to be set aside and the purchaser entitled to refund of the amount deposited by him.[84]

2.8.8 Arbitration Clause in a Contract May Be Valid

Where a contract, entered into without free consent is voidable (on account of undue influence, fraud, misrepresentation or mistake of fact), the arbitration clause contained therein may be valid. The issue about non-existence of a concluded contract, however, can be left open to be decided by arbitrator.[85]

2.8.9 The Doctrine of Fairness and Statutory Contract

The doctrine of fairness is nothing but a duty to act fairly and reasonably and not arbitrarily. It is a doctrine developed in the administrative law field to ensure rule of law and to prevent failure of justice where an action is administrative in nature. It certainly cannot be invoked to amend, alter, or vary an express term of the contract between the parties. This is so even if the contract is governed by a statutory provision, i.e. where it is a statutory contract. The Supreme Court of India has observed: "It is one thing to say that a statutory contract or for that matter, every contract must be construed reasonably, having regard to its language. But to strike down the terms of a statutory contract on the ground of unfairness is entirely different."[86]

---

[84] Mahendra Kumar Roongta v. Official Liqudator, AIR 1996 Cal. 146.

[85] Enercon (India) Ltd. v. Enercon GMBH, (SC): 2014(1) ArbiLR 257: 2014(3) JT 49: 2014(5) SCC 1: 2014(2) SCR 855:: 2014(2) R.A.J. 112: 2014(2) Scale 452:: 2014 AIR (SCW) 3513: 2014 All SCR 2317 2014 AIR (SC) 3152.

[86] Mary v. State of Kerala, 2015(1) SCC (Civil) 295: 2014(14) SCC 272: 2013(9) SCR 1126: 2013 AIR (SCW) 6082: 2013(6) Recent Apex Judgments (R.A.J.) 565: 2014 AIR (SC) 1: 2013(13) JT 538.

2.9 Contingent Contract

Section 31 of the Indian Contract Act defines contingent contract in these words:

> A contingent contract is a contract to do or not to do something, if some event, collateral to such contract, does or does not happen.

ILLUSTRATION

A contracts to pay B Rs.10,000/- if B's house is burnt. That is a contingent contract.

The contract of life insurance and similar other contracts are contingent contracts within the meaning of this section. Whether a contract is of the kind specified in the Act may be a question of fact or construction. For example, the Supreme Court has held that where a contract requires a No Objection Certificate (NOC) from a concerned authority, it was incidental. Such a condition did not render the contract contingent and much less void being impossible of performance.[87]

Sections 32 to 36, of the Indian Contract Act deal with the enforcement of contingent contracts. The said sections are self-explanatory. Section 32 reads:

> 32. *Enforcement of contracts contingent on an event happening*
>
> Contingent contracts to do or not to do anything if an uncertain future event happens cannot be enforced by law unless and until that event has happened. If the event becomes impossible, such contracts become void.

ILLUSTRATIONS

(a) A makes a contract with B to buy B's horse if A survives C. This contract cannot be enforced by law unless and until C dies in A's lifetime.

(a) A makes a contract with B to sell a horse to B at a specified price, if C, to whom the horse has been offered, refuses to buy him. The contract cannot be enforced by law unless and until C refuses to buy the horse.

(a) A contracts to pay B a sum of money when B marries C. C dies without being married to B. The contract becomes void.

In building and engineering contracts, an example of a contingent contract is when a main contractor, before submitting his tender, enters into a contract with a subcontractor, whose contract is dependent upon the future event of the main contractor succeeding in securing the contract or a contract may be conditional upon the obtaining of permission to sublet the contract.

---

[87] J.P. Builders v. A. Ramadas Rao (SC): 2011(1) SCC 429: 2011(2) MLJ 222: 2011 AIR SC (Civil) 230; 2011(1) Recent Apex Judgments (R.A.J.) 88: 2010(15) SCR 538; 2010(12) Scale 400: 2011(1) Apex Court Judgments (SC) 307.
Also see Asman Investments Ltd. v. K.L. Suneja (Delhi), 2011(181) DLT 156: 2011(124) DRJ 693: 2011(38) R.C.R. (Civil) 590.

EXAMPLE FROM CASE LAW

A tenderer, having purchased tender documents for a project, was approached by another company for joint participation. The tenderer sent a draft agreement for signature by the second company; the draft contained an arbitration clause. The second company returned the agreement duly signed but with material variations converting the joint liability of the parties to unilateral liability of the tendering company. The tenderer did not communicate with the other company but unilaterally submitted tender and subsequently withdrew the same. The second company claimed damages and invoked the arbitration clause. The tenderer filed an application under Section 33 in Delhi High Court. The High Court held the agreement had come into force the moment it was signed and sent and the tenderer, though he did not sign it acted on it, which amounted to indirect acceptance. On appeal, a three-judge bench of the Supreme Court had to decide if the submission of the unilateral tender amounted to acceptance of the counter-offer to bring into existence a concluded contract. It was held that the material alterations in the contract made a world of difference to draw an inference of a concluded contract. The joint liability of the parties was made unilateral liability of the parties. A clause containing material part of the terms for performance of the contract was deleted. Thereby there was no consensus *ad idem* on the material terms of the contract. It was further held that even if proposal and counter-proposal were assumed to have concluded into an agreement, it was a contingent contract and, by operation of Section 32 of the Contract Act, the counter-proposal could not be enforced.[88]

The provisions of Sections 33 to 36 are self-explanatory and reproduced below for ready reference.

33. *Enforcement of contracts contingent on an event not happening*

Contingent contracts to do or not to do anything if an uncertain future event does not happen can be enforced when the happening of that event becomes impossible and not before.

ILLUSTRATION

A agrees to pay B a sum of money if a certain ship does not return. The ship is sunk. The contract can be enforced when the ship sinks.

34. *When event on which contract is contingent to be deemed impossible if it is the future conduct of a living person*

If the future event on which a contract is contingent is the way in which a person will act at an unspecified time, the event shall be considered to become impossible when such a person does anything which renders it impossible that he should so act within any definite time, or otherwise than under further contingencies.

ILLUSTRATION

A agrees to pay B a sum of money if B marries C. C marries D. The marriage of B to C will now be considered impossible, although it is possible that D may die and that C may afterwards marry B.

---

[88] U.P. Rajkia Nirman Nigam Ltd. v. Indure Pvt. Ltd., AIR 1996 SC 1373. Also see: Union of India v. Rail Udyog (Delhi), 2002(1) Arb. LR 58.

35. *When contracts become void which are contingent on happening of specified event within fixed time*

Contingent contracts to do or not to do anything if a specified uncertain event happens within a fixed time become void if, at the expiry of the time fixed, such event has not happened, or if, before the time fixed, such event becomes impossible.

When contracts may be enforced which are contingent on specified event not happening within fixed time: Contingent contracts to do or not to do anything if a specified uncertain event does not happen within a fixed time may be enforced by law when the time fixed has expired and such event has not happened, or, before the time fixed has expired, if it becomes certain that such event will not happen.

ILLUSTRATIONS

(a) A promises to pay B a sum of money if a certain ship returns within a year. The contract may be enforced if the ship returns within the year, and becomes void if the ship is burnt within the year.

(b) A promises to pay B a sum of money if a certain ship does not return within the year or is burnt within that year. The contract may be enforced if the ship does not return within the year, or is burnt within the year.

36. *Agreements contingent on impossible events void*

Contingent agreements to do or not to do anything if an impossible event happens, are void, whether the impossibility of the event is known or not to the parties to the agreement at the time when it is made.

ILLUSTRATIONS

(a) A agrees to pay B Rs. 1,000, if two straight lines should enclose a space. This agreement is void.

(b) A agrees to pay B Rs. 1,000 if B will marry A's daughter C. C was dead at time of the agreement. The agreement is void.

2.10 Collateral Contracts

Collateral means "side by side; parallel". A collateral contract is a subsidiary or an additional supportive contract entered into by the parties either prior to or side by side of the main contract; its existence depends on the main contract. For example, there may be a contract, the consideration for which is the making of some other contract. An agreement between the main contractor and a subcontractor for executing a part of a work prior to submission of tender by the main contractor for the work is an example of a collateral contract.

Where a formal contract is a result of a series of oral or partly written or partly oral distinct and separate, earlier or contemporaneous agreements, it is possible that not all the terms of the final formal contract need necessarily be found in any one collateral document. However, if the final contract is not reached and formal agreement is not signed, it may still be possible to deduce an agreement or agreements between the parties from the terms of collateral contracts. The oral evidence rule will not be attracted in the absence of a formal written contract. Where, however, a formal written contract is executed and its terms are inconsistent with the collateral contracts, the collateral contracts may be deprived of any effect due to the oral evidence rule.[89]

2.11 Implied Contract

An express contract is one whose terms are declared by parties in so many words, either orally or in writing, at the time the contract is entered into. As against this in an implied contract, the law implies one or more terms from the conduct of the parties. Shortly stated an implied contract is an actual contract, circumstantially proved. Terms may be implied into a contract if they are necessary for the efficacy of the contract. For example, there may be a duty on an employer to cooperate with the contractor, if the work cannot be executed without such cooperation.[90] Similarly, there could be an implied term not to prevent a contractor from carrying out the contracted work in an orderly manner.[91]

In a suit for compensation for work and labour, the claimant may recover on the basis of an implied contract on *quantum meruit* (fair and reasonable value) basis, if the evidence fails to show an express contract but does support an implied contract.

2.12 Performance of Reciprocal Promises

Sections 51 and 52 of the Act are relevant and read as follows:

51. *Promisor not bound to perform, unless reciprocal promisee ready and willing to perform*
When a contract consists of reciprocal promises to be simultaneously performed, no promisor need perform his promise unless the promisee is ready and willing to perform his reciprocal promise.

52. *Order of performance of reciprocal promises*
Where the order in which reciprocal promises are to be performed is expressly fixed by the contract, they shall be performed in that order; and where the order is not expressly fixed by the contract, they shall be performed in that order which the nature of the transaction requires.[92]

The contents of the section are self-explanatory and the Act also gives illustrations to further make clear the provisions of the section. The illustration below is an example of the application of Sections 51 and 52 to a construction contract.

EXAMPLE FROM CASE LAW

The appellant-tenant in a case agreed to pay rent at a new rate after reconstruction of premises with toilet facilities and though the premises were reconstructed, the toilet was not constructed. The Supreme Court held that the appellant was liable to pay rent at the new rate only after the toilet was constructed.[93]

---

[89] For oral evidence rule see paragraph 4.10.12 in Chapter 4 of this volume.
[90] Merton London Borough v Stanley Hugh Leach Ltd., (1985) 32 BLR 51.
[91] Allridge (Builders) Ltd v Grand Actual Ltd., (1996) 55 Con. LR 91.
[92] This section is considered in paragraph 10.5.1 Chapter 10 while discussing breach of contract.
[93] Mohammed v. Pushpalatha, 2008(2) CTLJ 92 (SC).

2.13 Quasi-Contracts – Work Done Without Contract

Occasionally, it so happens that a contractor executes the work without there being a written or oral contract. The question in such cases arises as to what relief, if any, the contractor is entitled to in the absence of an express agreement. Sections 68 to 70 of the Indian Contract Act make provisions to govern such situations. The said sections are self-explanatory and read:

68. *Claim for necessaries supplied to person incapable of contracting, or on his account*
If a person, incapable of entering into a contract, or any one whom he is legally bound to support is supplied by another person with necessaries suited to his condition in life, the person who has furnished such supplies is entitled to be reimbursed from the property of such incapable person.

ILLUSTRATIONS

(a) A supplies B, a lunatic, with necessaries suitable to his condition in life. A is entitled to be reimbursed from B's property.

(b) A supplies the wife and children of B, a lunatic, with necessaries suitable to their conditions in life. A is entitled to be reimbursed from B's property.

69. *Reimbursement of person paying money due by another, in payment of which he is interested*
A person who is interested in the payment of money which another is bound by law to pay and who therefore pays it is entitled to be reimbursed by the other.

ILLUSTRATION

B holds land in Bengal, on a lease granted by A the zamindar. The revenue payable by A to the government being in arrear, his land is advertised for sale by the government. Under the revenue law the consequences of such sale will be the annulment of B's lease. B, to prevent the sale and the consequent annulment of his own lease, pays to the government the sum due from A. A is bound to make good to B the amount so paid.

70. *Obligation of person enjoying benefit of non-gratuitous act*
Where a person lawfully does anything for another person, or delivers anything to him not intending to do so gratuitously, and such other person enjoys the benefit thereof, the latter is bound to make compensation to the former in respect of, or to restore, the thing so done or delivered.

ILLUSTRATIONS

(a) A, a tradesman, leaves goods at B's house by mistake. B treats the goods as his own. He is bound to pay A for them.

(b) A saves B's property from a fire. A is not entitled for compensation from B, if the circumstances show that he intended to act gratuitously.

EXAMPLES FROM CASE LAW

1.   A state government superintending engineer sanctioned the estimated cost of the work. That work was allotted and done by the petitioner. From this, it follows that petitioner had performed his part and what remained was the part on behalf of state government. In the supplementary counter affidavit while not disputing the aforesaid fact, it was contended that as no formal agreement was entered into and no formal work order issued, payment cannot be made to the petitioner. It was observed:[94] "as the work was done for the Government and for the benefit of public, the petitioner is entitled to payment." The court applied Section 73. However, it is respectfully submitted that Section 70 is more appropriate in such cases as was held in Example (2) below.

2.   A contractor had undertaken the work of constructing godowns for the Civil Supplies Department of the state. When the contract work was in progress, the contractor, at the request of the Sub-Divisional Officer, Virambagh, submitted their estimate for the construction of a katcha road, guard room, kitchen and office room for the use of the department. The Additional Deputy Director of Civil Supplies, during his visit to Virambagh requested the contractor to carry out the work. The contractor executed both the works. He then submitted his bills for payment. The state declined to pay the bills, whereupon the contractor filed a suit in the Calcutta High Court. The state in defence alleged that under Section 175(3) of the Government of India Act, 1935 (analogous to Art. 299(1) of the Constitution) the request for the construction works were invalid, unauthorized and did not constitute a valid contract. In the absence of binding contracts the state was not liable to pay the contractor's bills. The High Court held that having regard to Section 175(3) of the Government of India Act, there was no valid contract between the contractor and the state. However, the contractor's claim against the state was justified under Section 70 of the Indian Contract Act. The Supreme Court while dismissing the state's appeal with costs confirmed both the above findings.[95]

2.13.1 Section 70 Analysed

It is worthwhile to consider the requirements to be satisfied to invoke the provisions of Section 70 of the Indian Contract Act. The three conditions laid down by Section 70 are:

   (i)   A person should lawfully do something for another person;

   (ii) The other person for whom something is done must enjoy the benefit thereof;

   (iii) The thing done must not be done fraudulently or dishonestly nor must it be done gratuitously.

When these conditions are satisfied, Section 70 imposes upon the latter person the liability to make compensation to the former in respect of the things so done. When claim for compensation is made by one person against another under Section 70, it is not on the basis of any subsisting

---

[94] Ram Pravesh Prasad v. State of Bihar, AIR 2007 Pat. 26. Also see Municipal Committee, Pundri v. Bajrang Rao Nagrath, AIR 2006 P. & H. 142. Also see: Birampur Construction Pvt. Ltd. v. State of Bihar, (Patna) 2017(4) PLJR 885. State of Himachal Pradesh v. Shamsher Chauhan, (Himachal Pradesh): 2016(3) SimLC 1350; Reserve Bank of India v. M/s A.B. Tools Pvt. Ltd. (Himachal Pradesh) (D.B.): 2015(2) SimLC 1197: 2015(8) R.C.R.(Civil) 510: 2015(3) Him. L.R. 1808.

[95] M/s B.K. Mondal and Sons v. State of West Bengal, AIR 1962 SC 779; Municipal Committee, Pundri v. Bajrang Rao Nagrath, AIR 2006 P. & H. 142.

contracts between the parties. It is on the basis of the fact that something was done by the party for another and the said work so done has been voluntarily accepted by the other party.

Governments are subject to the provision of Section 70 and if a government has accepted the things delivered to it or enjoyed the work done for it, such acceptance and enjoyment would afford a valid basis for claims for compensation against it.

It must be noted that where the parties are bound by a contract and have acted under it, Section 70 will have no application. It is the contract alone which governs their respective obligations.[96] Section 70 deals with quasi-contract, i.e. where there is no contract. In cases where the contract agreement does not specifically provide for compensating reasonably for the work done by the contractor, the principle of *quantum meruit* (fair and reasonable cost) would be applicable.[97] Where admittedly extra work has been done and accepted by the other party without raising any dispute, the presumption that the work has been accepted is implied.[98]

Sections 71 and 72 are self-explanatory and the provisions thereunder may also be noted.

71. *Responsibility of finder of goods*
A person who finds goods belonging to another and takes them into his custody, is subject to the same responsibility as a bailee.

72. *Liability of person to whom money is paid, or thing delivered by mistake or under coercion*
A person to whom money has been paid, or anything delivered, by mistake or under coercion, must repay or return it.

ILLUSTRATIONS

(a) A and B jointly owe Rs. 100 to C. A alone pays the amount to C, and B, not knowing this fact, pays Rs. 100 over again to C. C is bound to repay the amount to B.

(b) A railway company refuses to deliver up certain goods to the consignee, except upon the payment of an illegal charge for carriage. The consignee pays the sum charged in order to obtain the goods. He is entitled to recover so much of the charge as was illegally excessive."

2.14 Contract of Guarantee and Indemnity

2.14.1 Contract of Guarantee

In big construction projects large sums of money have to be invested in mobilization of men and machinery. Although this is primarily a responsibility of the contractor undertaking such projects, provisions are made in contracts whereunder the owner advances specified sums to the contractor for mobilization. Invariably such sums are advanced against the security of bank guarantees. The law applicable to such guarantees is embodied in Chapter VIII of the Indian Contract Act. Relevant sections read as under:

---

[96] Dabur India Ltd. v. Hansa Vision Ltd., (Delhi) (D.B.): 2012(191) DLT 335: 2012(3) BC 809: 2014(12) R.C.R.(Civil) 671.

[97] The NHPC Limited v. Oriental Engineers, (Gauhati): 2016(3) GauLR 793: 2016 AIR (Gauhati) 93: 2016(3) GauLT 235: 2016(4) ArbiLR 31: 2016(4) GauLJ 16: 2016(6) R.A.J. 636: 2016(2) NEJ 239; Gopal Chandra Bhui v. Bankura Zilla Parishad, (Calcutta) (DB) 2015 AIR (Calcutta) 124: 2015(5) Cal. H.C.N. 356: 2015(19) R.C.R.(Civil) 18.

[98] Board of Trustees for the Port of Cal. v. Royal Constructions, (Calcutta) (DB) 2013(4) ICC 31: 2013(2) Cal. H.C.N. 641: 2012(20) R.C.R.(Civil) 813. Also see AIR 1977 SC 329.

126. *"Contract of guarantee", "surety", "principal debtor" and "creditor"*

A "contract of guarantee" is a contract to perform the promise, or discharge the liability of a third person in case of his default. The person who gives the guarantee is called the "surety"; the person in respect of whom the guarantee is given is called the "creditor." A guarantee may be either oral or written.

Further aspects of enforcement of bank guarantee are considered in Chapters 7 and 12.

2.14.2 Contract of Indemnity

A contractor is required under certain provisions incorporated in the construction contracts to indemnify the owner against any loss caused by the contractor's men or machinery to the property of others. The relevant provisions of the Indian Contract Act applicable read as under:

---

124. *Contract of indemnity defined*

A contract by which one party promises to save the other from loss caused to him by the conduct of the promisor or any other person, is called a "contract of indemnity".

ILLUSTRATION

A contracts to indemnify B against the consequences of any proceedings which C may take against B in respect of a certain sum of Rs. 200. This is a contract of indemnity.

EXAMPLES FROM CASE LAW

1. A plaintiff appointed the defendant as storage and handling contractor for handling export consignments. The defendant executed indemnity and custody bond by which he agreed to remain liable for all losses, damages and shortages occurring in respect of transaction. The plaintiff established that various consignments were sent to the defendant by railways. The defendant claiming short supply had lodged claims in respect thereof with the railway authorities. Held: the defendant was obliged to make good the value of the short supply in terms of the indemnity and custody bond.[99]

2. A plaintiff, relying on the terms of an insurance contract, claimed for shortage in the goods received at their plant; this was certified by surveyor of the plaintiff and its commercial officers. The claim for loss in transit based on a contract of insurance was, however, subject to a notice of claim to the insurer and a certificate by insurers' surveyors. The insurer rejected the claim on the ground that notice of shortage and claim was not served on them or their agents. It was held that the defendants could not be expected to accept the claim for the shortage without notice to them to verify and survey them. The suit was dismissed on the ground that the loss was not proved.[100]

---

[99] Steel Authority of India Ltd. v. T.D. Kumar & Bros. Pvt. Ltd., (Calcutta): 2014 AIR (Calcutta) 184: 2014(43) R.C. R.(Civil) 843.

[100] Hindustan Copper Ltd. v. New India Assurance Co. Ltd., (Bombay) 2010(5) BCR 575: 2010(6) AIR Bom.R 597: 2010(6) ALL MR 83: 2010(5) Mh.LJ 551: 2010(46) R.C.R.(Civil) 390.

Rights of the indemnity-holder when sued are the subject matter of Section 125, the contents of which are self-explanatory and read as follows:

125. *Rights of indemnity-holder when sued*
The promisee in a contract of indemnity, acting within the scope of his authority, is entitled to recover from the promisor:

(1) all damages which he may be compelled to pay in any suit in respect of any matter to which the promise to indemnify applies;

(2) all costs which he may be compelled to pay in any such suit if, bringing or defending it, he did not contravene the orders of the promisor, and acted as it would have been prudent for him to act in the absence of any contract of indemnity, or if the promisor authorized him to bring or defend the suit.

(3) All sums which he may have paid under the terms of any compromise of any such suit, if the compromise was not contrary to the orders of the promisor, and was one which would have been prudent for the promisee to make in the absence of any contract of indemnity, or if the promisor authorized him to compromise the suit.

2.15 Contracts Which Need Not Be Performed

In addition to Sections 64 to 67 dealt with earlier, the following provisions of the Indian Contract Act dealing with this aspect are self-explanatory.

62. *Effect of novation, rescission and alteration of contract*
If the parties to a contract agree to substitute a new contract for it, or to rescind or alter it, the original contract need not be performed.

ILLUSTRATIONS

(a) A owes money to B under a contract. It is agreed between A, B and C that B shall henceforth accept C as his debtor instead of A. The old debt of A to B is at an end, and a new debt from C to B has been contracted.

(b) A owes B Rs. 10,000. A enters into an agreement with B and gives B a mortgage of his (A's) estate for Rs. 5,000 in place of the debt of Rs. 10,000. This is a new contract and extinguishes the old.

(c) A owes B Rs. 1,000 under a contract. B owes C Rs. 1,000. B orders A to credit C with Rs. 1,000 in his books, but C does not assent to the arrangement. B still owes C Rs. 1,000, and no new contract has been entered into.

63. *Promisee may dispense with or remit performance of promise*
Every promisee may dispense with or remit, wholly or in part, the performance of the promise made to him, or may extend the time for such performance, or may accept instead of it any satisfaction which he thinks fit.

ILLUSTRATIONS

(a) A promises to paint a picture for B. B afterwards forbids him to do so. A is no longer bound to perform the promise.

(b) A owes B Rs. 5,000. A pays to B Rs. 1,000, and B accepts them in satisfaction of claim on A. This payment is a discharge of the whole claim.

(c) A owes B under a contract, a sum of money, the amount of which has not been ascertained. A without ascertaining the amount gives to B, and B, in satisfaction thereof, accepts, the sum of Rs. 2,000. This is a discharge of the whole debt, whatever may be its amount.

(d) A owes B Rs. 2,000, and is also indebted to other creditors. A makes an arrangement with his creditors, including B, to pay them a composition of eight annas in the rupee (50%) upon their respective demands. Payment to B of Rs. 1,000 is a discharge of B's demand.

2.16 Contract to Be in Writing

It is not a legal necessity that a building or engineering contract, except when it is with a corporation or a government, should be in writing. However, owing to the complex nature of building and engineering contracts, it is advisable that they should be expressed in writing. Written agreements help to eliminate many grounds of future disputes between the parties. Examples of agreements that must necessarily be in writing are listed below, along with the relevant provision of the law:

1. Agreements without consideration made on account of natural love and affection. (Section 25 of the Indian Contract Act.)

2. An agreement to refer to arbitration any question between parties (to the agreement), which has already arisen. (Section 28 exception 2 of the Indian Contract Act.)

3. Memorandum of Association, Articles of Association of a company and contracts by companies (Sections 4, 10 and 46 of the Indian Companies Act of 2013 albeit that where the law does not require a contract in writing if it is made by a natural person, a company need not make such a contract in writing).

4. Creation of a trust (Section 5 of the Indian Trusts Act 1882).

5. A sale, a mortgage, a lease and a gift governed by the provisions of the Transfer of Property Act 1882.

6. Agreements made with a corporation, which must contract in writing or under seal in regard to the subject matters of such agreements.

2.17 Contract Documents

Parties to a building or engineering contract should exercise great care in naming the documents which form the basis of their agreement. A typical construction contract may include any or all of the following documents.

1. Tender notice.
2. General instructions and directions for the use of contractors.
3. Letters of acceptance, work order or other letters.
4. Form of contract.

5. Conditions of contract.
6. List of materials, if any, agreed to be supplied to the contractor by the owner, and the conditions of supply.
7. Bill of Quantities.
8. Specifications: General and Particular.
9. Drawings.
   The letters, if any, exchanged between the parties while the contract was being negotiated, should be included in the list, as shown above, if their contents have not been included in the written agreement. It often happens that the parties seek clarifications of some points or impose or accept special conditions prior to entering into a written agreement. Failure of the parties to reduce all such terms and conditions in writing and to include them in the written agreement may put a party to a great loss.

2.17.1 PPP Contracts – Documents

In addition to the documents listed above, contract documents in Public Private Partnership (PPP) contracts (discussed in more detail in paragraph 2.18.4) generally include the following additional documents.

1. **Cash Flow Projections** giving details of how the contractor intends to generate finance and use it till the recovery of the capital, investment cost and profit during the concession period.
2. **Several Schedules** describing the project site, project facilities, site delivery schedule, design requirements, construction requirements, operation and maintenance requirements, performance security, etc.
3. **State Support Agreement**: The implementation of the agreement requires extensive and continued support and grant of certain rights and authorities by the government. For this purpose the State Support Agreement is signed and forms part of the contract documents.
4. **Substitution Agreement**: The contractor, called the concessionaire, invariably obtains large sums of money from financial institutions such as consortium of banks, i.e. the lenders. In case the concessionaire fails to implement the project, the lenders have a right to propose a person called a selectee to be approved by the employer to substitute the concessionaire to implement the contract. To this end a tri-party agreement is signed by the employer, the contractor and the financial institutions.
5. **Handback Requirements**: Under this head are included the minimum requirements of an itemwise inventory of all assets to be handed over to the employer at the end of the concession period.

2.18 Types of Contract

Basically building and engineering contracts can be divided into the following groups.

1. Contract between an owner and a contractor.
2. Contract between a contractor and a subcontractor.
3. Engineer's or architect's contract (with the owner) for engineering and architectural services.

Contracts of the second and the third types are dealt with in Chapters 14 and 15, respectively. Legal aspects of different types of contracts between the owner and the contractor are briefly considered below.[101]

2.18.1 Lump Sum Contracts

In a lump sum contract, the contractor agrees to perform the work as shown on the drawings and described by the specifications including all contingencies for a specified lump sum. The form must have been evolved to meet the need of the owner to know in advance the total cost of the project that he may be required to incur. It is suitable for the projects where the drawings and designs are fully prepared before inviting bids. Experience, however, shows that no matter how much care and caution is taken in carrying out preliminary investigations and explorations, site conditions actually encountered do require some changes to be made. If the scope or nature of the work changes the cost must also change accordingly. At that stage it is difficult to get the contractor to agree to a reasonable cost. To avoid this problem, for building works, this form is sometimes used for the work in superstructure (above plinth) and the foundation work is executed on an item rate basis by use of measurement contracts discussed below.

Since a construction contract can rarely be completed without some changes being made, over the period the form was improved with a provision made whereby the specified fixed sum is subject to adjustments. The lump sum contracts in present day practice contain Bill of Quantities (lump sum with BOQ form) to regulate the amount to be added to or deducted from the fixed sum on account of additions and alterations not covered by the contract. These provisions do not change the intrinsic nature of the contract.

One of the main advantages of this form is that detailed measurements of the work done are not required to be recorded except in respect of additions and alterations.[102] The deductions to which the employer is entitled as per contracts are clearly recorded and attended to while issuing final certificates.

An Entire Contract

A lump sum contract is, generally speaking, an "entire" contract. An entire contract is one where one party has to fulfil his part of the agreement fully before the obligation of the other party to the contract arises. Thus, in a strict lump sum contract, the contractor may not receive under the contract, the cost of the work done, however substantial it may be, if he has failed to complete the contract fully in accordance with the agreed terms. In one case it was held that a lump sum contract which stipulates periodical payment to the contractor during the progress of the work, nevertheless remained an entire contract.[103] It was, however, held in the same case that the contractor, who stopped the work and claimed the entire balance because of the owner's failure to pay the instalment as agreed, could collect only in *quantum meruit* (i.e. fair and reasonable value). Thus, the contractor could recover the contract price less payments made and less the cost of completion. Even if a contract provides for fluctuation or price-adjustment clauses by reference to price indices, the contract may remain a lump sum contract.[104]

Another essence of a strict lump sum contract is that the contractor on completion of work gets the entire contract price although he may not have to do certain things.

---

[101] Also see Chapter 1, paragraph 1.4.1 to 1.4.10

[102] R. C. Thakkar v. Gujarat Housing Board, AIR 1973 Guj. 34.

[103] New Era Homes Corporation v. Forster, 299, N. Y. 303, 86 NE 2nd 757 (1947).

[104] Mascareignes Sterling Co. Ltd. v. Chang Cheng Esquires Co. Ltd., (Mauritius) [2016] UKPC 21.

EXAMPLE FROM CASE LAW

A contract contained a condition that the contractor should submit a priced schedule with his tender. In the priced schedule submitted by the contractor, whose tender was accepted, there was a calculation mistake. One of the items in the schedule of quantities was 420 cubic yards of clinker. The contractor priced at 5 s per cubic yard. However, the total cost of that item was shown as £150 instead of £105. The discrepancy was noticed during the progress of the work. On completion of the work, the owner sought to deduct £45 from the agreed contract sum. It was held that the contract was a lump sum contract and £45 could not be deducted.[105]

The parties would be presumed to have entered into a lump sum contract if they have reached an agreement by regarding the various items as a whole. In other words, the total price agreed was a single consideration for the whole of the work. Such pure form lump sum contract is suitable only when the planned scope of the work, designs and specifications are standardized and not likely to change. It may be noted that the basic criterion in determining whether a contract is a lump sum one or not is the intention of the parties.

EXAMPLE FROM CASE LAW

A contract for filling up and consolidating the earth in hollows along the side of a certain drain was entered into with a municipality. The rate tendered and sanctioned was stated to be Rs. 29,696/-. This rate was entered into on the basis of the BOQ showing the quantity of earthwork to amount to 2,400,180 cubic feet. The contract included a condition that

> on the completion of the work the contractor shall personally make a final measurement with the officer appointed to superintend the work and sign his final bill within one month, failing which the measurements of the work shall be considered as correct and payments made accordingly.

The quantity of work actually done by the contractor was only 1,624,920 cubic feet. The contractor sought to recover the entire sum of Rs. 29,696/- on the plea that the contract was a lump sum. It was held by the Privy Council, that the true meaning of the contract was that the contractor was to be paid for doing the specified work according to the measurements of what he had done and not a lump sum of Rs. 29,696/- irrespective of such measurements.[106]

2.18.2 Measurement Contracts

Measurement contracts also known as schedule rate contracts are those in which the contractor undertakes to execute the work at fixed rates, the sum he is to receive depending on the quantities and kind of work done or material supplied. Contracts with BOQs are grouped under this heading. The main types are:

(i) Item rate or unit price contract.

(ii) Percentage rate contract.

---

[105] Mitchell v. Magistrates of Delkeith, (1930), S.L.T. 80.
[106] Municipality of Hyderabad v. Mangharam Hukumutrai, AIR 1950 PC. 62.

In these types of contract, the work to be executed is divided into a number of items. The unit of measurement of quantities under each item is agreed upon. The BOQ sets out the estimated quantities under each item, a brief description of the work to be carried out under it and the unit in which the quantities are to be measured. In an item rate or unit price contract, the tenderer quotes his rates separately for each item, both in figures and in words. In the percentage rate contract, estimated rates are also set out by the engineer and the tenderer has to quote a common percentage figure by which the estimated rates are to be increased or decreased for making payments to him, should his tender be accepted. Thus, it is only a modified form of the first type. It claims an advantage that there is no scope for a tenderer to submit an unbalanced tender. This type, thus, recommends itself for those works where the quantities are likely to vary considerably from the estimated quantities. Apart from the mode of tendering the essential characteristics of both the types are the same.

It is a matter of common experience that in executing a work, the contractor may have to perform hundreds of operations. However, a typical building contract may not include more than 50 or 60 items. In other words, for all the work that the contractor is required to execute for the completion of an item of work, he will be paid on the basis of actual quantities executed under that item multiplied by the agreed unit rate. The basic presumption, which is obvious from these considerations, is that the unit rate quoted by the contractor includes cost of executing all ancillary or contingent work necessary for completion of that item as called for by the agreement.

To avoid disputes as to which ancillary or contingent work is included and which falls beyond the scope, the contract includes specifications. These are instructions to the contractor and set out, besides quality of materials and workmanship, contingent and ancillary works that are clubbed with the item and for which no separate payment will be made.

Traditional methods in vogue in India called for too much clubbing, making the task of an estimator and a tenderer difficult. The Indian Standards Institution of India has come out with the Standard Mode of Measurement of Building and Civil Engineering Works, I.S. 1200, which eliminates this drawback. Unfortunately, the Standard lacks legal sanction. It is, however, in the best interest of parties to follow the Standard Mode of Measurement as the same may be treated as an evidence of custom.

A properly worded description of the items coupled with scientifically prepared specifications would give no scope for disputes. For, if the work requires something to be done which neither the description of the item nor the specifications spell out, a contractor may have a proper claim for such work. Where a rate has been agreed upon, for any described work, the rate can only include such ancillary or contingent work. The said rate can never cover something which calls for a totally different rate, looking to the nature of the work involved.[107]

Another point needs to be noted with reference to these types of contracts. Measurement contracts may be classified as "severable", as opposed to an "entire contract". In a severable type contract, a breach by one party in respect of any one item does not justify the other in repudiating the contract. Any loss sustained by one party because of a breach by another can be recovered from the defaulting party. The doctrine of part performance and substantial performance are applicable to measurement contracts. These doctrines are discussed in detail in Chapter 10.

2.18.3 Cost-Plus Contracts

Apart from the types of contracts considered above which are adopted for public as well as private works, there is one more group of contracts, which may be adopted for private works only. By their nature they appear to be unsuitable for public works. These have already been discussed at length in Chapter 1. As stated therein, in all such contracts care must be exercised in properly defining the "cost". Generally cost includes the cost of materials, labour, and plant and

---

[107] Gujarat Electricity Board v. S.A. Jais & Co., AIR 1972 Guj. 192 (193); see Ch. 5.

machinery, use of which is normally warranted, but excludes the overheads and profit, which must be included in "fee" payable.

In an agreement that provides for all construction work to be done on a "time and material basis", the contractor's compensation will be limited to the cost of labour and materials only. He cannot charge for overheads or profit, as is customary in cost-plus contracts.[108] Thus, while agreeing to undertake a work on a time and material basis the contractor must be careful to include his overhead charges and profit margin in the costs of materials and labour.

In a cost-plus contract, can the owner avoid the contract on a plea that the final cost of the work is excessive? This question arose before a court in relation to a cost-plus contract entered into for the manufacture of fuel-oil tanks. At the time the contract was entered into, because of steel shortage, fuel-oil tanks were difficult to obtain. When the tanks were manufactured, the party for whom they were made refused to accept and pay for the same on the ground that the cost of manufacturing the tanks was excessive. The contractor filed a suit to recover compensation for the tanks on a cost-plus basis. In allowing the contractor a recovery, the court held:[109] "Though the price was higher than either of the parties expected, the defendant, in the absence of fraud or misrepresentation cannot now be heard to claim that it was unreasonable."

In another case, the cost-plus contract contemplated that changes would be made in the work as it progressed. The changes made by the architect were, however, so substantial that the cost of the work was increased to a figure far in excess of what the owner had planned to spend. It was held that this ground did not justify the owner's abandonment of the work.[110] In a cost-plus type contract, compensation cannot exceed the amount fixed by the contract. In other words, the contractor would be entitled to the cost on the basis of *quantum meruit* (fair and reasonable value) for the works, if he is to bring an action to recover the cost of work performed up to the time of the abandonment of the same. It will include cost of his labour and materials finished plus the amount stipulated in the contract for profit and overheads.

2.18.4 PPP Contract

PPP can be described as a government service or a private business venture which is financed and operated through a partnership of government/public sector undertaking and one or more private sector companies. The first systematic programme Private Finance Initiative (PFI), aimed at encouraging PPPs was introduced in the UK in 1992. The focus was on reducing public sector borrowing. A number of Australian state governments adopted PFI. The first and the model for others, is Partnerships Victoria. In Canada, the vanguards for PPP have been provincial organizations supported by the Canadian Council for PPP established in 1993 with representatives of both the public and the private sectors. The Federal Conservative Government in Canada solidified its commitment to PPP by creation of a Crown Corporation (P3 Canada Inc.) in 2009 with an independent Board of Directors reporting through the Minister of Finance to the Canadian Parliament. This enables the one-off and naive clients to work with an experienced central procurement organization. Each province within Canada also adapts the P3 for its needs as, for example, the cheapest project (Alberta) or the best value for money (British Columbia). The typical Canadian short list of bidders has three tenderers after an open prequalification round. The documents produced in Canada are more standardized than those used in the UK thus reducing the time taken to close a deal from approximately two years (in the UK) by half in Canada, thus reducing the market uncertainty. Canada does not prioritize the same objective as in the UK of avoiding government borrowing, as public borrowing is a better deal for the taxpayers than private borrowing. Yet, the Canadian model is not beyond controversy. A 2013 academic study of 28 P3 projects, carried out by the University of Toronto, showed that the PPP

---

[108] Doughtly v. Irdale, 108 N.E. Court of Appeal of Ohio (1952) 2nd 754.

[109] Continental Copper and Steel Industries v. Bloom, Supreme Court of Errors of Connecticut 139 Conn. 700 (1953).

[110] Oliver L. Taetz Inc. v. Groff et al., 253, S.W. 2nd ed, 824, Supreme Court of Missouri (1953).

costs were, on average, 16% higher than those of a traditional contract. P3 Crown Corporation was wound down after about seven years although the government claimed that there was a saving of $1.7 billion across 25 large infrastructure projects.[111]

In India, PPPs have been successfully implemented in developing infrastructure, particularly road projects under the National Highway Authority of India. Recently, there has been a shift toward encouraging private participation in the government works and promoting of PPPs. The Ministry of Housing and Urban Poverty Alleviation in its National Urban Housing Habitat Policy, 2007 specifically mentions the participation of the private sector as one of its aims. It envisages that the state government and the central government shall act as facilitators and enablers. The Maharashtra State Housing Policy, dated 23rd July 2007, provides for private participation. Pursuant to the declared policy by the central and state governments, the Maharashtra State Housing Board and Maharashtra Housing and Area Development Authority (MHADA) are well within their rights to apply the Swiss Challenge Method (see below) with respect to the MHADA lands that were lying undeveloped, since the same was being applied only on trial basis as a method of encouraging private participation.

The Development in Law in India

It is a well-settled principle that in the matters of government contract, the scope for judicial review is very limited and that the court cannot substitute its own decision for that of the government. Even as early as in 1986 the Supreme Court of India held that when the state government is granting licence for putting up a new industry, it is not at all necessary that it should advertise and invite offers for putting up such industry. The state government is entitled to negotiate with those who have come up with an offer to set up such industry.[112] In 2009, the Supreme Court of India observed:[113]

> It is by now well settled that non-floating of tenders or absence of public auction or invitation alone is no sufficient reason to castigate the move or an action of a public authority as either arbitrary or unreasonable or amounting to mala fide or improper exercise or improper abuse of power by the authority concerned. Courts have always leaned in favour of sufficient latitude being left with the authorities to adopt their own techniques of management of projects with concomitant economic expediencies depending upon the exigencies of a situation guided by appropriate financial policy in the best interests of the authority motivated by public interest as well in undertaking such ventures.

2.18.5 Swiss Challenge Method

As per the Swiss Challenge Method, the developer who has given the original proposal has the opportunity that is first right of refusal. However, the said developer has to match/raise his bid (rate) with the highest proposal tendered. The original proposer shall have the opportunity to take up the project on highest offer, and in the event that he refuses, then the highest bidder shall have the right to implement the project. However, if such highest bidder refuses the offer then the earnest money deposited by that bidder will be forfeited.

It is noteworthy that the Swiss Challenge Method is adopted in Chile, Costa Rica, Guam (U.S. Territory), Indonesia, Korea, Philippines, South Africa, Sri Lanka, Taiwan (China), Virginia (U.S.) and also in India by Andhra Pradesh, Rajasthan, Madhya Pradesh, Chhattisgarh, Gujarat,

---

[111] Retrieved from www.canada.ca/en/office-infrastructure/news/2017/11/government_of_canadaannounceswind-downofpppcanadacrowncorporatio.html on 22nd October 2018.

[112] State of M.P. and Others vs. Nandlal Jaiswal and Others, (1986) 4 SCC 566.

[113] Ravi Development v. Shree Krishna Prathisthan, AIR 2009 SC 2519.

Uttaranchal, Punjab States and Cochin Port Authorities. The law seems to be settled by the Ravi Development case[114] wherein it was held: "The decision to apply Swiss Challenge Method clearly fell within the realm of executive discretion."

Controversy

A common problem with PPP projects is the higher rate of return to private investors than the government's bond rate, though the most or all income risks were borne by the public sector. As regards the first objection, it has to be admitted that the government can sustainably fund projects at a cost of finance equal to its risk-free borrowing rate, based on the weighted average cost of capital (WACC), which can be less than the rate of return expected by a private party. However, there exist constraints because the government borrowing must ultimately be funded by the taxpayer and therefore has to be kept within prudent limits. The borrowing levels are currently not too low in most countries. PPPs are bound to be important in the economic activity of several countries for timely solutions to development problems.

The second objection has some basis as a number of Australian studies showed that in most cases the schemes proposed were inferior to those based on competitive tendering of public owned assets.[115] However, this is surmountable by strict formal procedures for the assessment of PPP. It is noteworthy that nowadays a new model is also being discussed: PPCP (Public Private Community Partnership) aimed at social welfare schemes eliminating the prime focus on profit.

---

[114] Ravi Development v. Shree Krishna Prathisthan, AIR 2009 SC 2519.
[115] Economic Planning Advisory Commission, (EPAC) 1995a, b, etc.

3

Contracts by Government

3.0 Introduction

Contracts by government need to be specially discussed because they raise some problems which do not or cannot possibly arise in the case of contracts entered into by private persons. In order that public funds may not be depleted by clandestine contracts made by any and every public servant, there is often a specific procedure according to which governments' contracts must be made. The law that governs the procedure to be followed is embodied in Art. 299 of the Constitution of India. In addition, the provisions of Arts. 14 and 226 of the Constitution and relevant provisions of the Indian Contract Act are considered below with examples.

3.1 Provisions of the Law

The following provisions of the law are dealt with in this chapter.

Provision of the Law	Paragraph No.
Constitution of India, Art. 299	3.2–3.6
Constitution of India, Arts. 14 and 226	3.11
Indian Contract Act, Section 70	3.7
Indian Contract Act, Sections 230, 235	3.8

3.2 Provisions Made in the Constitution of India

Art. 299 reads as follows:

1. All contracts made in exercise of the executive power of the Union or of a State shall be expressed to be made by the President, or by the Governor of the State, as the case may be and all such contracts and all assurances of property made in the exercise of that power shall be executed on behalf of the President or the Governor by such persons and in such manner as he may direct or authorise.

2. Neither the President nor the Governor shall be personally liable in respect of any contract or assurance made or executed for the purpose of this Constitution, or for the purpose of any enactment relating to the Government of India heretofore in force, nor shall any person making or executing any such contract or assurance on behalf of any of them be personally liable in respect thereof.

3.2.1 The Provisions of Art. 299 Are Mandatory

Where a statute requires that a thing shall be done in the prescribed manner or form but does not set out the consequences of non-compliance, the question whether the provision is mandatory or directory has to be answered in the light of the intention of the legislature as disclosed by the object, purpose and scope of the statute. If the provision is mandatory, the thing done, not in the manner or form prescribed, can have no effect or validity; if it is directory, penalty may be incurred for non-compliance, but the act or thing done is regarded as good. Placing reliance upon Maxwell on the Interpretation of the Statutes,[1] the Constitutional Bench of the Supreme Court of India held:[2]

> It is clear that the Parliament intended in enacting the provision contained in S.175(3) that the State should not be saddled with liability for unauthorized contracts and with that object provided that the contracts must show on their face that they are made on behalf of the State, i.e. by the Head of the State and executed on his behalf and in the manner prescribed by the person authorized. The provision, it appears, is enacted in the public interest, and invests public servants with authority to bind the State by contractual obligations incurred for the purpose of the State.
>
> It is in the interest of the public that the question whether a binding contract has been made between the State and private individual should not be left open to dispute and litigation; and that is why the legislature appears to have made a provision that the contract must be in writing and must on its face show that it is executed for and on behalf of the Head of the State and in the manner prescribed. The whole aim and object of the legislature in conferring powers upon the head of the State would be defeated if in the case of a contract which is in form ambiguous, disputes are permitted to be raised whether the contract was intended to be made for and on behalf of the State, or on behalf of the person making the contract. This consideration by itself would be sufficient to imply a prohibition against a contract being effectively made otherwise than in the manner prescribed. It is true that in some cases, hardship may result to a person not conversant with the law and who enters into a contract in a form other than the one prescribed by law. It also happens that the Government contracts are sometimes made in disregard of the forms prescribed; but that would not in our judgment be a ground for holding that departure from a provision which is mandatory and at the same time salutary may be permitted."

The apprehension of hardship likely to result, expressed by the Supreme Court is illustrated by the case below.

EXAMPLE FROM CASE LAW

A contractor executed certain works but some parts of the work were in progress. The government invited fresh tenders in respect of the incomplete works. In a writ petition filed by the contractor challenging the action of the authority, it was held that since the acceptance letters were not issued in the name of the governor, the contracts signed were not legally valid. Holding that the petitioner had an alternate remedy available to pursue the matter before the civil courts, the writ was dismissed.[3]

3.3 Formality of Contracts on Behalf of the Government

The provision incorporated in Art. 299 was considered in a series of cases. The following interpretation is now generally accepted.

---

[1] *Maxwell on Interpretation of Statutes*, 10th Ed. p. 376.
[2] Bhikraj v. Union of India, AIR 1962 SC 113 (119).
[3] Abdul Aziz Sheikh v. State of J. & K., AIR 2000 J. & K. 113.

3.3.1 Formal Written Contract

The words "expressed to be made" and "executed" suggest that there must be a deed for a formal written contract. A contract by correspondence or an oral contract is, accordingly, not binding upon the government.[4] The use of the word "shall" in making the provision is intended to make the provision itself obligatory and not directory.[5] It is now well settled that the provisions of Art. 299 of the Constitution which are mandatory in character require that the contract made in exercise of the executive power of the Union or of a state government must satisfy three conditions, namely:

(i) it must be expressed to be made by the president or by the governor of the state, as the case may be;

(ii) it must be executed on behalf of the president or the governor, as the case may be, and

(iii) its execution must be by such person and in such manner as the president or governor may direct or authorize.

Failure to comply with these conditions nullifies the contract and renders it void and unenforceable.[6]

3.3.2 The Contract Must Be Expressed to Be Made by the President or Governor

After the first requirement of a formal written agreement is satisfied, the second essential requirement is that the agreement "shall be expressed to be made by the President or by the Governor of the State, as the case may be". In the absence of such an expression, the agreement is likely to be held void. For example, wherein the agreement was expressed to be made by the "Government of the State of Kerala" and not by the governor, the Kerala High Court held that the agreement was not valid.[7] Thus, even if the correspondence shows that the formalities necessary for a concluded contract have been satisfied and the parties were *ad idem* by the time the letter of acceptance was written, there is no valid or binding contract if the letter of acceptance is not issued by a person authorized to execute the contracts, for and on behalf of the governor of the state or the President of India, as the case may be.[8]

EXAMPLES FROM CASE LAW

1. A letter of acceptance stated that the tender had been accepted on behalf of the President of India but it omitted to say as to what person or officer had accepted it and the signatory of the letter neither purported to accept the tender, nor did he sign it for and on behalf of the president and the letter looked like a communication of acceptance of the president rather than an acceptance having been made according to law. It was held by the Delhi High Court that no valid contract resulted.[9]

---

[4] Karamshi v. State of Bombay, AIR 1964 SC 1714; State of Haryana v. M/s O. P. Singhal & Co., AIR 1984 P. & H. 358 Shri Narayan Gosain v. Collector, Cuttack, AIR 1986 Ori. 46. See, however, paragraph 3.5 in this chapter.

[5] M/s B. K. Mondal and Sons v. State of West Bengal, AIR 1962 SC 779.

[6] State of Bihar v. M/s Karam Chand Thapar and Brothers Ltd. (1962) I, SCR 827; (AIR 1962 SC 110); Bikhraj Jaipuria v. Union of India, (1962) 2, SCR 880: (MR 1962 SC IB) and State of West Bengal v. B.K. Mondal and Sons, (1962) Supp. I SCR 876: (AIR 1962 SC 779); Samir Mohanty v. State of Odisha, (Orissa) (DB): 2013(sup) Cut LT 392: 2013(2) ILR (Cuttack) 16: 2012(73) R.C.R.(Civil) 369: 2013 (supl.) *Ori. Law Rev.* 449.

[7] K.N. Vidyadharan v. State, AIR 1980 Ker. 212.

[8] Union of India v. N.K. Pvt. Ltd, AIR 1972 SC 915.

[9] Union of India v. U.S. Dugal & Co., AIR 1972 Delhi 110.

2. In a case decided by the Supreme Court of India[10] the facts were as follows. The executive engineer, the appellant, invited tenders and the respondent submitted his tender in response to the invitation. The tenders were recalled and the respondent re-submitted his tender. The executive engineer informed the respondent by a telegram that his tender was accepted. A letter followed the telegram. Neither the telegram nor the letter expressed that the Governor of Punjab accepted the tender. The respondent subsequently withdrew his offer. No formal contract was signed by and between the parties. The executive engineer levied a penalty under Clause 2 of the agreement. The matter was referred to arbitration by the appellant and the respondent refused to admit the existence of a valid and binding contract between the parties. The respondent filed an application under Section 33 of the Arbitration Act, 1940 contending that there was no valid acceptance of the contract. The trial court dismissed the application with costs. In revision, the High Court came to the conclusion that there was no valid contract. The Supreme Court confirmed the judgment of the High Court. It was observed, "The acceptance letter, at least, must conform to the requirement of Art. 299(1) of the Constitution."

3.3.3 Acceptance by Telegram

A valid acceptance by telegram must comply with Art. 299 of the Indian Constitution. The requirements are that the telegraphic message must be expressed to be made by the president or the governor and that the telegram must be executed by such person and in such a manner as the president might have authorized. In the normal course, therefore, it is advisable to wait till a formal confirmation of the telegraphic message duly signed by the authorized person is received before commencing work or taking any steps in performance of the contract.

EXAMPLES FROM CASE LAW

1. In a case where the acceptance of the contractor's tender was communicated to him by the proper government authority under the ministry's abbreviated telegraphic address, it was held that the acceptance by telegram was valid.[11]

2. In another case,[12] however, it was held to be invalid on the ground that it did not meet the requirements of Art. 299 of the Indian Constitution. In that case the telegraphic message at the end had the abbreviated telegraphic address of the ministry but no evidence was led to prove that the telegram emanated from the President of India or was signed by the President of India or by any person authorized by any general or special order made by the president.

3.3.4 Deed Must Be Executed by an Authorized Person

The third requirement of Art. 299 is that the deed must be executed by a person duly authorized to do so by the president, in the case of the central government and by the governor in the case of the state government.[13] The contract also fails if it is executed by a person who is not authorized in that behalf under Art. 299 by the president or governor, as the case may be.[14] The article does not prescribe any

---

[10] State of Punjab v. M/s Om Prakash Baldev Krishnan, AIR 1988 SC 2149.

[11] M/s Chiranji v. Union of India, AIR 1963 Punj. 372.

[12] A. K. N. Vidyadharan v. State, AIR 1980 Ker. 212.

[13] New Marine Coal Co. v. Union of India, AIR 1964 SC 152.

[14] Karamshi v. State of Bombay, AIR 1964 SC 1714; see AIR 1984 P. & H. 358 AND AIR 1986 Ori. 64.

particular mode in which the authority must be conferred by the president or the governor. The authority may be conferred either by a general order or by an ad hoc order upon a particular officer for the purpose of a particular contract. Such order may be notified in the Official Gazette or established by other evidence. However, the Himachal Pradesh High Court has held that the fact that the contracts under Art. 299(1) have to be in writing, and that there can be no dispute with that proposition. But, the Supreme Court has not laid down, so far as we can see that the direction or authorization contemplated by Art. 299(1) must be in writing.[15] It is advisable to insist on a written order or authorization to avoid leading evidence to prove oral order.

Formality of contracts by government is also applicable to contracts by municipalities or corporations in accordance with the respective Acts. The first of the following examples will make the point clear.

EXAMPLES FROM CASE LAW

1. A bidder participated in an auction and his bid was accepted, whereupon he paid the earnest money. The acceptance of the bid contained signatures of the chairman of the board and the bidder. The rules contemplated that in case of acceptance of the bid, an agreement on stamp paper would be executed, but no such agreement was executed. The Municipal Rules in the case, contained the general provision that every contract entered into by or on behalf of the Municipal Corporation shall be signed by the chairman and executive officer and shall be sealed with the common seal of the Municipality. Relying upon earlier decisions of the Supreme Court and the Delhi High Court,[16] it was held that there was no concluded contract.[17]

2. The deputy commissioner of a district entered into a contract with A for putting up of certain additional structures to a hospital. A sued the government alleging a breach of those contracts. The government repudiated liability on the ground that the deputy commissioner had no authority to enter into those contracts on its behalf. It was held, that in the absence of a rule, regulation or notification being brought to the notice of the court by which the deputy commissioners were generally authorized to enter into contracts of the kind, the government would not be bound by the contract.[18]

3. An agreement was entered into on behalf of the Dominion Government by the additional chief engineer and not by the executive engineer. The contractor then wrote a letter to the executive engineer to confirm the contract but it was held that it would have no effect. It was held:

> In any case, a person cannot be bound by a one-sided offer which is never accepted, particularly when the parties intend that the contract should be reduced to writing. ... a letter written to the Executive Engineer subsequent to the acceptance of the tender would have no effect inasmuch as the Government can only be bound by contracts that are entered into in a particular way and which are signed by a proper authority.[19]

4. A state government invited tenders for the collection of toll tax for a new bridge. The petitioner's bid was the highest. He deposited 10% amount within 24 hours as required. As

---

[15] State of Himachal Pradesh v. M/s Crown Timber and Food Ltd. & Others, AIR 1978 H.P. 1

[16] State of U.P. v. Kishorilal Minocha, AIR 1980 SC 680 and Union of India v. Uttam Singh Dugal & Co. (Pvt) Ltd., AIR 1972 Del. 110.

[17] Municipal Board Khumber v. Yadu Nath Singh, AIR 2004 Raj. 79.

[18] State of Bombay v. Pannalal, AIR 1958 Bom. 56; Chandradhan v. State of Bihar, AIR 1976 Pat. 15.

[19] Thawardas v. Union of India, AIR 1955 SC 468. Also see AIR 1976 M.P. 199; see M/s Smart Chip Ltd. v. State of U.P., AIR 2003 All. 80.

per the conditions of the contract, the petitioner was to be given certain amenities like providing electric and water connections, construction of room with a window for collection, etc. The authorities failed to provide the said amenities but insisted on the respondent depositing further instalments. It was the case of the petitioner that the old road and old bridge continued to be in existence, so that there was less tendency to use the toll bridge. The petitioner requested for agreed amenities and asked that the old road be closed. A number of meetings were held but did not resolve the dispute. Subsequently, the petitioner was informed that the old road could not be closed and he was directed to deposit a further amount with interest. The petitioner challenged the action and sought refund of the amount already deposited. It was urged on behalf of the petitioner that there was no agreement as required by Art. 299 of the Constitution, a fact not denied by the other side. It was held that the action of the state was arbitrary. The writ was allowed with the direction to the state to refund the amount deposited within three months failing which the amount would carry an interest of 12% p.a. from the date of receipt till the date of payment.[20]

5. A contract was executed in the prescribed form. During the course of the execution of work some disputes arose between the parties resulting in a rescission of the contract by the state government. The contractor's request to appoint an outside arbitrator was accepted by the state and a communication was signed by the deputy secretary of the state. The arbitrator's award on the dispute was challenged by the state government on the ground that the subsequent agreement regarding the appointment of an outsider as an arbitrator was null and void as the deputy secretary was not duly authorized to execute the said contract on behalf of the governor. It was observed:

> The document ex facie has been executed on behalf of the Governor. Having been executed in course of the official business and in the discharge of executive functions ostensibly it is to be taken to conform to the requirements of Art. 299 of the Constitution ... The appellant wants to be relieved of the burden under the contract. In such a situation it is for the appellant to prove to the satisfaction of the Court that due to lack of proper authorization the contract became void.[21]

In another case it was argued that the original contract containing the arbitration clause was superseded by a collateral agreement entered into between the Railways and a firm of contractors. The collateral agreement was not signed by the Railways in accordance with Art. 299 of the Constitution of India. It was held that the subsequent collateral unsigned contract was not binding on the parties.[22]

3.4 No Implied Contract under Art. 299

In view of Art. 299(1) of the Constitution there can be no implied contract between the government and any other person. The reason being that if such implied contracts between the government and another person were allowed, they would in effect make Art. 299(1) superfluous, for then a person who had contract with the government which was not executed at all in the manner provided in Art.

---

[20] Tilak Raj v. State of Rajasthan, 2000(1) WLC 521.
[21] State v. R.B. Ojha, AIR 1977 Pat. 258.
[22] Union of India v. M/s Sohoun Constructions, AIR 1989 A.P. 350.

299(1) could contend that an implied contract may be inferred on the facts and circumstances of a particular case.[23] It was held by the Rajasthan High Court that the recording of minutes of a discussion between a private party and a state minister did not amount to a valid agreement between the private party and the state in accordance with Art. 299(1) of the Constitution.[24]

3.5 Contract by Correspondence May Be Valid

The position that the provisions of Art. 299 are made mandatory is now well settled. However, the question remains if the provisions of Art. 299 spell out clearly a need for a formal contract. If it is found that those provisions clearly spell out necessity of a formal deed, it would result in great injustice in a number of cases. In construction contracts there are several instances of the works having been executed without the parties signing a formal deed of contract. In all such cases if the contractors were to be paid for the work done not on the basis of agreed rates but on a *quantum meruit* basis, almost every case would go into litigation. As a matter of fact, the Supreme Court of India, as early as in 1954, took note of this factor and observed:[25]

> It would, in our opinion, be disastrous to hold that hundreds of Government officers who have daily to enter into a variety of contracts, often of a petty nature, and sometimes in an emergency, cannot contract orally or through correspondence and that every petty contract must be effected by a ponderous, legal document couched in a particular form.

It was held that there would be nothing to prevent ratification. But if the original agreement is void, it is difficult to see how a void contract remains capable of ratification. This decision, therefore, only helps in showing the gravity of the problem and its awareness by the Supreme Court but does not suggest any answer to the problem, it is respectfully submitted. Perhaps, it may indirectly help those cases in which the parties are either nearing the end of the completion of a contract or after the work is completed sign the formal document. The difficulty in such a solution is that the proper government officer should show his willingness to comply with the requirements though at a late stage. Experience shows that such willingness comes only after the contractor is required to compromise his claims in many ways. This may be unjust. Simple justice demands that an agreement otherwise valid, except for executing a formal deed, must be held to be valid and binding under Art. 299 of the Constitution and, indeed, a just solution is given by the Supreme Court itself in the Rallia Ram case decided in 1963. It is especially noteworthy that this decision duly refers to the decision of the Supreme Court in an earlier case.[26] In the Rallia Ram case,[27] after citing with approval the decision of the Supreme Court in Bhikraj Jaipuria v. Union of India,[28] it was held:

> Section 175(3) does not in term require that a formal document executed on behalf of the Dominion of India and the other contracting party, alone is effective. In the absence of any direction by the Governor-General under Section 175(3) of the Government of India Act prescribing the manner, a valid contract may result from correspondence if the requisite conditions are fulfilled.

In a construction contract, tenders are invariably invited in a bulky blank tender form, which not only includes all the conditions of the contract but a formal deed as well. These blank tender

---

[23] K. P. Choudhary v. State of M.P., AIR 1967 SC 203 (206).

[24] Kotah Match Factory v. State, AIR 1970 Raj. 118.

[25] Chaturbhuj Vithaldas v. Moreshwar Parashram, AIR 1954 SC 236 (243).

[26] Reported in AIR 1962 SC 113.

[27] Union of India v. Rallia Ram, AIR 1963 SC 1685.

[28] Bhikraj Jaipuria v. Union of India, AIR 1962 SC 113.

forms are filled in by the tenderers and signed at relevant places. When the duly authorized government officer sends a communication of acceptance by any mode such as telegram, telex or post, a binding contract comes into existence under the law at the time when and at the place where the letter of acceptance is posted. Although such a letter of acceptance may mention signing of a formal deed at a later stage on payment of the requisite security deposit the validity of the contract would not depend upon execution of a formal deed subsequently. The only exception to this will be the cases in which the letter of acceptance incorporates certain important conditions that would require acceptance by the successful tenderer. Even in such cases, if the successful tenderer communicates his acceptance of such conditions, a binding and enforceable contract under Art. 299 of the Constitution will result. In such cases, however, the agreement would be made at the place where and at the time when such acceptance is posted by the contractor.[29]

The above can be taken as a well-settled position in law as a result of the decision of the Supreme Court in the Union of India v. Rallia Ram case. This proposition finds support in a subsequent decision of the Supreme Court[30] wherein it was held:

> It is now settled by this Court that though the words "expressed" and "executed" in Article 299(1) might suggest that it should be by a deed or by a formal written contract, a binding contract, by tender and acceptance can also come into existence if the acceptance is by a person duly authorised on this behalf by the President of India. A contract whether by a formal deed or otherwise by persons not authorised by the President cannot be binding and is absolutely void.

The following cases will make the above proposition clear.

EXAMPLES FROM CASE LAW

1. Where in an auction sale, the highest bid was provisionally accepted subject to final approval through resolution of competent authority in terms of conditions of auction sale; it was held that the provisional acceptance and payment of half the bid amount by the bidder did not bring about the concluded contract. The highest bidder in the said case had withdrawn the offer before he was communicated the acceptance by the competent authority. The 50% bid amount was forfeited by the authority. It was held by the Karnataka High Court that in the absence of a concluded contract, the forfeiture of the bid amount, being very much part of the sale consideration, was in the nature of penalty, for imposition of which, Section 74 of the Contract Act was not attracted and was unjustified. However, it was held that the forfeiture of the earnest money was justified.[31]

2. The plaintiff in a case admitted that his bid was accepted and the acceptance intimated by a letter directing him to pay the bid amount in four instalments. He declined to do so on the ground that the formal agreement duly signed by the competent authority had not been made over to him. It was held that the attitude of the plaintiff was thoroughly misconceived and a termination of the contract by the respondent was upheld though forfeiture of security deposit was held untenable because under the rules the said action could only be taken by the government and not by the authority competent to execute the agreement.[32]

---

[29] Murthy and Bros. v. State, AIR 1971 Mad. 393; also see AIR 1982 Ori. 147 and AIR 1980 Ori. 40.

[30] Union of India v. N. K. Pvt. Ltd., AIR 1972 SC 915; also see: M/s Bhagwati Enterprises v. Raj. State Road Transport Corpn., AIR 2006 Raj. 233.

[31] Hubli-Dharwad Municipal Corporation v. Chandrashekar Shetty, AIR 2009 Kar. 41. Also see Punya Coal Road-lines v. Western Coalfields Ltd., 2008 (1) CTLJ 117 (Bom.) (DB) (SN) wherein it was held that earnest money was rightly forfeited because the bidder withdrew his tender during its validity, irrespective of the fact that subsequently the tender was cancelled. Also see: Harminder Singh v. Punjab and Sind Bank, AIR 2008 P. & H. 39.

[32] State of Orissa v. Ganeswar Jena, AIR 1994 Ori. 94.

3.6 Execution of a Formal Contract: When a Condition Precedent

If the correspondence exchanged between the parties does not conclusively establish an agreement between the parties, the execution of a formal contract may become necessary. For example, a contract was under negotiations for more than 17 months, the original tender lapsed by sheer passage of time, and the subsequent additions, subtractions and modifications of the terms during the course of more than 17 months changed its shape beyond recognition. Even the letter of acceptance mentioned a large number of conditions. In view of these facts, the Delhi High Court held that the execution of a formal contract was an inevitable condition precedent.[33]

3.6.1 Provision in Agreement Contrary to Directions by the Governor

If a contract contains two provisions apparently opposed to each other, the provision which is contrary to the stipulations contained in directions issued by the governor cannot be enforced. A question to be decided by the Kerala High Court was if the arbitration Clause 73 of the Madras Detailed Standard Specification which was to be read as part of the contract could survive the specific deletion, by a government order, of the arbitration clause for disputes involving sums below Rs. 2 lacs. Relying upon the earlier full bench decision of the same court,[34] it was held that not only the arbitration clause but the entire arbitration process was consciously "annihilated". The Kerala High Court held, "A, provision of a contract which is contrary to the stipulations contained in directions issued by the Governor under the above Article cannot be enforced against the executive Government."[35]

3.7 Effect of Non-Compliance with Requirements of Art. 299(1)

The discussion above makes it clear that non-compliance with the requirements of Art. 299(1) renders a contract void, and as such the pleas of implied contract or estoppel or ratification or specific performance or damages for breach of contract are not permitted. The question then remains for consideration as to what relief, if any, a party to such a void contract will be entitled if such a party has carried out any work or provided services or supplied materials to or received any advance from the government. In such situations Sections 65 or 70 of the Indian Contract Act will be attracted.

3.7.1 Applicability of Section 70 of the Act

In view of the foregoing discussion, it is true that there is no liability to pay under a void contract, but the aggrieved party can nevertheless claim compensation against the government under Section 70 of the Indian Contract Act. The said section reads as follows:

> 70. *Obligation of person enjoying benefit of non-gratuitous act*
> Where a person lawfully does anything for another person or delivers anything to him, not intending to do so gratuitously and such other person enjoys the benefit thereof, the latter is bound to make compensation to the former in respect of, or to restore, the thing so done, or delivered.

---

[33] Union of India v. U.S. Dugal and Co. (Delhi), AIR 1972 Del. 110.

[34] State v. C. Abraham, AIR 1989 Ker. 61 (F.B.).

[35] M/s Leo Construction Contractors v. Govt. of Kerala, AIR 1989 Ker. 241.

ILLUSTRATIONS

(a) A, a tradesman, leaves goods at B's house by mistake. B treats the goods as his own. He is bound to pay A for them.

(b) A saves B's property from fire. A is not entitled to compensation from B, if the circumstances show that he intended to act gratuitously.

When a claim for compensation is made by one person against another under Section 70 it is not on the basis of a subsisting contract between the parties but on a different kind of obligation namely, a quasi-contractual or restitution obligation. Section 70 may be applicable in the case of a void contract if three conditions set out below are satisfied:[36]

1. A person should lawfully do something for another person.
2. The other person for whom something is done must enjoy the benefit thereof.
3. The thing done must not be done fraudulently or dishonestly, nor must it be done gratuitously.

In a case covered by Section 70, the person doing something for another or delivering something to another cannot sue for the specific performance of the contract, nor ask for damages for the breach of the contract, for the simple reason that there is no contract between him and the other person.[37]

3.7.2 Section 70: When Not Applicable

Section 70 cannot be relied upon against persons who are incompetent to contract. The circumstances contemplated by the section are those in which the law implies a promise to pay. Thus, a minor who cannot be sued on an express contract cannot be sued on an implied contract. Where, however, a corporation receives money or property under an agreement which turns out to be *ultra vires* or illegal, it is not entitled to retain the money or the property, and restitution and compensation can be ordered independently of the express contract. It is well established that a person who seeks restitution has a duty to account to the defendant for what he has received in the transaction from which his right to restitution arises. In other words, an accounting by the plaintiff is a condition of restitution from the defendant.

The section will not be applicable if any one of the three requisite conditions is not satisfied. If a person delivered something to another, the latter may refuse to accept the thing or return it. In that case, Section 70 would not come into operation. It is only when he voluntarily accepts the thing or enjoys the work done that the liability under Section 70 arises.

Applying the test of restitution under Section 62[38] to the case under Section 70, the Calcutta High Court held:[39]

> A person whose contract is void for non-compliance with Article 299(1) of the Constitution is entitled to compensation under Section 70 of the Contract Act. It is, however, his duty to account to the other party for what he has received in the transaction before his right to restitution arises. If he fails to do so his claim for compensation or refund for deposit will be dismissed.

---

[36] Mulamchand v. State of M.P., AIR 1968 SC 1218. Also see: Union of India v. R.S. Venkataraman (Madras), 2011(4) MLJ 269: 2011(6) R.C.R.(Civil) 2712. B. K. Mondal v. State of West Bengal, AIR 1962 SC 779; Nirvik Printers P. Ltd. v. Orissa Primary Education Programme, 2010(1) CTLJ 78 (Orissa) (DB).

[37] Nanalal Madhavji Varma, plaintiff v. State of A.P., AIR 1982 Cal. 167.

[38] For Section 62 see paragraph 3.7.4 in this chapter.

[39] Nanalal Madhavagi v. State of M.P., AIR 1982 Cal. 167.

The Patna High Court has made a similar decision.[40]

3.7.3 Absence of Alternative Case under Section 70 Not Fatal to the Suit

It is obvious that the first plea of the party claiming money due would be under the alleged valid contract. Normally if the validity of the contract is doubtful, the plaintiff should set out in the plaint an alternative case that he is entitled to get compensation. However, mere absence of setting up the alternative case will not disentitle him to such relief. The Supreme Court in a case observed:[41]

> The Respondent claimed under oral agreement compensation at prevailing market rates for work done by him even if he failed to prove an express agreement in that behalf, the court may still award him compensation under S.70 of the Contract Act. By awarding a decree for compensation under the Statute and not under the oral contract pleaded, there was in the circumstances of this case no substantial departure from the claim made by the respondent.

Relying upon the above observations it was held by the Allahabad High Court:

> In my view, this case therefore very clearly lays down that it is not necessary that an alternative case must necessarily be pleaded. I do not think that the mere absence of a formal alternative case under S. 70 of the Contract Act is fatal to the maintainability of the suit in question.[42]

3.7.4 Agreed Rates May Be Reasonable Compensation

The following case is a typical example of developments in cases where work is done without a formal agreement. The superintending engineer rejected all the tenders and ordered that until fresh tenders were invited and finalized, for urgent cases, the lowest rates for the carriage tender decided in respect of Drainage and Sewerage Division should be adopted and the lowest or any other contractor might be asked to execute the works. Accordingly the plaintiff executed the work for three months on the verbal orders of the officers of the Division, as no written order was issued at any level. Seven out of 19 bills submitted by the plaintiff were paid. The matter was, in the meantime, referred to the Chief Engineer, who decided that the carriage charges should be paid at the declared rates or scheduled rates whichever is less. The contractor was to execute the agreement and recoveries were to be made from the pending bills. The contractor refused to do so. In a suit filed by the contractor both the trial court and the High Court held that the requisite conditions contemplated under Art. 299 were fulfilled except that formal deed of contract was not executed. The Orissa High Court[43] held that obviously the above decision of the Chief Engineer was a deviation from the terms of the contract agreed upon by the parties for carriage of materials. Once the plaintiff agreed to execute the work at a particular rate and got the work done to the satisfaction of the defendant, it is no longer open for the defendant or any of its officers to vary the rate to the prejudice of the plaintiff. It was further held that the plaintiff is entitled to reasonable compensation under Section 70 of the Contract Act and the rates approved are reasonable. This decision, it is respectfully submitted, lays down a good guiding principle for deciding reasonable compensation under Section 70 in the cases of its kind.[44] This approach will restrict the scope of enquiry. If the court is to uphold that the agreed rates are reasonable, there will be no serious error to justify interference in appeal.[45]

---

[40] Manoharlal v. Union of India, AIR 1974 Pat. 56.

[41] Subamanium v. Thayappa, AIR 1966 SC 1034.

[42] Union of India v. Saheb Singh, AIR 1977 All 277(279).

[43] State v. Ananda Prasad, AIR 1985 Ori. 142; M/s Jain Mills & Electrical Stores v. State, AIR 1991 Ori. 117.

[44] Mir Abdul Jalil v. State, AIR 1984 Cal. 200.

[45] Subramanyam v. Thayappa, AIR 1966 SC 1034.

EXAMPLE FROM CASE LAW

A plaintiff submitted a tender for recovery of toll tax on a bridge. Earnest money was deposited. The offer of the plaintiff was the highest. It was duly accepted. However, due to some reasons, an agreement was not entered into nor the work of toll collection started. The earnest money was forfeited and compensation was sought to be recovered as arrears of land revenue by issuing an RRC (Revenue Recovery Certificate). This order was challenged and the trial court declared that the state had no right to recover the amount by way of RRC. The High Court upheld the said judgment subject to modification about the direction of payment of *ad valorem* court fees.[46]

The facts in another case decided by the Orissa High Court are worth noting. In that case, the state accepted the quotation of a firm for the supply of electrical goods. The goods were dispatched in time by rail, but, in transit, there was some delay and the goods reached the destination after the due date stipulated for delivery. The state retained the goods but did not pay for the same on ground of the delay. On these facts, it was rightly held by the Orissa High Court that Section 70 of the Contract Act would have no application to the case as the original contract subsisted and even if Section 70 had application, as no evidence was adduced on the side of the plaintiffs as to what was the market price of the cables by the date of their supply, the contractual rate could be assumed to be the market value of the goods on the relevant date.[47]

3.7.5 Section 65: When Applicable

Section 65 of the Indian Contract Act enshrines the principle of restitution. It provides that any person who has received any advantage under a void contract is bound to restore it, or to make compensation for it, to the person from whom he received it.

EXAMPLE FROM CASE LAW

A contractor entered into an agreement with the government to construct a godown cum inspection room and received advance payment for the same. He failed to complete the work. The government filed a suit to recover the amounts advanced due to the failure of the contractor to complete the work. It was found that the contract was void from its inception, not being in conformity with Art. 299(1) of the Constitution of India. However, the government could recover the amount under Section 65 of the Indian Control Act, 1872.[48]

3.8 Personal Liability of Government Officer Executing the Contract

A plain reading of Clause (2) of Art. 299 makes it clear that neither the executive heads nor the officials executing contracts for and on behalf of the government are personally liable under the contract. A very important question nevertheless remains to be considered. Where the government is not bound for want of due compliance with Art. 299(1), is the officer who executed the contract

---

[46] State of M.P. v. Siyaram Verma, 2004(1) CTLJ 356(M.P.) (DB).
[47] M/s Jain Mills & Electrical Stores v. State, AIR 1991 Ori. 117.
[48] State of Orissa v. Rajballav, AIR 1976 Ori. 19.

personally liable? The answer to this question can be sought in provisions of Sections 230 and 235 of the Indian Contract Act. Section 230 reads as follows:

> 230. *Agent cannot personally enforce, nor be bound by, contracts, on behalf of principal*
> In the absence of any contract to that effect, an agent cannot personally enforce contracts; entered into, by him on behalf of his principal, nor is he personally bound by them.
> **Presumption of contract to contrary** Such a contract shall be presumed to exist in the following cases:
>
> 1. Where the contract is made by an agent for sale or purchase of goods for a merchant residing abroad;
> 2. Where the agent does not disclose the name of his principal;
> 3. Where the principal, though disclosed, cannot be sued.

It is a fact that government officers act as agents on behalf of the government to the knowledge of everybody. Thus, only Clause (3) of the above Section is relevant for the discussion and is analysed thus. The essential ingredients that must be satisfied for applicability of this Section, in the opinion of the authors, are:

(i) There must be a contract entered into by the agent on behalf of his principal.

(ii) The principal, though disclosed, cannot be sued.

Now, in cases within the principle of Art. 299 and this section, the second condition may be fulfilled. It may be shown that the government in whose service the officer was employed is the principal disclosed and in view of Art. 299, the government cannot be sued. However, is the first condition satisfied? Section 230 of the Contract Act clearly implies an existence of a contract for its application. It is now well settled that a contract entered into without complying with the conditions laid down in Art. 299(1) is void and cannot be ratified. Under the circumstances the first condition is not fulfilled and Section 230(3) is not applicable.[49]

In view of the above, the decision of the Calcutta High Court[50] which was given by placing reliance upon the decision of the Supreme Court in Chaturbhuj's case[51] has also to be regarded as not laying down the law correctly.

It now remains to be seen whether the officer will be personally liable under Section 235 of the Indian Contract Act. The section reads as below:

> 235. *Liability of pretended agent*
> A person untruly representing himself to be the authorized agent of another, and thereby inducing a third person to deal with him as such agent, is liable, if his alleged employer does not ratify his acts, to make compensation to the other in respect of any loss or damage which he has incurred by so dealing.

The essential ingredients for applicability of the section can be stated thus:

1. A person represents himself to be the authorized agent of another.
2. He thereby induces a third person to deal with him as such agent.
3. His representation is untrue.
4. His alleged employer does not ratify his acts.

---

[49] The State of U.P. and another, v. Murari Lal and Brothers Ltd., AIR 1971 SC 2210.

[50] Union of India v. B.M. Sen, AIR 1963 Cal. 456.

[51] See note 25 and paragraph 3.5 above.

If the above ingredients are satisfied such a person is liable to make compensation to the other in respect of any loss or damage which the other has incurred by so dealing.

It is to be noted that the word "contract" which is used in Section 230 is not used in this section. If the officer has no authority from the president or the governor to enter into a transaction on behalf of the government, it is obvious that the first three ingredients could easily be satisfied. The fourth ingredient, therefore, assumes importance. The section uses the wording "if his alleged employer does not ratify his acts". A careful reading of this brings out the basic underlying presumption, namely, that the act of the agent is capable of ratification. Thus, where an act of the agent is not capable of ratification, the fourth ingredient is absent. It is, therefore, submitted that a government officer will not be personally liable even under this section because the contract in violation of Art. 299(1) cannot be ratified. The Supreme Court of India in a case observed:[52]

> But it seems that S. 235 also can become applicable only if there is a valid contract in existence. This appears to follow from the words "if his alleged employer does not ratify his acts". The contract should thus be such that it is capable of ratification.

The attention of the reader is invited to the fact that the Supreme Court has kept the question open for future determination. This is obvious from the following observations in the same paragraph of the judgment:

> However, we do not wish to express any final opinion on the applicability of Section 235 of the Contract Act to cases where the contract suffers from the infirmity that requirements of Art-299(1) of the Constitution have not been complied with.

The reason was that the appeal before the Supreme Court was from the decision of the High Court based on Section 230. It is significant to note that the observations on the applicability of Section 235 were made with reference to the judgment in the trial court in that case. The trial court had held that the defendant government officers had no authority to enter into a contract on behalf of the state government but still they purported to do so. There was an implied warranty of authority which had to be presumed and the plaintiff was entitled to receive compensation for breach of that warranty under Section 235 of the Contract Act. Under the circumstances, it is submitted that their Lordships of the Supreme Court have more than indicated the way Section 235 may be interpreted.

3.9 Constitutional Rights and Obligations of Parties to Government Contracts

What are the constitutional obligations of the state when it takes action in exercise of statutory or executive power? Is the state entitled to deal with its property in any manner it likes or award a contract without any constitutional limitations upon it? What are the parameters of its statutory or executive power in the matter of awarding a contract in dealing with its property? Can a party aggrieved by the action taken by the state seek legal redress? These questions fall in the sphere of both administrative law and constitutional law. They assume special significance in a modern welfare state, which is committed to egalitarian values and dedicated to the rule of law. For example, it has been held by the Apex Court that disposal of public properties owned by state or its instrumentalities should be by public auction or by inviting tenders. In exceptional cases, a contract may have to be granted by private negotiation, but clearly that should not be done without adequate reasons as it shakes the public confidence.[53]

---

[52] The State of U.P. v. Murari Lal and Brothers Ltd., AIR 1971 SC 2210.
[53] Brihan Mumbai Electric Supply v. Laqshya Media P. Ltd., 2010(1) CTLJ 1 (SC).

The above questions cannot be decided in the abstract. They can be determined only against the background of facts. A contract is a commercial transaction and evaluating tenders and awarding contracts are essentially commercial functions. In such cases, principles of equity and natural justice stay at a distance. If the decision relating to award of contracts is *bona fide* and is in public interest, courts will not exercise the power of judicial review and interfere, even if it is accepted for the sake of argument that there is a procedural lacuna.[54] The Supreme Court has aptly observed: "We must reiterate here that it is not for this Court to award the contracts by accepting or rejecting the tender bids. It is exclusively for the appellant herein to do that."[55]

A tender is an offer. It is something which is communicated and invites acceptance. Broadly stated, it must be unconditional; must be in the proper form; and the person by whom tender is made must be able to and willing to perform his obligations. The terms of the invitation to tender cannot be open to judicial scrutiny because the invitation to tender is in the realm of contract. However, a limited judicial review may be available in cases where it is established that the terms of the invitation to tender were so tailor-made to suit the convenience of any particular person with a view to eliminate all others from participating in the bidding process. The bidders participating in the tender process have no other right except the right to equality and fair treatment in the matter of evaluation of competitive bids offered by interested persons in response to notice inviting tenders in a transparent manner and free from hidden agenda. One cannot challenge the terms and conditions of the tender except on the above stated ground, the reason being the terms of the invitation to tender are in the realm of the contract. No bidder is entitled, as a matter of right, to insist the authority inviting tenders should enter into further negotiations unless the terms and conditions of notice so provided for such negotiations.

The authority has the right not to accept the highest bid and even to prefer a tender other than the highest bidder, if there exist good and sufficient reasons, such as, the highest bid not representing the market price, but there cannot be any doubt that the authority's action in accepting or refusing the bid must be free from arbitrariness or favouritism.[56]

In recent times, judicial review of administrative action has become expansive and is becoming wider day by day. The traditional limitations have been vanishing and the sphere of judicial scrutiny is being expanded. State activity, too, is fast becoming pervasive. As the state has descended into the commercial field and giant public sector undertakings have grown up, the stake of the public exchequer is also large, justifying larger social audit, judicial control and review by opening up to public scrutiny: these necessitate recording of reasons for executive actions including cases of rejection of highest offers. That very often involves large stakes, and availability of reasons for action on the record assures credibility to the action, disciplines public conduct and improves the culture of accountability. Looking for reasons in support of such action provides an opportunity for an objective review in appropriate cases both by the administrative superior and by the judicial process.[57] For example, in one case, in which the tender was rejected not only on the ground that the rate was less than 15% from the ceiling rate, a division bench of the Jharkhand High Court declined to interfere with the order refusing to grant tender in favour of the appellant taking into consideration various factors.[58]

---

[54] Siemens Public Communication Pvt. Ltd. v. Union of India, AIR 2009 SC 1204.

[55] THDC India Ltd. v. Voith Hydro GMBH CO., (SC) 2011 AIR (SCW) 2243: 2011(4) SCC 756: 2011(3) BC 6: 2011 AIR SC (Civil) 1022: 2011(5) R.C.R.(Civil) 189: 2011(3) R.A.J. 76: 2011(5) SCR 618: 2011(3) Scale 677: 2011(3) JT 592: 2011(2) SCC (Civil) 470.

[56] Meerut Development Authority v. Association of Management Studies; Pawan Kumar Agarwal v. Meerut Development Authority and Anr., AIR 2009 SC 2894.

[57] M/s State Enterprises etc. v. The City and Industrial Development Corporation of Maharashtra Ltd., (1990) 2 JT SC 401; see State of Orissa v. Tata Iron & Steel Co. Ltd., 2008(1) CTLJ 380 (SC).

[58] Sandeep Kumar v. Bihar State Food and Civil Supplies Corpn., AIR 2007 Jhar. 91.

3.9.1 Scope of Judicial Review – Constitutional Provisions

Judicial review is invariably in the light of Art. 14 and Art. 226 of the Constitution of India. These provisions read:

> **Art. 14.** The State shall not deny to any person equality before the law or the equal protection of the law within the territory of India.

> **Art. 226.**

> (1) Notwithstanding anything in Article 32 every High Court shall have power, throughout the territories in relation to which it exercises jurisdiction, to issue to any person or authority, including in appropriate cases, any Government within those territories directions, orders or writs, including writs in the nature of habeas corpus, mandamus, prohibition, quo warranto and certiorari, or any of them, for the enforcement of any of the rights conferred by Part III and for any other purpose.

> (2) The power conferred by clause (1) to issue directions, orders or writs to any Government authority or person may also be exercised by any High Court exercising jurisdiction in relation to the territories within which the cause of action, wholly or in part, arises for the exercise of such power, notwithstanding that the seat of such Government or authority or the residence of such person is not within those territories.

> (3) Where any party against whom an interim order, whether by way of injunction or stay or in any other manner, is made on, or in any proceedings relating to, a petition under Clause (1), without:

>> (a) furnishing to such party copies of such petition and all documents in support of the plea for such interim order, and

>> (b) giving such party, an opportunity of being heard, makes an application to the High Court for the vacation of such order and furnishes a copy of such application to the party of whose favour such order has been made or the counsel of such party, the High Court shall dispose of the application within a period of two weeks from the date on which it is received or from the date on which the copy of such application is so furnished, whichever is later, or where the High Court is closed on the last day of that period, before the expiry of the next day on which the High Court is open; and if the application is not so disposed of, that period, or, as the case may be, the expiry of the said next day, stands vacated,

> (4) The power conferred on a High Court by this Article shall not be in derogation of the power conferred on the Supreme Court by Clause (2) of article 32.

Art 32(2) referred to above reads:

> (1) The right to move the Supreme Court by appropriate proceedings for the enforcement of the rights conferred by this part is guaranteed.

> (2) The Supreme Court shall have power to issue directions or orders or writs, including writs in the nature of habeas corpus, mandamus, prohibition, quo warranto and certiorari whichever may be appropriate, for the enforcement of any of the rights conferred by this part.

At the outset, the question arises as to what are the rights and liabilities of the parties to a tender invitation? This question has been briefly discussed in Chapter 1 and is further considered below under the heading Doctrine of Legitimate Expectation (paragraph 3.10.2).

3.10 Types of Cases under Art. 226

It is trite that if an action on the part of the state is violative of the equality clause contained in Art. 14 of the Constitution of India, a writ petition would be maintainable even in the field of contracts. A distinction indisputably must be made between a matter which is at the threshold of a contract and a breach of contract; whereas in the former the court's scrutiny would be more intrusive, in the latter the court may not ordinarily exercise its discretionary jurisdiction of judicial review unless it is found to be in breach of Art. 14 of the Constitution. While exercising contractual powers also, the government bodies may be subjected to judicial review in order to prevent arbitrariness or favouritism on its part. Indisputably, inherent limitations exist, but it would not be correct to opine that a writ will never lie in contractual matters.

When there is a contractual dispute with a public law element, and a party chooses the public law remedy by way of a writ petition instead of a private law remedy of a suit, he will not get a full-fledged adjudication of his contractual rights, but only a judicial review of the administrative action. The question whether there was a contract and whether there was a breach may, however, be examined incidentally while considering the reasonableness of the administrative action. The issue whether there was a concluded contract and breach thereof becomes secondary. In exercising writ jurisdiction, if the High Court finds that the exercise of power in passing an order was not arbitrary and unreasonable, it should normally desist from giving any finding on disputed or complicated questions of fact as to whether there was a contract, and relegate the petitioner to the remedy of a civil suit.[59] It is so because where serious disputed questions of fact are raised requiring appreciation of evidence and, thus, for determination thereof, examination of witnesses would be necessary; it may not be convenient to decide the dispute in a proceeding under Art. 226 of the Constitution of India.

The types of cases in which breaches of alleged obligation by the state or its agents can be set up are divisible into four main types as follows:[60]

(i) Where a petitioner makes a grievance of breach of promise on the part of the state in cases where on assurance or promise made by the state he has acted to his prejudice and predicament, but the agreement is short of a contract within the meaning of Art. 299 of the Constitution;

(ii) Where the contract entered between the person aggrieved and the exercise of a statutory power under Acts or Rules framed thereunder and the petitioner alleges a breach on the part of the state;

(iii) Where the contract entered into between the state and the person aggrieved is non-statutory and purely contractual and the rights and liabilities of the parties are governed by the terms of the contract, a petitioner complains about breach of such contract by the state; and

(iv) Where the contract entered into between the state and the person aggrieved is non-statutory and purely contractual, but such contract has been cancelled on a ground, *de hors* any of the terms of the contract which is per se violative of Art. 14 of the Constitution.

Insofar as the exercise of power under Art. 226 of the Constitution, the Patna High Court has held in the first two and the fourth category above, applications under Art. 226 of the Constitution are maintainable.[61] It was held that where the contract entered into between the state and the person

---

[59] Kisan Sahkari Chini Mills Ltd. v. Vardan Linkers WITH State of Uttaranchal v. Vardan Linkers and Ors.; Doiwala Sugar Company Ltd. and Ors. v. Vardan Linkers and Ors., AIR 2008 SC 2160.

[60] Radhakrishna Agarwal v. State of Bihar, AIR 1977 SC 1496.

[61] M/s Pancham Singh v. State, AIR 1991 Pat. 168 (F.B.).

aggrieved is non-statutory and purely contractual but such contract has been cancelled on a ground *de hors* any of the terms of the contract, and which is per se violative of Art. 14 of the Constitution, the High Court in such case can exercise its jurisdiction under Art. 226 and the writ petition under Art. 226 by the aggrieved person would be maintainable. Where a contractor's application for an extension of time for one month to complete the balance work worth Rs. 40 lacs was not allowed by the authorities and the contract was terminated with the work being awarded to another agency at Rs. 80 lacs, the Patna High Court allowed the writ and directed the authorities to extend the time by one month at the old rates.[62]

Thus, in an appropriate case, a writ petition against the state or an instrumentality of the state even arising out of a contractual obligation is maintainable. Merely, because some disputed questions of fact arise for consideration, that by itself cannot be a ground to refuse to entertain a writ petition in all cases as a matter of rule. Further, a writ petition involving a consequential relief of monetary claim is also maintainable.[63]

These categories can be further subdivided into nine sub-categories in respect of the procedure to be followed in inviting and accepting tenders which may attract Art. 226 as follows:

1. Eligibility of tenderers
2. Prequalification of tenderers
3. Reclassification of tenderers
4. Essential conditions of tender notification
5. Negotiations prior to acceptance
6. Acceptance/rejection of tenders
7. Nomination of or preference given to a tenderer or a class of tenderers
8. Blacklisting and/or disqualification of tenderers
9. Breach/termination of contract after the tender is accepted

Illustrative cases under each of the above nine categories are given below to help highlight the principles applied by courts. Knowledge of these principles may be of help to the State authorities responsible for administrative contracts and also to a tenderer seeking redress, if he feels aggrieved by any action of the State authority. European law contains public procurement principles that have recently been enforced in successful actions for damages including an award for loss of profit.[64]

3.10.1 Eligibility of Tenderers

Prescribing of Conditions in the Tender – When Amenable to Judicial Review?

Normally, terms of the invitation to tender may not be open to judicial scrutiny, but the courts can scrutinize the award of contract by the government or its agencies in exercise of their power of judicial review to prevent arbitrariness or favouritism. The court may refuse to exercise its jurisdiction, if it does not involve any public interest.[65]

The employer is the best judge in the matter of contract. Evaluating tenders and awarding contracts being essentially commercial functions, principles of natural justice and equity are not likely to be attracted. Prescribing of conditions in the tender for contract can be challenged and interfered with by a court only on limited grounds of being arbitrary, unreasonable, unfair or *mala*

---

[62] State of Bihar v. Ram Binod Choudhary, AIR 2009 Pat. 115.

[63] Shimnit Utsch India Pvt. v. State of Karnataka, (Karnataka) (DB): 2012 ILR (Karnataka) 4073: 2012(36) R.C.R. (Civil) 178: 2013(1) KCCR 544.

[64] Treumer, S., "Damages for Breach of the EC Public Procurement Rules – Changes in European Regulation and Practice", PPLR 2006, 4, 159–170.

[65] Directorate of Education and Others v. Educomp Datamatics Ltd. and Others, AIR 2004 SC 1962 = 2004 AIR SCW 1505 = (2004) 4 SCC 19.

fide. A condition is amenable to judicial review if it fails to satisfy the test of reasonableness and fairness in action. No hard and fast rule can be laid down in this regard. Impugned condition of average annual turnover fixed in view of the total cost of the project and overall public interest, it was held, cannot be said to be arbitrary, unreasonable, unfair or *mala fide*.[66] The legal position on this subject was summed up after a comprehensive review of principles of law applicable to the process for judicial review by the Supreme Court of India in the following words:

19 From the above decisions, the following principles emerge:

(a) The basic requirement of Article 14 is fairness in action by the State, and non-arbitrariness in essence and substance is the heartbeat of fair play. These actions are amenable to the judicial review only to the extent that the State must act validly for a discernible reason and not whimsically for any ulterior purpose. If the State acts within the bounds of reasonableness, it would be legitimate to take into consideration the national priorities;

(b) Fixation of a value of the tender is entirely within the purview of the executive and courts hardly have any role to play in this process except for striking down such action of the executive as is proved to be arbitrary or unreasonable. If the Government acts in conformity with certain healthy standards and norms such as awarding of contracts by inviting tenders, in those circumstances, the interference by Courts is very limited;

(c) In the matter of formulating conditions of a tender document and awarding a contract, greater latitude is required to be conceded to the State authorities unless the action of tendering authority is found to be malicious and a misuse of its statutory powers, interference by Courts is not warranted;

(d) Certain preconditions or qualifications for tenders have to be laid down to ensure that the contractor has the capacity and the resources to successfully execute the work; and

(e) If the State or its instrumentalities act reasonably, fairly and in public interest in awarding contract, here again, interference by Court is very restrictive since no person can claim fundamental right to carry on business with the Government.

It was further held:

20. Therefore, a Court before interfering in tender or contractual matters, in exercise of power of judicial review, should pose to itself the following questions:

(i) Whether the process adopted or decision made by the authority is *mala fide* or intended to favour someone; or whether the process adopted or decision made is so arbitrary and irrational that the court can say: "the decision is such that no responsible authority acting reasonably and in accordance with relevant law could have reached"; and

(ii) Whether the public interest is affected. If the answers to the above questions are in negative, then there should be no interference under Article 226.[67]

---

[66] Yummy Bites v. State of Punjab, (Punjab and Haryana) (DB): 2018(2) R.C.R.(Civil) 16; also see M/s Real Mazon India Ltd. v. State of Punjab, (Punjab and Haryana) (DB): 2018(2) R.C.R.(Civil) 12.

[67] Michigan Rubber (India) Ltd. v. State of Karnataka and Ors., (2012) 8 SCC 216(229), paragraph 19. Also see: Sundar v. Jodhpur Vidhyut Vitaran Nigam Limited, (Rajasthan) (Jodhpur Bench): 2017(2) WLC (Raj.) (UC) 350; Raunaq International Ltd. v. I.V.R Construction Ltd. & Ors., reported in 1999 (1) SCC 492: Sh. Lalfakawma v. State of Mizoram, (Gauhati) (DB): 2017 AIR (Gauhati) 201: 2017(6) GauLJ 643; M/s. Kelelhounei Angami v. State of Nagaland, (Gauhati) (Kohima Bench): 2017(3) GauLT 774: 2017(4) GauLJ 176; MRP Enterprise v. The Union of India, (Gauhati): 2017(2) GauLT 192: 2017(3) GauLR 335: 2017(2) GauLJ 617: 2017(5) NEJ 77: 2018(1) BC 65.

In 2007, the Apex Court summarized some principles while exercising the right of judicial review in respect of award of contracts thus:[68]

> We are also not shutting our eyes towards the new principles of judicial review which are being developed; but the law as it stands now having regard to the principles laid down in the aforementioned decisions may be summarized as under:
>
> i)   If there are essential conditions, the same must be adhered to.
>
> ii) If there is no power of general relaxation, ordinarily the same shall not be exercised and the principle of strict compliance would be applied where it is possible for all the parties to comply with all such conditions fully.
>
> iii) If, however, a deviation is made in relation to all the parties in regard to any of such conditions, ordinarily again a power of relaxation may be held to be existing.
>
> iv) The parties who have taken the benefit of such relaxation should not ordinarily be allowed to take a different stand in relation to compliance of another part of tender contract, particularly when he was also not in a position to comply with all the conditions of tender fully, unless the court otherwise finds relaxation of a condition which being essential in nature could not be relaxed and thus the same was wholly illegal and without jurisdiction.
>
> v)   When a decision is taken by the appropriate authority upon due consideration of the tender document submitted by all the tenderers on their own merits and if it is ultimately found that successful bidders had in fact substantially complied with the purport and object for which essential conditions were laid down, the same may not ordinarily be interfered with.
>
> vi) The contractors cannot form a cartel. If despite the same, their bids are considered and they are given an offer to match with the rates quoted by the lowest tenderer, public interest would be given priority.
>
> vii) Where a decision has been taken purely on public interest, the Court ordinarily should exercise judicial restraint.

Although the scope of judicial review or the development of law in this field has been noted earlier, it needs to be remembered that each case, however, must be decided on its own facts. Public interest may be one of the factors to exercise power of judicial review. In a case where a public law element is involved, judicial review may be permissible.[69] The following examples based on the latest case law will help the reader understand the principles applied in judicial review

EXAMPLES FROM CASE LAW

1. A petitioner's bid was rejected for non-submission of 90 days bid validity affidavit and financial statements of five years. The petitioner was granted an opportunity to approach the tender review committee for compliance of tender requirements. Further, the committee was directed to pass fresh orders whether to reject or accept the bid of petitioner.[70]

---

[68] B.S.N. Joshi and Sons Ltd., M/s. v. Nair Coal Services Ltd., AIR 2007 SC 437.
[69] Noble Resources Ltd. v. State of Orissa, AIR 2007 SC 119. Binny Ltd. v. Sadashivqn, 2005 AIR SCW 3774; G.B. Mahajan v. Jalgaon Municipal Council, AIR 1991 SC 1153.
[70] S. Thanmi v. State of Manipur, (Manipur): 2017(3) GauLT 771.

2. It was for authorities issuing tender to decide the period of tender and to further decide whether there should be a composite tender or a tender in parts. Neither the participants nor court could dictate terms and conditions of tender.[71]

3. A petitioner's bid, although the highest, was rejected as being defective on ground of non-deposit of earnest money in the form of a call deposit. There was no indication/ stipulation in the terms and conditions of the NIT or in form supplied, that the earnest money had to be furnished only by way of a call deposit and in no other form/method. The prescribed form supplied by the respondent corporation, in which the offer had to be submitted by the tenderer, also did not stipulate that such earnest money has to be furnished in the form of the call deposit though it required furnishing certain information relating to the call deposit. Such information as sought for in the form could not be treated as "terms and conditions of the NIT". The petition was allowed and the order rejecting the tender was held to be arbitrary.[72]

4. A tender notice for the development, operation and management of a hospital on "public-private partnership" (PPP) basis was published. In response, the petitioner submitted a bid without deviating from the essential terms of the NIT. The court found that the petitioner's tender was beneficial to the government. It was held that merely because the government reserved their right to reject any tender, they did not possess any unfettered discretion of ignoring a tender which was more beneficial to the government than the other tenders by stamping it as non-responsive. The rejection, instead of seeking a clarification, suffered from *mala fides* in law. Accepting a less beneficial offer is contrary to public interest. The decision of the authority was liable to be set aside.[73]

5. The Patna High Court directed the respondent authority to give an opportunity to the appellant to submit its tender after the last date. Instead, the respondent issued a fresh advertisement inviting tenders for the contract from the bidders other than those who had already submitted their tenders. It was held that by expanding the scope of competition, by giving opportunity to other leftover bidders along with the appellant, no prejudice was caused to the appellant.[74]

6. A rejection of a bid on the ground that the bank guarantee was short by seven days was challenged by a writ petition. It was held that the earnest money deposit furnished by the petitioner was short by a period of seven days but that the error was rectified immediately on bid opening and prior to the scrutiny, evaluation and consideration of bids of various bidders. The discrepancy and error had occurred on account of amendments of the bid document by the respondents from time to time. The petitioner had not gained an unfair advantage in this process. Further, it was observed that the deviation was not a material deviation of an essential condition. Therefore, it was held that it would be hyper technical and not in public interest if the petitioner's bid were to be rejected. A direction was issued to consider it along with other bids.[75]

---

[71] Messrs B. Himmatlal Agrawal v. Maharashtra State Power Generation Co. Ltd., Nagpur (Bombay) (DB) (Nagpur Bench): 2015(1) R.A.J. 143: 2014(3) AIR Bom.R 697: 2014 AIR (Bombay) 108: 2014(4) BC 612: 2014(32) R.C.R. (Civil) 399.

[72] Enjil Choudhury v. Assam Fisheries Development Corporation, (Gauhati): 2011(5) GauLT 49: 2011(6) GauLJ 311: 2011(11) R.C.R.(Civil) 359.

[73] Shalby Limited, v. The State of Goa, (Bombay) (Goa Bench) (D.B.): 2011(6) AIR Bom. R 242: 2011(6) BCR 866: 2012(1) Mh.LJ 533: 2012(6) R.C.R.(Civil) 1348: 2011(2) Goa L.R. 337: 2012(2) R.A.J. 519: 2011(6) ALL MR 154.

[74] Ashraf Khan v. State of Bihar, (Patna) (DB): 2011(1) PLJR 218: 2010(66) R.C.R.(Civil) 266.

[75] PES Installations Pvt. Ltd. v. Union of India, (Delhi) (DB): 2015 AIR (Delhi) 108: 2015(24) R.C.R.(Civil) 564.

7. Where three companies were found eligible and two of them had quoted an identical rate, the allegation that the two companies had formed a cartel was not accepted.[76] Reliance was placed upon the Supreme Court's judgment wherein it was held that the mere question of identical price and an offer of a further reduction by themselves would not entitle the two companies automatically to corner the entire market by way of monopoly since the final allotment of the quantities was vested in the authorities who at their discretion could distribute the same to all the manufacturers. It was observed that what was only a suspicion was strengthened by the post-tender attitude of the said manufacturers who quoted a much lower price. It could not positively be concluded on the basis of these two circumstances alone that the two companies had formed a cartel; there had to be further evidence of the existence of the alleged cartel.[77]

8. A tenderer (appellant) did not comply with an essential condition of eligibility and the state respondents rejected their tender documents. The appellant challenged the state's action on the ground that the tender that was selected had not complied with the essential conditions of the tender notice. It was observed that as the appellant himself had not complied with the terms and conditions of the tender, the appellant did not have *locus standi* to challenge the selection by the state respondents. On this aspect, the appeal was dismissed by placing reliance on a decision of the Supreme Court[78] that giving judicial relief at the instance of a party who does not fulfil the requisite criteria of a tender is misplaced.[79] *Prima facie*, this decision might seem a little unfair or arbitrary. The selected tenderer's sample of the geomembrane was of Indian Standard Specification IS 15351:2008, but as per the Gazette Notification, the same should have conformed to IS 15351:2015. The appellant had failed to submit its Tax/VAT Clearance Certificate along with his tender; the appellant was, in fact, a Tax/VAT Registered Dealer but the documents did not show that the appellant was a registered dealer of geomembrane. The respondent had to be fair in its selection of tenders, but could choose if the non-compliance with a condition by the tenderers was a minor condition or an essential condition. Reliance was also placed on an earlier decision of the Supreme Court about the employer's discretion over prioritization of the tender selection criteria. The Supreme Court had observed:[80]

> [W]hether a term of NIT is essential or not is a decision taken by the employer which should be respected. Even if the term is essential, the employer has the inherent authority to deviate from it provided the deviation is made applicable to all bidders and potential bidders as held in Ramana Dayaram Shetty. However, if the term is held by the employer to be ancillary or subsidiary, even that decision should be respected. The lawfulness of that decision can be questioned on very limited grounds, as mentioned in the various decisions discussed above, but the soundness of the decision cannot be questioned, otherwise this Court would be taking over the function of the tender issuing authority, which it cannot.

9. On the perusal of the records produced by the government advocate, the Gauhati Court found that when the respondent made the allotment of work orders in respect of 57 schools, no

[76] ACE India Transport Pvt. Ltd. v. Rajasthan State Mines and Minerals, AIR 2005 Raj. 307.

[77] Union of India v. Hindustan Development Corporation, AIR 1994 SC 988.

[78] Raunaq International Ltd. v. I.V.R Construction Ltd. & Ors., 1999 (1) SCC 492.

[79] Sh. Lalfakawma v. State of Mizoram, (Gauhati) (DB): 2017 AIR (Gauhati) 201: 2017(6) GauLJ 643; See also. S. Price Hightech Pvt. Ltd. v. National Thermal Power Corpn. Ltd., (Andhra Pradesh): 2012(1) ALT 721: 2012(1) Andh LD 46: 2011(16) R.C.R.(Civil) 436.

[80] Central Coalfields Limited & Anr. v. SLL-SML (Joint Venture Consortium) & Ors., reported in 2016 (8) SCC 622, paragraph 48.

specific or cogent reasons were given when rejecting the recommendations made in a comparative statement. There were no contemporaneous notes in the records. It was held that there was no transparency and fair play on the part of the respondents while taking a decision to award the work orders to a particular tenderer. The impugned work order was set aside and the respondents were asked to reconsider the case of the petitioner tenderer in terms of the recommendation made in the comparative statement.[81]

10.   A writ petition was filed in the High Court by a petitioner whose contract was illegally terminated in order to award a bid in respect of the work to respondent no. 4. It was challenged on ground that this decision was in breach of a mandatory condition of the tender. Ignoring the interim order passed by the High Court that the petitioner should not be compelled to hand over the plant to respondent no. 4, the respondent-state handed over the work to respondent no. 4 without seeking prior permission from the High Court. The High Court rejected the contention that since respondent no. 4 had satisfactorily executed work for more than one year, that court's interference in the proceeding might not be in the public interest. The petitioner whose contract was illegally terminated was entitled to an award of the work. The bid of respondent no. 4 was accepted in breach of mandatory conditions of the tender. It is in the public interest that the government adheres to the conditions of the contract and does not act in breach of the same. It was also held that there is no absolute principle that in cases where only a single bid is found acceptable, the work under a tender cannot be awarded to that tenderer.[82]

11.   The first part of a clause in a contract was mandatory and required the contractor to rectify defects/damages at his risk and cost during the defect liability period of one year from the date of completion of the construction work. To supplement this condition, the second part of the clause stipulated that the contractor was also required to submit an affidavit/undertaking in this regard. Would the submission of such an affidavit/undertaking be considered mandatory or not? The trial judge held that the affidavit/undertaking by the contractor for rectifying defect of construction during the defect liability period should have been submitted at the time of submitting the tender and that it was a mandatory condition. This did not appeal to the Gauhati High Court which held:[83]

> The second part of clause 16 does not say that such affidavit/undertaking is required to be submitted along with the tender documents. ... Thus, from the above it is clear that the first limb of clause 16 is an essential requirement whereas the second limb is only ancillary to the object of the first requirement."

The clause appears to be strange because if it was a condition of the contract that the defects/damage should be rectified at the contractor's cost, an undertaking, that too on an affidavit, should be unnecessary.

3.10.2 The Doctrine of Legitimate Expectation

Doctrine means a rule, principle, theory or tenet of the law.[84] Legitimate used as an adjective according to *Black's Law Dictionary* means that "which is lawful, legal, recognized by law or according to law". Thus, for legal purpose, expectation cannot be the same as anticipation. It is

---

[81] M/s. Kelelhounei Angami v. State of Nagaland, (Gauhati) (Kohima Bench): 2017(3) GauLT 774: 2017(4) GauLJ 176.

[82] VA Tech Wabag Limited District Dhanbad v. State of Jharkhand, (Jharkhand): 2015(2) AIR Jhar R. 616: 2015(2) J. C.R. 679: 2015(22) R.C.R.(Civil) 404.

[83] Khedi Trade & Development Agency (M/S) v. Z. Rulho, (M/S), (DB) 2015(2) GauLT 1064: 2015(57) R.C.R.(Civil) 604.

[84] *Black's Law Dictionary*, 6th Ed., p. 481.

different from a hope, a wish or a desire, nor can it amount to a claim or demand. The doctrine of legitimate expectation means a rule of law that recognizes expectation of a person to be lawful or legal. This concept first entered English Law in 1969[85] and is currently explained by Laws LJ in terms of "fairness in public administration" so that "a change of policy which would otherwise be legally unexceptional may be held to be unfair by reason of prior action, or inaction, by the authority".[86] This would lead to a duty to consult (the public or a specific section of the public) and/or a substantive duty to keep its promise because acting against "legitimate expectations" would be tantamount to an abuse of power. Legitimate expectations may arise from unqualified and unambiguous representations or an established course of conduct. The EU law has recognized this principle of protection under the umbrella of "rule of law" since 1965.[87] The EU courts expect public authorities not to interfere with vested rights unless it is absolutely necessary in the public interest.

A person may have a legitimate expectation of being treated in a certain way by a public authority even though he may have no legal right in private law to receive such a treatment. The expectation may arise from a representation or promise made by the authority including an implied representation, or from consistent past practice. It may have a number of different consequences including:

(i) It may give locus standi to seek leave to apply for judicial review,

(ii) It may mean that the authority ought not to act so as to defeat the expectation without some overriding reason for public policy to justify its doing so, or

(iii) It may mean that, if the authority proposes to defeat a person's legitimate expectation, it must afford him an opportunity to make representation on the matter.

The Supreme Court of India has held that

"legitimate expectation may arise:

(a) if there is an express promise given by a public authority; or

(b) because of the existence of regular practice which the claimant can reasonably expect to continue; but

(c) such an expectation must be reasonable.

However, if there is a change in policy or in public interest, the position is altered by a rule or legislation, no question of legitimate expectation would arise."[88]

A party who has been granted a licence may have a legitimate expectation that it will be renewed unless there is good reason not to do so and may therefore be entitled to greater procedural protection than a mere applicant for a grant.[89]

---

[85] Schmidt v. Secretary of State for Home Affairs, (1969) 2 Ch. 149.

[86] R. (on the application of Niazi) v. Secretary of State for the Home Department, [2008] EWCA Civ 755, paragraph 50. Also see Paponette v. Attorney General of Trinidad and Tobago, [2010] UKPC 32 in which the legitimate expectation of self governance of a taxi association under the Constitution of Trinidad and Tobago was protected by holding that it was an abuse of power to subject them to the control and management of its competitor. This could have been justified if the government showed an overriding public interest to justify this course of action.

[87] Case 111/63 Lemmerz-Werke v. High Authority of the ECSC, [1965] ECR 677; Eugénio Branco, L^da v. Commission of the European Communities, Judgment of the Court of First Instance (Fifth Chamber), 30 June 2005, II 2560.

[88] Madras City Wine Merchants' Association v. State of T.N., 1994 (5) SCC 509. Also see: A.C. Roy v. Union of India, AIR 1995Cal. 246.

[89] *Halsbury's Laws of England*, 4th Ed., Vol. I (I) 151 cited with approval. It is submitted in Union of India v. Hindustan Development Corporation, AIR 1994 SC 988 (1013).

The Kerala High Court, however, clarified that the doctrine of legitimate expectation has no application if the authority's action is in the public interest, and upheld the action by an authority of inviting tenders for the grant of licence to medical shops which were held previously under a three-year licence.[90] Similarly, in a case where a contract is concluded by acceptance of an offer, the doctrine of legitimate expectancy has no application. As soon as a contract is concluded, the expectation, if any, comes to an end; thereafter the parties will be bound only by the terms of the contract.[91]

A question arises: can the lowest tenderer for a works contract or the highest bidder in an auction sale of public property, seek the benefit of this doctrine when his tender or bid is not accepted, all tenders are rejected and/or negotiations are held with all bidders with a view to obtaining an offer beneficial to the state? The answer to this question is not a straightforward "yes" or "no". For example, in an auction, the highest bid was highly inadequate and shockingly low and the Punjab & Haryana High Court upheld the government's refusal to grant approval. It was further observed that mere participation and deposit of earnest money did not confer any vested right in a bidder to seek confirmation of sale.[92]

Even though the authority inviting tenders or bids has the right to reject any tender, including the lowest tender (or the highest bid in an auction), if this right is exercised arbitrarily, without reasonable care and caution, the credibility of the procedure of inviting sealed tenders or bids would be lost. In such a case, the answer to the question will be in the affirmative. On the other hand, without the test of reasonableness, an absence of arbitrariness or the presence of a cogent reason, the principle, simply stated, that once tenders have been invited and the lowest tenderer (or highest bidder) has come forward to comply with the conditions stipulated in the tender notice, it is not permissible to switch over to negotiations with all tenderers and thereby reject the lowest tender (or highest bid) is too wide to be acceptable.[93]

EXAMPLES FROM CASE LAW

1. Where the tender notice did not indicate that the supply order would be divided among different suppliers or that a selected supplier would not be permitted to make the supply of the entire requirement, the authority could not restrict the supply order by the tenderer recommended by the purchase board to permit the balance supplies to be made by non-recommended suppliers. Such a decision was held to be legally unsustainable and quashed by applying the doctrine of legitimate expectation.[94]

2. Where cogent reasons were given in evaluation sheets pertaining to each and every tenderer, it was held that the decision rejecting the petitioner's bid was not arbitrary.[95]

The terms of the invitation of tenders cannot be open to judicial scrutiny because the invitation is in the realm of contract. However, a limited judicial review may be available in cases where it is established that the terms of the invitation to tender were so tailor-made as to suit the convenience of any particular person with a view to eliminating all others from participating in the bidding process. The bidders participating in the tender process have no other right except the right to equal and fair treatment in the matter of evaluation of competitive bids in a transparent manner and free from hidden agenda.[96] Where

---

[90] M.S.N. Medicals, Thiruvalla v. K.S.R.T. Corpn., Thiruvananthapuram, AIR 1995 Ker. 119.

[91] D. Wren International Ltd. v. Engineers India Ltd., AIR 1996 Cal. 424.

[92] Laxmi Narain v. State of Haryana & Ors., AIR 2009 (NOC) 566 (P. & H.). Also see Harjit Singh v. State of U.P., (Allahabad) (DB) (Lucknow Bench): 2017(5) ADJ 272: 2017(3) All WC 3060.

[93] Food Corporation of India v. Kamdhenu Cattle Feed Industries, AIR 1993 SC 1601 (1605).

[94] Y. Yeangpong Konyak v. State of Nagaland, 2008 (2) CTLJ 376 (Gauhati).

[95] GCS Computer Tech Pvt. Ltd. v. State of Haryana, 2008 (1) CTLJ 110 (P. & H.) (DB) (SN).

[96] Meerut Development Authority v. AMS, 2009(1) CTLJ 212 (SC). Also see: Nex Tenders (I) Pvt. Ltd. v. Ministry of Commerce & Industries, 2009(1) CTLJ 309 (Delhi) (DB).

the tender invitation contained a condition that the bidder should have completed the work of National Highway/Expressway and that experience in the construction of state highways was not to be considered relevant, the condition was not found to be unreasonable or arbitrary.[97]

A few examples will help further clarify the above viewpoints.

EXAMPLES FROM CASE LAW

1. Tenders were invited for the sale of stocks of damaged food grains by Food Corporation of India (FCI) in accordance with the terms and conditions contained in the tender notice. K's bid was found to be highest on opening of the tenders. FCI not satisfied with the adequacy of the amount offered in the highest tender, instead of accepting any of the tenders, invited all the tenderers to participate in the negotiations. FCI then received offers totalling to Rs. 1.10 crore from a bidder as against Rs. 90 lacs offered by K who refused to increase rates in the negotiation. Upon refusal to accept K's tender, a writ petition was submitted on the ground that FCI's action was arbitrary and, therefore, in breach of Art. 14 of the Constitution. This contention found favour with the P. & H. High Court. On appeal, the Supreme Court observed:

> Procuring the highest price for the commodity is undoubtedly in public interest. Accordingly, inadequacy of the price offered in the highest tender would be a cogent ground for negotiating with the tenderers giving them equal opportunity to revise their bids with a view to obtain the highest available price. Retaining the option to accept the highest tender, in case the negotiations do not yield a significantly higher offer would be fair to tenderers besides protecting public interest – this procedure involves giving due weight to the legitimate expectation of the highest bidder to have his tender accepted unless outbid by a higher bidder would be a reasonable exercise of power for public good.

2. Global tenders were invited for implementing a scheme for[98] moderation of Air Traffic Control Service at Bombay and Delhi Airport by NAA on turnkey basis in four parts:

    i. Prequalification bid;
    ii. Technical bid;
    iii. Commercial bid; and
    iv. Financial bid.

Three tenders were received. All of them met the prequalification criteria. After seeking clarifications, an external experts' committee was appointed by the tender committee to look into the technical evaluation. The external experts shortlisted two bidders. Thereafter the financial bids of the said two bidders were opened. Negotiations followed with the two and final offers submitted by the two were opened. One of the two unsuccessful bidders and the third bidder who was not shortlisted challenged the award of the tender by filing writ petitions on the ground that the decision to award the contract was arbitrary and actuated by fraud and *mala fides*. A preliminary objection was raised to the maintainability of the writ petition on the ground that the petitioners were foreign companies and therefore not entitled to the protection of fundamental right conferred by Art. 19(1)(g) of the Constitution. The Delhi High Court held that having invited tenders from all over the world and having taken the position that the tenders would be considered on their respective merits, the petitioners were entitled to legitimate expectation of fair consideration. Like a citizen, a foreigner is also entitled to avail personal rights enshrined in Art. 14 of the Constitution. In any case, the foreign companies could challenge the decision by availing principles of

---

[97] G. R. Infraprojects Ltd. v. Airports Authority of India, 2009 (1) CTLJ 375 (Delhi) (DB).
[98] Food Corporation of India v. Kamdhenu Feed Industries, AIR 1993 SC 1601(1605).

administrative law without recourse to Arts. 14 and 19 of the Constitution. The writ petitions were considered and dismissed on merits.[99]

3.   Every year the Indian Railways Board enters into a contract with a manufacturer for supply of cast steel bogies for building wagons. It invites tenders from about a dozen established sources of proven ability. A limited tender notice was issued for the procurement of 19,000 cast steel bogies from 1–4–92 to 31–3–93. On opening of the tenders it was found that the top three manufacturers quoted an identical price of Rs. 76,660/- per bogie while others quoted between Rs. 83,000/- and Rs. 84,500/- per bogie. Before finalization of the tenders, the government announced two major concessions including reduction of customs duty. After taking into account the effect of the concessions, the tender committee concluded that Rs. 76,000/- per bogie would be a reasonable rate. It was concluded that the three large manufactures had formed a cartel and quoted lower rates. Counter-offers were given to the three big ones at Rs. 65,000/- per bogie and to the others at Rs. 76,000/-. This was challenged in the Delhi High Court on the ground of discrimination. The High Court allowed the writs and directed that all the suppliers should make the supply at Rs. 67,000/- per bogie. It was also observed that the allegation that the three big manufacturers had formed a cartel was based on extraneous considerations. Aggrieved by this judgment, the Union of India filed a Special Leave Petition. The Supreme Court held that there was not enough material to conclude that the three manufacturers formed a cartel. However, there was scope for entertaining suspicions by the tender committee that they formed a cartel since all the three quoted identical prices. The Railways Board had taken the stand that the three big manufacturers offered a low price with the hope of getting the entire or larger quantity allotted, resulting in a monopoly extinguishing the smaller manufacturers. It was also held that dual pricing was reasonable under the circumstances of this case. The plea of legitimate expectation advanced on behalf of the three manufacturers was considered and taking a review of the judicial pronouncement from various countries, the Supreme Court rejected it on the ground that "the protection of such legitimate expectation does not require the fulfilment of the expectation where an overriding public interest requires otherwise". However, it was observed that the doctrine gives the applicant sufficient *locus standi* for judicial review and that it is to be confirmed mostly to the right of a fair hearing before decision. It does not give scope to claim relief straightaway from the administrative authority as no crystallized right as such is involved.[100] It needs to be noted that the above judgment was given prior to the enforcement of the Competition Act, 2002 which declares price fixing agreements void.[101]

4.   Flats were allotted by a Housing Board subsequent to the sanction of a layout plan in which an area was earmarked for community facility. A writ petitioner asked the court to restrain the Board from converting that area for residential or commercial uses. It was held that the allottees had a legitimate expectation that they would be provided with community facilities.[102]

5.   Where the government invited tenders for the sale of a certain quantity of G.I. pipes and the tender notice stipulated that the highest quoted rate would be accepted and yet, the state allotted a certain quantity to a non-tenderer after accepting the highest tender, it was held that the doctrine was applicable and the action was arbitrary.[103]

6.   An invitation to tender was given to a bidder without stipulating a minimum requirement to provide four trains per hour service throughout the day in an area undergoing regeneration. This was challenged by another bidder on the ground that email exchange with the respondent had

---

[99] Thompson – CSF v. National Airport Authority of India, AIR 1993 Del. 252.

[100] Union of India v. Hindustan Development Corporation, AIR 1994 SC 988.

[101] Excel Crop Care Limited v. Competition Commission of India, AIR 2017 SC 2734=2017(8) SCC 47.

[102] Bhagat Sing Negi v. H.P. Housing Board, AIR 1994 60.

[103] Mogo Nagi v. State of Nagaland, AIR 1995 Gau. 6.

led to a legitimate expectation of it being given an opportunity to make further representation on the essential need for a four trains per hour service. The Court of Appeal was not convinced that this was the effect of the email exchange and that even if such an expectation were raised, the respondent would have been entitled to depart from it in the broader public interest. In a massive project it was "simply unreal" that two emails would have generated expectations upon which it was reasonable for the bidder to rely.[104]

3.10.3 Procedure to Be Followed for Inviting and Accepting Tenders

In a state governed by rule of law, every tender process by the state or its instrumentalities should be transparent, fair and open. The state cannot adopt a tender process which does not notify the procedure that will be followed for the acceptance or rejection of tenders. The process in such a case is not transparent, fair or open. In such a case, a writ will lie and a court will interfere.[105] For example, a registered Class I contractor could contend that the issuance of a tender notice by the state did not have sufficiently wide circulation to invite the participation of eligible bidders, as a result of which he was not aware of such notice inviting tender. The action of the state denied him an opportunity to participate in the tender process and infringed his right, giving him a locus standi to maintain a writ petition.

EXAMPLES FROM CASE LAW

1. The Constitution mandate requires that there should be a specific authorization by the governor to invite tenders for contracts. The contract documents were signed by the respondent on behalf of the governor but there was no authorization in that regard as required under Art. 299 of the Constitution. Rules of Business of the State required a cabinet approval for contracts valued at more than Rs. 5 crores. Though the contract value was more than Rs. 5 crores, there was no approval of the cabinet. In the absence of a cabinet approval, the contract could not be considered to be a concluded contract, despite the parties having executed formal contract documents. Further, the non-publication of the notification inviting tender in the *Indian Trade Journal* as required under Rule 10 of the Karnataka Transparency in Public Procurement Rules was held to have vitiated the entire tender processing. Under these facts, the Karnataka High Court held that the contract became void and unenforceable.[106]

2. By a memo dated 20th November 1973, an acceptance regarding the tender submitted by the appellant-contractor was conveyed by the Executive Engineer, but this very communication required the parties to sign an agreement pertaining to the work in question. Another communication by the Executive Engineer dated 25th January 1974 required the signing of an agreement. The admitted position of fact between the parties was no such agreement was signed. On the contrary, the communications dated 20th November 1973 and 25th January 1974 issued by the state made it clear that the agreement between the parties had yet to be entered into.[107]

---

[104] The Queen (on the application of London Borough of Enfield) v. Secretary of State for Transport v. Abellio East Anglia Limited, and others, [2016] EWCA Civ 480, paragraph 53.

[105] Dutta Associates Pvt. Ltd. v. Indo Merchantiles Pvt. Ltd., 1996 (10) JT 419; 1997 (1) SCC 53. Also see Ashwin S. Shah v. Municipal Corporation of Greater Mumbai, 2010 (1) CTLJ 302 (Bombay) (DB).

[106] Shimnit Utsch India Pvt. v. State of Karnataka, (Karnataka) (DB): 2012 ILR (Karnataka) 4073: 2012(36) R.C.R. (Civil) 178: 2013(1) KCCR 544. Samir Mohanty v. State of Odisha, (Orissa) (DB): 2013 (sup) CutLT 392: 2013(2) ILR (Cuttack) 16: 2012(73) R.C.R.(Civil) 369: 2013 (sup1) Ori. Law Rev. 449.

[107] State of Haryana v. Chaman Lal Mukhija and anr., (Punjab and Haryana): 2012(5) R.A.J. 178: 2012 AIR (Punjab) 104: 2012(4) BC 417: 2013(6) R.C.R.(Civil) 1008.

3. Due to urgency, tender notices were displayed on notice boards of PWD and were also dispatched to the concerned division for information to interested bidders but no notice inviting tenders was published in any media. It was held that the state could not allege that there were time constraints over the publication of a notice inviting tenders when the necessary sanction from the central government was still awaited.[108]

The rule is that an executive authority must be vigorously held to the standards by which it professes its action to be judged and it must scrupulously observe those standards on pain of invalidation of an act in violation of them. However, it cannot follow that the executive authority cannot at all deviate from or relax the standards. That would depend upon the nature of the act, the necessity for making the deviation or relaxation, and the potential effects of such a deviation or relaxation.

A distinction must be made between those terms and conditions which are essential terms and the conditions of eligibility, and those which are incidental or inconsequential in nature. If the relaxation is made in the essential conditions such as a condition relating to turnover, fairness and equal treatment requires that the process should be carried out afresh.[109]

EXAMPLES FROM CASE LAW

1. The respondent's invitation to offer was for a contract to collect toll for 90 weeks. The offer/proposal given by the petitioner was also to collect toll for 90 weeks. However, the letter of acceptance given by the respondent to the petitioner was for a period of only 52 weeks without a change in the amount offered. The period of contract itself was varied in terms of acceptance. It was held that the variation in terms of acceptance was not trivial, minor or immaterial. The variation was substantial and to the prejudice and detriment of the petitioner. Such a qualified acceptance would not convert an offer into a promise. Forfeiture of the petitioner's earnest money was not justifiable.[110]

2. In response to a request for prequalification, the petitioner submitted details and secured third highest eligibility points. He was informed by the authority that his application had not been shortlisted. No clarification was sought in terms of the invitation to facilitate evaluation before coming to the conclusion. It was held that the authority could not claim that it has got unfettered discretion with which courts have no right to interfere. The matter was remitted back for reconsideration.[111]

The authority or tendering committee has the power to relax conditions, to extend the deadline for submission of bids, or to issue an amendment. For example, in a case, the bidding date was deferred as certain bidders were prevented/obstructed from submitting bids, an extension given was held justified.[112]

---

[108] M/s. K.K. Chire and Sons v. State of Nagaland, (Gauhati) (Kohima Bench): 2012 AIR (Gauhati) 20: 2012(2) R.A.J. 664: 2012(4) BC 150: 2012(2) GauLJ 730: 2011(4) NEJ 767: 2012(3) GauLR 504: 2011(30) R.C.R.(Civil) 815: 2016(5) GauLT 332.

[109] Konark Infrastructure Pvt. Ltd. v. Commissioner Ulhasnagar Municipal Corpn., AIR 2000 Bom. 389. Also see footnote 101 above.

[110] M/s. Abhay Construction v. State of Maharashtra, (Bombay) (DB) (Aurangabad Bench): 2014(21) R.C.R.(Civil) 741: 2014(4) AIR Bom. R 132: 2014(4) Mh.LJ 829: 2014(6) BCR 82.

[111] PSA Ennore Pte. Limited v. Union of India, 2010(1) CTLJ 84 (Madras).

[112] Pioneer Construction v. The State of West Bengal, (Calcutta) (DB): 2011 AIR (Calcutta) 199: 2011(25) R.C.R. (Civil) 602.

EXAMPLES FROM CASE LAW

The Petitioner tenderer complied with the essential qualifying criteria, and deposited earnest money by way of a demand draft (DD). The DD expired on 3rd October 2014. The decision to accept a tender could not be finalized up to 7th October 2014 "due to procedure in vogue". The tenderers were informed on 8th October 2014 to extend the validity of their offers as well as renew their DDs. The petitioner was not able to renew its DD due to reasons beyond its control. The respondent concluded that the petitioner had failed to file a valid DD within time. It was held that this was not proper and that the respondent could have negotiated a non-essential term of contract with petitioner. The respondent was directed to pay Rs. 2,00,000/- as costs to the petitioner.[113]

2.   A tender invitation notice provided that for an existing licensee to be eligible, he ought to have cleared past dues relating to licence fee. An affidavit format to be submitted along with the tender referred to all existing licences. The contention of a petitioner that the licence fee related to stalls for which tenders were invited and not all stalls was rejected and the writ petition dismissed.[114]

3.   A tender invitation for request for proposal stipulated that a consortium agreement should provide for "joint and several liability" whereas the proposal submitted by a consortium included the words "joint and several responsibility". It was held by the Kerala High Court and confirmed by the Supreme Court that objective words can be interpreted subjectively. It was pointed out that the words liability and responsibility were used interchangeably by the Project Advisor/Sponsor. It was held that the requisite words could be inserted at the stage of writing the final licence agreement. The bid submitted by the consortium was held valid and the authority was directed to consider it.[115]

4.   Where bank guarantees submitted by the petitioner did not satisfy the requirements of conditions of bidding (e.g. it did not mention that it was being given in lieu of deposits, etc.), it was held that the rejection of the tender cannot be said to be arbitrary.[116]

5.   A bid of the respondent was condemned only on the ground that the certificate furnished by its chartered accountant mentioned "financial year" instead of "assessment year". The documents produced had to support the claims of the annual turnover achieved in the past three years by the bidders. The cumulative total turnover for the past three years was to be taken into account to arrive at average annual turnover instead of looking at turnover of a particular year in those three years. The petitioner's objections to the respondent's bid were, therefore, wrong.[117]

Condition Requiring the Tenderer to Submit Documents if Condition Precedent?

If there is a condition in the tender notice requiring the tenderer to submit documents along with the tender, such a condition may not be termed a condition precedent, but it could be characterized as a condition subsequent. If it is a condition subsequent, it can be relaxed by the concerned authority and a tenderer can be asked to submit documents later. For example, shortlisted tenderers were asked

---

[113] Ratnesh Srivastav v. Union of India, (Calcutta): 2016 AIR CC 2601.

[114] Deepak & Co. v. Indian Railways Catering & Tourism Corpn. Ltd., 2009(1) CTLJ 115(Delhi) (DB).

[115] State of Kerala v. Zoom Developers Pvt. Ltd., 2009(1) CTLJ 295 (SC).

[116] Tama Fabrication Works v. State of Arunachal Pradesh, 2010(1) CTLJ 63 (Gauhati) (DB).

[117] S. Price Hightech Pvt. Ltd. v. National Thermal Power Corpn. Ltd., (Andhra Pradesh): 2012(1) ALT 721: 2012(1) Andh. LD 46: 2011(16) R.C.R.(Civil) 436.

to submit their reports on energy saving, "positively" by 1st August 2007. One tenderer sought a day's extension of time to submit the report; the report was submitted on 2nd August 2007. His request seeking extension of time was rejected. A writ petition was filed. It was held that the action of the respondent authority was arbitrary, and opposed to larger and overwhelming public interest because the authority had not set a deadline for the finalization of tenders and no additional advantage was claimed that would not have been available to other remaining firms.[118]

In another case, a tender condition required tenderers to have one year's experience of transportation of controlled food grains under doorstep delivery in the State of Rajasthan. The said condition of production of experience certificate, it was held, could not be said to be a condition precedent before accepting the tender, but it could be called a condition subsequent. In such a situation, if the concerned authority, on the request of the tenderer, had asked him to submit the requisite experience certificate, it had committed no illegality nor exceeded its jurisdiction in doing so, but on the contrary, it had the power to relax such type of condition of production of document. It was held:

> Furthermore, such type of condition can only be relaxed in favour of such person whose tender was found lowest and valid in all respects and in this case, as per the tender proceedings, the tender of the respondent No. 2 was found [to be the] lowest one and it was valid in all respects.[119]

Ordinarily, a court refuses to substitute its discretion in the place of the authority's discretion to examine the comparative merits of bidders, but it will not abdicate its jurisdiction and authority to examine the merit of any allegations of arbitrariness in the procedure or the decision to give someone an unfair advantage or to keep eligible persons out of competition.

EXAMPLES FROM CASE LAW

1. Furnishing of income-tax returns was a condition precedent for the opening of the price bids but not the technical bids. Further, a copy of the Permanent Account Number (PAN) card and copy of the latest income-tax return could be submitted one day before the opening of the price bids. Those also had to be uploaded online. No stipulation or criterion was set out in the tender documents except that a copy of the latest income-tax clearance certificate had to accompany the tender document. Admittedly, a bidder had not submitted the latest income-tax return along with his tender but had furnished the PAN card and income-tax clearance certificate for the year 2002–2003 online. His tender was disqualified on the basis of failure to provide a copy of the latest income-tax return. On these facts, it was held that since the technical bid of the tenderer contained the qualification data prescribed in paragraph (10) of the government order, it had to be opened.[120]

2. The State of Assam invited bids for the settlement of a river fishery. The highest bidder failed to deposit security deposit and other required amounts after his bid was accepted. The state thereafter issued a notice for resale at the cost of the bidder after warning him that his earnest money would be forfeited. In the resale, the highest bidder was asked to produce some documents. Pending scrutiny, the state cancelled the resale notice. Deciding the writ petition filed by the highest bidder in the resale process, the Gauhati High Court held that "every power is coupled with duty and discretion is unfettered. The impugned notice – not backed by acceptable logic and rationale and is a product of mechanical exercise of power". The notice of cancelling resale was set aside and quashed and the state was directed to complete the resale process.[121]

---

[118] S. R. Comm. Systems v. Greater Hyderabad Municipal Corpn., 2009 (1) CTLJ 138 (AP).
[119] Kesulal Mehta v. R.T.A.D. Co-op. Federation Ltd., AIR 2005 Raj. 55.
[120] K. Sudarshan Reddy v. Govt. of A. P., AIR 2005 AP 228.
[121] Bijon Kumar Das v. State of Assam, AIR 2003 Gau. 164.

If the scrutiny shows no arbitrariness, courts may refuse to grant any relief.[122] Wherein a tender invitation required the proof of turnover of the firm over the last two relevant years with supporting documents and also required the tenderer to produce proof of work experience for the last two years with full details and supporting documents, and the checklist had specifically mentioned that the production of proof of turnover with latest profit and loss account duly certified by a chartered accountant was a mandatory requirement, the Supreme Court held:

> In our opinion, the High Court was not justified in coming to the conclusion that production of the documents mentioned herein above along with the tender form was not mandatory and the High Court was also not justified in coming to the conclusion that neither the rules and conditions governing the tender nor the advertisement calling for tender made it mandatory for an intending tenderer to produce those documents and specially proof of turnover for the relevant year.[123]

In another case, the tender notice categorically provided a three-tier eligibility test and invited offers in three covers. On opening the first cover, the petitioner tenderer was found ineligible. The other two covers were not opened. The A.P. High Court declined to interfere in the decision.[124]

3.11 Acceptance/Rejection of Tender – Interference by Court – When Justified – Guiding Principles

In the case of a state or public body, there could be, in a given case, an element of public law/public interest involved. The elements of public interest include:

1. Public money;
2. Public works and utility services;
3. Expeditious completion;
4. Quality of goods supplied or works executed.

The High Court will interfere with the acceptance or rejection of tenders by the government or other public authorities when a decision is influenced by extraneous considerations which ought not to have been taken cognizance of. Even if the action was taken in good faith, it is bad in law and cannot be upheld.

EXAMPLES FROM CASE LAW

1. An invitation of tenders required that the tenders should remain valid for a period of four months and the rates should remain in force for four months. A tender was accepted after the date of validity of the rates had elapsed. It was held that the

---

[122] M/s Continental Construction Ltd. v. Tehri Hydro D.C. Ltd., AIR 2002 SC 3134; 2002(3) Arb. LR 255(SC). Also see: M/s Bharat Construction Co. v. State of Rajasthan; AIR 2002 Raj. 279; M/s Doshi Ion Exchange & Chemical Inds. Ltd. v. Union of India, AIR 2001 Raj. 276.
   Madhu Construction Co. v. National Aluminium Co. Ltd., AIR 2001 Ori. 169.
   M/s Ritz Construction Co. v. Union of India, AIR 2001 J. & K. 7.
[123] Laxmi Sales Corporation v. M/s. Bolangir Trading Co., AIR 2005 SC 1962 = 2005 AIR SCW 1337.
[124] Neelambar Ropeways v. A.P. Tourism Develop. Corpn. Ltd., 2002(3) Arb. LR 150 (AP); N.D. Grover v. State of M. P., 2002(3) Arb. LR 224 (MP). Also see: Jay Bee Energy Services Pvt. Ltd. v. Oil India Ltd., 2002 (2) Arb. LR 374 (Gauhati); Sharda Floor Pvt. Ltd. v. State of Chhattisgarh, AIR 2005 Chhat. 12.

acceptance was bad and stood quashed and the earnest money was ordered to be refunded with interest.[125]

2. A successful tenderer refused to execute an agreement and perform its part on the ground that the acceptance was not in terms of the invitation. The trial judge accepted that contention but on close scrutiny of the terms of the invitation, the High Court found no error in the action of forfeiting the earnest money.[126]

3.11.1 Guiding Principles

The Supreme Court of India had, from time to time, indicated the principles to be followed by High Courts when a writ petition is filed in the High Court challenging the award of a contract by a public authority.[127] First and foremost is that in order to avail of prerogative remedy, the conduct of the party seeking the remedy would be borne in mind. It is, therefore, of utmost necessity that the petitioner should approach the court with clean hands.[128] If there is no candid disclosure of relevant and material facts or the petitioner is guilty of misleading the court, his petition may be dismissed at the threshold without considering the merits of the claim.[129]

A modern state which is committed to egalitarian values and dedicated to the rule of law, has to act while awarding a contract, under the constitutional mandate of Art. 14, as also the judicially evolved rules of administrative law. It must, therefore, be taken to be the law that where the government is dealing with the public, whether by way of giving jobs or entering into contracts or issuing quotas or licences or granting other forms of largesse, the government cannot act arbitrarily at its sweet will and, like a private individual, deal with any person it pleases, but its action must be in conformity with a standard or norm which is not arbitrary, irrational or irrelevant. In the well-known case of Ramana Dayaram Shetty v. The International Airport Authority of India, it was observed:[130]

> It is indeed unthinkable that in a democracy governed by the rule of law the executive Government or any of its officers should possess arbitrary power over the interests of the individual. Every action of the executive Government must be informed with reason and should be free from arbitrariness. That is the very essence of the rule of law; and its bare minimal requirement ... The State cannot, therefore act arbitrarily in entering into relationship, contractual or otherwise with a third party, but its action must conform to some standard or norm which is rational and non-discriminatory.

In the case of M/s. Kasturi Lal Lakshmi Reddy v. State of Jammu & Kashmir,[131] it was reiterated:

> Whatever be its activity, the Government is still the Government and is subject to restraints inherent in its position in a democratic society. The constitutional power

---

[125] Jain Electricals and Contractors v. RIICO, 2008(1) CTLJ 485 (Rajasthan).

[126] State of Bihar v. Bhawani Industries Ltd., 2008(2) CTLJ 420 (Patna) (DB).

[127] Several principles have been referred to in earlier paragraphs in this chapter.

[128] G. Jayashree v. Bhagwandas S. Patel, 2009(1) CTLJ 10 (SC).

[129] K. D. Sharma v. Steel Authority of India Ltd., 2008(2) CTLJ 11 (SC); see Tata Power Co. Ltd. v. Union of India, 2009(1) CTLJ 357; Nex Tenders (I) Pvt. Ltd. v. Ministry of Commerce & Industries, 2009(1) CTLJ 309 (Delhi) (DB).

[130] Ramana Dayaram Shetty v. The International Airport Authority of India, AIR 1979 SC 1628.

[131] M/s. Kasturi Lal Lakshmi Reddy v. State of Jammu & Kashmir, AIR 1980 SC 1992.

conferred on the Government cannot be exercised by it arbitrarily or capriciously or in an unprincipled manner.

In the case of Mahabir Auto Stores v. Indian Oil Corporation, the petitioners' firm was carrying on the business of sale and distribution of lubricants for 18 years obtaining its supplies from the Indian Oil Corporation. Abruptly and without any notice, the Indian Oil Corporation stopped the supply of lubricants to the firm. In that connection it was observed as follows:

> Where there is arbitrariness in State action of this type of entering or not entering into contracts, Article 14 springs up and judicial review strikes such an action down. Every action of the State executive authority must be subject to rule of law and must be informed by reason. ... Even though the rights of the citizens are in the nature of contractual rights, the manner, the method and motive of a decision of entering or not entering into a contract, are subject to judicial review on the touchstone of relevance and reasonableness, fair play, natural justice, equality and non-discrimination in the type of the transactions and nature of the dealing as in the present case.[132]

This case was mentioned in a later decision of the Supreme Court, where it was observed that the courts' jurisdiction is wider at the threshold of formation of a contract than in the cases involving the enforcement of its terms and conditions. However, it was also made clear that the court would interfere if the terms of the contract were against public policy or the state acted arbitrarily, unfairly, or unreasonably or in a discriminatory manner.[133]

In Tata Cellular v. Union of India, the Supreme Court observed:

> It cannot be denied that the principles of judicial review would apply to the exercise of contractual powers by Government bodies in order to prevent arbitrariness or favouritism. However, it must be clearly stated that there are inherent limitations in exercise of that power of judicial review. Government is the guardian of the finances of the State. It is expected to protect the financial interest of the State. The right to refuse the lowest or any other tender is always available to the Government. But, the principles laid down in Article 14 of the Constitution have to be kept in view while accepting or refusing a tender. There can be no question of infringement of Article 14 if the Government tries to get the best person or the best quotation. The right to choose cannot be considered to be an arbitrary power. Of course, if the said power is exercised for any collateral purpose the exercise of that power will be struck down.[134]

In connection with the termination of the appointment of the District Government Counsel by the State Government of Uttar Pradesh the same question was considered by the Supreme Court in the case of Kumari Shrilekha Vidyarthi v. State of U.P.[135] wherein it was held:

> In our opinion, it would be alien to the constitutional Scheme to accept the arguments of exclusion of Article 14 in the contractual matters. The scope and permissible grounds of judicial review in such matters and the relief which may be available are different mailers but that does not justify the view of its total exclusion. This is more so when the modern trend is also to examine the unreasonableness of a term in such contracts where the bargaining power is unequal so that these are not negotiated contracts but standard form contracts between parties having unequal bargaining power.[136]

[132] Mahabir Auto Stores v. Indian Oil Corporation, AIR 1990 SC 1031.
[133] United India Insurance Co. Ltd. v. Manubhai Dharamsinhbhai Gajera, AIR 2009 SC 446.
[134] Tata Cellular v. Union of India, AIR 1996 SC 11; also see: Bhupendra Engg. & Construction Pvt. Ltd., 2004 (1) CTLJ 450 (Jharkhand).
[135] Kumari Shrilekha Vidyarthi v. State of U.P., (1990) 3 SCJ 336.
[136] Bhupendra Engg. & Constn. Pvt. Ltd. v. State of Jharkhand, 2004(1) CTLJ 450 (Jharkhand).

In conclusion, it can be said that the law is well settled that in respect of proceedings and decisions taken in administrative matters, the scope of judicial review is confined to the decision-making process and does not extend to the merits of the decision taken.[137] The other principles can be summed up as follows:

1. The court must be satisfied that there is some element of public interest involved. The court must weigh conflicting public interests before intervening. Only when the court concludes that there is an overriding public interest involved, should that court intervene.

2. A mere difference in prices offered by two tenderers may or may not be decisive.

3. The proposed project may be considerably delayed thereby escalating the cost far more than the saving which may result by the court's decision. The High Courts do not interfere in a given case only because it would be lawful to do so; the High Court may refuse to issue any writ in the event it is found that substantial justice has been done to the parties or that in the larger interest it would not be prudent to issue such a writ. The right of an individual sometimes has got to give way to the right of the public at large. In the interest of the public, the works allotted to a tenderer should be allowed to continue in view of the fact that thereby the public exchequer would benefit to a huge extent.[138]

EXAMPLES FROM CASE LAW

1. Tenders were invited by the T.N. Electricity Board for the construction of a hydroelectric project. The tenderer who had failed to comply with the tender conditions was successful and started the work. The acceptance of the tender was held to be illegal. However, in view of the fact the work was started and nine months' time had already elapsed, the Madras High Court held that it would not be in the public interest to put the hands of the clock back and to direct the authorities to reconsider the tenders. It was therefore held that the aggrieved party should approach the appropriate court for an action in damages.[139]

2. The M. S. Electricity Board invited tenders from registered contractors, but accepted the tender of an unregistered contractor. The Bombay High Court held that it was an essential condition and that the decision of the Board was liable to be quashed as the successful tenderer had not fulfilled an essential condition. However, the work had already been commenced by the tenderer and the High Court thought it fit not to upset the decision of the Board and to declare the contract void. The decision not to declare the contract void was taken in the peculiar facts and circumstances of the case.[140]

In the case of a public interest litigation (PIL) petition challenging the award of contract, it needs to be seen that the petitioner is litigating *bona fide* for public good and that the PIL is not a cloak for attaining private ends of a third party or the petitioner. The court can examine the previous record of public service rendered by the organization bringing PIL.

When two tenders are found to be the lowest, the authority can accept the tender of the bidder who is senior with regard to the date of registration. An order to divide the work between the two renderers, where the rules do not permit division of the work is likely to be struck down.[141]

---

[137] K. Vinod Kumar v. Palanisamy and ors., 2004(1) CTLJ 436 (SC); NIIT Ltd. v. Bihar State Electronics Corpn. Ltd., 2008(1) CTLJ 500(Patna) (DB).

[138] Also see: Y. Swamidhas v. Thamil Nadu Anaithu Vagai Oppan., Sangam, 2002(1) Arb. LR 54 Madras (DB).

[139] M/s Chinnamman & Co., Kadamparai v. T.N.E. Board, Madras, AIR 1986 Mad. 302.

[140] Haribai Velhi v. M.S.E.B., AIR 1988 Bom. 114.

[141] Khursheed Alam v. State of Bihar, AIR 1994 Pat. 20.

When a decision has been taken *bona fide* and the choice has been exercised on legitimate considerations and not arbitrarily, there is no reason why the court should entertain a petition under Art. 226.

Where there is an allegation of *mala fides* and the court, based on the material before it, is satisfied that the allegation needs further investigations, the petition should be entertained.

When the decision-making procedure has been structured and tender conditions set out the requirements, the court is required to examine if these conditions have been considered.[142]

If any relaxation in the tender requirements has been granted, the court should examine if it was for *bona fide* reasons, whether the tender conditions permit such relaxation, and if the decision has been arrived at after a fair consideration of all offers.

If an evaluation committee of experts is appointed to evaluate offers, the expert committee's knowledge plays a decisive role and price is only one of the criteria. The court should not substitute its own decision for the decision of the expert evaluation committee.

When interim orders are being passed, balance of convenience would play a major role in shaping the interim relief. For example, where neither any *mala fides* were alleged nor any allegation of collateral motive was present, except that the tenderer whose bid was accepted did not fulfil the qualifying criteria, and the petitioner too did not fulfil the qualifying criteria, it was held by the Supreme Court that the High Court had seriously erred in granting the interim order.[143]

It is not a rule that merely because some disputed questions of fact arise for consideration; the court should refuse to entertain a writ petition in all cases.

A writ petition involving a consequential relief of monetary claim is also maintainable.[144] After holding that the disputes in question could not be agitated in a writ petition, the dismissal might have result in a miscarriage of justice on account of lapse of time resulting in the foreclosure of all other remedies, the Supreme Court refused to dismiss the writ petition.[145]

If the government acts fairly, though faltering in wisdom, the Court should not interfere.[146]

EXAMPLES FROM CASE LAW

1. Where, at the instance of the court, tenders were to be recalled but while issuing a fresh notice for recall, new conditions for eligibility were inserted, the court refused to interfere at the instance of the party which did not fulfil the requisite criteria. It was held that if by inserting certain terms the state wanted the best to come forward, such incorporation could not be said to be arbitrary.[147]

2. When the lowest tenderer is found to lack the requisite experience and all the tenderers agree to execute the work at the rates tendered by the lowest tenderer; who should be selected? The Orissa High Court upheld the decision under the provisions of Orissa PWD

---

[142] Also see: M/s Kesar Enterprises Ltd. v. State of U.P., AIR 2001 All. 209.

[143] Raunaq International Ltd. v. I.V.R. Construction Ltd., AIR 1999 SC 393.
      Also see: Asst. Collector of Central Excise, West Bengal v. Dunlop India Ltd., AIR 1985 SC 330.

[144] ABL International Ltd. v. Export Credit Guarantee Corporation of India Ltd., 2004 (1) CTLJ 1 (SC); (2004)3 SCC 553, referred to in Karnataka State Forest Industries Corporation v. M/s Indian Rocks, AIR 2009 SC 684.

[145] Kerala State Electricity Board v. Kurien E. Kalathi, 2000(6) SCC 293; 2000 (2) Arb. LR 652 (SC). Also see: Ganga Retreats & Towers Ltd. v. State of Rajasthan, 2004(1) CTLJ 104 (SC).

[146] Fertilizer Corporation Kamgar Union (Regs.), Sindri v. Union of India, (1981) 1 SCC 568; AIR 1981 SC 344; Ravinder Singh v. Indian Oil Corporation Limited, 2009(1) CTLJ 178 (Delhi) (DB).

[147] M/s Dredge & Dive v. State of Jharkhand AIR 2000 Jha. 124; M/s Riaz Construction Co. v. Union of India, AIR 2001, J. & K.7. The protection of legitimate expectations was held compatible with the discontinuance of a free beauty contest procedure to allocate digital frequencies (which would have ensured the grant of a tender to a bidder) in favour of a fee-based tendering procedure so long as there had been no precise assurances from an authorized and reliable source: Europa Way Srl and another v. Autorità per le Garanzie nelle Comunicazioni (Case C-560/15) [2018] 1 CMLR 31.

Code of awarding the contract to the lowest eligible tenderer and only if he refuses to extend such offer to the next lowest tenderer.[148]

3. The division bench of the Calcutta High Court offered a public officer an opportunity to justify his non-acceptance of the lowest rates and the acceptance of the tenders with high rates for supply of dietary articles to district hospitals; the officer gave no reason. The division bench ordered an invitation of fresh tenders and also directed investigations by a competent police officer as to why, without any reasons, the higher rates were accepted and to forward the report to the Vigilance Department for taking steps against erring officers, if it was found that the officers shunned public duty.[149]

4. An invitation of tenders was issued for the construction of earthen dam and spillway. After scrutiny of the tenders received it was found that the lowest tenderer did not possess the requisite experience and machinery. The second lowest tender was rejected because of quotes of abnormally low rates without submitting an analysis of the rates quoted. The third lowest tender was recommended for acceptance. In the meantime, the second lowest tenderer appealed to the government, submitted his analysis of rates and stated that he had no work in hand and that he could, therefore, commence the work at once. The government directed a committee to reconsider the issue. The committee initially rejected the second lowest tenderer on the ground that he did not possess a valid licence. The licence was valid when the tender was submitted and was subsequently renewed. Thereafter it was decided that the work be awarded to second lowest tenderer if he was willing to execute the work at the cost of the lowest tenderer, if not give the same offer to the third lowest. The second lowest accepted the offer. It was held that the authorities had not acted illegally and the action did not suffer from the vice of arbitrariness or malice.[150]

In its pursuit of revenue, the state or its instrumentalities cannot ignore the essential conditions of the NIT or act contrary to public policy.[151] In an appropriate case, a writ petition against a state or its agency is maintainable even in matters relating to contracts.[152]

Where malice is attributed to the state, it can never be a case of personal ill-will or spite on the part of the state. It is malice in legal sense, an act which is taken with an oblique or indirect object. For example, an action of the state can be described as *mala fide* if it seeks to acquire land "for a purpose not authorized by the Act", i.e. other than for a statutory purpose.[153]

A couple of terms appearing in the principles stated above are briefly explained below.

Malice in Fact and Malice in Law – Meaning

The term malice in law means ill-will or spite towards a party or any indirect or improper motive in taking an action.[154] This is sometimes described as "malice in fact". "Legal malice" or "malice in law" means "something done without lawful excuse" in which the doer's state of mind is irrelevant.[155] In other words, it is an act done wrongfully and wilfully without reasonable or probable cause, and not necessarily an act done from ill feeling and spite.

---

[148] Debendranath Balbantaray v. Commr.-cum-Secretary Govt. of Orissa, AIR 2002 Ori. 142.

[149] Adhir Ghosh v. State of West Bengal, AIR 1998 Cal. 317.

[150] Debendranath Balabantaray v. Commr.-cum-Secretary Govt. of Orissa, AIR 2002 Ori. 142.

[151] Bijulibari Multipurpose Development Society v. State of Assam, 2004(1) CTLJ 365 (Gauhati) (DB).

[152] See e.g. Omjee Finance Pvt. Ltd. v. Amravati Municipal Corporation, 2008(1) CTLJ 176 (Bombay) (DB).

[153] State of Andhra Pradesh v. Goverdhanlal Pitti, AIR 2003 SC 1941.

[154] Paragraph 651, *Halsbury's Laws of England*; retrieved from www.lexisnexis.com November 2018.

[155] Paragraph 651, *Halsbury's Laws of England*; retrieved from www.lexisnexis.com November 2018.

Doctrine of Public Policy

The doctrine of public policy is contained in a branch of common law. It is governed by precedents. What is "opposed to public policy" would be a matter depending upon the nature of the transaction. The pleadings of the parties and the materials brought on record would be relevant so as to enable the court to judge the concept as to what is for public good or in public interest or what would be injurious or harmful to public good or public interest at any relevant point in time as distinguished from the policy of a particular government.[156] Section 23 of the Indian Contract Act provides that where consideration and object are not lawful, the contract would be void. However, for Section 23 to apply, the consideration or object must be forbidden by law or it must be of such a nature that it would defeat the provision of any law or it is fraudulent or it involves or implies injury to the person or property of another or the court regards it as immoral or opposed to public policy.[157]

A contract being "opposed to public policy" is a defence under Section 23 of the Indian Contract Act and the court while deciding the validity of a contract has to consider:[158]

a. Pleadings in terms of Order VI, Rule 10 of the Code of Civil Procedure;

b. Statute governing the case;

c. Provisions of Part III and IV of the Constitution of India;

d. Expert evidence, if any;

e. Materials brought on record of the case; and

f. Other relevant factors, if any.

A party against whom illegality is pleaded also gets an opportunity to defend himself. Hence, this essential function to decide on what is public policy cannot be delegated by the legislature to the executive through a subordinate legislation. The legislature of a State, however, may lay down as to which acts would be immoral being injurious to the society. Such a legislation being substantive in nature must receive the legislative sanction specifically and not through a subordinate legislation or executive instructions. The phraseology "opposed to public policy" may embrace within its fold such acts which are likely to deprave, corrupt or be injurious to the public morality and, thus, essentially should be a matter of legislative policy.

3.11.2 No Tender Submitted – Locus Standi to Maintain a Writ Petition?

Can a person challenge the action of acceptance of a tender if he has not submitted a tender because he did not fulfil the requisite qualification? He can, if his grievance is that if it were known that non-fulfilment of the condition of eligibility would be no bar to the consideration of a tender, he also would have submitted a tender and competed for the award of a contract. But he was precluded from submitting a tender and entering the field of consideration by reason of the condition of eligibility.

EXAMPLE FROM CASE LAW

An NIT for the transportation of coal required tenderers to own one-third of the total number of trucks required for the tender work. A person who did not fulfil the condition of owning the requisite number of trucks was allotted the work. It was held by the Patna High Court that the person who had not submitted a tender because of

---

[156] Meerut Development Authority v. AMS, 2009(1) CTLJ 212 (SC).

[157] Zoroastrian Co-op. Hsg. Socy. Ltd. v. Dist. Registrar, Co-op. Societies (Urban), AIR 2005 SC 2306= 2005 AIR SCW 2317.

[158] State of Rajasthan v. Basant Nahata, AIR 2005 SC 3401 = 2005 AIR SCW 4456.

not owning one-third number of trucks had locus standi to file a writ petition and that the petition had not become infructuous merely because tenders of others were already accepted.[159]

3.12 Conditions of Eligibility

Where, for public works, tenders are invited from tenderers with certain qualifications, the object is to restrict competition to a certain category or kind of contractors with some credentials such as registration. This is to invite tenders from persons with some experience, ability and infrastructure who also own the required machinery and equipment with technically competent personnel available to complete the work.[160]

What would be the eligibility criteria for the participation in a tender is a matter to be decided by the state authority and this being a policy decision the court, in exercise of its discretionary powers, will interfere with such a decision only if it is arbitrary, unreasonable and irrelevant. For example, the condition that a tenderer has to be a permanent resident of the concerned block was held to be wholly without any authority of law and in violation of Art. 19 of the Constitution of India.[161]

The rule is well settled that the state cannot accept the tender of a person who does not fulfil the requisite qualification and, as such, by accepting such a tender, the state would be guilty of discrimination between the person whose tender was accepted and the person who might have, but on account of being unqualified, did not submit his tender.

EXAMPLES FROM CASE LAW

1. The original NIT for the grant of a contract for raising, calibration and transport of iron ore for a three-year period included the eligibility criteria that the agency will not be allowed to participate in the tender if the work in hand of the said agency in the same mine was not likely to end within six months of the date of issue of the NIT. The purpose was to avoid the possibility that the work tendered for and the agency's work in hand would operate concurrently. After a couple of cancellations of the NIT, the amended NIT eliminated the overlapping margin of six months. On merits, the High Court held that there was a perfectly good reason for doing away with the six months' margin for a pre-existing contract in the same mine. On appeal the Supreme Court expressed complete agreement with the High Court's view.[162]

2. In an NIT, one of the conditions for eligibility was three years' experience of manufacturing the required item. One of the bidders, an Indian company, did not satisfy this condition, but assured that its foreign collaborator had the experience. It was held that the experience and expertise of the foreign collaborator could be of no avail to the Indian company. It was further held that the fact that the foreign company held 10% of the share capital would not advance the case of the Indian company any further. It was further observed that if two

---

[159] Jai Bharat Transport Co. Central Coal Fields Ltd., AIR 1989 Pat. 170.

[160] M/s The Indian Hume Pipe Co. Ltd. v. Bangalore Water S. & S. Board, AIR 1990 Kant. 305. Also see M/s Promuk Hoffman International Ltd. etc. v. State of Rajasthan & Ors., AIR 2009 (NOC) 567 (Raj.).

[161] Mahendra Kumar Mohanty v. Collector Khurda, 2008(1) CTLJ 442 (Orissa) (DB).

[162] S. S. and Company, M/s. v. Orissa Mining Corpn. Ltd., AIR 2009 SC 461.

manufacturing units are merged into one, and one of them has the requisite experience and expertise, the new entity after the merger might be given the benefit.[163]

3.   The eligibility criteria in an NIT required the shipbuilder to have built and delivered tugs of 15 ton bollard pull and above. One tenderer had built and delivered tugs of 13 ton capacity. Higher capacity tugs were being built, but not delivered, at the time of submission of the tender. The Special Tender Committee found that the work of building tugs of higher capacity was in progress and declared that the tenderer was qualified. By the time the tender came to be accepted after clarifications, negotiations with all parties and receiving revised finance bids, the tugs of higher capacity were delivered by the tenderer whose bid incidentally was found to be the lowest. The Orissa High Court rejected the plea that the tenderer ought to have been disqualified in the first instance and his offer not considered at all.[164]

4.   An NIT for the contract for transportation of essential commodities to ration shops stipulated that the tenderers should possess at least two lorries for carrying out the said contract. Rejection of the tender of the lowest tenderer who had, on the date of submission of his tender, only one lorry and the award of the contract to the next tenderer, who besides satisfying the conditions also had huge experience in transportation, was held valid and proper.[165]

5.   An NIT included a prequalification criterion of having successfully executed drilling work of 5,000 m and grouting to the extent of 100 tonnes in drainage gallery of masonry dam/tunnel or similar structures. In the subsequent notice, the words "tunnel or similar structures" were eliminated. A writ contending that due to the change the whole process was vitiated was dismissed.[166]

6.   A condition prohibiting defaulting contractors from tendering was upheld by observing that the question of contractual obligations cannot be gone into in the writ jurisdiction.[167]

7.   A condition of submitting a good moral character certificate of no more than three months was part of an NIT for participating in the auction of the right to collect toll on a bridge. The Allahabad High Court held that this was mandatory and could not be relaxed.[168]

8.   An NIT for printing, binding and supply of Yellow Pages telephone directories contained a stipulation that the tenderers should substantiate their experience with documentary proof by furnishing credentials in the field of having compiled, printed and supplied telephone directories to large telephone systems with capacity of more than 50,000 lines. A company, which had no experience and wanted to depend upon the experience of its shareholders, found its offer rejected, though it offered the highest royalty. In a writ petition, it was held by the Delhi High Court that the approach of the authorities in arriving at a conclusion that the petitioner did not satisfy the eligibility condition was not arbitrary.[169]

---

[163] M/s. Continental Pump and Motors Ltd. Gaziabad v. State of Bihar, AIR 1995 Pat.183.

[164] Bharati Shipyard Pvt. Ltd. v. Paradeep Port Trust, AIR 1995 Ori.147; PIP Associates v. The chairman, Tamil Nadu Electricity Board & Ors., AIR 2009 (NOC) 864 (Mad.).

[165] L.V. Basavraj v. Dy. Commr., Chikmagalur Dist., AIR 2002 Kant. 81.

[166] Reddy Brothers & Company v. State of M.P. and others, 2002(3) Arb. LR 347 (MP).

[167] Mohammad Ashraf Gilkar v. State of J. & K., AIR 2000 J. & K. 30.

[168] Bashishtha Narain Pandey v. Commissioner, Basti Divn., AIR 2002 All. 280. Also see in respect of Sales Tax Clearance Certificate: Sushila Devi Jhawar v. State of Tripura, 2008(2) CTLJ 433 (Gauhati).

[169] M/s New Horizons Ltd. v. Union of India, AIR 1994 Delhi 126= 1995 (1) SCC 478 = 1995 AIR SCW 275. Also see Ganpati-RV-Talleres Algeria Track Pvt. Ltd. v. Union of India, 2008(2) CTLJ 137 (Delhi) (DB). Bipson Surgical Private Limited v. Govt. of NCT of Delhi, 2008(2) CTLJ 500 (Delhi) (DB).

9. A tenderer was denied an award of a contract on the ground that he was not registered with the Building Construction Division of the PWD. PWD contended that after bifurcation of the PWD into Road Construction Department and Building Construction Department, the state had made the registration in the bifurcated Building Construction Department compulsory. Although the tenderer renewed his registration, albeit after the deadline for submission of tenders, he was found ineligible to take up the work. The tenderer contended that he was the most senior contractor duly registered in the PWD and that his renewal request for registration was approved for both the departments. The court examined the register of the Building Construction Department which showed that all the contractors including the tenderer were registered for Road and Building Departments (by indicating R and B). It rejected PWD's contention by quashing the work order in favour of another tenderer.[170]

10. Where a fresh auction was ordered on the ground that an important condition was not inserted in the earlier auction notice, the M.P. High Court, placing reliance upon a decision of the Supreme Court,[171] held that the fresh auction could not be said to be illegal, arbitrary or *mala fide*.[172]

11. The State Government of Meghalaya invited tenders for operating a weighbridge. It accepted the highest bid at Rs. 1.21 crores and entered into a contract with the highest bidder. The tender accepting committee had assessed the value at Rs. 2 crores. The action was challenged by an unsuccessful bidder on the ground that the bid amount was speculatory and predatory. Both the single judge and the division bench allowed the writ and fixed the approximate value of the contract at a much lower amount of Rs. 40,29,600/-. Allowing an appeal, the Supreme Court set aside the judgment and the order of the High Court by holding that the reasonable assessment made by the High Court was only a tentative expression of opinion and that the state authorities had a right to differ. It was held that the offering of bids after knowing the commercial value of the contract was a matter left to the business acumen or prudence of the tenderers and that the principle of predatory pricing was wholly alien to this type of contract. It was further held that principle of monopoly also did not come into play in this type of contract.[173]

12. An NIT for a materials handling contract specified an eligibility condition of having experience of handling of steel materials of at least 28,000 MT during any of the last five financial years. It was challenged in a writ petition under Art. 14 on the ground that it was discriminatory and excluded the petitioner having experience but not fulfilling the eligibility condition. Dismissing the writ it was held by the Allahabad High Court:

> Whenever any condition is imposed with an objective of increasing the efficiency even if it restricts the field of eligibility [it] could not be said to be discriminatory or violative of Art. 14 of the Constitution of India. Discrimination under this is discrimination among the same class of persons similarly placed not among different class of person dissimilarly placed.[174]

13. A Public Sector Undertaking (PSU) invited tenders for the transportation of food grains. Out of the seven tenders received, the tender of the second lowest tenderer was

---

[170] Dhanpat Prasad v. State of Bihar, AIR 2004 Pat. 80.

[171] State of Orissa v. Harinarain Jaiswail, AIR 1972 SC 1816.

[172] Arunkumar v. Chief Engineer, P.W.D., Bhopal, AIR 1989 M.P. 288.

[173] Jespar I. Slong v. State of Meghalaya, 2004 (1) CTLJ 527 (SC); AIR 2004 SC 3533.

[174] Vijay Kumar Ajay Kumar v. Steel Authority of India Ltd., AIR 1994 ALL 182.

recommended for acceptance by the tender committee. The lowest tenderer filed a writ petition in the High Court. The acceptance was then revoked and the second lowest tenderer was persuaded in negotiations to accept the rates less than those quoted by the lowest tenderer. The lowest tenderer was not even called for negotiations. The defence of his poor financial position was rejected but the High Court observed that no standard for fiscal fitness was laid down and that the plea was trotted out to reject the lowest tender. It was held that the action of accepting the second lowest tender was arbitrary. The authorities were directed to invite fresh tenders.[175]

14.   A tender condition stipulated the eligibility criteria of having satisfactorily completed at least two runways/national highways, preferably rigid pavement works involving considerable earth filling each valued Rs. 400 lacs and one work of Rs. 600 lacs during the last five years and having annual turnover of Rs. 500 lacs in each of the last three years. Having failed to get a blank tender form on the ground of non-eligibility, a contracting company filed a writ in the High Court of Madhya Pradesh. The petitioner had two projects of the required value ongoing but they were not yet completed and he had produced certificates to the extent that the progress on the two projects was satisfactory. The High Court, by an interim order, directed to issue a blank tender form and the petitioners duly submitted their tender. However, at the final hearing, the single judge and also the division bench dismissed the writ. An appeal was filed in the Supreme Court. It was held that the word "completed" would indicate that as on the date of the application for the tender, the tenderer should have completed at least two works. In other words completion of the two works was a precondition. It was held: "Under these circumstances, the view taken by the High Court cannot be said to be unwarranted".[176]

In another case, the Delhi High Court accepted the plea that "satisfactorily completed" meant commissioning of the project work.[177] On appeal, the Supreme Court confirmed the said decision.[178] The division bench of the Calcutta High Court in another case held, and rightly so, it is respectfully submitted, that completion certificate does not necessarily mean performance certificate which is issued after the maintenance period is over.[179]

15.   The lowest tenderer as a firm or through its partners, had no previous experience of the type of work required and could not produce the necessary certificate. This called for rejection of the lowest tender. The contract was awarded to the second lowest tenderer after negotiating the rates to be on par with those quoted by the lowest tenderer. It was held that there was no arbitrary or illegal action on the part of the respondent.[180]

16.   Tenders were invited for clearing and unloading bitumen from a railway station and for the transportation of the same. One of the qualifications for eligibility was registration as a Transport Carrier under the Indian Banker's Association. It was held by the Gauhati High Court that the condition was a symbol of denoting credibility and could not be said to be superficial.[181]

17.   A tender notification fixed the eligibility criteria of the bidder having executed a single contract of value not less than Rs. 72 lacs for supply and installation of an ash handling system. The said criteria were challenged as arbitrary and irrational and to prefer one

---

[175] Pawan Kumar v. Food Corporation of India, AIR 1999 P. & H. 76.

[176] M/s Shapers Constructions (P) Ltd. v. Air Port Authority of India, 1996 (2) Arb. LR 612.

[177] Electrical Mfg. Co. Ltd. v. Power Grid Corpn. of India Ltd., 2008(1) CTLJ 533 Delhi (DB).

[178] Electrical Mfg. Co. Ltd. v. Power Grid Corpn. of India Ltd., 2009(1) CTLJ 290 (SC).

[179] Nicco Corporation Ltd. v. Cable Corporation of India Ltd., 2008(2) CTLJ 321 (Calcutta) (DB).

[180] M/s Adilaxmi Constructions v. C E (R&B), Buildings, Hyderabad, AIR 1999 AP 437.

[181] Anal Roy Choudhary v. State of Tripura, AIR 1999 Gau.5.

contractor to the others. Relying on the Tata Cellular case and taking cognizance of the fact that four eligible tenderers had already put in their papers, it was held that the eligibility criteria were not irrational or arbitrary.[182]

18.   A petitioner tenderer had challenged an NIT that imposed a condition of submitting a security of one-third of the amount of consideration and gave only seven days for the submission of tender. It was held that the condition of security was to safeguard the interest of the state and not to oust small contractors and that it was not arbitrary or illegal. Similarly, taking cognizance of the facts that five tenders were submitted and prior contract was coming to an end, it was held that the seven days' time allowed for the submission of tenders was proper, being equal for all.[183]

19.   In an NIT for the supply, erection, testing and commissioning of a power station, the eligibility condition required a performance certificate from certain specified authorities for similar work. On opening of the tenders, the lowest tenderer was found to have fully complied with all the required qualifications including techno-commercial considerations. One tender was rejected on the opening day because of non-production of a performance certificate. All the engineers concerned recommended the acceptance of the lowest valid tender. The authority competent to accept tender, based on the advice given by the Law Secretary, not only declared a rejected tender as valid but accepted the same at the rate negotiated with the lowest tenderer. The lowest tenderer filed a writ petition. The Law Secretary had given the advice based on the decision of the Supreme Court of India in New Horizons Ltd. v. Union of India.[184] It was held that as per the law interpreted by the Apex Court in its above said decision, the deciding authorities must rigorously hold on to the standards specified in the NIT. The High Court noted that the decision of first rejecting the tender and afterwards accepting it was wrong. In result, the High Court allowed the writ petition and set aside the letter of intent issued in favour of the tenderer whose tender was first rejected and subsequently accepted in preference to the lowest tenderer. The High Court further referred the matter back to the authorities to take a decision afresh.[185]

20.   Open tenders for leasing of supply, installation and commissioning of computer systems in various government/government-aided senior secondary and middle schools were invited. The expenditure was to the tune of Rs. 100 crores per annum. In consultation with the Technical Advisory Committee, and in the light of experience of failure by the tenderers in earlier years, the director took the decision to invite tenders from firms having turnover of more than Rs. 20 crores over the last three years. Aggrieved by this condition, a writ petition was filed in the Delhi High Court. The division bench of the High Court allowing the petition held that the financial turnover of the bidder had nothing whatsoever to do with computer education and, as such, the term was arbitrary and irrational. Aggrieved against the aforesaid judgment, appeals were filed in the Supreme Court of India. Relying upon the decisions in the earlier cases[186] it was held "that the terms of the invitation to tender are not open to scrutiny the same being in the realm of contract. ... The courts would interfere with the administrative policy decision only if it is

---

[182] Mecgale Pneumatics v. Bhillai Electrical Supply Company Ltd., AIR 2004 Chhat. 5; 2004 (1) CTLJ 147 (Chatt.).

[183] Vinod & Associates v. State of M.P., 2004(1) CTLJ 549 (MP).

[184] New Horizons Ltd. v. Union of India, 1995(1) SCC 478. See paragraph 3.10.3 in this chapter.

[185] P.S.C. Engineers Pvt. Ltd. v. State of Tripura, AIR 2000 Gau. 198.

[186] Tata Cellular v. Union of India, AIR 1996 SC 11; Air India Ltd. v. Cochin International Airport Ltd., AIR 2000 SC 801 and Monarch Infrastructure Pvt. Ltd. v. Commissioner, Ulhasnagar Municipal Corporation, AIR 2000 SC 2272; also see: Aggarwal & Modi Enterprises (Cinema Project) Pvt. Ltd. v. NMDC, 2009(1) CTLJ 191 (Dlhi) (DB); R.S. Daikho v. State of Manipur, (Manipur) (Imphal Bench): Law Finder Doc Id # 973,462 2018(1) BC 166.

arbitrary, discriminatory, malafide or actuated by bias". Appeal was allowed and the decision of the High Court was set aside.[187]

21.   The Bombay High Court placing reliance on the decision in Example (20) above, dismissed petitions challenging the validity of tender conditions that required (a) the payment of "earnest money of Rs. 2,00,18000/- (Rupees Two Crore Eighteen Lakh only) (*sic*), through pay order or DD of nationalized or Scheduled Bank", (b) experience of minimum of one year of collection of octroi of 50% of amount of the estimated offer of realization, and (c) that the bidder should have executed the tender for one full year.[188]

22.   Where an intending tenderer was denied an opportunity to obtain a blank tender form on account of utter chaos and confusion due to acts of some miscreants, the Madras High Court recalled the tender process and ordered re-tender only for the process which was affected due to the acts of miscreants and not for the time prior to the happening of chaos and confusion.

3.12.1 Disqualification of an Intending Tenderer Whose Near Relative Is Working in the PSU, if Valid?

A petitioner was denied a tender form on the ground that his brother was working as a clerk with BSNL and as such he could not give an affidavit stating that none of his near relatives was working in the BSNL unit. A division bench of the Punjab and Haryana High Court held the said condition of filing an affidavit was clearly unconstitutional and violated Art. 14 of the Constitution. It was pointed out that the petitioner's brother could not have influenced the decision of the General Manager. It was further observed that under Rule 4(3) of the CCS (Conduct) Rules, a government servant is restrained from dealing with any matter or giving sanction to any contract in favour of his relatives requiring reference of such matters to his superior for his decision. It was held that the clause was contrary to the mandate of Rule 4(3) of the CCS (Conduct) Rules.[189] The division bench of the Himachal Pradesh High Court had struck down similar provision.[190]

   In another case, a firm's tender for a transport contract was rejected; an officer's father was a partner in the firm. The tender was in breach of a stipulation that a near relative of an employee was not permitted to tender. It was held that the rejection was valid.[191] The Delhi High Court upheld the ban and observed:[192] the decision not to enter into contract with the relative of employees of public undertaking is neither arbitrary nor unreasonable. The controversy appears to have been settled by the Supreme Court decision wherein, expressing the doubt as to if a person belonging to Class III or IV can be in a position to affect the decision-making process, it was suggested that the provision can certainly be made to the effect "that other things being equal, preference will be given to those whose relatives are not in employment in any unit."[193]

---

[187] Directorate of Education v. Educomp Datamatics Ltd., AIR 2004 SC 1962; 2004(1) CTLJ 339 (SC). Also see State of Kerala v. Nup Kumar, AIR 2005 Ker. 276.

[188] Mega Enterprises v. State of Maharashtra, AIR 2007 Bom. 156 =2008 (1) CTLJ 120 (Bom.) (DB).

[189] Tarsem Singh v. Bharat Sanchar Nigam Ltd., AIR 2004 P. & H. 156.

[190] Narinder Kumar v. Union of India and another, CWP No. 33 of 1995.

[191] M/s. Rajendra Road Lines v. Indian Oil Corpn., AIR 2003 All. 77.

[192] S.N. Engineering Works v. Mahanagar Telephone Nigam Ltd., 1996 (37) DRJ 446; followed in Ashwani Garg v. N. D.M.C., 2008(1) CTLJ 530 (Delhi) (DB).

[193] B.S.N.L. Ltd. v. Bhupender Minhas, 2008(1) CTLJ 240 (SC).

3.13 Modification of Eligibility Conditions

Once the authority decides upon and lays down the eligibility conditions in an NIT, no major modification in the said condition should be attempted subsequent to the date of issue of the NIT. If absolutely necessary, any modification should be published by way of addendum to the NIT and communicated by post to the tenderers who had already been supplied with blank tender forms. However, the Supreme Court of India has held that the authority inviting tenders is the best judge of its interests and needs, and that it is always open to it to suitably modify or change the eligibility criteria. It was further observed:

> Whenever a change is introduced in the eligibility criteria either by introducing some new conditions restricting or altogether doing away with certain previous concessions it might hurt the interests of someone or the other but for that reason the change(s) made in the eligibility criteria cannot be labelled as mala fide.[194]

The well-established position in the law is illustrated by the following cases.

EXAMPLES FROM CASE LAW

1. Tenders were invited for supply of a product named "Bionol". By an amendment the product was specified to be Cardol/Cardonol. A bidder who had deposited three samples of Bionol was selected. Another tenderer filed a petition seeking declaration that the acceptance of the tender was illegal and arbitrary. It was found that the bidder whose tender was accepted was neither the manufacturer nor the authorized distributor of the product as claimed. The impugned work order was cancelled.[195]

2. A delay in the submission of samples, it was held, did not disentitle a tenderer from being considered from tendering, particularly when there was no mandatory requirement to submit samples.[196]

3. Purchase orders were to be placed on bidders selected from the listed tenderers' ratings. A change in the ratings after the date of opening of the tenders was not permissible. The authority was directed to consider the tenders on the basis of the original ratings and place additional purchase orders from the balance supplies.[197]

4. A tender notice for the transportation and loading of iron ore lump and fines into railway wagons specified eligibility criteria of 8 lacs MT. Tenders were cancelled on four occasions. On the fifth occasion, the qualification was reduced to 2.5 lacs MT and the work was awarded to a party. The unsuccessful party filed a writ petition that favour was shown to the party whose tender was accepted. The writ was dismissed. After the judgment was received, the writ petitioner found that for the fourth tender, three tenderers had qualified but the authority had presented that none had qualified. When this important fact was discovered, a review was allowed and the authorities were directed to consider the case of the petitioner and the opposite party.[198]

---

[194] M/s S.S. & Company v. Orissa Mining Corpn. Ltd., AIR 2009 SC 461; 2008(1) CTLJ 223 (SC).

[195] Mysore Agro Chemical Co. Pvt. Ltd. v. Union of India, 2008(1) CTLJ 267 (Gauhati).

[196] Spaceage Switchgears Ltd. v. Govt. of NCT of Delhi, 2008(2) CTLJ 361 (Delhi) (DB).

[197] Telephone Cables Ltd. v. Bharat Sanchar Nigam Ltd., 2004(1) CTLJ 537 (Delhi) (DB).

[198] Ores India Pvt. Ltd. v. Steel Authority of India, AIR 2003 Ori. 134.

5. A petitioner purchased tender documents on 21st April 2007, and formed a consortium on 30th April 2007. However, after a pre-bid meeting, the respondent altered the profile of the tender invitation documents by inserting a clause to the effect that bids could not be submitted by a special purpose vehicle or by a bidding consortium. The petitioners challenged the said provision as amounting to a denial of fundamental right to trade. The High Court rejected the said contention.[199]

Some further interesting questions that are likely to arise are discussed below.

3.13.1 Experience of Partners, If Can Be Considered as That of the Firm?

An interesting question, not easy to answer, may arise in some cases, where the tender notice stipulates conditions of eligibility of minimum experience of, say, 3 or 5 years and turnover of a particular amount in any year during the previous 3 to 5 years. Can individual entities who do not meet the requisite criteria form a partnership or a joint venture (JV) for the purpose of tendering and be considered eligible? In other words, the question to be decided is: can the experience of individual partners be considered as the experience of partnership firm itself? For proper appreciation of the above question one more question needs to be raised and answered.

Is a Partnership Firm an Entity or a "Person" in Law?

At law, a limited company is a separate and distinct juristic person from its shareholders.[200] However, the Supreme Court has held that a firm has no legal existence and that its property vests in its partners.[201] However, it is true that the law recognizes a partnership firm as a distinct personality for the purpose of income tax by virtue of the specific provisions of the Income Tax Act. Under the Limited Liability Partnership Act 2000, an English limited liability partnership (LLP) is considered a body with separate legal personality. A Scottish limited partnership is also considered to have separate legal personality from its members. A similar entity LLP is a creation of the Indian Limited Liability Partnership (LLP) Act 2008.

The Indian Partnership Act defines an "act of a firm" to mean any act or omission by all the partners, or by any partner of agent of the firm which gives a right enforceable by or against the firm. Section 4 of the said Act defines "partnership" as a relationship between persons called "partners" who have entered into a partnership and collectively "affirm" to share the profits of a business carried on by all or any of them acting for all.

From the provisions of the above two sections, if an act of a firm means an act of partners – "experience" of the firm can be said to mean the experience of the partners of the firm. Indeed, in New Horizons Limited v. Union of India, it was held, "[i]n respect of a joint venture company, the experience of the company can only mean the experience of the constituents of the venture".[202]

Relying on the above decision, the Andhra Pradesh High Court held: "though a firm is distinct and separate from a company and since it has no personality of its own, the experience of the partners can be treated as the experience of the firm".[203] Where, however, an NIT specifically excluded JVs/consortiums/MOUs from submitting tenders and further provided that the

---

[199] Bhaskar Industries Ltd. v. Maharashtra State Elect. Distn. Co. Ltd., 2008(1) CTLJ 194 (Bombay) DB.

[200] Bacha F. Guzdar v. Commissioner of IT., Bombay, AIR 1955 SC 74.

[201] Narayanappa v. Bhaskara Krishnappa, AIR 1966 SC 1300.

[202] New Horizons Limited v. Union of India, 1995(1) SCC 478, 1995 AIR SCW 275.

[203] Avula Constructions Pvt. Ltd. v. Sr. Divn. Electrical Engineer, T.D, AIR 1999 AP 318; C.K. Asati v. Union of India, AIR 2005 M.P. 96.

credentials of the firm submitting tender alone would be considered, a tender submitted by the firm which relied on the experience of two of its partners as partners of another firm, was rejected as not valid. It was held by the Allahabad High Court that there was no error in the decision.[204]

Where an NIT specified that not more than one bid be submitted by one bidder, it was held that consortium members would constitute one bidder and its constituent members could not form part of another consortium.[205]

Where a tender was rejected merely on the ground that the income-tax return filed was in the name of the firm and not in the personal name of the petitioner, the action was held as bad on the ground that there is nothing in law which separates the partner from the firm as the firm has no legal entity.[206]

The above statement lays down the correct position in law, it is respectfully submitted. The earlier decision of the same High Court in another case can be distinguished on the facts peculiar to the said case and/or not having laid down good law, in so far as it held: "the experience of partners of the firm prior to the constitution of the partnership firm cannot be treated as the experience of the partners of the said firms".[207]

The question of considering the experience of partners as that of the firm may not be easy to answer. The facts of a given case will be decisive.

EXAMPLES FROM CASE LAW

1. An NIT stipulated a two package system: technical bids and commercial bids. One of the conditions of eligibility was that the tenderers should be ISO 9001 certified and approved by the Research Design and Standards Organisation (RDSO). The appellant in the Supreme Court had received a letter informing him that its commercial bid would be opened but subsequently it was not opened and the appellant was informed that it did not meet the eligibility criteria. No reason was specified. A writ filed in the High Court was dismissed. An appeal was filed in the Supreme Court. The appellant had attached ISO 9001 certification of its JV partners and the ISO certificate in respect of the JV Company was in existence prior to the date of opening of the tenders. One of the JV partners was also an RDSO approved manufacturer. Appeal was allowed and the tender Evaluation Committee was directed to consider the bid of the appellant.[208]

2. The respondent company did not have the requisite experience and qualification to submit a tender by itself, but it had taken over an existing partnership firm having experience and financial capacity. It was held that the respondent company could not be disqualified only on the ground that it had no experience in its own name.[209]

3. A tender notification for running ferry boat service stipulated that the bidders should have at least three years' experience of management of ferry *ghat* service. The authority accepted the tender of a person having experience in a society running such service. It was held by the Calcutta High Court that the said person had no requisite qualification to bid.[210]

4. An NIT invited bids along with documentary proof showing the requisite experience. The petitioner was not a JV but there was only a certain amount of equity participation by

---

[204] Vijay Stone Products v. Union of India, 2010(1) CTLJ 192 (Allahabad) (DB).

[205] Shapoorji Pallonji Co. Ltd. v. MSGPCL, 2010 (1) CTLJ 201 (Bom.) (DB).

[206] M.V.V. Satyanarayana v. Engineer-in-Chief (R & B), Hyderabad, 2008 (2) CTLJ 207 (AP).

[207] M/s Margadarsi Borewells v. Singareni Colleries Co. Ltd., AIR 1997 AP 188.

[208] Ganpati RV – Talleres Algeria Track Pvt. Ltd. v. Union of India, 2009(1) CTLJ 1 (SC).

[209] G. R. Engineering Works Ltd. Mumbai v. Oil India Ltd. & Ors., AIR 2009 (NOC) 1067 (Gau.).

[210] Birendra Prasad Singh v. State of West Bengal, AIR 2003 Cal. 142.

a foreign company. It was held that a company is an independent legal person and distinct from its members. Experience of its shareholders could not be the experience of the company. The plea that the authorities should have looked behind the façade of corporate identity of the petitioner was rejected. That was not part of the authority's duty.[211]

3.13.2 Where One Partner Is Left Holding the Responsibility for JV, Parameter for Judging Experience

It is not uncommon in the construction industry in India that a tender is submitted by a JV but after the work is awarded, the entire or balance work is carried out only by one member of the JV. In such an event, what should be the parameters for judging experience? This question arose before the Delhi High Court. An Indian company entered into a JV agreement with a foreign company with the share of the foreign company at 75% and the Indian company at 25%. The performance of the foreign company was not satisfactory and it opted out of the JV, leaving the Indian company to complete the work. Accordingly, a modified JV agreement was signed and the employer was informed. A certificate of completion was duly issued. However, in a subsequent tender invitation, the Indian company was not considered qualified by giving it the benefit of 25% of the cost of the work completed by it on behalf of the JV. On these facts, the Delhi High Court held:

> The expressions used by clause 3.3.1 are "experience", "substantially completed", "successfully completed", "worked on", and "participated" all implying performance, execution or positive achievements/attainments. These expressions, coupled with the objects of the tender documents, leave no manner of doubt that actual experience gathered by a bidder/applicant has to be examined. ... Another reason why we cannot subscribe to the interpretation canvassed by NHAI [National Highways Authority of India] is that the clause deals with a working joint venture, and not one that has failed, where the partners go their own ways, leaving one or the other to complete the entire project, as in this case ... For the above reasons, we hold against NHAI on the issue of the correct interpretation of clause 3.3.1, and find in favour of the petitioners; they are entitled to reckon actual experience gathered in "the working" of the Surat Manor contract.[212]

3.13.3 Experience of Collaborator, If to Be Considered as Experience of Tenderer

When an NIT stipulates that a tenderer should possess certain experience, would it mean the experience of the tenderer itself or that of its collaborator? It appears that the expression would mean the experience of the tenderer itself and not that of the collaborator. An award of a contract to a public sector undertaking without the qualifying experience but in collaboration with a foreign company having the experience was liable to be quashed and the authority was directed to accept the tender of the lowest tenderer at a price reduced on account of the changed parameters offered to the public sector undertaking alone.[213]

---

[211] M/s New Horizons Ltd. v. Union of India, AIR 1994 Delhi 126.
[212] Patel Engineering Ltd. v. National Highways Authority of India, AIR 2005 Del. 298.
[213] P.C.T. Ltd. v. Bangaigaon Refinery and Petrochemicals Ltd., AIR 1994 Delhi 322.

EXAMPLES FROM CASE LAW

1. One of the conditions for eligibility was three years' experience of manufacturing the required item. One of the bidders, an Indian company did not satisfy this condition, but assured that its foreign collaborator had the experience. It was held that the experience and expertise of the foreign collaborator could be of no avail to the Indian company. Further, the foreign company held 10% of the share capital of the Indian company. It was further observed that if two manufacturing units were merged into one, and one of them had the requisite experience and expertise, the new entity after the merger might be given the benefit of it.[214]

2. A condition of eligibility for tendering provided:

> The tenderer should have designed, installed and constructed at least three sets of equipment/machinery of treatment plants of type/types specified in appendix "A" which are in successful commercial operation for at least two years. The tenderer must indicate in the tender details including capacities, date of start of operation etc. of the Plants.

The tenderer whose tender was accepted had collaborated with an outside firm for technical know-how which firm fulfilled the qualifications. The petitioner whose tender was not accepted had also availed of the facility and tendered in another case on the basis of a collaboration with an outside firm. On these facts, the Punjab and Haryana High Court refused to interfere with the decision to accept the tender. The High Court distinguished its earlier decision[215] on the facts of the case.[216]

3.13.4 Eligibility Condition by Registration – Classification – Reclassification

It is customary to avoid stipulating elaborately the conditions of eligibility in the tender advertisement by mentioning the eligibility by class in which the contractors are registered with different public authorities. The list of registered contractors is generally renewed periodically by each department or authority. The conditions of renewal may contain fresh conditions and contractors already listed cannot claim to be exempted from fulfilling the said conditions.[217]

The PWDs of the central as well as the state governments maintain a list of contractors registered under different categories. Care must be taken while revising the list not to make new classification applicable with retrospective effect. If it is so made, it is likely to be declared invalid.[218]

3.14 Prequalification of Tenderers

In the case of major projects involving the deployment of considerable funds, sophisticated machinery/equipment and substantial expertise, it is quite common to invite applications for prequalification of bidders. A notice inviting such applications must clearly give full particulars of expected qualifications. The authority must have a free hand in setting the terms of a tender and is further entitled to pragmatic adjustments, which may be called for by the particular circumstances.

---

[214] M/s. Continental Pump and Motors Ltd. Gaziabad v. State of Bihar, AIR 1995 Pat.183.

[215] M/s Driplex Water Engineering Ltd. v. The Punjab State Electricity Board, (P. & H.) (DB) AIR 1991 P. & H. 38.

[216] M/s Flow Treatment Incorporate v. Punjab W.S. & Sewerage Board, AIR 1993 P. & H. 66.

[217] J.P. Aggarwal v. Director General of Works, 2002(3) Arb. LR 426 (Delhi); Paradise Hotel Restaurant v. Airport Authority of India, 2002(3) Arb. LR 105 (Gauhati) (DB).

[218] B.G. Ahuja v. State of Maharashtra, AIR 1991 Bom. 307 (309).

Courts would interfere with the administrative policy decision only if it is arbitrary, discriminatory, *mala fide* or actuated by bias. As already stated in Chapter 1, there must be a rational nexus between the demand made and conditions imposed.[219] For example, a prequalification notice imposing a condition of minimum turnover of Rs. 100 crores was likely to be oppressive, and the authority offered to lower it to Rs. 25 crores. The P&H High Court held that the revised condition was neither harsh nor oppressive, and answered the petitioner's legitimate concerns.[220]

A tenderer who fulfilled the prequalification requirements was purported to be kept out of the next phase of submission of price bids by the authority introducing a provision: "the prequalified applicants 'shall be ranked on the basis of respective aggregate experience score and shortlisted for submission of bids'." The list was also limited to the top five or six scorers. The provision to limit the competition to a few top scorers when there were many more tenderers satisfying the prequalification requirements was held to be clearly objectionable. The Delhi High Court upheld the decision of the Government of India to prospectively delete the said clause in respect of highway projects and to retain the same for 60 road projects in respect of which the shortlisting was duly completed.[221]

3.14.1 Joint Venture

It is not uncommon in the case of large projects that, where one contractor is not qualified on his own to submit a tender, he enters into a JV agreement with another so that the JV partners become eligible to submit their tender. Care must be taken to see that the JV's application for prequalification is properly scrutinized and that if there is any doubt, a clarification should be sought. In the absence of such a procedure, the decision not to prequalify is liable to be questioned.[222]

EXAMPLES FROM CASE LAW

1. A tender condition stipulated, "annual turnover of Rs. 20 crores (to authenticate more than 20 crores annual turnover, last three years statements duly authenticated by C.A ... is required)". The question arose: was the turnover exceeding Rs. 20 crores required for one year or for three years? It was held that the three years' figures were called for to ascertain that there was a progressive increase in the turnover and that the last year's figure was not exaggerated. The department's contention that it had called for Rs 20 crores' turnover for three consecutive years was rejected and the petitioner's tender was directed to be considered.[223]

2. An order disqualifying a tenderer for non-submission of his labour contract licence was set aside by a division bench of the Orissa High Court. The licence could be obtained only after award of contract. The licence was not required to be submitted in terms of the tender notification. The technical bid called for the submission of profile and experience, and the tenderer was not required to prove its experience.[224]

3. The NIT for prequalification stipulated a condition that the applicant must have annual turnover in any one financial year out of the last five years of Rs. 10 crores and total turnover of Rs. 20 crores or more in the said five years. A tenderer whose maximum

[219] See Chapter 1, paragraph 1.6.
[220] Nair Coal Services Ltd. v. Punjab State Electricity Board, 2009(1) CTLJ 119 (P. & H.) (DB).
[221] National Highway Builders Federation v. NHAI, 2009(1) CTLJ 86 (Delhi) (DB).
[222] Laxmi Sales Corporation v. M/s. Bolangir Trading Co., AIR 2005 SC 1962 = 2005 AIR SCW 1337. Also see Asia Foundations & Constructions Ltd. v. State, AIR 1986 Guj. 185 (206–7).
[223] Sterling Lab. v. UOI & Ors., 2004 (1) CTLJ 154 (Delhi) (DB).
[224] M/s. Trident Softech Pvt. Ltd. v. State of Orissa and others, AIR 2004 NOC 170 (Orissa).

turnover was Rs. 8 crores and a total turnover of Rs. 15.99 crores was not allowed to participate. The writ petition challenging this decision was dismissed.[225]

4. In response to separate invitations issued by the state to prequalify prospective bidders for the construction of a concrete dam and excavation, tunnelling and other works for an underground riverbed powerhouse and appurtenant works for the Sardar Sarovar project, the petitioners submitted their prequalification information and documents, as a JV consortium. When other parties were prequalified and the petitioners did not hear anything from the government, they repeatedly asked the state government to advise them on the prequalification and, on getting no response from the state, they approached the Gujarat High Court for an appropriate writ, order or direction. Thereafter they received replies from the government that they were not prequalified either for the dam or for the powerhouse work on the basis that the information given by the petitioners in form "C" annexed to their application for prequalification showed that the first petitioner company had no experience in dam construction and that the contract work was to be executed by the first petitioner company under the guidance and supervision of the second petitioner. It was argued that no exception could be taken to the decision rejecting the petitioners' applications when the responsibilities of the partners of the JV were not clearly demarcated. However, the JV agreement furnished by the petitioners clearly stated that the responsibilities under the contract were to be assumed by them jointly and severally. The Gujarat High Court observed that the authorities did not consider the most relevant information contained in the basic JV contract and proceeded on the basis of some imprecise statements in form "C". It was also observed that if there was some doubt due to the statements in form "C" they could have sought a clarification. It was held that the respondents had failed to apply their minds to the correct criteria to be adopted in deciding the question of prequalification in case of JV consortia applications. The decision rejecting the application for prequalification was set aside as bad in law and also on the ground of application of wrong criteria.[226]

5. A petitioner, a Class I contractor, had applied for prequalification for the work of construction of commercial complexes and parking lots. His application was rejected. The petitioner insisted before the Andhra Pradesh High Court that the respondents' action was in breach of the principles of natural justice. If the respondents had given him a notice, he alleged, he would have satisfied the authorities that he was capable of executing the works and had gained experience though not in the same building constructions. The High Court observed that the rules as well as the published notices required the contractors to furnish the specified particulars and details. Where a contractor is found to be relatively unqualified, the rejection to enter into a contract with him does not involve the forfeiture of any pre-existing rights or interest, nor does it defeat his legitimate expectations. The High Court held that a tenderer is not entitled to a hearing before the rejection of his claim at the stage of prequalification. It was, however, further observed that the decision should not be arbitrary and that insistence on reasons is a valuable safeguard against the exercise of discretionary power. The record must establish that the claims were in fact considered and the rejection must be supported by germane and relevant reasons. It need not be like a judicial order; the reasons may be brief and if demanded by an aggrieved person, they may be supplied at his cost.[227]

---

[225] M/s. Bharat Construction Co. v. State of Rajasthan, AIR 2002 Raj. 279. Also see Hira Lal S/o of Shri Suraj Bhan v. Food Corporation of India, AIR 2004 NOC 173 (Punj. and Har.).

[226] Asia Foundations & Constructions Ltd. v. State, AIR 1986 Guj. 185 (206–7).

[227] S. M. Quadri v. Spl Officers, Hyderabad Municipal Corporation, AIR 1987 A.P. 6.

3.15 Modification of Tender Requirements Prior to Submission of Bids

If the original terms of the tender notice are changed, all the tenderers should be given an opportunity to resubmit their tenders in conformity with the changed terms.[228] There may be situations where there are compelling reasons necessitating departure from the rule, but then the reasons for the departure must be justified by compulsion and not by compromise. The justification should be for compelling reasons and not just for convenience.[229]

EXAMPLES FROM CASE LAW

1. A Water Users Association challenged the decision of the state to invite tenders by national competitive bidding rather than awarding the work exclusively to the Association. It was held by the Andhra Pradesh High Court, placing reliance on a Supreme Court decision,[230] that it is a settled principle that the court will *prima facie* presume that the state policy is based on all reasonable factors. The writ was dismissed.[231]

2. The Madhya Pradesh High Court did not find fault with the following procedure.[232] After the tenders were submitted a meeting was held in which all the tenderers participated. Thereafter, on the basis of information received through the tender forms, a rational and reasonable criterion was laid down. A list of five eligible tenderers was prepared from whom the selection was made. It was held that there was nothing unfair in this procedure.

3. A tender invitation stipulated that the employer may amend the bidding documents and notify in writing to all prospective bidders and that the bidders would be required to take into account such an amendment. The employer amended the bid documents by inserting an additional condition. One tenderer requested for the withdrawal of the said amendment but nevertheless submitted his tender. His tender was the lowest and yet rejected. He filed a writ petition in the Orissa High Court. It was held that the amendment was permissible and the writ petition was rejected.[233]

3.16 Opening of Tenders

Tenders must be opened in the presence of the tenderers or their representatives for achieving transparency, fairness and so as not to provide room for manipulation to suit the whims of the state agencies.[234] That is the reason why most tender notices publish the date, time and place of opening of tenders. If tenders are opened behind the back of tenderers due to lack of communication of the date and time of the opening of tenders, the action is likely to be held illegal.

---

[228] Harminder Singh v. Union of India 61, AIR 1986 SC 1527.
[229] Sachidanad Pandcy v. State of West Bengal, AIR 1987 SC 1109 quoted in Haji T.M.: Hasan's case.
[230] Ram Krishna Dalmia and others v. Shri Justice S. R. Tendolker and others, AIR 1958 SC 538.
[231] Margam Anjaneyulu, President WUA v. S.E. Irrigation Circle, 2002(3) Arb. LR 462 (AP).
[232] M/s Mal Ca Constructions (I) Pvt Ltd. v. M.P. Housing Board, AIR 1990 M.P. 49.
[233] M/s. EMCO Ltd. v. Grid Crpn. of Orissa Ltd., AIR 2003 Ori. 168.
[234] West Bengal Electricity Board v. Patel Engg. Co, Ltd., AIR 2001 SC 682.

EXAMPLES FROM CASE LAW

1. The technical bids were scrutinized after 18 days of opening of the bids and the financial bid was opened the very next day, allegedly by telephonic intimation to the bidders whose technical bids were found in order. Only two tenderers were present at the time of opening of the financial bids. A tenderer filed a writ alleging that he did not receive any intimation about the date and time of opening of the financial bids. It was held that the decision-making process was not fair, *bona fide* and transparent. The tender notice and subsequent actions taken thereon were set aside.[235]

2. The Union of India invited tenders for organizing tourism related expositions. There were 12 applicants. Out of the 12, four were shortlisted. The selection committee found that all the four tenderers appeared to be equal in experience and other factors. The main criterion for selection was the guarantee of upfront payment. The highest bidder was appointed as an event manager. The third lowest filed a writ petition, which was dismissed by the single judge but upheld by the division bench. The successful tenderer and the Union of India appealed to the Supreme Court. The main ground that appealed to the division bench was that according to the rules framed, the minister in charge had to issue directions to the secretary to appoint an event manager, which procedure was not followed. The Supreme Court found, after scrutiny of the documents, that the secretary had permitted due to urgency and the minister being otherwise being busy. However, the file was put up before the minister with a note stating that event manager was appointed for three years. The minister endorsed "file returned" and signed. It was held by the Supreme Court that the decision of the division bench was erroneous on factual grounds and set it aside.[236]

3.16.1 Revoked Tender – If and When to Be Considered?

A tenderer who revokes his tender and demands refund of his earnest money has no right to insist on his bid being opened. However, where the earnest money was not refunded and yet the tender was not opened and contract was awarded to the party whose tender was found to be higher than the unopened tender, it was held that something had gone wrong in the entire tendering process, calling for interference by the court. The court, holding that the judicial review is not concerned with reviewing the merits of the decision involved but the decision-making process itself, cancelled the tender process and directed to invite fresh tenders.[237]

3.16.2 Tender Submitted after the Time Fixed

Would a tender submitted after the time fixed for the purpose be a valid tender for being accepted for consideration? If not considered valid, would it amount to an arbitrary decision on the part of the administrative authority and could the administrative authority relax the time limit and entertain a tender submitted beyond the time limit? These are important questions likely to arise in cases involving alleged delay of a few minutes, each party referring to the time shown by its own watch.

The authors submit that tenderers who had submitted tenders are likely to disclose the tendered cost or price to others on the presumption that the time limit for submission of the tenders is over. If a tenderer is permitted to submit a tender late, he is likely to exploit this knowledge to the

---

[235] M/s Tafcon Projects (I) Pvt. Ltd. v. Union of India, AIR 2004 SC 949.

[236] M/s. Tafcon Projects (I) Pvt. Ltd. v. Union of India, AIR 2004 SC 949.

[237] Rana Construction and Engineers v. Food Corporation of India, 2008(1) CTLJ 143 (Gauhati).

detriment of the person who has disclosed his offer. Strict adherence to the condition of time limit for submission, therefore, deserves to be followed. The conditions and stipulations in a tender notice have two types of consequences. The first is that the party issuing the tender has the right to scrupulously and rigidly enforce them. The second is that it is not that the party inviting tenders cannot deviate from the guidelines at all in any situation, but that any deviation, if made, should not result in arbitrariness or discrimination.[238] A division bench of the Gauhati High Court in deciding writ appeals followed the above principles. It was held:

> By applying the above principles also it appears that if an administrative decision was taken or administratively the authorities acted in a particular manner, namely, in this case by not opening the tenders submitted by the petitioner-respondents, it would not be open for the court to sit over that administrative action or non-action on the part of the authorities, unless it suffered from the vice of discrimination or arbitrariness.[239]

The above case involved yet another question. Two tenders were opened on another day without notice to the tenderers whose tenders were opened earlier on the scheduled date and time. It was held that it was incumbent upon the authorities to send individual notice to all the tenderers to be present, if they so chose, at the time of the second opening of tenders. It was further observed that this fact assumed importance because the respondent petitioners had not impleaded the other tenderers as parties in their writ petitions.

The Allahabad High Court refused to interfere with the decision of the government not to open a tender submitted 15 minutes late.[240] In another case the tenders were opened 50 minutes before time and in the absence of one of the members of committee and the lawful tenderers. The government decided to re-tender. The action of the government could not be considered arbitrary and no right accrued to the highest bidder.[241]

3.17 Fulfilment of Requisite Conditions of Tender Notice

Tender notices invariably incorporate several conditions to be fulfilled by the intending tenderers. The requirements can be classified into two main categories: those which lay down essential conditions of eligibility and the others which are merely ancillary or subsidiary. In the first case the authority issuing the tender may be required to enforce them rigidly. In the other cases, it must be open to the authority to deviate from and not to insist upon the strict literal compliance of the condition in appropriate cases.[242] As a matter of general proposition it cannot be said that an authority inviting tenders is bound to give effect to every condition mentioned in the notice in meticulous detail, and is not entitled to waive even a technical irregularity of little or no significance. For example, in a case where revised bid became necessary in view of the changed situation, namely, alterations, amendments and clarification in the schedules and the drawings, resulting in alterations and amendments both in respect of part-A, i.e. supply of materials, and part-B, i.e. service, the Supreme Court held that it could not be said that such offers were liable to be rejected on the ground that they varied from the

[238] Tata Cellular v. Union of India, (1994) 6 SCC 651: AIR 1996 SC (11); G.J. Fernandez v. State of Karnataka, (1990) 2SCC 488: AIR 1990 SC 958.

[239] Sailen Konwar Dutta v. M/s Satya Capital Services (P) Ltd., AIR 2000 Gau. 152; also see: M/s S. Ali v. Union of India, 1995 (1) Gauhati LT 458.

[240] Sailesh Kumar Rusia v. State of U.P., All. 237. Also see Nokia India Pvt. Ltd. v. Mahanagar Telephone Nigam Ltd., AIR 2003 Del. 474.

[241] State of J. & K. v. Qazi Nazir Ahmad, AIR 2000 J. & K. 73.

[242] Poddar Steel Corpn. v. Ganesh Engineering Works, AIR 1991 SC 1579; also see Mahabir Auto Stores v. Indian Oil Corporation, AIR 1990 SC 1031 reversing the decision in AIR 1989 Delhi 315.

guidelines. The tenderer whose tender was subsequently modified reducing the cost and was accepted was directed to pay Rs. 1 crore to the lowest non-successful tenderer.[243]

The following cases illustrate which conditions are essential and which are not.

EXAMPLES FROM CASE LAW

1. The non-submission of a Loan Clearance Certificate from government and Sales Tax Clearance Certificate along with the tender was held to be a breach of an essential condition and the tender could not be accepted.[244]

2. In case of an "online tender", a tenderer did not upload a scanned copy of Demand Draft (DD)/Bank Guarantee (BG) and his tender was not opened on that count. The tender condition stipulated that a bid would be rejected if original BG/DD were not submitted before the opening of price bids. The bidder had duly complied with this stipulation and submitted the original BG/DD. It was held that the authorities were not justified in excluding the price bid submitted by the said bidder from the field of consideration.[245]

3. A petitioner did not possess the requisite experience or financial status, and did not submit his PAN. The non-supply was held fatal to his eligibility.[246]

4. A division bench of the Gauhati High Court held that the tender invitation conditions requiring a tender to be signed in the presence of a Gazetted Officer and to be accompanied by Bakijai Clearance Certificate were essential conditions and that the non-fulfilment of the same rendered a tender invalid.[247]

5. A tender was not accompanied by documents evidencing the possession or holding on lease of the requisite machinery and labour licence. It was rejected. The tenderer's writ petition was dismissed as having no merit.[248]

6. Two cover bid systems usually require the technical and price bids to be submitted in separate covers. An NIT did not require the submission of tenders in two covers. Subsequently by letter the tenderers were asked to submit the tenders in two covers. A tenderer did not submit his tender in two covers and his tender was not considered. He filed a writ petition. It was held that the condition of two covers not being a part of the original tender notification, the non-consideration of his tender was not justified. The authority was directed to reconsider the petitioner's tender.[249]

7. Where the NIT specifically required the tenderers to submit their bids in a two cover system and, admittedly, the petitioner had not submitted his tender application in two cover system, it was held that the rejection of the tender was proper for non-compliance of the basic requirement of submission in two cover system.[250]

---

[243] Subhash P. and M. Ltd. v. W. B. Power Devpt. Corpn. Ltd., AIR 2006 SC 116 = 2005 AIR SCW 5579.

[244] Bikash Bora v. The State of Assam, AIR 2004 NOC 79 (Gauhati); 2004 (1) CTLJ 66 (Gauhati) (DB).

[245] Manisha & Mulay (JV) v. A. Krishna Reddy, AIR 2007 (NOC) 2248 (AP).= 2008 (1) CTLJ 59 (AP) (DB).

[246] Hemkund Transport Service v. Union of India, 2008(1) CTLJ 115 (J&K) (SN). also see: P. H. Transport v. State of Gujarat, AIR 2005 Guj. 326.

[247] Bijulbari Multipurpose Development Society v. State of Assam, 2004 (1) CTLJ 365 (Gauhati) (DB).

[248] M/s. Singh Electrical and Constructions v. State of Jharkhand, AIR 2004 Jhar.13.

[249] M.K. Hegde Constructions Pvt. Ltd., M/s. v. Karnataka University, AIR 2002 Kant. 41.

[250] T.V. Rajakumar Gowda v. State of Karnataka & Ors., (Karnataka) 2009(1) Kant LJ 388.

3.17.1 Qualification Certificate Issued by Officer or Authority – Care to Be Taken

A tender notification specified eligibility criteria and further stipulated that tenderers should produce documentary proof in support of meeting the requirements of the said criteria. A tenderer produced a certificate of a private party. An unsuccessful tenderer challenged the correctness and validity of the certificate submitted by that tenderer. The authorities made a reference to the party issuing the certificate and confirmed the facts underlying the certificate. A division bench of the Orissa High Court dismissed the writ filed by the unsuccessful tenderer whose tender was found invalid but observed that there is no difficulty when the certificate is issued by a government official. The tenderer can, at any point of time, verify whether the officer concerned has issued the certificate. Problem arises when a certificate is issued by any non-government entity. A careful approach has to be adopted and a verification has to be done to find out the authenticity of the document and acceptability thereof.[251]

Letters asking the tenderers to produce certain documents were delayed in dispatch so that some tenderers were unable to produce those documents. The tenders of the tenderers who had submitted the documents were opened and the petitioner was found to be the lowest. When the other tenderers pointed out the mistake of the department, an opportunity was given to those tenderers to submit the requisite documents. Those who qualified technically and whose tenders were not opened were called for the opening of the tenders along with the petitioner who was the lowest when tenders were first opened. The respondent no. 7 was found to be the lowest tenderer and his tender was accepted. The petitioner filed a writ petition. After scrutiny of relevant documents, the division bench of the Jharkhand High Court held that there was no arbitrariness or unfairness in the awarding of the contract.[252]

Another petitioner was the lowest tenderer in a tender invitation, in which the NIT mentioned the earnest money as Rs. 5.09 lacs instead of Rs. 5.90 lacs. The mistake was amended prior to the opening of the bids and the petitioner paid the balance amount. The case was recommended to the head office for award of the contract but the respondent invited fresh tenders. It was held the action of the respondent was not justified and a direction was given to award the contract to the petitioner.[253]

3.17.2 Reference to Tender Schedule for Full Details of Conditions Necessary

It is important to note that tender advertisements are generally incomprehensive and may not be treated as a code. One is expected to look into the tender schedule to have full details and knowledge of the terms and conditions of the tender. When the clauses in the tender invitation are absolutely clear, categorical and unambiguous, nothing can be read into the same.[254]

EXAMPLE FROM CASE LAW

A tender schedule provided that the tender covers which did not contain the superscription of the important particulars including an acceptance of the terms and conditions would not be opened. A tender which did not contain the superscription was returned unopened. In the writ filed by the tenderer it was contended that the tender notice did not contain the said stipulation and that the tenderer had signed all the pages inside the cover including those relating to the acceptance of conditions. The Andhra Pradesh High Court held that the petition was bereft of any merits and dismissed it.[255]

---

[251] Santilata Sahoo v. State of Orissa, AIR 1999 Ori. 199.

[252] Mahto Automobiles v. Union of India, 2004(1) CTLJ 519(Jharkhand) (DB).

[253] Shri Mateshwari Indrani Constn. P. Ltd. v. State of Rajasthan, 2009(1) CTLJ 510 (Raj.).

[254] Nair Coal Services Ltd. v. State of M.P., 2009(1) CTLJ 518 (M.P.) (DB).

[255] APNPD Shilpa Engineering Contractors and Suppliers v. CL., 2004 (1) ctlj 30(ar).

3.17.3 Payment of Earnest Money/Security Deposit

Tender notifications invariably specify for the payment of earnest money in one form or another with the tender to be submitted or for the payment of a security deposit for the auction sale, etc. This kind of a provision is an essential condition and a tenderer is advised to follow it carefully or else face the rejection of his tender. For example, where the invitation of tender stipulated "demand draft of any associated bank of State Bank of India shall not be accepted" and a tenderer submitted an account payee pay order issued from Dena Bank as earnest money, it was held by Gauhati High Court that there was a clear violation of the terms of the NIT and that the tender was liable to be rejected.[256] However, as a matter of general proposition it cannot be said that an authority inviting tenders is bound to give effect to every condition mentioned in the notice scrupulously. It may waive a technical irregularity of little or no significance.

A tenderer failed to submit a performance guarantee in another contract, and further failed to demonstrate his financial capacity to furnish the security. His tender, though the lowest, was rejected and a decision was taken to call for fresh tenders. The authority had lost confidence in the tenderer. The Delhi High Court declined to interfere in the authority's decision.[257]

EXAMPLES FROM CASE LAW

1.   A tender notification required the intending tenderers to supply a security deposit in the form of a Banker's Guarantee from a Scheduled Bank in a given form or in cash for the named sum. A tenderer failed to comply with this condition. His tender was therefore rejected. A writ was filed in the Karnataka High Court. Placing reliance upon a ruling of the Bombay High Court,[258] it was argued that not every condition of the tender could be considered to be essential, and that, therefore, the rejection of the tender must be held to be arbitrary. The Karnataka High Court, placing reliance upon a decision of the Supreme Court, held that the condition was an essential one and that non-compliance with it enabled the authority to reject the tender. The Supreme Court had held that where an essential condition is not complied with, it is certainly open to the person inviting the tender to reject the same. The decision of the Bombay High Court was distinguished on the ground that in that case the condition related to earnest money and it was released not only in favour of one but of all the tenderers.[259]

2.   Diesel Locomotive Works' NIT Clause 6 required the tenders to be accompanied by earnest money, cash or by a DD drawn on the State Bank of India. A tenderer had sent a cheque of the Union Bank of India drawn on its own branch and not on the State Bank. On these facts, reversing the decision of the High Court, the Supreme Court held:

> In the present case the certified cheque of the Union Bank of India drawn on its own branch must be treated as sufficient for the purpose of achieving the object of the condition, and the tender committee took the abundant caution by a further verification from the bank. In this situation it is not correct to hold that the Diesel Locomotive Works had no authority to waive the technical compliance of Clause 6, especially when it was in its interest not to reject the said bid which was the highest.[260]

---

[256] Rafique Ahmed Mazumdar v. Hindustan Paper Corpn. Ltd., 2008(1) CTLJ 107 (Gauhati) (SN).

[257] Ramunia Fabricators SDN BHD v. ONGC Ltd., 2008 (2) CTLJ 346 (Delhi) (DB).

[258] M/s B.D. Yadav and N.R. Meshram v. Administrator of the City Nagpur, AIR 1984 Bom. 351.

[259] N.O. Shetty v. K.S.R.T. Corpn., Bangalore, AIR Kant. 94.

[260] Poddar Steel Corpn. v. Ganesh Engineering Works, AIR 1991 SC 1579; also see Mahabir Auto Stores v. Indian Oil Corporation, AIR 1990 SC 1031, reversing the decision in AIR 1989 Delhi 315.

3.   The state deviated from an earlier policy of collecting earnest money at Rs. 5000/- for participating in an auction for tendu leaves and on the basis of separate tenders for each unit. According to the revised policy each participant was required to deposit gate money at the rate of Rs. 25/- per bag through a bank draft. First lot was for as many as 65,479 bags requiring the prospective bidder to deposit Rs. 16,36,975/-. Bidders challenged the said order by a writ petition. In defence it was stated that tendu leaves are perishable in nature and need to be disposed of before the arrival of the fresh crop. The past experience showed that the petitioner and other purchaser had formed a cartel and boycotted the auction. Besides, the increase in the amount of security deposit would also help to exclude illegitimate and fake bidders. The Allahabad High Court, holding that there were special reasons for resorting to distress/clearance sale, dismissed the writ.[261]

4.   The lowest tender was rejected on the ground that the earnest money in a Fixed Deposit Receipt was wrongly drawn in the name of "Dy. FA and CAO", "instead of FA and CAO". Subsequently, the bank clarified to read it in the correct name. It was held that a decision might also be arbitrary and discriminatory where it is unduly oppressive and unjustifiably inflicts excessive hardship on a citizen. The decision of rejecting the lowest tender and accepting another tender was set aside.[262]

5.   The lowest tenderer submitted an original of a National Savings Certificate (NSC) by way of earnest money. His tender was rejected on the ground that the said certificate showed on its face that it had been pledged to the state. The evidence showed that the state had released the certificate from the pledge. It was observed by the Kerala High Court that the concept of pledge under Sections 172 and 179 of the Indian Contract Act is that there is a transfer of physical or constructive possession of goods. What was required was the production of an NSC or another bond. An inference can be drawn that the tenderer intended to create a pledge also. The writ was allowed and direction to accept the tender was given.[263]

3.17.4 Sealing of Tender with Adhesive Tape and Not by Red Sealing Wax – If Makes the Tender Invalid?

The object and purpose of a tender condition that the tender document must be sealed with red sealing wax is that the confidentiality and secrecy of the document should be strictly maintained. It is the substance and not the form that matters for compliance of a condition such as this. In a case, of the three tenders received one was sealed with an adhesive tape and not with red sealing wax as notified by the tender notification. The tenders were opened. However, in the next meeting, the Purchase Committee decided that the tender of the petitioner should be ignored for non-compliance with the tender condition and, since the minimum three valid tenders were not received, fresh tenders should be invited. It was held that admittedly the secrecy and confidentiality of the tender document had not been breached or impaired, so that the refusal to consider the petitioner's tender with the other two tenders was wholly illegal and unreasonable.[264]

---

[261] U.P. Biri Evam Patta Udyog Samaiti v. U.P. Forest Corpn., AIR 2004 All. 21. Also see: The Times Travels v. Chief Managing Director, 2008 CTLJ 503 (Delhi) (DB).

[262] Bibhu Bhushan Choudhury v. Union of India, AIR 2000 Gau. 192.

[263] Haneefa v. Pathanamthitta Municipality, 2008(1) CTLJ 287 (Kerala).

[264] M/s Antex Printers v. Ram Manohar Lohia Avadh University; AIR 2002 All. 237.

3.18 Negotiations Prior to Acceptance

Where the procedure for the prequalification of tenderers is not followed, a question sometimes arises as to the procedure to be adopted while accepting a tender higher than the lowest or accepting a bid lower than the highest for the disposal of government property. Invariably, NITs include a provision entitling the accepting authority to accept any tender or to reject all tenders including the right to accept a tender higher than the lowest. The law on the point is well settled that the government, for good and sufficient reasons, has the right not to accept the lowest tender or all the tenders. In Ramana v. International Airport Authority,[265] the Supreme Court has held that no one has any right to enter into a contract but all who offer tenders or quotations are entitled to an equal treatment. The government, while granting a contract, is not free like an ordinary individual to deal with any person it pleases. It is obligatory upon the government to act fairly and at any rate it cannot act arbitrarily. The government may not deal with anyone, but if it does so, it must do so fairly, without discrimination and unfair procedure. These principles and guidelines were also referred to and relied upon by the Gauhati High Court.[266]

In another case, the tender condition prohibited variations in the offer made. However, the circumstances justified a variation. All tenderers were allowed to submit revised offers. The lowest offer based on the revised price bid was rejected and the tender of the party allegedly the lowest was accepted after the said party reduced the offer to the lowest bid. The Supreme Court observed:

> [T]his was a case where in any event, all the tenderers should have been invited and given an opportunity to reduce their bids before accepting the most competitive of them in public interest and in the interest of the project. Unilateral action of secret negotiations with one of the tenderers behind the backs of the others is arbitrary and not supported by usually incorporated condition of contract, conferring absolute right to accept or reject any tender.[267]

This is so, even where the governmental policy requires preference to be given to a certain category such as a Small Scale Industrial Unit.[268]

EXAMPLES FROM CASE LAW

1. Tenders were invited publicly from registered contractors from category S-V and others. The tenders received were processed in accordance with the M.P. PWD Manual. The tenderers were called for negotiations and to offer revised tenders with a view to reducing the cost and to extend the validity period. The petitioner filed a writ petition contending that although its tender was the lowest, the contract was not given to the petitioner. The court held that the invitation to the tenderers for negotiation was not in consonance with the PWD Manual. The order inviting the tenderers for negotiation was quashed and the state was directed to consider the offers received on merit. Thereafter, the state rejected all the tenders on the basis of the recommendation of an expert committee. Tenders were re-invited. Again the petitioner filed a writ petition. The state placed a comparative statement before the court. Another petitioner intervened. Relying upon the decision of the Supreme Court,[269] the High

---

[265] Ramana v. International Airport Authority, AIR 1979 SC 1628.

[266] M/s. Noble Sales Agency v. State of Assam, AIR 1992 Gau. 46.

[267] Subhash P. and M. Ltd. v. W.B. Power Devpt. Corpn. Ltd., AIR 2006 SC 116 = 2005 AIR SCW 5579.

[268] M/s The Indian Hume Pipe Co. Ltd. v. Bangalore Water S & S Board, AIR 1990 Kant. 305.

[269] AIR 1986 SC 1527.

Court rejected the petition mainly on the ground that the re-invitation of the tenders reduced the cost of the project.[270]

2.   The FCI took a policy decision to abolish contract labour system for handling food grains at various godowns in India and instead decided to offer guaranteed employment of perennial and regular nature to the erstwhile workmen of the then existing contractors by engaging a labour co-operative of such workmen. The FCI was supposed to encourage the formation of labour co-operative societies for this purpose. Despite having taken this policy decision, FCI invited tenders from independent contractors who used to tender under the older policy and negotiated with them. Work order was given to such a tenderer. The Karnataka High Court directed that a co-operative society should be formed of labourers who were working at those godowns and that such societies should be awarded contracts without calling for any tenders in the first instance for two years. This would ensure that there would be no break in continuity of work for the labourers. It was further held that inviting opinion and recommendation of the Registrar of Co-Operative Societies was not necessary.[271]

3.   Where the state government decided to get a project work completed by a PSU due to the difficulty of allotting the work to the private sector, the Patna High Court held that the decision was not arbitrary, unreasonable, discriminatory or violative of the mandates of law.[272]

3.18.1 Circular Issued by Central Vigilance Commission

It is established practice of public authorities, whenever the tenders received are not acceptable, to undertake negotiations with the persons who responded to the tender invitation. Care should be taken while doing so, to invite all the participants and give an equal opportunity to them. The decision-making process by giving an opportunity to lower the rate to one tenderer only, ignoring others, is not fair and it would violate Art. 14 of the Constitution.[273] Any selective negotiations do not fit into the scheme and any action of an authority based on selective negotiations is likely to be held illegal and quashed by the court.[274]

It is necessary to take cognizance of a circular issued by the Central Vigilance Commission (CVC) directing all the departments of the country to the effect that only the lowest tenderer should be called for negotiations. This direction may not be binding on the state government departments. For example, the State of Madhya Pradesh had prepared a manual in accordance with which all tenderers were invited for negotiations. The petitioner in a case himself participated in negotiations and submitted rates. However, he submitted a petition challenging the action of the state to enter into negotiations with all tenderers and sent negotiated rates in sealed envelopes in preference to his lowest tender. It was held the circular issued was applicable to central government departments and corporations, and cannot override the provisions of the manual prepared by the state government. The action of the state was strictly within the rules.[275]

The administrative instructions which have no statutory force do not give rise to any legal right in favour of the aggrieved party and cannot be enforced by any court of law against the administration.

---

[270] Precision Tecnofab and Engineering Co. Ahmedabad v. State, AIR 1990 M.P. 55.

[271] Karnataka C.L. & T. Co-op Society Ltd. v. FCI Madras, AIR 1994, Kant 147.

[272] State of Bihar v. National Project Construction Corpn. Ltd. 2008(1) 494 (Patna) (DB).

[273] M/s Noble Sales Agency v. State of Assam, AIR 1992 Gau. 46.

[274] M.V. Krishna Reddy v. Govt. of A.P., AIR 2003 A.P. 81; 2003(1) Arb. LR 446 (AP).

[275] Shrikishan & Co. v. State of Chhattisgarh, AIR 2003 Chhat. 18.

Their breach may subject the subordinate authority to disciplinary or appropriate action, but they cannot be said to be statutory rules having force of law.[276]

In practice, it has been observed that certain authorities of the central government try to implement the directions contained in the circulars issued even in cases of concluded contracts during the performance of the contract, though such circulars do not form a part of the contract. The authorities forget that the terms of the concluded contract cannot be unilaterally altered, much less to the disadvantage of the other party to the contract. The law can be said to be well settled that the circular letters cannot ipso facto be given effect to unless they become part of the contract. Any novation in the contract is required to be done on the same terms as are required for entering into a valid and concluded contract.[277]

3.18.2 Single Tender

There are cases when only a single tender or a single valid acceptable tender is received even against open invitation, thus indicating a lack of competition. There may be cases where the lack of competition is due to a restrictive specification which does not permit many vendors to participate. The authority in such cases may review the specification of the item to facilitate wider competition. If not, there is no harm in processing the single tender in the normal way. It should be noted that in the case of a single tender, CVC has issued guidelines advising retendering. This may happen when there is a single tender because the invitation was issued to one firm only. It can also happen when only one valid quote is received in open competition. However, when several prequalification bids are received, and only one among them is responsive, the CVC guideline for retendering is not attracted. The Supreme Court of India observed:

> Regard being had to the facts and circumstances as stated above, the action of the respondents seems to be quite arbitrary, as in similar situation, instances of which have been given herein before, the same Tender Committee having found only one bidder as responsive has processed further the matter of tender, whereas in the instant case, decision has been taken to cancel the tender and to go for re-bidding. In a situation, which seems to have been admitted by the State, that on account of revision of the schedule rate of the articles, the estimated amount would go higher and, thereby, decision taken by the authority is against the public interest, therefore, I do consider it to be a fit case for interference.[278]

3.18.3 Rules of the CPWD Manual Regarding Negotiations

Rules of the CPWD Manual in respect of negotiations prior to acceptance have been dealt with in Chapter 1.[279] The rule stipulating that negotiations can only be had with the lowest tenderer need not be carried out to the letter. Judicially, what is expected is that all tenderers are treated equally on par in the matter of negotiations as well. What is the position of the rules in the Code and if violation of the same will be prejudicial to the action taken, or in short if the rules have statutory force or not was considered by the Supreme Court in a case. The Supreme Court of India held:

---

[276] Pradeep Kumar Sahoo v. State of Orissa, 2010(1) CTLJ 72 (Orissa) (DB).

[277] Bharat Sanchar Nigam Ltd. v. BPL Mobile Cellular Ltd., 2008 (1) CTLJ 537 (SC)= 2008 (8) SCALE 106=2008 AIR SCW 6743.

[278] CWE-SOMA Consortium v. State of Jharkhand, (Jharkhand): 2015(1) JBCJ 521: 2014(4) J.C.R. 471: 2015(4) R.A. J. 553: 2014(3) AIR Jhar R. 741: 2015 AIR (Jharkhand) 40: 2015(1) JLJR 34: 2014(46) R.C.R.(Civil) 181.

[279] See paragraph 1.14 in Chapter 1.

Taking first the contention with respect to the Code not being followed in the matter of tenders, the question that arises is whether this Code consists of statutory rules or not ... If they have no statutory force they confer no right on anybody and tenderer cannot claim any rights on the basis of these administrative instructions. If these are mere administrative instructions it may be open to Government to take disciplinary action against its servants who do not follow these instructions but non-observance of such administrative instructions does not in our opinion confer any right on any member of the public like a tenderer to ask for a writ against the Government by petition under Article 226. The matter may be different if the instructions contained in the Code are statutory rules. ... In the view we take it is unnecessary for us to consider this, for we are of the opinion that no claim for any relief before a court of law can be founded by a member of the public, like the appellant, on the breach of mere administrative instructions.[280]

EXAMPLE FROM CASE LAW

The Gujarat High Court found that the rates quoted by one party were not unconditional and coupled with more than one "ifs and buts" while the rates quoted by the other party were straightforward, unconditional and more favourable as a whole. It was held that the competing parties whose tenders were found to be valid could not be subjected to a differential treatment. It was held that one party could not be excluded from negotiations and both the parties should have been called for negotiations.[281] Also see Example (1) under 3.18 above.[282]

In conclusion, it is submitted that when the lowest eligible tenderer is established, the other tenderers go out of the picture and the authority may be justified in negotiating with the lowest tenderer alone to bring the rates further down.[283]

3.19 Acceptance/Rejection of Tenders

The government may enter into a contract with any person but in so doing the state or its instrumentalities cannot act arbitrarily. The tenders should be adjudged on their own intrinsic merits in accordance with the terms and conditions of the NIT. The general principle of law is that a contract with the government does not stand on a different footing from a contract with a private party. Merely because one person is chosen in preference to another, it does not follow that there is a violation of Art. 14 of the Constitution. Normally, the lowest bidder for works contracts or the highest bidder in an auction ought to be awarded the contract, but this is not an absolute rule and the governmental authority can deviate from this for good and valid reasons.[284]

Where the NIT provided that tenders received below a minimum bid amount would be rejected and the highest tender received, though below the minimum bid amount, was accepted, the Bombay High Court held that the tender condition did not expressly bar the acceptance of a tender below the minimum bid amount and that the decision was *bona fide*, neither arbitrary nor discriminatory.[285] Tenders were re-invited time and again but to no avail. Thus the government is

---

[280] G.J. Fernandez v. State of Mysore, AIR 1967 SC 1753.

[281] Siddhi Travels v. Indian Airlines Ltd., AIR 2000 Guj. 102.

[282] Precision Tecnofab and Engineering Co. Ahmedabad v. State, AIR 1990 M.P. 55.

[283] Kumar Transports v. Central Ware Housing Corporation, 2002(3) Arb. LR 119 (Madras); Koratemjen v. State of Nagaland, 2002(3) Arb. LR 622 (Gauhati).

[284] Era Infra Engineering Ltd. v. Delhi Development Authority, 2010 (1) CTLJ 21 (Delhi) (DB).

[285] Kishore Chandrakant Shah v. State of Maharashtra, 2008(2) CTLJ 52 (Bombay) (DB).

necessarily entitled to make a choice. But the choice cannot be arbitrary or fanciful. The choice must be dictated by public interest and must not be unreasoned or unprincipled.[286]

It is not generally advisable to cancel a tender merely because only one valid tender has been received. But when there is intrinsic evidence that the single tender is financially depressed and there is a real prospect of getting a much higher bid, the authority can set aside the tendering process and recall fresh tenders.[287] The Madras High Court refused to relax a condition for eligibility[288] by relying upon a decision of the Supreme Court.[289]

If the state or its instrumentality chooses to invite tenders, it must abide by the result of the tendering process. If the tender was to be given to the lowest bidder according to the NIT, the authority cannot arbitrarily or capriciously accept the bid of a party which is high and to the detriment of the state.[290] Fairness demands that the authority should notify in the NIT itself the procedure which they propose to adopt while accepting the tenders.

EXAMPLES FROM CASE LAW

1. A petitioner's tender, though found the lowest, was not accepted for some items. It was contended that the Ministry of Defence had issued a letter laying down a condition to reject the tenders quoting below 20% of reasonable rates as fictitious. The petitioner's rates in respect of the items in question were found to be far lower than the reasonable rates fixed by the authority. The P&H High Court took cognizance of CVC's letters to the effect that the post-tender negotiations form the main source of corruption, and hence are banned except in case of negotiations with the lowest tenderers. It was held that the post-tender negotiations were not only permissible but mandatory in case of a tenderer whose rates were lower than 20%. In other words, no tender could be rejected outright on the ground of the rates being below the threshold 20%. Only after negotiations, if the rates tendered were found unworkable, the tender could be rejected.[291]

2. A municipal corporation invited tenders for the settlement of markets and after receipt of the tenders, decided not to accept any tender at any amount more than double the estimated value. This was in view of allegations of extortion during the previous term of the lease. Applying the said basis, all higher offers were ignored and finally the tender of the eighth tenderer was accepted. Three writ petitions were filed challenging this action. Recognizing the power of the authority to refuse to accept the highest or a higher bid, it was held that if the authority felt that any bid in excess of double the estimated value would be unreasonable, the tenderers should have been put on notice by due publication in the NIT. The procedure adopted was held to be in derogation of the rights of the petitioners to fair treatment. As a result, the order accepting the tender was quashed and set aside.[292]

3. Where the government discriminated between persons similarly situated by excluding from consideration the petitioners' tender on the basis of some undisclosed criteria, the action was held to be wholly unreasonable, arbitrary and violative of Art. 14 of the Constitution.[293]

---

[286] Ramana Dayaram Shetty v. International Airport Authority of India, AIR 1979 SC 1628.

[287] Virendra Kapoor v. Airports Authority of India, 2009(1) CTLJ 154 (Delhi) (DB); Umesh Kumar Paswan,etc. v. Union of India, 2010(1) CTLJ 324 (Patna).

[288] Taxi Owners-cum-Drivers Asso. v. Reg. Dir., SRBC, ONGC, 2004 (1) CTLJ 304 (Madras).

[289] West Bengal State Electricity v. Patel Engineering Co., (2001) 2 SCC 451; 2001 (1) Arb. LR 540 (SC).

[290] Harminder Singh Arora v. Union of India, AIR 1986 SC.1527. M/s. Kesar Enterprises Ltd. v. State of U.P., AIR 2001 All. 209.

[291] R.R. Co-op, Labour and Construction (LOC) Ltd. v. State of Punjab, AIR 1999 P. & H. 244.

[292] Saurabh Das v. Gauhati Municipal Corpn, AIR 2004 Gau. 38; 2004(1) CTLJ 456 (Gauhati).

[293] Golam Mohammad v. Supdt., N.R.S. Medical College and Hospital, AIR 2001 Cal.5.

4. A division bench of the Allahabad High Court held that a foreign company, while competing with Indian companies for the supply and installation of electronic meters, could not complain of violation of rights under Art. 19 when a tender of an Indian company, though second lowest, was accepted in preference to the foreign company. It was held that a foreign company cannot claim any relief for violation of Art. 14.[294]

5. Where the essential conditions of tender documents were deviated from while accepting the tender, it was held that the award on the basis of varied terms of contract was bad. It is noteworthy that during the pendency of the petition, the tender with deviated condition was accepted and by the time the petition was heard part performance was completed. The learned single judge refused to consider this aspect and held that delay could not deprive the petitioner of its right.[295] An appeal was filed before the division bench, which reversed the above finding on the ground that no *mala fide* was shown.[296]

3.19.1 Conditional Tenders

There is no concept of conditional offer in the law of contract. Offer is always on certain terms. The acceptance thereof should be absolute and unconditional to make it a concluded contract. Conditional tender, therefore, means the tenderer proposing certain conditions in addition to or in variance with the conditions contained in the NIT. Where a tender invitation contains a stipulation that conditional tenders were liable for summary rejections, rejection of a tender containing conditions will *be valid*.[297] However, if the conditions are contained in a forwarding letter which clearly states that the bid submitted was not subject to the said assumptions/ deviations, it may not amount to a conditional offer. For example, a tender invitation provided that the bidder must specify what statutory charges it would not bear. In other words, by default the bidder would be liable to bear charges. The petitioner quoted Rs. 51.57 crores inclusive of sales tax. The other tenderer quoted Rs. 47.35/- subject to applicability of forms C and D and if not applicable 12.5% sales tax would be added. The authority added 12.5% to the petitioner's tender also and awarded the contract to the other bidder. The Delhi High Court held that the terms were misinterpreted and directed the authority to award the contract to the petitioner.[298]

The government contracts, as already stated, impose predetermined conditions and tender invitations expressly stipulate that any deviations from the said conditions would render the tender as non-responsive offer. The nature of deviation proposed would decide if the tender is responsive or not.

EXAMPLES FROM CASE LAW

1. A covering letter contained certain assumptions and asked if the authority could agree to these assumptions/deviations. The tenderer withdrew the assumptions and made an endorsement to that effect on the same covering letter withdrawing all the conditions before opening of the tender. The tender was thereafter opened with the consent of the other bidders, as per the minutes of meeting. However, on the next day, a protest was lodged

---

[294] Power Measurement Limited v. U.P.P. Corpn. Ltd., AIR 2003 All. 153.

[295] Aristocraft International Pvt. Ltd. v. Union of India, AIR 2001 M.P. 99.

[296] P.S. System (India) Ltd. v. Aristocraft International Pvt. Ltd., AIR 2001 M.P. 135.

[297] M/s. Asian Techs. Ltd. v. State of Kerala, AIR 2001 Ker. 388.

[298] Arraycom (India) Ltd. v. Union of India, 2009(1) CTLJ 184 (Delhi) (DB).

by one of the tenderers stating that these assumptions were nothing but conditions attached to the tender and therefore the tender should have been rejected as non-responsive. In a writ petition filed, the Delhi High Court held that only a material deviation or reservation of the nature specified in an NIT clause would make the bid substantially non-responsive. It was further held that the minutes of the meeting demonstrated the transparency in the whole process of opening tenders, which was accepted by all tenderers who signed the Record Form. It was held that the action did not lack fairness. The bids were allowed to be evaluated and considered.[299]

2. The respondent invited tenders for the work of construction of a road. The construction had to be done by an automatic power unit. Conditional tenders were liable to be rejected. Four tenders were received, out of which the petitioner's tender was the lowest. His tender was not accepted on the grounds that he had no experience of road work and that he did not own or possess an automatic plant and machinery. His tender was also conditional inasmuch as he demanded 5% mobilization advance. The second lowest tender was also rejected on similar grounds. The respondents, therefore, called respondents nos. 3 and 4, the other two tenderers, for negotiations. Both these respondents offered to reduce the tendered sum. The tender committee decided to divide the work equally in two parts and awarded the same to respondents nos. 3 and 4. It is pertinent to note that the tender submitted by respondent no. 4 was also conditional inasmuch as he had demanded 10% mobilization advance to be adjusted in five equal instalments among other conditions and yet his tender was not rejected on that ground. The petitioner approached the Madhya Pradesh High Court challenging the award of contract to respondents nos. 3 and 4. While holding that it was a fit case for interference, the High Court also observed that there was no condition in the tender notice that the tenderer must possess experience of road work or that he must own or possess an automatic power unit for doing the tarring work. In the absence of such clear declaration of conditions in the tender notice, the petitioner and the second lowest tenderer could not have been outright excluded from consideration and negotiations by the respondent. The authority was held to have clearly departed from the declared mode of awarding the contract. Reliance was placed upon a decision of the Supreme Court.[300] Testing the action of the authority on the touchstone of Art. 14 of the Constitution, the High Court set aside the award of contracts in favour of the respondents nos. 3 and 4.[301]

Occasionally, a review of the decision to accept a tender based on the recommendations of a Tender Advisory Committee, or a higher authority may be justified in the absence of any *mala fide*.[302] An unbalanced tender, though the lowest, can be rejected by the state on that count and the work be awarded to the other contractor.[303]

The lowest tender can be rejected for lack of experience and the tender of the next lowest can be accepted at the rates quoted by the lowest tenderer on the ground that he had the experience of construction of the type of work.[304]

---

[299] Gammon India Ltd. v. Union of India, 2003(1) Arb. LR 353(Delhi) (DB).

[300] K.N. Guruswamy v. State of Mysore, AIR 1954 SC 592; also see AIR 1979 SC 1628 & AIR 1989 SC 592 1642.

[301] B. Shukla v. Chairman, S.A.D.A Singrauli Distt. Sindhi, AIR 1990 M.P. 365.

[302] M/s. Nestor Pharmaceuticals (Pvt.) Ltd. v. Union of India, AIR 1995 Delhi 260.

[303] Lalzawmliana v. State of Mizoram, AIR 2001 Gau. 23.

[304] Arwish Marak v. State of Meghalaya, 2004(1) CTLJ 467 (Gauhati).

3.19.2 Arithmetical Errors in Tender

The question as to what action should be taken when a tenderer, after opening of the tender but before its acceptance, informs the authority that he has committed a mistake and desires to allow him to rectify his mistakes or withdraw the tender is a vexed one and cannot be answered to meet all situations. If the mistake is genuine, *bona fide* and promptly brought to the notice of the authority, the tender should not be considered as it is, if the tenderer has sought permission to correct or revoke his tender.

EXAMPLES FROM CASE LAW

1.   Where the lowest tenderer intimated that his tender rate Rs. 23,76,000/- against an item be read as Rs. 32,76,000/-, which resulted in an overall increase in the tender cost from Rs. 32 crores to Rs. 41 crores, the contention that he had not withdrawn the tender but corrected the error was rejected and forfeiture of earnest money was held as proper and justified.[305]

2.   A tenderer committed an error in submitting his offer. A percentage development share was mentioned of 81.04% but the figure written was 74,477.00 instead of 2,32,139.00. As a matter of fact 74,477 sq. yards represented the minimum guarantee share any bidder was required to offer. The tenderer submitted a representation that there was an error which needed to be corrected. The authority permitted to correct the error. Placing reliance on a couple of Supreme Court decisions,[306] it was held by the Andhra Pradesh High Court that the overwhelming public interest lay in not interfering with the decision taken by the authority.[307]

The above two examples show how the judicial response is likely to be different. The authors are firmly of the opinion that at the stage of opening of tender allowing the tenderer to withdraw the tender on the ground of error is, no doubt, in the interest of the authority inviting tender. Allowing him to do so would not cause any prejudice to the interest of the public authority. This again cannot be considered as a rule because it may not be followed if a tenderer is found to be habitually indulging in such practice. The question as to his right to revoke the tender under the provisions of the Indian Contract Act is independent and its answer will be governed by the terms and conditions of tender invitation. The tendency of the courts of late to forfeit earnest money if a tenderer revokes his offer is indicated by a few decisions considered in Chapter 12. At this stage the discussion is limited to what is a proper and legal action that the authority should take. The question as to whether he should be allowed to correct his mistake is a difficult one. If the mistake is of the nature that would change the complexion of the tender it may not be allowed to be corrected. The legal position that emerges is briefly summed up as follows.

Can Court Direct to Permit Corrections of Errors in a Tender?

It would be beyond the scope of the power of the writ court to direct the authority to allow the tenderers to correct arithmetical errors, particularly when the errors are of such nature and

---

[305] Villayati Ram Mittal P. Ltd. v. Union of India, 2009(1) CTLJ 113 (Delhi) (DB); also see Siemens Aktiengesellschaft v. Delhi Transco Limited, 2009(1) CTLJ 147 (Delhi) (DB).

[306] Raunaq International Ltd. v. I.V.R. Construction Ltd., AIR 1999 SC 393; Air India Ltd. v. Cochin International Airport Ltd., AIR 2000 SC 801.

[307] G. Chinna Babu v. Visakhapatanam Urban Development Authority, 2008(2) CTLJ 337 (AP).

magnitude that if permitted, it would give a different complexion to the bid. It is likely to encourage and provide scope for discrimination, arbitrariness and favouritism.

The question that needs to be addressed is whether, in equity, a tenderer is entitled to relief of correction of mistakes. In this respect, the following reference to American Jurisprudence is relevant:

> As a general rule, equitable relief will be granted to a bidder for a public contract where he has made a material mistake of fact in the bid which he submitted, and upon the discovery of that mistake acts promptly in informing the public authorities and requesting withdrawal of his bid or opportunity to rectify his mistake, particularly where he does so before any formal contract is entered into.[308]

The Supreme Court of India did not apply the above rule of equity to the first two cases in the examples below.

EXAMPLES FROM CASE LAW

1.   In an original bid invitation, no quantity was mentioned in respect of an item of work in the column of "Quantity". A bidder indicated "As Required" and the column of "Total Price Euro" was left blank; the unit prices were, however, quoted by all bidders. The quantity was made known to bidders after the commercial bids were opened. At that stage, the bidder inserted the quantity "1" and inserted the unit rate and the total cost "8,977.34." The authority accepting the tender multiplied the unit rate by the quantity required and which was subsequently notified. The same method was uniformly adopted for the other two bidders. It was held accepting the interpretation as sought by the bidder would amount to rewriting the entries in the bid document and reading into the bid documents terms that did not exist therein.[309]

2.   A tenderer filled in rate in Indian rupees per unit of excavation at Rs. 148.08/- and in the next line where it was to be quoted in US dollars, the same rate in Indian rupees was repeated. The rate was sought to be corrected to US$3.38. Similar corrections were sought to be made in respect of other items. The amount column, however, showed the correct amount in US dollars. The rules of tender invitation and consideration stipulated that the rate would be treated as correct and the amount would be corrected. The tenderer sought correction by treating the first rate for 50% quantity and the second to be for the balance quantity. This was not in compliance with the tender invitation. The Supreme Court of India set aside the order of the High Court directing the authority to consider the bid after correction along with other bids.[310]

3.   A tenderer submitted the highest bid offer of Rs. 51 lacs. The second bid was for a sum less than Rs. 27 lacs. The highest bidder made a subsequent offer of Rs. 29 lacs. The authority accepted the second lowest offer originally received. The highest bidder filed a writ stating that the authority ought to have accepted its original bid amount Rs. 51 lacs. The division bench of Gauhati High Court dismissed the appeal and held that the action of the authority was neither arbitrary nor illegal.[311]

---

[308] 43 *American Jurisprudence, Public Works and Contracts*, Sec. 63, p. 805. Cited in James T. Taylor etc. v. Arlington Ind. School Dist. 335, S.W.2d 371 (1960).

[309] Siemens Public Communications Networks Pvt. Ltd. v. UOI, 2008 (2) CTLJ 445 (SC).

[310] West Bengal State Electricity v. Patel Engineering Co., (2001) 2 SCC 451; 2001 (1) Arb. LR 540 (SC).

[311] Bijulbari Multipurpose Development Society v. State of Assam, 2004 (1) CTLJ 365 (Gauhati) (DB). Also see: Spartan Carriers Pvt. Ltd. v. Central Coal Fields Ltd., 2008(1) CTLJ 470 (Jharkhand).

3.19.3 Splitting of Work between Two Tenderers

If a tender invitation contains a stipulation that the tenderer is agreeable to accept a part order against the tender if the employer so desires, the employer can accept the lowest tender for part work. And he can award the balance part to the next lowest tenderer, at the rate quoted by the lowest tenderer.[312]

EXAMPLE FROM CASE LAW

In a tender notification for construction of flyover/bridge/rail-over bridge in packages, package 1 included three flyovers at different locations. The tender notice included a condition reserving the right to the state to delete any flyover from the package. The petitioner's tender was found to be the lowest considering package 1 as a whole, however, a comparative statement showed that in respect of one flyover it was higher than the 2nd and the 3rd lowest. The authorities excluded that flyover from package 1 and ordered that work to the 3rd lowest tenderer. In the process, the state saved Rs. 1.0 crore. In a writ petition, the lowest tenderer sought a direction to the authorities to allot all items of package 1 to the petitioner. Placing reliance on a couple of Supreme Court judgments,[313] it was held: "it cannot be said that the process of deleting Item – C of Package – I and awarding it to the 3rd Respondent, suffers from any illegality, arbitrariness, irrationality, unfairness or procedural impropriety and not in public interest, as such, no interference is called for at the hands of this Court, accordingly, the writ petition is devoid of merits."[314]

3.20 Rejection of the Lowest Tender

It has been well established by the cases discussed hereinabove that the authority has discretion to reject the lowest tender for valid reasons. Ordinarily the authority rejecting all tenders and the lowest in particular, must record reasons for the rejection and communicate the same to the concerned parties unless there is a justification not to do so.[315] If a clarification is sought from any tenderers at all, the authority must seek it from all the tenderers.[316] However, mere non-communication of reasons for rejection may not be sufficient to interfere with the decision in all cases, particularly, if there was every reason to believe that the petitioners were aware of the reasons.[317]

EXAMPLES FROM CASE LAW

1. The petitioner's tender, though found to be the lowest and recommended by both the Superintending and the Chief Engineer, the Commissioner of Tenders, the final authority in the finalization of tenders, did not scrutinize the tender and the matter was transmitted to

---

[312] M/s Nair Coal Services (P) Ltd. v. Maharashtra State Electricity Board, AIR 1994 Bom. 163.

[313] Krishna Murari Prasad v. Mitar Singh, AIR 1994 SC 488 and Tata Cellular v. Union of India, AIR 1996 SC 11.

[314] M/s VSN Benarji, Engineers & Contractors v. State of AP and others, AIR 1998 A.P. 29.

[315] M/s Star Enterprises v. CIDCO of Maharashtra Ltd., (1990) 3 SCC 280.

[316] SAG ELV Slovensko and Others, Case C-599/10 followed in Clinton v. Department for Employment and Learning, [2012] N1QB2.

[317] Nagesh M. Daivajna v. State of Goa, AIR 1998 Bom. 166.

the state government. The state government rejected the said tender without assigning any reasons. In reply to the petition before the High Court, the state averred that on receipt of credible information, investigations were carried out and it was found that fraudulent practices of a grave and serious nature were indulged in, which necessitated cancellation of all bids with the permission of the World Bank. The record showed the permission was given by the World Bank reluctantly and there was no material to prove fraudulent practices. It was held that "viewed from any angle, the inescapable conclusions are that the process adopted by the Government in rejecting the bid of the Petitioner cannot be said to be fair and reasonable."[318] The government was directed to transmit the bids and connected documents to the Commissioner of Tender, for evaluating the tenders and taking appropriate decision uninfluenced by the allegations, noting and opinions recorded by various officers.

2. The Punjab and Haryana High Court set aside the decision not to accept the petitioner's lowest tender and allot the work to another agency without the permission of the higher authority as required by the manual of the authority. The work carried out by the other agency was ordered to be measured and paid for with the balance work ordered to be allotted to the concern found most suitable and competent, not necessarily the petitioner. It was held that since the case had been dealt with by the officials in a casual manner, in violation of rules and instructions, the cost of Rs. 5000/- was ordered to be borne by the officers responsible for the lapses.[319]

3. Four tenders were received. Respondent no. 4 was the lowest, the petitioner was the second lowest, and respondent no. 3 was the third lowest. Respondent no. 4 did not fulfil the eligibility criteria and after a complaint by the petitioner and upon enquiry, his contract was cancelled. The work was awarded to respondent no. 3, albeit on a temporary basis, but not to the petitioner. It was held that the award to respondent no. 3 was not reasonable and the authority was directed to award the contract to the petitioner.[320]

4. A tenderer who did not submit his Sales Tax Clearance Certificate was awarded the contract in preference to another tenderer who had quoted the same price and also attached a Sales Tax Clearance Certificate. The person whose tender was accepted had applied for registration but was not even registered under the Sales Tax Act. It was held that the award of the contract was wholly illegal. However, major work had already been completed by the tenderer whose tender was accepted. Under the circumstances the petitioner was awarded compensation of Rs. 1 lac to be recovered from the salary of the person or persons held responsible.[321]

When the lowest tenderer is found to lack the requisite experience and all the tenderers agree to execute the work at the rates tendered by the lowest tenderer, the provision made in the Orissa PWD Code providing for awarding the contract to the lowest eligible tenderer and only if he refuses, to extend such offer to the next lowest tenderer, recommends itself to all similar cases.[322]

---

[318] Lanco Constructions Ltd., v. Government of AP, AIR 1999 AP 371.

[319] M/s Pritam Singh v. State of Punjab, AIR 1997 P. & H. 194.

[320] Airogo Travel & Cargo Pvt. Ltd. v. Union of India, 2004 (1) CTLJ 220 (Delhi).

[321] A.K. Constructions v. State of Jharkhand, 2004(1) CTLJ 487(Jharkhand).

[322] Debendranath Balbantaray v. Commr.-cum-Secretary Govt. of Orissa, AIR 2002 Ori. 142.

Art. 14 Is Attracted in Case of Limited Competition

Addressing letters to a few selected tenderers, inviting them to submit their offers with earnest money etc., amounts to a tender invitation in a limited competition, and selection of a tenderer will attract the Provision of Art. 14.[323]

3.20.1 Rejection of the Highest Bid or the Lowest Tender – When Not Arbitrary?

The purpose of holding an auction or inviting bids in open competition is to get maximum benefit to the public exchequer. If the highest bid itself falls below expectations, it is rightly not accepted. Such a decision of a Development Authority is justified and reasonable. The fact that a subsequent auction resulted in much more than the earlier bid establishes that expectations of respondents were not be belied. A direction was issued to act upon a second auction.[324]

EXAMPLES FROM CASE LAW

1. **Non-workable rates**: The lowest tenderer had quoted 0.043% above the estimated cost and the second lowest 20.67% above the estimated cost. The lowest tender was rejected on the ground that the rates were non-workable because the department's estimate was based on a schedule of rates two years prior to the tender notification. The department had worked out the justified estimation cost, which worked out to 5.66% above the original estimate. The CPWD Manual stipulates that contractors cannot be allotted work whose rates work out the cost 5% below the justified estimated cost. It was held that the action inviting the second lowest to negotiate and accept the cost at 1.579% above justified estimated cost was not illegal.[325]

2. The Guwahati High Court upheld a rejection of the lowest tender for supply of goods on the ground that the rates tendered were below the market rates.[326]

3. In yet another case decided by the Orissa High Court, all tenders received were found to be defective. The owner gave an opportunity to all the tenderers to rectify the defects. The lowest two tenders were rejected by an expert committee because the prices quoted were lower by over 50% and 30% of the estimated cost and, as such, considered non-workable. The second lowest tenderer filed a petition. It was held that the petitioner could not have a legitimate expectation of award of contract and that the owner had the right to choose the right person with requisite experience and ability to do the work. While adjudicating the constitutional validity of executive decision concerning economic matters, certain freedom to the executive should be allowed. On these grounds the writ petition was dismissed.[327]

4. **Weighted average of financial and technical bid**: Tenders were invited in two separate bids, namely a technical bid along with experience certificates, turnover certificates and plan of dimensions and locations of sites for appointment as distributor/dealer of cellular service. The petitioner's tender in respect of the technical bid was accepted. The petitioner's financial bid was

---

[323] Larsen & Toubro Ltd., Calcutta v. Neyveli Lignate Corporation Ltd., AIR 1999 MAD. 306.

[324] Harjit Singh v. State of U.P. (Allahabad) (DB) (Lucknow Bench): 2017(5) ADJ 272: 2017(3) All WC 3060.

[325] Saikhom Raghumani Singh v. Chief Engineer I, PWD, AIR 1999 Gau 143.

[326] Rubul Chandra Deka v. State of Assam, AIR 2003 Gau. 169.

[327] NU Calcutta Construction Co. v. National Aluminium Co. Ltd., AIR 2000 Ori. 186. Also see: Jasbir Singh Chabra v. State of Punjab, 2010(1) CTLJ 232(SC); Optel Telecommunications Ltd. v. Union of India, AIR 2001 M.P. 161; M/s. Bishal Enterprises v. State of Orissa, AIR 2003 Ori. 207. Also see National Radio & Electric Co. Ltd. v. Union of India, AIR 2003 Del. 308; Suresh Goel & Associates v. Union of India, 2009(1) CTLJ 162 (Delhi) (DB).

found to be the lowest. However, she was denied the dealership. She filed a writ petition. In defence, it was contended that 75% weighting was given to the financial bid and 25% was given to the technical bid. After adding the two weightings, the petitioner ranked second. Relying on the decision of the Supreme Court of India in Tata Cellular case, the Allahabad High Court declined to interfere with the decision on the ground that to do so would amount to sitting in appeal over the decision of the High Power Committee appointed for evaluation of the bids. The petition was dismissed.[328]

5. **Commercial transaction of a complex nature**: Acceptance of the tender of Air India (fifth lowest) in preference to the lowest tenderer, by negotiations with the fifth lowest alone, was held to be arbitrary, illegal and opposed to the principles of natural justice by a division bench of the Kerala High Court.[329] This decision in respect of tenders for the contract for ground handling service at Cochin International Airport Ltd. (CIAL) was set aside by the Supreme Court of India by holding that CIAL did not commit any wrongdoing by considering the fact that Air India is an airline and a national carrier and could be in a position to bring more traffic of Air India and other domestic lines to the airport. It was held:

> in a commercial transaction of a complex nature what may appear to be better, on the face of it, may not be considered so when an overall view is taken. In such matters the Court cannot substitute its decision for the decision of the party awarding the contract.[330]

6. **When tenderer did not comply with the mandatory condition**: In another case decided by the Allahabad High Court, the tender condition required the tenderer to submit a character certificate not more than three months old along with the tender. When the tenderer did not comply with this condition, it was held that consideration of the tender not complying with the mandatory condition was bad and the tenderer in question was not eligible to participate. The second lowest tenderer had not challenged the decision and hence re-auction was ordered.[331]

7. **When the lowest tenderer declines to accept new time schedule**: The state invited tenders for the construction of godowns with a construction period of one year. The lowest tender was accepted subject to change of the completion period to five months. The lowest tenderer declined to accept the reduced period of completion. The lowest tenderer in another phase agreed to complete within five months. His tender was accepted and he was entrusted with the work. It was held that the decision neither suffered from any infirmity nor was it based on extraneous considerations calling for interference with the court.[332]

---

[328] Balaji Coal Linkers, U.P. v. Bharat Sanchar Nigam Ltd., AIR 2004 All. 141; 2004(1) CTLJ 582 (Allahabad) (DB).

[329] Cambatta Aviation Ltd., v. Cochin International Airport Ltd., AIR 1999 Ker. 368.

[330] Air India Ltd. v. Cochin International Airport Ltd., AIR 2000 SC 801. Deepak Builders v. State of Uttarakhand (Uttarakhand): 2014 AIR (Uttaranchal) 2: 2014(3) R.A.J. 750: 2013(2) U.D. 517: 2013(2) UAD 746: 2013(39) R.C. R.(Civil) 534. Gammon India Ltd. v. Union of India (Delhi) (D.B.): 2003(102) DLT 141: 2003(1) ArbiLR 353: 2003(1) AD(Delhi) 841.

[331] Bashishtha Narain Pandey v. Commissioner, Basti Divn., AIR 2002 All. 280.

[332] Jayaprakash Nanda v. G.M., Orissa State Warehousing Corpn., AIR 2002 Ori. 199. Also see Sudarshan Marketing v. Chief Commercial Manager, W. Rly., AIR 2004 Bom. 114.

3.20.2 *Lowest Tenderer Not Entitled to Hearing?*

The question as to whether the lowest tenderer or highest bidder in an auction sale is entitled to a hearing if his tender/bid is not acceptable needs to be considered. The acceptance or rejection of a tender is based not on any statute or statutory rules and the authority concerned does not pass any order vested with certain statutory power. Second, mere submission of a tender does not create any right to property in favour of the lowest tenderer. In any event, therefore, the petitioner cannot claim any hearing before cancellation of such tender; particularly when the right to accept or reject any tender without assigning any reason is reserved in the tender invitation.[333] In this connection the following observations of the Supreme Court of India need to be noted. It was held:

> Government certainly has a right to enter into a contract with a person well known to it and specially one who has faithfully performed his contracts in the past in preference to an undesirable or unsuitable or untried person. Moreover, the Government is not bound to accept the highest tender but accept a lower one in case it thinks that the person offering the lower tender is on an overall consideration to be preferred to the higher tenderer.[334]

The above observations of the Supreme Court, it is respectfully submitted, do not amount to giving unrestricted power to the government. If the government acts arbitrarily in rejecting the lowest tender for a works contract or a highest bid for disposal of public property, such an act would be open to scrutiny by courts and if found arbitrary, the court may correct the action. For example, the Andhra Pradesh High Court set aside the order of cancellation of a tender which order was made even before the tenderer had complied with the required formalities and without giving an opportunity to explain what irregularity, if any, was committed.[335]

The Punjab and Haryana High Court upheld the action of the Collector of rejecting the highest bid at an auction on the ground that the amount offered was much lower than the value of the property under auction. It was held that the highest bidder did not acquire any vested rights in respect of the sale unless the Collector approved the bid.[336]

A court did not interfere where, after scrutiny, it was found that the petitioner's tender was not ignored or rejected arbitrarily. It was duly considered and an analysis of the rates was invited and only after due consideration of the analysis was the tender submitted by the petitioner rejected.[337]

In another case, the Supreme Court reversed a decision of the Patna High Court and held that there was nothing wrong with not awarding a contract on the ground of the tenderer having been blacklisted earlier. It was held that there was no need to issue a show cause notice inasmuch as the authority was taking note of an existing blacklisting order and not sitting in judgment over such an order or itself issuing such order.[338]

3.21 Rejecting All Tenders

The power to reject all the tenders cannot be exercised arbitrarily and must depend for its validity on the existence of cogent reasons for such action. For example, a tender to build, operate and transfer (BOT) an Inter State Bus Terminus was cancelled when the Development Authority

---

[333] Jaidev Jain & Co. v. Union of India, AIR 1972 Cal. 253.

[334] Trilochan Mishra v. State of Orissa, 1971(3) SCC153 (160,161).

[335] M/s Sathi Devi Mahila Mandali v. Medical Superintendent, AIR 2004 NOC 97 A.P.

[336] M/s Swadesh Rubber Industries v. Sardar Singh, AIR 1994 P. & H. 306. Also see K. Dashratha v. Mysore City Municipal Corporation, AIR 1995 Kant. 157.

[337] Omprakash v. Union of India AIR 1982 Cal. 340.

[338] Patna Regional Development Authority v. M/s Rashtriya Pariyojana Nirman Nigam, AIR 1996 S.C. 2074.

agreed to new development control norms allowing greater utilization of space for commercial purposes. It was held that no fault could be found in the decision taken.[339]

In the contractual sphere, as in all state's actions, the state and all its instrumentalities have to conform to Art. 14 of the Constitution of India of which non-arbitrariness is a significant facet. This imposes the duty to act "fairly and to adopt a procedure which is fair play in action". As such the unfettered power to reject or accept any bid without informing any ground conferred on employer is *ultra vires* Art. 14 of the Constitution of India.[340] However, unless a work order is issued, no right accrues to the tenderer and the internal note sheets of an authority have no face value.[341]

3.22 Nomination/Preference to a Class of Tenderers

Negotiating with one party alone and awarding a contract to it, or distributing the work among a few contenders without rejecting or accepting tenders cannot be said to be per se bad actions. The wider concept of equality before the law and the equal protection of laws is that there shall be equality among equals. Even among equals there can be unequal treatment based on an intelligible differentia having a rational relation to the objects sought to be achieved. For example, Consumers' Co-operative Societies form a distinct class by themselves. Benefits and concessions granted to them ultimately benefit persons of small means and promote social justice in accordance with the directive principles.[342]

Preference to government company: It is open to the government to favour a PSU as a contractor where the risk of breach of contract is less and there is no risk of financial loss to the government, even if the contract amount is higher than the price quoted by a private contractor. For example, where a contract for supply of machines was granted to the respondent, a fully owned central government undertaking located within the state, instead of the appellant private contractor with office in a different state, it was held that the decision of the government was not arbitrary.[343]

Although a citizen has a fundamental right to carry on a trade or a business, he has no fundamental right to insist upon the government or any other individual doing business with him. Where a right is conferred on a particular individual or a group of individuals to the exclusion of others, reasonableness of restrictions has to be determined with reference to the circumstances relating to the trade or business in question. This should be carried out in an objective manner and from the standpoint of the interest of the general public and not from the standpoint of the interest of the persons upon whom the restrictions are imposed.

EXAMPLES FROM CASE LAW

1. A circular of the Kerala Government in respect of a scheme for the supply of pump sets to farmers imposed a restriction that pump sets must be purchased from dealers approved by the government. The Supreme Court held that the restriction was not violative of Arts. 14 and 19 of the Constitution of India.[344]

---

[339] Suncity Projects Pvt. Ltd. v. Government of N.C.T. of Delhi, 2009(1) CTLJ 168(Delhi) (DB).

[340] M/s. Gouranga Lal Chatterjee v. State of West Bengal, AIR 2003 Cal. 44. Also see Aroma Enterprise v. Murshidabad Zilla Parishad, AIR 2003 Cal. 251; K.M. Pareeth Labba v. Kerala Live Stock Development Board Ltd., AIR 1994 Ker. 286.

[341] Dipak Kumar Sarkar v. State of W.B., AIR 2004 Cal. 182.

[342] Madhya Pradesh Ration Vikreta Sangh Society v. State of Madhya Pradesh, AIR 1981 SC 2001.

[343] Biodigital (P) Ltd. v. State of Kerala, (Kerala) (DB): 2011 AIR CC 486: 2010(4) ILR (Kerala) 462: 2010(27) R.C.R. (Civil) 198.

[344] Krishnan Kakkanth v. Govt. of Kerala, AIR 1997 SC128.

2. The State Government of Punjab issued a notification for giving preference to the Co-operative Labour and Construction Societies in the matter of allotment of works by its departments and local bodies, etc. By the said notification unskilled works up to any value and skilled work up to Rs. 10 lacs were to be allotted to these societies only by way of tenders within common schedule of rates fixed by PWD. If more than one society offered tenders, the lowest rate accepted could be made applicable to the other societies for doing that work up to their capacities. The leftover work alone could be executed by inviting open tenders. A writ petition was filed challenging the above notification. Reliance was placed upon several decisions of the Supreme Court of India[345] and it was held that the notification did not suffer from any constitutional or legal infirmity.[346]

A direction by PWD that all unskilled works up to any value, and skilled works up to the limit of Rs. 2 lacs for each work, should be allotted to the Co-operative Labour and Construction Societies by way of tenders within the ceiling rates fixed by the competent authority, is not violative of Art. 14 of the Constitution on the ground of discrimination. It was held that private contractors were not altogether excluded from consideration. The classification between private contractors and co-operative societies was held to be reasonable.[347]

3.22.1 Change in Policy of Government

It cannot be disputed that the government has a right to change its policy from time to time, according to the demands of the time and situation, and always in the public interest. If the government has the power to accept or not to accept the highest bid and if the government has also the power to change its policy from time to time; it must follow that a change or revision of policy subsequent to the provisional acceptance of the bid but before its final acceptance is justification enough for the government's refusal to accept the highest bid at an auction.[348]

3.22.2 Public-Private Partnership (PPP) Contract: "Swiss Challenge Method"

Infrastructure development projects in the past were thought of as a way of providing infrastructure at no extra cost to the public. For this reason, the projects used to be generally investigated, designed and financed by the PWDs of the state or central government. The projects needed several years for realization from the stage of conception to becoming operational. After the need for a project would be felt, a project report based on preliminary survey and investigations used to be prepared and, based on the approximate estimates, its feasibility determined and administratively approved by the concerned department. Following administrative approval, a provision would be made in the government's annual budget for sanctioning the funds for the preparation of detailed investigation, designs and drawings and estimates to be given approval or technical sanction. The project would then wait for the requisite finance to be provided in the annual budget. By the time the budget provision was made, several years used to pass, rendering the estimates outdated and requiring further revision and further sanctions. Even ongoing projects were required to be suspended or slowed down

---

[345] Indian Drugs and Pharm. Ltd. v. Punjab Drugs Manufacturers Assocn., AIR 1999 SC 1626; Krishnan Kakkanth v. Govt. of Kerala, AIR 1997 SC128; P.T.R. Exports (Madras) Pvt. Ltd. v. Union of India, AIR 1996 SC 3461; Indian Railway Service of Mechanical Engineer Association v. Indian Railway Traffic Service Association, 1993 AIR SCW 2342.

[346] M/s Maya Construction Co. v. State of Punjab, AIR 2004 P. & H. 35.

[347] K. Moideenkutty Haji v. Superintending Engineer, P.W.D., AIR 2001 Ker. 294.

[348] State of Uttar Pradesh v. Vijay Bahadur Singh, AIR 1982 SC 1234 (1236); Danya Electric Company v. State, AIR 1994 Madras 180. Also see Shankarlal v. Indore Development Authority, AIR 1995 M.P. 182. Also see M/s. Babu Ram Gupta v. Mahanagar Telephone Nigam Ltd., AIR 1995 Delhi 223.

for want of funds. This resulted in claims and litigations, and further delays and escalation in cost. Since the public utility projects earned little or no revenue, private parties did not venture into the field of construction of such projects. When it was realized in recent times that the users of the facilities created by such projects would not mind paying for the facility, the traditional scenario changed dramatically. The law and the courts in India too had to take cognizance of this development.

PPPs are emerging as an important means for implementation of long-term infrastructure assets and related services such as roads, under the National Highway Authority of India. Recently a new model is also being discussed: PPCP (Public-Private Community Partnership) aimed at social welfare schemes, eliminating the prime focus on profit.

3.22.3 The Development in Law in India

Even as early as in 1986, the Supreme Court of India held that when the state government is granting licence for putting up a new industry, it is not at all necessary that it should advertise and invite offers for putting up such industry.[349] The state government is entitled to negotiate with those who have come up with an offer to set up such industry. In a later case, in 2009, the Supreme Court of India observed:

> Recently, there has been shift towards encouraging private participation in the government works and promoting of public-private partnership. The Ministry of Housing and Urban Poverty Alleviation in its National Urban Housing Habitat Policy, 2007 specifically mentions participation of private sector as one of its aims. It envisages that the State Government and the Central Government shall act as facilitators and enablers. A private party can submit the proposal under Swiss Challenge Method.[350]

3.22.4 Swiss Challenge Method

By the Swiss Challenge Method, the developer who has given the original proposal has the opportunity (first right of refusal) to carry out the project. However, the said developer has to match/raise his bid (rate) with the highest proposal tendered. The original proposer has the opportunity to take up the project on the highest offer, and in the event that he refuses, then the highest bidder shall have the right to implement the project. However, if such highest bidder refuses the offer, then the amount deposited shall be forfeited.

The Swiss Challenge Method is adopted in Chile, Costa Rica, Guam (U.S. Territory), Indonesia, Korea, Philippines, South Africa, Sri Lanka, Taiwan (China), Virginia (U.S.) and also in India by Andhra Pradesh, Rajasthan, Madhya Pradesh, Chhattisgarh, Gujarat, Uttaranchal, Punjab states and Cochin Port authorities.

The Bombay High Court, relied on a decision of the Supreme Court[351] and held that the Swiss Challenge Method suffered from the vice of arbitrariness and unreasonableness. The Supreme Court did not agree and held on appeal:

> The decision to apply Swiss Challenge Method clearly fell within the realm of executive discretion ... It is not possible to reject the claim of State of Maharashtra and MHADA, in view of shortage of land, increasing cost in housing sector, the Central and State Governments recommended strongly for public-private joint ventures and in the said category Swiss Challenge is the acceptable democratic method as compared to other options.[352]

---

[349] State of M.P. and Others v. Nandlal Jaiswal and Others, (1986) 4 SCC 566.

[350] Ravi Development v. Shree Krishna Prathisthan, AIR 2009 SC 2519.

[351] Monarch Infrastructure Pvt. Ltd. v. Commissioner, Ulhasnagar Municipal Corporation, AIR 2000 SC 2272.

[352] Ravi Development v. Shree Krishna Prathisthan, AIR 2009 SC 2519.

3.22.5 Apex Court Suggestions to State Government

The effort of public-private participation can only be possible when private entities are aware of such a scheme. Also in the scheme of availing a new system, thorough rules and regulations are needed to be followed otherwise unfairness, arbitrariness or ambiguity may creep in. In order to avoid such ill-effects the state governments should consider the following aspects suggested by the Supreme Court:

1. The state/authority shall:

    (a) Publish in advance the nature of Swiss Challenge Method and particulars
    (b) the nature of projects that can come under such method
    (c) Mention/notify the authorities to be approached with respect to the project plans
    (d) Mention/notify the various fields of the projects that can be considered under the method
    (e) Set out rules regarding time limits on the approval of the project and respective bidding

2. Set out the rules that are to be followed after a project has been approved by the respective authorities to be considered under the method

3. All persons interested in such developmental activities should be given equal and sufficient opportunity to participate in such venture and there should be healthy inter se competition among such developers.

It was further clarified that these suggestions are not exhaustive and the state is free to incorporate any other clauses for transparency and proper execution of the scheme. The state government should frame regulations/instructions on the above lines and take necessary steps in future.

A right to carry out balance work after a concession agreement was terminated was the subject of an NIT. It was held that the right to match the bid of the lowest tenderer (L-1) or to exercise right of first refusal would come into play only if the concessionaire (whose agreement was terminated) was to participate in the tender process pursuant to the NIT. Having failed to participate in the tender process and, more so, despite express terms in tender documents, validity whereof was not challenged, the concessionaire could not be heard to contend that it had acquired any right whatsoever.[353]

3.22.6 Price Preference for Public Sector Undertakings

Price preference for PSUs would depend upon the government policy at the time of finalization of tenders.

EXAMPLES FROM CASE LAW

1. The NIT provided for a pre-bid meeting during which it was disclosed, in answer to a query, that the price preference for PSUs would depend upon the government policy at the time of finalization of tenders. Thereafter, the owner intimated to all tenderers that the price preference to PSUs shall be applicable and, as such, the parties were given the option to re-quote their price if they so desired. The petitioner did not submit a revised bid. When the tender of a PSU was accepted, the petitioner filed a writ petition on the ground that the price

---

[353] National Highways Authority of India v. Gwalior Jhansi Expressway Limited, AIR 2018 SC 3380.

preference was not included in the NIT but only intimated subsequently by a letter. It was held that there was no arbitrariness or preferential treatment. Incidentally in the case the price quoted by the PSU was lower by Rs. 50 crores and, as such, the award to the said PSU was held to be in the public interest.[354]

2. In pursuance of a policy decision, the state government invited proposals for private sector participation for setting up Mini Hydro plants on canal drops at ten sites. Tenders were received for one of the sites which were evaluated on the basis of uniform criteria for all the sites. According to the said criteria, ranking was given to the tenderers. However, in the meantime, the government changed its policy. A letter of intent was issued to the petitioner on the basis of build operate and own (BOO) subject to the terms stipulated, where upon Rs. 8 lacs were deposited as non-refundable processing fees. Subsequently, the government reviewed the matter and decided to cancel the letter of intent issued to the petitioner and to award the contract, in the public interest, to respondent no. 3 – a PSU, which was not entitled to submit proposals in terms of the advertisements. Placing reliance on the decision of Ramana's case,[355] it was held that after having invited proposals only from the private sector, approved and allocated a site to the petitioner, the government could not have arbitrarily reviewed the decision in the name of public interest. It was observed that public interest might have been better served if the government had allowed public sector participation in the advertisements inviting proposals. The writ petition was allowed and the decision of the government was set aside.[356]

3. A tender for the construction of a township for Ardha Kumbha Mela was accepted. A part of the work concerning the Police Department was allotted to another agency, leading to a challenge by the contractor. The Allahabad High Court held that the petition involved an issue relating to a breach of contract and the contract contained an arbitration clause and that, therefore, the petitioner was not entitled to any relief from the court.[357]

4. A government had been giving contracts for maintenance of lifts on tender basis for the years 1983–85. The tenders were given to the tenderer offering the lowest bid, the petitioner, for the two years. After the expiry of the period of two years on 31st March 1985, the Chief Engineer, PWD decided to give maintenance work to the 4th respondent who had offered 35% higher rates than the ones offered by the petitioner, on a nomination basis instead of by inviting tenders. It was contended on behalf of the respondents that with a view to effect quality, better service and to avoid risk to the users of the lift, it was decided that the maintenance work of the lifts be given to the 4th respondent. It was also admitted that there was no complaint against the petitioner with regard to the work done by him during 1983–85. It was held that mere assurance of better service by the 4th respondent was only a ruse to extend arbitrary favour to him. The action was held to be discriminatory and *ultra vires* Arts. 19(1)(g) and 14 of the Constitution.[358]

---

[354] Mitsui Babacock Enery (India) Ltd. v. Union of India, AIR 2000 Ori. 170.

[355] Ramana's case, AIR 1979 SC 1628.

[356] Bhoruka Power Corpn. Ltd. v. State of Haryana, AIR 2000 P. & H. 245.

[357] M/s. Lalloo Ji & Sons v. State of U.P., AIR 1995 All. 142; State of Kerala v. K.P.W.S.W.L.C. Co-op. Socy. Ltd., AIR 2001 Ker. 60.

[358] A.E. Services v. Electrical Engineer, (GL)PWD & R.&B., AIR 1986 A.P. 358; Nex Tenders (I) Pvt. Ltd. v. Ministry of Commerce & Industries, 2009(1) CTLJ 309 (Delhi) (DB).

5. A circular issued by government directing to get the works executed through the nominees of Panchayat, members of the legislative assembly, or of the chief minister, was held to be violative of Art. 14.[359]

6. The government invited tenders in respect of used X-ray plates, etc. to be sold to Small Scale Units. Later on the government passed a resolution purporting to distribute these two items among three parties including the petitioner, also indicating the areas in which each of the parties would operate. The petitioner's tenders were the highest for both the items and the petitioner claimed that he ought to have been awarded the items exclusively. The Bombay High Court held that the said resolution was clearly bad in law. The resolution was quashed by the High Court.[360]

7. Five star hotels were to be established in Calcutta and this intention of the Government was well publicized. In view of the requirement for expertise and sound financial position only two parties came forward. The government entered into negotiations with one of them instead of inviting tenders or holding public auctions. It was held that the government was justified in doing so.[361]

8. On 11th December 1984, the State Government of Maharashtra passed a resolution laying down a new policy to ensure rapid construction of reliable and good quality tenements in Maharashtra for mass housing and such other requirements. Under Section 28(1)(a)(xii) of the Maharashtra Housing and Area Development Act, 1976, it was one of the duties of MHADA to undertake and promote pre-fabrication and mass production of houses. The government came to the conclusion that with the view to encouraging the industrialized prefab construction technology, the name of Shirke-Siporex Consortium needed to be considered in the overall policy decision. It was decided and laid down in the resolution that out of over Rs. 200 crores worth of building projects executed each year by the state government and semi-government bodies, the building construction work of about 5 lacs square feet of built-up area at a cost of Rs. 16 to 18 crores be entrusted every year to M/s. B. G. Shirke & Co. by the semi-government bodies mentioned in the resolution to the extent of the value indicated against each name of the semi-government bodies. The resolution was challenged by M/s. CIDCO Contractors' Association and others. It was held that the action of the state[362]

> in the circumstances, is bona fide, reasonable and in the best interests of the society. There is not a tinge of arbitrariness, unreasonableness or malafideness in the said action. I am fortified in my conclusion in this behalf of the Supreme Court judgment in the case M/s. Kasturi Lal v. The State of J. & K.[363]

This decision was confirmed in an appeal.

---

[359] Tek Nath Sapkota v. State, AIR 1995 Sikkim 1.
[360] Nitin Industrial Associates, Khamgaon v. State, AIR 1986 Bom. 298.
[361] Shri Sachidan and Pande v. State of W.B., AIR 1987 SC 1109.
[362] Writ Petition No. 538 of 1985.
[363] M/s. Kasturi Lal v. The State of J. & K., AIR 1980 SC 1992.

3.23 Declared Defaulter vis-à-vis Blacklisting – Meaning of

The expression "declaration" has a definite connotation. It is a statement of material facts. It may constitute a formal announcement or a deliberate statement. A declaration must be announced solemnly or officially. It must be made with a view "to make known" or "to announce".[364] When a person is placed in the category of a declared defaulter, it must precede a decision. The expression "declared" is wider than the words "found" or "made". Declared defaulter should be an actual defaulter and not an alleged defaulter.

When "declared defaulter" status is proclaimed or published, so far as the affected person is concerned, its effect would be akin to blacklisting. When a contractor is blacklisted by a department, he is debarred from obtaining a contract, but in terms of the NIT, when a tenderer is declared to be a defaulter, he may not get any contract at all. He may have to wind up his business. The same would, thus, have a disastrous effect on him. Whether a person defaults in making a payment or not would depend upon the context in which the allegations are made as well as the relevant statute operating in the field. When a demand is made, if the person concerned raises a *bona fide* dispute in regard to the claim, so long as the dispute is not resolved, he may not be declared to be defaulter.[365]

3.24 Blacklisting or Disqualification of Tenderers

Blacklisting of a tenderer has the effect of preventing a person from the privilege and advantage of entering into a lawful, profitable relationship with the government. The first question that arises is whether a person who is put on the blacklist by the state government is entitled to a notice to be heard before the name is put on the blacklist. The second question then would arise as to whether an order of blacklisting, which is made after affording an opportunity of being heard, can be quashed on the ground that it is arbitrary.

The law is well settled that nobody should be blacklisted without being given an opportunity of being heard.[366] An order blacklisting a person cannot be made without following the principle of natural justice. An authority must give a show cause notice against proposed blacklisting, followed by an opportunity of hearing. Recommendations made by the Central Bureau of Investigations (CBI) are not binding upon the authority which is called upon to take decision in the matter. It is important that the authority does not proceed with any preconceived notions. A fair and impartial approach has to be adopted in the matter by independent assessment of the facts and circumstances of each case. Moreover, an order of blacklisting cannot be for an indefinite period but for only a specified period.[367] The Supreme Court of India has held:

> The fact that a disability is created by the order of blacklisting indicates that the relevant authority is to have an objective satisfaction. Fundamentals of fair play require that the person concerned should be given an opportunity to represent his case before he is put on the blacklist.[368]

---

[364] Prativa Pal v. J.C. Chatterjee, AIR 1963 Cal. 470 at 472.

[365] B.S.N. Joshi and Sons Ltd., M/s. v. Nair Coal Services Ltd., AIR 2007 SC 437.

[366] Dandapani Roula v. State, AIR 1986 Ori. 220; M/s Quality Traders v. Union of India, AIR 2005 All. 3.

[367] Brite Aricon v. Airports Authority of India (Delhi): 2013(203) DLT 408: 2014(1) AD(Delhi) 483: 2014(13) R.C.R. (Civil) 2308.

[368] E.E. & C. Ltd. v. State of W.B., AIR 1975 SC 266. Also see T.P. Yakoob v. Kerala State Civil Supplies Corporation Ltd., AIR 2004 NOC 38(Kerala); BSBK (P). Ltd. v. Delhi State Industrial Development Corporation Ltd., 2004(1) CTLJ 217 (Delhi).

As regards the second question, the answer is that a mere formal compliance with the requirement of law of giving an opportunity cannot offer real justice to the petitioner if the order which is made is arbitrary.[369] Such an order is liable to be set aside. Deregistration from the list of contractors amounts to blacklisting and the principles enunciated by the Supreme Court in the above case would be applicable.[370]

EXAMPLES FROM CASE LAW

1. In performing a contract for the supply of galvanized steel pipes, a contractor indulged in delaying tactics by raising non-contentions pleas because, due to increased steel prices, he was not willing to stick to his offer. The Department acted fairly in not cancelling his tender immediately but granted him an extension of time. The contractor failed to perform his part of the obligation. An order was made forfeiting his earnest money and blacklisting the contractor. This was upheld.[371]

2. An original NIT specified a two-month time limit to complete the work, but by an amendment the time was reduced to 25 days. The petitioner's was the only tender and all correspondence was sent to the address provided by the petitioner. The petitioner refused to commence the work on the ground of non-awareness of the reduced time limit. The petitioner was debarred from tendering for the period of two years after issue of a show cause notice and being given reasonable opportunity. His writ was dismissed. The appeal against the order was also dismissed by the division bench of the Delhi High Court.[372]

3. A contractor registered in Class "A" was removed from the "approved list" because he had participated in only three tenders and had failed to secure a contract though a large number of tenders were floated during the said period. It was held that there was no illegality in removing the petitioner's name from the list.[373]

4. An order blacklisting a contractor for three years on account of a failure to eliminate deficiencies pointed out in a rate contract, which contract was not formally executed, was held as illegal. It was observed that till due execution of the rate contract, there was no relationship binding the parties.[374]

5. Where a petitioner's name was removed permanently from the list of enlisted contractors for his failure to execute a couple of petty works, it was held that the action taken was highly disproportionate to the petitioner's default. The order removing the petitioner's name from enlisted contractors was quashed.[375]

6. A petitioner challenged the decision of an authority not to open his tender because a First Information Report (FIR) had been filed against him and others by the CBI, alleging the commission of certain offences in connection with earlier works contracts executed by the petitioner. The petitioner argued that the action amounted to virtual blacklisting of the petitioner without affording to him an opportunity of being heard. The single judge made an interim order directing the authority to give the petitioner an

---

[369] Ramana Dayaram Shetty v. The International Airport Authority of India and others, AIR 1979 SC 1628. Also see: Bhim Sain v. Union of India, AIR 1981 Del. 260.

[370] S.P. Timber Industries v. Union of India, AIR 1990 Del. 312.

[371] State of Bihar and Ors. v. Bhawani Tries Ltd., AIR 2008 Patna 121.

[372] B.S. Construction Co. v. Commissioner of MCD, 2008(1) 257 (Delhi) (DB).

[373] Hotel Ridge View, New Delhi v. Union of India, 2004(1) CTLJ 565(P. &H.).

[374] Manjit Sales Corporation v. State of Punjab, 2002(3) Arb. LR 430 (P. & H.).

[375] Narendra H. Chandwani v. M.C.D., 2009(1) CTLJ 270 (Delhi).

opportunity. The authorities rejected the petitioner's representation but the single judge dismissed his writ. In the course of hearing of the petitioner's appeal, the report submitted by the CBI indicated that the appellant did not commit any criminal act nor make any misrepresentation. Declining to accept an act on the basis of the CBI report alone, the division bench directed the authority to make a fresh decision in respect of the tender strictly in accordance with law and also to inform the petitioner of the materials upon which it proposes not to consider the petitioner's tender.[376]

7. A director of a contracting company physically assaulted an employee of the authority with a sharp weapon causing an injury near the right eye. An FIR was lodged. The director also submitted a letter to the Commissioner of Police making allegations against the officials. An Enquiry Officer submitted a report observing that the employee's version was correct and that the director's allegations appeared to be in retaliation. A show cause notice was issued to the company as to why it should not be blacklisted. A reply was submitted. In the meantime, the company's tender was not considered even though it was found to be the lowest. A writ petition was filed. The High Court held that the director of the company and the company itself were two separate legal entities and even if any unbecoming act was committed by the director, that should not stand in the way of the contract being awarded to the company. The Contractor Registration Board debarred the company and its director for five years. The single judge allowed a writ petition. However, the authority was allowed to issue a detailed and reasoned order. The authority issued a fresh order debarring the company and its director for five years. The writ filed against the said order was allowed by the High Court. On the authority's appeal to the Supreme Court, it was held that in a legalistic sense, an incorporated body like a company and its directors are separate entities for certain purposes, though in practice, many directors act as alter egos of the companies. There would be vicarious and constructive liabilities for the director's acts. The acts of the companies are carried out through its directors or employees. The Supreme Court recognized that a strained relationship between the contractor and the contractee could have implications in the execution of the contract. In the absence of any proved *mala fide* or irrationality of the authority, and in view of the background facts, the Supreme Court upheld its discretion not to accept the tender; that exercise of its discretion was not open to judicial review.[377]

8. A clause in the works contract stipulated as follows:

> 7.1 The contractor may subcontract any portion of work, up to a limit specified in Contract Data with the approval of Engineer but may not assign the contract without the approval of the Employer in writing. Sub-contracting does not alter the Contractor's obligations.

The Contract Data provided the limit of subcontracting up to 50% of initial contract price with prior approval of the engineer. The contractor had engaged a subcontractor for executing a portion of the work. The engineer on the site was aware that a subcontractor was executing part of the work. Neither the engineer nor the employer was satisfied with the progress of the work and issued a notice asking if prior permission was sought for the subcontracting. The letter warned that the employer would not agree for a time overrun and directed to make efforts to make up the time loss. Subsequent to the complaint, the subcontractor was removed. Subsequently, when the contractor was found to be the lowest

[376] M/s Navayuga Engineering Company Limited v. Visakhapatnam Port Trust, AIR 1998 A.P. 222.
[377] D.D. Authority v. UEE Electricals Engg. (P) Ltd., AIR 2004 SC 2100; 2004(1) CTLJ 345 (SC).

tenderer for another package, a show cause notice was issued as to why the contractor should not be debarred from further award of work for one year. The contractor replied taking the stand that the agency was given the work on a trial basis and if its performance was found satisfactory, permission would have been obtained to subcontract part of the work. The engineer was fully aware of this arrangement and initially had raised no objection. Only when the employer's representative objected to the said arrangement, was the subcontractor removed. No action was taken thereafter and even an extension of time was recommended. However, the contractor was debarred from participating in any bidding process and the entire bidding process of the other package was cancelled. A writ petition was submitted by the contractor in respect of the orders cancelling the tendering process and his blacklisting. The single judge approved both the employer's orders. On appeal, the division bench did not accept the employer's contention that prior approval of the employer ought to have been taken. The provisions of the clause speak about "the approval of the Engineer" for subcontracting and the approval of the employer in writing. The subcontracting did not alter the contractor's obligations. As a result, the blacklisting order was set aside. However, the relief in respect of the contractor's bid for the other package was declined on the ground that the employer had the right to accept or reject any tender.[378]

9. A contractor was debarred for three years on the ground that he had gone back on an oral promise to lower the rates during negotiations and subsequently insisted on the tendered rates as the correct price. All of this took place before the acceptance of the tender. The debarring order was set aside by the P. & H. High Court on the ground that it was not a reasoned order and that the petitioner had clearly admitted that its representative had made a mistake which was corrected at the first possible opportunity and that thereafter the contract was awarded to another party so that the respondent had not suffered in any manner whatsoever.[379]

10. A contractor, after negotiating with a municipal corporation for the supply of iron, obtained the acceptance of the corporation but failed to enter into a contract or to supply the material, and even did not reply to a show cause notice. The corporation issued an order blacklisting the contractor for two years. The Delhi High Court declined to interfere with the said order. The Supreme Court supported the decision of the Delhi High Court.[380]

11. A contractor was debarred from tendering in future for alleged failure to commence a work awarded to him. The order was given considerable publicity by sending copies to other departments and organizations. It was held that that was a misuse of power with an ulterior motive; the order amounted to blacklisting the contractor without giving him a notice and also, the punishment was disproportionate to the offence alleged. The debarring order was quashed.[381]

12. The Gujarat High Court upheld an order blacklisting a contractor for ten years for having submitted fake bank guarantees. The contractor was issued a show cause notice and given an opportunity to present his case.[382]

---

[378] M/s. P.T.S. Umber Mitra Jaya v. N.H. Authority of India, AIR 2003 Mad. 221.

[379] Ives Drugs (India) Pvt. Ltd. v. State of Haryana, AIR 2004 P. & H. 250; 2004 (1) CTLJ 622 (P. & H.) (DB).

[380] M/s Nova Steel India Ltd. v. M.C.D., AIR 1995 S.C. 1057.

[381] M/s V.K. Dewan & Co. v. Municipal Corporation of Delhi, AIR 1994 Delhi 304.
      Also see: M/s Unibros v. All India Radio, AIR 1995 Delhi 368.

[382] Asif Enterprises v. O.N.G.C. Ltd., AIR 2002 Guj. 264.

13.   Where the government terminated a contract of an approved contractor who had failed to complete a work within the stipulated time limit, forfeited his security deposit and blacklisted his name without hearing him, the Madras High Court held that the blacklisting order did not relate to any particular contract and hence could not have been ordered without hearing the contractor.[383]

14.   In a case where indisputably no notice was given to the appellant of the proposal to blacklist him, the state contended that there was no requirement in the rule of giving any prior notice. The Supreme Court held: "Even if the rules do not express so, it is an elementary principle of natural justice that parties affected by any order should have the right of being heard and making representations against the order. In that view of the matter, the last portion of the order in so far as it directs blacklisting of the appellant in respect of future contracts, cannot be sustained in law."[384] It was clarified that after following due process the state would be at liberty to take any appropriate action.[385]

15.   Registration of a contracting firm was cancelled because a complaint by one of its directors against the chief engineer was found to be incorrect. The evidence showed that the concerned authorities had decided to teach a lesson to the contracting firm. The chief engineer against whom the complaint was made attended a meeting of the Board of Chief Engineers to decide on the action to be taken. The Andhra Pradesh High Court held that such a course was not permissible in law.[386]

16.   DDA floated tenders for three works. The petitioner quoted the lowest rates. It was stated that the petitioner was verbally instructed to commence the work by the then super-intending engineer (SE). However, another person replaced that SE; the second SE asked the petitioner to extend the validity of his tender and later on negotiated with the second lowest tenderer before awarding the work to him. Aggrieved by this, the petitioner filed a writ in the Delhi High Court. The High Court held that the mere fact that a particular contract was not awarded to a party did not mean that the said party had been blacklisted and that there was no bar placed on the way of the petitioner submitting tenders in future.[387]

17.   When an authority called the second lowest tenderer to negotiate and to reduce his rates, thus rejecting the petitioner's lowest tender, a writ petition was filed. It was alleged that the officer concerned had always been antagonistic towards the petitioner. The High Court observed that it is not possible to hold that every refusal to award a particular contract would amount to blacklisting. It was held:

> The acceptance of the second lowest tender on comparative considerations does not violate the principles of natural justice and, therefore, no notice and opportunity to the party, whose lowest tender was rejected, is necessitated. What has to be examined in such circumstances in accepting any tender other than the lowest is whether the lowest tender has been treated fairly and honestly in rejecting the same.[388]

---

[383] Preetam Pipes Syndicate v. T.N. Slum Clearance Board, AIR 1986 Mad. 310.

[384] Raghunath Thakur v. State of Bihar, AIR 1989 SC 620.

[385] Also see: M/s Saraswati Dynamics Pvt. Ltd. v. Union of India, AIR 2003 Del. 146; M/s Southern Painters v. Fertilizers & Chemicals Travancore Ltd., AIR 1994 SC 127; M/s D.C. Bhura and Sons v. Hindustan Paper Corpn. Ltd., AIR 1996 Gau. 86; M/s Elite Engineering Co. v. Bihar State Electricity Board, AIR 2000 Pat. 170.

[386] Sri S.S. Constructions Pvt. Ltd. v. Eng-in-Chief I and CAD Department, Hyderabad (AP), AIR 1999 A.P. 270.

[387] R.K. Aneja v. Delhi Development Authority, AIR 1989 Del. 17.

[388] M/s. Amarchand Shanna v. Union of India, AIR 1988 A.P. 45.

3.24.1 Reasons for Blacklisting Need to Be Recorded

In view of the expanding scope of the active role played by the state in the construction sector, the requirement to record reason is one principle of natural justice which must govern the exercise of power by administrative authorities. Therefore, except in cases where the requirement has been dispensed with expressly or by necessary implication, an administrative authority exercising judicial or quasi-judicial functions is required to record the reasons for its decision.[389]

Relying upon the above principle, the Punjab and Haryana High Court quashed a non-speaking (i.e. unreasoned) order blacklisting a firm. It was also held that its sister concern could not be blacklisted without giving a notice, although some of the partners in both the firms were common.[390]

3.25 Cancellation of Contract

Once an offer is validly accepted, the result is a concluded contract. A cancellation of such a concluded contract would amount to a breach of contract. A person committing a breach of contract is liable to pay compensation to the other party for loss, if any, incurred by the innocent party. Can a public authority avail of a freedom which a private party has of terminating a concluded contract and assuming the risk of its consequences? Any provision made in the contract empowering one party to cancel the concluded contract may not be valid in law. For example, a clause in a contract empowering an authority to cancel a contract without giving reasons in case of complaints would be held illegal. An order of a cancellation using such a clause was held unsustainable particularly when no enquiry was held nor an enquiry report given by the competent authority. Under such facts, the order would be in violation of the principles of natural justice.[391] The examples below will further clarify the matter.

EXAMPLES FROM CASE LAW

1. A contract executed on behalf of the governor in accordance with Art. 299 of the Constitution of India, was terminated by the Additional Secretary of Transport Department, Government of West Bengal without following stipulations under Art. 166(2) of the Constitution of India. The state government did not produce any evidence to show that the Additional Secretary was duly authorized to issue a letter of termination. It was held that the termination could not stand the test of law.[392]

2. An allotment of work was cancelled where no agreement was executed, no security money deposited and the petitioner had written a letter to the respondent stating that the work-site was not fit for work and that it would execute the agreement only after the site was fit for commencing the work. The petitioner filed a writ seeking a direction to the respondents for executing the agreement in favour of the petitioner. The petitioner failed as he had failed to execute the agreement within 21 days as per the terms of the tender. The

---

[389] S.N. Mukherjee v. Union of India, AIR 1990 SC 1984 (1996).

[390] M/s Pritam Singh and Sons v. State of Punjab, AIR 1996 P. & H. 260.

[391] Object Technologies, Bangalore v. State of Karnataka (Karnataka): 2017(4) Air Kar R 652: 2017(6) KantLJ 310. Also see: West Bengal Motor Vehicles Weighbridge Corporation Ltd. v. State of West Bengal (Calcutta): 2016(4) Cal. L.T. 205; M/s Ranjeet Construction Bihar v. State of Jharkhand (Jharkhand): 2015(2) AIR Jhar R. 602: 2015(22) R.C.R.(Civil) 400.

[392] West Bengal Motor Vehicles Weighbridge Corporation Ltd. v. State of West Bengal, (Calcutta): 2016(4) Cal. L.T. 205.

challenge to the forfeiture of his earnest money also failed. The writ petition was dismissed.[393]

3.   A petitioner was allotted a contract of operating a parking slot. The petitioner delayed the deposit of advance licence fees. The respondents started a fresh tendering process to re-advertise the parking slot. The petitioner participated in the second process as well. On these facts it was held that the forfeiture of the earnest money deposited by the petitioner vis-à-vis the earlier contract would not only be against the principles of equity, fair play and good conscience but a harsh decision on the face of it. The single judge ordered a refund of the earnest money to the petitioner for the first tender. However, the division bench held that the petitioner had to bear reasonable costs which were to be deducted from earnest money to be refunded.[394]

4.   A tender awarded to a petitioner was cancelled allegedly because the Chief Engineer who executed the agreement was not authorized to do so. That explanation was given for the first time in a counter affidavit in court but the bidder was not apprised of any such fact prior to the cancellation of his tender. The petitioner had already commenced the work and made a huge investment in material and labour. It was held that the state's actions had to stand to reason and ought to maintain transparency. The petitioner's accrued right was considerably jeopardized and he was deprived of a valuable right. The order of cancellation of tender was quashed.[395]

3.26   Breach of Contract – Writ Petition Not Maintainable?

As already stated in the first two chapters, once a tender is validly accepted it results in a binding contract. Subsequent breach of contract by the public authority will not attract Art. 226 or Art. 14 of the Constitution of India. In commercial contracts, the breach of contract can be properly adjudicated upon in a civil suit or arbitration if the contract contains an arbitration clause; such adjudication is not proper in a writ petition. However, this does not mean that a writ petition ought not to be entertained unless the arbitration remedy is exhausted. The existence of an alternative remedy is not a bar to filing of a writ petition for the enforcement of a fundamental right or where there has been a violation of principles of natural justice.[396] It is a settled principle of law for a writ of *mandamus* that the petitioners must have a legal right to enforce the performance of an alleged duty by the respondent.[397] A writ petition is not maintainable for the

---

[393] M/s Atri Construction v. State of Jharkhand, (Jharkhand): 2015(2) AIR Jhar R. 718: 2015(3) J.C.R. 212: 2015(22) R.C.R.(Civil) 423.

[394] Sr Divisional Commercial Manager S.E. Railway Chakradharpur v. Sunita Singh, (Jharkhand) (D.B.):2015(2) AIR Jhar R. 49: 2015(1) JLJR 296: 2015 AIR (Jharkhand) 49: 2015(2) JBCJ 404: 2015(3) BC 454: 2015(2) J.C.R. 432: 2014(22) R.C.R.(Civil) 940: 2015(5) R.A.J. 549.

[395] Arjun Yadav v. State of Jharkhand, (Jharkhand): 2011(4) AIR Jhar R. 115: 2012 AIR (Jharkhand) 2: 2012(6) R.C.R.(Civil) 751: 2011(106) AIC 777: 2011(4) J.C.R. 236: 2011(4) JLJR 258.

[396] Also see E. Venkatkrishna v. Indian Oil Corporation Ltd. Bombay, AIR 1989 Kant. 35; Ramit Enterprises v. Hindustan Petroleum Corporation Ltd., 2004 (1) CTLJ 202 (Delhi). Also see: Prem Chand v. Union of India, 2004 (1) CTLJ 200 (Delhi). Also see: R.K. Machine Tools Ltd. v. Union of India, 2010(1) CTLJ 172 (Delhi). Also see: Whirlpool Corporation v. Registrar of Trade Marks, Mumbai, (1998) 7 SCC243= 1998 (Suppl.) Arb. LR 553 (SC); Harbanslal Sahnia v. Indian Oil Corpo. Ltd., JT 2002 (10) SC 561; Popcorn Entertainment v. CIDCO, (2007) 9 SCC593=2007 (1) CTLJ 137 (SC).

[397] C.K. Achutan v. State of Kerala, AIR 1959 SC 490; M/s Radhakrishna Agarwal v. State of Bihar, AIR 1997 SC 1496; Life Insurance Corporation of India v. Escorts Ltd., 1986 SC 1370; M/s Mahabir Auto Stores v. Indian Oil Corp. Ltd., AIR 1989 Del. 315.

enforcement of a contract. Such a case may not attract Art. 226.[398] Also, a petition under Art. 32 of the Constitution would also be wholly misconceived. No fundamental right is involved. At best, a party should take the matter to a civil court, if so advised, or arbitration, if available, and claim damages for the alleged breach of contract.[399]

3.26.1 When Termination of Contract for Reasons De Hors the Contract – Writ May Be Maintainable

Every rule has some exceptions. Thus, where a tender is accepted but the acceptance is invalid, a writ will be maintainable if an attempt is made to forfeit the earnest money or security deposit. Similarly, a full bench of the Patna High Court held that when a termination of contract is for reasons *de hors* the contract, an application under Art. 226 is maintainable. The State of Bihar had invited tenders for the construction of a spillway of a reservoir project. The petitioner's tender was duly accepted and an order to commence the work was given. A formal agreement was also signed by the parties. By a subsequent communication, the work order was cancelled on the basis of orders of the Engineer-in-chief and the Chief Engineer. The state's case as pleaded in the High Court was that the original estimate was proposed to be modified, bringing down the cost from Rs. 31.40 crores to Rs. 15.2 crores because the possible discharge itself was changed from 8,217 cusecs to 4,600 cusecs as advised by the Central Water Commission. Because of these later developments, in the public interest and in order to safeguard the public revenue, the work order and the agreement were cancelled. The petitioner referred to some of the terms of the agreement to show that the respondents were at liberty to revise the design or drawing of the project, including the estimated cost in terms of the contract itself and, as such, there was no reason to cancel the contract. On these facts, the High Court observed, "even according to the respondents, the ground for cancellation of the work order and the agreement is not referable to any one of the terms of the agreement but is *de hors* the said agreement".[400] As to the exercise of power under Art. 226 of the Constitution, a reference was made to the case of M/s Radhakrishna Agarwal v. State of Bihar.[401] Where the contract entered into between the state and the person aggrieved is non-statutory and purely contractual but such contract has been cancelled on ground *de hors* any of the terms of the contract, and which is per se violative of Art. 14 of the Constitution, even in such cases applications under Art. 226 of the Constitution are maintainable. It was further observed that the High Court, in its exercise of its discretionary jurisdiction under Art. 226 of the Constitution could not issue a writ of *mandamus* in the nature of "a decree for specific performance of a contract", when a doubt was raised about the original drawing, design and the estimated cost of the project. The writ application was allowed and the order in question was quashed. It would be open to the state government to inform the grounds for cancellation to the petitioner and to furnish to the petitioner the revised drawing, design and estimated cost in respect of the project in question as approved by the Commission. Thereafter, the respondents could proceed in accordance with the law.[402]

EXAMPLE FROM CASE LAW

A tender for the collection and purchase of gum in lot mentioned some estimated quantity. After the tenders were accepted and during the period of the contract, gum in excess of the

---

[398] G. Ram v. Delhi Development Authority, AIR 2003 Del. 120.

[399] C.K. Achutam v. State of Kerala, AIR 1959 SC 490.

[400] M/s Pancham Singh v. State, AIR 1991 Pat. 168 (F.B.).

[401] M/s Radhakrishna Agarwal v. State of Bihar, AIR 1977 SC 1496.

[402] See also paragraph 5.15 in Chapter 5.

estimated quantity became available. The agreement contained a provision that if the production of gum exceeded the notified quantity, the purchaser was bound to take the increased quantity at the agreed price. However, the relevant government department refused to allow the contractors to lift the additional quantity. In a writ petition, the High Court allowed the contractors to lift the additional quantity during the relevant period at the contract price. In the meantime, the government had sold the excess gum. The Supreme Court directed, on appeal, that the state should supply the excess quantity from a fresh lot.[403]

---

[403] State of Madhya Pradesh v. Ramswaroop Vaishya, AIR 2003 SC 1067.

4

Interpretation of Contract

4.0 Introduction

What does interpretation of a provision in a contract mean? Interpretation is the method by which the sense or meaning of a word, phrase or a provision is understood. The question of interpretation can arise only if two or more possible meanings are sought to be placed on a contractual provision – one party suggesting one construction and the other, a different one.[1] It is, therefore, in the best interest of all the parties to a contract that they express their intention clearly, and without ambiguity. However, for various reasons the meaning of a contract or some of its terms is often obscure. More often than not this becomes a source of dispute between the parties. In order to settle such a dispute, the disputed term or provision has to be interpreted or given a proper meaning.

In this chapter we outline the distinction between construction and interpretation, and also describe some commonly used rules of interpretation. This is followed by a brief discussion on the oral evidence rule and conflict of laws.

4.1 Provisions of the Law

The following provisions of the law are dealt with in this chapter.

Provision of the Law	Paragraph No.
Indian Evidence Act, Sections 91–99	4.10.12
Indian Evidence Act, Section 115	4.10.9
The Indian Contract Act, Sections 23, 24	4.10.2
Provisions in Standard Form Contracts	
FIDIC,[2] Clause 1.15	4.10.5
APDSS/MDSS, Clause 59	4.10.1

4.2 What Is Interpretation?

The dictionary meaning of the word "interpretation" is "an explanation" or a "way of explaining".[3] *Black's Law Dictionary* gives the meaning of interpretation as: "The art or process

---

[1] State J. & K. v. K.V. Ganga Singh, AIR 1960 SC 356 (359).
[2] Unless expressly stated, all references are to the FIDIC Red Book.
[3] *Oxford English Dictionary* – retrieved from www.en.oxforddictionaries.com/definition/interpretation.

of discovering and ascertaining the meaning of a statute, will, contract or any other written document, that is, the meaning which the author designed it to convey to others."[4] But this alone does not help. The word "meaning" may mean the literal meaning of the words used or, if we consider the question "what does the writer mean?", it also means the purpose or the intention he has in mind which is expressed through the words and phrases. So the interpretation of the contract almost always involves ascertaining the intention of parties to the contract – as conveyed by the words set out in writing or by the spoken words the parties used in the case of an oral contract.

The general rule in respect of contracts that have been reduced to writing is that one must look only to that writing for ascertaining the terms of the agreement between the parties. However, it does not follow that it is only what is set out expressly in a document that can constitute the terms of the contract between the parties. It is useful to read a contract as a whole to deduce its meaning. However, the terms of a contract do not always give complete guidance. This is where lawyers' and judges' skills are useful. For example, most cases on contracts that reach the UK's Supreme Court tend to concern issues of interpretation or remedial questions.[5] Despite the growth in boiler plate clauses and standard form contracts, what parties mean in a contract may sometimes have to be implied from what has been expressed and/or omitted. Sometimes parties write correspondence on the side that may throw light on their intentions. The meaning of a contract is, in the ultimate analysis, a question of construction of its provisions and it is well established that in construing a contract it would be legitimate to take into account surrounding circumstances.[6]

4.3 Distinction between Interpretation and Construction

Interpretation and construction represent two classic approaches to finding the meaning of a written constitution. The Supreme Court of India referred to the American jurist Cooley who explains the difference between them by suggesting that interpretation involves finding the true sense of any form of words that their author intended to convey, and construction is the drawing of conclusions that go beyond the direct expression of the text. The latter may be drawn from elements known from and given in the text or conclusions which are in the spirit, though not within the letter of the text.[7]

It is thus clear that in the strict usage, the term "construction" is wider in scope than interpretation. While the latter is a linguistic approach and is concerned with ascertaining the sense and meaning of the subject matter, the former may also be directed to explaining the legal effects and consequences of the document in question, including giving guidance on what should be done where the contract is silent on any matter. The interpretation, as such, must necessarily stop at the written text. The two are not the same. The profession, however, has not accepted this distinction, and the two expressions are, in practice, synonymous. The more common term is construction.

4.4 Kinds of Interpretation

Interpretation could be categorized as follows:

---

[4] *Black's Law Dictionary Online* – retrieved from www.thelawdictionary.org/interpretation/.
[5] McMeel, Gerard, "Language and the Law Revisited: An Intellectual History of Contractual Interpretation", *Common Law World Review*, 34(3), 256.
[6] Khardah Company Ltd. v. Raymon Co. (India) Pvt. Ltd., AIR 1962 SC 1810: (1963) 3 SCR 183.
[7] In re Sea Customs Act, AIR 1963 SC 1760 (1794).

1. **Authentic interpretation**: This applies to legislation. It is conferred by the legislator and is obligatory on the courts. For example, most statutes comprise an interpretation or a definition clause, which defines the meaning of certain words occurring frequently in the legislation.

2. **"Customary" or "usual" interpretation:** It is that which arises from successive or concurrent decisions of the court on the same subject matter, having regard to the spirit of the law, jurisprudence and usages.

3. **"Close", "strict" or "literal" interpretation:** This is adopted when the words are taken in their narrowest meaning without consulting anything outside the written document.

4. **"Extensive" or "liberal" interpretation:** This adopts a more comprehensive significance of the words used.

In addition to the above commonly used terms, there are others like: "extravagant interpretation" which takes the meaning beyond that indicated *prima facie* by the words used, and "free or unrestricted interpretation" which proceeds simply on the general principles of interpretation in good faith.

Depending on the nature of duty, even if the words used in a statute are *prima facie* enabling, a court shall infer a duty to exercise such power which is invested in aid of enforcement of a citizen's right. The word "may" used in Section 93 of the Punjab Municipal Act 1911 and Section 90 of the Himachal Pradesh Act 1968 was construed as "compulsory" in light of its object of giving effect to the residents' rights.[8] Similarly, while considering the NIT requirements regarding experience of a contracting firm, it has to be borne in mind that the said requirement is contained in a document inviting offers for a commercial transaction. The construction of such a term must take into account the standpoint of a prudent businessman. This was recognized by Lord Collins SCJ in the following words: "In complex documents ... there are bound to be ambiguities, infelicities and inconsistencies. An over-literal interpretation of one provision without regard to the whole may distort or frustrate the commercial purpose."[9]

When a businessman enters into a contract whereunder some works are to be performed, he seeks to assure himself about the credentials of the person who is entrusted with the performance of the work. Such credentials are to be examined from a commercial point of view, which means that if the contract is to be entered with a company he will look into the background of the company and the persons who are in control of the same in their capacity to execute the work. He would not go by the name of the company but by the persons behind the company. While keeping in view the past experience he would also take note of the present state of affairs, and the equipment and resources at the disposal of the company. When one company or entity is not capable to satisfy the eligibility criteria, a joint venture (JV) company is formed to pool together the resources and experience of two or more entities. Such a JV company formed for the purpose of tendering, by itself, cannot have the experience called for. But the JV company can rely on the experience of the constituents of the venture. Thus, experience of a JV company can only mean the experience of the constituents of the venture.[10]

4.5 Authority with Whom Interpretation Rests

A construction contract by its very nature and scope provides fertile ground for disputes. Resolution of disputes by the process of arbitration or litigation in court will obviously hinder the smooth progress of the work. It is for this reason that the construction contracts invariably

---

[8] Jagannath and others v. The Kullu Municipal Committee and others, AIR 2003 HP 5.

[9] Re Sigma Corp. (in administrative receivership) [2009] UKSC 2; [2010] 1 All ER 571 at paragraph 35. Also see, Rainy Sky SA v. Kookmin Bank [2011] UKSC 50; [2011] 1 WLR 2900.

[10] New Horizons Limited v. Union of India, 1995(1) SCC 478, 1995 AIR SCW 275.

provide that decisions of an architect or engineer as to questions arising in respect of certain matters shall be final and conclusive. Such subject matters include:

1. The interpretation of drawings and specifications.
2. Quality of materials and workmanship.
3. Subletting of the contract.

Where a construction contract includes a stipulation such as the above, the decisions of the engineer of architect have the effect of the award of an arbitrator. It has been held that in the absence of fraud or such gross mistakes as imply bad faith or failure to exercise an honest judgement, the decisions of the engineer or architect are conclusive and binding upon the parties.[11]

All disputes or differences other than those in respect of which the engineer's or architect's decision is expressed to be final and binding are referred for adjudication to an arbitrator, if the contract contains an arbitration clause, or to the court. The interpretation of the contract then rests with the arbitrator or the court.

4.6 Nature of Rules of Interpretation

We have to understand the principles of interpretation as judicially noted. Historically, these rules have evolved with the passage of time. These are not rules of law. These are only guiding principles. There are a number of rules of construction. Many of them are called golden rules. Many of these rules are artificial, some are contradictory, some are uncertain, so much so that no less a person than His Lordship Hon'ble Mr. Justice Krishna Iyer observed, "the golden rule is that there is no golden rule".[12]

4.7 Rules of Interpretation

It is useful to acquaint the reader with some of the general rules that have evolved and been followed in construing a contract. Many times aid is sought by referring to English court decisions. A word of caution at this stage will not be out of place.

4.7.1 English Law and Court Decisions – How Far and When Relevant?

Before considering specific rules, it is worthwhile noting that while interpreting the provisions of the Indian Contract Act, it is not always permissible to import the principles of English law. When a rule of English law receives statutory recognition by the Indian legislature, the language of the Act determines the scope of it, uninfluenced by the manner in which the principles may have been laid down in English law. For example, can a plaintiff in an Indian court base a cause of action on common law and equity as opposed to the express provisions of Section 27 of the Contract Act? This question arose before the Supreme Court of India which ruled that while the Contract Act, 1872, does not profess to be a complete code dealing with the law relating to contracts, to the extent the Act deals with a particular subject, it is exhaustive upon the same and it is not permissible to import the principles of English law *de hors* the statutory provision, unless the statute is such that it cannot be understood without

---

[11] Dandakaranya Project v. P.C. Corporation, AIR 1975 MP 152.
[12] C.J. Vaswani v. State of W.B., AIR 1975 SC 2473 (2476).

the aid of the English law.[13] In this case the doctrine of restraint of trade was the subject matter. The Supreme Court determined that under Section 27 of the Contract Act, a service covenant extended beyond the termination of the service was void.[14]

It is useful to summarize here the current position in English law as there are several elements of English legal interpretation or construction of contract that can apply to an Indian case under the Contract Act. Lord Justice Mance's position that the construction of contracts is an "iterative process"[15] is a good starting point when a court or an arbitrator has to consider a contract that is not clear or unambiguous. Different potential meanings of each clause in question need to be considered against the whole of the contract in its commercial context.

The classic summary of English legal principles of interpretation of contracts was set out by Lord Justice Hoffman[16] as follows:

1. Ascertain the meaning which the contract would convey to a reasonable person having all the background knowledge which would reasonably have been available to the contracting parties at the time of the contract.

2. Lord Wilberforce's reference to the background as the "matrix of fact",[17] is an understated description of what the background may include. It can include anything which would have (a) been available to the contracting parties at the time of the contract, and (b) affected the way in which the language of the document would have been understood by a reasonable man.

3. However, for reasons of practical policy, the admissible background excludes (a) the pre-contract negotiations of the parties, and (b) the parties' declarations of subjective intent. Those are admissible only in an action for rectification of the contract.

4. The meaning conveyed by a document to a reasonable man and the meaning of its words are not the same. The latter can be worked out from dictionaries and principles of grammar; the former is what the parties using those words against the relevant background would reasonably have been understood to mean. The background may enable the reasonable man to choose between the possible meanings of ambiguous words but also to conclude that the parties might, for whatever reason, have used the wrong words or syntax.

5. The rule that provides that we should give words their "natural and ordinary meaning" reflects our reluctance to accept that people have made linguistic mistakes, particularly in formal documents. If the background shows that something must have gone wrong with the language, the law does not require a court to attribute to the parties an intention which they plainly could not have had. Lord Diplock emphasized this when he said that if detailed semantic and syntactical analysis of words in a commercial contract leads to a conclusion that flouts business common sense, it must be made to yield to business common sense.[18]

---

[13] Superintendence Co. of India v. Krishan Murgai, AIR 1980 SC 1717.

[14] The Supreme Court decision in AIR 1980 SC 1717 was followed in Sandhya Organic Chemicals P. Ltd. v. United Phosphorous Ltd., AIR 1997 Guj.177.

[15] Re Sigma Corp. (in administrative receivership) [2009] UKSC 2; [2010] 1 All ER 571 at paragraph 12.

[16] Investors Compensation Scheme Ltd. v. West Bromwich Building Society (No.1); Armitage v. West Bromwich Building Society; Alford v. West Bromwich Building Society Investors Compensation Scheme Ltd. v. Hopkin & Sons, [1998] 1 W.L.R. 896 at 912–913. Followed in Mt Højgaard A/S v. E.ON Climate and Renewables UK Robin Rigg East Limited; Court of Appeal (Civil Division); [2015] EWCA Civ. 407; [2015] 4 WLUK 686.

[17] Prenn v. Simmonds, [1971] 1 WLR 1381.

[18] The Antaios Compania Neviera S.A. v. Salen Rederierna A.B., [1985] 1 A.C. 191, 201.

4.7.2 Conflict between Written and Printed Parts of Contracts

Where some parts of the contract are written and the other parts are printed, in case of conflict, the written parts of the contract will prevail.[19] If printed words on a form are rendered inconsistent by the written words and the former are not struck out, they may be rejected.[20] Words deleted and initialled in a working draft cannot be read again for the purpose of construing the final contract.[21] The Supreme Court of India has held that where a contract consists of a printed form with cyclostyled amendments, typed additions and deletions and handwritten corrections, an endeavour must be made to give effect to all the provisions. However, in the event of apparent or irreconcilable inconsistency, the following rules of construction will normally apply:

(i) The cyclostyled amendments will prevail over the printed terms;

(ii) The type-written additions will prevail over the printed terms and cyclostyled amendments;

(iii) Handwritten corrections will prevail over the printed terms, cyclostyled amendments and typed written additions.[22]

The above rules have evolved from the well-known maxim of construction that written, stamped or typed additions, when inconsistent with the printed terms, would normally prevail over the printed terms. This maxim proceeds on the assumption that the printed form contained the original terms, and changes thereto were incorporated by the cyclostyled amendments, followed by changes by type-written additions and lastly the handwritten additions. The logical explanation for such assumption is this: the printed form contains standardized terms to suit all contracts and situations. It is not drafted with reference to the special features of a specific contract. When such a standard form is used with reference to a specific contract, it becomes necessary to modify the standard/general terms by making additions/alterations/deletions, to provide for the special features of that contract. This is done either by way of an attachment of an annexure to the standard printed form incorporating the changes, or by carrying out the required additions, alterations, or deletions in the standard form itself by typing/stamping/handwriting. Lord Ellenborough's following observations with reference to printed forms of contract with handwritten additions are still valid:

> that the words superadded in writing (subject indeed always to be governed in point of construction by the language and terms with which they are accompanied) are entitled nevertheless, if there should be any reasonable doubt upon the sense and meaning of the whole, to have a greater effect attributed to them than to the printed words, inasmuch as the written words are the immediate language and terms selected by the parties themselves for the expression of their meaning, and the printed words are a general formula adapted equally to their case and that of all other contracting parties upon similar occasions and subjects.[23]

Another parallel principle here is that where a contract has several annexures or attachments prepared at different points of time, unless a contrary intention is apparent, the latter in point of time would normally prevail over those earlier in point of time.

---

[19] Southland Royalty Co. v. Pan Am Petroleum Corp., 378 S.W.2d. 50,57 (Tex 1964).

[20] Western Assurance Co. of Toronto v. Poole, (1903) 1. K. B. 376.

[21] A&J Inglis v. Buttery, (1878) 3 App. 552.

[22] M.K. Abraham & Co. v. State of Kerala, (SC): 2009(5) Recent Apex Judgments (R.A.J.) 691: 2009(4) R.C.R.(Civil) 366: 2009(7) SCC 636: 2010 AIR (SC) 1265: 2009(3) ArbiLR 130: 2010 AIR (SCW) 574: 2009(9) Scale 288.

[23] Robertson v. French, [1803] 102 ER 779, 782.

4.7.3 Ignorance of Contract Terms

Ignorance through negligence will not relieve a party from his contract obligations. He who signs or accepts a written contract, in the absence of fraud or other wrongful act on the part of another contracting party, is presumed to know its contents and to assent to them. This doctrine arises from the well-settled principle that affixing a signature to a contract creates a conclusive presumption, except as against fraud, that the signatory read, understood and assented to its terms. Ignorance of the terms of the contract does not mean that there was an absence of intent.[24] Similarly, a party advised by lawyers in the formation of the contract would be presumed to have the knowledge and understanding of their lawyers; the knowledge of the lawyers is imputed to the party they advise.

4.7.4 Words Given Their Plain Meaning

The words that are free from any ambiguity are usually construed according to their natural meaning. A court will not construe a written contract in such a way as to modify the plain meaning of its words under the guise of interpretation. Technical words used in a contract should be given their accepted technical meaning.[25] Oral evidence as to their meaning may be given, usually by an expert in the field. ·

4.7.5 Conduct of Parties

The contracting parties' conduct after signing the contract can be taken into account in determining its proper interpretation, especially where such conduct pre-dates any controversy arising out of the contract. The subsequent conduct may be admissible as evidence to prove the nature of the agreement or the legal relationship of the parties, even though this may vary or add to the written instrument.[26] However, unilateral conduct of one party submitting a tender may not amount to the acceptance of the counter-offer made by the other party.[27] It is the mutual conduct that can be useful to construe the meaning of the contract.

4.7.6 Intent of Parties

Construing contracts has as its objective the ascertainment of the intent of the parties at the time the contract was entered into and to give effect to that intention, if it can be done consistently with legal principles. The intent of the parties is to be gathered from the contract as a whole. Where the parties to a contract have expressed their intention without ambiguity, the court may not alter their agreement although the bargain is hard or unwise.[28]

EXAMPLES FROM CASE LAW

1.   A clause in a contract provided for compensation for a party's failure to supply or delayed supply of materials by the other party. The Supreme Court of India held that the clause was never intended to cover refusal to deliver materials. As such, in the event of such a refusal, the liquidated damages clause would not apply. The arbitrator's award of damages to the aggrieved party, in excess of liquidated damages, was held proper and upheld.[29]

---

[24] Timeplan Educational Group Ltd. v. NUT, [1997] 1RLR 457.

[25] Paragraph 12–052 of *Chitty on Contracts*, 31st Edition (2012).

[26] AG Securities v. Vaughan, [1990] 1 AC 417 at 469 and 476.

[27] U.P. Rajkia Nirman Nigam Ltd. v. Indure Pvt. Ltd., AIR 1996 SC 1373.

[28] FSHC Group Holdings Limited v. Barclays Bank Plc, [2018] EWHC 1558 (Ch); [2018] 6 WLUK 448.

[29] Steel Authority of India Ltd. v. Gupta Brother Steel Tubes Ltd., 2010(1) R.C.R.(Civil) 101: 2010(1) Recent Apex Judgments (R.A.J.) 1: 2009(10) SCC 63: 2009(4) SCC (Civil) 16: 2009(12) Scale 393: 2010(1) W.L.C. 62: 2009(4) R.A.

2. A contract construed on the basis of correspondence contained a letter stipulating as follows: "Mode of measurement: We have based our price on the total built-up area of one floor [four flats] including stair-case and common corridor but excluding balconies only. Hence work should be measured on the built-up area, excluding balcony areas." It was contended that the said plans were modified later and that the flats, as finally constructed, did not have any balconies so that no question of excluding the balconies' area could arise. However, no agreed or sanctioned plan modifying the plan attached to the tender notice was produced. The claim for balcony area was allowed by the arbitrator. On appeal, the Supreme Court observed and held as follows:

> 7. The above stipulation clearly says that total built-up area of a floor shall include staircase and common corridor but shall exclude balconies. It expressly provides that "work should be measured on the built-up area excluding balcony area". It is undisputed that in the plan of flats attached to the Tender notice, balconies are provided. ... The appellant could not have constructed flats except in accordance with the plans attached to the Tender notice, unless of course there was a later mutually agreed modified plan – and there is none in this case. We cannot, therefore, entertain the contention at this stage that there are no balconies at all in the flats constructed and that, therefore, the aforesaid stipulation has no relevance. We must proceed on the assumption that the plans attached to the Tender notice are the agreed plans and that construction has been made according to them and that in the light of the agreed stipulation referred to above, the areas covered by balconies should be excluded. In this view of the matter we agree with the Division Bench that the arbitrators over-stepped their authority by including the area of the balconies in the measurement of the built-up area. It is axiomatic that the arbitrator being a creature of the agreement, must operate within the four corners of the agreement and cannot travel beyond it. More particularly, he cannot award any amount which is ruled out or prohibited by the terms of the agreement. In this case, the agreement between the parties clearly says that in measuring the built-up area, the balcony areas should be excluded. The arbitrators could not have acted contrary to the said stipulation and awarded any amount to the appellant on that account. We, therefore, affirm the decision of the Division Bench on this score.[30]

4.7.7 Every Word and Provision to Be Given Effect to

It is an elementary rule in interpretation of contracts that whenever reasonably practical, every term in a contract should be given effect to.[31] A contract should not be construed in a manner that would render meaningless any provisions therein. However, it may not always be possible to give effect to every term of a contract, especially when the contract is in a standard form.[32] Where there is an inconsistency between a specific and a general provision, the specific provision takes precedence.[33]

J. 172: 2009(3) ArbiLR 466: 2009 AIR (SCW) 7191: 2010(2) AIR Jhar R. 73: 2009(4) All WC 4100: 2009(3) Law Herald 2547: 2009(4) CivilLJ 884: 2009(3) Apex Court Journal 609.

[30] New India Civil Erectors (P) Ltd. v. Oil & Natural Gas Corpn., AIR 1997 SC 980.

[31] James Orr v. Alexander Mitchell and Others, (1892) 19 R 700.

[32] Macquarie Internationale Investments Ltd. v. Glencore UK Ltd., [2010] EWCA Civ 697, [2010] All ER (D) 175 (Jun).

[33] Woodford Land Ltd. v. Persimmon Homes Ltd., [2011] EWHC 984 (Ch), [2011] All ER (D) 205.

EXAMPLE FROM CASE LAW

A clause in the works contract in a case provided: "7.1 The contractor may subcontract any portion of work, up to a limit specified in Contract Data with the approval of Engineer but may not assign the contract without the approval of the Employer in writing. Subcontracting does not alter the Contractor's obligations." The Contract Data provided the limit of subcontracting up to 50% of the initial contract price with prior approval of the Engineer.

The contractor engaged a subcontractor for executing a portion of the work. The engineer on the spot was aware that subcontractor was executing a part of the work. Neither the engineer nor the employer was satisfied with the progress of the work and in the notice issued sought whether prior permission was taken for subcontracting. Subsequent to complaint, the main contractor removed the subcontractor. However, on the said cause, the contractor was debarred from participating in any bidding process. A writ petition by the contractor followed in respect of the orders cancelling the tendering process and black-listing. The single judge accepted the defence of the employer and upheld both the orders. On appeal, the division bench interpreted the provisions as follows:

1. In the first place, there is no user of the words "prior approval" either in respect of the sub-contractor in respect of the assignment of contract.

2. Then, there is a definite difference in the concept of sub-contracting of the portion of the work and assigning the original contract itself. While for sub-contracting only the approval of the Engineer is required, for assignment of the contract itself, the approval of NHAI that too in writing, is necessary. There is undoubtedly a dichotomy in the concepts of "sub-contracting" and "assignment of contract". It is therefore, obvious that for such sub-contracting, there would be no need of the approval of NHAI though unfortunately, the parties have understood otherwise.

As a result the blacklisting order was set aside.[34]

4.7.8 Law Will Not Make a Better Contract

The law will not (a) make a better contract for the parties than they themselves have seen fit to enter into, (b) alter it for the benefit of one party to the detriment of another, or (c) supply a term with respect to which the contract is silent. The court's duty is always interpretation: not so much to find out the intention of the parties, but to find the meaning of the words used by them.

4.7.9 Erasures and Alterations

Erasures and alterations made prior to the signing of a contract are given effect to, but those made by one party, after signatures by both parties do not affect the contract in any way. To avoid disputes, erasures and alterations, where they are really unavoidable, should be initialled by both parties before signing the contract.

---

[34] M/s. P.T.S. Umber Mitra Jaya v. N.H. Authority of India, AIR 2003 Mad. 221.

4.7.10 Ambiguity in Phrase – Evidence of Subsequent Conduct of Both Parties Is Relevant

Extrinsic evidence to determine the effect of a contract is permissible where there remains a doubt as to the true meaning of the contract. Evidence of the acts done under it is a guide to the parties' intention, particularly when acts are done after the date of the instrument.[35]

4.7.11 Intention – Surrounding Circumstances When Considered

In construing a document, whether in English or vernacular, the fundamental rule is to ascertain the intention from the words used; the surrounding circumstances are to be considered but that is only for the purpose of finding out the intended meaning of the words which have actually been employed.[36]

4.7.12 Interpretation of a Document in One Case – How Far Relevant in Another Case?

Unless the language of the two documents is identical, an interpretation put upon one document is no authority for the proposition that a document differently drafted, though using partially similar language, should be similarly interpreted.[37] Besides, it is obvious that in construing documents, the usefulness of precedents is usually of a limited character; after all, courts have to consider the material and relevant terms of the document with which they are concerned; and it is on a fair and reasonable construction of the said terms that the nature and character of the transaction evidenced by it has to be determined.[38]

4.7.13 Two Constructions Possible

It is a settled rule of interpretation that if there are two possible constructions of a document, one of which will give effect to all the clauses therein while the other will render one or more of them nugatory, it is the former construction that should be adopted on the principle expressed in the maxim *"ut res magis valeat quam pereat"*.[39]

4.7.14 Conflict between Earlier and Later Clauses

If there is a conflict between an earlier clause and a latter clause in a contract, and it is not possible to give effect to both of them, then the rule of construction is that it is the earlier clause that must override the later clause and not "vice versa". If the latter provision of the deed were introduced by the word "but" or the words "provided always", "nevertheless" or the like, there would be no difficulty. However, there is no necessity to find any such words, if a latter clause says in so many words or as a matter of construction that an earlier clause is to be qualified in a certain way, effect can, and must, be given to both the clauses.[40]

EXAMPLE FROM CASE LAW

A contract contained two clauses, one of which provided that "the price shall be firm and without any escalation on any ground whatsoever until the completion of the work", and the second clause stipulated that the prices were to be given in the currency of the bidder

---

[35] Abdulla Ahmed v. Animendra Kissen, AIR 1950 SC 15 (21).
[36] Ram Gopal v. Raw Cotton Co., AIR 1955 SC 376.
[37] Abdulla Ahmed v. Animendra Kissen, AIR 1950 SC 15.
[38] Triveni Bani v. Smt. Lilabai, AIR 1959 SC 620 (626).
[39] Radha Sunder v. Mohd. Jahadur Rahim, AIR 1959 SC 24 Para 11.
[40] Forbes v. Git and others (PC.), AIR 1921 PC 209 (211).

and that "if the bidder expects to incur a portion of this expenditure in currencies other than those stated in his bid, and so indicates in his bid, payment of the corresponding portion of the prices as so expended will be made in these other currencies". The Supreme Court of India held that the two conditions must be construed in such a manner that effect may be given to both of them. The bid had two components, namely, an Indian currency and a US dollar component. It was observed that the second component involving fluctuation did not amount to an escalation of price and did not violate the first clause. The claim was for the foreign exchange component and no more.[41]

4.7.15 Equitable Principle – When Applicable?

A person cannot in equity or in law assign what has no existence. However, a person can contract to assign property which is to come into existence in the future, and when it has come into existence, equity, treating as done that which ought to be done, fastens upon that property, and the contract to assign thus becomes a complete assignment.[42] The equitable principle only implements or effectuates the agreement of the parties. Equity does not, however, take upon itself the task of making any new agreements for the parties either by filling up a lacuna or gap in their agreement or otherwise. If, therefore, there is no agreement between the parties to transfer a future decree, the equitable principle cannot come into play at all.[43]

4.7.16 Departure from Literal Interpretation of Statutes – When Permissible

The cardinal rule for the construction of statutes is to read them literally, that is by giving to the words used by the legislature their ordinary, natural and grammatical meaning. If, however, such a reading leads to absurdity and the words are susceptible of another meaning, the court may adopt the same. But if no such alternative construction is possible, the court must adopt the ordinary rule of literal interpretation.[44]

The rule of interpretation is well established that:

> Where the language of a Statute, in its ordinary meaning and grammatical construction leads to a manifest contradiction of the apparent purpose of the enactment, or to some inconvenience or absurdity, hardship or injustice, presumably not intended, a construction may be put upon it which modifies the meaning of the words, and even the structure of the sentence.[45]

4.7.17 Construction Tending to Make Part of a Statute Meaningless

It is well settled that in construing the provisions of a statute, courts should be slow to adopt a construction which tends to make any part of the statute meaningless or ineffective; an attempt must always be made so to reconcile the relevant provisions as to advance the remedy intended by the statute.[46]

---

[41] Pure Helium India Pvt. V. Oil & Natural Gas Commission, AIR 2003 SC 4519.

[42] Collyer v. Isaacs, (1881) 19 Ch D 342 at p. 351.

[43] Jugalkishore Saraf v. M/s. Raw Cotton Co. Ltd., AIR 1955 SC 376.

[44] Jugal Kishore v. Raw Cotton Co., AIR 1955 SC 376.

[45] Tirth Singh v. Bachittar Singh, AIR 1955 SC 830 (833).

[46] Siraj-ul-Haq v. SC Board of Waqf, AIR 1959 SC 198 (204).

4.7.18 Severance of the Good from the Bad

The fact that a clause in a deed is not binding on the ground that it is unauthorized cannot *ipso facto* render the whole deed void unless it forms such an integral part of a transaction so as to render it impossible to sever the good from the bad.[47]

4.7.19 Use of Marginal Notes or Paragraph Headings – When Permissible

Where a section of a statute is unambiguous, the marginal note may not be used as an aid to its interpretation.[48] It is also true that a heading cannot control the interpretation of a clause if its meaning is otherwise plain and unambiguous, but it can certainly be referred to as indicating the general drift of the clause and affording a key to better understanding of its meaning.[49]

4.7.20 The Living Approach Theory

It is worthwhile to conclude the discussion on various doctrines and rules of interpretation of contracts by referring to the observations of His Lordship Hon'ble Mr. Justice Krishna Iyer, who observed:

> Judicial interpretation is a not bloodless and sterile exercise in spinning subtle webs, sometimes cobwebs, out of words, and phrases otherwise simple, but to unfold the scheme of the legislation insightfully, sense its social setting and read the plain intendment. This living approach can do justice to law.[50]

Every decision given by a court of law is based on the factual matrix of the case before the court. A three judge bench of the Supreme Court has observed:

> Courts should not place reliance on decisions without discussing as to how the factual situation fits in with the fact situation of the decision on which reliance is placed. Observations of Courts are not to be read as Euclid's theorems nor as provisions of the statute. These observations must be read in the context in which they appear. Judgments of courts are not to be construed as statutes. To interpret words, phrases and provisions of a statute, it may become necessary for judges to embark into lengthy discussion, but the discussion is meant to explain and not to define. Judges interpret statutes, they do not interpret judgments. They interpret words of statutes, their words are not to be interpreted as statutes.[51]

The living approach theory is equally applicable to the interpretation of contracts. If interpretation were to consist only of ascertaining the dictionary meaning of the words used, it could be done by a computer. But a human judge can discover the intention of the parties and discover the essence of a document.

4.8 Construction – Strict v. Liberal and Literal v. Logical Construction

Historically, courts used to be strict about the written words of a contract. They went by grammatical meaning. They refused to look at outside aids and refused to fill in any gaps. It was

---

[47] K.D. Co. v. K.N. Singh, AIR 1956 SC 446 (459).
[48] Shakuntala v. Mahesh, AIR 1989 Bom. 353.
[49] Union of India v. Raman Iron Foundry, AIR 1974 SC 1265.
[50] Bar Council of India v. M.V. Dhabolkar & Others, AIR 1975 SC 2029.
[51] Haryana Financial Corpn. v. M/s. Jagdamba Oil Mills, AIR 2002 SC 834= 2002 AIR SCW 500.

for the parties to anticipate every contingency that might befall them and to provide a term to protect from such contingencies. If the parties did not do so, they were bound by the written words. Courts would not write in any exception or implication to protect a party. Those were the days when contracts were few and were indeed the outcome of negotiations and somewhat equal bargaining power. Under those circumstances, the process followed by the court – known as "strict" construction, "grammatical meaning", "literal construction" – could be said to be justifiable. But with the advent of the industrial revolution, expansion of commerce and trade, advancement of science, revolutions in the means of transport and communication, etc., the number of contracts entered into multiplied manifold. The idea of an agreement freely negotiated between the parties has given way to the necessity of a uniform set of printed conditions, which can be used time and again. Such contracts are called, quite appropriately, the contracts of adhesion wherein all the terms and conditions are prepared, printed and handed over by one party to the other. The role of the other party is reduced to merely adhering to these terms and conditions. The freedom of contract exists only to sign the agreement containing these terms or to go without. Nevertheless, courts were forced to apply to this situation the ordinary principles of the law of contract, which was not entirely capable of providing a just solution for a transaction in which freedom of contract exists on one side only. In particular, the party preparing and delivering the printed documents was permitted to exempt himself unfairly from certain of his liabilities at the law and thus to deprive the other party of the compensation which he might reasonably have expected to receive for any loss or damage arising out of the transaction. Under these changed circumstances, the attitude of the court, being strict about the written words, going by the grammatical meaning, etc., was not capable of providing a just solution. Therefore, as observed by Lord Denning:

> Simple justice demanded that the buyer or the consumer should be protected, even though he did not insert an express term on his own behalf. So the courts filled in the gaps. They did this by means of the doctrine of "implied terms." A number of such doctrines were evolved to find a just solution. These doctrines gave rise to a "liberal" interpretation.[52]

Before we study these doctrines and their application to the standard form contracts in use in the construction industry, it is interesting to note that in one case, the High Court adopted the strict construction rule and the Supreme Court the liberal construction rule. The High Court set aside an award made by an arbitrator appointed by "the Secretary in Ministry of Food and Agriculture" on the ground that at that time there were two secretaries in the ministry and that the use of the article "the" made the agreement uncertain and therefore invalid. On appeal, it was observed by the Supreme Court: "Though this argument appears attractive at first sight, a little scrutiny will reveal that it is unsound. It is based on a highly technical and doctrinaire approach and is opposed to plain common-sense. ... The Secretary in charge of the Department of Food filled the description 'Secretary in the Ministry of Food and Agriculture' given in clause."[53] From the judgments of the Supreme Court of India, the following rules emerge:

1. A contract must be interpreted in such a manner as to give efficacy to the contract rather than to invalidate it.

2. It would not be right to apply strict rules of construction to a contract entered into between two lay parties, which are ordinarily applicable to a conveyance and other formal documents.

3. The meaning of such a contract must be gathered by adopting a common-sense approach and it must not be allowed to be thwarted by a narrow pedantic and legalistic approach.

---

[52] *The Discipline of Law*, London, Butterworth, 1979, p. 33.
[53] Union of India v. D.N. Revri & Co., AIR 1976 SC 2257.

4. The answer to the question "why this provision" would be permissible to be looked into to gather the intention of the parties to the contract. In other words "object and reason" of the provision may be ascertained and taken note of while interpreting a contract.

5. The observation that "[t]his is, in our opinion, hyper-technical argument which seeks to make a fortress of the dictionary and ignores the plain intendment of the contract" stands confirmed by a subsequent pronouncement by the Supreme Court that while literal meaning can be ascertained from a dictionary, the court is not subject to the dictatorship of the dictionaries.[54]

6. Post-contract conduct of the parties, which may throw light on the interpretation put upon by the parties themselves, is relevant and can be looked into.[55]

7. That if two alternatives are possible, one each under strict or literal construction and "liberal" or "living approach" construction, respectively, the option to be exercised, because it is most likely to find favour with an arbitrator or court in a judicial proceeding, will be to adopt the construction under the liberal rule of interpretation.

4.9 *"Force Majeure"* Clause

The French expression *"force majeure"* literally means superior force. When performance of a part of the contract is not possible due to causes which are outside the parties' control, and which could not be avoided by the exercise of due care, this provision protects the parties. For example, in a case the *force majeure* clause read as follows:

> Neither party shall be held responsible for any loss or damage or delay in or failure of performance of the contract, if any, to the extent that such loss or damage or such delay in or failure of performance is caused by FORCE MAJEURE including but not limited to acts of God, restraint of State/Central Government, devastating fires, major accidents, declared or undeclared hostilities, riots, rebellion, explosion, stripes, epidemic, severe inclement weather like cyclone, gale etc. compliance with any request, ruling orders or decree or local or any Indian Government authorities and any other similar cause or causes which cannot with reasonable diligence be controlled or provided against by the parties hereto.

The arbitrator stated in his award that a generator breakdown could not be considered as covered under the *force majeure* clause which clearly included any cause(s) which could not, with reasonable diligence, be controlled or provided against by the contracting parties. In the arbitrator's opinion, the respondent could diligently control the situation by providing standby generator sets to avoid production loss. The respondent could not avoid its liability to compensate the claimant for the loss suffered due to the generator breakdown. It was held by the Calcutta High Court that in view of the reason given by the arbitrator, the High Court had no occasion to interfere with his decision by substituting its own view.[56]

Usually the burden of proving that a *force majeure* clause applies is on the party that asserts reliance on it.[57]

---

[54] Subhash Chandra v. State of U.P., AIR 1980 SC 800.
[55] Abdulha Ahmed v. Animendra Kissen, AIR 1950 SC 15 (21).
[56] Oil & Natural Gas Commssion Ltd. v. M/s Dilip Constructions, AIR 2000 Cal. 140.
[57] Channel Island Ferries Ltd. v. Sealink UK Ltd., [1988] 1 Lloyd's Rep. 323.

4.10 Doctrines of Interpretation

Some important doctrines of interpretation that are often required to be used in interpreting contract provisions in the standard forms are explained below with examples. At the outset, however, it is pointed out that in the process of interpretation, the question "what does the provision in the agreement or agreement itself mean?" may not necessarily be the same as "what did the parties intend when they executed the document?" They are presumed to have intended to say that which they have in fact said. In effect, a contract is what the court or arbitrator says it is.

4.10.1 Contra Proferentum *Rule*

In *Anson's Law of Contract*,[58] this doctrine is summarized in these words:

> The words of written documents are construed more forcibly against the party using them. The rule is based on the principle that a man is responsible for ambiguities in his own expression and has no right to induce another to contract with him on the supposition that his words mean one thing, while he hopes the Court will adopt a construction by which they would mean another thing, more to his advantage.

The drafter of the contract should make clear his meaning and if he does not, the document will be interpreted against the drafter. This rule is old[59] but it has come into its own after the growth of standard form contracts. Under these clauses, as already stated earlier, the party drafting the agreement tended to make provisions which exempted that party from a legal liability to pay compensation to the other party. When interpretation of these exemption clauses came before courts and the judges felt that such clauses were a challenge to judicial conscience and had to be so interpreted as to keep them within reasonable limits, *contra proferentum* gained prominence. In England, the rule has a limited scope[60] especially since the Unfair Contract Terms Act 1977. English courts tend to rely on the words of contracts, commercial sense, and the context, both documentary and factual.[61] Further, in purely commercial transactions, there is very little scope for the application of this doctrine in England.[62]

There are two things worth noting about this rule: first, it has to be called a rule of the last resort to be pressed into service where all other rules of construction fail to yield, and second it is applicable only in cases of doubts and ambiguity. In simple words, the rule is that where two alternatives are possible as a result of considering and invoking all permissible rules, then that interpretation will be preferred which is against the person using or drafting the words or expressions which have given rise to the difficulty in construction. The Supreme Court of India has pressed this rule into service even when admittedly there was no ambiguity in the expression used by the parties. The first of the following examples is the case in question.

EXAMPLES FROM CASE LAW

1. An insurance policy for a property contained a clause stipulating *inter alia* that if the property insured or any part thereof be destroyed or damaged by "the following:

---

[58] 24th Edition at page 150.

[59] Bramall & Ogden Ltd. v. Sheffield City Council, 29 BLR 73.

[60] Persimmon Homes Ltd. v. Ove Arup & Partners Ltd., [2017] EWCA Civ 373.

[61] K/S Victoria Street v. House of Fraser (Stores Management) Ltd., [2011] EWCA Civ 904.

[62] K/S Victoria Street v. House of Fraser (Stores Management) Ltd., [2011] EWCA Civ 904; Transocean Drilling UK Ltd. v. Providence Resources PLC, [2016] EWCA Civ 372; and Persimmon Homes Ltd. and others v. Ove Arup Partners Ltd. and others, [2017] EWCA Civ 373.

... Impact by any rail/road vehicle or animal." It was not in dispute that the damage to the building and the machinery was caused on account of a road construction bulldozer having been driven close to the building. The insurance company resisted the claim on the ground that the damage was not caused by any direct collision of a road vehicle. The claim was dismissed by the District Consumer Forum but allowed by the State Commission. On appeal, the National Commission accepted the decisions of the State Commission but reduced the quantum of damages. Against the said decision, the insurance company filed an appeal in the Supreme Court. The Supreme Court observed that the only point for consideration was whether the word "impact" contained in the policy covered the damage caused or not. It was held:

> In order to interpret this clause, it is necessary to gather the intention of the parties from the word used in the policy. If the word "impact" is interpreted narrowly the question of impact by any rail would not arise as the question of any rail forcibly coming to the contact of a building or machinery would not arise. In the absence of specific exclusion and the word "impact" having more meanings in the context, it cannot be confined to forcible contact alone ... Although there is no ambiguity in the expression "impact", even otherwise applying the rule of contra proferentum, the use of the word "impact" ... must be construed against the appellant.[63]

The appeal was dismissed.

2. Clause 59 in **MDSS** or **APDSS** standard form contract used in the states of Tamil Nadu and Andhra Pradesh for public works, reads: "59. *Delays and Extension of Time:*- No claim for compensation on account of delays or to the work from any cause whatever shall lie, except, as hereinafter defined. Reasonable extension of time will be allowed by the Executive Engineer or by the officer competent to sanction the extension." The facts of a case dealt with by the author involving the above provisions were as follows.

A project involved the manufacture and laying of a pipeline for a water supply project. It stipulated a certain time limit for its completion. The contract included a provision whereby the government had undertaken to make available to the contractor foreign exchange for importing "know-how" and machinery. The foreign exchange could not be made available in time so much so that the entire work including the erection of machinery, etc. had to be carried out beyond the stipulated period. The contractor claimed damages for the losses caused by the delay. The state government rejected the claim on the ground that the contract contained a provision which precluded the contractor from claiming damages due to delay in completion. The clause provided: "No claim for compensation on account of delays or hindrances to the work from any cause whatever shall lie, *except* as *hereinafter defined* ..." (emphasis supplied).

The interpretation of this provision attracted the rule of *contra proferentum*. The opening sentence suggests that the agreement defines circumstances under which claim for compensation "shall lie" but neither the relevant clause nor any provision subsequent to it stipulated the circumstances "when the claim for compensation shall lie". One of the rules of interpretation is that "every word or phrase must be given some meaning" which is based on the presumption that the drafter of a formal document does not waste words.[64] The words, "except as hereinafter defined" cannot be reduced to silence. It only means that the subsequent provisions in the said clause make exceptions and allow "compensation" to the contractor. The subsequent provisions only allow "extension of time". That means the word

---

[63] United India Insurance Co. Ltd. v. M/s Pushpalaya Printers, AIR 2004 SC 1700.
[64] Union of India v. Raman Iron Foundry, AIR 1974 SC 1265.

"compensation" did not mean "money compensation". The dictionary meaning of the word "compensation" is "compensating". To compensate means "counter-balance, make amends for things"; in other words, to counter-balance the loss of anything. The loss of time can be counter-balanced or amended by granting an "extension of time". Therefore, the entire provision could be said to deal with extension of time and not refer to money compensation. So interpreted, the clause would read: "No claim for extension of time on account of delays or hindrances to the work from any cause whatever shall lie, except as hereinafter defined. Reasonable extension of time will be allowed ...". So read the words "except as hereinafter defined" used by the drafter immediately find their rightful place and utility.

Reference can be gainfully made to the Supreme Court decision where the Supreme Court dealt with Clause 18 with its heading "Recovery of sum due". It observed, "It is true that a heading cannot control the interpretation of a clause if its meaning is otherwise plain and unambiguous, but it can certainly be referred to as indicating the general drift of the clause and offering a key to better understanding of its meaning. The heading of clause 18 clearly suggests that this clause is intended to deal with the subject of recovery of sum due".[65]

Relying on the above observations, if one looks at the heading of Clause 59, it reads "*Delays and Extension of Time*". It is evident that the clause suggests that it intended to deal with delays and extension of time and that monetary compensation does not figure in it. It is arguable that the state drafted the clause in such a manner as to make a contractor sign it believing that it only deals with delays and extension of time while believing that an arbitrator would construe it to mean "money compensation". This calls for adopting that interpretation which favours the contractor and is against the drafter, the state government.

The arbitrator made an award in favour of the contractor which was not challenged in the court of law. This clause was, in subsequent cases, interpreted differently by the High Courts and the Supreme Court.[66] The interpretation so given by the courts is likely to result in injustice in some deserving cases. A brief reference to the facts of these cases will make this self-evident, it is submitted.

In State of A.P. v. Associated Engineering Enterprises, Hyderabad,[67] Mr. Justice Jeevan Reddy, speaking for the division bench again referred to Clause 59 and observed at paragraph 26 as follows:

> Applying the principle of the above decision to the facts of the case before us it must be held that Clause 59 bars a claim for compensation on account of any delays or hindrances caused by the department. In such a case, the contractor is entitled only to extension of the period of contract. In such a case, the claim for compensation is clearly barred by Clause 59 of the APDSS which is admittedly a term of the agreement between the parties.

In a later decision in State of A.P. v. S. Shivraj Reddy,[68] a different view was taken. The claim in this case pertained to the work done beyond the contract period. It was held that the contractor must be paid as per the standard specification rates (SSRs) for the reason that the state had defaulted in handing over the site at the agreed time. It was further held that the claim was not barred by Clause 59 of APDSS. An earlier decision[69] was distinguished on the basis that the claim in that case was for payment of compensation for loss of profit or damage. The contractor was

---

[65] Union of India v. Raman Iron Foundry, AIR 1974 SC 1265.

[66] Ch. Ramalinga Reddy v. Supdt. Engineer, (1994) 2nd December 1994, (1994) 5 Scale 67.

[67] State of A.P. v. Associated Engineering Enterprises, Hyderabad, AIR 1990 Andh Pra 294.

[68] 1988 (2) APLJ 465.

[69] CMA No. 677/81 and CRP No. 385/82 dated 19th April 1982.

not claiming escalation of rates but contract rates and most of the site was handed over after the original contract period had elapsed.

In Govt. of A.P. v. P.V. Subba Naidu,[70] Jeevan Reddy, J. speaking for the division bench referred to the Supreme Court decision in P.M. Paul v. Union of India,[71] and distinguished that decision on the ground that the cases involving a breach of contract by the government stood on a different footing. In P.M. Paul's case it was pointed out that once it was held that the arbitrator had jurisdiction to find that there was a delay in the execution of the contract due to the respondent's conduct, the respondent was liable for the consequences of the delay, namely, an increase in prices. The award on escalation charges was held to be incidental to the finding of breach of contractual obligation by the government. Therefore, it was held that the arbitrator has jurisdiction to go into the question.

In the case of Ch. Ramalinga Reddy v. Superintending Engineer,[72] the Supreme Court approved the view taken by the division bench and rejected the claim for escalation. The Supreme Court considered the claim for payment of extra rates for the work executed beyond the agreed time at the schedule of rates prevailing at the time of execution and held that Clause 59 of the APDSS made the claim impermissible. It was observed that the claim made in that case fell outside the defined exceptions in Clause 59. P.M. Paul's case above was distinguished on the following two grounds: "There was in P.M. Paul's case no clause in the contract which provided that the respondent would not be liable to pay compensation on account of delay in the work from any cause nor was it stipulated, when extension of time was granted to the appellant to complete the work, that no claim for compensation would lie." Then, it was observed that the jurisdiction of the arbitrator was limited by the terms of the contract, and that where the contract plainly barred the appellant from making any claim, it was impermissible to make an award in respect thereof and the Court was entitled to intervene.

The Andhra Pradesh High Court posed the following question in a case involving different factual matrix:[73]

9. The question is whether in the light of the aforementioned decisions, especially the latest decision of the Supreme Court, Clause 59 precludes claim No. 6 being raised and allowed by the arbitrator and whether the arbitrator can be said to have committed an error of jurisdiction or a patent error of law in granting extra rate based on Standard Schedule of Rates in force during the relevant year i.e., 1985–86. Before dealing with the question of applicability of Ramalinga Reddy's case (1994 (5) Scale 67) (SC) (supra) to the facts of the present case, we would like to advert to what the arbitrator has said vis-a-vis Clause 59. The arbitrator drawing support from the decision of this Court in V. Raghunadha Rao v. State of A. P.[74] opined that clause 59 was totally inequitable and unreasonable. The learned arbitrator observed, "there ought to be some sense of proportion; the contract period of one year cannot be extended for two more years without paying any compensation for the extra cost involved." As the execution of the contract within the time-limit stipulated was clearly frustrated by a fundamental breach or failure on the part of the Department, Clause 59 cannot be put against the contractor. The arbitrator further commented, "payments were made at the increased rates for the work done in 1984 season after accepting a supplemental agreement. By the same argument, the respondent ought to pay at further increased rates based on the Standard Schedule of Rates (SSR) of 1985–86 for the work done after May, 1985 since water was again released in July, 1984 irregularly without giving adequate time for the completion of the work. The argument of the respondent invoking Clause 59 is invalid in the light of his own acceptance of the

---

[70] (1989) 2 APLJ 362: AIR 1990 NOC 90.

[71] AIR 1989 SC 1034.

[72] Ch. Ramalinga Reddy v. Superintending Engineer, (1994 (5) Scale 67) (SC).

[73] Govt. of A.P. and others v. V. Satyam Rao, AIR 1996 A.P. 28; C.R.P. No. 2460 of 1993 and C.M.A. No. 891 of 1993, D/- 8 - 12 - 1995.

[74] V. Raghunadha Rao v. State of A. P., (1988)1 Andh LT 461.

breach in contract necessitating payment at increased rate and the claim as such cannot be resisted.

It was further observed and held:

It is an undisputed fact that the appellants themselves made extra payment based on the Standard Schedule of Rates of 1984-85 for the work done in 1984 season pursuant to a supplemental agreement entered into between the parties. ... The appellants must, therefore, be deemed to have waived their right to insist on the adherence to Clause 59 which purports to exclude the liability to pay compensation in the circumstances specified therein. ... We, therefore, see no-error of jurisdiction or an error of law apparent on the face of the award in allowing claim No. 6A partly.

Similar to the above case, the courts will have to distinguish the decisions of the Supreme Court in Continental Construction Company v. State of Madhya Pradesh[75] and Associated Engineering Co. v. Government of Andhra Pradesh,[76] which are likely to be cited in support of exemption clauses.

Before concluding the discussion on the interpretation of Clause 59, it is necessary to have a look at the provisions found in the construction contracts containing exemption clauses and answer the question: is construction contracting gambling or science? These provisions, to cite a few by way of examples, include:

No compensation for alteration in or restriction of work to be carried out.

The Contractor will have to proceed with the work as and when drawings are released. At times it may be necessary for the contractor to retard his work. The contractor shall have absolutely no claim ... on this account and will not be entitled to any compensation whatsoever on account of delay in release or issue of drawings ....

The unit rates quoted shall be firm and shall remain applicable during the entire period of execution of work up to the completion and no escalation in rates will be permitted due to increase in prices of materials, rise in labour wages, railway freight or due to any other reasons.

If, for some reason, the purchaser is unable at any time to issue materials as mentioned hereafter, and the work has to be stopped for want of materials, the Contractor shall not be entitled to any monetary claim arising out of such circumstances.[77]

There is no scientific basis to estimate with a reasonable degree of accuracy the monetary effect of the above clauses; in the absence of definite answers to questions like: how much curtailment in the scope, what extension of time may be involved, how much will be the price rise during the extended period and so on. These uncertainties mean that what is to be included in the cost of the works to be tendered is more of a gamble with the risk of even the government's breaches of contract being taken by the contractors. The facts of the above cases and the decisions thereon suggest that if justice is to be rendered to the parties, the interpretation put on Clause 59 of APDSS by the Hon'ble Supreme Court of India deserves to be reviewed, it is respectfully submitted. There are many decisions of the Supreme Court which can be said to be directly applicable for such review.

It is heartening to note that the courts have developed new doctrines, which seem to offer some escape from even the most carefully drafted exemption clauses. These doctrines are considered as follows:

---

[75] Continental Construction Company v. State of Madhya Pradesh, AIR 1988 SC 1166.

[76] Associated Engineering Co. v. Government of Andhra Pradesh, AIR 1992 SC 232.

[77] U.P. State Electricity Board v. M/s. Om Metals and Minerals (Pvt.) Ltd., AIR 1995 All 246, paragraph 9.

4.10.2 Unfair and Unreasonable Contract and Interpretation of Exemption Clauses

Many building and engineering contracts are standard form contracts entered into with the state or large corporations on a "take it or leave it" basis. These do not afford the contractors a realistic opportunity to bargain or negotiate the terms of the contract. These contracts are called contracts of adhesion and the principles of interpretation of such contracts are based on the fact that the weaker party to such contracts did not have much choice but to accept one-sided and seemingly unfair terms of such contracts. Such unfair clauses generally seek to exempt a party from liability to compensate the other under the law, or to limit the liability thereunder, thus giving such clauses the names exemption or exclusion or limitation clauses.

One rule of construction of exemption clauses propounded by Lord Denning, M. R. is that the court would not allow a party to a contract to exempt himself from his liability at common law when it would be unconscionable for him to do so.[78] In England, the Unfair Contract Terms Act, 1977, governs the interpretation of exemption clauses. This Act applies to contracts entered into after 1st February 1978. In respect of contracts made on or after 1st October 2015, the 1977 Act does not offer protection to consumers who will be protected by the Consumer Rights Act 2015. To the extent that the Unfair Contract Terms Act 1977 applies, it lays down guidelines and limitations over the use of exemption clauses in contracts in relation to liability in both tort and contracts. English law does require a strict interpretation of an exemption clause.[79] If it is drafted with clarity, an exemption clause may be upheld even in relation to a breach of a fundamental term of a contract,[80] but not in the case of a fraud by the person seeking to rely on the exemption clause.[81]

The Supreme Court of India has laid down a principle of far-reaching effect in a landmark case decided in 1986. Although the case relates to a service contract, the legal principle recognized and acted upon by the Supreme Court is of general application. By reference to the development in the laws of several countries with respect to the exemption clauses, the Supreme Court deduced the following principle:

> This principle is that the courts will not enforce and will, when called upon to do so, strike down an unfair and unreasonable contract, entered into between parties who are not equal in bargaining power. It is difficult to give an exhaustive list of all bargains of this type. ... One can only attempt to give some illustrations. For instance, the above principle will apply where the inequality of bargaining power is the result of the great disparity in the economic strength of the contracting parties. It will apply where the inequality is the result of circumstances whether of the creation of the parties or not, it will apply to situations in which the weaker party is in a position in which he can obtain goods or services, or means of livelihood only upon the terms imposed by the stronger party or go without them. *It will also apply where a man has no choice, or rather no meaningful choice, but to give his assent to a contract or to sign on the dotted line in a prescribed or standard form or to accept a set of rules as part of the contract, however unfair, unreasonable and unconscionable a clause in that contract or form or rules may be.* ... The court must judge each case on its own facts and circumstances.[82]

> (Emphasis supplied)

The Supreme Court also laid down that the principle might not apply in the following circumstances:

---

[78] Gillespie Brothers & Co. Ltd. v. Roy Bowles Transport Ltd., (1973) Q.B. 400.

[79] Glynn v. Margetson & Co., [1893] AC 351; Canada Steamship Lines Ltd. v. King, The [1952] AC 192.

[80] Photo Production Ltd. v. Securicor Transport Ltd., [1980] AC 827; George Mitchell (Chesterhall) Ltd. v. Finnery Lock Seeds Ltd., [1981] 1 Lloyd's Rep. 476.

[81] S. Pearson & Son Ltd. v. Dublin Corp., [1907] AC 351.

[82] Central Inland Water Transport Corp. Ltd. v. Brojo Nath, AIR 1986 SC 1571 (1610–11); followed in National Insurance Co. Ltd. v. M/s Boghara Polyfab Pvt. Ltd., AIR 2009 SC 170, (182).

a. Where the bargaining power of the contracting parties is equal or almost equal; or

b. Where both parties are businessmen and the contract is a commercial transaction.

It is significant to note that the Supreme Court recognized the fact that in today's complex world of giant corporations with their vast infrastructural organizations and with the state through its instrumentalities and agencies entering into contracts in almost every branch of industry and commerce, there can be myriad situations which may result in unfair and unconscionable bargains between parties possessing wholly disproportionate and unequal bargaining power.

The Supreme Court has further elucidated the principle by observing as follows:

> The types of contracts to which the principle formulated by us ... applies are not contracts which are tainted with illegality but are contracts which contain terms which are so unfair and unreasonable that they shock the conscience of the court. They are opposed to public policy and require to be adjudged void.[83]

The Supreme Court has applied Section 23 of the Indian Contract Act while declaring such a contract void. Sections 23 and 24, which are self-explanatory read as follows:

> 23. *What considerations and objects are lawful, and what not*
> The consideration or object of an agreement is lawful, unless it – is forbidden by law; or is of such nature that, if permitted, it would defeat the provisions of any law; or is fraudulent; or involves or implies injury to the person or property of another; or the Court regards it as immoral or opposed to public policy.

> 24. *Agreement void, if considerations and objects unlawful in part*
> If any part of a single consideration for one or more objects, or any one or any part of any one of several considerations for a single object, is unlawful, the agreement is void.

As an example, a printed condition on the reverse of laundry receipts issued by the proprietor of a laundry purported to restrict his liability for quantum of loss to 20 times the laundering charges or half of the value of unreturned articles, whichever was less. It was held that the condition was, *ex facie*, opposed to public policy and fundamental principles of law of contract. Such a stipulation could not constitute a valid defence in an action by the bailor against the bailee.[84]

4.10.3 Forum Selection Clauses – When May Not Be Enforced?

Forum selection clauses purport to oust the jurisdiction of otherwise competent courts in favour of a foreign jurisdiction. It comes under the wider category of exemption clauses. To balance parties' contractual freedom with public good in having local courts adjudicate certain claims, courts have developed a test to determine whether such clauses should be enforced. In common law provinces, a forum selection clause cannot bind a court or interfere with a court's jurisdiction. As the English Court of Appeal recognized long ago, "no one by his private stipulation can oust these courts of their jurisdiction in a matter that properly belongs to them".[85] Instead, where no legislation overrides the clause, courts apply a two-step approach to determine whether to enforce a forum selection clause and stay an action brought contrary to it.

---

[83] AIR 1986 SC 1571.

[84] R.S. Deboo v. Dr. M.V. Hindlekar, AIR 1995 Bom. 68; Vinayakappa v. Dulichand, AIR 1986 Bom. 193.

[85] The Fehmarn, Cargo Lately Laden on Board the Fehmarn (Owners) v. Fehmarn (Owners); [1958] 1 All E.R. 333, at p. 335. [1958] 1 WLR 159.

Two-Step Approach

At the first step, the party seeking a stay based on the forum selection clause must establish that the clause is valid, clear and enforceable, and that it applies to the cause of action before the court.[86] Once the party seeking the stay establishes the validity of the forum selection clause, in the second step the onus shifts to the plaintiff. At this second step of the test, the plaintiff must show strong reasons why the court should not enforce the forum selection clause. In exercising its discretion at this step, a court must take into consideration all the relevant circumstances, including the convenience of the parties and fairness. For example, unequal bargaining power of the parties and the rights that a consumer relinquishes under the contract, without any opportunity to negotiate, may provide compelling reasons for a court to exercise its discretion to deny a stay of proceedings, depending on the other circumstances of the case.[87] The authors are in respectful agreement with the law laid down in these Canadian cases.

4.10.4 What Is an Unconscionable Contract?

An unconscionable contract is such an agreement as no reasonable and prudent man would make. Mere pecuniary inadequacy of consideration will not generally make the terms of a contract seem too unfair for enforcement unless the degree of inadequacy is extreme. The inadequacy must be so extreme as to call for application of equity. In construing such a contract, one must start with the presumption that neither party intends to abandon any remedies for its breach arising by operation of law, unless clear express words are used in order to rebut this presumption.[88] In other words, "one cannot say that the parties can, in a contract, have contemplated that the clause should have so wide an ambit as in effect to deprive one party's stipulations of all contractual force. To do so would be to reduce the contract to a mere declaration of intent. Thus, it may be correct to say that there is a rule of law against the application of an exception clause to a particular type of breach."[89]

EXAMPLE FROM CASE LAW

An agreement for the extraction of mineral sand contained a provision that the contractor could not seek relief in payment on the basis of non-extraction of sand. The contract further provided that the contractor had to offer to pay compensation to an occupier or owner of land wherefrom minor mineral was to be extracted and had to approach the Mining Engineer to get the amount of compensation determined by the Collector in case the occupier refused his consent. The contractor had to and did approach the Collector for getting the compensation fixed. The Collector failed to do so. Sand could not be extracted for even one day. The agreement was terminated for failure to pay the contract money. The security deposit of over Rs. two crores was forfeited. An appeal was partly allowed by an appellate authority converting the security money into contract money and seeking the balance to be paid by the contractor. A writ petition filed in the High Court was partly allowed by reducing the contract money. On appeal, the Supreme Court observed that the High Court had committed an error in not going into the principal issue involved in the matter namely:

---

[86] See, Preymann v. Ayus Technology Corp., 2012 BCCA 30 and The Eleftheria, [1969] 2 WLR 1073.

[87] Deborah Louise Douez v. Facebook, Inc. & Another, CDJ 2017 Canada 011 Supreme Court of Canada Case No. 36616. Also see: Straus v. Decaire, 2007 ONCA 854, at para. 5 (CanLII). See Expedition Helicopters Inc. v. Honeywell Inc., 2010 if th ONCA 351, 100 O.R. (3d) 241, at para. 24; Stubbs v. ATS Applied Tech Systems Inc., 2010 ONCA 879, 272 O.A.C. 386, at para. 58.

[88] Modern Engineering (Bristol) Ltd. v. C. Miskin & Son Ltd., [1981] 1 Lloyd's Rep. 135.

[89] Suisse Atlantique Societe d'Armement SA v. NV Rotterdamsche Kolen Centrale [1967] 1 AC 361,432. Also see AstraZeneca UK Ltd. v. Albemarle International Corporation, [2011] 2. C.L.C. 252.

"... whether Clause 18A of the agreement would remain enforceable despite the fact that the appellant allegedly could not extract any sand by reasons of omission and commission on the part of the concerned respondents." Similarly the appellant's plea that the contract became impossible of performance was not considered. The High Court had merely proceeded on the basis of the appellant having entered into contract with eyes wide open. It was observed: "... but, same would not, in our opinion, mean that they were bound to pay the contract amount, get its security amount forfeited, as also pay interest at the rate of 24 per cent, although it could not, by reason of acts of omission and commission on the part of the respondents, carry out the mining operation as per the terms of the agreement." It was further observed: "Whether in such a situation the doctrine of frustration will be invoked or not should have been considered by the High Court." In result, the High Court judgment was set aside and the matter was remitted back to the High Court for consideration of the matter afresh in the light of the observations made.[90]

The exclusion clauses included and interpreted in the examples in para. 4.10.5 below commence with the words: "*Neither party shall be liable ...*" which can be said to make the clause somewhat fair and it will be interpreted as per its terms. It may not be declared as easily or wholly invalid as a clause which excludes the liability of only one party to the contract but not the other.

4.10.5 Consequential Damages

Exemption clause from payment of damages will not exempt from liability to pay direct damages but only consequential damages. A clause excluding "responsibility for consequential damages" was held by the Court of Appeal not to exclude liability for damage occurring naturally or directly.[91]

In a construction contract both the parties are aware that the cost agreed is worked out by ascertaining the cost of five elements namely, materials, labour, machinery, overheads and profit, and that the total cost is dependent on the time allowed for completion. If a breach is committed by the employer resulting in delay, not only the cost of first four heads would escalate resulting in actual loss, but the contractor would also suffer from the loss of profit expected from the deployment of resources for the extended stay at site, and it is a direct loss. It needs to be remembered that profit is the only remuneration the contractor gets for his stay at the work site. Compensation on the first four heads would be reimbursement for actual out of pocket expenses incurred. Whether loss of profit is a direct loss or an indirect one may be controversial in relation to other business contracts such as supply of goods, machinery or technology contracts but not for construction contracts unless the contract expressly so stipulates. For example, under FIDIC Red Book, Clause 1.15, contract loss that can be recovered is direct loss and expenses falling under the definition of cost, which includes all expenses including escalation due to inflation, exchange rate, overheads, loss of productivity, etc., but it excludes profit. The provision for compensation mentions amount of such costs, which shall be added to the contract price, calls for reference to definition of cost. It is suggested that to be fair, the FIDIC provision needs to be amended to include loss of expected profit, for the reason stated hereinabove.

The exclusion clauses can be differently interpreted to include or exclude profit as seen from the following examples.

---

[90] Jay Durga Finvest Pvt. Ltd. v. State of Haryana, AIR 2004 SC 1484.

[91] Millar's Machinery Co. Ltd. v. David Way and Son, [1934] 1 WLUK 6; (1935) 40 Com. Cas 204; also see Saint Line v. Richardsons Westgarth & Co. Ltd., [1940] 2 K.B. 99.

EXAMPLES FROM CASE LAW

1. An exclusion clause in a subcontract read: *"Neither party shall be liable to the other under this sub contract for loss of profits, revenue, business goodwill, indirect consequential loss of damage."* The court had to decide if this clause excluded the defendant's liability for all loss of profits that is direct and indirect or for only indirect loss of profit. It was held that the clause covered all loss of profits. It was observed that if the parties intended to exclude indirect profit only, the parties were expected to make this clear.[92] This interpretation was not commercially unreasonable or inconsistent with business sense.

2. In a contract based on a minimum spend, the plaintiff Polypearl claimed loss of profit on the shortfall. The contract included a clause reading:

> Neither party will be liable to the other for any indirect or consequential loss (both of which include, without limitation, pure economic loss, loss of profit, loss of business, depletion of goodwill and like loss) howsoever caused (including as a result of negligence) under this Agreement, except in so far as it relates to personal injury or death caused by negligence.

The court noted that words in the clause made the meaning of the clause ambiguous. It was held that the clause excluded liability for indirect/consequential loss of profits, but not direct loss of profit.[93] Loss of profits, it was rightly noted, could rise as a direct or an indirect loss. It was held the most likely and often the only damage that Polypearl would suffer to meet the minimum spend commitment would be a loss of profit. It was unlikely that a business person would exclude this direct loss.

In addition to the above doctrines of interpretation there is one more doctrine that is relevant to consider in this *chapter, namely the doctrine of fundamental breach of contract.*

4.10.6 Fundamental Breach of Contract

There are in every contract certain terms, which are fundamental, the breach of which amounts to complete non-performance of the contract. A fundamental term was conceived to be something more basic than a warranty or even a condition. It formed the "core" of the contract and therefore could not be affected by any exemption clause. In other words, no party to a contract could exempt itself from responsibility for a fundamental breach. Though the limits of this doctrine were not precisely defined it was said that a party could claim the protection of an exemption clause "when he is carrying out his contract, not when he is deviating from it or is guilty of breach which goes to the root of it".[94]

In construction contracts, for the contractor's obligation to complete the work, the consideration payable by the owner is the cost of the work, either lump sum or on the basis of agreed rates. A time-cost graph, if plotted, shows that the cost of the work changes with the time for its completion. As such, it can be presumed that the sum or rates tendered by the contractor bear a definite relationship with the time limit for completion agreed to by the parties. Though it has been held by the Supreme Court of India that the time is not of the essence,[95] an injured party's right to claim damages due to delay is not affected by the said decision. In cases where the delay is abnormal, the agreed sum or rates become an inadequate consideration. In such cases, to get

---

[92] Fujitsu v. IBM UK Ltd. [2014] EWHC 752 (TCC).

[93] Polypearl Limited v. E.ON. Energy Solutions Limited, [2014] EWHC 3045 (QB); [2014] 10 WLUK 62.

[94] Spurling (3) Ltd. v. Bradshaw, (1956), WLR 461 per Denning L.J. at p. 465.

[95] Hind Construction Contractors v. State of Maharashtra, AIR 1979 SC 720.

justice, this doctrine can be pressed into service. The exemption clauses would not be operable in cases where the delay is unreasonable, it is submitted.

EXAMPLE FROM CASE LAW

The work was to be completed within a period of one year, but due to financial difficulties, a smaller budget having been provided in the said year, the Department requested the contractor to spread the work over two more years, i.e. to complete the same in three years. In the same case besides a 20% increase in the rates, the contractor also claimed interest on the balance amount due to him which the Department had wrongly retained. The contract contained an exemption clause worded as follows: "The contractor shall not be entitled to interest upon any payments in arrears or upon any balance which may on final settlement be found due to him". The following observations of the Supreme Court of India clearly lay down the doctrine of fundamental breach, it is respectfully submitted. Their Lordships observed:

> In our view the reliance on this clause is of no avail to the appellant for the simple reason that this clause will be applicable *provided the work was completed according to the specifications and time-schedule fixed in the original contract.* Moreover in the instant case the plaintiff had issued notice claiming interest under the Interest Act and the High Court, has, in modification of the trial Court's decree awarded interest from the date of notice till payment. The claim for interest, therefore, was rightly allowed.[96]

(Emphasis supplied)

Thus, where excessive delay is caused on account of a failure on the part of the owner to fulfil one or more of the following obligations, this doctrine may be useful to obtain justice for the contractor.

i. Handing over the site of work in full within a reasonable time.

ii. Supplying working drawings and designs, which are the basis for execution of the work by the contractor.

iii. Supplying materials such as cement, steel, if agreed to be supplied by the owner and such materials are essential ingredients and without which substantial or essential parts of the work cannot be carried out.

iv. Supply of special equipment or release of foreign exchange to purchase some machinery or equipment and the work cannot be carried out without the machinery or the equipment.

v. Material changes in the designs or specifications of the work from those originally agreed upon, made very late so that the work could not be completed within a reasonable extended period also.

vi. Any other similar default which causes a delay in completion of the work.

A Word of Caution

A word of caution is necessary for those who enthusiastically invoke this doctrine. The doctrine of fundamental breach is a rule of construction. There is no rule of substantive law to the effect that a party cannot contract out of or limit his liability for a fundamental breach. There is a new approach to this in English law. On a question of principle, the House of Lords appears to conclude that there is no

---

[96] Hyderabad Municipal Corporation v. M. Krishnaswami, AIR 1985 SC 607.

difference in operation between a fundamental breach and a breach of a condition. The fundamental breach, simply, gives to the innocent party an option to affirm or repudiate the contract. The innocent party, when faced with a fundamental breach, can either repudiate on the spot or he must affirm.[97] If the innocent party affirms the contract, the proposition is that the whole contract, including the exclusion clause, continues to apply and continues to govern the legal relationship between the parties.

It follows, therefore, that in a case where the contract is affirmed and yet the exclusion clause is still held to be inapplicable, this is because the exclusion clause, on its proper construction, does not cover the breach in question. There may be circumstances wherein the exclusion clause covers the breach, and yet, it may be held incapable because it may be found to be unreasonable. The Patna High Court held a clause that exempted the Railway authority from its liability to pay interest in *any* situation as arbitrary, unjust, unreasonable and violative of Arts. 14 and 19 of the Constitution.[98] The test of reasonableness will be governed by the provision of Sections 3 and 11 of the Unfair Contract Terms Act, 1977, in England. Section 3 of the Unfair Contract Terms Act 1977 reads as follows:

(1) this section applies as between contracting parties where one of them deals[99] on the other's written standard terms of business.

(2) As against that party, the other party cannot by reference to any contract term –

    (a) when himself in breach of contract, exclude or restrict any liability of his in respect of the breach; or

    (b) claim to be entitled –

        (i) to render a contractual performance substantially different from that which has reasonably expected of him, or

        (ii) in respect of the whole or any part of his contractual obligation, to render no performance at all, expect in so far as (in any of the cases mentioned above in this sub-section) the contract term satisfies the requirement of reasonableness."

It is not that the doctrine is dead and buried by the decision of the House of Lords. The Unfair Contract Terms Act, 1977 has made a difference to its application. In India, the Supreme Court would continue to follow the trend set in the absence of amendment of the Indian Contract Act on the lines of the Unfair Contract Terms Act 1977. In India, Section 23 of the Indian Contract Act applies. There are circumstances, the principle notwithstanding, where an exclusion clause on its construction does not apply. They include:

    i. Where the clause is so worked as to be held not wide enough to cover the kind of breach committed;

    ii. Where the terms of the clause are so wide that it is unreasonable to apply them literally;

    iii. Where the clause leads to absurdity;

    iv. Where there is a "strong though rebuttable presumption" that the parties did not intend the clause to cover fundamental breach.[100]

There is an advantage in pleading a breach of a fundamental term. The burden lies on the party inserting the exclusion clause in the agreement to prove that the loss was not due to fundamental breach, if the innocent party has specifically pleaded a fundamental breach.[101]

---

[97] Photo Production Ltd. v. Securicor Transport Ltd. [1980] A.C. 827.

[98] Hari Om Apartment v. East Central Railway, AIR 2010 (NOC) 185 (PAT.).

[99] The words "as consumer or" deleted by the Consumer Rights Act 2015.

[100] Suisse Atlantique Societe d'Armement SA v. NV Rotterdamsche Kolen Centrale [1967] 1 AC 361.

[101] Woolmer v. Delmer Price Ltd. (1955) 1 QB 291; J. Spuring Ltd. v. Bradshaw [1956] 1 WLR 461 (466).

4.10.7 The Doctrine of Implied Terms

When one party alleges a breach of contract in respect of the breaches mentioned in paragraph 4.13 above, the potential response from the respondent may include a contention that the contract does not include any terms such as:

1. Full and complete possession of the site will be handed over within a reasonable time, or
2. Working drawings will be supplied at proper time or times, or
3. Material will be supplied as and when needed/demanded by the contractor, or
4. Change orders will be given at such time as not to affect the planned progress of the contractor's work, etc. and therefore, there is no breach committed.

The very fact that the contract provides for completion of the work within the stipulated period, which is one of the factors on which the contractor's rates are based, the above or similar terms which reason and justice demand, require to be written in the contract. Therefore, though there is no express term in the contract, the law itself, which means the court/arbitrator, will imply a necessary term. This is the doctrine of "implied terms".

The doctrine of "implied terms" does not involve asking whether the parties implicitly agreed on a term; courts recognize that they never agreed on it at all because they never envisaged that a particular situation would arise. In such cases, courts seek to find the parties' presumed intent, that is, what they presumably would have agreed if they had envisaged the situation. A court presumes that the parties would have agreed upon a fair and reasonable solution and it then declares what that fair and reasonable solution is. This whole process is part of construction of contracts. The court construes the contract so as to give effect to presumed intent. The doctrine can be useful in rendering justice in cases of exemption clauses also. Lord Denning, on page 46 of *The Discipline of Law*, observed[102]:

> It is important to notice that, in order to decide whether the exemption or limitation clause applies, you must construe the contract, not in the grammatical or literal sense, or even in the natural and ordinary meaning of the words but in the wider context of the "presumed intention" of the parties – so as to see whether or not, in the situation that has arisen the parties can reasonably be supposed to have intended that the party in breach should be able to avail himself of the exemption or limitation clause. ... In other words, in order to ascertain the "presumed intention" of the parties, you must ask this question: "If the parties had envisaged the situation which has happened, would they, as reasonable persons, have supposed that the exemption or limitation clause would apply to protect the wrongdoer?" The spokesman of the reasonable man is, and must be the court or the arbitrator.

The current position of English law, which has changed since the above words of Lord Denning, is summarized in the following manner by Lord Hoffman:

- A court has no power to improve upon the document which it is called upon to construe, whether it be a contract, a statute or articles of association of a company.
- A court cannot introduce terms to make the contract fairer or more reasonable.
- Court's concern is only to discover what the document means.
- The meaning it seeks is not necessarily or always what the authors or parties would have intended, but rather it is the meaning which the document would convey to a reasonable

---

[102] Lord Denning, *The Discipline of Law*, London, Butterworths, 1979.

person having all the background knowledge which would reasonably be available to the persons to whom the instrument is addressed.[103]

It is thus an objective meaning which an English court would seek when construing a contract which is silent on a particular scenario. This is what, Lord Hoffman acknowledged, used to be called the intention of the parties, or the intention of Parliament, or the intention of whatever person was the author of the document. In substance, what the court or arbitrator will arrive at should be very similar, whether the traditional or more modern English approach is used.

EXAMPLE FROM CASE LAW

The Supreme Court of India had to deal with the challenge to an arbitral award on the ground that the arbitrators could not have gone into the question of construction of the contract. Briefly stated the facts involved were as follows: A contract for the supply of helium diving gas was awarded on the basis of a global tender invitation. Three different categories of rates were to be quoted by the tenderers, both foreign and Indian. The appellant before the Supreme Court of India had opted for a bid quoting Indian price with a foreign exchange component. The appellant's bid was found to be the lowest in that the appellant had bid a price of Rs. 150/- per cubic metre out of which US$0.5 was to be the foreign exchange component. On increase in the price of the US dollar, the appellant claimed the difference in price of US dollar as on the date of contract and the date of supply. One of the conditions of the contract read: "Bidder shall quote firm price without any escalation for any ground whatsoever until they complete the work against this tender or any extension thereof." The claim was rejected by the authority. In the ensuing arbitration proceedings, the arbitrator gave a non-speaking award in favour of the appellants over the exchange rate fluctuations. The division bench of the High Court of Bombay set aside the award holding that the award was without jurisdiction. On appeal the Supreme Court of India held:

> It is trite that the terms of the contract can be express or implied. ... The appellant quoted the foreign exchange component in its bids in terms of the notice inviting tenders. The same was asked for by the respondent itself for a definite purpose. A contract between the parties must be construed keeping in view the fact that the fluctuation in the rate of dollar was required to be kept in mind by the respondent having regard to the fact that the tender was global in nature and in the event the respondent was required to pay in foreign currency, the same would have an impact on the cost factor ... The arbitrators were called upon to determine a legal issue, which included interpretation of the contract. The arbitrators, therefore, cannot be said to have travelled beyond jurisdiction in making the award. Award passed by arbitrator is sustainable.[104]

4.10.8 Ejusdem Generis

The *ejusdem generis* rule is usually called upon with reference to construction of statutes, namely, that where several words preceding the general word point to a narrow meaning, the general word shall not extend in its effect beyond subjects *ejusdem generis* (of the same class). It applies to the construction of contracts as well. According to the Supreme Court, "[t]he true scope of the rule of *ejusdem generis* is that words of a general nature following specific and particular words

---

[103] Attorney General of Belize v. Belize Telecom Ltd., [2009] UKPC 10; [2009] 1 WLR 1988.

[104] Pure Helium India Pvt. Ltd. v. Oil & Natural Gas Commission, AIR 2003 SC 4519 = 2003 (8) SCC 593.

should be construed as limited to things which are of the same nature as those specified and not its reverse, that specific words which precede are controlled by the general words which follow."[105]

In construction contracts, for example, a provision is made exempting the owner from his liability to pay compensation to the contractor on account of delay in the supply of materials, etc., where such delay is caused by "force majeure, act of God, act of enemies of the Republic of India or any other reasonable cause". This provision will attract application of this rule for interpreting the scope and nature of "any other reasonable cause". This rule is an important aid in interpretation of contracts and can be useful in a number of situations as follows.

EXAMPLES FROM CASE LAW

1. A construction contract contained a commonly found *force majeure* clause providing that neither party would be liable to the other for any loss or damage occasioned by or arising out of an act of God, such as unprecedented rains, etc. The contractor claimed compensation on account of loss caused due to flooding of the work area. The arbitrator allowed the claim holding that the flood was not caused by unprecedented rain and the award was upheld by the High Court. The Supreme Court dismissed the appeal on the ground that no evidence was produced by the appellant state to prove that the damage was caused by unprecedented rains.[106]

2. The defendants had pledged their stock of food grains with a bank to obtain loans from time to time. In reply to the demand of the bank to liquidate the accounts, the defendants repeatedly requested the bank to sell the stocks and release the outstanding balance. The bank did nothing and the stock became worthless. The trial court dismissed the suit filed by the bank for the recovery of dues with interest on the ground that the bank was not in a position to redeliver the goods pledged. On the first appeal to the High Court the bank submitted that the dismissal of the suit was illegal because Clause 9 of the agreement granted total exemption to the bank from liability in the event of loss, deterioration or damage to the pledged goods "whether caused by theft, fire, rain, flood, earthquake, lightning or any other cause whatever". A number of cases were cited in support of the proposition that it is open to a bailee to contract himself out of any liability due to his own negligence. The Madhya Pradesh High Court raised the question whether the bank had contracted out of its liability due to the negligence of its servants. Clause 9 of the pledge agreement exonerated the bank from any liability for the loss, deterioration or damage whether caused by theft, fire, rain, flood, earthquake, lightning or any other cause whatever. This clause nowhere exempted the bank from liability for the negligence of its servants. The cases enumerated in Clause 9 were all natural causes without human intervention and which could not be prevented by any amount of foresight or care. The contract also contemplated loss of goods on account of theft, an act of a third party. Relying on Cooch Bihar Commercial Co. v. Union of India,[107] the bank, the bailee in this case, was held liable for the damage caused by the negligence of his servants.[108]

---

[105] Amar Singji v. State of Rajasthan, AIR 1955 SC 504.

[106] State of U.P. v. Allied Constructions, AIR 2004 SC 586; 2003 (7) SCC 396; 2003 (3) Arb.LR 106. AIR 2003 SCW 6075.

[107] Cooch Bihar Commercial Co. v. Union of India, AIR 1960 Cal 455.

[108] Central Bank of India v. M/s Grains and Gunny Agencies, AIR 1989 M.P. 28.

In English law, the principle is not a rigid rule of construction but it is used as a flexible aid to construction and reflects the current English focus on commercial common sense and the need to construe the contract as a whole in its commercial context. Ward, a development company, purchased a property from Burrows, an investment company, in 2007. Ward was permitted to dispose of or transfer part of the land under Clause 1.1(c) of the sale agreement in the following words: "the transfer/dedication/lease of land for the site of an electricity sub-station, gas governor kiosk, sewage pumping station and the like, or for roads, footpaths, public open spaces or other social/community purposes." Ward sold five completed buildings to a social housing provider. The sale agreement allowed buildings to be sold in open market; that sale was not negotiated in the open market. Ward relied on Clause 1.1(c) to suggest that the sale was for social or community purposes. The Court of Appeal concluded that it would have been very strange to describe the transfer of a completed dwelling as a "transfer of land", particularly when regarding the specific instances of transfers of land itemized in Clause 1.1(c). Land for the site of an electricity sub-station, gas governor kiosk or sewage pumping station, or for use as a road or footpath, or as a public open space, is unlikely to have any buildings on it at the date of transfer, and would certainly not have had a dwelling house on it. The words "or other social/community purposes" had to be read in the light of the specified purposes preceding them. This decision was made despite the Court of Appeal's agreement with the general statement that provision of affordable housing units to a registered social landlord clearly achieved both social and community purposes and that the national and local planning policy was in favour of encouraging such a provision.[109] Burrows sought compensation from Ward for the loss of the opportunity to negotiate a reasonable price for releasing Ward from its contractual obligations on the basis that the benefit of the contractual restriction was a potentially valuable property, and that Burrows was deprived of the opportunity to exploit it for what it was worth. The Court of Appeal agreed. Burrows alleged that the hypothetical negotiation could take into account the fact that it would in the event have suffered no actual loss, and that affordable housing policy was foreseen at the time of the sale agreement.

4.10.9 Doctrine of Estoppel

Estoppel is a rule of evidence. It is a personal disqualification laid upon a person peculiarly circumstanced from proving peculiar facts. It is embodied in Section 115 of the Indian Evidence Act. The section reads:

> When one person has, by his declaration, act or omission, intentionally caused or permitted another person to believe a thing to be true and to act upon such belief, neither he nor his representative shall be allowed, in any suit or proceeding between himself and such person or his representative, to deny the truth of that thing.

ILLUSTRATION

A intentionally and falsely leads B to believe that a certain piece of land belongs to A, and thereby induces B to buy it and pay for it.

The land afterwards becomes the property of A, and A seeks to set aside the sale on the ground that, at the time of the sale, he had no title. He must not be allowed to prove "his want of the title".

---

[109] Burrows Investments Limited v. Ward Homes Limited, [2017] EWCA Civ 1577; [2018] 1 P&CR 13.

Before a party is precluded from denying the truth of his declaration, act or omission under the section, all the three conditions mentioned below must be satisfied[110]:

(i) There must be a representation by a person to another;

(ii) The other shall have acted upon the said representation; and

(iii) Such action shall have been detrimental to the interests of the person to whom the representation had been made.

It is necessary to mention yet another problem situation. Many times promises are given or assurances are made to a contractor, either orally or in writing, which the government officers subsequently find difficulty in fulfilling. In such a situation, it should be checked whether promissory estoppel or estoppel is applicable.

4.10.10 The Doctrine of Promissory Estoppel

The Supreme Court has explained and summed up this doctrine in a couple of cases. The process of refinement of law in this branch commenced in Anglo Afghan Agency case, where the Supreme Court held:

> We are unable to accede to the contention that the executive necessity releases the Government from honouring its solemn promises relying on which citizens have acted to their detriment. Under our constitutional set-up, no person may be deprived of his right or liberty except in due course of and by authority of law: if a member of the executive seeks to deprive a citizen of his right or liberty otherwise than in exercise of power derived from the law – common or statute – the Courts will be competent to, and indeed would be bound to, protect the rights of the aggrieved citizen.[111]

In the M. P. Sugar Mill case, the Supreme Court observed:

> The law may, therefore, now be taken to be settled as a result of this decision, that where the Government makes a promise knowing or intending that it would be acted on by the promisee and, in fact, the promisee, acting in reliance on it, alters his position, the Government would be held bound by the promise and the promise would be enforceable against the Government, at the instance of the promisee, notwithstanding that there is no consideration for the promise and the promise is not recorded in the form of a formal contract as required by Art 299 of the Constitution. It is elementary that in a republic governed by the rule of law, no one, however high or low, is above the law. Everyone is subject to the law as fully and completely as any other and the Government is no exception. It is indeed the pride of constitutional democracy and rule of law that the Government stands on the same footing as a private individual so far as the obligation of the law is concerned. The former is equally bound as the latter. It is indeed difficult to see on what principle can the Government, committed to the rule of law, claim immunity from the doctrine of promissory estoppel? Can the Government say that it is under no obligation to act in a manner that is fair and just or that it is not bound by considerations of "honesty and good faith"? Why should the Government not be held to a high standard of rectitude while dealing with its citizens?[112]

In a subsequent decision the doctrine was summed up in these words:

---

[110] See Navayuga Exports Ltd. v. Mineral Development Corporation, AIR 1998 A.P. 391.

[111] Union of India v. M/s. Anglo Afghan Agencies, (SC) AIR 1968 SC 718; 1968(2) SCR 366.

[112] M. P. Sugar Mills v. State of U. P., AIR 1979 SC 621.

(1) "The plea is not available against the exercise of the legislative functions of the State.

(2) The doctrine cannot be invoked for preventing the Government from discharging its functions under the law.

(3) When the officer of the Government acts outside the scope of his authority, the plea of promissory estoppel is not available. The doctrine of ultra vires will come into operation and the Government cannot be held bound by the unauthorized acts of its officers.

(4) When the officer acts within the scope of his authority under a scheme and enters into an agreement and makes a representation and a person acting on that representation puts himself in a disadvantageous position, the Court is entitled to require the officer to act accordingly to the scheme and the agreement or representation. The officer cannot arbitrarily act on his mere whim and ignore his promise on some undefined and undisclosed grounds of necessity or change the conditions to the prejudice of the person who had acted upon such representation and put himself in a disadvantageous position.

(5) The officer would be justified in changing the terms of the agreement to the prejudice of the other party on special considerations such as difficult foreign exchange position or other matters which have a bearing on the general interest of the State."[113]

The examples below will help further clarify the above concept.

EXAMPLES FROM CASE LAW

MSTC, SAIL's agent, invited offers for the disposal of a wagonload of Cast Iron Skull Rolls (CIS Rolls), which the original purchaser did not take delivery of. The petitioner purchased the said quantity. After some deliveries were made, MSTC and SAIL neglected to make further deliveries of CIS Rolls and rejected steel rolls. Rather than honouring the commitment in terms of MSTC's sale order, further attempts were made to auction the contracted goods. On these grounds, the petitioner sought intervention of the Calcutta High Court under Art. 226 of the Constitution on the ground that the action of MSTC and SAIL was in breach of Art. 14 of the Constitution. The Calcutta High Court observed:

> The governmental agency must act in a commercial venture to earn more profit but that does not obviously clothe the governmental agency to affect persons who have already entered into transactions with the Government on the basis of the declared policy. Law Courts shall not permit affectation of an individual by reason of an attempt to be wiser after the event. It must act on the basis of a declared policy. ... Change of policy cannot affect the present transaction though it may have its due effect on the future transaction. ... While it is true that the doctrine of promissory estoppel may not be said to be strictly applicable in the fact and circumstances of the matter under consideration but the principles akin thereto cannot, however, altogether be ignored.

As a result the writ petition was allowed.[114]

---

[113] M/s Jit Ram Shiv Kumar v. State of Haryana, AIR 1980 SC 1285.
[114] Steel Crackers v. M.S.T.C., AIR 1992 Cal. 86 (91). Also see: Subhash Chand v. State of Haryana 2008(1) CTLJ 42 (P. & H.) (FB). Also see: Gujarat State Financial Corporation v. M/s. Lotus Hotels Pvt. Ltd., AIR 1983 SC 848; 1983(3) SCC 379.

Promissory Estoppel – When Not Attracted

The doctrine of promissory estoppel is primarily an equitable relief. The applicability of this doctrine depends upon the wording of the alleged promise made. The government acquired and handed over land to the Industrial Corporation, a public sector company for development of industry. The Industrial Corporation earmarked some area of focal point for residential purpose and allotted residential plots by draw after proper advertisement. The government, however, refused to allow change of land use from industrial to residential. It was stipulated in the application form for the allotment of the plots and also in the advertisement that acceptance of applications and earnest money would not put the Industrial Corporation under an obligation to allot a plot to the applicant. It was held that no promise, much less an enforceable promise, was made by the Industrial Corporation to the prospective applicants that by making an application pursuant to the advertisement and on being declared successful in the draw of lots, they would get residential plots.[115]

Once a contract is entered into by the parties, the parties are to act in terms of the contract only. The doctrine of promissory estoppel cannot have any application there. If there is any breach of the contract terms, the parties would be at liberty to exercise their remedies for the breach of the contract. The doctrine of promissory estoppel would not in any way help the complainant.[116] This is similar to the Lord Sumption's observation that the scope of estoppel cannot be so broad as to destroy the entire advantage of certainty provided in a contract which included a "no oral modification" clause.[117]

In English law, in order to establish a defence of promissory estoppel, the following elements need to be established:

1. Parties have a legal relationship giving rise to rights and duties to each other;
2. One party promises or represents (by words or conduct, but not inactivity/omission) to the other that he will not enforce against the other his strict legal rights arising out of that relationship;
3. The promisor intends that the promisee will rely on the representation;
4. Actual reliance by the promisee on that representation; and
5. It is inequitable for the promisor to go back on his promise.

Under English law, there is no requirement to show detriment by the promise; it is enough if the promisee has changed his position in reliance on the promise so that it would be unjust or inequitable to allow the promisor to act inconsistently with his promise. Of course, if no detriment can be shown, it is a relevant factor in the determination of the injustice or inequity of the situation. The formulation does leave scope for taking into consideration unjust enrichment by the promisor.

4.10.11 The Doctrine of Public Policy

Public policy connotes some matter which concerns public good and public interest. It indicates a general and well-settled opinion relating to man's duty to his fellow men. The concept of what is for the public good or in the public interest or what would be injurious or harmful to the public good or the public interest has varied from time to time.[118] The concept of public policy is, therefore, illusive,

---

[115] Jasbir Singh Chhabra v. State of Punjab (SC): 2010(2) JT 637: 2010(4) SCC 192: 2010(3) ICC 369: 2011(1) MLJ 607: 2010(2) Law Herald (SC) 1688: 2010(2) R.C.R.(Civil) 474: 2010(2) Recent Apex Judgments (R.A.J.) 380: 2010(sup) AIR (SC) 682: 2010 AIR (SCW) 1804: 2010(1) W.L.C. 481: 2010(2) Scale 754: 2010(1) L.A.R. 215: 2010(2) SLT 492: 2010(2) Apex Court Judgments (SC) 93.

[116] Kalpana Das v. Contai Co-op. Bank Ltd., AIR 2005 Cal. 95; also see: C.V. Enterprise (AIR 1984 Cal. 306).

[117] Rock Advertising Limited v. MWB Business Exchange Centres Limited, [2018] UKSC 24; [2018] 2 WLR 1603.

[118] See Paragraph 93 of Central Inland Water Transport Corporation Ltd. v. Brojo Nath, AIR 1986 SC 1571.

varying, and at times vague, unsatisfactory and uncertain, being capable of taking on different meanings. It has also been described as an "untrustworthy guide", "unruly horse", etc.[119] The question is who should decide it, or in other words should it be allowed to be a ground of judicial decision? As it is, it is the province of the judge to expound the law by sound reason and just inference; not to speculate upon what is the best, in his opinion, for the advantage of the community. Different Indian High Courts have had occasion to express their views on this concept in their judgments.[120] It is well settled that the term public policy is not capable of a precise definition. Whatever tends to injustice such as a restraint of liberty, commerce and natural or legal rights, whatever tends to the obstruction of justice or to the violation of a statute, and whatever is against good morals can be said to be against public policy. In a judicial sense, public policy does not mean simply sound policy, or good policy, but the policy of a state established for the public whether by law, by courts, or by general consent.[121]

The need for applying the touchstone of public policy was explained by Sir William Holdsworth in terms that a body of law like the common law, which has grown up gradually, necessarily acquires some fixed principles, and if it is to maintain these principles it must be able, on the ground of public policy or some other like ground, to suppress practices which, in ever new forms, seek to weaken or negate them.[122]

Since the doctrine of public policy is somewhat open-textured and flexible, two positions have emerged over it – a narrow and a broad view. According to the narrow view, courts cannot create new heads of public policy whereas the broad view countenances judicial law-making in some areas. The earlier trend of the decisions in India favoured the narrow view when the Supreme Court observed that

> though the heads are not closed and though theoretically it may be permissible to evolve a new head under exceptional circumstances of a changing world, it is admissible in the interest of stability of society not to make any attempt to discover new heads in these days.[123]

In later decisions, the Supreme Court has, however, leaned toward the broad view.[124]

Application to Construction Contracts

Since invariably construction contracts include arbitration as a mode of dispute resolution, the applicability of the doctrine of public policy frequently comes up in the challenges to arbitration awards. Where in Indian statutes the words "public policy" are used without indicating whether they refer to public policy of India or not, the words must be taken to refer to Indian public policy. The words "public policy" used in the Arbitration and Conciliation Act 1996 expressly refer to the public policy of India in respect of the recognition and enforcement of domestic and New York Convention awards. An award of an arbitral tribunal cannot be questioned on the ground that it is contrary to the public policy of another country such as the USA. This is similar to the view of the English Court of Appeal that it is for the English court enforcing an award to decide the matter of English public policy.[125]

---

[119] Gherulal Parakh v. Mahadeodas, AIR 1959 SC 781 (793).

[120] Bhagwant v. Gangabishan, AIR 1940 Bom. 369; Mafizuddin v. Habibuddin, AIR 1957 Cal. 336; Colapatri v. Colapatri, AIR 1964 Andh. Pra. 465; and Ratan Chand v. Ashkar, AIR 1976 Andh. Pra. 112.

[121] P. Rathinam/Nagbhusan Patnaik v. Union of India, AIR 1994 SC 1844; Clough v. Gardiner, 182 NYS 803, 806, 111 Misc. 244.

[122] Holdsworth, W., *A History of English Law*, (London: Methuen, 1952), Vol. III, p. 55.

[123] Gherulal Parakh v. Mahadeodas Maiya, 1959 Suppl. (2) SCR 392(440): AIR 1959 SC 781(795).

[124] See Murlidhar Agarwal v. State of U.P. (1975) 1 SCR 575 at p. 584: (AIR 1974 SC 1924 at p. 1930): Central Inland Water Transport Corporation v. Brojo Nath Ganguly (supra) (1986)(2) SCR 278 at p. 373: (AIR 1986 SC 1571 at p. 1612); Rattanchand Hira Chand v. Askar Nawaz Jung (1991)3 SCC 67 at pp. 76–77: (1991 AIR SCW 496 at pp. 502–503).

[125] Anatolie Stati and others v. The Republic of Kazakhstan, [2018] EWCA Civ 1896; [2018] 8 WLUK 107.

In the field of private international law, courts refuse to apply a rule of foreign law or recognize a foreign judgment or a foreign arbitral award if it is found to be contrary to the public policy of the country in which it is sought to be invoked or enforced.[126] A distinction is drawn while applying the said rule of public policy between matters governed by domestic law and a matter involving conflict of laws. The application of the doctrine of public policy in the field of conflict of laws is less likely when a purely municipal legal issue is involved.[127]

The approach of the American courts to the doctrine of public policy in its application to recognition and enforcement of foreign arbitral awards under the New York Convention is reflected in the decision of the US Court of Appeal for the Second Circuit which held that an expansive construction of public policy could vitiate the New York Convention's basic effort to remove pre-existing obstacles to enforcement.[128] It was suggested that the Convention's public policy defence should be construed narrowly. The state's most basic notions of morality and justice are to be the guiding factors. It was observed, in the context of arbitration agreements in international business transactions, that one cannot have trade and commerce in world markets and international waters exclusively on one's own terms and laws, and resolved in domestic courts.[129] Concerns of international comity, respect for the capacities of foreign and transnational tribunals, and sensitivity to the need of the international commercial system for predictability in dispute resolution requires the enforcement of parties' agreement, even if a contrary result might be the outcome of the dispute in a domestic context.[130]

Donaldson MR suggested approaching public policy considerations with extreme caution and sought some element of illegality or clear injury to the public good or where the enforcement of the award would be "wholly offensive to the ordinary reasonable and fully informed member of the public".[131] In Germany, public policy (a breach of *ordre public*) is anything that contravenes fundamental rules and ideas of justice, and in France, a distinction is made between international public policy and the national public policy.[132] In Islamic law the concept looks to the general spirit of *Shari'a* and its sources. As recognized by the International Law Association's Committee on International Commercial Arbitration, courts and laws are increasingly referring to "international public policy" in the context of international arbitration awards.[133] Any such international or transnational public policy must be given a narrow construction and be approached with even more caution than that suggested by Donaldson MR in the context of domestic public policy. In fact, actual examples of the application of such a policy are rare and it is difficult to see whether the concept "international" adds anything to the consideration of a state's public policy, unless the state happens to have very outdated laws.

The defence of "public policy" which is permissible under Section 48(2)(b) of the Indian Arbitration and Conciliation Act, 1996 should be construed narrowly. It should be noted that under Section 34(2)(b)(ii) of the Arbitration and Conciliation Act, 1996, the same phrase can be given a wider meaning, as held by the Supreme Court of India.[134] When the statute has couched

---

[126] Halsbury's Laws of England, *Conflict of Laws* (Volume 19 (2011)), paragraph 428.

[127] Vervaeke v. Smith, (1983) 1 AC 145 at p. 164; Dicey and Morris, *Conflict of Laws*, 11th Ed., Vol. I, p. 92; Cheshire and North, *Private International Law*, 12th Ed., pp. 128–129.

[128] Parsons and Whittemore Overseas Co. Inc. v. Société Générale De L'Industrie Du Papier (Rakta) and Bank of America, (1974) 508 F 2d 969.

[129] Fritz Scherk v. Alberto-Culver Co., (1974) 41 L.Ed. 2d, 270 at pp. 279 and 281.

[130] Mitsubishi Motors Corporation v. Soler Chrysler-Plymouth Inc., (1985) 87 L.Ed. 2d, 444.

[131] D.S.T. v. Rakoil, [1987] 2 Lloyd's Rep. 246.

[132] See Redfern and Hunter, *Law and Practice of International Commercial Arbitration*, 2nd Ed. p. 445.

[133] Articles 1498 and 1502 of Title V of the New Code of Civil Procedure (1981) referring to "ordre public international". Portuguese law has Article 1096(f) of the Code of Civil Procedure (1986) that includes a similar thing. The Milan Court of Appeal considers that the public policy referred to in Article V.2(b) is international public policy: (1997) XXII Yearbook 725, decision of 4th December 1992.

[134] Oil and Natural Gas Corporation Ltd. v. SAW Pipes Ltd., 2003(2) Arb. LR 5 SC; AIR 2003 SC 2629. Also see: Associate Builders v. Delhi Development Authority (SC): 2015(3) M.P.L.J. 368: 2015(3) SCC 49: 2014(13) Scale

the grounds of challenge in respect of both foreign and domestic awards using the very same words, there seems to be no justification to interpret the identical provisions in the same enactment differently, it is respectfully submitted. The said judgment, by necessary implication, widens the scope and ambit of challenge to an award under Section 34.

4.10.12 Oral Evidence Rule

Where all the terms of a contract have been reduced into writing, the document is assumed to express the full intent of the parties concerned. This is obvious because any evidence tending to change such expressed intent would defeat the purpose of writing. Once a contract is reduced to writing, by operation of Section 91 of the Evidence Act, it is not open to any of the parties to seek to prove the terms of the contract with reference to some oral or other documentary evidence to find out the intention of the parties. Under Section 92 of the Evidence Act, where the written instrument appears to contain the whole terms of the contract then parties to the contract are not entitled to lead by oral evidence to ascertain the terms of the contract. It is only when the written contract does not contain the whole of the agreement between the parties and there is any ambiguity that oral evidence is permissible to prove the other conditions which also must not be inconsistent with the written contract.

Under Section 91, when the terms of a contract have been reduced to the form of a document, or where any matter is required by law to be reduced to the form of a document, the document itself or secondary evidence of its contents must be put in evidence. Explanation 1 to the section further states that the section is applicable to cases in which the contracts are contained in one document or more than one document.

Section 92 declares that when the terms of any such contracts have been proved according to Section 91, that is either by the production of the document itself, or by the production of secondary evidence to it, "no evidence of any oral agreement or statement shall be admitted, as between the parties to any such instrument or their representative in interest, for the purpose of contradicting varying, adding to, or subtracting from this terms."

In view of the above provisions, it is to be noted that conversations held between the parties, or declarations made by either of them, whether before, or after, or at the time of the completion of a contract will be rejected and the written contract will be given effect to. Care must, therefore, be taken to see that all material terms settled in negotiations are expressly included in the written contract. Where terms have been negotiated by correspondence between the parties and a formal document (particularly when it is a standard form) is to be signed, the terms should be incorporated in the formal document or the letters should be included to form a part of the contract.

EXAMPLE FROM CASE LAW

The plaintiffs, along with their tender, attached a letter inserting certain terms by writing in ink to establish the case that the acceptance of the plaintiffs' tender would be tantamount to the acceptance of the terms contained in the letter in which there were insertions in writing to the effect that it was on a multi-slab basis. There was no signature either by the persons submitting the tender or by the persons receiving the same on the handwritten portion of the letter. The question whether such handwritten portion was originally there or was subsequently inserted assumed great significance. The trial judge noticed that the certified copy of the letter issued by the Board on 11th July 1978, clearly contained the handwritten portion

226: 2015 AIR (SC) 620: 2015(124) CorLA 318: 2015(2) SCC (Civil) 204: 2015(1) JBCJ 57: 2015(4) Mh.LJ 576: 2014(6) R.A.J. 693: 2015(1) Recent Apex Judgments (R.A.J.) 641: 2014(4) ArbiLR 307: 2014(215) DLT 204: 2014 AIR (SCW) 6861: 2015 All SCR 375: 2015(5) Cal. H.C.N. 92: 2015(1) Apex Court Judgments (SC) 542: 2015(33) LCD 345.

and concluded that the handwritten portion was there at the time of submission of the tender. The division bench of the High Court allowed variance of the terms of the written contract relying upon the statement of witnesses and granted a decree on a multi-slab basis. On appeal, this conclusion was set aside by the Supreme Court of India on the ground that the tender itself was submitted on 12th July 1978 and the Board could not have granted a certified copy of the letter on 11th July 1978 when the plaintiffs' case itself was that the letter was attached to the tender. The Supreme Court held that the handwritten portion in the letter was not there at the time of submission of the tender but that it was subsequently inserted, obviously with the connivance of the officers of the Board. The division bench of the High Court was held to have committed an obvious error in allowing variance of the terms of the written contract relying upon oral evidence.[135]

Oral Evidence – When Admissible

To the general rule, which forbids the admission of oral evidence, there are six exceptions. These are embodied in Section 92 itself. These provisions are reproduced below:

> PROVISO (1)
> Any fact may be proved which would invalidate any document, or which would entitle any person to any decree or order relating thereto, such as fraud, intimidation, illegality, want of due execution, want of capacity in any contracting party, want or failure of consideration, or mistake in fact or law.

Under this proviso, evidence is admitted to prove that there was no agreement between the parties, and therefore no contract.

> PROVISO (2)
> The existence of any separate oral agreement as to any matter on which a document is silent, and which is not inconsistent with its terms, may be proved. In considering whether or not this proviso applies, the Court shall have regard to the degree of formality of the document.

Under this proviso, a contemporaneous oral agreement to the effect that a written contract was to be of no force or effect and that it was to impose no obligation at all until the happening of a certain event, may be proved.

> PROVISO (3)
> The existence of any separate oral agreement, constituting a condition precedent to the attaching of any obligation under any such contract grant or disposition of property, may be proved.

The proviso is self-explanatory.

> PROVISO (4)
> The existence of any distinct subsequent oral agreement to rescind or modify any such contract, grant or disposition of property, may be proved, except in cases in which such contract, grant or disposition of property is by law required to be in writing, or has been registered according to the law in force for the time being as to the registration of documents.

---

[135] Tamil Nadu Electricity Board v. N. Raju Reddiar, AIR 1996 SC 2025; 1996(8) SCC400.

Under this proviso, a prior written contract may be modified by a subsequent written or oral agreement and this modification may be shown by writings, by words or by the conduct of the parties, or by all three. Where the original contract is of such a nature as that the law requires it to be in writing, subsequent modification of it must also be in writing.

> PROVISO (5)
> Any usage or custom by which incidents not expressly mentioned in any contract are usually annexed to contracts of that description, may be proved: Provided that the annexing of such incident would not be repugnant to, or inconsistent with, the express terms of the contract.

Under this proviso, oral evidence may also be accepted to confirm the existence of custom if custom gives a special meaning to special words in the contract document. It will not, however, be admitted to vary or contradict the written contract.

> PROVISO (6)
> Any fact may be proved which shows in what manner the language of a document is related to existing facts.

Where the terms of the document themselves require explanation, oral evidence can be led, under this proviso, to show in what manner the language of the document is related to existing facts.

EXAMPLE FROM CASE LAW

The description of an excavation item in a contract included the words "in any soil, murum, sock, etc". It was argued on behalf of the owner that the expression "sock" which admittedly has no meaning with reference to an engineering contract was a typing mistake for the term "rock" and that it could be so corrected by invoking the proviso to Section 92 of the Indian Evidence Act. This contention was upheld by the Gujarat High Court, and confirmed by the Supreme Court, although the correction itself was not allowed because there was no oral evidence pleaded to that effect.[136]

Interpretation of Documents by Oral Evidence

Sections 91 and 92 of the Indian Evidence Act define the cases in which documents are exclusive evidence of transactions which they embody. Sections 93 to 99 deal with rules for the construction of documents by oral evidence. The rules are summarized below:

1. *Section 93* deals with patent ambiguities. It prohibits the admission of oral evidence to make the language of a document certain, if on its face, the language is ambiguous or defective. For example, if in a written contract, blank spaces have been left, the court will not admit oral evidence to determine how such blanks were intended to be filled up.
2. *Section 94* forbids the admission of oral evidence to show that common words, whose meaning is plain, not appearing from the context to have been used in a peculiar sense, have been in fact so used.
3. *Section 95* reads: "When language used in a document is plain in itself, but is meaningless with reference to the existing facts, evidence may be given to show that it was used in a peculiar sense." This section is illustrated thus: "A sells to B, by deed, 'my house in

---

[136] Gujarat Electricity Board v. S.A. Jais & Co., AIR 1972, Guj. 192 (193); confirmed in AIR 1988 SC 254.

Calcutta'. A has no house in Calcutta, but it appears that he had a house in Howrah, of which B had been in possession since the execution of the deed. These facts may be proved to show that the deed related to the house at Howrah."

4. *Section 96* modifies the rule laid down in Section 94 by providing that where the language of a document correctly describes two sets of circumstances but could not have been intended to apply to both, evidence may be given to show to which set it was intended to apply. The section gives two illustrations, one of which reads: "A agrees to sell to B, for Rs. 1000/-, 'my white horse'. A has two white horses. Evidence may be given of facts which show which of them was meant."

5. *Section 97* reads: "When the language used applies partly to one set of existing facts, and partly to another set of existing facts, but the whole of it does not apply correctly to either, evidence may be given to show to which of the two it was meant to apply."

ILLUSTRATION

"A agrees to sell to B 'my land at X in the occupation of Y'. A has land at X, but not in the occupation of Y, and he had land in the occupation of Y, but it is not at X. Evidence may be given of facts showing which he meant to sell."

Provisions of this section apply to a case, where land within boundaries is sold and is wrongly described as containing a certain area. The error in area is regarded as a mere mis-description and does not vitiate the deed.

---

6. *Section 98* permits evidence to be given as to the meaning of illegible characters or of foreign, obsolete, technical, local and provincial expressions, and of words used in a peculiar sense. For example, evidence was allowed in a case to define meaning of the term "working days". The contract in that case provided for completion of the work in 125 working days. Evidence can be given of the meaning which the words "working days" or the like carry in construction contracts because such evidence only explains the meaning of expressions and cannot be said to vary the written contract.

7. *Section 99* provides that persons who are not parties to a document may give evidence tending to show a contemporaneous agreement varying the terms of the document. It is clear from this section that the principle of Section 92 does not apply to third persons. The illustration given below the section helps in understanding this provision. It reads:

> A and B make a contract in writing that B shall sell a certain type of cotton, to be paid for on delivery. At the same time they make an oral agreement that three months credit shall be given to A. This could not be shown as between A and B, but it might be shown by C, if it affected his interest.

4.10.13 Conflict of Laws/The Doctrine of Proper Law

With the increase in the number of construction contracts involving Indian companies undertaking construction works in other countries and vice versa, it is essential that parties to the contract understand the doctrine of proper law. In cases of international contracts to be performed in India or where an Indian party is involved in contracts to be performed outside India, most commercial parties tend to provide expressly for the selection of proper law and jurisdiction in the contracts.

Definition

Proper law is the law which the court is to apply in determining the obligations under a contract. It must, however, be appreciated that not all the matters affecting a contract are necessarily governed by one law, and the circumstances sometimes require different questions to be submitted to different laws. For example, the question whether an agreement has been reached, competency of the parties, validity of the contract and interpretation to be put upon a particular clause in the contract may not necessarily fall to be governed by the same law. Nevertheless, the court will not readily and without good reason split a contract in this respect,[137] and it can be said that in all cases there is a primary system of law called the proper law, which usually governs most matters affecting the formation and substance of the obligations.

The problem of ascertaining the proper law can be more perplexing in the case of contracts than in almost any other topic. In a given case, one can visualize three possibilities, namely:

1. Express choice by the parties.
2. Inferred choice.
3. Absence of any choice, either express or inferred.

Many commercial contracts are cross-border and interdependent. Not all would necessarily be governed by the same proper law even if they are interlinked. When common law applies to a case, the English courts will likely take an expansive view of the jurisdiction agreements and take into account the whole relationship between the parties on the reasoning that commercial parties would not intend their claims to be covered by different jurisdictions.[138] Their approach is unlikely to be decided simply by the labels "exclusive" or "non-exclusive" attached by parties to their choices of jurisdiction.[139] Parties' agreement may be enforced where their intention is clear even if this results in fragmentation of a dispute;[140] one-stop shop for dispute resolution is only a presumption. However, if a construction is possible to avoid two or more clauses competing, it would be preferred. When dealing with a standard form contract used globally, the English court's approach is going to focus on promoting consistency, predictability and certainty in commercial transactions.[141]

(A) **Express Choice by the Parties**

The determination of the proper law of contract may not normally involve any difficulty if the parties have recorded expressly which legal system is to apply to their agreement.[142] The difficulty in this case, however, arises due to the fact that parties are not free to submit the validity of their contract to any law of their own choosing. Thus, the express intention of the parties is subject to certain limitations, designed, in the main, to prevent the misuse of the discretion conferred on the parties to select the law to govern their contract. If the express choice is *bona fide* and legal and there is no reason for avoiding the choice on the ground of public policy, the choice will be conclusive. Within the European Union, the questions of conflict of laws are decided by reference to the EU Regulation 1215/2012, Brussels I (recast) and this (unlike Art. 23 of Regulation 44/2001) applied where one or more of the parties were domiciled in a Member State.

---

[137] Kahler v. Midland Bank, [1950] AC 24 (42).

[138] Fiona Trust & Holding Corp. v. Privalov, [2007] EWCA Civ 20 affirmed in [2007] UKHL 40. Also see, Monde Petroleum SA v. Westernzagros Ltd., [2015] 1 Lloyd's Rep. 330.

[139] Citibank NA, London Branch v. Oceanwood Opportunities Master Fund and Others, [2018] EWHC 305 (Ch).

[140] Deutsche Bank AG v. Comune Di Savona, [2018] EWCA Civ 1740.

[141] BNP Paribas SA v. Trattamento Rifiuti Metropolitani SPA, [2018] EWHC 1670 (Comm).

[142] Acrow (Automation) Ltd v. Rex Chainbelt, Inc., [1971] 1 WLR 1676.

(B) **Inferred Choice of the Proper Law**

Where the parties have failed to lay down expressly in their contract the law applicable to it, the court may be able to infer the law which the parties intended to apply. Two approaches have evolved over ascertaining the intention of the parties. Some courts emphasize the presumed intention of the parties and declare that the task of the court is to infer from the terms and circumstances of the contract what the parties' common intention would have been had they considered the matter at the time when the contract was made.[143] Others say that the court must determine for the parties what they would have intended, had they considered the matters. There is a clear difference between these two views. In the first view, the doctrine of implied term is used, while in the second view it conjectures no probabilities, but applies the external standard of a reasonable man. The second approach is more reasonable and recommends itself. According to this theory, the proper law is the legal system with which the contract has the most substantial connection.[144]

On this view of the matter, every term of the contract, every detail affecting its formation and performance, every fact that points to its natural seat is relevant. No one fact is conclusive. The matters which need to be taken into account include:

1. The form of the documents made with respect to the transaction.[145]

2. The style and terminology in which the contract is drafted.[146] As, for instance, the language may be appropriate for one system of law, but inappropriate for another.

3. The use of a particular language.[147] This, however, is a factor of minor importance.[148]

4. The currency in which payment is to be made.[149]

5. The nature and location of the subject matter of the contract.[150]

6. The residence of the parties and their nationality.[151]

7. A connection with a preceding transaction.[152]

8. The fact that one of the parties is a government.[153]

9. If the contract contains a clause whereby the parties agree that any dispute shall be submitted to arbitration in a particular country, there is a powerful, though not conclusive, inference that the parties have selected the law of the country of arbitration as the proper law.[154]

10. Whether any inference can be made as to the intention of the parties from the fact that the contract, or one of its terms, is valid under one relevant system of law but not under another. If yes, the parties may be taken to have intended that their contract should be governed by the system of law by which it was valid.[155] However, such a fact is only evidence and not conclusive evidence as to the intention of the parties.

---

[143] Lloyd v. Guibert (1865), L.R., IQB., 115 (120); Rex. v. International Trustee, [1937] AC 500 (529).

[144] United Railways of Havana and Regla Warehouses Ltd., Re: [1960] 2 WLR 969.

[145] Compagnie Tunisienne de Navigation S. A. v. Compagnie d' Armenment Maritime S. A., [1971] AC 572, 583.

[146] Whitwort Street Estates (Manchester) Ltd. v. James Miller and Partners Ltd., [1970] AC 583; 603, 608, 611, 612.

[147] St. Pierre v. South American Stores (Gath and Chaves) Ltd., (1937) 2 ALL ER 349.

[148] Coast Lines Ltd. v. Hudig and Veder Chartering N.Y., (1972) 2, QB 34 47 50.

[149] Sayers v. International Drilling Co. N.V., [1971], 1, WLR 1176, 1183, 1186.

[150] British South Africa Co. v. De Beers Consolidated Mines Ltd. (1910) 1 Ch. 354, 383.

[151] Keiner v. Keiner, [1952] 1 All ER 643.

[152] The Metamorphosis [1953] 1 WLR 543.

[153] R. v. International Trustee for the Protection of Bond Holders A.G., [1937] AC 500.

[154] Tzortzis v. Monark Line A/B, [1968] 1 WLR 406 411; International Tank & Pipe SAK v. Kuwait Aviation Fuelling Co. KSC (1975) QB 224.

[155] Coast Lines Ltd. v. Hudig and Vender Chartering N.Y., (1972) Q.B. 34, 47, 50.

11. The place where the contract is made.[156]

12. The place where the contract is to be performed.[157]

(C) **Neither Express nor Inferred Choice of Proper Law**

It is probable in some cases that the possibility of a conflict of laws was absent in the minds of the parties. When the court is faced with the problem of deciding the proper law of the contract, in such cases the older authorities indicated that the court should adopt a subjective approach, viz., the doctrine of implied term. More recently, however, the courts have come to accept an objective approach, viz., the doctrine of presumed intent.

[156] The St. Joseph (1933) P. 119; [1933] 45 Lloyd's Rep. 180.
[157] Kremezi v. Ridway [1949] 1 ALL E.R. 662.

5

Variations and Deviations

5.0 Introduction

A construction contract differs from other business contracts, inasmuch as its subject matter invariably involves an element of uncertainty. No matter how much precaution and trouble is taken during the investigation, planning and preparation of contract documents, a large and complex work is rarely completed exactly in accordance with the original drawings and specifications. Additional work may be ordered by the employer and executed by the contractor under the original or a separate agreement, either made orally or in writing. However, it is common practice to make a suitable provision in the agreement itself. The contract price in such agreements having been generally agreed on the estimates based on such drawings and designs, changes invariably necessitate adjustments in the agreed cost no matter in which form the agreement is made. As a matter of fact, the entire or lump sum form of contract may require elaborate provisions for valuation of additions and alterations generally called variations as stated in Chapter 2.

Construction contracts, therefore, invariably include a stipulation empowering the engineer or architect to effect changes in the drawings, specifications and quantities, often resulting in omissions, alterations or additions to the original work, and further evaluation of the changes for the purpose of adjusting the contract sum. Probably no clause in a contract gives rise to disputes as often as the one relating to additions and alterations. This is obvious because the owner as well as the contractor may try to exploit this provision in the contract for their own respective benefit. Disputes may involve the basic question as to whether the change made amounts to an extra, which justifies an adjustment of the price. This chapter contains the analysis of provisions in commonly adopted standard form contracts, illustrated with decided cases.

5.1 Provisions in Standard Form Contracts

The relevant provisions of the General Conditions of some standard form contracts are considered, including the following (and see table over page).

The FIDIC Form of the 2017 edition deals with variations under a paragraph heading: "Variations and Adjustments" whereas the NITI Aayog Model Form used the heading "Change of Scope". Whatever the paragraph heading might be, the provisions deal with additions, alterations and deletions to the originally agreed scope of work and effect thereof on the contract price and the time limit for the completion of the works. The variations may result in extra work or additional work.

5.2 Extra Work and Additional Work Distinguished

Additions and alterations made by an engineer or architect generally result into either of two distinct categories of extras or omissions. The first category of variations includes: the works which are not expressly or impliedly included in the original contract, and therefore are not

Provisions in Standard Form Contracts	Paragraph Nos.
FIDIC,[1] 1987 edition, Clauses 1, 51, 52, 68, 70	5.2, 5.4, 5.6, 5.21.2
FIDIC, 2017 edition, Clauses 4.1, 8.3, 12, 13	5.2, 5.4.3
FIDIC Red Book, 1999 1st edition, Clauses 3.3, 12, 12.4, 13, 14, 20	5.4, 5.4.3, 5.4.4, 5.11.1, 5.21.2
FIDIC Yellow Book, 1999 1st edition, Clause 1.1.6.8	5.21.1
Ministry of Statistics and Programme Implementation, Government of India (MOS & PI)	5.5
NITI[2] Aayog Model Form 2017, Clauses 4, 8.3.3, 13.1–13.5, 17	5.6
PWD B-1 and B-2 Standard Forms of State Government, Clause 14	5.7, 5.9.2, 5.11.2
CPWD Form, Clauses 12, 12A, 13*	5.4.4, 5.7, 5.8.1, 5.9.1, 5.9.2, 5.17

* There are similar provisions in the Military Engineering Service and Railway Engineering Department forms.

included in the original contract price, provided the work is done within the framework of the original contract.[3] It is the work arising outside of and entirely independent of the contract and not required in its performance. Such a work is called "extra work". The second category of variations includes work necessarily required in the performance of the contract, not intentionally omitted from the contract and evidently necessary for the completion of the work – this is "additional work".

The distinction between extra and additional work is made by the courts because extra work, as defined above, when ordered by the engineer or architect and accepted and carried out by the contractor, forms a new contract. Such a new contract remains binding, even if, for any reason, the original contract is unenforceable. As against this, additional work is a part of the original contract and is subject to the original terms and conditions. However, in practice, the distinction between extra work and additional work is often not easy to draw. A contractor may always raise the question whether deviations ordered resulting in extras or omissions on a large scale do not radically change the original nature of the contract so as to demand total novation of the contract.

The FIDIC Standard Form 1987 edition also recognized the distinction between the extra work and additional work to some extent. Clause 51.2 provided that the contractor would not make any variation "without an instruction of the Engineer". The proviso to the said clause stipulated that no instruction would be required if there was a decrease or increase in the quantity of any work if it was the result of quantities exceeding or being lower than those stated in the Bill of Quantities (BOQ) and not as a result of the engineer's instructions under Clause 51. It is suggested that the 1987 (reprint 1992) clause 51 under the heading "Alteration, Additions and Omissions" was a comprehensive provision dealing with all aspects without requiring a reference to any other clauses. It provided for variation limit of additions and deductions taken together at 15% of the contract price.

Art. 13.1 of the FIDIC 2017 edition stipulates that the contractor shall not make any variation "unless and until the Engineer instructs a variation under sub-clause 13.3.1 (Variation by instruction)". It also provides: "Each variation may include changes to the quantities of any item of work included in the contract (however, such changes do not necessarily constitute a variation)". In this respect reference has to be made to Clause 14.1(c) which provides that the quantities set out in the BOQ or other schedules are estimated and not to be taken as actual or

---

[1] Fédération Internationale des Ingénieurs-Conseils.
[2] National Institution for Transforming India, a Government of India think tank.
[3] Food Corporation of India v. Vikas Majdoor Kamdar Sahkari Mandli Ltd., (SC): 2008(1) R.C.R.(Civil) 284: 2007(6) Recent Apex Judgments (R.A.J.) 583: 2007(12) JT 517: 2008(1) ICC 663: 2007(13) Scale 126: 2007(11) SCR 1038: 2007(13) SCC 544: 2008(1) W.L.C. 309: 2008(1) DNJ 183: 2008(1) R.A.J. 140: 2008(1) All WC 495: 2008(2) M.L.W. 25: 2007 AIR (SCW) 6999: 2008(2) MLJ 857: 2008(1) JLJR 149: 2008(1) PLJR 149. State of U.P. v. Chandi-a Glipta & Co., AIR 1977 All 28 (32).

correct. It makes a reference to clause 12.3(b) which stipulates that a permissible increase or decrease of more than 10% in the quantities of an item of work constitutes a variation. The engineer will have to be on watch so as to order a variation when the quantity of any item is likely to change from the quantity in the contract by more than 10%. There is no limit specified for overall increase of contract price. Before these provisions are discussed in detail, a basic question needs to be raised and answered.

5.3 What Constitutes a Variation?

Apart from the problems caused by the extent of the architect's or engineer's authority to order variations and the effect of any contractual requirements of form, there is little doubt that most frequent and difficult questions of construction that arise in construction contracts relate to whether particular items of work are, in fact, a variation of the work undertaken. The description of the item in a BOQ and the detailed description of the work in the specifications have to be examined. Whether a particular work described in contract documents such as BOQ or specifications includes ancillary work is a question of construction which must often depend upon the circumstances of the case. A carefully drawn BOQ prepared by the owner's advisors sometimes omits to make any mention of ancillary items, or mentions some ancillary items and not others. It is possible to construe that the ancillary work, though described in the specifications but not having been mentioned in the BOQ, may amount to extras. To avoid this problem situation, some authorities, like the National Highways Authority of India, incorporate a reference to the standard specifications in the description of the item, and the specifications include a mode of measurement and payment that lists the ancillary activities which are required to be carried out but which will not be separately payable. However, there is a remarkable lack of authority on this kind of a problem. A contractor will be advised to include in his price ancillary works which are mentioned in the specifications of a given item while quoting his price against that item.

The FIDIC Form recognizes the above difficulty and empowers the engineer to hold that if the nature and amount of any varied work is such that the agreement rate(s) are rendered inappropriate or inapplicable, the parties shall agree on a suitable rate or price. Some modified conditions in use also extend this power to cover the time of variation ordered as an additional ground to render the agreement rate/price inapplicable.

5.4 FIDIC Form

FIDIC is a French acronym[4] for the International Federation of Consulting Engineers. Founded in 1913, FIDIC's global membership includes more than 60 countries. The standard form of contracts in use in India for many major projects is the adapted version of the FIDIC Form 4th edition, 1987 and amended reprint, 1992. A revised form was published in 1999. The second edition of 2017 is also published. The provisions of all the editions are considered below. FIDIC forms discussed in this book mainly relate to the Red Book which deals with civil engineering contracts designed by the employer.

FIDIC Form 4th edition 1987 and amended reprint 1992 contains provisions 51.1, 51.2 and 52.1 to 52.4 under the main heading "Alterations, Additions and Omissions".

Clause 51.1 empowers the engineer to make any variation of the form, quality or quantity of the works or any part thereof that may, in his opinion, be necessary. This general power enables the engineer to increase or decrease the quantity of any work, omit any such work, change the character or quality or kind of any such work, change levels, positions and

---

[4] Fédération Internationale des Ingénieurs-Conseils (FIDIC).

dimensions of any part of the work, order additional work necessary for the completion of the work, and change any specified sequence or timing of construction of any part of the works. Omission of an item of work should not be done for the purpose of awarding it to another contractor but for the reason that the work is no longer necessary. The provision further stipulates that the ordering of variation would not vitiate or invalidate the contract but the effect would be valued in accordance with the provisions of Clause 52. It contains a proviso that if variations were necessitated by some default or breach of contract by the contractor, the additional cost, if any, would be borne by the contractor. It further provides that no such instruction would be necessary if increase or decrease resulted from the quantities deviating from the quantities stipulated in the BOQ.

The notable feature of this provision is that, *prima facie*, the variations need not necessarily be ordered in writing. Clause 51.2 stipulates that the contractor shall not make any variations without an instruction of the engineer. It does not say "instruction in writing". However, this provision needs to be read with provisions of Clause 68 under the heading "Notices". The said provisions require all certificates, notices or instructions to be given to the contractor by the employer or the engineer under the terms of the contract to be sent by post, cable, telex or fax. Therefore, although the provisions of Clause 51 do not per se require the instructions to be given in writing, contractors are well advised to insist on written instructions. If, in spite of a demand, written instructions are not issued, a contractor should place on record the variations orally ordered, which if not contradicted, would be deemed to have been issued in writing. This suggestion stands expressly incorporated in the 1999 edition of the FIDIC Form under Clause 3.3.

The right of the contractor to be paid for variations orally ordered would not be prejudiced if the contractor can establish by evidence the fact of variations having been orally ordered and confirmed in writing by the contractor. It is further pertinent to note that the right to be paid for varied work is subject to the condition contained in the proviso under Clause 52.2 below which requires a notice to be given within 14 days of the date of such instruction (other than in the case of omitted work) before the commencement of the varied work, by the contractor to the engineer of his intention to claim extra payment or a varied rate or price or by the engineer to the contractor of his intention to vary a rate or price.

Clause 52.1 specifies the procedure for the valuation of variations as follows:

1. Varied work should be valued at the contract rates if the engineer declares that the said rates are applicable.
2. If the contract does not contain any rates or prices for the varied work, new rates should be worked out from the rates of prices of similar items in the contract, if reasonably possible.
3. The new rates should be mutually agreed between the engineer and the contractor after consulting the employer.
4. If no agreement is possible, the engineer should fix the rates or prices and notify to the contractor with a copy to the employer. Should finalization of the rates by mutual consultation or otherwise be likely to take time, the provision empowers the engineer to determine provisional rates or prices for the purpose of issuing interim payment certificates.

Clause 52.2 carves out exceptions to the above rules by way of two provisos. The first proviso stipulates that if the nature or amount of any varied work, when compared with the nature or amount of the whole work, is such that in the opinion of the engineer originally agreed contract rates or prices are rendered inappropriate or inapplicable, new rates would be agreed between the engineer and the contractor in consultation with the employer. Failing such agreement, power is given to the engineer to fix the rates finally or provisionally similar to the procedure under Clause 52.1.

The second proviso requires the contractor to protect his right to be paid extra at a varied rate or price by giving notice within 14 days of the date of such instruction. Such a notice is not required if the engineer, at the time of instructing to vary the work, gives notice of his intention to vary a rate or price. The notice contemplated by this provision must also be in the form and manner provided under Clause 68.

Clause 52.3 provides for addition to or deduction from the contract price such a sum as may be agreed between the contractor and the engineer after due consultation of the engineer with the employer and the contractor. This sum is payable in case additions to or deductions from the contract price under all the varied work valued under Clauses 52.1 and 52.2 taken together are in excess of 15% of the effective contract price. The effective contract price for this purpose is defined to mean the contract price excluding provisional sums and allowance for day works, if any. Clause 1.1(e)(i) defines the contract price to mean the sum stated in the letter of acceptance.

It may be noted that the provisions of Clause 52.3 are attracted only after all the varied work under Clauses 52.1 and 52.2 are given effect. As such the contractor is entitled to claim additional payments on account of variation over and above the payments received by him under Clauses 52.1 and 52.2. The provision for additional payment is generally restricted to compensate "the Contractor's Site and general overhead costs of the Contract" on account of variations. It is significant to note that the agreement expressly provides for escalation under FIDIC conditions Clause 70 for increase or decrease of the cost of labour and materials, which generally include fuel oil and lubricants, but there is no provision made for compensating the loss, if any, in respect of overheads on account of variations. This provision expressly stipulates that the additional amount is to be worked out having regard to the contractor's site and general overhead costs of the contract. This additional amount can be worked out after ascertaining the provision made by the contractor in his tender for overheads including site and head office overheads. The percentage provision will be applicable only on the amount by which such additions or deductions are in excess of the 15% of the Effective Contract Price. For example, if the additions and deductions exceed 15% of the Effective Contract Price by, say, Rs. 10 million and the contractor's overheads are ascertained at 15% of the contract price (that is to say 17.65% of the basic cost of men, material and machinery), the additional amount would be Rs. 15,00,000/-.

Clause 52.4 empowers the engineer to issue an instruction that any varied work shall be executed on a day work basis. In such an event, the contractor is entitled to payment under the terms set out in the daywork schedule included in the contract and at the rate and prices affixed thereto by him in the tender. The provision further lays down the procedure for the substantiation of deployment of material, labour, etc. by the contractor.

5.4.1 Disputes and Differences Regarding the Interpretation of Clause 52

The above provisions are self-explanatory. However, in practice, the provisions of Clause 52.2 give rise to dispute and differences regarding the interpretation of "if the nature or amount of any varied work relative to nature or amount of the whole of the works or any part thereof, is such that, "the rate or price contained in the contract is rendered inappropriate or inapplicable." The points on which difference of opinion may arise include:

1. *The increase in the quantum of work vis-à-vis the original quantity under a given item.* The authors are of the opinion that the variation in the quantity of an item of work in excess of 20 to 25% of the originally agreed quantity may be used as a basis for rendering the agreement rate inapplicable. These percentages are generally provided in other popular forms of standard form contracts in India. However, the new edition of FIDIC eliminates this difficulty by expressly limiting the excess to 10% of the originally agreed quantity subject to other conditions.

2. *Whether the new rate to be decided is to be applicable to the quantity in excess of the originally agreed quantity or whether for the entire quantity executed under the item in*

question. Invariably the employer puts forth the plea that the quantity in excess alone will attract the new rate, the contractor having been bound to execute to the originally agreed quantity at its tendered rate. The authors believe that the provisions of Clauses 51.1 and 51.2 together entitle the contractor to payment for the entire quantity at the new rate. In this respect, Clause 51.1, which empowers the engineer to increase or decrease the quantity, provides that the "effect, if any, of all such variations shall be valued in accordance with Clause 52".

Clause 52.1 expressly provides for additions to be valued at the rates and prices set out in the contract. The proviso in Clause 52.2 carves out an exception to the above rule. The wording used, "a suitable rate or price shall be agreed" if the varied work "rendered inappropriate or inapplicable the rate or price contained in the contract", leaves no doubt as to the intention of the parties. Once it is held that the variation renders the contract rate inappropriate or inapplicable, it is for the full quantity and not only for the quantity in excess of the stipulated quantity. Unlike other contract conditions, this provision does not stipulate any limit of variation in respect of individual items. Under the circumstance it is difficult to interpret that the new rate would be applicable for quantity in excess of agreed quantity alone. It is stated that there is no scope to interpret the provisions in any manner other than as suggested by the authors.

3. *Whether a contractor is bound to execute quantities before the parties mutually agree equitable rates.* Failing mutual agreement in respect thereof, the rates would be decided in arbitration. FIDIC provisions, as already seen above, empower the engineer to fix provisional rates for the purpose of interim payments. The question may still arise in cases where (i) the engineer unreasonably or under the pressure from the employer holds that the agreed rates are applicable to the varied quantities or items of additional works and/or (ii) the provisional rates fixed by the engineer are not workable. It is submitted that the answer to the question is not certain and free from doubt. It is advisable to keep the work in progress pending arbitration and if the adverse effect on the cash-flow affects the progress of the work, its consequences will be on the party which is unsuccessful in the arbitration.

Whether the case will be governed by the decision of the Supreme Court considered below,[5] in which it was held that a contractor need not execute the work if rates for the varied work were not mutually agreed, is also debatable in view of the arbitration clause. The arbitration clause clearly provides that the obligations of the employer, the engineer and the contractor would not be altered by reason of an arbitration conducted during the progress of the works. Some contracts expressly prohibit stoppage of work by the contractor due to an ongoing arbitration. It is suggested that the contractor should keep the work in progress pending settlement of the rates if the engineer has fixed the rates provisionally for interim payments or the contractor is not sure that the final decision would go in his favour. Only if the contractor is sure of his entitlement for new rates and the rates are not agreed or the rates are not fixed by the engineer for interim payment, can the ratio of the said decision be attracted to the case.

5.4.2 Abnormal Variation in Quantity – Power to Fix Rate in Absence of Engineer's Instructions

It is customary in construction contracts to stipulate that the quantities shown in the BOQ are approximate and likely to vary and that the contractor is bound to execute the varied work at the agreed rate if the variation is within the limit prescribed in the contract. A mere discrepancy between the approximate estimated quantity in the schedule of rates and that actually executed

---

[5] Bombay Housing Board v. Karbhase Naik & Co., AIR 1975 SC 763.

would not necessarily amount to a variation.[6] Where the quantities increase or decrease without any order by the engineer, the question may arise, under FIDIC standard conditions, whether the rate of the item in question is open for renegotiation.

Clause 51 which opens with the sentence "[t]he Engineer shall make any variation" contemplates an order by the engineer. It further provides that all such variations shall be valued in accordance with Clause 52. Clause 51.2 opens with the statement prohibiting the contractor from making "any such variation" without the order of the engineer. The words "any such variation" clearly mean the variations under Clause 51.1, sub-clause (a) of which stipulates increase or decrease in the quantity resulting from the engineer's order. Clause 51.2 contains a proviso which provides that no instruction is required for an increase or decrease in the quantity of any work where such increase or decrease is not the result of an instruction given under the clause, but is the result of the quantities exceeding or being less than those stated in the BOQ. Clause 51.2 does not contain the provision for the valuation of variation as under Clause 52. The Clause 52 stipulation that "[a]ll variations referred to in Clause 51 and any additions to the contract price which are required to be determined in accordance with Clause 52", clearly is not limited to Clause 51.1, but includes both Clauses 51.1 and 51.2. Thus, the intention of the parties to the agreement containing the above provisions is clear that a variation in excess of the limit stipulated by the agreement, not resulting from an order by the engineer, will also be valued under Clause 52.

If there is an abnormal variation in quantities, the time limit also would be extended. If during the extended time the prices go up, a contractor cannot be bound to the rates originally quoted by him for the quantities shown in the BOQ plus or minus the agreed limit for variation. Similarly, if the rate quoted is higher than normal, the employer also cannot be deprived of the benefit likely to result on account of "economies of scale", that is to say decreased per unit cost as the output increases. Thus, there is no reason why variations in quantity, both positive and negative, beyond the limits set in the contract, whether or not instructed by the engineer, should not lead to a renegotiation of the rates at the instance of any party. An elaborate discussion and interpretation of the above terms is to be found in the case decided by the Division Bench of Delhi High Court in the example below.[7]

EXAMPLE FROM CASE LAW

For an estimated quantity of 6,08,000 m^2 of geogrid/geotextile reinforcement fabric in RE walls, the respondent contractor quoted a rate of Rs. 300/- per m^2. The actual quantity increased to 19,58,105 m^2 of geogrid as compared to 6,08,000 m^2 as estimated in the BOQ. The dispute referred to arbitration concerned the consequences of the geogrid/geotextile material exceeding the estimated quantity as indicated in the BOQ contained in the contract. While the appellant claimed that, pursuant to the contract, the engineer was entitled to seek a renegotiation of the rates since the quantity of geogrid required to execute the contractual work exceeded the BOQ quantity by nearly three times, according to the respondent No. 1 the rates did not call for renegotiation under the contractual terms. The relevant contractual terms, which are contained in the General Conditions of Contract (GCC) and Conditions of Particular Application were FIDIC 1992 Reprint with amendments. The Delhi High Court held that there was no reason why the variations in the quantity beyond the limits set in the contract, whether or not instructed, should not lead to a renegotiation of the rates at the instance of either party, and that it would be the only fair

---

[6] Grinaker Construction (Transvaal) Pty Ltd. v Transvaal Provincial Administration, Supreme Court, Appellate Division (South Africa), [1983] 1 WLUK 217; 20 BLR 30.

[7] NHAI v. Som Datt Builders-NCC-NEC, (JV), 2010(1) Arb. LR 185 (Delhi) (DB).

reasonable and equitable way to work the contract. Whether the variation in quantity is in the positive or in the negative direction, in either case, if the variation exceeds the tolerance limits set in the contract, renegotiation of rates would be called for.[8]

In the High Court's view, Clause 52.1 clearly and categorically states that "all variations referred to in Clause 51" are open to renegotiation provided the conditions stated in the contract are satisfied. There is no reason to limit the meaning of the words "all variations referred to in Clause 51" to only those variations which are the result of an instruction. There was, it held, no logic behind the interpretation given to the contract by the arbitral tribunal and accepted by the single judge. That interpretation appeared to be contrary to the express language used by the parties in their contract.[9]

The decision of the Appellate Division of the South African Court,[10] interpreting a somewhat similar contractual clause, was relied upon by the trial court. In that case, automatic increases or decreases in quantities were held not to constitute a variation. The Delhi High Court distinguished it on the basis of absence of the words "all variations referred to in Clause 51" in the clause. In a very studied judgment, and after elaborate discussion, it was held that the interpretation adopted by the arbitral tribunal and upheld by the single judge was patently implausible, unfair, unreasonable, erroneous and illegal. The said interpretation in the award, shocked the "conscience" of the High Court. The order of the single judge as well as the award passed by the arbitral tribunal was set aside.[11]

5.4.3 FIDIC 1999 Edition

In the 1999 edition, Clause 13 under the heading "Variations and Adjustments" includes provisions similar to Clauses 51 and 52 of the earlier edition. However, certain provisions of Clauses 3.3 and 12 are also relevant insofar as giving oral instructions and the evaluation of the varied work is concerned. Clause 3.3 is considered in paragraphs 5.4 and 5.11.1 below. The other provisions are briefly considered below.

Clause 13.1

This provision empowers the engineer to initiate variations at any time prior to issuing the Taking Over Certificates for the works. There are two methods for initiating variations: first, by issuing an instruction and/or second, by requesting the contractor to submit a proposal. The scope of variation is more or less the same as covered by Clause 51.1 of the earlier edition including the stipulation that the contractor shall not make any alteration or modification until the engineer instructs or approves a variation. The question of approval of a variation would arise in the case where proposal to vary is submitted by the contractor under Clause 13.1 and/or 13.2.

Clause 13.2

This clause empowers the contractor, *suo moto*, at any time, to submit to the engineer a written proposal which, in the opinion of the contractor will, if adopted, accelerate completion, reduce

---

[8] NHAI v. Som Datt Builders-NCC-NEC (JV), 2010(1) Arb. LR 185 (Delhi) (DB).

[9] Ibid.

[10] Grinaker Construction (TVL) (Pty) Ltd. v. Transvaal Provincial Administration, 1982(I) AD 78.

[11] National Highways Authority of India v. Som Datt Builders-NCC-New, (Delhi) (DB): 2010(1) Arb. LR 185 (Delhi): 2010(4) R.A.J. 352.

the cost of work either of executing or maintaining and operating the works, improve the efficiency or value of the completed works or otherwise be of benefit to the employer. The cost of preparation of the proposal is to be borne by the contractor and it is to include the items listed in Clause 13.3. The provision further stipulates that if the proposal is approved by the engineer and involves a change in the design of part of the permanent works, and then unless otherwise agreed by both the parties, the contractor is to design that part of work under his general obligations listed in Clause 4.1(a) to (d). If the change results in a reduction in the contract value the engineer is to determine a fee, which would be included in the contract price. The quantum of the fee shall be half (50%) of the difference between the following amounts:

i. A reduction in the contract value as a result of the variations, excluding adjustments for changes in legislation and escalation payable under Clauses 13.7 and 13.8, respectively, and

ii. A reduction in the value to the employer of the varied works on account of reduction, if any, in quality, anticipated life or operational efficiencies.

It is further stipulated that in case amount (i) is less than amount (ii), no fee shall be payable. In short, the fee payable would be 50% of the net saving effected in the original cost on account of changes suggested by the contractor after deducting a reduction in value on account of quality, life, efficiency, etc. due to the changes so made.

Clause 13.3

This provision prescribes the procedure of variation in case of the engineer requesting the contractor to submit a proposal under Clause 13.1. On receipt of such a request, the contractor may promptly give notice to the engineer if the contractor cannot comply with the request. In the notice so issued the contractor must give reasons with supporting particulars such as the contractor's inability readily to obtain the supplies of materials or machinery/equipment required for the variations, etc. In such an event, the engineer has the power to cancel, confirm or vary the instruction under Clause 13.1.

If the contractor accepts the request, he should promptly submit (a) a description of the proposed work, (b) a programme for its execution and, if necessary, a modification to the programme submitted under Clause 8.3, and (c) an extension for time for completion, if warranted. The contractor must include in his proposal, an evaluation of the variation. The engineer should respond to this proposal with approval, disapproval or comments. The contractor must not delay any work while waiting for the engineer's response.

Each instruction to execute a variation, with any requirements for the recording of costs, must be issued by the engineer to the contractor, who should acknowledge its receipt. Each variation is to be evaluated in accordance with Clause 12 unless the engineer instructs or approves otherwise in accordance with this clause. In short, the engineer acting under the provisions of Clause 13 is empowered to decide the cost of variations independently of Clause 12 or in accordance with Clause 12. The provisions of Clauses 12.3 and 12.4 are relevant for this purpose.

Clauses 12.3 and 12.4 provide the following steps for the evaluation of the contract price "except as otherwise stated in the contract":

a. For each item of work, the appropriate rate or price shall be the rate or the price of item specified in the contract, or

b. If there is no such item, the rate or price specified for similar work included in the contract.

c. A new rate is to be agreed in the case of two contingencies contemplated in sub-clause (a) and/or (b). Sub-clause (a) provides for excessive variation in quantities in respect of items included in the contract but not specified as a "fixed rate item". An item in the BOQ qualifies for a new rate, if it is not a fixed rate item and further fulfils the following three conditions:

  i. The measured quantity of the item is changed by more than 10% from the quantity of the said item in the BOQ or other schedule.

 ii. This change in the quantity multiplied by the agreed rate exceeds 0.01% of the accepted contract amount, and

iii. This change in the quantity directly changes the cost per unit quantity of this item by more than 1%.

The above provisions limit the power to order variations to a maximum of 10% of the original quantity of any item of work at the originally stipulated rate and entitle the contractor to claim payment at new rates if the change in quantity multiplied by the specified rate for the item in question exceeds 0.01% of the accepted contract amount.

This clause is unduly complicated and not very well drafted. The second condition above can cause confusion even if the overall objective of Clause 12 is clearly to provide for a new rate in the case of significant, as opposed to minor, variations, whether upward or downward. The words "this change in quantity" can suggest that the consideration of 0.01% of contract sum is to be computed by applying the specified rate to (a) the total measured quantity, or (b) the difference between the total measured quantity and the BOQ quantity. It is more likely that (a), the total measured quantity is indicated by this phrase. Otherwise, it would have used an expression such as "the additional" or "difference in quantity". "Change" means "instead of" or "substitution" according to plain dictionary meaning.[12] A related question is whether the new rate to be determined under Clause 12.3 applies to the additional quantity or the total quantity and FIDIC's guidance in this respect is that it applies to the total quantity.[13] This view supports the authors' contention that the second condition refers to the total measured quantity, and not the difference between the total measured quantity and the BOQ quantity.

The words "such specified rate for this item" in the second condition obviously mean the agreed rate shown in the BOQ, because that is the only rate specified at that stage to determine if the new rate is required to be worked out. It appears that the first and third provisions would be satisfied if the quantity of a given item differs by more than 10% from the quantity of the item in the BOQ, if "the Cost per unit quantity" is also to mean the originally agreed rate. The drafters in such a case would not have used words different from the words used in the second condition. It has to be presumed that the third of the above provisions stipulates that a new rate will be payable only if the unit cost of the item in question due to changed quantity will cost more than 1% of the accepted unit rate. But at this stage it is not decided if the new rate is at all attracted. One possible view is that if the first two conditions are satisfied, the new rate should be worked out and it will be payable only if it is found to be more than 1% of the originally agreed rate. The other view involves dividing the tendered cost by the new quantity to work out the rate and compare it with the agreed rate. If the difference is more than 1%, the new rate is to be worked out.

FIDIC 2017 edition eliminates the doubt raised in the earlier edition of this book by stipulating "the rate or price specified in the BOQ" in place of the "agreed rate". However, the uncertainty over the meaning of "this change in quantity" continues. This latest edition adds two more conditions. As a result, to qualify for an extra payment on account of excessive variation beyond agreed limit, Clause 12.3(b) stipulates conditions, all of which must be satisfied, it is submitted. These are:

  (i) Measured quantity changed by more than 10% of the quantity stated in the BOQ;

 (ii) The change in the quantity multiplied by the BOQ rate exceeds 0.01% of the contract sum;

---

[12] *Oxford English Dictionary*, accessed online at https://en.oxforddictionaries.com.
[13] Retrieved from http://fidic.org/node/923.

(iii) The change in the quantity directly changes the cost per unit of the item by more than 1%;

(iv) The item in question is not specified as fixed price or not subject to adjustment for any change in quantity; and/or

(v) The work is instructed under Variations and Adjustment Clause 13 and sub-paragraph (a) or (b) above applies.

The question whether the last condition (v) is in addition to or alternative to all the earlier conditions is left to the reader's interpretation by the use of words "and/or" at the end of sub-clause (iv). The intention appears to be that if the work is instructed under Variations and Adjustment Clause 13, this condition alone is sufficient to qualify the work to be valued at the new rate. "Or" thus stands as an alternative for conditions (a) and (b). If sub-clause (a) applies variation order under Clause 13 is a must. Besides, sub-clauses (b)(i) to (iv) will have no application. If (b) deals with a variation in excess of the limit, sub-clauses (b)(i) to (iv) are relevant.

The question arises if by use of and/or at the end of (iv) it is obligatory to obtain a variation order even if the quantity alone increases, qualifying for payment at a new rate. From practical considerations, the need for a written variation order in the case of excessive variation in quantity alone should not be insisted upon. However, it is necessary to issue the order fixing the rate or price for variations and adjustments ordered under Clause 13.

Clause 12.3(c) of the 2017 edition also deals with fixing of the rate or price for variations and adjustments ordered under Clause 13. It incorporates the usual condition found in all standard form contracts to derive a new rate based on the BOQ rates with suitable adjustments. In cases where no rate or price is specified in the contract and no specified rate or price is appropriate because the varied item is not of similar character or is not executed under similar conditions as any item in the contract, the new rate or price should be derived from the reasonable cost of executing the work including applicable percentage of profit stated in the contract data and in its absence of it, at 5%. The provision empowers the engineer to determine a provisional rate or price for the purposes of interim payment certificates until the matter is agreed or determined under Clause 3.7 subject to the dispute resolution provision in the agreement.

5.4.4 Omission in Scope of Work

Clause 12.4 of the FIDIC Form 1999 and 2017 editions and Clause 13 of the CPWD Form deal with deletions/omissions of works from the scope of the contract.

Clause 12.4 of FIDIC 2017 Edition

This clause deals with evaluation of omissions. It opens with the phrase, "Whenever the omission of any work forms part or (or all) of a Variation …". It means a variation order may be part addition, part omission or a full order consisting of only omission. The contractor has to submit a proposal for price adjustment subject to four conditions set out below:

(a) The value of omission which has not been agreed;

(b) The contractor will incur or has incurred the cost which if the work would not have been omitted, would have been deemed to be covered by the accepted contracted amount;

(c) The omission of the work will result (or has resulted) in the sum not forming part of the contract price, and

(d) The cost is not deemed to be included in the evaluation of any substituted work.

In respect of omissions, the contractor is entitled to claim extra for the cost of omitted work if the omitted work was such that its cost was included in the rate/price forming part of the contract price provided that the cost is not deemed to be included in the evaluation of any substituted work. The addition to which the contractor will be entitled will be in the form of compensation for the loss of overheads, expected profit besides cost of preliminary works/part of work, if executed and rendered valueless. For the purpose of evaluating the addition to the contract sum on account of omissions in the work, due credit should be given to the cost of any substituted work.

The contractor will be entitled to the additional amount only when all the three conditions are satisfied and the contractor has given notice to the engineer accordingly, with supporting particulars. The provision empowers the engineer to determine the cost which is to be included in the contract price under the power vested in him by the provisions of Clause 3.5. The said provisions require the engineer to consult each party and endeavour to reach an agreement by the parties. If such an agreement is not achieved, the engineer shall make a fair determination and give notice to both the parties with supporting particulars. The parties are to give effect to each such agreement or determination unless and until it is revised under clause 20 pertaining to claims, disputes and arbitration.

The deletion of part or whole of the work and its consequences are provided for in Clause 13 of the CPWD Form. The difference in the two conditions dealing with the same subject matter are noteworthy.

5.4.5 FIDIC Form 2017 Edition – Noteworthy Features

1. The right to order a variation can be exercised at any time before the issue of Taking Over Certificate.
2. The contractor is empowered "promptly" to serve notice to the engineer giving details of reasons why accepting the variation order would be difficult for him.
3. The engineer shall promptly, after receipt of notice from the contractor, respond by giving notice to the contractor "cancelling, confirming or varying" the order issued under Clause 13.3.1.
4. Under the heading "Value Engineering", Clause 13.2 enables the contractor, at any time, to submit a written proposal, at his cost, for the issue of a variation order under Clause 13.3.1 which will accelerate completion, reduce the price, improve efficiency or otherwise be of benefit to the employer.
5. The engineer shall, at the sole discretion of the employer, but as soon as practicable, respond to the proposal submitted by the contractor. If he gives his consent with or without comments, he shall issue a variation.
6. Clause 13.3.1 lays down the procedure for issuing a variation by the engineer by instruction by giving a notice, but does not specify a written notice. If an oral order is issued, it is advisable that the contractor confirms it in writing.
7. On receipt of a variation notice, the contractor is to submit his proposal of modification of the programme for completion of the works and any adjustment to the contract price within 28 days or such other time mutually agreed by the contractor and the engineer.
8. The engineer shall proceed under Clause 3.7 to determine issues of an extension of time and/or adjustment to the contract price. The provision calls for the engineer to initiate a consultation with both the parties and provide record to both the parties. If the parties do not reach an agreement, the engineer shall proceed under Clause 3.2 to determine the issues and give a notice within 42 days. If either party is dissatisfied with the engineer's determination, a notice of dissatisfaction would be issued stating the reasons for the same within 28 days. The matter would then be referred to Dispute Adjudication Board or arbitration as provided in the agreement.
9. Clauses 13.4 and 13.5 deal with Provisional Sums and Daywork, respectively. The provisions are self-explanatory.

10. Clause 13.6 under the paragraph heading: "Adjustment for Changes in Laws" provides for an adjustment to the contract price, an increase or a decrease resulting from any legislation including the judicial or official government interpretation or implementation of the law. It also contemplates an extension of time, if required, entitling the contractor to seek it. The procedure for this is set out.

11. Clause 13.7: "Adjustment for Changes in Cost." This sub-clause is attracted only if cost indexation is included in the contract. It specifically provides that until such time that current cost index is available, the price adjustment will be worked out on the basis of a provisional index for the issuance of an Interim Payment Certificate. The provision also stipulates working out a price adjustment favourable to employer, in case the contractor fails to achieve completion within the stipulated for it.

5.5 Government of India – Planning Commission Form

The relevant provisions are self-explanatory and read as follows:

CLAUSE 4 – Variations, Extras/Substituted Items

A. Variation permitted should be ± 25% in quantity of each individual item, and ± 10% of the total contract price. Within 14 days of the date of instruction for executing varied work, extra work or substitution, and before the commencement of such work, notice shall be given either (a) by the contractor to the owner of his intention to claim extra payment or a varied rate or price, or (b) by the owner to the contractor of his intention to vary a rate or price.

B. For items not existing in the Bill of Quantities or substitutions to items in the Bill of Quantities, rate payable should be determined by methods given below and in the order given below:
   1. Rates and prices in Contract, if applicable.
   2. Rates and prices in the Schedule of Rates applicable to the Contract ± tendered percentage, where appropriate.
   3. Market rates of materials and labour, plus 10% for overheads and Profits of contractor.
   4. Escalation to be paid as admissible.

C. If there is delay in the owner and the contractor coming to an agreement on the rate of an extra item, provisional rates as proposed by the owner should be payable till such time as the rates are finally determined.

D. For items existing in the Bill of Quantities but where quantities have increased beyond the variation limits, the rate payable for quantity in excess of the quantity in the Bill of Quantities plus the permissible variation should be:
   1. Rates and prices in contract, if reasonable, failing which
   2. Market rates of material and labour, plus 20% for overheads and profits of contractor.

5.6 NITI Aayog Model Form – Noteworthy Features

This form incorporates the best features of many standard form contracts including FIDIC Form 2017 edition. For example, under Change of Scope Clause 13.1.1, the authority reserves the right to require a contractor to make modifications or alterations to the works ("Change of Scope") before the issue of the Completion Certificate (as against Taking Over Certificate in FIDIC Form). The option to the contractor to submit a proposal for Change of Scope involving additional cost or reduction in cost is also provided. Any such Change of Scope shall be made and valued in accordance with the provisions of this Article 13. The said provision is, in many ways, different from and an improvement to the FIDIC Form as can be seen from the following.

Art. 13.1.2 defines Change of Scope to mean:

> (a) change in specifications of any item of Works; (b) omission of any work from the Scope of the Project except under Clause 8.3.3; provided that, subject to Clause 13.5, the Authority shall not omit any work under this Clause in order to get it executed by any other entity; or (c) any additional work, plant, materials or services which are not included in the Scope of the Project, including any associated tests on completion of construction.
>
> 13.1.3 If the Contractor determines at any time that a Change of Scope will, if adopted, (i) accelerate completion, (ii) reduce the cost to the Authority of executing, maintaining or operating the Project, (iii) improve the efficiency or value to the Authority of the completed Project, or (iv) otherwise be of benefit to the Authority, it shall prepare a proposal with relevant details at its own cost. The Contractor shall submit such proposal, supported with the relevant details including the amount of reduction in the Contract Price, if any, to the Authority to consider such Change of Scope. The Authority shall, within 15 (fifteen) days of receipt of such proposal, either accept such Change of Scope with modifications, if any, and initiate proceedings therefor in accordance with this Article 13 or reject the proposal and inform the Contractor of its decision. For the avoidance of doubt, the Parties agree that the Contractor shall not undertake any Change of Scope without a Change of Scope Order being issued by the Authority, save and except any Works necessary for meeting any Emergency.

The rest of the self-explanatory Clause 13 reads as follows.

13.2 *Procedure for Change of Scope*

13.2.1 In the event of the Authority determining that a Change of Scope is necessary, it may direct the Authority's Engineer to issue to the Contractor a notice specifying in reasonable detail the works and services contemplated thereunder (the "Change of Scope Notice").

13.2.2 Upon receipt of a Change of Scope Notice, the Contractor shall, with due diligence, provide to the Authority and the Authority's Engineer such information as is necessary, together with preliminary documentation in support of: (a) the impact, if any, which the Change of Scope is likely to have on the Project Completion Schedule if the works or services are required to be carried out during the Construction Period; and (b) the options for implementing the proposed Change of Scope and the effect, if any, each such option would have on the costs and time thereof, including the following details: (i) break down of the quantities, unit rates and cost for different items of work; (ii) proposed design for the Change of Scope; and (iii) proposed modifications, if any, to the Project Completion Schedule of the Project. For the avoidance of doubt, the Parties expressly agree that, subject to the provisions of Clause 13.4.2, the Contract Price shall be increased or decreased, as the case may be, on account of Change of Scope.

13.2.3 The Contractor's quotation of costs for the Change of Scope shall be determined on the following principles: (a) For works of similar nature compared to the Works being executed, the quotation shall be based on the rate for the work inclusive of all labour, Materials, equipment, incidentals, overheads and profit derived in accordance with the provisions of Clause 17.3; and the price adjustment in accordance with Clause 17.8 shall apply to the rates so worked out. (b) For works not similar in nature to the Works being executed, the cost of work shall be derived on the basis of [MORTH Standard Data Book and] the applicable schedule of rates for the relevant circle, as published by the respective State Government, and such rates shall be indexed with reference to the WPI[14] once every year at the commencement of the financial year, with the base being the month and year of the publication of the said schedule of rates; provided, however, that for any item not included in the schedule of rates, the prevailing market rates as determined by the Authority's Engineer shall apply, and for any item in respect of which [MORTH Standard Data Book] does not provide the requisite details, the Authority's Engineer shall determine the rate in accordance with Good Industry Practice.

13.2.4 Upon reaching an agreement, the Authority shall issue an order (the "Change of Scope Order") requiring the Contractor to proceed with the performance thereof. In the event that the Parties are unable to agree, the Authority may: (a) issue a Change of Scope Order requiring the Contractor to proceed with the performance thereof at the rates and conditions approved by the Authority till the matter is resolved in accordance with Article 24; or (b) proceed in accordance with Clause 13.5.

13.2.5 The provisions of this Agreement, insofar as they relate to Works and Tests, shall apply mutatis mutandis to the works undertaken by the Contractor under this Article 13.

13.3 *Payment for Change of Scope*

Payment for Change of Scope shall be made in accordance with the payment schedule specified in the Change of Scope Order.

13.4 *Restrictions on Change of Scope*

13.4.1 No Change of Scope shall be executed unless the Authority has issued the Change of Scope Order save and except any Works necessary for meeting any Emergency.

13.4.2 Unless the Parties mutually agree to the contrary, the total value of all Change of Scope Orders shall not exceed 10% (ten per cent) of the Contract Price.

13.4.3 Notwithstanding anything to the contrary in this Article 13, no change arising from any default of the Contractor in the performance of its obligations under this Agreement shall be deemed to be Change of Scope, and shall not result in any adjustment of the Contract Price or the Project Completion Schedule.

13.5 *Power of the Authority to Undertake Works*

13.5.1 In the event the Parties are unable to agree to the proposed Change of Scope Orders in accordance with Clause 13.2, the Authority may, after giving notice to the Contractor and considering its reply thereto, award such works or services to any person on the basis of open competitive bidding from amongst bidders who are pre-qualified for undertaking the additional work; provided that the Contractor shall have the option of matching the first ranked bid in terms of the selection criteria, subject to payment of 2% (two per cent) of the bid amount to the Authority, and thereupon securing the award of such works or services. For the avoidance of doubt, it is agreed that the Contractor shall be entitled to exercise such option only if it has participated in the bidding process and its bid does not exceed the first ranked bid by more than 10% (ten per cent) thereof. It is also agreed that the Contractor shall provide assistance and cooperation to the person who undertakes the works or services hereunder, but shall not be responsible for rectification of any Defects in works carried out by other agencies. For the avoidance of doubt, the Authority acknowledges and agrees that it shall not undertake any works or services under this Clause 13.5.1 if such works or services cause a Material Adverse Effect on the Contractor.

13.5.2 The works undertaken in accordance with this Clause 13.5 shall conform to the Specifications and Standards and shall be carried out in a manner that minimises disruption to the Project. The provisions of this Agreement, insofar as they relate to Works and Tests, shall apply mutatis mutandis to the works carried out under this Clause.

---

[14] Wholesale Price Index.

13.5.3 Notwithstanding anything to the contrary in this Article 13, no change arising from any default of the Contractor in the performance of its obligations under this Agreement shall be deemed to be Change of Scope, and shall not result in any adjustment of the Contract Price or the Project Completion Schedule.

5.7 Provisions in Standard Form Contract of Central and State Public Works, Railways, Military Engineering Services and Other Departments of the Government of India

The provisions made in contract by the Government of India and public bodies in India include provision reading somewhat as Clause 12 of the Standard Form below:

> The Engineer-in-charge shall have power to make any alterations in, omissions from, additions to or substitution for, the original specifications, drawings, designs and instructions, that may appear to him to be necessary or advisable during the progress of the work, and the contractor shall be bound to carry out the work in accordance with any instructions which may be given to him in writing signed by the Engineer-in-charge, and such alterations, omissions, additions or substitutions shall not invalidate the contract; and any altered, additional or substituted work which the contractor may be directed to do in the manner above specified as part of the work shall be carried out by the contractor on the same conditions in all respects on which he agreed to do the main work and at the same rates as are specified in the tender for the main work. The time for the completion of the work shall be extended in the proportion that the additional or substituted work bears to the original work, and the certificate of the Engineer-in-charge shall be conclusive as to such proportion. And if the altered, additional or substituted work included any class of work for which no rate is specified in this contract, then such class of work shall be carried out at the rates entered in the schedule of rates of the CPWD Schedule of Rates of the year on which the estimated cost shown in the tender is based provided that when the tender for the original work is a percentage above the schedule rates the altered, additional or substituted work required as aforesaid shall be chargeable at the said schedule rate plus the same percentage deduction/addition and if such class of work is not entered in the said schedule of rates, then the contractor shall within seven days of the date of the receipt of the order to carry out the work inform the Engineer-in-charge of the rate which it is his intention to charge for such class of work, and if the Engineer-in-charge does not agree to this rate he shall by notice in writing be at liberty to cancel his order to carry out such class of work and arrange to carry it out in such manner as he may consider advisable provided always that if the contractor shall commence work or incur any expenditure in regard thereto before the rates shall have been determined as lastly hereinbefore mentioned, then and in such case he shall only be entitled to be paid in respect of the work carried out or expenditure incurred by him prior to the date of the determination of the rate as aforesaid according to such rate or rates as shall be fixed by the Engineer-in-charge. In the event of a dispute the decision of the Superintending Engineer of the Circle shall be final.

Under this clause the engineer in charge is apparently empowered to:

1. Increase the quantum of work under each item of the BOQ to any extent without fixing a new rate.

2. Order additions and alterations in the work for which no rate is specified in the contract at the mutually agreed rate but not exceeding the rates entered in the schedule of rates of the concerned division.

3. Make any additions to the original specifications that may appear to him to be necessary or advisable during the progress of the work and the contractor is bound to carry out the work in accordance with any instructions given to him in writing signed by the engineer-in-charge.

4. As regards the payment for the additional work which the contractor may be directed to do it provides that if such class of work is not entered in the said schedule of rates then the contractor should inform the engineer-in-charge within seven days of the receipt of the order the rate he wants to charge for such class of work and the engineer-in-charge, if he does not agree to the said rate, may cancel the order for such additional work, and if the contractor has, commenced the work or incurred expenditure in regard thereto before the determination of the rates the contractor shall be paid in respect of work carried out or expenditure incurred by him prior to the determination of the rates according to such rates or rates as shall be fixed by the engineer-in-charge and in the event of a dispute the decision of the Superintending Engineer of the Circle would be final.

The above and similar provisions give rise to the following questions:

1. Is the power of the engineer to order additional work under the above stipulation unlimited?
2. If the work is executed by the contractor after submitting rates expected by him, the engineer does not cancel the order but after completion the engineer decides rates different from those claimed/demanded by the contractor, which rates the contractor would be entitled to?
3. When may a contractor be not bound to execute variations?

Answers to the above questions and the shortcomings of the above standard form provisions are considered below:

5.8 Right to Change Is Limited

The concept of variation of the quantity of work is a common feature of works contracts. This is because in contracts relating to major works, the estimates of work at the time the tenders are invited can only be approximate. A variations and deviations clause, though it apparently allows the engineer to order changes, does not authorize him to change radically the original nature of the contract. The law on this point, simply stated, is that the engineer has no power to change the essential nature or the main purpose of the contract, but he may make changes incidental to the primary object of the contract. The construction of such clauses is not only in accordance with their obvious purpose, but is also strongly supported by public policy. If the government were empowered by such clauses to alter materially the object of the contract after construction has started, all bidders would have to take such a possibility into consideration and materially raise their bid in anticipation of such losses, thus increasing the cost of public work.[15]

In this respect, the following observations of McCardie J.[16] are still noteworthy almost a century since the decision: "It is essential to remember, however, that words, even though general, must be limited to circumstances within the contemplation of the parties."

A contract literally suggested that after £300,000 profit was earned by the contractor, he could be compelled to work without reward or limit, on any further "extras" which the commissioners

---

[15] Boomer v. Abbett, California Court of Appeals, Civ. 15408, 24th November 1953.
[16] Naylor, Benzon and Co. v. Krainische Industrie Gesellschaft, [1918] 1 KB 331 at p. 335.

might order, only the most compelling language would induce a court to construe the instrument as placing one party completely at the mercy of the other. Where the language of a contract is capable of a literal and wide, but also of a less literal and a more restricted, meaning, all the relevant circumstances can be taken into account in deciding whether the literal or a more limited meaning should be ascribed to it.[17]

Thus, the law can be said to be well settled that the power of the employer to vary the terms relating to the quantum of work cannot be unlimited. The passage in *Hudson's Building and Engineering Contracts*[18] that such power "although apparently unlimited, must in fact be limited to ordering extras of a certain value and type" was approved by the Supreme Court of India in the case of the appellant's claim for payment for the additional work of cutting hard rock. The appellant had claimed a higher rate of Rs. 200/- per thousand cft. for this additional work. Under the agreement the appellant was required to execute hard rock cutting to the extent of 7,54,530 cft. but actually he was required to execute such cutting to the extent of 18.15 lacs cft. The extent of the additional work was about 10.60 lacs cft., i.e. about 140%. While undertaking the execution of the additional work of hard rock cutting, the appellant wrote a letter to the executive engineer, requesting revision of the rate for hard rock cutting and stating that the minimum working rates for this item were 52% above the tendered rates. On the executive engineer's request, the appellant submitted his analysis of rate for hard rock cutting. The executive engineer also had an analysis of rates done on the basis of the data collected on actual observation and he arrived at a figure of Rs. 237/- per thousand cft. All the concerned engineers recommended the payment at the rate of Rs. 200/- per thousand cft. However, the government did not agree to pay at a rate in excess of the rate of Rs. 129/- per thousand cft. plus 2% stipulated under the agreement. The dispute was, therefore, referred to arbitration. The arbitrator partly allowed the claim.

Before the single judge, the respondent submitted that Clause 12 of the agreement between the parties provided that additions to the contract work had to be carried out by the contractor on the same conditions in all respects on which he agreed to do the main work and at the same rates as specified in the tender for the main work. The single judge rejected the said contention and held that the arbitrator was determining only the value of the additional work at the rate of Rs. 200/- which had been agreed by the engineer-in-charge and the superintending engineer of the circle as contemplated by Clause 12 and the scope of the inquiry before the arbitrator was only the quantity of work which was additional to the quantities specified in the agreement.

The division bench took the view that under Clause 12 of the agreement, the provision with regard to the fixing of the rate by the engineer-in-charge and the superintending engineer of the circle comes into play only when the additional item of work does not form part of the main work and the rates for such work are not specified in the schedule of rates. The division bench held that since the additional hard rock cutting job done by the appellant was part of the main work and the agreement provided a rate for the said item, there was no occasion for the engineer-in-charge or the superintending engineer to fix the rate for the extra quantity of hard rock cutting and that the action of the arbitrator in allowing the rate to the contractor in excess of the agreed rate for the job of hard rock cutting was against Clause 12 of the agreement, and that thereby the arbitrator had exceeded his jurisdiction.

To decide the issue, a three-judge bench of the Supreme Court[19] cited with approval, the following passage from *Hudson's Building and Engineering Contracts*:[20]

---

[17] Parkinson (Sir Lindsay) & Co. Ltd. v Commissioners of His Majesty's Works and Public Buildings, [1949] 2 K.B. 632, p. 662.

[18] 1979, Sweet & Maxwell Ltd., 10th ed., p. 549.

[19] S. Harcharan Singh v. Union of India, (1990) 4 SCC 647. Also see: Food Corporation of India v. Vikas Majdoor Kamdar Sahkari Mandli Ltd., (SC): 2008(1) R.C.R.(Civil) 284: 2007(6) Recent Apex Judgments (R.A.J.) 583: 2007(12) JT 517: 2008(1) ICC 663: 2007(13) Scale 126: 2007(11) SCR 1038: 2007(13) SCC 544: 2008(1) W.L.C. 309: 2008(1) DNJ 183: 2008(1) R.A.J. 140: 2008(1) All WC 495: 2008(2) M.L.W. 25: 2007 AIR (SCW) 6999: 2008(2) MLJ 857: 2008(1) JLJR 149: 2008(1) PLJR 149.

[20] 8th Edn., pp. 294, 296.

It may be that it can be inferred from the terms of the contract that the power to order extras, although apparently unlimited, is in fact limited to ordering extras up to a certain value and, in such a case, extras ordered in excess of that amount, although work of a kind contemplated by the contract, may yet be quite outside the terms of the contract. If the extra work ordered is outside the contract the terms of the contract have no application.

The Apex Court also thought it relevant to take note of a decision of the Court of Appeal in England in Parkinson (Sir Lindsay) and Co. Ltd. v. Commissioners of His Majesty's Works and Public Buildings,[21] and observed:

> Here also the question has often arisen whether the contractor under the variation clause is liable to execute the extra or additional quantities of the tendered items at the tendered rates to an unlimited extent. In some awards given by the arbitrators in the Central Public Works Department of the Government of India the variation of the tendered quantities under the variation clause in the contract has been restricted to 10% beyond which the contractor was entitled to claim as extras and awards have been accepted and implemented by the Government. It appears that the standard form of contract of the Central Public Works Department has been amended and now it specifically permits for a limit of variation called "deviation limit" up to a maximum of 20% and up to such limit the contractor has to carry out the work at the rates stipulated in the contract and for the work in excess of that limit at the rates to be determined in accordance with Cl. 12-A under which the Engineer-in-charge can revise the rates having regard to the prevailing market rates.[22]

It was held that while considering the claim of the appellant, the arbitrator was required to consider the terms of the contract and to construe the same. It was, therefore, permissible for the arbitrator to consider whether Clause 12 of the contract enabled the engineer-in-charge to require the appellant to execute additional work without any limit or a reasonable limit should be placed on the quantity of the additional work which the appellant might be required to execute at the rate stipulated for the main work under the contract. The award was upheld.[23]

EXAMPLES FROM CASE LAW

1. The contractors agreed with His Majesty's Commissioners of Works and Public Buildings to erect an ordnance factory according to the general conditions and specifications and BOQs and drawings annexed for the contract sum of £3,500,000. Under the general conditions of contract, the Commissioners had power, at their absolute discretion, to modify the extent and character of the work or to order alterations of or additions to the works and it was the duty of the contractor to comply with the architect's instructions in this respect. In the contract it was also provided that it is probable that further work to the value of approximately £500,000 would be ordered on a measured basis under the terms of the contract. The contract was amended by a deed of variation and it was provided that exceptional methods should be used to hasten the work and that a system of uneconomic working should be introduced to bring about the completion of the factory by the date fixed by the contract. The Commissioners ordered work to be executed greatly in excess of the amount contemplated although not different in character from that covered by the varied contract. The works could not be completed until a year later than the time anticipated and the actual cost of the contracts was £6,683,056 which amount had been

[21] Parkinson (Sir Lindsay) and Co. Ltd. v. Commissioners of His Majesty's Works and Public Buildings, [1949] 2 K.B. 632.

[22] S. Harcharan Singh v. Union of India, (1990) 4 SCC 647: Paragraph 78.

[23] S. Harcharan Singh v. Union of India, (1990) 4 SCC 647: (AIR 1991 SC 945); 1990(2) Arb LR 243.

paid to them along with £300,000, the maximum profit under the deed of variation. During the progress of the work the contractors had complained to the Commissioners that they were being called on to execute more work than was contemplated by the varied contract and claimed that they were entitled to extra remuneration for the work in excess of that contemplated, but they proceeded with the work at the request of the Commissioners leaving the issue to be subsequently decided by arbitration. The arbitrator found that the estimated cost of the work under the varied contract was £5,000,000 and awarded £90,298 as proportionate or reasonable profit or remuneration to the contractors for the additional work. The award was upheld by the Court of Appeal on the view that a term must be implied in the varied contract that the Commissioners should not be entitled to require work materially in excess of the sum of £5,000,000 and that such excess work having been executed by the contractors, the Commissioners were liable to pay the contractors reasonable remuneration therefor. The Commissioners relied on Condition 33 of the original contract which gave the Commissioners an unlimited power of ordering extras even to the extent of altering the character of the work. The contractors, on the other hand, placed reliance on the observations of McCardie J.,[24] "that words, even though general, must be limited to circumstances within the contemplation of the parties". The Court of Appeal upheld the sum awarded by the arbitrator to the contractor as a proportionate or reasonable profit to the contractors for the additional work.[25]

2. A contract for the construction of additional runways on a city-owned airport empowered the engineer to make such changes in the work as he might consider necessary or advisable to complete the work provided that such changes did not change the estimated cost of the work by more than 25%. It was held that the engineer could not eliminate part of the work linking the existing runways and the proposed runways. This created a gap between the existing and proposed runways; the gap made access to the proposed work more difficult and greatly increased the cost of the work. The contractor brought an action to recover the increased cost. The contractor's claim was resisted on the ground that the engineer had the right to effect changes to the extent of 25% of the estimated cost; whereas the estimated cost of the part of the work that was eliminated amounted to only 20%. The court, allowing the contractor a recovery, stated:

> The power vested in the engineer to effect changes in the quantities of the work is not so extensive as to enable him to abrogate or change the contract which the parties executed ... nor does it authorize the defendant to employ such right to defeat the object of the contract which is reasonably deducible from its terms.[26]

The changes which could be ordered had to be viewed against the background of the work described in the contract and the language used in the specifications; they had to be directed either to the achievement of a specific improvement or the elimination of work not integrally necessary to the project. The purpose of such powers of change was to maintain a degree of flexibility in adapting conditions of the contract to the objective sought. However, the discretion of the engineer had to be exercised within the framework of the contract and for the purpose of implementing the work as originally intended, not in an arbitrary manner.

---

[24] Naylor, Benzon and Co. v. Krainische Industrie Gesellschaft, [1918] 1 KB 331, p. 335.

[25] Parkinson (Sir Lindsay) and Co. Ltd. v. Commissioners of His Majesty's Works and Public Buildings, [1949] 2 K.B. 632.

[26] Hansler v. City of Los Angeles, 268, p. 2d, 72 District Court of Appeals of Calif., 23rd March 1954.

3. Where the original design of a building showed a staircase from the second floor to the third floor level, and the design was later altered by the employer cancelling the staircase of the second and third floor levels, the contractor had to use extra manpower and scaffolding to work on the roof. The contractor's claim for extra cost was partly awarded by the arbitrator. It was held that there was no infirmity in the award.[27]

The changes which may be ordered have to be, thus, examined against the background of the work described in the contract and the details in the specifications. The changes must be:

(a) Within the overall contractual framework
(b) With an objective of completing the work as originally intended
(c) Aimed at improving the contractual work or removing from the contractual scope the work that is not fundamentally necessary for the completion of the original project
(d) Used in a fair (not arbitrary) manner.

The language and the spirit of the alteration and omission clause deserve to be considered. For example, in the first example, the contract for the construction of a subway granted an engineer sweeping unrestricted power "to omit any portion of work without constituting grounds for any claim by the contractors for payment or allowances for damages", and the work deleted did not detract from the ultimate purpose of the contract. In the second example, a contractor undertook the work of building a section of a lock which was to be built at a particular point and the state had reserved the right to make such additions or deductions from work or changes in the plans and specifications covering the work as may be necessary. When it was found that the soil conditions made it impossible to construct the lock at the designated site, the state deleted it from the contract and relocated the lock to an area where another contractor was performing some work. In both these cases, the contractors were denied recovery of loss of profit because the deletion served the purpose of the contracts and came within the contractual term for variation.[28]

5.8.1 Limit on Variation in Quantities

There is a good reason why, in modern works contracts, a limitation of 20% or even up to 25% is placed on the power of alteration, whether plus or minus. Such a limitation up to 20% or 25% now imposed under Clause 12A of the standard terms of CPWD contracts is considered below at length. A typical representative provision is contained in "Standard Contract Clauses for Domestic Bidding Contracts" published by the Ministry of Statistics and Programme Implementation, Government of India, 2001. Clause 40.1 of the said conditions limits the right to change the quantities of each individual item to plus/minus 25% and of the total contract price to 10%.

Clauses 40.2 and 40.3 provide for payment for items not in the BOQ or substituted items, and for quantities in excess of the BOQ quantities plus the variation limit. Three alternatives are given and the rates are to be worked out in the order mentioned, namely, (i) contract rates plus escalation, or (ii) rates and prices in the schedule of rates applicable to the contract plus ruling percentage, or (iii) market rates of materials, labour, hire charges of plant and machinery used plus 10% for overheads and profit. Clause 40.3 makes it clear that the new rate is to apply to

---

[27] Shiv Kumar Wasal & Co. v. Delhi Development Authority, (Delhi) 2016(154) DRJ 420: 2016(4) R.A.J. 157: 2015(54) R.C.R.(Civil) 717.
[28] N.E. 2d 559. 15. Kinser Construction Co. v. State, 204, N.Y. 381,97, N.tE. 871; Dc Belso Construdion Corp. v. City of New York, 278, N.Y. 154.

quantities in excess of the variation limit, that is for the quantities in excess of 125% of BOQ quantities.

However, there are still some contracts entered into wherein the engineer's power to order extra quantities is seemingly unlimited by the use of words such as "any variation". In some contracts, it is stipulated that the limit on variation will not apply to foundation items. In both these cases, it is necessary to put an implied limitation such that one party is not put completely at the mercy of the other.[29] Where the language of the contract is capable of a literal and wide but also of a less literal and a more restricted meaning, all relevant circumstances can be taken into account in deciding whether the literal or a more limited meaning should be ascribed to it.[30]

5.8.2 Exclusion of Foundation Items from Variation Limit

Some contracts include provisions which state that the "variation limit" agreed between the parties will not apply to foundation items which can be altered to "any variation". It is possible to argue that foundation items being more difficult to evaluate due to unknown and uncertain working conditions, the parties agree that even 25% extra quantity will not be governed by the original rate and any extra beyond a reasonable limit of the originally estimated quantity may attract new rate(s). In such a case, the reasonable limit may be 10%. The decision of the Supreme Court of India[31] supports this view so long as the court or the arbitrator considers some "limited" excess quantity over and above the agreed quantity being payable at the originally agreed rate(s). This is so because the conditions normally stipulate that the quantities shown in the BOQ are approximate and likely to vary. It can be said that the contractor, relying upon this provision, incorporated in his price build-up some relief to cover excessive variation of quantities.

5.8.3 Variation of Varied Quantity

The above principles deserve application to cases even where the variation limit is agreed by the parties but the extent to which the varied quantity can be ordered is not agreed. Occasionally, in some contracts involving huge amounts, it has been observed that the power to order variation in excess of the limit gets abused and varied quantities are again varied and that, too, without mutually agreeing to the rates/prices of varied items. The dispute resolution mechanism does not yield a result to generate cash-flow for executing the work during its progress. Under the circumstances, the contractor's right to refuse to execute varied work, unless the rates/prices are agreed either amicably or by a dispute resolution mechanism, must be upheld, it is submitted. The difficulty is that often the nature of the extra work is such that unless it is completed, the balance work cannot be proceeded with. The contractors, therefore, do not take the risk to refuse to carry out the varied work and accept provisional payment based on unworkable rates. Within reasonable limits this works, but if the financial burden is unbearable it results in termination of the contract.

It needs to be noted that under the general law of contracts, once the contract is entered into, any clause giving absolute power to one party to override or modify the terms of the contract at his sweet will or to cancel the contract even if the opposite party is not in breach, will amount to interfering with the integrity of the contract.[32]

---

[29] *Hudson's Building and Engineering Contracts*, 10th ed. p. 549.

[30] Sir Lindsay Parkinson & Co. Ltd. v. Commissioners of Works, Court of Appeal, [1949] 2 K.B. 632.

[31] S. Harcharan Singh v. Union of India, (1990) 4 SCC 647: (AIR 1991 SC 945).

[32] National Fertilizers Ltd. v. Puran Chand Nangia, (SC) AIR 2001SC 53, 2000(8) SCC 343; Maddala Thathiah v. Union of India, AIR 1957 Madras 82. Also see: AIR 2001 SC 53.

5.9 Valuation of Cost of Extra Work

5.9.1 Clauses 12 and 12A of the CPWD Form

Clause 12 of the CPWD form stipulates that the rates for additional, altered or substituted work shall be worked out in accordance with the five sub-clauses in their respective order as follows:

(i) Rates stipulated in the contract

(ii) Rates to be derived from the contract rates

(iii) Rates entered into in PWD schedule of rates plus/minus the tender percentage above or below the estimated cost

(iv) Rates to be derived from schedule of rates plus/minus the tender percentage above or below the estimated cost of the sub-work

(v) Rates to be submitted by the contractor within seven days.

The clause at the end provides:

> Except in case of items relating to foundations, provisions contained in sub-clauses (i) to (v) above shall not apply to contract or substituted items as individually exceed the percentage set out, in the tender documents (referred to herein below as "deviation limit"), subject to the following restrictions.

The clause then provides four restrictions:

(a) The deviation limit referred to above is the net effect of (algebraical sum) of all additions and deductions ordered.

(b) In no case shall the additions and deductions (arithmetical sum) exceed twice the deviation limit.

(c) The deviations ordered on items of any individual trade included in the contract shall not exceed plus/minus 50% of the value of that trade in the contract as a whole or half the deviation limit, whichever is less.

(d) The value of additions of items of any individual trade not already included in the contract shall not exceed 10% of the deviation limit.

After defining individual trade to mean trade sections such as excavation and earthwork, concrete, wood work, etc. the clause ends thus:

> The rates of any such work except the items relating to foundations which is in excess of deviation limit shall be determined in accordance with the provisions contained in clause 12A.

The form contains Clause 12A for determining the rates for the work in excess of the deviation limit, which reads as follows:

> Clause 12A
>
> In the case of contract or substituted items which individually exceed the quantity stipulated in the contract by more than the deviation limit, except the item relating to foundation work, which the contractor is required to do under Cl. 12 above, the contractor shall within 7 days of the receipt of the order, claim revision of the rates supported by proper analysis of items for quantities in excess of the deviation limit, notwithstanding the fact that the rates for such item exist in the tender for the main work or can be derived in accordance with the provisions of sub-cl. (ii) of Cl. 12 and Engineer-in-charge may revise their rates, having regard to the prevailing

market rates and the contractor shall be paid in accordance with the rates so fixed. The Engineer-in-charge shall, however, be at liberty to cancel his order to carry out such increased quantities of work by giving notice in writing to the contractor and arrange to carry it out in such a manner as he may consider advisable. But under no circumstances, the contractor shall suspend the work on the plea of non-settlement of rates of items falling under this clause. All the provisions of the present paragraph shall equally apply to the decrease in the rates of the items for quantities in excess of the deviation limit, notwithstanding the fact that the rates for such items exist in the tender for the main work or can be derived in accordance with the provisions of sub-clause (ii) of the precedent Cl. 12 and the Engineer-in-charge may revise such rates having regard to the prevailing market rates.

A similar provision is to be found in forms in use by the Military Engineering Services of Government of India and in Indian Railway contracts. This provision is also not free from doubt and invariably gives rise to a dispute.

Items in BOQ Amount to Different Items or Single Item

Variation clauses in some contracts stipulate two conditions to make a new rate applicable for the varied work. The conditions are that the item in question must be priced more than 1% of the contract price and the varied quantity be in excess of the limit agreed, namely 20% or 25%. Clause 12A of the CPWD Form above serves as a typical example.

Sometimes while interpreting such provisions of the contract a question arises as to whether certain items in the BOQ amount to different items or a single item. For example, do cement concrete of Grade M-150, M-200 and M-250 constitute separate individual items or one single item of concreting? The interpretation given by arbitrators, one of whom was a technical expert, that each item is separate was upheld by the H. P. High Court.[33] The following examples will highlight some of the problem areas.

EXAMPLES FROM CASE LAW

1. The total contract sum was exceeded by 20%. The refusal of the contractor's claim was based on the ground that Clause 12A is attracted only when the quantities of individual items increase and not the overall cost. The dispute was referred to arbitration. The arbitrator allowed the relief. It was held by the Supreme Court that the contract did visualize the contractor raising a claim for revision of rates. The dispute was as to when such a claim could be raised and clearly related to the interpretation of the terms of the contract. The construction placed upon the contract by the contractor could not be said to be an implausible one and the decision of the arbitrator was held to be final and binding.[34]

2. The controversy in this case was whether cement concreting M-150 grade and cement concreting M-200 grade were separate items or these along with cement concreting M-250 grade constituted one item. The arbitrators referred to Clause 12A, CPWD Form and the schedule of quantities and rates to hold:

A cursory glance of this clause shows that the revised rates are admissible in respect of the item(s) which "individually exceeds the quantity stipulated in the contract by more than twenty per cent." It is also evident that in the "Schedule of Quantities and Rates" ... and items (iii) and (iv) i.e. cement concrete Grade 150 and M-200 have been shown as

---

[33] M/s. Hydel Construction Ltd. v. H.P. State Electricity Board, AIR 2000 HP 19. See Example (2) below for more details.

[34] Himachal Pradesh State Electricity Board v. R.J. Shah & Co., 1999(2) Arb. LR 316 (S.C.).

"individual items" with different rates and in fact the respondent itself invited separate rates for all the items listed therein (including the items under dispute). The Items (iii) and (iv) in the Table ... are distinct and separate from each other and qualify for revised rates in terms of Clause 12-A as the deviation in each case exceeds twenty per cent.

The above findings of the arbitrators were set aside by the single judge who held:

> On the basis of the agreement between the parties, the contention put forth on behalf of the Board appears to be correct. At this stage, Item No. IV of the Schedule of Quantity and Rates attached with the Agreement can safely be referred. This item deals with the term "Concreting". Thus, Item No. IV is an individual Item of Concreting comprising of various grades of concreting. In those various grades, as per table put in by the Board, the deviation was only to the extent of 8%.

On appeal against the above decision, reliance was further placed upon Clause 5.01 "Scope of Work" which inter alia stipulated

> [t]he items of concrete will have to be split up into several items according to the class of concrete to be used and its location, and will be measured and paid for accordingly. The general specifications described hereinafter shall, however, in relevance apply to all concrete items.

It was held by the H.P. High Court:

> In view of the law laid down in Municipal Corporation of Delhi v. M/s Jagan Nath Ashok Kumar; Jagdish Chander v. Hindustan Vegetable Oils Corpn. and S. Harcharan Singh v. Union of India ... we are of the view that it was not for the learned single Judge to interfere with the interpretation given by the Arbitrators, one of whom was a technical expert, that cement concrete Grade M-150, M-200 and M-250 are separate individual items and come to his own independent conclusion that these constitute single individual item of concreting. Therefore, we set aside the findings of the learned single Judge in respect of Claim No. 1 by rejecting the Objections of the Board in this regard and accept the award of the Arbitrators for Rs. 23,38,840.73 against Claim No. 1 as correct and make it rule of the Court.[35]

3. An item in a contract read: "Reinforcement for RCC work including cutting, bending and binding and placing in position etc. complete (b) Cold Twisted Bars." The quantity as per the agreement worked out to 2,64,453 kg and the rate was Rs. 7.50/- per kg. Drawings were to be supplied by the respondent. On the basis of the drawings, it transpired that the quantity of the said agreement as provided for in the contract would be inadequate and the same exceeded the deviation limit provided for in the contract. The deviation limit provided under the contract was 25%. As per the terms of the contract, the rates for individual item executed beyond 25% were to be paid in accordance with Clause 12A of the agreement reproduced above.

It was the case of the petitioner that as per Clause 12A, the petitioner requested the respondent to revise the rates for this item of work executed beyond the deviation limit. Rates claimed were Rs. 8.68/- per kg. The rate quoted by the petitioner for the excess work executed beyond the deviation limit was according to the prevalent market rate at that point of time. It was further contended by the petitioner that the respondent never denied the rate of Rs. 8.68/-

[35] M/s. Hydel Construction Ltd., Appellant v. H.P. State Electricity Board, AIR 2000 HP 19.

per kg as the market rates, nor denied the claim of the petitioner under Clause 12A of the agreement. The respondent's only defence was that the petitioner had not fulfilled other conditions required under the said clause for making this claim. What those conditions were, the respondent never specified. It was the grievance of the petitioner that the arbitrator, behind the back of the petitioner/objector, took into consideration the rate of other alleged contemporary agreements. The arbitrator stated as follows:

> I uphold the contentions of the respondent in this regard – in fact keeping in view that the reinforcement was being issued by respondents at a fixed cost of Rs. 6.50 per kg. Comparing with the agreement rates of other contemporary agreements, I am of the view that the agreement rate of Rs. 7.50 per kg of the item already reflected its market rate.

The Delhi High Court held:

> From the reading of these observations it is apparent that the arbitrator did consider some contemporary agreements. There is nothing on the record to show that these agreements were relied [upon] by the respondent or were shown to the petitioner by the arbitrator. Hence, reliance by the arbitrator on contemporary agreements to arrive at market rates behind the back of the petitioner was tantamount to misconduct of the proceedings. The respondent has not been able to show even today that contemporary agreements were placed before the arbitrator to prove that Rs. 6.58 per kg was the prevalent market rate. Therefore, ... the award against Claim No. 1 is liable to be set aside. Order accordingly."[36]

Though the case was decided on technicalities, the right to get paid for at a new rate, under similar circumstances, was vindicated. Clause 12 imposes the condition of demanding the new rate within seven days of receipt of order duly supported by rate analysis, which it is advisable to satisfy for a valid claim.

4. **Interpretation of plus minus 25%.** An arbitration award interpreted a "variation" clause in the contract to allow the appellant to vary the extent of the contract work, both upwards and downwards up to 25%. The award was challenged under the Indian Arbitration Act, 1940 concerning the interpretation of the question whether (as contended by the appellant) the said 25% was to be arrived at by taking into account the net overall increase in the work, i.e. by adding up the increases in work and deducting therefrom the decreases in work, or whether (as contended for the respondent-contractor) the 25% was to be computed by adding up the total variations, both involving the increase in the work and the decrease in the work. The importance of the point is that if the variations exceeded 25% of the contract price, the contractor was not confined to the contract rates but could claim market rates.

The relevant clause read as follows:

> The contract price has been arrived at on the basis of your quoted rates in your tender and the enclosed schedule of quantities, your quoted rates shall hold good for a variation of ± 25% (plus/minus twenty five percent) of the contract price stated in this work order, beyond which your quoted rates will be suitably revised subject to mutual agreement.

It was contended that the arbitrator had acted without jurisdiction in granting extra amount or higher rates for the work done up to the extended date 30th October 1986. This was prohibited by several clauses of the NIT, Special and General Conditions and under annexure attached to the work order. The variation limit of plus/minus 25% of the contract

---

[36] M/s. Kirpal Singh Khurana and Sons v. Union of India, 59(1995) DLT 259.

price was applicable on the "total contract price" and not on any individual quantities or items. Any revision of rates would be permissible only after the total contract price stood increased or decreased beyond 25% on actual execution and completion of the contract project. In any event, the arbitrator could not have allowed a uniform increase of 50% for all items.

On the other hand, the respondent-contractor contended that the question was not one of increase or decrease in the total contract value. If the sum totals of the variations, i.e. both plus and minus exceeded 25%, the contract rates were no longer binding and market rates had to be paid. The trial court had found, as a fact, that the sum total of the additions and deletions in the work exceeded 100%.

The rival contentions were summed up by the Supreme Court of India thus:

> The contention of the appellant is that the above exception is applicable only to the net difference between the increases and decreases and if it works out to more than 25% of the contract value, then rates can be revised. For example if the contract value is Rs. 50 lakhs, the increases are of a value of Rs. 15 lakhs and the reductions are of a value of Rs. 10 lakhs, the net difference according to the appellant, in the overall contract value is only Rs. 5 lakhs and being 10% of Rs. 50 lakhs, there can be no escalation in rates.
>
> On the other hand, the respondent contends that one has to add up the total variations both plus and minus and hence, in the above example, the value of total variation, both plus and minus amounts to Rs. 25 lakhs which works out to more than 25% (in fact 50%) of the contract price and the enhanced rates will be applicable.

It was further observed:

> The point raises certain important issues concerning integrity of the contract. The concept of variation of the question of work is no doubt a common feature of works contracts. This is because in contracts relating to major works, the estimates of work at the time the tenders are invited can only be approximate. But, it was also realised that the power of the employer to vary the terms relating to the quantum of work cannot be unlimited. In Hudson's Building and Engineering Contracts (8th ed.) (pp. 294–296) it has been pointed out that this power "although unlimited, is in fact limited to ordering extras up to a certain value."

Reliance was placed on the decision in S. Harcharan Singh v. Union of India,[37] and it was held:

> We are of the view that the above said clause "25%±" was understood by the arbitrator in a reasonable manner as being applicable to a case where the value of the sum total of the additions and deletions exceeded 25% of the contract price. That construction, in our view, cannot be said to be vitiated by any serious error of law. The following are our reasons.
>
> When a contractor bids in a contract, he has to offer reasonable rates for the works which are both difficult to perform and other works which are not that difficult to perform. Every contractor tries to balance his rates in such a manner that the employer may consider his offer reasonable. In that process the contractor tries to get a reasonable margin of profit by balancing the more difficult (and less profitable items) and the less difficult (and more profitable items). His bid is, normally, a package. If the employer is permitted in law to make variations upwards and downwards – even if it be up to a limit beyond which market rates become payable – then the interpretation of the clause must be one which balances the rights of both parties. For example, if the plus and minus variations go beyond 25% and are made in a manner increasing the less profitable items

---

[37] (1990) 4 SCC 647: (AIR 1991 SC 945).

and decreasing the more profitable items, and if the net result of the contract is to be the basis, as contended by the appellant, then it may work out that the contractor could be made to perform a substantially new contract on the same contracted rates. In fact, if the said reasoning of the appellant is accepted and if, in a given case, the value of the increases in unprofitable items is 50% of the contract value and the value of the reductions of the remaining more profitable items is 50% of the contract value, it could still be contended for the appellant that the net variation was nil, even though that was a situation where the contract had been substantially modified and was almost a different contract from the one stipulated. Such an unreasonable construction is to be avoided and was rightly avoided by the arbitrator.

It was further held:

> The additions and decreases in work are, in our opinion, therefore both independent for the purpose of finding out the ± 25% variation and have to be pooled together. The arbitrator was right in thinking that the case fell within the exception. Obviously, he must have felt that the plus and minus variations are more than 25% and that the contract rates are no longer binding. His construction of the clause appears to be rational and just and cannot be said to be unreasonable. In result, the appeal was dismissed.[38]

5. The extra claim was based on the fact that the quantity of excavation of hard rock was abnormally high and much beyond the anticipated quantity indicated in the agreement and even much in excess of the so-called 25% of the work as per the GOMS No. 2289 dated 12th June 1968. The claimants claimed a separate rate for such extra excavation and the arbitrators rejected it. The trial court remitted the claim to arbitrator for reconsideration. The decision was challenged in the High Court which referred to Clause 59 of the agreement on delay and extension of time, and various letters of the superintending engineer rejecting the claim, and concluded that the contractor-claimant would not be entitled to be paid at any higher rate for such additional excavation work and accordingly set aside the order of the trial judge remitting the claim to the arbitrator.

Clause 59 of APDSS has absolutely no application to the monetary claim and deals with extension of time as already submitted in Chapter 4.[39] As regards the absence of a provision similar to Clause 12 of the CPWD form, reliance was placed upon the GOMS No. 2289, dated 12th June 1968. Even otherwise, a provision giving unrestricted power to vary the quantity has to be limited to a reasonable excess. Variation up to 20% to 25% of the originally stipulated quantity can be considered to be reasonable if not as a custom or usage in the construction industry. The decision of the High Court, it is respectfully submitted, did not lay down good law. However, the Supreme Court upheld the said decision. Before the Supreme Court the case was presented by placing reliance on Clause 63 of the agreement.

> Clause 63. *Payment for additions and deductions for omissions*: (A) The contractor is bound to execute all supplemental items that are found essential, incidental and inevitable during the execution of the work, at the rates to be worked out as detailed below:
>
> (a) For all items of work in excess of the quantities shown in schedule A of the tender the rates payable for such items shall be either the tender rates or the

[38] National Fertilizers v. Puran Chand Nangia, AIR 2001 SC 53.
[39] See paragraph 4.10.1 in Chapter 4.

standard schedule of rates for the items plus or minus the overall tender percentage accepted by the competent authority whichever is less.

(b) For items directly deducible from similar items in the agreement, the rates shall be derived by adding to or subtracting from the agreement rate of such similar items, the cost of difference in quantity of material or labour between the new items and the similar items in the agreement, worked out with reference to the Schedule of rates adopted in the sanctioned estimate plus or minus the overall tender percentage.

(c) For new items which do not correspond to any items in the agreement, the rates shall be standard schedule rate plus or minus the overall tender percentage. The terms "standard schedule of rates" used in the above sub-clauses (a), (b) and (c) means the schedule of rates on which the sanctioned estimate was prepared.

(d) In the event of the Executive Engineer and the Contractor failing to agree on a rate for such additional work, the Executive Engineer may, at his option either:

(i) employ other parties to carry out the additional work in the same manner as provided for under clause 48, or

(ii) the Contractor shall execute the work upon written orders from the Executive Engineer and the cost of labour and materials plus 10 per cent thereon shall be allowed therefor, provided that the vouchers for the labour and materials employed shall have been delivered to the Executive Engineer or his representative within seven days after such work shall have been completed. If the Executive Engineer considers that payment for such work on the basis of the vouchers presented is unduly high, he shall make payment in accordance with such valuation as he considers fair and reasonable and his decision to the matter shall be final, if the amount involved in additional payment is Rs. 1000 or less, for each occasion on which such additional works shall have been authorised. If such amount exceeds Rs. 1000, the contractor shall have the right to submit the matter to arbitration under the provisions of the arbitration clause 73.

(e) If, in the opinion of the Executive Engineer a rate for the additional work is not capable of being properly arrived at prior to execution of work, or if the work is not capable of being properly measured, then the cost and payment thereof shall be dealt with as provided for in the preceding sub-clause (d)(ii).

This is what the Supreme Court observed:[40]

The ... contention in fact centres round the question as to whether for the additional quantity of excavation work, the contractor would be entitled to at a higher rate in accordance with Clause 63 of the agreement. ... this contention had never been raised either before the arbitrator or before the subordinate Judge or even before the High Court. In fact the claim petition filed before the arbitrator is rather cryptic and absolutely vague, not indicating on what basis the additional claim is made, though the foundation for the claim was there, namely there had been an increased amount of excavation work beyond the agreement.

Referring to the decision in. S. Harcharan Singh v. Union of India,[41] it was observed:

---

[40] Ramachandra Reddy and Co. v. State of A.P., AIR 2001 SC 1523.
[41] (1990) 4 SCC 647: (AIR 1991 SC 945).

We fail to understand how the aforesaid decision will be of any assistance to the claimant in the present case, where there is no clause like Clause 12A nor is there any letter from the competent authority agreeing to payment at a higher rate for the additional work beyond the limit of 25% as provided under the GOMS No. 2289 dated 12th June 1968. Arbitrator being a creature of the agreement, unless agreement either specifically or inferentially provides for a higher rate to be awarded for any additional or excess work done by the contractor, it would not be permissible for the arbitrator to award for the so-called additional work at a higher rate. ... But such recommendation of the Executive Engineer, who was not competent to decide the question of awarding a higher rate for the excess quantity of excavation will not clothe any jurisdiction on the arbitrator to award the contractor at a higher rate nor would it entitle the contractor to get a higher rate for the claim in question on the basis of agreement. Now coming to the very clause, ..., we find that the said clause relates to supplemental item, which have been found essential, incidental and inevitable during the execution of the work. The excavation of hard rock cannot be held to be a supplemental item and on the other hand, is an item of work tendered and accepted, and as such clause 63 will have no application to the claim item No. 1 ... The appeal fails and is dismissed but in the circumstances there will be no order as to costs.[42]

This case is an excellent illustration of how the interpretation or construction of a business contract neglected to follow the dictates of the Supreme Court itself that meaning of such contracts must be gathered by adopting a commonsense approach and not be allowed to be thwarted by a narrow pedantic and legalistic interpretation,[43] it is respectfully submitted. The authors, with great respect, submit that this judgment does not lay down a good law. The provisions of Clause 63(a) read as follows:

(a) *For all items of work in excess of the quantities shown in schedule A of the tender* the rates payable for such items shall be either the tender rates or the standard schedule of rates for the items plus or minus the overall tender percentage accepted by the competent authority whichever is less.

(Emphasis supplied)

Clearly, the opening words of the provision show that the provisions of Clause 63 are not restricted to supplemental items alone but *"For all items of work in excess of the quantities shown in schedule A of the tender."* Supplemental items of work are not anticipated at the time of signing the contract but likely to arise during the construction of the work and their quantities do not reflect in the schedule of quantities. The paragraph heading is also general and reads: *"Payment for additions and deductions for omissions"*. Eliminating an item of work from the schedule of quantities resulting in omissions is not the same thing as ordering the execution of supplemental items *"that are found essential, incidental and inevitable during the execution of the work"*. In fact, for all practical purposes the said clause is a provision similar to Clause 12 of the CPWD contract form, it is respectfully submitted.

Clause 63, as the heading gives the general drift of the intention of the parties, could be pressed to support the claims of quantities *"in excess of the quantities shown in schedule A of the tender"* without stipulating any percentage limit unless the above referred to circular is incorporated in the contract. Under the circumstances relying on the earlier decisions and observations of the Supreme Court a reasonable limit is required to be impliedly put upon in construing the said provision, it is submitted.

---

[42] Ramachandra Reddy and Co. v. State of A.P., AIR 2001 SC 1523.
[43] Union of India v. D.N. Revri & Co., AIR 1976 SC 2257.

6. In a case decided by the Bombay High Court, provisions of Clause 14 in a contract were interpreted. Clause 14 clearly provided that,

> the Engineer-in-charge shall have power to make any alteration in or additions to the original specifications, drawings, designs and instructions that may appear to him to be necessary or advisable during the progress of the work, and the contractor shall be bound to carry out the work in accordance with any instructions in this connection, which may be given to him in writing signed by the Engineer-in-charge and such alteration shall not invalidate the contract and any additional work which the contractor may be directed to do in the manner above specified as part of the work shall be carried out by the contractor on the same conditions in all respects on which he agreed to do the main work and at the same rates as are specified in the tender for the main work.[44]

It was held that the clause clearly showed that *when the extra work in contemplation at the time of tender is done, for which the rates are provided in the original tender, the contractor cannot claim any extra rate.* He can claim extra rate only if that class of work is not provided in the tender (emphasis supplied).

The words emphasized in the above judgment show that the originally agreed rates would be binding *when the extra work was in contemplation at the time of submission of tender.* If the work were in contemplation of the parties at the time of tendering, it would have been reflected in the agreed scope of the contract and the contract sum. It would cease to be extra. The work is extra because when the contract was signed it was not within the contemplation of the parties. This judgment needs to be seen in the light of the facts of the case wherein the work of construction of a bridge was involved and the depth of foundation was increased, thereby increasing the quantities of the items of work. It appears that the clause in question did not provide for any limit on variation.

5.9.2 Work Executed by the Contractor after Submitting Rates, the Engineer Does Not Cancel the Order under Cl. 12A – Effect?

The plain objective of Clause 12A in the CPWD Standard Form Contract[45] is that for the variation ordered, if there is neither agreed rate nor rate entered in the schedule of rates of the division, the contractor is to inform the engineer the rate he intends to charge for the type of work ordered within seven days of his having received the variation order. A question arises: if the engineer neither accepts the rate nor cancels the variation order by a written notice, and the contractor executes the work, can the contractor be paid for at his quoted rate or in some other manner?

The way the language is used, it is reasonable to conclude that the engineer by his silence can be deemed to have accepted the rate.[46] However, the Supreme Court of India in another case involving a similar provision in the contract held differently.[47] It was observed that until the rates were settled by agreement, the contractor was under no obligation to carry out the additional or altered work. He could legitimately have said that in the absence of scheduled rates in the division for the type of work in question or an agreement in regard to the rates, he was not bound to carry out the additional or altered work. The contention that since the engineer-in-charge did not exercise his liberty to cancel the order, there was a concluded contract

---

[44] State of Maharashtra and others, Appellants v. Saifuddin Mujjaffarali Saifi, AIR 1994 Bom. 48.

[45] See paragraph 5.9.1 for the wording of the clause.

[46] Union of India v. Khetra Mohan Banerjee, AIR 1960 Cal. 190 confirmed in appeal by the Supreme Court in CA No. 206 of 24th November 1962.

[47] Bombay Housing Board v. Karbhase Naik & Co., AIR 1975 SC 763.

between the parties was rejected. It was held that in the absence of some positive act on the part of the engineer-in-charge agreeing to the rate, "there was no agreement as to the rate and that *the respondent was not bound to carry out the work*" (emphasis supplied).

The Supreme Court has duly laid down the law, it is respectfully submitted, that failure of the contractor to execute the extra work until there is an agreement in respect of the rates would not amount to a breach of the contract on his part. However, this interpretation of the contract provision is easily said but difficult to follow. The extra work ordered may be of small quantum compared to the subsequent work, which cannot be executed till completion of the varied work. In such a situation it may amount to a suspension of the whole work under the contract, which could lead to huge losses to both the parties. Invariably, under pressure, the contractor executes the extra work without the rates having been agreed and ultimately lands himself in dispute and litigation. In the interest of justice, the interpretation earlier upheld by the High Courts and the Supreme Court deserves to be restored, it is respectfully submitted.

Clause 13 of the CPWD Form

Clause 13 reads:

> *No compensation for alterations in or restriction of work to be carried, out-* If at any time after the commencement of the work the President of India shall for any reason whatsoever not require the whole thereof as specified in the tender to be carried out, the Engineer-in-charge shall give notice in writing of the fact to the contractor who shall have no claim to any payment of compensation whatsoever on account of any profit or advantage which he might have derived from the execution of the work in full, but which he did not derive in consequence of the full amount of the work not having been carried out, neither shall he have any claim for compensation by reason of any alterations having been made in the original specifications, drawings, designs and instructions which shall involve any curtailment of the work as originally contemplated.
>
> Provided that the contractor shall be paid the charges on the cartage only of materials actually and bona fide brought to the site of the work by the contractor and rendered surplus as a result of the abandonment or curtailment of the work or any portion thereof and then taken back by the contractor, provided however, that the Engineer-in-charge shall have in all such cases the option of taking over all or any such materials at their' purchase price or at local current rates whichever may be less. In the case of such stores having been issued from Government stores and returned by the contractor to the Government stores credit shall be given to him by the Engineer-in-Charge at rates not exceeding those at which they were originally issued to him, after taking into consideration any deduction for claims on account of any deterioration or damage while in the custody of the contractor and in this respect the decision of the Engineer-in-Charge shall be final.

Clause 13 Interpreted

An understanding of the proper interpretation of the above clause is necessary. First, it must be remembered that this provision, being an "exemption clause" in a printed form, will be strictly construed against the employer. As such, for the provisions of this clause to be applicable the following conditions must be satisfied.

i. The notice in writing should have been given by the engineer-in-charge. This stipulation is mandatory.[48]

ii. Such a notice should have been given by the engineer-in-charge "after the commencement of work". Thus, where after an issue of the work order but before commencement of the

---

[48] Bombay Housing Board v. Karbhase Naik & Co., AIR 1975 SC 763(767).

work, the contract is cancelled, the department cannot avoid liability to pay compensation to the contractor for that breach of contract.

iii. The contents of the first paragraph clearly point out two contingencies:

    a. "The full amount of the work not having been carried out", and

    b. Alterations in the work "which shall involve any curtailment of the work as originally contemplated."

The above two contingencies can occur only after the commencement of the work and, therefore, the words in the para-heading "restriction of work" can mean restriction at the stage executed and, therefore, can encompass within its sphere total abandonment of the whole of the balance work, it is submitted. This could result on the one hand, in the abandonment of the whole project in certain contingencies only, for example, a project such as construction of dams, canals, roads or rails, etc. is dropped, and on the other hand it may result in "restriction of the work" to the stage already executed, for example, not constructing the top storey or two of a multi-storied building project.The words "materials … rendered surplus as a result of the abandonment or curtailment of the work or any portion thereof" in the second para. lend support to this interpretation.

iv. The words "not require the whole thereof as specified in the tender to be carried out." can mean: the part of it (work already executed included) may be carried out as specified in the tender and the balance work is to be carried out with "altered specifications, drawings, designs and instructions." In other words, a possible view to be taken can be stated thus: "The President of India shall for any reason whatsoever not require the whole work to be carried out as specified in the tender; but can modify, alter or restrict its scope and have it executed in any manner without abandoning the whole work."The use of the words "as specified in the tender" following the words "the whole thereof," it is submitted, makes the dual interpretation possible. If this indeed is the case, the words of a paragraph heading can control the meaning, and "restriction" of work cannot by itself mean total abandonment of the work or project. But the use of words "abandonment or curtailment of the work or any portion thereof" in the second paragraph makes this interpretation weak, if not untenable. The doubt can be eliminated by amending the contract by deleting the words "as specified in the tender" in the opening sentence without causing any misunderstanding in the remaining part, if the intention of the parties is to cover total abandonment of the balance work as a whole or in part.[49]

v. Lastly, the words "The President of India shall for any reason whatsoever not require the whole (work) as specified in the tender", make this provision inapplicable to cases where the deleted work is not, in fact, totally abandoned.

Under the terms of most contracts, an employer who exercises the power to omit work must genuinely require the work not be done at all, and cannot exercise such a power with a view to having the work carried out by someone else.[50]

EXAMPLES FROM CASE LAW

1. The state terminated a contract with a view to getting the work executed through another agency, with changed designs and specifications, and relied on Clause 3.11 of the contract, which was similar to Clause 13 above. The Patna High Court held:

---

[49] For example, see the wording of the clause in Example (1) below.
[50] See: *Hudson's Building and Engineering Contracts*, 10th ed. pp. 340 and 533.

In view of the aforesaid clause 3.11 the Government for any reason whatsoever may decide to abandon or to reduce the scope of the work. The expression 'abandon' means, the abandonment of the project itself. In other words, the Government does not want to proceed with the construction of the project. In the present case it is not the stand of the State that the project itself has been abandoned. In the present case the State is very much anxious to proceed with the construction of the project but on the basis of the different drawing and design. As such the action of the State is not referable to clause 3.11 of the agreement. There is no escape from the conclusion that the ground for cancellation of the contract does not flow from any of the terms of the agreement, it is de hors any terms of the contract and as such can be the subject matter of judicial review.[51]

2.   A contract for erecting a factory required the contractor to follow the architect's instructions to fabricate the steel which was to be supplied by the owner at no cost to the contractor. The architect subsequently wrote to the contractor that the owner had awarded the contract for fabrication of structural steel to another company. The contract contained a provision empowering "The Architect in his absolute discretion and from time to time to issue written instructions or directions ... in regard to the omission ... of any work" and making it obligatory on the part of the contractor, who "shall forthwith comply with all the instructions of the Architect". It was held that the provisions of the contract would authorize the architect, doubtless within certain limits, to direct that particular item of work shall not be carried out, but that the provisions do not authorize him to say that particular items so included shall be carried out not by the contractor with whom the contract is made but by some other builder or contractor.[52] Such a power would be most unreasonable and very clear words would be required to confer it on the architect. An exercise of power to issue a variation order such as the one issued was thus outside the power conferred by the variation clause and, therefore, amounted to a breach of contract entitling the contractor to terminate the contract.

3.   A contract for roadwork included an item of moving and spreading topsoil. The specifications provided that if topsoil available at the site of work was insufficient, the engineer could order the contractor, in writing, to obtain topsoil elsewhere for which the agreement prescribed a rate. The engineer, rather than asking the contractor to bring soil from elsewhere, instructed a third party to bring more soil from elsewhere on to the site. It was held that this was a breach of contract on the part of the employer. The power to omit the work from the contract, did not entitle the owner to take away part of the contract work so that it might be given to another contractor.[53]

5.10   Decision Regarding Rate – When Not Open to Arbitration?

A dispute over the rate for extra work cannot be referred to an arbitrator appointed under an arbitration clause if it opens with the words, "Except where otherwise provided in the contract all questions and disputes ... shall be referred to ... arbitration." Where Clause 13A of a contract provided for the Engineer-in-Charge to have the power to make any alternation in, addition to, etc. as per normal provisions and ended with the wording as above, it was held by the Madhya Pradesh High Court that the dispute regarding rate could not be referred to an arbitrator

---

[51] M/s. Pancham Singh v. State (FB), AIR 1991 Pat. (168).
[52] Carr v. Berriman (J.A.) Pty. Ltd., (1953) 27 ALJ 273. Australia.
[53] Commissioner for Main Roads v. Reed & Stuart Pyt. Ltd., High Court of Australia (1974) 12 BLR 55.

appointed under Clause 14.[54] The Supreme Court of India upheld this decision of the Madhya Pradesh High Court.[55]

5.11 Change Orders Must Be Written

The clause relating to additions and alterations generally makes it clear that the contractor is bound to execute the extra or additional work in accordance with any instructions, which may be given to him in writing signed by the engineer-in-charge.

It is a common experience that engineers on site give a number of instructions and orders, including change orders, orally. A question arises as to whether the contractor is duty bound to follow oral orders or instructions which amount to a variation or deviation of the contract.

In ordinary language, the use of "may" implies an option. But, legally, "may" might mean "must". The meaning will have to be construed in the context in which the term is used. It has been pointed out in Chapter 3 that, in India, a valid contract with a government must be in writing. As such, any amendments, modifications or additions in respect of that contract must also be in writing. Not only that but the amendments, modifications or additions must be made by a person who has the authority from the government to that effect. If the engineer-in-charge is empowered by the government to make changes in the contract, it is necessary that all orders which make changes resulting in additions and alterations must be in writing and signed by him. He may not be able to delegate the authority to his subordinate. Thus, "may" in this clause means "must".

Many a time a contractor writes to the engineer for a confirmation of the oral order given to him. In a great number of cases, he receives no reply to his letter. Should the contractor carry out the extra work? The answer suggested is no, unless he expects no extra payment. The Supreme Court of India has held that to allow a claim on account of a deviation/alteration, it is essential that there is a prior permission in writing from the authority.[56] Where a contract provided: "The contractor shall not in any way alter the works or any part thereof in respect of designs, quality, materials or specifications without the previous permission in writing of the Engineer", the Supreme Court set aside an arbitration award allowing a contractor's claim for excavation at the original site which was abandoned, holding that admittedly there was no written order though he was orally promised payment.

The above decision was held inapplicable in a case decided by the Delhi High Court in which a number of extra items were ordered and executed. The engineer had proposed for approval rates for eight items which were not accepted by the contractor and in respect of the other items there was no decision of the authority placed on record to show that the decision had attained finality and as such the case was considered on merits by the arbitral tribunal. The condition in the contract made it obligatory on the part of the contractor to submit every month an account giving particulars of the extra or the additional work. A breach of this provision by the contractor was also not urged. The award was upheld.[57]

5.11.1 Absence of Written Variation Order: Contractor When Not Bound to Execute the Extra Work

The provisions made in the contract, as considered above, empower the employer through his architect or engineer to order extra or additional work. These provisions expressly limit the power of the architect/engineer to order the extra without requiring the prior consent of the employer.

---

[54] Dandakaranya Project v. P.C. Corporation, AIR 1975 M.P. 152.

[55] M/s. P.C. Corporation Ltd. v. Chief Administrator, Dandakaranya Project, AIR 1991 SC 957.

[56] State of J. & K. v. Dev Dutt Pandit, AIR 1999 SC 3196; Dinesh Prasad v. State of Jharkhand, (Jharkhand) 2012 AIR (Jharkhand) 57: 2012(112) AIC 397: 2012(1) J.C.R. 288: 2012(4) R.A.J. 511: 2012(1) JLJR 69: 2011(41) R.C.R. (Civil) 384; K.R. Anand v. Delhi Development Authority, (Delhi) 1997(68) DLT 143: 1997(4) AD (Delhi) 1006: 1997(2) ArbiLR 109: 1997(4) CurCC 269.

[57] D.S.A. Engineers v. Housing & Urban Dev. Corporation, 2004(2) Arb. LR 33 (Delhi).

When a variation is ordered, with or without consent of the employer, the agreements generally provide for a written order. Under these circumstances the contractor need not execute the extra work unless he is served with a written order. Some agreements do provide that if any variation is orally ordered, the contractor should confirm the oral order in writing and if the engineer/architect does not contradict the same within a stipulated time limit, the variation order would be deemed to have been issued in writing. For example, FIDIC Conditions, 1999 edition contains Clause 3.3, which stipulates that if oral instructions are issued to the contractor by the engineer or delegated assistant, who receives a written confirmation from the contractor within two working days and does not reply by a written rejection and/or instruction within two working days after receiving the confirmation, then the confirmation issued by the contractor constitutes the written instruction of the engineer or delegated assistant, as the case may be.

Example (1) below is an excellent illustration of what happens when work is executed without a written order.

EXAMPLES FROM CASE LAW

1. The entire dispute between the parties related to the second additional work which was allegedly allotted to the respondents by the petitioner's officers orally. The petitioners relied on Clauses 41 and 42 of the standard general conditions of contract. The High Court held that Clauses 41 and 42 ought to be read and interpreted in the light of Art. 14 of the Indian Constitution. The petitioner, as the state, was bound by the guarantee of Art. 14 of the Constitution. As a result of a series of judgments of the Supreme Court, it is settled law that the state and its instrumentalities cannot enter into contracts without giving an opportunity to every person who may be eligible to compete for that contract. The High Court held that even the works which can be awarded additionally to existing contractors in exercise of powers under Clauses 41 and 42 have to be given pursuant to a written contract. The provision in Clause 41 which permits allotment of work by verbal orders, which would undoubtedly have to be reduced to writing, have to be restricted to work of an urgent or marginal nature. Insofar as the present case was concerned, even if it is assumed that the work worth about Rs. 25 lacs could have been allotted by verbal orders, nevertheless, those orders were required to be reduced to writing in the form of a formal instrument by an officer who was competent to do so. Even if the requirement of a formal instrument was not mandatory, the requirement of a sanction being given to such a work by a competent officer was held to be mandatory. That power was vested in the engineer, and not the assistant engineer who was supposed to have signed the second and final variation statement. The petitioners would be liable for making payment for the work as per the terms of the contract only upon a written document signed by a competent authority.[58]

2. **Silence does not amount to acceptance – an illustration**: During the course of the execution of a contract between the State of Rajasthan and a contractor for construction of drainage siphons on a canal, the assistant engineer wrote to the contractor that it was proposed to increase the number of reinforced cement concrete (RCC) pipes and the steel content in the RCC work in the bank portion of the canal. He therefore asked the contractor to let him know whether the contractor would like to take up "the additional work at the same rates and the applicable rates of the schedule" as provided in the agreement. The contractor wrote back to the assistant engineer that he would execute the extra work other than that agreed upon in the agreement at 40% above the Chambal Schedule of Rates. The assistant engineer sent no reply to this letter. It was argued that the silence on the part of the Assistant Engineer amounted to implied acceptance of the rate quoted by the contractor. Relying on the decision of the Supreme Court of India, holding that

---

[58] Union of India v. M/s. Arora Associates, AIR 2003 Bom. 477.

mere silence could not be deemed to be acquiescence,[59] the Rajasthan High Court declined to accept the argument.[60] It was held that under the circumstances the contractor was not entitled to get increased rates on the basis of *quantum meruit*, i.e. just and reasonable value of the work done.

The plea of an implied acceptance in absence of a written approval thus fails in the case of government contracts, and contractors are well advised to insist on written orders of the authorized engineer before executing the extra or additional work. Verbal assurances are, it is pointed out, of no consequence. What happens if the contractor fails to get a written order? The dispute between the parties would then be whether the alleged work is covered in the contract by the rates agreed upon or not. Both the sides would rely on their rival interpretations and the contractor would be in a precarious position. If he does not carry out the oral order or leaves work at his own risk and cost, he could be subject to all the penalty clauses. On the other hand, if he executes the work under protest, he can expect justice at the hands of an arbitrator or in the absence of an arbitration clause in the contract, by taking the matter to the court. It would be the arbitrator or the court, as the case may be, which would resolve the dispute. For example, a contract contains a standard provision stipulating that no works beyond those included in the agreement would be paid for without a written order from the employer and the architect. During the execution of the works the employer needs certain work to be done which he insists is included in the contract. The contractor believes that the work is not within the scope of the contract and had to be paid for as an extra. It would be open to the arbitrator, if he concluded that the works in question were not in fact included in the contract, to hold that the employer had implicitly promised to pay for the said works, except that in contracts with governments, provisions of the Indian Constitution's Art. 14 would need to be kept in mind as discussed above.

There are two possibilities: either a particular work is covered by a contract and is to be paid for at the agreed rates, or it is outside the scope of the contract for which no rates are fixed under the contract. In the first case the claim would arise under the contract pursuant to its terms. In the second case the claim would be outside the contract on the principle of *quantum meruit* or Section 70 of the Indian Contract Act. Where the claim is of the first category, that is, arising out of a contract, it is subject to the terms of the contract. One of them usually stipulates that extra work will not be paid for unless a claim in writing is made within a specified time to a particular person. It is necessary to ensure that the contractor submits a written claim as specified in the contract because the condition in the contract is valid and enforceable.[61] If the authorities do not accept his claim, and the arbitrator or the court concludes that the work did arise out of the contract, the contractor may not be able to get higher rates than agreed. For, under the terms of the contract, a written order of the engineer-in-charge had to be obtained before executing the work involving additional expenses. In the absence of any such written order, it is not open to the court to hold the government liable for extra expenses by applying some principles or rule analogous to estoppel.

5.11.2 Contractor Need Not Execute the Work Unless and Until the Rates are Finalized?

If the variation order is issued in writing duly signed by the person authorized by the terms of the contract to do so, and if the terms of the agreement require that the parties mutually agree upon the rates/cost of the varied work, the contractor need not execute the work unless and until the rates are finalized.[62] The Supreme Court has observed:

---

[59] Union of India v. Watkins Mayor and Co., AIR 1966 SC 275.

[60] State v. Motiram, AIR 1973 Raj. 225 (228).

[61] Abercrombie and Williams v. Vandiver, 126 Ala, 513 So 491.

[62] Bombay Housing Board v. Karbhase Naik & Co., AIR 1975, SC 763.

We do not think that the respondent was bound to carry out the additions and alterations as there was no reply to the notice stating the rates it intended to charge. But it was free to commence and complete the work on the basis that since the rates quoted by it were not accepted, it would be paid at such rates to be fixed by the Engineer-in-charge and that if it was dissatisfied with the rate or rates fixed by the Engineer-in-charge, it could raise a dispute before the Superintending Engineer and that the time limit for completion would be extended in all cases of additions or alterations as stated in the last sub-para of clause 14.

We think that until the rates were settled by agreement the respondent was under no obligation to carry out the additional or altered work. The respondent could legitimately have said that in the absence of scheduled rates in the division for the type of work in question or an agreement in regard to the rates, it was not bound to carry out the additional or altered work. We are not satisfied that since the Engineer-in-charge did not exercise his liberty to cancel the order, there was a concluded contract between the parties. The failure to cancel the order for additional or altered work on receipt of the notice specifying the rate would not result in an agreement as to the rate to be charged. ... Failure of the contractor to execute the extra work till there is an agreement in respect of the rates would not amount to a breach of the contract on his part. The same cannot, however, be said to be the case in respect of additional work, and also in respect of extra work for which the price is stipulated in the contract.

And sometimes, it so happens that the engineer issues directions which, in the opinion of the contractor, are not only unwarranted by the facts of the case but amount to an extra, if executed. A contractor is bound to comply with the instructions. Both the engineer and the owner make it clear that for executing the work in the manner directed no extra payment would be made. What should the contractor do? The Supreme Court of India has held, as stated above, that the contractor is not bound to do additional or altered work unless rates for the same are mutually agreed upon. The Supreme Court of Canada in a majority decision held a similar view in the case in which the appellant, the main contractor for a bridgework, subcontracted the piling work to the respondent. During the execution of the piling work the employer's engineer insisted that many of the piles be driven to a greater depth than the respondent considered was required by the subcontract. The subcontractor had tendered for the construction work and the conditions of construction were amended by adding the words "bottom of timber piles to be below bottom of sheet piling". The subcontractor noted this addition at once but merely mentioned it to the head contractor and thereafter continued to complain of it, but nonetheless proceeded to attempt to carry out the work. This led to larger costs than if the work had been performed in accordance with the contract as first drafted. No additional compensation could be obtained unless it was proved that there was a new contract. The respondent complained to the appellant that the work was not required but was told to comply with the instructions of the engineer. Both the engineer and the employer made it clear that they would not authorize extra payment. The respondent claimed extra payment both from the employer and the appellant. Its action against the employer was dismissed. The action against the appellant also failed in the trial court, but on appeal it was held that the respondent's view of the meaning of the contract was correct. The Supreme Court of Canada took the view (by majority) that if the engineer's instructions were not justified, the respondent's correct course was to refuse to carry them out without a variation order.[63] This case was distinguished in a 1975 decision where there was no change in the original contract leading to extra costs; rather the extra costs were caused by a breach of contract by the owner that led to adverse work conditions.[64]

The authors are in full agreement with the dissenting view that this was an unrealistic course of action to propose. It must be remembered that many a time the terms of the contract are so difficult to construe that there can be a difference in judicial opinion as to its true meaning. If the contractor, who holds one view, is to stop the work, he could be under the threat of virtual

---

[63] Peter Kiewit Sons Co. of Canada Ltd. v. Eakins Construction Ltd., Supreme Court of Canada (1960) 22 DLR(2d) 465.

[64] Penvidic Contracting Co. Ltd. v. International Nickel Co. of Canada Ltd., 53 DLR (3d) 748.

financial ruin if his view turns out to be not acceptable to the judiciary. To say that because under these circumstances the contractor was not prepared to stop work and so risk a ruinous loss which would have befallen him if his view turned out to be erroneous, the employer will retain the benefit of all the additional work without paying for it, would be to countenance an unjust enrichment. Consider as an example the case discussed below.

The plaintiff alleged that during the execution of the work, he found several defects in putting up the foundation both for abutment and piers. The excavation had to be done deeper than the designed level and this necessitated further dewatering in the riverbed. The plaintiff, therefore, wrote a letter to the defendant pointing out difficulties in respect of excavation of foundations and dewatering. The letter also pointed out that the rates in the tender for this work were far less than the cost of carrying out this work. The appellant/defendant argued that under the stipulated conditions of tender, the plaintiff was not entitled to an additional amount or extra rates for excess excavation and dewatering. The trial court allowed the claims. On appeal it was held:

> It would be also important to observe that unexpected difficulty or expense is in general, no excuse for non-performance. The contractor in such a case cannot rely on his ignorance of such matters as defects in the soil nor any implied warranty by the employer that the bills of quantities, plans and specifications are accurate or that the work is capable of performance in the manner set out in the invitation to tender or limit his liability to exactly excavated stated quantities of work. More particularly, when there is a clause that the contractor shall himself acquaint with the site, with the locality and all matters relating to the contract, he cannot back out from this condition and say that the quantities mentioned therein may not be complete or that nature of soil is unknown. He should visualize all probable contingencies and also all other matters, which could in any way influence tender or the contract. This risk in such contracts has to be taken by the contractors. He cannot resile and say that he was unaware of the soil or local conditions.

The High Court referred to Clause 14 of the contract which showed that when extra work contemplated at the time of tender was done for which rates were provided in the original tender, the contractor could not claim any extra rate. He could claim extra rate only if that class of work was not provided in the tender. The judgment of the trial court was set aside.[65]

The judgment contained observations of far reaching consequences putting the total risk of uncertainties on the contractor in investigating the site conditions, preparation of drawings and estimate of cost, etc. It is submitted, with respect, that the said observations do not lay down good law. Usually, risk allocation is fair and efficient if it is borne by a party that can foresee conditions and take steps to reduce their impact on the costs. Given that the state agencies usually have access to the site conditions and do the preparation of the initial drawings and specifications, they are better placed to bear the risk of uncertainty than the contractor. On the other hand, the contractor might be better placed to bear the risk of fluctuations in, say, the cost of labour. The above High Court decision shows how judicial opinions can vary and an outright refusal by a contractor to carry out what is perceived to be extra work may prove disastrous to the contractor.

Many standard form contracts in India place an obligation on the contractor to carry out all extra items not included in the schedule and found necessary during execution of the work, and an obligation on the government department to pay for such extra items. A contractor claimed that the work tendered for was to be executed by "pile driving" but it was actually done by "jetting" having regard to the nature and condition of the soil at the sites, and that this was an extra item of work, for which he was entitled to extra remuneration. The state (employer) took the view that "jetting" was not an extra item of work but only a revised and improved technical method of executing the work tendered for, and that, therefore, the contractor was not entitled to any extra remuneration. The matter was referred to arbitration in accordance with the arbitration clause included in the contract. The arbitrator decided that jetting was not an extra item of work. The award of the arbitrator was set

---

[65] State of Maharashtra and others, Appellants v. Saifuddin Mujjaffarali Saifi, AIR 1994 Bom. 48.

aside by the trial court under the provisions of the Arbitration Act. The state appealed to the High Court.[66] The Kerala High Court allowed the appeal and set aside the decision of the trial court. The Supreme Court considered the decision of the Kerala High Court in a civil appeal and held that the two letters which admitted jetting as an extra item, both internal correspondence one written by the executive engineer to the chief engineer and the other written by the superintending engineer to the chief engineer, were material documents to arrive at a just and fair decision to resolve the controversy and that it was incumbent upon the arbitrator to get hold of all the relevant documents including these two letters, even if the respondent authority did not produce these documents. The Supreme Court found that although the arbitrator had held that jetting was not an authorized extra covered by the agreement, he also observed:

> The Chief Engineer had rejected the claims of the contractor on grounds of non-inclusion of this (Jetting) in the agreement which was executed subsequent to the direction issued by the department to adopt jetting. The Chief Engineer's decision totally ignores the next sentence in that letter "Meanwhile you may execute the agreement". By this sentence the issue of extra payment for jetting is left open even after the execution of the agreement.

This conclusion of the arbitrator rejecting the claim on the ground that jetting was not an authorized extra was rationally inconsistent. The award, therefore, suffered from a manifest error apparent *ex facie*. As a result, the judgment of the High Court was set aside and that of the trial court restored. The award was quashed. The arbitrator was directed to complete the proceedings after considering all the relevant documents, including the internal correspondence, after giving an opportunity to the parties to make submissions.[67]

5.12 Recovery in the Absence of Written Order – When Possible?

Where the terms of a contract provide for a written variation order and the extra work is found to be outside the contract, it follows that the terms of the contract have no application, and although the production of an order in writing may be a condition precedent to the recovery of payment for an extra work done under the contract, it is not a condition for payment for extra work which is outside the contract. It is not necessary to produce the contract in evidence to recover payment for that work except, in the event of the defence being raised, to prove that the work in question is outside the contract and payment is to be made for it not at contract rates but upon a *quantum meruit* or otherwise in accordance with the separate contract or request relied on. This provision has found a statutory recognition under Section 70 of the Indian Contract Act. The Supreme Court of India has held, as mentioned earlier, that once it is shown that the government took the benefit of extra work which was not gratuitously done, and which created a quasi-contractual obligation, the claim for compensation under Section 70 of the Contract Act could be enforced against the government. Art. 299 of the Indian Constitution is no bar to application of Section 70 under the circumstances embodied in that section.[68]

In case of contracts for private works, which by law are not required to be in writing, oral orders would be valid. Even where such a contract provides that deviation orders must be in writing, a contractor can, in certain circumstances, recover payment in absence of a written order. For example, where oral change orders were frequently handled without a protest by either party,

---

[66] State v. K.P. Poulose, AIR 1973 Ker. 242. 22. AIR 1973 Ker. 242.
[67] Poulose v. State of Kerala, AIR 1975 SC 1259.
[68] State of W.B. v. M/s. B. K. Mondal and Sons, AIR 1962 SC 779; Food Corporation of India v. Vikas Majdoor Kamdar Sahkari Mandli Ltd., 2007(13) Scale 126: 2007(11) SCR 1038: 2007(13) SCC 544:: 2007 AIR (SCW) 6999.

during the course of prosecution of the work, it may be held that such a conduct constitutes the waiver of the contract provision.[69]

A clause requiring a variation order to be in writing may be drafted such that an order is not a condition precedent to the contractor's right to payment. For example, a clause may provide that "the contractor shall execute such alterations as the architect may direct in writing" and may not exclude any claim for work not so ordered. In such a case there is nothing to prevent the owner being liable under the law on a separate contract, express or implied. Also, instructions can be said to have been given through drawings or sketches. Where a respondent argued that the special conditions of the contract required the plaintiff to obtain an order before execution of the work, and that without such an order the plaintiff was not entitled to get paid for the same, on the basis of the facts of the case, the Allahabad High Court held:

> The submission made cannot be accepted for two reasons. The first reason being that the ... plaintiff ... had been orally asked to do the R.C.C. work. ... Reference may also be made to Condition No.5 of the Detailed Specifications, which requires that "all work shall be carried out in accordance with the detailed drawing to be supplied or as directed by the Engineer-in-charge from time to time". As the work was done by the plaintiff in accordance with the oral instruction received by him from the Engineer-in-charge, therefore, it cannot be said that the plaintiff was not entitled to get the price for R.C.C. work. Ext.2 ... in the second drawing also shows alterations and changes in the nature of work, i.e. reinforced concrete work. Therefore, this exhibit itself can be treated as an order in writing. Accordingly, the submission of the counsel for the State that as the plaintiff had not obtained any order in writing, he could not get the price for this item, is liable to fail.[70]

5.13 Instruction to Assist Contractors in Difficulty

Sometimes, due to unforeseen difficulties, the contractor suggests a few alternatives to execute the work and the engineer, with a view to assisting the contractor faced with a difficulty, orders changes. These would be difficulties that the contractor is obliged to surmount by whatever methods he chooses to adopt at his cost. Such an action by the engineer may be of benefit to the owner as well as to the contractor, since (i) the delay which might be costly to the owner may be avoided, (ii) the difficulty may throw doubt on the long-term suitability of the permanent work after completion, and hence in the owner's interest may call for a design change which incidentally will assist the contractor. Such a situation may give rise to a difficult problem. Two contentions are likely to be advanced by the owner: (i) that the change was agreed solely to assist the contractor faced with a difficulty and not for any interest or advantage of the owner; or (ii) the work concerned was not a variation at all, but only contingent work which on true construction of the contract was the obligation of the contractor.

In this connection it should be noted that if a contractor is already bound, on true construction of the contract, to do a certain work for a certain price, there will be no consideration for any promise by the owner to pay for it under the mistaken impression that it is a variation. It may be so even if the engineer gives an instruction to execute certain work expressly stating it to be a variation, if, on the true construction of the contract the work is included in the contract price. Even where payments have been released, there is no estoppel.[71]

[69] Brookhaven Landscape & Grading Co. v. J. F. Barton Contracting Co., 676 F. 2d 516 (11th Cir. 1982); V.L. Nicholson Co. v. Transcon Inv. & Fin. Ltd., 595 S.W. 2d 474 (Tenn. 1980); Union Building Corp. v. J&J Building & Maintenance Contractors, Inc., 578 S.W. 2d 519 (Tex. Ct. Ap.1979).

[70] State of U. P. v. Chandra Gupta, AIR 1977 All 28.

[71] T.N. Electricity Board and another v. N. Raju Reddiar, AIR 1996 SC 2025.

To avoid doubt, architects or engineers should be careful in the use of language, particularly when authorizing such an alteration, and they should expressly disclaim any intention to give an instruction or an order in the matter under the terms of the variation clause.

5.14 Extras Caused by Misrepresentation

It has already been pointed out that a contract to which consent is obtained by a misrepresentation or fraud is voidable at the option of the party whose consent was so caused. Thus a contractor, who submitted his tender on a belief that certain conditions as indicated by plans and specifications actually exist, and has relied upon them in submitting the tender, is entitled to recover the value of such extra work as was necessitated by conditions being other than those represented.

If a contract contains positive representations as to the conditions, character, or nature of the work amounting substantially to a warranty, there may be recovery based on such representation. This may be true even though the contractor was required to make investigations and satisfy himself upon these matters. Where, however, the intention of the parties is such that the contractor was to rely upon his own investigations and examinations, he cannot recover upon the representation. But a recovery may be possible where the representations made are fraudulent, even though the contractor is required by the contract to make independent investigations.

Thus, the ultimate guide in determining whether or not there can be a recovery for the extra cost of doing a work under a contract is the contract itself. It is always a question of the intention of the parties, and that has to be gathered from the provisions of the contract.

5.15 Extras Caused by Legal Provision

If a change is caused due to some legal requirement overlooked in the specifications, the contractor can recover the increased cost of the work even though the general provisions of the contract called for compliance with the requirement of the law.

EXAMPLE FROM CASE LAW

The specifications gave a contractor an option to use either plasterboard or plaster over lath in the bathrooms of the building. The contract contained the provision that the contractor would comply with rules and regulations of the state affecting work of this character. The contractor decided to use plasterboards. The law, however, required the use of plaster over lath and, therefore, a change order was delivered to the contractor. Subsequently the contractor brought an action to recover the increased cost on account of the change order. The trial court denied him the recovery. The Appellate Court holding that the contractor was entitled to recover the increased cost of the work caused by the change order, reversed the judgment. The Appellate Court observed:

> The city prepared the plans and specifications and asked for competitive bidding on the basis thereof. It would not be expected that the bidders would examine the various laws and building codes as to each item specified to see if the codes required some different method of construction. Such procedure would make the bidders' interpretation of the law rather than the specifications controlling it. It would make the specifications so indefinite and uncertain as to destroy the validity of any contract awarded pursuant thereto.[72]

---

[72] Green Construction Company v. City of New York, 283 AD 485 (1954), Appellate Division, First Dept. N.Y. at p. 487.

5.16 When Contract Based on "Approximate Estimate"

It has already been pointed out in Chapter 1 that an estimate as an offer may be binding. If, in such a case, the cost of the work exceeds the cost indicated by the "approximate estimate" submitted by the contractor, can he recover the excess cost? In one case it was held that he could. The court observed that to make an estimate ordinarily means to calculate roughly or to form an opinion as to the amount, from imperfect data. The use of the word "estimate" and especially "approximate monthly estimate" precludes accuracy.[73] A Canadian contractor had entered into an agreement for the construction of a postal terminal including the supply and installation of mail-handling equipment of varying complexity. The specification for the equipment was not available at the time of the contract and it was expected that the equipment part of the work would be executed by a subcontractor. The contractor was asked to tender for all the costs of overheads, supervision and profit for that work on the basis of an estimate provided to the contractor. In fact, the actual cost of that work was in excess of the estimate. The contractor asked for a variation order and an increase in payment using an escalation formula. The respondent contended that the estimate in the contract was simply an estimate and that if it was inaccurate, the risk was taken by the contractor. The Canadian Supreme Court held that the estimate was actually a representation as the contractor could not have worked out the cost of that equipment without a specification.[74]

5.17 Extra Claims beyond the Scope of Agreement

Where an agreement expressly prohibits extra payment in certain contingencies, a very good case is required to be made to overcome the contractual stipulation, if indeed it is possible to do so. The courts are inclined to invalidate an award which allows extra payment in spite of contractual stipulations.[75] The following cases will serve as illustrations of the above well-established position.

EXAMPLES FROM CASE LAW

1. A contractor claimed extra for dewatering in respect of two items. He was denied recovery. In respect of one item, the schedule of quantities and the bids clearly provided that dewatering was included in the work required to be done and there was no mention of dewatering at all in the technical specifications. It was held that the fact that the technical specifications were silent did not mean that the same were contrary to what was stated in the schedule of quantities and bids. The fact that in the cost analysis also there was no mention of dewatering in respect of this item, did not improve the situation. As regards the other item, the schedule of quantities and bids specifically included dewatering in the work required to be done and there was also a specific provision in one para. of the technical specifications that nothing is to be paid extra for dewatering but there was no mention in the technical specification relating to rates about it.[76]

---

[73] A. Beeler v. Mitter, 254, S.W. 2d, 986 Kansas City Court of Appeals, 1953.

[74] Cana Construction Co. v. Queen, The, [1983–1] WLUK 66; (1983) 21 DLR (4th) 12.

[75] FCI v. Surendra, Devendra & Mahendra Transport Co., 2003 (1) Arb. LR 505 SC; State of Kerala v. Mathai, 2003 (3) 28 Kerala; Associated Engineering Co. v. Government of Andhra Pradesh, AIR 1992 SC 232. Also see: Delhi Jal Board v. Kaveri Infrastructure Pvt. Ltd., (Delhi) 2014(206) DLT 136: 2014(1) BC 580: 2014(2) R.A.J. 286: 2013(39) R.C.R.(Civil) 280.

[76] State of U.P. v. M/s. Allied Construction Engineers and Contractor, AIR 1996 All. 295.

2. A contract included the following conditions:

> 3.15. Rates quoted by the contractor shall hold for work at all heights and depths. ... 3.19. Centering and shuttering required for double height slab shall be done by the contractor as per approved drawing issued by the Engineer in-Charge.

The contractor contended that the increase in the height of the shuttering was necessitated due to the change in drawing appended with the bid documents and any change in the original tender drawing drastically affecting the quoted rate could not be changed without payment of an extra amount. Further, that the relevant clause was not Clause 3.15 which provided that the rates quoted by the contractor should apply for all heights and depths, but Clause 3.19 which stated that shuttering and centring should be as per the approved drawing issued by the engineer-in-charge. The above contention had found favour with the arbitrator whose award was set aside by the trial court on the ground that the arbitrator did not even refer to Clause 12 of the conditions. On appeal, it was held:

> A perusal of the counter statement clearly shows that the respondent had not relied on Clause 12 at all. ... Clause 12 was not referred to by respondent in their objections filed in this Court. Accordingly the learned Single Judge's judgment to the extent it relies upon Clause 12 cannot be sustained.[77]

It is not clear from the decision what Clause 12 provided.

3. Generally, NITs contain a stipulation showing the quarry chart from which the required materials can be obtained. Many times these stipulations do not guarantee either the quality or adequacy of the materials from these sources. The contracts, however, contain provisions such that the contractor shall make his own arrangements for quarries or for supply of water and no assurance is given regarding the availability of materials from quarries or water from the sources indicated. Yet, the claims made by a contractor for extra lead or extra expenditure incurred may not be tenable. The relief can be denied even if an assurance to that effect was available on the record.[78] Similarly, if an agreement makes a provision for the payment of escalation for certain limited materials, escalation in respect of other materials or matters may not be payable. The Indian Supreme Court case in example 4 below highlights the need to exercise caution on the part of tenderers as also arbitrators/umpires.

4. The terms of the agreement included an escalation provision which was limited to matters such as, diesel oil, labour, etc. but not to napa slabs used for canal lining. The agreement expressly prohibited claims for price adjustment other than those provided in the agreement. It further stipulated that the contractor should make his own arrangements to obtain the napa slabs as per the standard specifications. The government department did not accept any responsibility either in handing over the quarries or producing the napa slabs or any other facilities. The contractor was not entitled to any extra rate due to change in selection of quarries. The Supreme Court held these provisions to amount to a prohibition against price adjustment or award for escalated cost in respect of any matter falling outside the price adjustment provision in the contract. In the same case the Supreme Court set aside the claim for extra lead for water allowed by the umpire and observed:[79]

---

[77] Guru Flehar Constructions v. DDA, Arb. LR 2002(2) 254 Delhi (DB).
[78] State of Karnataka v. Stellar Construction Co., 2003 (1) Arb. LR 40 (Karnataka) (DB).
[79] Associated Engineering Co. v. Government of Andhra Pradesh, AIR 1992 SC 232.

The contract specifically stated that it was the responsibility of the contractor to make its own arrangements for the supply of water. The Government gave no assurance to the contractor regarding the availability of water or the prices payable therefor. The umpire, therefore, had no jurisdiction to allow the claim.

From the brief facts available from the judgment it appears that the agreement provided for the government department to widen banks to five and three metres' width to facilitate transport of materials. The contractor had to maintain the haul roads. The contractor's claim for extra expenditure incurred due to the flattening of canal slopes and the consequent reduction in the top width of banks used as roadways, that was allowed by the umpire, was set aside by the High Court and the Supreme Court. The Supreme Court observed:

> In the absence of any provision to pay for extra expenditure and in the light of specific provision placing the sole responsibility for the maintenance of the haul roads on the contractor, the arbitrator had no jurisdiction to award ... extra rate.[80]

The representation that canal banks would be widened to facilitate the transportation of materials was made at the time of tendering the rates, thereby inducing the contractors not to account for the cost of providing "any other haul roads required by the contractor and not specified in the plan to be carried out by the contractor at his cost" in his tendered rates. If the representation proved false the contractor deserved to be compensated for the extra expenditure. The award being non-speaking, the basis for it was not revealed and in the process it got struck down as a bad award, it is submitted.

5.   The above case and the decision of the Supreme Court in it were referred to and relied upon by the full bench of the Kerala High Court. The facts of the Kerala High Court case were as follows. In a roadwork contract to be completed within 15 months, there were delays on the part of the government in handing over the site, supplying cement and steel, etc. For the purpose of the extension of time, a supplemental agreement was executed between the parties in which the contractor agreed not to claim compensation for the extra work or expenditure and not to claim higher rates for labour, materials, etc. The said supplemental agreement stipulated that payment would be according to the terms and conditions in the original agreement, the only benefit being an extension of time. In spite of the above facts, the contractor raised disputes referring the claims for compensation and higher rates to arbitration. The state relied in its defence on the supplemental agreement. The arbitrator did not even care to look into the objection of the state and to call for the supplemental agreement. He made a non-speaking award allowing compensation. The trial court rejected the objection to the award on the ground that the award was a non-speaking award and there was no error apparent on the face of the award or misconduct on the part of the arbitrator. On appeal, the division bench referred the main point to the full bench as to whether the arbitrator had acted in excess of authority and whether a non-speaking award could be attacked on any of the grounds mentioned in Sections 16 and 30 of the Act. It was held:

> In our view, the latest decision of the Supreme Court in Associated Engineering Co. v. Govt. of Andhra Pradesh, (1991) 4 SCC 93: (AIR 1992 SC 232), clinches the issue in favour of the appellants. That decision has now clearly laid down that if an arbitrator while giving a non-speaking award, acts in contravention of the clear, obvious or patent terms of the main contract which deal with the rights and

---

[80] Ibid.

obligations of the parties, such action will be without jurisdiction. It is also held that for the purpose of finding out if the arbitrator has so acted, it is open to look outside the award, including affidavits, pleadings and terms of the main contract. Such conduct also amounts to legal misconduct.[81]

Accordingly the High Court set aside the award for the compensation and remitted the said part of the award for the reconsideration.

A claim for extra lead for procuring metal from faraway places was allowed by an arbitrator on the basis that the extra lead was not contemplated by the parties at the time of the contract. It was also upheld by the Supreme Court on the ground that the arbitrator could be said to have taken a reasonably possible view. However, in respect of a couple of other claims which were awarded, ignoring the express provisions of the agreement, the same award was partly set aside.[82]

5.18 Owner's Liability to Pay for Changes Made

Where the contractor voluntarily, and without any request by the owner, executes extra work or provides better materials than those stipulated for, he can have no claim against the owner for more than the contract price.[83] Under similar circumstances a subcontractor too shall have no claim against a contractor. If the owner has consented to alterations from the works specified he would be liable to pay only if he has expressly been informed that additional expense might be incurred. If the change is allowed by way of a concession to the contractor, the owner is not liable to pay more than the contract price.

Where the contract provides notice requirements in relation to variations, these should be complied with so that the contractor is not disbarred from making a claim for an extension of time or payment for the additional work. If the contract gives the owner an absolute discretion to forfeit rights for failure to comply with notice clauses, the general contract law principle applies that the discretion should not be abused.[84] An implied term would expect that the discretion ought to be exercised within the limits of honesty, good faith and genuineness and without arbitrariness, capriciousness, perversity and irrationality.[85] It would be an abuse of discretion to reject a request for a variation or to seek to forfeit a contractor's right to additional payment or an extension of time, merely because the information was not given "without delay" or that some information was missing.[86]

EXAMPLE FROM CASE LAW

Contractors agreed to furnish and erect all the iron and general work for a lump sum and the contract made an express provision that no alteration in the specified work should be made by the contractor. The contractors found that they were unable to make certain girders owing to the thinness of the metal specified. The contractors made the girders with

---

[81] In State of Kerala v. V.P. Jolly, AIR 1992 Kerala 187.
[82] M/s. Shyama Charan Agarwala and Sons v. Union of India, AIR 2002 SC 2659.
[83] Wilmot v. Smith, (1828) 3 0 & P 453.
[84] Bluewater Energy Services BV v. Mercon Steel Structures BV, [2014] EWHC 2132 (TCC).
[85] Ibid., para. 1009.
[86] Ibid., para. 1011.

thicker metal than that stipulated in the contract, but with the oral consent of the engineer. It was held that there was not sufficient evidence to imply that there was to be a payment for that additional thickness.[87]

It would seem, by analogy, that if the owner consents to the contractor making use of less expensive materials than those specified, he cannot, unless there is a new contract, claim that the contractor should make a corresponding reduction in price.[88] Thus, the owner will be liable for any cost of additions and substitutions made by the contractor of his own accord only if the owner ratifies such changes. Merely retaining the extra work done or superior material used does not make the owner liable for its full or any value.

5.19 Basis of Compensation for Extra Work

Where a work is to be done on an item rate basis, payment for additional quantities of work will be made on the basis of the unit prices stipulated in the contract for an increase in the quantities within the variation limit.

If a contract is silent as regards the basis of compensation for extra work, payment will, as a general rule, be made on the basis of fair and reasonable value of the work done or material furnished.[89]

Generally, when a dispute is referred to an arbitrator for deciding the rate of an extra item, both the parties submit their respective rate analyses in support of the rates proposed by them. Evidence has to be led to prove the cost of materials, labour and machinery assumed in the rate analyses. The arbitrator determines the rate on the basis of the evidence. Occasionally, rather than leading evidence of the prevailing prices at the time the extra work was executed, the parties place reliance upon the price indices to show the variation in the cost that was prevailing at the time of signing of the agreement. Care must be taken in such cases to use the price index prevailing at the time of execution of the work. The Bombay High Court remitted back an award for reconsideration of the rates for which the arbitrator admittedly did not use the price indices for the period the extra work was done.[90] As to what is the fair and reasonable value of a work done, the following cases will serve as examples.

EXAMPLES FROM CASE LAW

1. A contractor had brought an action for the recovery of balance in which the only dispute to be settled was as to the value of the extra work executed. Three experts testified as to what they would consider to be fair and reasonable charges for the extra work. Their estimated cost varied. The lowest estimate was $1,149 and the highest estimate was $2,140. The trial court as to this item allowed $1,550, which was sustained on appeal. As to the value of labour, the estimates of the three experts varied from $1,960 to $2,140. The judgment of the trial court which allowed the highest estimated value was reduced to $2,000 by the Appellate Court.[91]

---

[87] Tharsis Sulphur & Copper Co. v. Mc Elroy & Sons, (1878) 3 App Cas 1040 (1054).

[88] *Halsbury's Law of England*, 2nd ed., Vol. 3, p. 268.

[89] Carr v. J.A. Berriman Pvt. Ltd. (1953) 27 A. L. J. 273 (Australia); see *Hudson's*, 10th ed., pp. 340 & 533.

[90] Municipal Corporaion of Greater Mumbai v. Prestressed Products (India) 2003(2) Arb. LR 624 (Bombay).

[91] Lindberg v. Brandt et al., 112 N.E. 2d. 746 Appellate Court of Illinois, 1953.

2. A contract provided for excavation in hard murrum. The contractor was, in fact, required to carry out the work of excavation in soft rock, payment for which was to be made on the basis of *quantum meruit*, that is, the fair and reasonable value of the work done. The contractor had, during the performance of the contract, demanded a rate of Rs. 45/- per 100 cu. ft. for the type of soil met with. In the cross-examination, one of the witnesses of the owner had admitted it as a proper rate. The geologist appointed by the contractor, an expert witness who had examined samples of the rocks in question, opined after seeing the other two rates for hard rock excavation and the working rate of the plaintiff that Rs. 80/- was, the proper rate. The Gujarat High Court allowed a rate of Rs. 45/- per 100 cu. ft. in view of the circumstances that the work had to be done in the seabed where it could be carried on only for a few hours when the site was not flooded due to high tide.[92]

It will be noticed that the best way to deal with pricing of variations and deviations is to include express stipulations in the contract. The provisions of the contract generally state the basis on which payment for extra work would be made, should it be required for the performance of the contract. When using standard form contracts, this option of negotiation is not always available. Such stipulations would avoid disputes between parties, which may otherwise arise at a later stage.

5.20 Absence of Measurements – Record Maintained by Contractor May Be Relied Upon

Where variations lead to the need for an extension of time, it is advisable to keep a record to show that the claiming party was actually delayed by the factors of which it complains. It does not follow as a matter of logic or practice, on a construction or fabrication project, that, simply because a variation is ordered or that information is provided later than programmed or that free issue materials are issued later than envisaged originally, the claimant is delayed. As Akenhead J. has said, if the real cause of the delay is, say, overwork or disorganization within the claimant, the fact that there have been variations, late instructions or information or late issue of materials may be simply coincidental.[93]

When an owner wants to deny a contractor's claim for "extra" work, he should ensure that the work is measured accurately by his representative. These measurements can be kept jointly, where agreement is possible. In the absence of measurements of such work having been recorded by the owner or his representative, it is possible that the measurements as recorded by the contractor would be relied upon in judicial proceedings. This would be useful in the event that the owner fails to establish the contention that the said work was not extra; by then it might have become impossible to measure the work if it was covered by subsequent work. The Calcutta High Court agrees that: "[a]s to measurement of works it was also the duty of the appellant to keep proper account in respect thereof and if they failed to maintain such record, the contractor could not be deprived of the legal dues."[94]

However, it needs to be noted that the record should be properly kept and entries proved or else it may not be accepted, as happened in the case below. Joint measurements, of course, are ideal so

[92] M/s. R.C. Thakkar v. (The Bombay Housing Board by its successors) now The Gujarat Housing Board, AIR 1973 Guj. 34.

[93] Cleveland Bridge UK Limited v. Severfield-Rowen Structures Limited, [2012] EWHC 3652 (TCC), paragraph 98.

[94] Union of India v. M.L. Dalmia & Co., AIR 1977 Cal. 266 (272, 273). Also see M/s Jagan Nath Ashok Kumar v. D.D. A., AIR 1995 Del. 87; K.C. Skaria v. Govt. of State of Kerala, AIR 2006 SC 811 = 2006 AIR SCW 265.

long as the parties note that the liability is not agreed. In this instance, please note the case discussed below in paragraph 5.21.1.[95]

EXAMPLE FROM CASE LAW

In respect of the record maintained by the contractor, the Gauhati High Court's observation is self-explanatory:

> Now let us have a look at the Exhibits 14, 15, 16 and 18. Ext. 15(1) is an exercise book containing some entries. Also it contains some slips of papers. There is no page number in the exercise book. Each and every entry was not proved by any of the witnesses. It is not understood as to why this book was admitted in the evidence without proving entries made therein. This Ext. does not come within the definition of Section 34 of the Evidence Act. It is not understood as to why both the courts below accepted this exercise book with some entries made therein along with some slip papers to thrust the liability on the plaintiff. No doubt, this book was exhibited without objection from the defendant. But that did not take away the right of the court to question and/or scrutinize the evidentiary value of the same, Section 61 of the Evidence Act provided that contents of documents may be proved either by primary or by secondary evidence. Here the contents of the exhibits were not proved either by primary or by secondary evidence. So these exhibits must be brushed aside. These Ext. 15(1) and 15(2) are mere scrap of papers. They were not kept in the regular course of business and the same could never be admitted as evidence under the provisions of law. The said measurement book did not bear any signature of the departmental officers and that were never seen or submitted prior to the same being exhibited in court. Ext. 14 is a typed copy which claims to be the work done from March, 16, 1975 up to 31st March 1976. It gives the nature of work carried out and engaging average labour strength of 150 numbers daily and it also gives the volume and value of work. This also has no evidentiary value at all. Ext. 16 is another sheet which claims that it is the work done from April, 1976 up to 31st March 1977. It shows the nature of work carried out and also volume and value of the work. It does not bear the signature of anybody. The typed copies were exhibited by the plaintiff. Exts.17 and 18 are other two typed copies of the same nature. All these exhibits have no evidentiary value whatsoever and they cannot thrust any liability on any person.[96]

5.21 Common Grounds on Which Extras are Claimed – Illustrative Cases

Illustrative cases to highlight further the above principles are grouped under the following heads:

Extra work in foundation excavation

Changes in design, drawings, specifications

Extra lead/lift

Additional difficulty in carrying out the work.

---

[95] Associated Engineering Co. v. Government of Andhra Pradesh, AIR 1992 SC 232.
[96] State of Meghalaya v. Joinmanick Nosmel Giri, AIR 1995 Gau. 23.

5.21.1 Extra Work in Foundation Excavation

Foundation work can give rise to unexpected conditions, especially when a contractor has not had an opportunity to conduct in-depth tests. The following examples highlight how the risk of such unforeseen conditions is dealt with.

EXAMPLES FROM CASE LAW

1. The facts of a case decided by the Gujarat High Court clarify the concept of what may amount to "extra" work with reference to excavation. The contract was for construction of a seabed tank. The dispute in the case was regarding the following three items: (i) excavation work, (ii) carting the excavated stuff, and (iii) dewatering from the bed of the tank. The observations and findings by the Supreme Court were as follows.[97]

5.21.1.1 Excavation Work

The dispute was quite narrow in its scope, namely, whether the item of rock-cutting was an item included in the contract and thus liable to be paid for at the contract rates only, namely, Rs. 8/- to Rs. 14/- per 100 cu. ft. in respect of excavation work at different levels. The excavation referred to in the contract was of murrum including hard murrum. It was common ground that there was no specific mention of excavation in rock or cutting of rock. The relevant term only spoke of "excavation for tank in any soil, murrum, *sock*, etc." (emphasis supplied). The only dispute was whether the word "sock" was intended to mean "rock" as contended by the defendant. Both the trial court and the division bench of the Gujarat High Court examined the evidence and concluded that the word "sock" could not refer to rock. The Supreme Court agreed.

The facts and circumstances considered by the High Court were as follows: The contractor completed the work as directed and was granted an extension of time for about 13 months without imposing any penalty on the ground that rock cutting was involved. The contractor's four letters demanding extra rate remained unreplied for more than eight months. On this background, commenting on the statement made by the engineer the High Court observed:

> In view of the silence for all these months and the extension being granted to the plaintiff because of the difficult work which he had to carry out with manual labour in these difficult circumstances in sea bed, the denial by this engineer can hardly be swallowed by anybody that there was no rock at all or that the plaintiff did not carry out any rock cutting by chiseling.[98]

5.21.1.2 Dewatering from Bed of Tank

This item was described in the contract thus: "Dewatering from the bed of the tank including strutting, shoring if necessary, including pump and all accessories for this job, etc. complete." The rate was Rs. 2/- per 100 cu. ft. of excavation. Power was to be supplied by the Board at the stipulated rate. The contractor in terms accepted the rate and power charges as stated by the Board. The contractor claimed payment by repeating this agreed rate three times for (i) excavation, (ii) laying concrete bed and (iii) completing the masonry walls so that for the entire work he would, in all get a rate of Rs. 6/- per cu. ft. The trial court upheld the claim of the contractor. Reversing this judgment, the Gujarat High Court held:

---

[97] S.A. Jais and Company v. Gujarat Electricity Board, AIR 1988 SC 254, 1987 (Sup) SCC 614.
[98] Gujarat Electricity Board v. S.A. Jais and Company, AIR 1972 Guj. 192.

The trial Court has ignored the material word in the abstract in item No.7, dewatering "from the bed of the tank". The term "from" is very relevant in showing that what was agreed to was not dewatering only during the process of excavation but the entire job of dewatering because water would be coming out from the tank, which was excavated during all the three processes. That is why the complete job was mentioned as dewatering complete.

On appeal, the Apex Court agreed with the High Court. The Supreme Court did not agree that the work of dewatering was to be paid for each time dewatering was done as the tank was flooded by tidal water and amounts for dewatering were liable to be paid each time dewatering had to be done; the respondent had made one payment for dewatering once. The trial court had awarded additional amounts for dewatering not for the repeated flooding by tides, but on the basis that dewatering was required for excavation for the tank, concreting the bottom of the tank as well as for making masonry walls, and dewatering had to be paid for separately each (time) it was done. The Supreme Court found that there was no justification for this.

In reply to the argument of the contractor that the quantity mentioned in the abstract against the item was 104,900 cu. ft. which was the quantity of the excavation and, therefore, that the dewatering must be only for the process of excavation, it was observed

that the quantity is mentioned because the measure which was adopted for fixing this rate was not the time for which the pump works or the water which is taken out but a lump sum of Rs. 2/- for dewatering from the bed of the tank by way of a complete job on the basis of 100 cu. ft. of this excavation work.

It was further held that the rates for the items of cement concrete and masonry were agreed for the complete job. These rates therefore covered everything including dewatering, without which these items could not be done. The Supreme Court confirmed the findings of the High Court.

5.21.1.3 Extra Bulkage in Transport of Excavated Stuff

This claim for an additional amount of Rs. 4,662.84/- was for the transport of that excavated material. The minor claim was accepted by the trial court but disallowed by the division bench of the Gujarat High Court. The basis of this claim is that there was an increase in the volume of murrum when it was excavated from the ground; when it forms part of the ground it is in a compact condition, whereas once excavated it is loose and thus larger in volume. The claim was on the estimate that there would be an increase of 25% in bulkage when murrum was excavated from the ground. The High Court held that the defendant had already paid an increased amount of cartage on the basis that when material is excavated from the ground, there would be an increase in volume of 9.80%. There was no reliable evidence to establish that there would be an increase in volume exceeding 9.80% which had been paid for. The Supreme Court concurred.[99]

1. Over Breakage: An agreement for the construction of a canal provided in the specifications that over breakage was permissible up to a maximum of 5 cm limited to an average 2.5 cm. It was the contention of the contractor that over breakages occurred due to adverse geological conditions not foreseen at the time of estimating the cost of the work; it was beyond the control of the contractor and occurred in spite of due care and caution exercised by him. The respondent's contentions were manifold, namely, that (a) the contractor should be responsible for extra excavation, (b) the contractor should have taken necessary precautions after studying the nature of the soil strata, (c) the report of the geologist relied upon was not accurate, and (d) the fact of measurements being recorded could not be taken as

---

[99] S.A. Jais & Co. v. Gujarat Electricity Board, AIR/ SC 254.

acceptance of the work. The arbitrators allowed the claim of the contractor. In a petition under Section 30 of the Arbitration Act, 1940 challenging the award, the Karnataka High Court, after reproducing the relevant provisions of the specifications, observed and held that the contractor would not be entitled for payment for extra breakages and their filling beyond the stipulated average depth of 2.5 cm if the extras were due either to inadvertence or were carried out for the convenience of the contractor. The claim was not rejected until the dispute was brought before the arbitrators. At no stage did the department inform the contractor that the over breakages in excavation were due to the contractor's inadvertence, but instead they informed the contractor that the claim was under consideration. Therefore, the arbitrators concluded that while the work was in progress at no time did the department warn or charge the contractors or allege that they were not resorting to proper methods of blasting in a manner so as not to cause over breakages or that they had been careless or indiscriminate in blasting which resulted in breakages. The arbitrators held that Clause 39 of the agreement which required the contractors to acquaint themselves of the site conditions before taking up the work did not militate against their claim. They also held that the recording of the measurements in the measurement book was not merely for the purpose of accounts, but that it disclosed tacit acceptance of the contractor's claim. In the circumstances, it was held that the over breakages occurred due to the peculiar geological strata. The Karnataka High Court refused to set aside the award.[100]

2. A contract for the design and construction of a road and tunnel in Gibraltar between Obrascon, a Spanish contractor, and the government of Gibraltar provided that a tunnel would run under an airport runway. Obrascon relied on the environmental statement and a site investigation report provided by the government. When the work was commenced, site contamination issues came up, leading to the problem of the use of that contaminated soil (or its storage). Obrascon also claimed that there were unforeseen issues with contaminated groundwater and rocks. It then proceeded to suspend work in order to carry out tests and a re-design. The government terminated its contract at a time when the two-year project was already six months late and only 25% of the work was executed. Various tenderers had raised questions about the site and the government had issued a tender bulletin that stated:

> Q1.7 – Could you tell us where the landfill is to tip the products from the tunnel excavation and demolitions? If there is none, could you tell us where there are possible storage areas for later use and the additional cost of this storage? A1.7 – Disposal of material is the Contractor's responsibility under the contract and no off-site storage areas have been identified.

Obrascon's tender was the lowest but also optimistic in its geotechnical parameters. The government had accepted another tender but on that tenderer's refusal to proceed, Obrascon had won the contract.[101] FIDIC Yellow Book (1st edition 1999) was incorporated into the contract conditions. Clause 1.1.6.8 stated, that "Unforeseeable" means "not reasonably foreseeable by an experienced contractor by the date of submission of the Tender". Akenhead J. observed that an experienced contractor would undoubtedly not limit itself to an analysis of the site investigation report made available at the tender stage. This observation may not necessarily be applicable if the contractor is quoting for work designed by the owner. Here, the contractor was undertaking a design and build contract. Many site investigations into the geotechnical aspects are based on boreholes which have obvious and known limitations that a sample taken from a 100–150 mm-wide hole may not be representative of material between

---

[100] State of Karnataka v. R. Shetty and Co. Engineers & Contractors, AIR 1991 Kant. 96(100–101).
[101] Obrascon Huarte Lain SA v. Attorney General for Gibraltar, [2014] EWHC 1028 (TCC).

that borehole and the next one. Trial pits, though larger, do not go to the same depth as boreholes. Given that the site in this case was a landfill site in proximity to a military site, the contaminants would have been randomly distributed. What was needed in relation to a brownfield site, Akenhead J. said, was "some intelligent assessment and analysis of why there was contamination there (namely the recent and less recent history)."[102] The presence of hydrocarbons and related derivatives would then have been expected, as well as contamination of the groundwater. The risk of contamination was thus not unforeseen. The Court of Appeal dismissed an appeal against the decision.[103]

5.21.2 Changes in Design, Drawings and Specifications – Illustrative Cases

The owner or the owner's representatives may introduce changes to the work by altering design, drawings or specifications. The examples below indicate how a dispute arising out of such changes may be resolved.

EXAMPLES FROM CASE LAW

1. The construction of the Hall of the Nations and the Hall of Industries for the Third Asian International Trade Fair was entrusted to the appellant construction company by the Union of India. The work consisted of construction of space frame structures in exposed concrete, a new venture in the field of construction technology and architecture. The designs could not be finalized prior to giving of the work order and substantial details were required to be given during the execution of the work which necessitated extra items and extra work. The value of the work went up from Rs. 91.57 lacs to 1.53 crore rupees. One of the disputed extra items pertained to the order substituting prefabricated structure by cast-in-situ for which a permanent steel staging became necessary. A huge quantity of additional steel was used for a continuous period of about nine months after which it had to be discarded. The other contested items included payment for additional platform of steel channels and kail wood set up for staging. The respondent's contention was that staging was included in the work and that it did not deserve separate payment. The contractor's argument was that the additional platform had to be set up on account of subsequent modification in the nature of the work and that it was different from the scaffolding put up for the movement of the labour force.

Clause 18 which relied upon by the respondent read:

> [t]he contractor shall supply and provide at his own cost all materials (except such special materials, if any, as may in accordance with the contract be supplied from the Engineer-in-charge's store), plant, tools, appliances, implements, ladders, cordage, tackle scaffolding and temporary works requisite for the proper execution of the work whether original, altered or substituted and whether included in the specification or other documents forming part of the contract or referred to in these conditions or not.

Undoubtedly, the wording of this clause was very wide, in particular "for the proper execution of the work whether original, altered or substituted and whether included in the specification or other document forming part of contract". But its interpretation had to be

---

[102] Ibid., para. 213.
[103] Obrascon Huarte Lain SA v. Attorney General for Gibraltar, [2015] EWCA Civ 712.

restricted to the changes made, which would not involve cost in excess of the cost originally contemplated by both the parties, that is by the owner while estimating the cost of construction prior to inviting tenders and by the contractor at the time of submission of his tender. Cost of such items of work, which were necessary for executing the main work but for which no separate payment was contemplated by the contract, was generally spread over or included in the rate or cost of executing the concerned item or items. As such the owner or his engineer/architect could not order substantial changes, which would involve cost in excess of that provided for at the time of signing of the agreement. Experienced engineers are fully aware of this and, as such, this defence is usually not raised before engineer arbitrators, but usually attempted to be raised in a court or before arbitrators who do not have engineering expertise. The Supreme Court of India, it is respectfully submitted, correctly upheld the award of the arbitrator allowing extra cost on the above accounts.[104]

2. The respondent-contractor entered into an agreement with the appellant for the construction of non-overflow and overflow sections with bridge spillway and other appurtenant works of Maudahe Dam in Hamirpur district in the State of Uttar Pradesh. In respect of two items of work, namely Items 13 and 15, it was alleged that the appellant changed the designs and drawings as a result of which the quantity of work became abnormally high compared to that estimated in the agreement. The contractor claimed a higher rate than what was agreed to in the agreement. The state refused to accede to the contractor's demand and the dispute was referred to a sole arbitrator. The contractor alleged that when the drawings and designs were changed, he had resisted and prayed for an alteration in the rate but the concerned authorities had assured him orally for such a change though ultimately they did not agree to it. It was also alleged that the agreement obliged the contractor to carry out the work as per directions of the concerned authorities and accordingly he had executed the work.

The appellant state denied its liability to pay at the revised rate while admitting that there had been a change in the drawings and designs relating to Items Nos. 13 and 15 and that on account of such changes, the quantity of work for the two items had increased. The appellant's allegation was that the variation in the quantity of work was covered by Clauses 11.25 and 13.11 of the agreement.

The arbitrator concluded, after analysing different clauses of the agreement, that (a) the contractor could not have refused the work in accordance with the alterations and modification in the drawings and designs, (b) there was a fundamental change in the drawings and designs which abnormally increased the quantum of work, and (c) under the agreement the contractor could not claim any excess rate for work up to 10% excess quantity, but beyond that he would be entitled to claim a higher rate. The court refused to interfere with the award.[105] However, it was held that in respect of the excess quantity of work executed by the claimant subsequent to the completion period indicated in the agreement when the claimant had made the claim at a higher rate and that claim was allowed by the arbitrator on the basis of analysis of rates given by him, then the amount already paid to him by the state in accordance with the escalation clause in the agreement had to be adjusted so that the contractor would not be entitled to a double benefit on that score. The award was accordingly modified.

3. **Change in RCC design: when it may amount to extra**: A contract provided for a 5-inch-thick RCC slab to cover an area of about 61,000 sq. ft. It was subsequently changed to 4½-inch-thick slab incorporating more steel than that required for a 5-inch-thick slab. The

[104] Puri Construction Pvt. Ltd. v. Union of India, AIR 1989 SC 777.
[105] State of U.P. Appellant v. M/s. Ram Nath International Const. Pvt. Ltd., AIR 1996 SC 782.

agreed rate was per unit area of the slab. Extra payment to the contractor was denied on the ground that the contract provided for changes in the design and specifications for which no compensation would be payable. The High Court held that the contractor had to be paid extra; the decision was approved by the Supreme Court. The provision in the contract that denied a payment of compensation was held to be inapplicable.[106] Such a dispute may not arise under a modern contract because the contracts nowadays provide for payment for reinforcement separately on a weight basis and for concrete on a volumetric basis.

4. **Introduction of reinforcement: when it may amount to extra**: Where a plaintiff was required to carry out cement concrete work but was subsequently ordered by the engineer-in-charge to provide reinforcement to connect different layers of 1:4:8 concrete at various places, it was alleged that the concrete work having been done in a ratio of 1:4:8 with brick ballast, it could not be considered as RCC. The fact that reinforcement was used was, however, admitted. It was held:

> Having considered the evidence given by the parties on the aforesaid question, we find that iron bars having been used by the plaintiff in the completion of the work mentioned in Schedule B, the work which the plaintiff did was R.C.C. and not cement concrete.[107]

5. **Changes requiring supply of panelled windows instead of glazed windows amounts to novation**: After the plaintiff company had manufactured glazed windows covering 2,000 square feet in accordance with the original design, the plan was revised and the plaintiff was required to manufacture and supply panelled windows. The cost of the material and the labour cost for panelling was to be borne by the plaintiff company. The plaintiff company asked the executive engineer for an enhanced rate, supporting it with a detailed rate analysis. The executive engineer recommended to the superintending engineer that the rate for new work be enhanced. It was alleged that the executive engineer advised the plaintiff to carry on with the supply like a good contractor. The plaintiff company consequently completed the supply as per the new design but the superintending engineer did not agree to the enhancement of the rate. The contract required the dispute to be referred to arbitration by the same superintending engineer. When the matter came before the trial court, it was held that there was no agreement between the parties as to the rates at which the plaintiff was to be paid for the revised work. The decision was later upheld by the High Court, observing that

> the subsequent requirement of the defendant, to supply panelled windows, instead of glazed windows, was in the nature of a novation; and the defendant was bound in law to compensate the plaintiff for the panelled windows; the obligation to supply which was de hors the original undertaking. In such a case, the plaintiff can claim a fair and reasonable price for the work done, or the goods supplied on the basis of quantum meruit that is so much as is deserved or merited. ... Here the original contract had been superseded by a new undertaking, and the new work was not complementary to the work originally contemplated, but outside its scope. The defendants cannot avoid payment of the extra cost involved in the new type of work which was required to be done.[108]

---

[106] Bombay Housing Board v. M/s. Karbhase Naik and Co., Sholapur, AIR 1975 SC 763.

[107] State of U.P. v. Chandra Gupta & Co., AIR 1977 All 28 (32).

[108] State of Punjab v. Hindustan Development Board Ltd., AIR 1960 Punj. 585.

6. **Extra cost for direction to use specific quality materials**: Where the owner or his architect/engineer insists upon the use of a specific quality material not originally specified, the contractor can claim extra cost. A contractor was directed to use new planks for shoring instead of old planks, and the arbitrator allowed the claim for extra payment. The award was upheld by the Calcutta High Court.[109]

5.21.3 "Extras" on Account of Additional Difficulties in Executing the Work

It is not infrequent in practice, particularly in engineering contracts for the construction of dams or bridges, or for the excavation for canals, sewers, etc., to encounter unexpected difficulties which may necessitate changes from the expected method of working. Most contracts contain express provision making such risk or contingencies the responsibility of the contractor, in the form of, for example, express disclaimers as to the state of the site, subsoil conditions, etc. The question as to whether the contractor in such a case can claim, and succeed in getting payment for, the extra expenditure occasioned by such an eventuality needs to be considered. The answer to this question is not simple and straightforward. It will depend upon the facts and circumstances of each case, the terms of the agreement including type of contract (i.e. lump sum or using bills), specifications, etc. The general trend of the court decisions is to hold the contractor liable for extra expenditure required to be incurred. The exceptions are made in cases where, for example, (i) the owner misrepresented the conditions; (ii) the difficulties encountered were in the variations ordered; or (iii) there was evidence to support that the change in design made by the engineer necessitated extra expenditure.

EXAMPLES FROM CASE LAW

1. A contractor was expected to excavate only hard soil but encountered hard soil mixed with pebbles and stones. Clause 11 of the agreement included a condition in that "Strata of E/W (Earthwork) as specified in the 'G' Schedule will only be payable to the contractor. In case strata other than the specified is met during the execution of work, no excess payment on this account will be allowed to the contractor." The arbitrator allowed the contractor's claim for extra rate under another clause but the award was set aside by the trial court on the ground that the arbitrator misconducted by ignoring the express provision of Clause 11. Upholding the appeal against the said decision, the Rajasthan High Court raised some very important points including that the state had not been honest with the contractor inasmuch as it was duty bound to examine the land on which the work was to be carried out before entering into the contract but failed to do so, and further denied the existence of pebbles and stones which the arbitrator confirmed by his site visit, and therefore the state could not take shelter under Clause 11. Even in a commercial contract, the state is supposed to be fair, just and reasonable. A contract has to be read holistically and not in piecemeal fashion. In case two clauses are inconsistent, the court should harmonize them.[110]

This judgment not only takes into account the doctrine of misrepresentation but also the doctrine of exemption clause, it is respectfully submitted. The other provision of agreement made price escalation payable only if the contract period exceeded 12 months. The original period was six months but it stood extended to more than 12 months due to no fault of the

---

[109] Calcutta Metropolitan W. & S. Authority v. Mis. Chakraborty Bros., AIR 1988 Cal. 423 (425–6).

[110] Heera Singh v. State of Rajasthan, AIR 2007 Raj. 213 =2008(1) CTLJ 97 (Rajasthan).

contractor. The award of payment of escalation was set aside by the trial court on this count. On appeal the High Court observed:

> [c]onsidering the fact that the period for completion of the project can be extended (and which was extended by the department), considering the fact that during the extended period the cost of material and cost of labour can increase, in order to compensate the contractor, Clause 45 was placed in the Agreement. Therefore, there is no cogent reason to limit the scope of the term "the stipulated period" to mean "the original period" for which the agreement was entered into. The term "the stipulated period" would have to include "the subsequently extended period" as well.[111]

2. A contractor encountered difficulties in the excavation of soil for the depths between 12 feet and 16 feet, but for structural stability of the bridge under construction, he needed to excavate to at least reach the sandy clay at a depth of 18 feet. Iron sheet piling was necessitated. Delay in completion resulted and no revision of rates was offered by the state, which instead called for fresh tenders. The Patna High Court held that the contractor was entitled to one-third of the cost of construction and refund of his earnest money.[112]

FIDIC conditions of contract for works of Civil Engineering Construction contain Clause 12.2 of 4th edition (Clause 4.12 of 1999 edition) which provides in such contingencies as under: (a) The contractor should give notice of physical obstructions or physical conditions not foreseeable by an experienced contractor to the engineer, with a copy to the employer. (b) The engineer, if satisfied that the obstruction/conditions could not have been reasonably foreseen by an experienced contractor, after the consultation with the employer and the contractor, should determine the extension of time and the additional amount of costs which the contractor may have to incur.

3. *Force majeure* **clause – compensation when allowed**: A contract contained the provision reading: "Neither party shall be liable to the other for any loss or damage occasioned by or arising out of act of God, such as unprecedented flood … ". An award for payment of compensation on account of flooding of the work area during progress was upheld by the trial court and also the High Court. The Supreme Court of India upheld the said decision for the reason that the appellant did not lead evidence that the rain was unprecedented and that, in fact, it was an act of God.[113]

In another case, with the contract containing a similarly worded clause, the claim was made on account of material washed away due to a breach of the protection bund which was also constructed by the same contractor who made the claim, the state sought protection under the *force majeure* clause. It was alleged that the unprecedented rains caused excessive discharge. The appeal against the award was dismissed by the High Court holding,

> [t]he State has not been able to substantiate its protection under force majeure clause as no evidence in support thereof has been brought on record as none has been pointed out by the learned Deputy Advocate General. There is no detail of discharge of water prior to the construction of the protection bund nor has any been mentioned after the construction of the protection bund. The State has also not pointed out the structural design of the protection bund and resultantly the strength for which it has been constructed. Thus, the argument in support thereof is totally misconceived and is therefore, rejected.[114]

---

[111] Heera Singh v. State of Rajasthan, 2008(1) CTLJ 97 (105)(Rajasthan).

[112] State of Bihar & Anr. v. S. Sgheyasuddin, AIR 2009 (NOC) 387 (Pat.).

[113] State of U.P. v. Allied Constructions, 2003 (3) Arb. LR 106 (SC).

[114] State of Punjab v. Parmar Construction Co., 2002(3) Arb. LR 32 (P. & H.).

It was further held:

> The argument that the material was not kept at the safe place, therefore, the contractor is liable on account of "volenti non fit injuria" is not sustainable. The learned Deputy Advocate General has not been able to point out any evidence or any communication by which the place for stacking the material by the contractor had ever been specified. He has also not been able to refer to any communication to the effect that after the material had been stacked by the contractor, it had ever been pointed out to him that the material should be removed from the place to avoid any possibility of the water coming in on account of excessive rains. Thus, the plea remains unsubstantiated. As such, the argument has no force and the same is rejected. In view of the above, the appeal is dismissed.

In yet another case decided by the Bombay High Court, a claim for murrum washed away by tidal waves was allowed by the arbitrator in spite of an exemption clause on the reasons recorded. The Bombay High Court upheld the award, stating that the construction which had been placed by the arbitrator was a possible view to take and that the court would not be justified in interfering with the award in view of the limitations placed on the exercise of its jurisdiction of Section 34 of the Arbitration and Conciliation Act, 1996.[115]

4. **Excess consumption of materials**: An arbitrator rejected a claim for recovery on account of excess consumption of steel and cement on the ground that the employer had not made any allegation that the contractor had diverted the steel and cement, and that the excess material could have been used in the work itself. It was held that the arbitrator had given good and sufficient reasons.[116]

5. In a subcontract for the construction of an autoclave, the design of a carbon and stainless steel autoclave was changed leading to an increase in the weight of the autoclave.[117] It was a one-off design so that standard specification would not apply to it. The price was agreed after negotiations on the basis of a weight of 542.2 tonnes. At this time the main contract had not been formalized. The main contractor was expected to form a consortium with the subcontractor which did not take place. FIDIC terms applied to the main contract but they were not incorporated into the subcontract. Had they been so incorporated, the main contractor would have had to appoint an engineer to administer the contract. As it was, the purchase order, when formalized, referred to a different weight. This also required fabrication on site rather than at the subcontractor's premises leading to extra expenditure. The subcontractor only saw the revised specification with the revised weight calculation, which provided for a different weight of 732.5 tonnes. The High Court proceeded on the basis that the spirit of FIDIC conditions would apply but they could not be unilaterally imposed by the main contractor on the subcontractor without having been properly incorporated into the subcontract. The High Court held that damages were payable for the variations along with an extension of time.

[115] Municipal Corporation of Greater Mumbai v. Prestressed Products (India), 2003(2) Arb. LR 624 (Bombay).
[116] Its Bhai Sardar Singh & Sons v. Delhi Development Authority, 2003(1) Arb. LR 387 (Delhi); also see M/s Jagan Nath Ashok Kumar v. D.D.A., AIR 1995 Del 87.
[117] Motherwell Bridge Construction Ltd. (t/a Motherwell Storage Tanks) v. Micafil Vakuumtechnik; Motherwell Bridge Construction Ltd. (t/a Motherwell Storage Tanks) v. Micafil AG, [2002] 1 WLUK 711; 81 Con. L.R. 44.

6

Time for Completion, Delay in Completion

6.0 Introduction

A construction project consists of many activities. Each activity requires some time for its completion. The cost of execution varies with the time required for its completion. The total cost of work is made up of two elements. The first is direct cost, which includes cost of materials, labour and plant. The second consists of overheads. These two bear a definite relationship with time. If a graph of time-cost relationship were plotted it would be clear that for every project there is an optimum time for its completion. In other words if a project were allowed to be completed in its optimum time the total cost would be minimum. If an attempt is made to complete the project earlier than the optimum time direct costs increase and indirect costs decrease but the overall cost is high. Similarly, if a project is planned to be completed in a period longer than the optimum time the direct costs may decrease but the indirect costs increase and the overall cost is high. Every attempt, therefore, must be made to plan a project in such a way that the contractor will get optimum time for its completion. There have been some instances where the time limit stipulated was a few months and the projects were not completed within four to five years. The cost of completion is bound to increase due to delay, on account of increases in both the direct cost of inputs and overheads.

If completion is desired in less than the optimum time, the owner should weigh the advantages of direct and indirect benefit which may thereby result, against the high cost of construction. It must, however, be remembered that no matter how much resources are put in, a project cannot be completed earlier than what is called a "crash time". Before inviting tenders for a work, careful thought must be given to the above aspects and a realistic duration fixed for completion of the work.

6.1 Provisions of the Law

The following provisions of the law are dealt with in this chapter.

Provision of the Law	Paragraph No.
Indian Contract Act, Section 46	6.17
Indian Contract Act, Sections 47–50	6.18
Indian Contract Act, Section 55	6.5, 6.15
Indian Contract Act, Section 63 and 64	6.12, 6.16.
Provisions in Standard Form Contracts	
FIDIC Form, 1992 edition, Clauses 41.4, 43.1, 44.1–44.3 1999 1st edition, Clauses 8.1–8.12 2017 2nd edition, Clauses 8.1–8.12	6.2, 6.8, 6.9, 6.11
FIDIC Form Clauses 7, 9 and 11	8.1.1
Ministry of Statistics and Programme Implementation, Government of India (MOS & PI), Clauses 27, 28, 29	6.2

(Continued)

(Cont.)

Provision of the Law	Paragraph No.
NITI Aayog Model Form, Clause 10.4	6.2, 6.11
PWD B-1 & B-2 Standard Forms of State Government, Clause 2	6.5.2, 6.11
CPWD, Clauses 2, 3, 6	6.2, 6.9 and 6.11
MDSS & APDSS, Clause 59	4.10.1
Military Engineering Services IAFW Form, Clauses 11A–11C	6.4

6.2 Mode of Specifying Time Limit

Once the time required for completion of a project is decided it may be stipulated in the contract in any one of the following four ways:

1. on a given date, or
2. within a definite number of calendar months, or
3. within a stipulated number of working days.
4. within a stipulated number of days.

In order that the contractor should be careful in avoiding delay in completion of a project the provisions in the standard form contracts usually empower the engineer to take certain actions during the performance of the project.

As an example, an age-old provision in Clause 2 of the B-2 form of contract used by the State of Maharashtra for executing public works is discussed here. A similar provision is found in the CPWD form as also forms adopted by other states and public undertakings.

The clause reads:

The time allowed for carrying out the work as entered in the tender shall be strictly observed by the contractor and shall be reckoned from the date on which the order to commence work is given to the contractor. The work shall throughout the stipulated period of the contract be proceeded with, with all due diligence (time being deemed to be of the essence of the contract on the part of the contractor) and the contractor shall pay as compensation an amount equal to one per cent, or such smaller amount as the Superintending Engineer (whose decision in writing shall be final) may decide, of the amount of the estimated cost of the whole work as shown by the tender for every day that the work remains uncommenced, or unfinished, after the proper dates. And further to ensure good progress during the execution of the work, the contractor shall be bound, in all cases in which the time allowed for the completion of any work exceeds one month to complete:

1/3 of the work in	*	the time
½ of the work in	*	the time
¾ of the work in	*	the time

*Note – The quantity of work to be done within a particular time to be specified above shall be fixed by the Officer competent to accept the contracts after taking into consideration the circumstances of each case, and inserted in the blank space kept for the purpose ...

In the event of the contractor failing to comply with these conditions he shall be liable to pay as compensation an amount equal to one percent or such smaller amounts as the Superintending

Engineer (whose decision in writing shall be final) may decide of the said estimated cost of the whole work for every day that the due quantity of work remains incomplete. "Provided always that the total amount of compensation to be paid under the provisions of this clause shall not exceed 10 percent, of the estimated cost of the work as shown in the tender."

This clause provides for the following:

i.   Time limit for completion, which is usually stipulated in calendar months
ii. Time shall be reckoned from the date of order to commence the work
iii. Time is deemed to be of the essence only on the part of the contractor
iv. It provides only four stages for review of progress and stipulates payment of liquidated damages if the progress is not maintained, as entered in the agreement
v.   The decision of the superintending engineer is made final as regards the levy of penalty
vi. The maximum amount of penalty is mentioned as 10% of the estimated cost as shown in the tender.

The first provision makes Section 55 of the Indian Contract Act applicable. When a contract stipulates a time limit, it is obvious, to avoid dispute, that it should also stipulate the date from which it shall be reckoned. Even after receiving the order to commence the work, the contractor would naturally require a week or two for the actual commencement of the work. FIDIC 1999 edition Clause 8.1 stipulates that the engineer shall give the contractor not less than seven days' notice of the commencement date. FIDIC 2017 edition provides for 14 days' notice. The provision further stipulates that if not specifically stated in the COPA (Conditions of Particular Application) the commencement date shall be within 42 days after the contractor receives the letter of acceptance of his tender. This form thus takes into account the fact that the mobilization period depends upon the nature and extent of the project and cannot be uniformly stated as 15 days or so. As such this provision is realistic, reasonable and fair and may be adopted for all contracts.

In the GOI Planning Commission Form the Start Date is given in the Contract Data. It is defined to mean the date when the Contractor shall commence execution of the works. It does not necessarily coincide with any of the Site Possession Date.

In the NITI Aayog Model Form, date of commencement is called "Appointed Date" and is defined as follows:

"Appointed Date" means that date which is later of:

(a) the 15th day of the date of this Agreement;
(b) the date on which the Contractor has delivered the Performance Security in accordance with the provisions of Article 7;
(c) the date on which the Authority has provided the Right of Way on at least 90% (ninety per cent) of the total land required for the Project in conformity with the provisions of Clause 8.2; and
(d) the date on which the Authority has provided to the Contractor the environmental and forest clearances for at least 90% of the total land required for the Project.

This is an ideal provision allowing the contractor the full time for performance and recommends itself for adoption for the major projects involving land acquisition problems.

The provision incorporated to ensure good progress in the old PWD forms is indeed outdated and needs to be drastically changed in view of developments in planning and scheduling of works. It is advisable that each tenderer be asked to submit along with his tender the construction schedule of the project based on the Critical Path Method (CPM) or Progress Evaluation Review Technique (PERT). Alternatively, the blank tender forms issued may include a construction schedule expected by the owner/engineer to which the contractor be directed to adhere in future planning and scheduling. On acceptance of a tender, the schedule should be included in the contract documents. It would be helpful in more ways than one to avoid disputes, which may arise later on during the performance of the contract, inasmuch as, with its use, the cause of delay can be pinpointed. The part of the clause dealing with liquidated damages is dealt with in Chapter 7.

The recently developed Standard Forms, mentioned hereinabove, provide for submission of a programme by the contractor within the time specified in the contract condition. For example, Clause 8.3 of FIDIC Form 2017 contemplates submission of initial programme for execution of the work within 28 days of the notified date of commencement under Clause 8.1. The provision further stipulates submission of revised programmes from time to time as detailed in the said provision.

6.3 Can Adequacy of Time Limit Be Questioned?

Volume 2 of the CPWD Manual stipulates the time limit that may be allowed for completion of a work depending upon the cost of construction. However, once a different time limit is stipulated and agreed to by the contractor he cannot question the adequacy of the time allowed relying on the provision in the CPWD Manual.[1] Once an undertaking is given, one cannot avoid liability under it merely because it is difficult to perform or not possible to perform except at an exorbitant cost.

6.4 Meaning of "Working Days"

Where a contract provides that the work shall be completed within a stipulated number of working days, dispute may arise as to the legal interpretation of the term "working days".

The term "working days" means the days on which it is possible to carry out the work or the work is carried out. The reason for not carrying out the work could be any, such as Sundays, holidays, bad weather, acts or omissions of the owner or his engineer/architect necessitating suspension of work, etc. Where the word working is not used, days will mean calendar days.

EXAMPLES FROM CASE LAW

1. In a case, the contract provided that the building was to be completed in 125 "working days". It further provided that should the contractor fail substantially to perform the work within the time limit allowed, the owner would be entitled to deduct $25 for each day of such delayed performance. However, in computing the amount of compensation for delayed performance Sundays and holidays were to be excluded from excess days of performance. The contractor failed to complete the work in 125 days excluding Sundays and holidays because of bad weather and acts of the owner. The owner, thereupon, deducted from the contract price compensation at $25 per day. In an action brought by the contractor to recover the sum so withheld, the owner claimed that under the contract all days except Sundays and holidays constituted "working days". The contention of the contractor was that the term "working days" had a well-defined meaning in the construction business and did not include, in addition to Sundays and holidays, days when the work could not be performed because of unusually bad weather or days when the work was delayed by the acts of the owner. He produced testimony of an expert in the building trade to support his contention. While sustaining the contractor's contention, the court observed that the owner and his architect, at the time the contract was

[1] Mohinder Singh v. Executive Engineer, AIR 1971 J. & K. 130.

entered into, must have known the meaning of the term "working days" when used in a construction contract.[2]

2. The Delhi High Court case in Simplex Concrete Piles v. Union of India,[3] raised an interesting question over the conflict between a contract clause "dis-entitling" a person from claiming damages to which he would be entitled under Sections 55 and 73 of the Indian Contract Act. The clauses before the court were 11A to 11C from the MES standard form contract for lump sum work. The arbitrator's award characterized the clauses as exemption clauses and held that they did not apply when the employer caused delays by its faults. The Delhi High Court was impressed by the reasoning in the decision. However, the government argued that even if the contractor was delayed by the government's faults, damages could neither be claimed by the contractor nor awarded by an arbitrator. In Ramnath International Construction (P) Ltd. v. Union of India,[4] the Supreme Court supported a contention that the clauses bar a contractor from receiving damages even for the government's own faults. In Asian Techs Ltd. v. Union of India and Others,[5] the Supreme Court held that the clauses only prevent the department from granting damages but not an arbitrator. The Delhi High Court decided that these two Supreme Court decisions could not be reconciled and were both of division benches. It decided to follow the guidance of the full bench of the Patna High Court in another case[6] that the High Court should follow that judgment of the Supreme Court which lays down the law "more elaborately and accurately". The Delhi High Court's own view was that Clauses 11A to 11C were "void by virtue of Section 23 of the Contract Act".[7] Public policy is a dynamic concept and the Delhi High Court upheld the arbitrator's award and observed,[8]

> A law which is made for individual benefit can be waived by an individual/private person, however, when such law includes a public interest/public policy element, such rights arising from the law cannot be waived because the same becomes a matter of public policy/public interest. ... Contracting out is permissible provided it does not deal with a matter of public policy. ... The issue therefore boils down to whether the rights which are created by Section 73 and 55 of the Contract Act can or cannot be contractually waived. ... Provisions pertaining to the effect of breach of contract, two of which provisions are Sections 73 and 55, ... are the very heart, foundation and the basis for existence of the Contract Act. This is because a contract which can be broken at will, will destroy the very edifice of the Contract Act. ... I therefore hold that the contractual clauses such as Clauses 11A to 11C, on their interpretation to disentitle the aggrieved party to the benefits of Sections 55 and 73, would be void being violative of Section 23 of the Contract Act.

This is a noteworthy view, especially relevant to the conditions of the Indian construction industry. In countries where purely commercial contracts of this nature may be entered into by parties advised by sophisticated lawyers, such issues may not arise. The Delhi High Court's view, albeit of limited value, being expressed when upholding an award rather than when deciding the case in the first instance, gives support to the view that exemption clauses will be interpreted very narrowly in India, especially where the party at fault who drafted the contract is attempting to rely on them.

---

[2] Lewis v. Jones, 251, S. W. 2d, 942, Court of Appeals of Texas, 1952.

[3] CS(OS) No. 614A/2002, decided 23rd February 2010.

[4] (2007) 2 SCC 453.

[5] (2009) 10 SCC 354.

[6] Amar Singh Yadav v. Shanti Devi and Others, AIR 1987 Pat. 191.

[7] Simplex Concrete Piles v. Union of India, CS(OS) No. 614A/2002 paragraph 10.

[8] Ibid., paras 12–16.

6.5 Meaning of "Time Shall Be Deemed to Be of the Essence of the Contract"

The stipulation in a typical construction contract making time to be of the essence of the contract is legally ineffective[9] and need not be incorporated. The reasons are as follows.

When time for performance is stipulated in the agreement and it is further stated to be of the essence of the contract, the parties to the contract understand that the performance of the contract at the precise time designated is of extreme and obvious importance. Section 55 of the Indian Contract Act, which deals with this matter, reads as follows:

> 55. *Effect of failure to perform at fixed time in contract in which time is essential*
>
> When a party to a contract promises to do a certain thing on or before a specified time, or certain things on or before the specified times, and fails to do any such thing on or before the specified time, the contract, or so much of it as has not been performed, becomes voidable at the option of the promisee, if the intention of the parties was that time should be of the essence of the contract.
>
> Effect of such failure when time is not essential.
>
> If it was not the intention of the parties that time should be of the essence of the contract, the contract does not become voidable by the failure to do such a thing on or before the specified time; but the promisee is entitled to compensation from the promisor for any loss occasioned to him by such failure.
>
> Effect of acceptance of performance at time other than that agreed upon.
>
> If, in case of a contract voidable on account of the promisor's failure to perform his promise at the time agreed, the promisee accepts performance of such promise at any time other than that agreed, the promisee cannot claim compensation for any loss occasioned by the non-performance of the promise at the time agreed, unless, at the time of such acceptance, he gives notice to the promisor of his intention to do so.

6.5.1 Promisee, Promissor

These words used in Section 55 are to be understood in respect of reciprocal promises to perform certain acts on or before specified times and not necessarily with reference to the entire transaction. For example, in a typical construction contract, to complete construction within the stipulated time, the contractor is the promissor and the owner is the promisee. However, to hand over the site for working or for making the payment, the owner is the promissor and the contractor is the promisee.

6.5.2 What Is the Effect of the Contractor's Failure to Complete the Work within the Time Specified in the Contract?

Section 55 provides two answers to this question. First, if the intention of the parties was that the time should be of the essence of the contract, the contract, or so much of it as has not been performed, becomes voidable at the option of the owner. That is, if the owner so desires he can terminate the contract by proper notice immediately after the lapse of due time and sue the contractor for recovery of damages for the breach of the contract. In the absence of a proper notice the option to avoid the contract will be deemed to have been waived and the contract subsisting.[10]

Second, if the time is not deemed to be of the essence of the contract, the contract does not become voidable at the option of the owner but only entitles him to compensation from the contractor for any loss occasioned to him by such failure. In such a case, the owner cannot, without making himself liable for breach of the contract, terminate the contract immediately after the time limit is over.

---

[9] See paragraph 6.7 in this chapter.
[10] Arun Prakash v. Tilsi Charan, AIR 1948 Cal. 510.

When the contract or part of it becomes voidable at the option of the owner and he does not annul it but agrees for an extension of time expressly or impliedly, he cannot claim compensation for the non-performance at the agreed time in the absence of notice to that effect. It does not, however, mean that the owner forfeits his right to claim damages if the contractor fails to complete the work at the extended time.[11]

If the owner is willing to extend the time of performance on the condition that such an extension will not prejudice his right to get the damages for the loss he sustained due to delay, he must give the contractor notice of his intention to recover damages at the time the extension is granted. Such a notice is not necessary if the time is not of the essence of the contract.

EXAMPLES FROM CASE LAW

1. The arbitral tribunal in a case found on scrutiny of the relevant documents that the employer, the National Highways Authority of India (NHAI), delayed in acquiring land and providing access to the site required for implementing the project and that the project was affected by numerous forced stoppages and interruptions of the construction works by local residents. The Scheduled Project Completion Date was specified as 26th February 2005. The actual completion date certified by an independent engineer was 20th June 2005, thus causing a delay of 114 days in the completion of the project. The arbitral tribunal also noted that the prolongation cost was claimed for the period of 114 days. There was an issue as regards the failure to give notice and whether for that purpose, the 3rd proviso to Section 55 of the Indian Contract Act 1872 would apply. The arbitral tribunal concluded that time was not the essence of the contract. Therefore, it was the second paragraph of Section 55 which would apply. Consequently, the arbitral tribunal held that the respondent would be entitled to compensation from NHAI for the loss occasioned due to prolongation without the requirement of a notice. The award was upheld.[12]

2. State v. Associated Engineering Enterprise, Hyderabad[13] raised certain important issues. It was decided under Section 30 of the Arbitration Act, 1940 and it still complicates the matter, inasmuch as the court admittedly did not enter "into the merits of the decision of the arbitrator". Brief facts of the case were as follows. An agreement was entered into between the State of Andhra Pradesh and the respondent contractor for executing the work of constructing approaches to the rail-cum-road bridge across the river Godavari at Rajahmundry. A period of 42 months was stipulated for completing the work, i.e. the work was to be completed on or before 21st December 1973. The respondent actually completed the work by 10th December 1974. Extension was granted up to the end of August 1974, subject to the imposition of penalty of Rs. 50/- per day after 1st September 1974. After the work was completed, disputes arose between the parties and were referred to arbitration in accordance with the agreement. The arbitrator made an award on 25th March 1981. The award was a non-speaking one. It was objected to on behalf of the state but was made a rule of the court by the trial court. Aggrieved, the state filed an appeal in the High Court. The High Court set aside the award under Claim No. 1 which decision raises an

---

[11] Mohammad Habib Ullah v. Bird & Co. (1921), LR 48 IA 175; also see: General Manager, Northern Railways v. Sarvesh Chopra, 2002 (1) Arb. LR 506 (SC).

[12] National Highways Authority of India v. Dhanoa Expressway Pvt. Ltd., (Delhi): 2017(4) AD(Delhi) 464, 2017(4) AD(Delhi) 464. Also see Union of India v. Tejinder Kumar Dua, 2014(10) R.C.R.(Civil) 828: 2013(200) DLT 60: 2013(5) R.A.J. 757.

[13] State v. Associated Engineering Enterprise, AIR 1990 A.P. 294; Amteshwar Anand v. Virender Mohan Singh, AIR 2006 SC 151; Hindusthan Construction Co. v. State of Bihar, AIR 1963 Pat. 254.

important issue as already stated. Claim No. 1 included compensation for delay in handing over part of the site. It was urged before the High Court that:

> Even if there is any delay in handing over the site, no claim for compensation can be made since any such claim is barred by clause 59 of the APDSS. It is also submitted that the contract does not provide for any such compensation, and hence the arbitrator, who had to operate within the four-corners of the contract, had no *power* to award any compensation on this account.

It was further observed:

> According to this Section, it was open to the respondent to avoid the contract on account of the Government's breach of promise to deliver the sites at a particular time, but, he did not choose to do so, and accepted the delivery of sites at a time *other* than what was agreed upon, between them earlier. If so, he is precluded from claiming compensation for any loss occasioned by such delay, unless, of course, at the time of such delayed acceptance of the sites, he had given notice to the Government of his intention to claim compensation on that account. It must be remembered that this provision of law was specifically referred to, and relied upon in the counter, filed by the Government to the respondent's claim before the arbitrator. But, it is not brought to our notice that the contractor had given such a notice (contemplated by the last sentence in Section 55). We must make it clear that we are not entering into the merits of the decision of the arbitrator. What we are saying is that such a claim for compensation is barred by law, except in a particular specified situation and inasmuch as such a particular specified situation is not present in this case, the claim for compensation is barred. It is well settled that an arbitrator, while making his award, has to act in accordance with the law of the land, except in a case where a specific question of law is referred for his decision.

The authors are in respectful disagreement with the view expressed by the High Court. It is submitted that agreement provided for both penalty for delay and extension of time under certain contingencies, thereby attracting the law laid down by the Supreme Court of India in Hind Construction Contractors v. State of Maharashtra,[14] that time ceased to be of the essence of the contract. As such, paragraph 2 of Section 55 of the Indian Contract Act is applicable. The third paragraph of the said section is applicable only when paragraph 1 of Section 55 applies, as is clear from the opening words of the third paragraph. Under the law, therefore, it was not open to the contractor to avoid the contract in the first instance and it was not obligatory on his part to give any notice contemplated under paragraph 3 of Section 55 of the Indian Contract Act in the second instance. The arbitrator did not misconduct on this count, it is respectfully submitted.

However, the High Court has further observed:[15]

> Even apart from Section 55, we are of the opinion that the arbitrator had no power to award compensation as claimed by the respondent. Clause 59 of the APDSS specifically bars such a claim. We have set out the clause in full hereinbefore. The meaning of the said clause was considered by a Bench of this Court of which one of us (Jeevan Reddy, J.) was

---

[14] Hind Construction Contractors v. State of Maharashtra, AIR 1979 SC 720. Also see Amteshwar Anand v. Virender Mohan Singh, AIR 2006 SC 151.

[15] State v. M/s. Associated Engineering Enterprises, Hyderabad (A.P.) (DB) AIR 1990 A.P. 294(302). Also see Konkan Railway Corporation Ltd. v. Oriental Construction Co. Ltd., 2013(3) BCR 140: 2013(21) R.C.R.(Civil) 244. Union of India v. K.R. Traders (Bombay) 2014(1) R.A.J. 310: 2013(37) R.C.R.(Civil) 746.

a member in A.A.O. No. 677/81 and C.R.P. No. 385 of 1982 disposed of on 19th April 1982. It was held:

> Coming to clause 59 of the preliminary specifications of "APDSS", it provides that neither party to the contract shall claim compensation "on account of delays or hindrances to work from any cause whatever." That the delays and hindrances contemplated by clause 59 include the stoppage, hindrances and delays on the part of the department as well, is clear from the following sentence, in the first part of the said clause, viz., "the Executive Engineer shall assess the period of delay or hindrances caused by any written instructions issued by him, at 25% in excess of the actual work period so lost". Indeed, the second para of the clause also contemplates delays and hindrances being caused on account of the failure of the Executive Engineer to issue necessary instructions. In such a case, the contractor has a right to claim the assessment of such delay by the Superintending Engineer of the Circle, whose decision is declared to be final and binding on the parties. But, any such claim has to be lodged in writing to the Executive Engineer within fourteen days of the commencement of such delay, or hindrance, as the case may be. We find it difficult, therefore, to say that clause 59 has no application to the present case. The words "from any cause whatever", occurring in clause 59, are wide enough to take in delays and hindrances of all types, caused by the department, or arising from other reasons, as the case may be. Thus, by virtue of clause 59, the contractor is precluded from claiming any compensation on account of delays or hindrances arising from any cause whatever, including those arising on account of the acts or omissions of the departmental authorities.

The other decisions of the same court were also referred to.[16]

The authors respectfully differ from the views expressed by the Andhra Pradesh High Court on the interpretation of Clause 59 of the APDSS. The said clause is given the heading: "Delays and Extension of Time". It nowhere refers to or speaks about money compensation. The word compensation in ordinary parlance means anything given to make things equivalent, to make amends for loss; it need not therefore be necessarily in terms of money.[17] The Supreme Court has held that compensation for one property may be by allotment of other property. Similarly, properly construed APDSS Clause 59 uses the word "compensation" for loss of time in the sense, "extension of time" and not monetary compensation. It does not even remotely hint of monetary compensation.

The view of the High Court that Clause 59 of APDSS clearly barred the claim for money compensation is not the correct view, it is submitted with respect. The clause reads: "No claim for compensation on account of delays or hindrances to the work from any cause whatever shall lie, *except as hereinafter defined*" (emphasis supplied). Thereafter the provisions speak of nothing except extension of time. The paragraph heading, though it cannot control the meaning of the text, certainly can be referred to for proper interpretation of the terms of the contract.[18] Until the Supreme Court reviews its decision in Ch. Ramalinga Reddy case,[19] the interpretation is bound to hold the field. This is demonstrated by the following example, in which case the court distinguished the said case under peculiar facts.

---

[16] Chief Engineer, Panchayat Raj Department v. B. Balaiah, (1985) I APU 224; State of A.P. v. S. Shivraj Reddy, (1988) 2 APU 465.

[17] State of Gujarat v. Shantilal, AIR 1969 SC 634.

[18] Union of India v. Raman Iron Foundry, AIR 1974 1974 SC 1265 (1270).

[19] Ch. Ramalinga Reddy v. Supdt. Engineer, (1994) 5 Scale 67, 1999(9)SCC610; 1999 (Supp) ArbLR 440.

EXAMPLE FROM CASE LAW

The respondent contractor was awarded a contract to execute the work of providing lining to a canal. The performance of the contract got prolonged far beyond the stipulated period, mainly because the PWD did not stop the flow of water in the canal. The respondent contractor sent a letter on 26th December 1986 intimating the fact that he was closing the contract. This was followed by another letter dated 14th October 1987 treating the contract as null and void and demanding settlement of his accounts, including compensation payable for the breach of contract. The department itself sent a communication to the petitioner on 29th January 1990 that the contract was being closed without levy of penalties. In the meanwhile, the disputes were referred to a panel of arbitrators. Later on the court appointed a retired chief engineer, PWD, as sole arbitrator in the place of the panel of arbitrators. The learned arbitrator awarded certain claims and rejected the counter-claims.

The department relied upon Clause 59 of the APDSS, which formed part of the contract to resist the claim of the contractor, and before the High Court also, reliance was placed on this clause and the decision of the Supreme Court in Ch. Ramalinga Reddy v. Supdt. Engineer.[20] It was observed:[21]

> The question is whether in the light of the aforementioned decisions, especially the latest decision of the Supreme Court, Clause 59 precludes Claim No. 6 being raised and allowed by the arbitrator and whether the arbitrator can be said to have committed an error of jurisdiction or a patent error of law in granting extra rate based on Standing Schedule of Rates in force during the relevant year, i.e. 1985–86. Before dealing with the question of applicability of Ramalinga Reddy's case (1994 (5) Scale 67) (SC) (supra) to the facts of the present case, we would like to advert to what the arbitrator has said vis-à-vis Clause 59. The arbitrator drawing support from the decision of this Court in V. Raghunadha Rao v. State of A. P., (1988) 1 Andh LT 461 opined that clause 59 was totally inequitable and unreasonable. The learned arbitrator observed, "there ought to be some sense of proportion; the contract period of one year cannot be extended for two more years without paying any compensation for the extra cost involved". As the execution of the contract within the time limit stipulated was clearly frustrated by a fundamental breach or failure on the part of the Department, Clause 59 cannot be put against the contractor. The arbitrator further commented: "payments were made at the increased rates for the work done in 1984 season after accepting a supplemental agreement. By the same argument, the respondent ought to pay at further increased rates based on S.S.R. of 1985–86 for the work done after May, 1985 since water was again released in July, 1984 irregularly without giving adequate time for the completion of the work. The argument of the respondent invoking Clause 59 is invalid in the light of his own acceptance of the breach in contract necessitating payment at increased rate and the claim as such cannot be resisted."

It was further held by the Andhra Pradesh High Court:[22]

> As far as the applicability of Clause 59 is concerned, we have already held that the appellants themselves waived their right to enforce that clause and entered into a revised agreement, agreeing to pay extra rate for the work done in 1984 season. No doubt, there was no similar agreement for the work done during 1985 season but the learned

---

[20] (1994) 5 Scale 67.

[21] Govt. of A.P. v. V. Satyam Rao, AIR 1996 AP 288. Also see: Simplex Concrete Piles (India) Ltd. v. Union of India, 2010(115) DRJ 616: 2010(5) R.A.J. 455: 2010(47) R.C.R.(Civil) 486: 2010(2) ILR (Delhi) 699.

[22] See Govt. of A.P. v. V. Satyam Rao, AIR 1996 AP 288, para. 9.

arbitrator considered it just and proper to adopt the same principle for the succeeding season also as the breach, which was undoubtedly fundamental in its nature,
persisted in the next season also. In doing so, the arbitrator did not go contrary to
the contractual terms but has only given effect to something that flowed as
a 'necessary concomitant' ... of what was agreed to between the parties in regard to
the execution of work beyond the contractual period.

6.6 When Is Time of the Essence of the Contract?

It is worthwhile to consider the situations under which the time of performance is deemed to be of the
essence of the contract. Mere mention of time during or before which something must be done will
not be enough.[23] Fixation of a period within which a contract is to be performed does not make
stipulation as to time being of the essence of the contract, nor does a default clause by itself evidence
intention to make the time of the essence of the contract.[24] A contract provision normally included to
meet this requirement is "the time shall be of the essence of this contract". However, such words will
not be effective if they are not in accord with all the other terms of the contract, such as provisions for
extension of time or penalty for delay. In the absence of an express stipulation like the one above, the
intention of the parties is to be inferred. It may be inferred from the antecedent conduct of the parties
but not from the subsequent conduct of the parties after the contract is made.[25]

When the parties to a contract desire to make the time to be of the essence of the contract, they
must do so by expressing their intention in explicit and unmistakable language. For example,
though it is a settled law that in respect of a contract relating to sale of immovable property, it
will normally be presumed that time is not of the essence of the contract,[26] where, in a contract
for sale of immovable property, a condition expressly stipulated time to be of the essence of the
contract and other stipulation read "Payment of the balance amount on or before ... is the
essence of the agreement" and further stipulated for forfeiture of the earnest money, the Supreme
Court held the time was of the essence of the contract.[27]

Time is always considered of the essence of the contract in the following cases:[28]

---

[23] Kisen Prasad v. Kunj Beharilal, 1925, 24 All L.J. 210, 91 IC. 790 (26) A.A. 278.

[24] Chand Rani v. Kamal Rani, AIR 1993 SC 1724; Gomathinayagam Pillai v. Palaniswami Nadar AIR 1967 SC 868;
Balasaheb Dayandeo Naik v. Appasaheb Dattatraya Pawar, 2008(1) CTLJ 299 (SC).

[25] Amn Prakash v. Tulsi Charan, AIR 1948 Cal. 510. Also see AIR 1974 Raj. 112; Vairavan v. K.S. Vidyanandam, AIR
1996 Mad. 353.

[26] Most Sita Devi v. Mahal Sahkari Grih Nirman Samiti Ltd., 2016 AIR (Patna) 191: Babulal Harchand Beldar
v. Sitaram Kathu Borse (Dhobi), 2016(5) Mh.LJ 816: 2016(5) AIR Bom. R 680: 2017(3) ALL MR 85; Mohammed
Abdul Azeem v. M/s. South India Prime Tannery Pvt. Ltd., (Telangana and Andhra Pradesh): 2016 AIR
(Hyderabad) 170; Radha Agrawal v. Krishna Sehgal; Rajasthan (Jaipur Bench) 2015(1) W.L.N. 344: 2014(67) R.C.
R.(Civil) 512; R.K.B.K. Fiscal Services Pvt. Ltd. v. Ishwar Dayal Kansal, (Delhi)(DB) 2014(210) DLT 66: 2014(34)
R.C.R.(Civil) 284: 2014(3) ILR (Delhi) 1671: 2014(9) AD(Delhi) 449; Azmat Ali (Md.) v. Abdul Sakkur Barbhuiyan
(Md.), (Gauhati) 2015(3) GauLT 223: 2016(1) NEJ 262: 2016(4) NEJ 523: 2015(54) R.C.R.(Civil) 345; Ram Kishan
v. Bhavi Chand, (Delhi)(DB), 1988(1) ArbiLR 133: 1988 AIR (Delhi) 20; Nanjachary v. P. Chennaveerachari,
2014(22) R.C.R.(Civil) 333: 2015 AIR (Madras) 73.

[27] A.K. Lakshmipathy v. Rai Saheb Pannalal H. Lahoti Charitable Trust, AIR 2010 SC 577. Also see I.S. Sikandar (D)
By Lrs. v. K. Subramani, 2014(1) Scale 1: 2013(15) SCC 27: 2014(1) R.C.R.(Civil) 236: 2013(6) Recent Apex
Judgments (R.A.J.) 487: 2014(1) CivCC 439: 2014(1) Law Herald (SC) 688: 2014(1) L.A.R. 259: 2015(3) Rajdhani
LR 339: 2014(118) CutLT 89: 2014(1) Apex Court Judgments (SC) 81: 2014(1) *Ori. Law Rev.* 504: 2013(6) All WC
6364.

[28] M/s Chemiplast Industries (Regd.) v. H.P. Agro Industries Corp Ltd., (H.P.): 1993(3) R.R.R. 266: 1993(3) LJR 210:
1995(1) BankCLR 11: 1993(2) *Simla Law Journal* 1185: 1994(sup) SimLC 398: 1993(1) BankCLR 11.

1. Where the parties have expressly agreed to treat it as of the essence of the contract.

2. Where delay operates as an injury; and

3. Where the nature and necessity of the contract require it to be so construed, for example, the object of making a contract.

Where time is of the essence of the contract and is extended, the extended date is also of the essence of the contract.

6.6.1 Time May Be Implied as Essential

Time may be implied as essential in a contract, from the nature of the subject matter with which the parties are dealing. For example, where in a case, the plaintiffs themselves admitted that value of suit property had increased by at least Rs. 1.50 crores within the period of three months, it was held: "Time was of the essence of the contract and since vendees were not ready and willing to perform their obligations under agreement they were not entitled to relief of specific performance."[29]

Whether or not time was of the essence of the contract would have also to be judged in the context and circumstances of the case.[30] The courts generally look at all the relevant circumstances, including the time limit specified in the agreement, and determine whether discretion to grant specific performance should be exercised. The time limit prescribed by the parties in the agreement has its own significance and value. It would not be reasonable to say that because the time is not made of the essence of the contract the time limit specified in the agreement has no relevance and can be ignored altogether. It has been rightly observed, it is submitted, that

> the rigor of the rule evolved by Courts that time is not of the essence of the contract in the case of immovable properties – evolved in times when prices and values were stable and inflation was unknown – requires to be relaxed, if not modified, particularly in the case of urban immovable properties. It is high time, we do so.[31]

In short, it must be kept in mind that when the parties agree and prescribe a certain time limit for taking steps by one or the other party, it must have some significance and that the said time limit cannot be ignored altogether on the ground that time has not been made of the essence of the contract. It is not necessary to state in the agreement itself that time is to be of the essence of the contract. The express covenants themselves in the contract may disclose the same, such as when clauses under an agreement provide the consequence in case if for any reason the requisite exemption/permission is not forthcoming by the agreed date or within such extended period as may be mutually agreed to, the agreement will become inoperative and unenforceable. In view of such provisions there remains no doubt whatsoever to hold that the parties have specifically agreed the time to be of the essence of the contract.[32]

EXAMPLES FROM CASE LAW

1. In a case, it was not only stipulated that time was of the essence of the contract but it was further stipulated that time was to be of prime importance. Held: it showed that intention of the parties was to make time of the essence of the contract.[33]

---

[29] Ritu Mercantiles Pvt. Ltd. v. Leelawati, 2014(7) R.C.R.(Civil) 2639: 2013 AIR CC 2438: 2013(201) DLT 301.

[30] Indira Kaur v. Sheo Lal Kapur, AIR 1988 SC 1074=1988(2) SCC 488.

[31] K.S. Vidyanadam v. Vairavan, (1997) 3 SCC 1: AIR 1997 SC 1751. Also see M/s. Ceean International Private Limited v. Ashok Surana, AIR 2003 Cal. 263; Kabirdass v. Vinothambal and others, 2002(3) Arb. LR 471(Madras).

[32] P. Purushotham Reddy v. M/s. Pratap Steels Ltd., AIR 2003 AP 141.

[33] A.K. Lakshmipathy (Dead) v. Rai Sai Saheb Pannala H. Lahoti Charitable Trust, 2010 AIR (SC) 577: 2010(1) SCC 287: 2009(4) ICC 625: 2010(1) R.C.R.(Civil) 922: 2010(6) R.C.R.(Civil) 20: 2010(1) Recent Apex Judgments (R.A.J.) 473: 2009(4) CivCC 650: 2009 AIR (SCW) 7144: 2009(13) Scale 457: 2010(1) All WC 317: 2010(2) AIR Jhar R. 27:

2. Where the entire consideration was to be paid by the plaintiff within six months from the date specified in the agreement, the plaintiff, instead of paying entire consideration, insisted upon delivery of possession and execution of sale deed in part proportionate to the amount paid by him. Held: it was tantamount to varying the terms of agreement. Since the time was of the essence of the contract the transaction failed primarily on account of non-payment of the entire amount of consideration within the specified period. The plaintiff was held not to be entitled to decree of specific performance. However, as there was no forfeiture clause in the agreement, the plaintiff was held to be entitled to refund of part consideration paid by him.[34]

3. In a contract for sale of immovable property, time was specified for payment of the sale price but not in regard to the execution of the sale deed. Held: time will become of the essence of the contract only with reference to payment of sale price but not in regard to execution of the sale deed.[35]

4. An agreement for the sale of land provided that the transaction would be completed within 30 days from the date of the agreement. Normally, in an agreement for the sale of immovable property, the presumption is against the time being of the essence. The plaintiff purchaser was willing to pay the complete purchase price within such time and the defendant had to do nothing except to execute a sale deed. Held: time was of the essence of contract.[36]

5. An agreement for the sale of urban land provided that within 12 months of the agreement, the plaintiff should construct a building. This would have required not only planning permission but also clearance from the government under the Urban Land Ceiling Act. The trial court granted a decree of specific performance directing the payment of the balance money due for the property. This was not paid for a few years and by the time the appeal was heard, the value of the property had increased considerably. The building was not constructed and nor was effort made to pay the balance sale consideration. Held: time being of the essence of the contract, the plaintiff was not entitled to specific performance.[37]

6. Appeal of the plaintiff was dismissed holding that even if time is not of the essence of the contract, the parties under the agreement have to perform their obligations within a reasonable time. The delay in filing suit for specific performance itself would operate as acquiescence or waiver on the part of the plaintiff and make it inequitable to grant relief of specific performance.[38]

7. An agreement of sale of immovable property stipulated that part consideration, Rs. 98,000/- only, was to be paid within ten days only of the execution of the agreement and the balance at the time of registration of the deed. It was further agreed that the vendor would redeem the property, which was mortaged, and also obtain an income tax clearance certificate.

2009(77) ALR 642: 2009(8) MLJ 693: 2010(1) OriLR 118: 2010(109) R.D. 27: Syed Aijaj Hasan v. Malik Mohammed Tahseen, (Delhi)(DB) 2014 AIR (Delhi) 104: 2014(34) R.C.R.(Civil) 125.

[34] Shankarlal Bijreja v. Ashok B. Ahuja, (Chhattisgarh)(DB) 2011 AIR (Chhattisgarh) 66: 2010(45) R.C.R.(Civil) 47: 2011(1) C.G.L.J. 498.

[35] Saradamani Kandappan v. S. Rajalakshmi, 2011 AIR (SCW) 4092: 2011 AIR SC (Civil) 1812: 2011(4) CivCC 271: 2011 AIR (SC) 3234: 2011(4) R.C.R.(Civil) 130: 2011(4) Recent Apex Judgments (R.A.J.) 524: 2011(6) MLJ 149: 2012(90) ALR 706: 2011(12) SCC 18: 2012(1) WBLR 786: 2011(8) SCR 874: 2011(6) Scale 768: 2011(2) W.L.C. 632: 2011(5) Andh LD 100: 2012(1) A.R.C. 39: 2011(8) JT 129: 2013(120) R.D. 564: 2011(4) CTC 640: 2011(4) LW 97: 2011(3) Apex Court Judgments (SC) 167.

[36] Savitramma v. Dr. M.P. Somaprasad, (Karnataka) 2011(1) Air Kar R 40: 2011 AIR CC 721: 2011(4) KantLJ 30: 2010(23) R.C.R.(Civil) 732.

[37] K.L. Sathyaprakash v. Kothari Motors Ltd., (Karnataka)(DB) 2011(1) Air Kar R 211: 2011(3) KantLJ 640: 2010(51) R.C.R.(Civil) 68.

[38] Jugraj v. P. Sankaran, 2010(3) CTC 297: 2010(4) MLJ 781: 2010(54) R.C.R.(Civil) 857: 2010(4) CivilLJ 542.

The agreement used the word "only" in respect of the amount and ten days time, it was held that the intention of the parties was to make time of the essence of the contract. When the purchaser was not ready and willing to pay the amount as agreed before possession and income tax clearance, it was held that the purchaser was not entitled for specific performance.[39]

8. In a case, the agreement was dated 21st June 1970 and the time prescribed for payment of balance was 2nd August 1970, i.e. about six weeks. It also contained a specific clause that in case the payment was not made, the earnest money of Rs. 2,500/- would stand forfeited. However, the plaintiffs made a part payment on 25th November 1970, the defendants accepted the same and an endorsement to that effect was made. Under the endorsement also it was specifically mentioned that the balance would be paid by 31st December 1970, i.e. within five weeks, and the registered document shall be obtained, failing which the agreement would stand cancelled. It was held:[40] "The above said terms ... clearly show that intention of the defendants to complete the sale transaction at the earliest point of time."

9. The plaintiff's case in brief was that he manufactured telescopic out-let shutters for hume pipes of various dimensions in his industrial unit and, coming to know that the State of Orissa (defendant No. 1) was in need of such shutters, he submitted his quotations before the executive engineer who, acting on behalf of the state, placed orders between 16th July 1990 and 20th July 1990 for supply of 133 shutters at the price quoted by him. When the executive engineer, defendant No. 4, placed the orders, the plaintiff represented to him that it may not be possible on his part to supply the shutters within the stipulated time on account of repeated load-shedding by the Orissa State Electricity Board and he was assured that if he faced any difficulty on account of load-shedding, on his application time for supply of the shutters shall be extended. On such assurance from defendant No. 4, the plaintiff started manufacturing the shutters. In course of manufacture there was frequent load-shedding. Therefore, on 28th July 1990 the plaintiff put in a written application to the executive engineer praying for extension of time, but received no reply. So he put in a second application on 10th August 1990 for extension of time and received a reply dated 25th August 1990 to the effect that the purchase order had been cancelled. But by then the plaintiff had completed manufacture of all the 133 shutters. Thereafter the plaintiff approached the defendants several times to take delivery of the articles and make the payment, but to no effect. According to him, two other firms with whom orders were placed for supply of the same type of shutters were given extension of time and they made the supplies during the extended periods and have been paid the price, but similar advantage was denied to the plaintiff. It was argued that time being not of the essence of the contract, and in view of the assurance given by the executive engineer, the plaintiff having manufactured the shutters and kept ready for delivery, he was entitled to a decree as prayed for. Defendant No. 4 alone filed written statement in the suit. He contended *inter alia* that in the notice calling for quotation it was stated that the materials were to be supplied within 10 days from the date of issue of the supply orders and no extension of time whatsoever would be allowed. The plaintiff accepted this condition and submitted his quotation but failed to supply the shutters within the time stipulated in the purchase orders and consequently the plaintiff could not legally claim any amount from the defendants. It was denied that the

---

[39] Chand Rani v. Kamal Rani, AIR 1993 SC 1742; also see Smt. Swarnam Ramachandran v. Aravacode Chakungal Jayapalan, AIR 2000 Bom. 410; Vairavan v. K.S. Vidyanandam, AIR 1996 Mad. 353.

[40] Dutta Seethamahalakshmamma v. Yanamadala Balaramaiah, AIR 2003 AP 430; Coromandel Indag Products (P) Ltd. v. Garuda Chit and Trading Co. P. Ltd., (SC) 2011(4) CivCC 330: 2011(4) R.C.R.(Civil) 677: 2011(5) Recent Apex Judgments (R.A.J.) 440: 2011(8) MLJ 112: 2012(1) Law Herald (SC) 862: 2011(12) SCR 115: 2011(9) Scale 139: 2011(10) R.A.J. 34: 2012(1) A.R.C. 68: 2011(8) SCC 601: 2011(4) SCC (Civil) 291.

plaintiff was given any assurance that time for supply would be extended on account of load-shedding. It was held:[41]

> On a consideration of the above material, it is held that time being the essence of the contract as revealed from the purchase orders, and the same having neither been waived nor extended, the Executive Engineer had the option to avoid the contract and no right under the contract accrued in favour of the plaintiff, because of non-performance of the contract by letter within the stipulated period.

If the plaintiff had obtained the executive engineer's assurance in writing before manufacturing, he might have been able to rely on a promissory estoppel.

6.6.2 Effect Where Parties Knowingly Give a Go-by to the Stipulation

It is to be noted that in the event that time is of the essence of the contract, the question of there being any presumption or presumed extension or presumed acceptance of a renewed date would not arise. The extension, if there be any, should and ought to be categorical in nature rather than being vague or in the anvil of presumptions. In the event the parties knowingly give a go-by to the stipulation as regards the time – the same may have two several effects: (a) parties name a future specific date for delivery and (b) parties may also agree to the abandonment of the contract. As regards (a) above, there must be a specific date within which delivery has to be effected and in the event there is no such specific date available in the course of conduct of the parties, then and in that event, the courts are not left with any other conclusion but a finding that the parties themselves, by their conduct, have given a go-by to the original term of the contract as regards the time being of the essence of the contract. In the event the contract comes within the ambit of Section 55, the remedy is also provided therein.[42]

6.7 When Time Ceases to Be of the Essence of the Contract

In spite of an express stipulation included in a contract making the time to be of the essence, the time shall not be of the essence if contradictory conditions are included in the contract. If the terms of the contract provided for a possible extension of time to complete the work, or for the payment of liquidated damages in the event of delay or a similar stipulation whereby the parties to the contract contemplate the possibility of delay in performance, the time ceases to be of the essence of the contract. In a case, the defendant had not issued any notice terminating the agreement immediately after the completion of the three-month period stipulated for payment of consideration. On the other hand, he continued to receive the balance sale consideration even after the expiry of the three-month period. Held: from the conduct of the defendant, he impliedly extended the period for making the balance payments. Time was not of the essence.[43]

This position is now very well settled by the decision dated 30th January 1979 of the Supreme Court of India.[44] The case involved questions of termination of the contract on the ground of

---

[41] State of Orissa and others v. M/s. Durga Enterprisers, AIR 1995 Ori 207.

[42] Arosan Enterprises Ltd., M/s. v. Union of India, AIR 1999 SC 3804= 1999 AIR SCW 3872.

[43] Chhandsi Abrol v. M. T. Rangaswamy, 2017(4) Air Kar R 717: 2018(2) ICC 58.Also see: M/s. Kailash Nath Associates v. Delhi Development Authority, 2015(1) Scale 230: 2015(1) JT 164: 2015(4) SCC 136:: 2015 AIR (SCW) 759.

[44] Hind Construction Contractors v. State of Maharashtra, AIR 1979, SC 720 (Example (1) below). Also see M. D., H. S.I.D.C. v. Hari Om Enterprises, AIR 2009 SC 218; Shambhulal Pannalal, Secretary of State, AIR 1940 Sind 1. Hindustan Construction Co. v. State of Bihar, AIR 1963 Pat. 254.

non-completion of the project work within the stipulated period, levy of penalty and forfeiture of security deposit, etc. which are very common in execution of building and engineering contracts. It is a landmark judgment and is considered in Example (1) below.

EXAMPLES FROM CASE LAW

1. A contract for the construction of an aqueduct[45] was granted to the appellant-plaintiff – by the respondent-defendant (the State of Maharashtra). The work order directing him to commence the work was given on 5th July, on the threshold of the rainy season. The period for completion of work was fixed at 12 months from the date stipulated for commencement of the work. The executive engineer rescinded the contract on the ground that the appellant-plaintiff had not completed the work within the stipulated time. The appellant-plaintiff filed a suit making a claim against the respondent-defendant alleging wrongful and illegal rescission of the contract. The appellant-plaintiff's case was that the initial fixation of 5th July as the date for commencement of the work was nominal, that the area where the work was to be done had usually heavy rainfall rendering it impossible to carry out any work from July to November and that, therefore, it was the practice of the PWD to deduct the period of monsoon in case of such type of works and that the appellant-plaintiff had been orally informed that this period would be deducted or not taken into account for calculating the period of 12 months under the contract and that on this assurance he had commenced the work towards the end of December. His case further was that in any event time was not of the essence of the contract, that on account of several difficulties, such as excessive rains, the lack of proper road and means of approach to the site, rejection of materials on improper grounds by the Government Officers, etc., over which he had no control, the completion of the work was delayed and that the extension of the time which was permissible under the contract had been wrongfully refused by the government officers. According to him none of these factors had been taken into account by the Government while refusing the extension and the contract was wrongfully rescinded and therefore, the respondent-defendant was liable in damages. The State of Maharashtra resisted the claim contending that time was of the essence of the contract. That the appellant-plaintiff knew the situation of the site and the so-called difficulties, that there was no excuse for him for not doing the work during the months of July to November, that the appellant-plaintiff had failed to carry out the proportionate work during the periods fixed in the contract and that since the appellant-plaintiff had rendered himself incompetent to complete the work in proper time it had to rescind the contract and the rescission was proper and for adequate reasons. It was further contended that the State was entitled to forfeit the security deposit, which it did, on the date when the contract was rescinded. The trial court decreed the suit. The High Court accepted the State's defence and set aside the decree of the trial court. On appeal, the Supreme Court observed:

> The first question that arises for our consideration, therefore, is whether time was of the essence of the contract. ... It cannot be disputed that the question whether or not time was the essence of the contract would essentially be a question of the intention of the parties to be gathered from the terms of the contract.

Reliance was placed on paragraph 1179 in the 4th Edition of *Halsbury's Laws of England* in regard to building and engineering contracts and the statement of law and it was held:

---

[45] Hind Construction Contractors v. State of Maharashtra, AIR 1979 SC 720.

It will be clear from the aforesaid statement of law that even where the parties have expressly provided that time is of the essence of the contract such a stipulation will have to be read along with other provisions of the contract and such other provisions, may, on construction of the contract, exclude the inference that the completion of the work by a particular date was intended to be fundamental, for instance, if the contract were to include clauses providing for extension of time in certain contingencies or for payment of fine or penalty for everyday or week the work undertaken remains unfinished on the expiry of the time provided in the contract such clauses would be construed as rendering ineffective the express provision relating to the time being of the essence of the contract.

The Supreme Court referred to Clause 2 and 6 of the conditions of the contract, which provided for penalty for the delay and extension of time under certain contingencies, and held:

[I]t would be difficult to accept the High Court's finding that the rescission of the contract on the part of the respondent-defendant was proper and justified on the basis that the same was neither shown to be mala fide nor unreasonable. It must be observed that it was never the case of the appellant-plaintiff that the rescission of the contract on the part of the respondent-defendant was mala fide. ...

It will thus appear clear that though time was not of the essence of the contract, the respondent-defendant did not fix any further period making time the essence directing the appellant-plaintiff to complete the work within such period. Instead it rescinded the contract straightaway ... Such rescission on the part of the respondent-defendant was clearly illegal and wrongful and thereby the respondent-defendant committed a breach of the contract, with the result that there could be no forfeiture of the security deposit. ... The appellant-plaintiff was also entitled to the payment of interest on the aforesaid sums and costs of suit as directed by the trial Court.

2.   A contract for the construction of an aqueduct was to be completed within 12 months from the date of the written order to commence it.[46] The work continued to be executed at a snail's pace. The department rejected the application seeking an extension of time and further informed the plaintiff that he had become liable to pay compensation under Clause (2) of the contract and called upon him to show cause why action should not be taken against him under Clause (3) thereof. There was not much progress and the plaintiff had stopped its execution when he was informed that the contract stood rescinded under Clause 3. The defendants also forfeited the plaintiff's security deposit.

The trial court found that the plaintiff had failed to prove that any sum was due to him for execution of the work or by way of damages, but further held that the forfeiture of the security deposit was illegal. On appeal, the High Court agreed with the findings of the trial court and certified the case to be a fit one for appeal by the defendant to the Supreme Court. The Supreme Court held:[47]

Clauses (2) and (3) have to be read together and interpreted with reference to each other and their provisions, read as one single whole, clearly mean that the contract was to continue to be in force till the completion of the work or its abandonment. The time was the essence of the contract only in the sense that if the plaintiff completed it within the original period of one year he would not be liable to pay any compensation but that in case he overstepped the said time limit he would have to compensate the defendants ... Till the time the contract was rescinded therefore, it was fully in force and the rescission

---

[46] State of Maharashtra v. Digambar, AIR 1979 SC 1339.
[47] State of Maharashtra v. Digambar, AIR 1979 SC 1339.

was consequently well-founded, being squarely covered by clause (3) of the contract, sub-clause (a) of which conferred on the Executive Engineer the right to forfeit the security deposit. Far from being illegal, the forfeiture was fully justified, and the High Court's finding to the contrary is liable to be reversed.

The above two decisions of the Supreme Court are very frequently cited and relied upon by two opposing parties in their cases. The two decisions do not lay down any conflicting propositions of the law, it is submitted. In both the decisions, it is very clearly laid down that time shall not be of the essence of the contract. In the latter decision, it is clarified that the wording to that effect in Clause 2 of the general conditions of the contract is only to identify the date from which the liquidated damages or penalty will be recoverable. The termination in the first case above was held to be wrongful and illegal as it was attempted to be supported only on the ground of time being of the essence of the contract and time was not extended, although there were factors justifying such an extension. On the other hand, the termination of the contract in the second case was supported by slow progress of the work resulting in virtual abandonment of the work by the contractor by stopping it. It is also clear that notices under Clause 2 were given, penalty imposed, an opportunity to show *bona fide* desire to complete the work by improving the progress was given and after all this when the contractor rather than improving the progress stopped the work, the contract was terminated under Clause 3. Thus the facts of the two cases are totally different justifying the differing verdicts of the Supreme Court in the two cases.

3.   The respondent held an auction sale on 28th November 1995. At this auction sale the appellant was the highest bidder. In terms of the conditions of the auction sale the appellant, on the same day, deposited a sum of Rs. 6,54,500/-. As per the terms of the auction sale the balance 75% of the amount had to be paid within 60 days from the issue of the demand letter. However, if sufficient cause was shown then the chairman could extend time up to a maximum period of 180 days, subject to payment of interest on the balance amount at the rate of 18% per annum. The demand letter was issued on 3rd January 1996. The balance sum of Rs. 1963,545/- had to be paid, in terms of the auction sale, within 60 days of the said letter. The appellant deposited money in instalments from time to time. The respondent, even though payments were made beyond time, accepted all these payments. The appellant had paid the entire amount including interest payable in respect of that plot. In spite of having accepted the delayed payments the plot was not delivered to the appellant. The appellant, therefore, sent legal notices. Only after receipt of the legal notices the respondent cancelled the allotment and forfeited the earnest money of Rs. 6,54,500/-. The appellant filed a writ petition, which came to be dismissed. The High Court held that payments made after the extended date were not valid and legal tender of money in accordance with law. The Supreme Court held:[48]

> In our view, the order of the High Court cannot be sustained. To be noted that by 27th September 1996 the entire amount payable for the plot had been deposited and delay in payment was less than 30 days. Thereafter in January, 1997 interest at the rate of 25% per annum, on delayed payments, was also paid. Both the delayed payments and the interest amount were accepted by the respondent. The moment those payments were accepted there was deemed extension of time. ... on the facts of this case, i.e. after accepting the delayed payments and interest respondent could not have cancelled the allotment.

---

[48] R.K. Saxena, v. Delhi Development Authority, AIR 2002 SC 2340. Also see: Bhagwan Singh v. Teja Singh, AIR 1995 P. & H. 64.

In the circumstances of the case, the impugned order was set aside. The letter of cancellation was quashed.

4.   In one case, the house and land, the subject matter of the contract, were required by the purchaser for immediate residential occupation. The condition of the contract named a day for completion of the purchase transaction. One more condition was included in the contract, which went on to provide that if, from any cause whatever, the purchase was not then completed, the purchaser should pay interest on the unpaid purchase money from that day until the actual completion of the purchase. In the suit that followed on this matter it was held: "inasmuch as parties to the contract evidently contemplated the possibility of the completion being postponed beyond the date named, time was not the essence".[49] The principle on which this case was decided is obviously applicable whatever the nature of the subject matter of a contract.

5. In a case where the completion of work was delayed by six months the court observed:

> But, looking to the whole of the deed we are of the opinion that the time of completion was not an essential part of the contract; first, because there is an express provision made for a weekly fine to be paid for every week during which the work should be delayed ...; and, secondly, because the deed clearly meant to exempt the plaintiff from the obligation as to the time for completion in case he should be prevented by fire or other circumstances admitted by the architect; and here, in fact, it is expressly found by the arbitrator that the delay was necessarily occasioned by the extra work.[50]

The case involved some additional factors which were as follows:

(i) The contract between the petitioner contractor and the CPWD was for construction of 27 km. of road costing Rs. 78,83,287/-.

(ii) The time limit for this work was stipulated as six months.

(iii) The contractor's tender was not accepted within 90 days, the limit stipulated for acceptance. The contractor informed the authorities by a letter prior to commencing the work that as the matter of final acceptance of the tender had been delayed by the authorities and the rainy season had set in, the work could not be completed in six months as originally stipulated.

(iv) Out of the 27 km length of the road, possession of 2 km of land was not given immediately to the contractor, but was given only after 3 months and 8 days.

(v) At the contractor's request, time for completion of the work was extended on and off in accordance with the provisions made in the contract.

(vi) Running account bills had to be paid every fortnight but only 29 running account bills were paid whereas 69 bills should have been paid.

(vii) The contract included an arbitration clause. On account of disputes between the parties an arbitrator was appointed to settle the same.

(viii) The work on certain unfinished items was given to other contractors. The final measurements of the work were claimed to have been taken in the presence of a representative of the contractor, as the contractor, it was alleged, did not turn up on the appointed date.

---

[49] Webb v. Hughes, L. R. 10 Ed. 281.
[50] Lamprell v. Billericay Union, (1849), 3 Ex. 283.

(ix) The contractor submitted a petition to the court requesting issue of an injunction restraining the authorities and other contractors from executing any item of work until the case of the contractor was finally decided.

(x) The authorities on the other hand urged that the road was very strategic and being a defence road no such injunction should be granted.

In view of the facts above-mentioned it was held that time was not of the essence of the contract. Further held:

> It would be a contradiction in terms that time was of the essence of the contract yet time being extended not once but a number of times. There is one more circumstance negativing this plea of time being the essence of this contract. In the original agreement running payments had to be made fortnightly. According to the petitioner only 29 running bills had been paid from 1967 to 1970 whereas 69 such bills should have been paid. If time was of the essence of the contract it should be for both the parties and not for only one of them. The respondents could not commit any defaults as far as their part of the performance of the contract is concerned, and insist on the performance of the contract by the other party within the time limit fixed.

The contractor's request for issue of an injunction was, however, turned down. Citing a similar case decided by the Calcutta High Court,[51] his Lordship observed:

> I do not think it is case in which I should exercise my discretion in favour of the petitioner in directing the respondents that further items of work entrusted to other contractors be held up indefinitely, till the Arbitrator decides the petitioner's case. Whether the road is a strategic one, a defence road or not, to hold up the construction and completion of the road would be against all canons of justice. Better roads are a source of convenience to everybody, it may be the army or the civil public.

6. A question that arose for consideration by the Bombay High Court was whether the respondent contractor could have refused to work, even if the petitioner had imposed conditions for not granting any compensation while granting extension of the original contractual period. It was observed:

> A perusal of clause ... makes it clear that the contractor was bound to carry out work even during the period when the dispute has arisen between the parties unless the petitioner would have otherwise directed the contractor to stop the work. The record indicates that the petitioner never directed the respondent to stop the work. On the contrary, the petitioner neither paid any compensation nor took any steps to stop the contractor from carrying out the work nor terminated the contract. The contractual period which was for 18 months, it took about 7 years to complete the work.

Held:

> In my view, the learned arbitrator has interpreted the terms of the contract and has dealt with the evidence produced by the parties and has rightly awarded the compensation in favour of the respondent during the prolongation period. Petition dismissed.[52]

---

[51] Ranjit Chandra v. Union of India, AIR 1963 Cal. 594.

[52] Union of India v. Bright Power Project, (Bombay) 2015(5) BCR 58: 2015(4) Mh.LJ 668: 2015(6) ArbiLR 428: 2015(6) R.A.J. 690: 2016(5) ALL MR 563: 2015(29) R.C.R.(Civil) 456.

Exemption clauses denying compensation for the increased cost and losses incurred cannot be valid when the time for completion stands extended to more than reasonable time.

6.8 Extension of Time for Delay

A construction contract usually stipulates circumstances under which the time for completion will be extended. A comprehensive list of valid grounds is as follows:

If the works be delayed by:

(i) Variation

(ii) *Force majeure*, or

(iii) Abnormally bad weather, or

(iv) Serious loss or damage by fire, or

(v) Civil commotion, local combination of workmen, strike or lock-out, affecting any of the trades employed in the work, or

(vi) Delay on the part of nominated subcontractors or nominated suppliers which the contractor has, in the opinion of the engineer-in-charge, taken all practicable steps to avoid or reduce, or

(vii) Delay on the part of other contractors or tradesmen engaged by the employer in executing work not forming part of the contract, or

(viii) Non-availability of stores which are the responsibility of the employer to supply, or

(ix) Non-availability or break-down of tools and plant to be supplied or supplied by the employer, or

(x) Suspension of work by the employer, or

(xi) Any other cause which, in the absolute discretion of the authority mentioned in the schedule attached to the contract, is beyond the contractor's control.

The provisions further stipulate that upon the happening of any such event causing delay, the contractor shall immediately give notice thereof in writing to the engineer-in-charge within the stipulated time if on that count he intends to seek an extension of time. The contractor shall nevertheless use constantly his best endeavours to prevent or make good the delay and shall do all that may be reasonably required to the satisfaction of the engineer-in-charge to proceed with the works.

Request for extension of time, to be eligible for consideration, under FIDIC Clause 20.1 of 1999 edition, shall be made by the contractor in writing not later than 28 days after the contractor became aware, or should have become aware of the happening of the event causing delay. The contractor may also, if practicable, indicate in such a request the period for which extension is desired.

The above provisions are ideal and in practice the provisions are included in other standard form contracts on the lines as above but none of the forms stipulate any time limit for an engineer to communicate extension of time after receiving the application from the contractor.

Early Warning Notice

Most of the recently developed standard forms include provision of Early Warning Notice to be given. The contractor is to warn the engineer/employer or his nominee at the earliest opportunity of specific likely future events or circumstances that may adversely affect the quality of the work, increase the contract price or delay the execution of works. The engineer or his nominee may require the contractor to provide an estimate of the expected effect of the event or circumstance on the contract price and completion date. The estimate is to be provided by the contractor as soon as reasonably possible. This is likely to be held as a condition precedent to the contractor's claim for extension of time or payment of compensation and hence advisable to be followed. For example, the provision of the NITI Aayog Model Form includes the provision reading as follows:

> 10.4.3 In the event of the failure of the Contractor to issue to the Authority's Engineer a notice in accordance with the provisions of Clause 10.4.2 within the time specified therein, the Contractor shall not be entitled to any Time Extension and shall forfeit its right to any such claims in future. For the avoidance of doubt, in the event of failure of the Contractor to issue notice as specified in this Clause 10.4.3, the Authority shall be discharged from all liability in connection therewith.

The provisions of such clauses are self-explanatory except, perhaps, the meaning of *"force majeure"*, which is discussed below.

6.8.1 *"Force Majeure"* and *"Vis Major"*

"Force majeure" is an expression taken originally from the Code of Napoleon. It means an absolute necessity or compulsion, circumstances beyond one's control. It covers a wider class of events than *"vis major"* an act of God, and includes man-made events such as strikes or wars, but it must be beyond the control of the person alleging it. This clause has to be construed in the light of the general background and the terms of the contract in question, so that differing decisions may be reached in different contracts.[53] This is so in the light of the fact that there are a number of *force majeure* clauses differing from one another.

The question arises whether the provision in the contract, which simply states that extension will be granted "if the work is delayed by *force majeure*", can be said to be vague? The question needs to be considered in the light of the fact that there are a number of *force majeure* clauses differing from one another. In this background, if an agreement is to include a provision "subject to *force majeure* clause", without any further particulars, it is likely to be held vague. However, if it were to mention "subject to usual *force majeure* clause", it may not be held as vague but the parties may be allowed to lead evidence to which clause the parties had in mind. In this respect the following observations of the Supreme Court of India need to be noted:[54]

> We were taken through the Encyclopedia of Forms and Precedents and shown a number of force majeure clauses, which were different. We were also taken through a number of rulings, in which the expression "force majeure" had been expounded, to show that there is no consistent or definite meaning. The contention thus is that there being no consensus ad idem, the contract must fail for vagueness or uncertainty. The argument, on the other side, is that this may be regarded as a surplus usage, and if meaningless, ignored. It is contended by the respondents that the addition of the word "usual" shows that there was some clause, which used to be included in such agreements. "The respondents also refer to S. 29 of the Indian Contract Act, which provides: "Agreements, the meaning of which is not certain, or capable of being made certain, are void" and emphasize the words "capable of being made certain", and contend that the clause was capable of being made certain, and ex-facie, the agreement was not void. "… An analysis of rulings on the subject, into which it is not necessary in this case to go, shows that where reference is made to "force majeure", the intention is to save the performing party from the consequences of anything over which he has no control. This is the widest meaning that can be given to "force majeure", and even if this be the meaning, it is obvious that the condition about "force majeure" in the agreement was not vague. …" Reliance was also placed upon a couple of cases.[55]

---

[53] Lebeaupin v. Crispin, (1920) 2 KB 714.
[54] Dhanrajamal Gobindram, M/s. v. M/s. Shamji Kalidas and Co., AIR 1961 SC 1285, 1961 (3) SCR 1020.
[55] Hillas & Co. v. Arcos Ltd., 1932 All ER 494; Adamastos Shipping Co. Ltd. v. Anglo-Saxon Petroleum Co. Ltd., 1959 AC 133.

The *force majeure* clause in an agreement stipulated that neither party to the contract shall be held responsible for any loss or damages or delay in or failure of performance of the contract, if any, to the extent that such loss or damage or delay in or failure of performance is caused by *force majeure*, or causes which cannot with reasonable diligence be controlled or provided against by the parties to the contract. On this backdrop the arbitrator contended that

> generator break down cannot be considered as covered under the force majeure clause, because the respondent could diligently control the situation by providing stand by generator sets to avoid production loss and as such cannot avoid its liability to compensate the loss the claimant suffered due to lack of diligence on the part of the respondent.

Deciding the validity of the award it was held: "As and when reason has been given by the Arbitrator, Court has no occasion to interfere with it by substituting its own view."[56]

6.8.2 Bad Weather, Strikes, etc.

Generally most contracts include an express provision for extension of time on account of inclement weather. For example, see Clause 8.5(c) of FIDIC Form 2017. However, in the absence of such a provision in the contract, delay caused by bad weather, strikes, etc. will not entitle the contractor for an extension of time. This is for the obvious reason that variableness of the climate or possibility of strikes was within the knowledge of the contractor when entering into the contract. A ship building contract contained a liquidated damages clause. It contained an exception for *force majeure* and/or strikes at the building yard or machinery workshops or at steel works supplying the ship or at the subcontractor's works. It was held that the universal coal strike of 1912 and a breakdown of machinery but not bad weather constituted *force majeure*.[57]

6.9 Written Request by a Contractor – a Condition Precedent

It must be remembered that non-fulfilment of a condition may deprive a contractor from getting an extension of time. For example, where a contract stipulated that all claims for extension of time for completion must be made in writing, it was held that the oral demand for an extension of time is not sufficient. While sustaining this proposition of law, the court, citing numerous cases, observed that: "a condition of this kind is a condition precedent compliance with which must be shown, because it must be assumed that the parties in inserting such provision attached both value and importance to it".[58] However, much will depend upon the wording of the contract provision and also the conduct of parties amounting to a waiver of the strict stipulation. Consider, for example, Clause 5 in the General Conditions of Contract of the Standard Form in use by the CPWD It reads:

> *If the contractor shall desire an extension for completion of work* on the grounds of his having been unavoidably hindered in its execution or on any other grounds he shall apply in writing to the Engineer-in Charge within 30 days of the hindrance.
>
> (Emphasis supplied)

The clause will apply "if the contractor shall desire an extension of time". The contractor may plead that he does not desire an extension unless the rates are revised, etc. and hence he did not

---

[56] Oil and Natural Gas Commission Ltd. v. M/s. Dilip Construction, AIR 2000 Cal. 140.

[57] Matsoukis v. Priestman, (1915) I K.B. 681.

[58] Austin – Griffith Inc. v. Goldberg et al., 79, S. E. 2d., 447 Supreme Court of South Carolina, 1953.

apply. The question arises whether or not the engineer-in-charge is competent to grant extension of time to the contractor in case he does not apply within 30 days of the hindrance occurring in execution of the work. It is submitted that where adequate and proper grounds exist the authority competent to grant an extension of time can do so even in the absence of an application from the contractor. This is particularly so while ordering extras or changes. Clause 12 of the same form includes a provision:

> The time for completion of the work shall be extended in the proportion the altered, additional or substituted work bears to the original contract work. Over and above this, a further period to the extent of 25% of such extension shall be allowed to the contractor.

The extension of time on this ground has to be granted even if there is no application from the contractor.

The provision in other standard form contracts is more or less similar. The older FIDIC Form uses the wording: "that the Engineer is not bound to make any determination ... of extension of time unless the contractor has notified the event causing delay in 28 days." The wording at the most may take away the right of the contractor to get an extension but not the power of the engineer to grant an extension, even in the absence of such notification by the contractor within 28 days. However, it is advisable that the contractor should adhere to the provisions of the contract in this respect to avoid unnecessary controversy and disputes. FIDIC Form 2017 edition makes elaborate provisions for the claims of extension of time and/or additional payment in Clause 8.8 read with Clause 20. The provisions are self-explanatory.

6.10 Conditions under Which Time Should Be Extended

To get an extension of time under the provisions normally incorporated in the standard forms, the following conditions must be fulfilled:[59]

  i.   The contractor must be hindered in execution of the work.
 ii.   The hindrance must be caused by any of the reasons included in the contract justifying an extension of time.
iii.   The hindrance must be of such a nature that it would necessitate an extension of time.
 iv.   The contractor under such circumstances must apply for an extension of time, in writing, to the engineer-in-charge within the time stipulated after the happening of the event causing delay.
  v.   In his application, the contractor must make out reasonable ground for such an extension; and
 vi.   At the same time make every effort to prevent or make good delay.

If all the above conditions are fulfilled, it is obligatory on the part of the engineer-in-charge or the competent authority to give a fair and reasonable extension of time. The Patna High Court's decision can be an example of when further extension is not justified. The High Court held:[60]

> On a consideration of the materials on record and the foregoing discussion, the irresistible conclusion is that the petitioner is absolutely incapable of taking up the work of the present magnitude. He is wholly unequipped and equally lacks the requisite seriousness to

---

[59] See for example, Clause 10.4.1 of Model Form of Government of India.
[60] Nevilal Rohita Construction Pvt. Ltd. v. State of Bihar, AIR 2005 Pat. 190.

execute a contract of this magnitude, particularly where time was practically the essence of the contract, as is obvious from the limited time of five weeks agreed upon between the parties for completion of the work. The fact that it was a contract for repair of embank-ment-cum-road, it was obviously an anti-flood project, and had to be completed within the stipulated period. It is in this background that I see that due to incompetence of the petitioner, the project which had to be concluded within a period of five weeks could not be completed even during the enlarged period of seven months and one week. The respondent authorities had, therefore, no option but to pass the order and the consequential order calling upon fresh tenders. The impugned order cannot be faulted.

6.11 Person Designated Should Grant Extension of Time

It is important to note that the person designated in the relevant contract provision should grant an extension of time. For example, according to Clause 5 of the CPWD Form of agreement, all letters of extension of time to be issued to the contractor should be signed by the engineer-in-charge as he is the only officer so empowered contractually to grant extension of time. The provision in the FIDIC Form empowers the engineer appointed by the employer, and named as such in Part II, to grant an extension of time in consultation with both the parties under Clause 3.7. In major projects, the engineer designated is generally a limited company and the engineer's representative is termed as team leader. The team leader is not competent to extend the time, it is submitted, unless the employer and the engineer authorize him to do so under the intimation to the contractor.

As per the NITI Aayog Model Form, the provision reads: "10.4.4 The Authority's Engineer shall, on receipt of a claim in accordance with the provisions of Clause 10.4.2, examine the claim expeditiously within the time frame specified herein". The clause goes on to specify a time of 30 days from the date of receipt of clarifications or supporting evidence, if sought.

6.12 Can Extension of Time Be Granted with Retrospective Effect?

It is observed in practice that formal extension of time is not granted during the originally stipulated/extended/formally extended period, but much later. In such cases the question arises as to whether extension so granted is legal. The question assumes importance because most construction contracts provide for recovery of liquidated damages. There are, it is submitted, three possible constructions of extension of time clauses in so far as the time for exercise of the power is concerned.

First the contract may contemplate that power should be exercised at once upon the occurrence of the event causing delay, for example, non-continuing causes, such as the ordering of extras.

Second the contract may contemplate that the power should be exercised when the full effect upon the contract programme is known, for example, continuing causes of delay, such as non-supply of drawings, decisions, failure to hand over site, strikes, etc.

Third the contract may contemplate exercise of the power at any time before issue of the final certificate.

As the ambit of most modern extension of time clauses usually comprehends delays due to causes of different kinds, in the absence of clear language to the contrary, the third alternative would normally prevail. Even where the contract contemplates the second alternative, the extension of time need not necessarily be granted before the contract date for completion. If, for example, the site is not handed over or the working drawings are not supplied until after the lapse of the contract date for completion, the extent of delay cannot be known or the necessary extension of time cannot be granted until after the stipulated time for completion has expired.

EXAMPLE FROM CASE LAW

In a case, the date stipulated for completion was 7th February 1949. The agreement provided for recovery of liquidated damages in the event of delay at the rate of £50 a week. On 19th Jaunary 1949, the contractor sought a 12-month extension. The architect merely gave a formal acknowledgement. The work was completed on 28th August 1950. The architect informed on 20th December 1950, extending the time of completion to 23rd May 1949. The contention that the architect must give the contractor a date which he can aim at, and that he cannot give a date which was passed was rejected. It was held that the retrospective extension of time was valid.[61]

This question can be viewed from another angle. As per the decision of the Supreme Court of India, the time limit stipulated for completion is not of the essence and as such it is obligatory on either party to agree for an extension of time.[62] If the contractor is permitted to carry out the work beyond the stipulated date it can be presumed that the extension of time is by implication granted and the contractor will be under an obligation to complete the work within a reasonable time. Any formal extension subsequently granted can be said to be confirmation of the implied extension. Section 63 of the Indian Contract Act supports this view. It reads:

> 63. *Promisee may dispense with or remit performance of promise*
>   Every promisee may dispense with or remit, wholly or in part, the performance of the promise made to him, or may extend the time for such performance, or may accept instead of it any satisfaction which he thinks fit.

The provisions of this section came for consideration by the Supreme Court of India wherein it was observed:[63]

> The true legal position in regard to the extension of time for the performance of a contract is quite clear under S. 63 of the Indian Contract Act. Every promisee, as the Section provides may extend time for the performance of the contract. The question as to how extension of the time may be agreed upon by the parties has been the subject matter of some argument at the Bar in the present appeal. There can be, no doubt, we think, that both the buyer and the seller must agree to extend time for the delivery of goods. It would not be open to the promisee by his unilateral act to extend the time for performance of his own accord for his own benefit. It is true that the agreement to extend time need not necessarily be reduced to writing. It may be proved by oral evidence. In some cases it may be proved by evidence of conduct. Forbearance on the part of the buyer to make a demand for the delivery of goods on the due date as fixed in the original contract may conceivably be relevant on the question of the intention of the buyer to accept the seller's proposal to extend time. It would be difficult to lay down any hard and fast rule about the requirements of proof of such an agreement. It would naturally be a question of fact in each case to be determined in the light of evidence adduced by the parties. Having regard to the probabilities in this case, and to the conduct of the parties at the relevant time, we think the appellants are entitled to urge that their oral evidence about the acceptance of the respondent's proposal for the extension of time should be believed and the finding of the learned trial judge on this question should be confirmed.

---

[61] ABC Ltd. v. Waltham Holy Cross V.D.C., [1952] 2 All ER 452; also see Fernbook Trading Co. Ltd. v. Taggort Supreme Court of New Zealand, [1979] I NZLR 556.
[62] Hind Construction Contractors v. State of Maharashtra, AIR 1979 SC 720.
[63] Keshavalal v. Lalbhai T. Mills Ltd., AIR 1958 SC 512.

Section 63 also embodies the doctrines of waiver and acquiescence. Waiver involves voluntary relinquishment of a known legal right, evincing awareness of the existence of the right and to waive the same. If a party entitled to a benefit under a contract, is denied the same, resulting in violation of a legal right, and does not protest, forgoing its legal right, and accepts compliance in another form and manner, issues will arise with regard to waiver or acquiescence by conduct.

Waiver by conduct was considered in P. Dasa Muni Reddy v. P. Appa Rao,[64] observing as follows:

> The doctrine which the courts of law will recognise is a rule of judicial policy that a person will not be allowed to take inconsistent position to gain advantage through the aid of courts. Waiver sometimes partakes of the nature of an election. Waiver is consensual in nature. It implies a meeting of the minds. It is a matter of mutual intention. The doctrine does not depend on misrepresentation. Waiver actually requires two parties, one party waiving and another receiving the benefit of waiver. There can be waiver so intended by one party and so understood by the other. The essential element of waiver is that there must be a voluntary and intentional relinquishment of a right. The voluntary choice is the essence of waiver. There should exist an opportunity for choice between the relinquishment and an enforcement of the right in question.

That waiver could also be deduced from acquiescence, was considered in Waman Shriniwas Kini v. Ratilal Bhagwandas & Co.,[65] observing:

> Waiver is the abandonment of a right which normally everybody is at liberty to waive. A waiver is nothing unless it amounts to a release. It signifies nothing more than an intention not to insist upon the right. It may be deduced from acquiescence or may be implied.[66]

But the matter does not end here. If any element of public interest is involved and a waiver takes place by one of the parties to an agreement, such waiver will not be given effect to if it is contrary to such public interest.[67] The test to determine the nature of interest, namely, private or public is whether the right which is renunciated is the right of party alone or of the public also in the sense that the general welfare of the society is involved. If the answer is the latter then it may be difficult to put estoppel as a defence. It is thus clear that if there is any element of public interest involved, the court steps in to thwart any waiver which may be contrary to such public interest.

The principle of waiver, though, is akin to the principle of estoppel. The difference between the two, however, is that whereas estoppel is not a cause of action (it is a rule of evidence), waiver is contractual and may constitute a cause of action; it is an agreement between the parties and a party fully knowing of its rights has agreed not to assert a right for a consideration. Whenever waiver is pleaded it is for the party pleading the same to show that an agreement waiving the right in consideration of some compromise came into being. Statutory right, however, may also be waived by his conduct.[68]

6.12.1 Novation/Alteration of a Contract

Under Section 62, apart from novation of a contract and rescission of a contract, alteration of a contract is mentioned. Alteration is understood generally in the sense of amendment. It is

---

[64] P. Dasa Muni Reddy v. P. Appa Rao, 1976 R.C.R. 58: (1974) 2 SCC 725.

[65] Waman Shriniwas Kini v. Ratilal Bhagwandas & Co., 1959 Supp (2) SCR 217.

[66] Kanchan Udyog Limited v. United Spirits Limited, 2017(3) R.C.R.(Civil) 433: 2017(4) Recent Apex Judgments (R. A.J.) 126: 2017(7) Scale 69: 2017(5) MLJ 616: 2017(3) BC 534: 2017(8) SCC 237: 2017(177) AIC 159: 2017(124) ALR 255: 2017(2) RJ 1473: 2017(6) WBLR 637.

[67] Lachoo Mal v. Radhey Shyam, 1971 R.C.R.(Rent) 320: (1971) 1 SCC 619, Indira Bai v. Nand Kishore, (1990) 4 SCC 668.

[68] Krishna Bahadur v. Purna Theatre, 2004 (4) S.C.T 137: (2004) 8 SCC 229.

settled law that an amendment to a contract being in the nature of a modification of the terms of the contract must be read in and become a part of the original contract in order to amount to an alteration under Section 62 of the Indian Contract Act.[69]

The effect of the alterations or modifications is that there is a new arrangement; a new contract containing as an entirety the old terms together with and as modified by the new terms incorporated.[70] The modifications are read into and become part and parcel of the original contract. The original terms also continue to be part of the contract and are not rescinded and/or superseded except in so far as they are inconsistent with the modifications. Those of the original terms which cannot make sense when read with the alterations must be rejected.[71]

6.13 Conditional Extension of Time

The extension of time for performance of the contract is, in a sense, a modification of the original contract and, therefore, must be agreed to and accepted by both the parties to the contract.[72] However, it does not require separate consideration. An extension of time by one party of its own accord given for its own benefit, but not accepted by the other party, may not result in a formal extension of time. In a case a party sought an extension of time for the performance of the contract subject to two conditions. It was held by the Supreme Court of India that unless both the conditions were agreed upon between the parties there would be no valid or binding extension of time under Section 63 of the Indian Contract Act.[73]

In view of the above position in law, it remains to be discussed as to what is the effect, in a construction contract, where either side may agree for an extension subject to certain conditions, which the other side is not willing to accept. What actually happens in practice is like this. The employer or engineer-in-charge, in case of public works, grants extension of time subject to payment of liquidated damages by the contractor. The contractor repudiates his liability to pay liquidated damages on the ground that the delay in completion was on account of defaults and delays on the part of the owner. On the other hand, the contractor seeks an extension of time subject to a number of conditions including revision of rates for the work done beyond the agreed time limit and payment of compensation. The employer agrees to an extension of time but refuses the liability to pay compensation or revision of rates.

The authors analyse these cases below.

In the first case, the imposition of penalty or recovery of liquidated damages for delay in completion being an agreed condition of the contract, the owner or his architect or engineer-in-charge, as the case may be, will be fully within his power to grant an extension subject to the condition of penalty/liquidated damages, if it is his contention that there was no default on the part of the employer.[74]

Similarly, in the second case, if it is the contention of the contractor that the delay in completion was on account of defaults on the part of the owner, he may seek conditional extension of time subject to payment of compensation, etc. The contract will remain valid but there would be, in both the cases, disputes between the parties about their rival claims, which would, in the end, be settled either amicably or through arbitration or in a court of law.

In other words, it is submitted that the cases of the above nature fall within the ambit of Section 55 of the Indian Contract Act and not under Section 63 of the said Act.

---

[69] Juggilal Kamlapat v. N.V. Internationale Crediet-En-Handels Vereeninging "Rotterdam", AIR 1955 Calcutta 65.

[70] Viscount Haldane in Morris v. Baron & Co., (1) (1918) Appeal Cases, 1 at 17.

[71] All India Power Engineer Federation v. Sasan Power Ltd., 2017(2) Recent Apex Judgments (R.A.J.) 386: 2016(12) Scale 553:: 2017(1) SCC 487: 2017(2) Rajdhani LR 91.

[72] Mahadeo Prasad v. Mathura Chowdhary, AIR 1931 All 589.

[73] Kesharlal v. Lalbhai T. Mills Ltd., AIR 1958 SC 512.

[74] Anand Construction Works v. State of Bihar, AIR 1973 Cal. 550.

It is observed, that in several cases in public works contracts, the application by the contractor listing causes for delay which include breaches of contract by the owner are accepted in words somewhat similar to "In view of the facts and circumstances stated by you in your application for an extension of time, the time for completion is extended up to", and yet the penalty/liquidated damages are shown recoverable. This is an abuse of power to levy penalty, resulting in unnecessary litigation wherein the right to recover penalty/liquidated damages is finally denied in arbitration or court proceedings.

6.14 If Extension of Time Not Granted, Even for Just Reasons

If the owner does not grant an extension of time, is a contractor legally within his right to stop work and claim damages? As already seen, it is obligatory on the part of the owner to grant an extension of time if the contractor has fulfilled all the necessary conditions for it. Refusal to grant an extension of time may, under appropriate circumstances, amount to the breach of the condition of the contract. The most important question to be decided would be: does the breach go to the root of the contract to enable the aggrieved party to treat the contract as at an end and claim damages? For, if it were found that the breach was so material, the contractor would be justified in declining to proceed with the work.

If it is assumed that the owner is willing to allow the contractor to complete the work without extending the time, the ultimate effect might be that the owner has set the time at large, and exonerated the contractor from liquidated damages for delay.[75] To recover compensation for delay is a right vested in the owner and he can waive it at any time. If the amount is deducted from the running bills, on this ground, it can also be adjusted in the final bill. The contractor can accept on account payments, under protest. If the owner deducts the amount from the final bill the contractor can claim recovery of the amount so deducted either in arbitration or through the court. As such, in the opinion of the authors, the breach does not go to the root of the contract and the contractor, under the circumstances, is well advised to fulfil his part of the contract, if the owner does not rescind it. The contractor should, however, be careful in protecting his interest by faithfully following the procedure laid down to get an extension of time.

6.15 Extension of Time – Effect on Compensation for Delay

In India, Sections 55 and 56 of the Indian Contract Act deal with consequences of breach of the contract provisions regarding time limit. Section 55 provides for situations where time is deemed to be of the essence of the contract and where it is not so deemed. In a construction contract, which provides for extension of time and/or levy of penalty, time is not deemed to be of the essence of the contract. As such, upon the contractor's failure to complete the work within the specified time the contract does not become voidable. As a result, the third paragraph of Section 55, which is applicable in case of a voidable contract, is not applicable. The second paragraph of Section 55 would apply.[76] The second paragraph of Section 55 entitles a party to compensation for any loss occasioned to it. Thus, the injured party can, without annulling the contract, ask for compensation in terms of the conditions of the contract and at the same time allow the contract to be performed.

If, under the provisions of a contract, extension of time were given without imposing penalty, the date of completion would stand extended to that extent. As such the contractor can validly claim exemption to pay compensation for delay, to the extent it is covered by such an extension. It

---

[75] Keshavlal Lallubhai Patel v. Lalbhai Trikumlal Mills Ltd., AIR 1958 SC 512.
[76] See paragraph 6.5.2 above; AIR 1973 Cal. 550.

is important that such an extension of time must be given, in case of public contracts, by a competent authority and in writing. The mere fact that the contractor was allowed and encouraged to continue the performance even after the due date may not suffice to claim exemption from payment of compensation for delay. Similarly, if the amount of compensation is not fixed, by the person competent to do it, soon after the due date, it might amount to a waiver of his right to fix the compensation and recover it under the contract.[77]

If, however, there is power to extend the time for delays caused by the owner, and such delays have in fact taken place but the power to extend the time has not been exercised due to failure to consider the matter within the time expressly or impliedly limited by the contract, the owner may have lost the benefit of the clause. The contract time has, in such a case, ceased to be applicable because of the owner's act of prevention, there is no date from which penalties/liquidated damages can run because any purported extension of time given is too late, and therefore, no liquidated damages can be recovered. By analogy it can be said that a purported extension of time that is too late can have an equally invalidating effect in cases where no element of breach by the owner is present and the cause of delay is a matter otherwise within the contractor's sphere of responsibility, such as bad weather.[78] Thus, unless there is a clear stipulation, it is not open to the owner, where the contract date has ceased to be applicable, to make out a kind of debtor and creditor account allowing so many days or weeks for delay caused by himself, and, after crediting that period to the contractor, to seek to recover from him damages at the liquidated rate for the remainder.[79]

EXAMPLES FROM CASE LAW

1. In a case, the contract permitted an extension of time for "extras or other causes beyond the control of the contractor". Liquidated damages were to be $1,000 per day. There was delay of 99 days and extension of time for 46 days. The owner, sued for $53,000. The trial judge found that 45 days delay had been caused by the owner's delivery of certain machinery in a defective condition requiring considerable repair work and that this was a breach of the contract by the owner. He accordingly awarded $8,000 liquidated damages. The Court of Appeal of British Columbia, applying Wells v. Army and Navy Society Ltd., held that on these facts no liquidated damages could be recovered.[80]

2. In the case of Wells v. Army and Navy, contractors undertook to erect buildings for a company within a year, unless delayed by alterations, strikes, subcontractors, "or other causes beyond the contractor's control". Another provision made the decision of the directors of the company in matters of time to be final; and liquidated damages were payable if the work was not completed within a time considered reasonable by them. There was a one year delay. The principal cause was the subcontractors, for which the directors were prepared to allow three months. The contractor contended that this was insufficient, and that the delay was caused by alterations, and also by the delay in giving possession and providing plans. The court found on the basis of evidence that there was substance in all the complaints, and it was impossible to say to what extent each one contributed to the delay, but the words "or other causes beyond the contractor's control" could not include the breaches of contract or other acts of the owner in not giving possession and failing to supply plans and drawings in due time, and consequently, it was held, liquidated damages could not be deducted.[81]

---

[77] State of Rajasthan v. Chandra Mohan, AIR 1971 Raj. 229.

[78] See *Hudson's Building and Engineering Contracts*, 10th ed., page 644.

[79] Dodd v. Churton, (1897) 1 Q.B. 562.

[80] Perini Pacific Ltd. v. Greater Vancouver Sewage, (1966), 57, DLR (2d), 307, Canada.

[81] Wells v. Army and Navy etc. Society, (1902) 86 LT, 764. *Hudson's Building and Engineering Contracts*, 10th ed., p. 630.

3. In a contract for the construction of a canal the time limit for completion was stipulated to be 30 months. The work remained incomplete due to non-fulfilment of the conditions of contract by the department. The contractor claimed compensation in the form of revised rates for the work done after the expiry of the original contract period. The arbitrators recorded a finding that there was delay in the execution of the work which was on account of non-fulfilment of the contract conditions for which the appellants were responsible and there were substantial changes in the working conditions and therefore the appellants were liable to pay damages/compensation towards the quantities of items that remained incomplete on the stipulated date of completion, and had to be completed thereafter. The arbitrators held that they had the power to grant damages under the provisions of the Contract Act notwithstanding the terms of the contract and it was stated that the appellants themselves accepted this position. The arbitrators found that the claims by the contractors were justified and awarded damages calculated on the basis of prevailing rates of labour, machinery, spares and petroleum, oil and lubricants 30 months after the commencement of the work, that is, during the extended period of execution of the items in question. The award was challenged in the court. In appeal, on behalf of the contractor, reliance was placed upon the decision of the Supreme Court in M/s. Tarapore Company's case.[82] The state placed reliance upon Continental Construction Co. Ltd. v. State of M.P.[83] to contend that specific escalation clause in the contract would bar award of extra cost. The Karnataka High Court held:[84]

> The effect of extension of time on damages claimed by the contractors has been exhaustively considered by several authors in several standard text books and a few of them are Halsbury's Law of England 4th edition, para 1281, Keating on Building Contracts at para 161 and Hudson on Building and Engineering Contracts, 10th edition at page 647, which view is reiterated in Emden on Building Contracts. All these text books are to the effect that when the time fixed by the contract ceases to be applicable on account of some act or default of the employer, a provision is generally inserted to extend the time. When the power to extend time has been properly exercised, the contractor will be liable to pay liquidated damages. In a true sense and on an examination of the matter in its proper perspective, what comes up for consideration in such a case is the determination of the question as to what are the rates applicable as a result of the extension of time granted and not awarding of damages as such. The enhancement in the rates itself will constitute the damages. When the contract itself does not bar such rates being given and the arbitrators in the case on hand on a consideration of the material on record, have arrived at the rate at which the contractors will be entitled for payment of extension of the contract time because of certain lapse on the part of the appellants and that rate being just and proper one, it cannot be said that the arbitrators have committed any error apparent on the face of the record calling for interference at the hands of this Court.

6.15.1 Consequences Where Time Is of the Essence of Contract

Where time is of the essence of contract, a failure to perform by the stipulated time will entitle the innocent party (i) to terminate the performance of the contract and thereby put an end to all the unperformed obligations of both the parties to the contract and (ii) claim damages for loss of the whole transaction on the basis of the fundamental breach of the contract. However, the innocent party, if it

---

[82] Tarapore and Company v. Cochin Shipyard Ltd., Cochin, AIR 1984 SC 1072.

[83] Continental Construction Co. Ltd. v. State of M.P., AIR 1988 SC 1166.

[84] State of Karnataka v. R.N. Shetty & Co., E. & C., AIR 1991 Kant. 96 (103).

accepts the belated performance instead of avoiding the contract, cannot claim damages for any loss occasioned by the non-performance of the reciprocal promise at the time agreed, unless at the time of such acceptance he gives notice to the promissor of his intention to do so.

6.15.2 Exemption Clauses and the Fundamental Breach

Thus, it appears that under the Indian law, in spite of there being a contract between the parties whereunder the contractor has undertaken not to make any claim for delay in performance of the contract occasioned by an act of the employer, still a claim would be entertainable in one of the following situations:

i.   If the contractor repudiates the contract exercising his right to do so under Section 55 of the Contract Act

ii. The employer gives an extension of time either by entering into supplemental agreement or by making it clear that escalation of rates or compensation for delay would be permissible

iii. If the contractor makes it clear that escalation of rates or compensation for delay shall have to be made by the employer and the employer accepts performance by the contractor in spite of delay and such notice by the contractor putting the employer on terms.

Thus, it may be open to the contractor, in the above situation, to prefer a claim touching an apparently excepted matter subject to a clear case having been made out for excepting or excluding the claim from within the four corners of "excepted matters". However, a party in its petition must show or *prima facie* suggest why such claims must be taken out of the category of excepted matters.[85]

6.16 Procedure of Terminating Contract after Expiry of Time Limit for Completion

Before discussing the procedure for termination of the contract, the conclusions and the principles very well established and considered above can be summed up as a check list to decide whether to terminate the contract or not.

1. In a construction contract mere stipulation of time limit for completion does not make that time the essence of the contract.

2. Even if the agreement clearly stipulates that the time shall be deemed to be of the essence of the contract, it shall cease to be of the essence if the contract contains stipulations regarding extension of time and imposition of penalty/liquidated damages.

3. Upon non-completion of the work either due to default(s) of the owner or of the contractor, the contract does not automatically come to an end on the expiry of the date of completion.

4. Either party to the contract is bound to grant extension of time, if sought for by the other party. The other party seeking extension, if it is guilty of having caused the delay, will be liable to make good the losses suffered by the injured party due to such delay either in accordance with the provision of the agreement or, in the absence of any such provision in the agreement, under Section 73 read with paragraph 2 of Section 55 of the Contract Act.

5. The extension to be agreed should be a reasonable extension. As to what is a reasonable extension, it is a question of fact and will depend upon facts and circumstances of each

---

[85] General Manager, Northern Railways v. Sarvesh Chopra, 2002(1) Arb. LR 515; Navayuga Engineering Co. P Ltd. v. Public Works Department, 2016(229) DLT 226: 2016(3) ArbiLR 6:: 2016(4) R.A.J. 387.

case. For example, if a contract involves construction work costing Rs. 50 million, originally agreed to be completed within ten months and if only half the work is completed within the stipulated time, the reasonable extension may be about four to five months.

6.16.1 The Procedure

If a defaulting party is likely to continue to default, every time the innocent party is not bound to grant an extension of time. In such an eventuality, although Section 55 does not so stipulate, it is open for the innocent party to give subsequent notice making the extended time of the essence of the contract. It is necessary that while giving such notice the intention of the party giving the notice should be very obvious from the notice itself inasmuch as the matter would be open to scrutiny by the court. The question whether the time prescribed in the notice is or is not of the essence of the contract would naturally depend upon the facts and circumstances of each case. The mere fact that the notice gave a certain time to perform the contract would not necessarily lead to the conclusion that the time prescribed was of the essence of the contract. In all such cases, the court has to look to the pith and substance of the notice and not at the letter of the notice and decide as to whether time was or was not essential to the subsistence of the contract. The real intention of the party who gives notice must be clear from the notice itself. It may in certain cases be necessary to rely upon surrounding circumstances. Nevertheless one has to largely look to the notice itself.[86]

After the time is extended and made of the essence, upon expiry of the extended time, which was made of the essence of the contract, the contract does not automatically come to an end. Section 55 makes such an agreement voidable. It must, however, be remembered that the only right which (the innocent party) gets in such a case is to avoid the contract. The contract does not automatically get determined. He has to further expressly or in unambiguous words determine the contract under Section 64 of the Contract Act.[87]

> 64: *Consequences of rescission of voidable contract*
>
> When a person at whose option a contract is voidable rescinds it, the party thereto need not perform any promise therein contained in which he is promisor. The party rescinding a voidable contract shall, if he has received any benefit there under from another party to such a contract, restore such benefit so far as may be, to the person from whom it was received.

In a construction contract, the contractor will be entitled to receive the value of the work done. That, however, will not prejudice the right to get compensation due to delay vested in the innocent party under Section 55 read with Section 73 of the Indian Contract Act.

EXAMPLES FROM CASE LAW

1. A.P. State Electricity Board – Appellants called for quotations from eight companies including the plaintiff for supply of materials. An order was placed which stipulated that the goods should be supplied before the specified date. The plaintiff sent the goods belatedly. The goods were received allegedly pending decision of the competent authority. The appellant's superintending engineer (operation) cancelled the purchase order and the plaintiff was requested to take back the goods. Instead of taking back the goods the plaintiff filed a suit seeking recovery of the cost of materials supplied. The trial court, on a consideration of the entire material placed before it, held in favour of the plaintiff and decreed the suit as prayed for. In the High Court it was

---

[86] T. Venkata Subrahmanayam v. V. Vishwanandharaju, AIR 1968 AP 190.
[87] Tandra Venkata Subrahmanayam v. Vegesana Viswanadharaju and another, AIR 1968 AP 190.

contended on behalf of the appellants that the contract being a commercial one time was of the essence of the contract and the delivery made by the plaintiff is no delivery at all in the eye of the law. The High Court took note of the following factors:

1. In the purchase order there was nothing to show that time was intended to be of the essence of the contract though the date was specified before which the supply of the goods was to be effective.

2. The conduct in the past established that in earlier transactions like the one in question, the board never intended time to be of the essence and goods were accepted by imposing penalty.

3. The board did not reject the goods when offered and the goods were accepted after the sample was approved and without putting any condition as to its verification and acceptance by superior officers.

4. Though the plaintiff's offer was valid for 30 days the board placed an order subsequent to the expiry of that period.

It was held:[88]

> For the foregoing reasons agreeing with the court below, I am of the opinion that time was never intended to be treated as the essence of the contract by the parties. If that is so, the Board is not entitled to cancel the order and the cancellation cannot be treated to be valid in the eye of the law.

2. In a case, the architect purported to give notice to the contractors under Clause 25(1) (of JCT 63 contract terms) stating that in his opinion they had failed to proceed regularly and diligently and that unless an appreciable improvement was shown within 14 days the contract would be determined. By a subsequent letter the employer purported to determine the contract. The contractor alleged that the notice of determination was wrongful and amounted to repudiation of the contract. The contractors refused to accept the repudiation and elected to proceed with the work. The employer claimed an injunction to restrain the contractors from remaining on the site. Observing that what is involved is the application of an uncertain concept to disputed facts the injunction was refused.[89]

6.17 Section 46 of the Indian Contract Act – When Attracted

If a contract does not specify the time for performance, the law will imply that the parties intended that the obligation under the contract should be performed within a reasonable time. Section 46 of the Contract Act provides that where, by a contract, a promissor is to perform his promise without application by the promisee, and no time for performance is specified, the engagement must be performed within a reasonable time and the question "what is reasonable time" is, in each particular case, a question of fact.[90]

The provisions are self-explanatory. It cannot be said that the time can be made of the essence of the contract only by express provision in the contract or by notice by one of the parties. The

---

[88] A.P. State Electricity Board v. Patel & Patel, AIR 1977 AP.

[89] London Borough of Hounslow v. Twickenham Garden Developments Ltd., Chancery Division (1970) 7 BLR 81.

[90] Rajinder Kumar v. Kuldeep Singh, 2014(2) Scale 135: 2014(104) ALR 741: 2014(2) JT 530: 2014(2) MLJ 496: 2014 AIR (SC) 1155: 2014(15) SCC 529.

law under Section 46, engrafts on the contract a condition that reasonable time in the absence of any specific time provided in the contract will be of the essence of the contract and if the contract is not performed within that reasonable time by any of the parties, the said party will be deemed to be guilty of breach of duty or breach of contract.[91]

When a contract for sale does not stipulate time it is expected that the same would be performed by the parties within a reasonable period. In a given case, the plaintiff is entitled to wait for a reasonable time and thereafter the suit can be brought within the general period of limitation, being three years. The decree for specific performance cannot be refused on the ground that there was no time stipulated for the same.[92]

Where the contract stipulated that the defendant shall execute a sale deed within one month after the receipt of permission of a competent authority, it was held that the time was not of the essence and the plaintiff could call the defendant to execute the sale deed even after the one month had elapsed.[93] Where, however, a suit was filed for specific performance after nine years, it was held that the plaintiff did not act reasonably and the suit was not maintainable.[94]

The question as to how long a plaintiff (even if he had performed the whole of his obligations under an agreement for sale, in which a time for performance is not fixed) could keep alive his right to specific performance was considered by the Apex Court.[95] In the said case the plaintiff came to the court after 29 years seeking to enforce the agreement. It was observed the question may have also to be considered by the court especially in the context of the fact that the relief of specific performance is discretionary and is governed by the relevant provisions of the Specific Relief Act. The Supreme Court further observed:

> But again, these questions cannot be decided as preliminary issues and they are not questions on the basis of which the suit could be dismissed as barred by limitation. The question of limitation has to be decided only on the basis of Article 54 of the Limitation Act and when the case is not covered by the first limb of that Article, normally the question of limitation could be dealt with only after evidence is taken and not as a preliminary issue unless, of course, it is admitted in the plaint that the plaintiffs had notice that performance was refused by the defendants and it is seen that the plaintiffs approached the court beyond three years of the date of notice.

Section 46 will be applicable in a case where the time fixed for completion has ceased to be applicable and the contractor is continuing the work under the contract.

EXAMPLE FROM CASE LAW

In a contract stipulating date for completion and a power to certify extension, no extension was either applied for or granted. Six months later the owner exercised power of forfeiture under the clause. The contractor brought an action for breach of contract contending that the clause ceased to have any application after the expiry of the time fixed by the contract for completion. It was held that the contract date having passed without any extension, the duty of the contractor was to complete the work within a reasonable time.[96]

---

[91] Sat Parkash v. Dr. Bodh Raj, (Punjab)(D.B.): 1958 PLR 97: 1958 AIR (Punjab) 111: 1958 ILR (Punjab) 97.

[92] P. Poppan and another v. Karia Gounder and others, 2002 (2) Arb. LR 666 Calcutta; also see: Lala Suymer Chand Goel v. Rakesh Kumar, 2002(Suppl.) Arb. LR 66 (Allahabad).

[93] Rakha Singh v. Babu Singh, 2003(1) Arb. LR 22 (P. & H.).

[94] Suryagandhi v. Lourduswamy, 2003(1) Arb. LR 234 (Madras).

[95] Gunwantbhai Mulchand Shah v. Anton Elis Farel, (SC): 2006(3) JT 212: 2006(2) SCC 634: 2006 AIR (SC) 1556: 2006(3) SCC 634: 2006(2) SCR 886: 2006(1) W.L.C. 741: 2006(1) A.R.C. 890: 2006(2) WBLR 423: 2006(3) Cal. H.C. N. 73: 2006 DNJ 514: 2006(2) M.P.L.J. 412: 2006(3) ALT 74: 2006 AIR (SCW) 1377.

[96] Joshua Henshaw & Sons v. Rochadale Corporation, (1944), K. B. 381 *Hudson's Building and Engineering Contracts*, 10th ed., p. 697. Also see Electronic Industries v. David Jones (1954) 91 C.L.R. 288 (Australia).

6.17.1 When Time Limit Is Not Specified – Oral Evidence – Whether Admissible?

Where a written contract is silent as to time of performance, a reasonable time is to be presumed without reference to parol (oral) evidence.[97] However, in a case, the court held that in order to determine what constitutes a reasonable time, the court should allow the parties to testify as to any conversation they had before the agreement was signed. In this case the trial court did not allow such a testimony on the ground that it tended to vary the terms of the written agreement. Reversing this ruling of the lower court the Appellate Court held that oral evidence was not offered to vary the terms of the written contract[98] but to show what the situation was when the parties entered into the contract and to determine what was a reasonable time for the performance of the contract.[99] The authors are in respectful agreement with this view.

6.18 Time and Place of Performance

The provisions of the Indian Contract Act other than those considered above, pertaining to time and place of performance include Sections 47 to 50, which are self-explanatory and reproduced below:

---

47. *Time and place for performance of promise, where time is specified and no application to be made*

When a promise is to be performed on a certain day, and the promisor has undertaken to perform it without application by the promisee, the promisor may perform it at any time during the usual hours of business on such day and at the place at which the promise ought to be performed.

ILLUSTRATION

A promises to deliver goods at B's warehouse on the 1st January. On that day A brings the goods to B's warehouse, but after the usual hour for closing it, they are not received. A has not performed his promise.

---

48. *Application for performance on certain day to be at proper time and place*

When a promise is to be performed on a certain day, and the promisor has not undertaken to perform it without application by the promisee, it is the duty of the promisee to apply for performance at a proper place and within the usual hours of business.

Explanation- The question "What is a proper time and place" is, in each particular case, a question of fact.

49. *"Place for performance of promise where no application to be made and no place fixed for performance"*

When a promise is to be performed without application by the promisee, and no place is fixed for the performance of it, it is the duty of the promisor to apply to the promisee to appoint a reasonable place for the performance of the promise, and to perform it at such place.

---

[97] Fiffels and Valet, Inc. v. Levy Co., 337 Mich. 177 (1953) 58 N.W.d2 899.

[98] Many contracts contain an entire agreement clause which, coupled with the rule prohibiting submission of oral evidence to vary a contract, may not permit a party to submit oral evidence. See Rock Advertising Ltd. v. MWB Business Exchange Centres Ltd., [2018] UKSC 24.

[99] Steinmn v. Olafson, 149 N.Y.S. 2d, 3d 31 Supreme Court, Appellate Term Second Dept. 1955.

ILLUSTRATION

A undertakes to deliver a thousand maunds of jute to B on a fixed day. A must apply to B to appoint a reasonable place for the purpose of receiving it, and must deliver it to him at such place.

50. *Performance in manner or at a time prescribed or sanctioned by promisee*

The performance of any promise may be made in any manner, or at any time which the promisee prescribes or sanctions.

ILLUSTRATIONS

(a) B owes A 2,000 rupees. A desires B to pay the amount to A's account with C, a banker. B, who also banks with C, orders the amount to be transferred from his account to A's credit, and this is done by C afterwards, and before A knows of the transfer, C fails. There has been a good payment by B.

(b) A and B are mutually indebted. A and B settle an account by setting off one item against another, and B pays A the balance found to be due from him upon such settlement. This amounts to a payment by A and B, respectively, of the sums which they owed to each other.

(c) A owes B 2,000 rupees. B accepts some of A's goods in reduction of the debt. The delivery of the goods operates as a part payment.

(d) A desires B, who owes him Rs. 100, to send him a note for Rs.100 by post. The debt is discharged as soon as B puts into the post a letter containing the note duly addressed to A.

7

Penalty/Liquidated Damages

7.0 Introduction

Breaches of contract are not uncommon in engineering and construction contracts. Damage, i.e. loss, may be pecuniary and/or non-pecuniary although in the context of this book, the focus is on the monetary loss. Damages mean money compensation claimed by the injured party for the injury done. Damages, which are ascertained with the fundamental consideration to place the innocent party in the position he would have occupied had the contract been performed according to its terms, are said to be compensatory. They may be liquidated or unliquidated. The amount of damages may have to be determined after the breach by a court or an arbitrator; these are unliquidated damages. Liquidated damages, the focus of this chapter, are the actual sums named in the contract, which the parties have themselves calculated would be a fair compensation for the breach of the contract. A penalty is the sum named in a contract by way of compensation; the amount recoverable is not necessarily the sum actually named, but the value of the precise damage suffered by the injured party not exceeding the sum named. In India, penalty is akin to unliquidated damages.

7.1 Provisions of the Law

7.1.1 FIDIC Form (Red Book) 1992 Edition, 1999 Edition and 2017 Edition

These forms (listed in the table over page) provide for payment of liquidated damages for delay in completion of the project at the rate "sum named in the Appendix or Contract Data" which shall be paid for every day of delay beyond the stipulated completion date up to the relevant date of completion. The delay damages are not to exceed the maximum amount stated in the contract data. This per day sum and the maximum limit is left to the parties to agree and the maximum amount named is generally not to exceed 10% of the contract sum.

The 2017 edition makes it clear that the delay damages shall be the only damages due from the contractor for failure to comply with Clause 8.2 (Time for completion). Exception is also made of damages resulting from termination of contract, which the contractor will be liable to pay for his obligation to complete the work. It also stipulates that the contractor's liability for delay damages shall not be limited in cases of fraud, gross negligence, deliberate default or reckless misconduct on his part.

The earlier forms empowered the employer to recover the damages from any monies due or to become due to the contractor without any certificate or decision of the engineer. The 2017 edition, however, subjects the employer's entitlement to claim delay damages to Clause 20.2.

Clause 20.2 provides for either party making a claim for additional payment to give notice to the engineer within 28 days of the party becoming or when it should have become aware of the event entitling it to make the claim. If the claiming party fails to give notice within 28 days, the engineer shall give notice of the said fact to the claiming party within 14 days. If the engineer fails to so notify, the notice belatedly served will be a valid notice. The other party has a right to challenge the deemed valid notice with details of the reasons for disagreement. In case the engineer gives notice within 14 days of receiving the claim notice, the claiming party has the right to justify delay and submit a full

Provisions of the Law	Paragraph No.
Indian Contract Act, Section 74	7.3.2
Provisions in Standard Form Contracts	
FIDIC Form, 1992 edition, Clauses 47.1, 47.2 1999 1st edition, Clause 8.7 2017 2nd edition, Clauses 8.8 and 20.2 read with 3.7	7.1.2, 7.9, 7.9.1 and 7.13
Ministry of Statistics and Programme Implementation, Government of India (MOS & PI), Clause 9 NITI Aayog Model Form, Clauses 1.2.1, 7.1.1, 7.1.3, 17.5	7.1.2 7.1.3
PWD B-1 & B-2 Standard Forms of State Government, Clause 2 CPWD, Clause 2	7.9
Military Engineering Services Form, Clause 50	7.9

detailed claim under Clause 20.2.4 within 84 days of the claiming party becoming or when it should have become aware of the event entitling it to make the claim. Clause 20.2.3 deals with contemporary records to be produced in support of the claim. If the claiming party fails to submit the detailed claim within the stipulated time of 84 days, the engineer by notice within 14 days after the elapse of the 84 days-period shall issue notice declaring the original claim notice no longer valid. The claiming party has a right to challenge the decision of the engineer and include its full response in the detailed statement.

Clause 20.2.5 stipulates that after receipt of the detailed statement, the engineer shall proceed in accordance with Clause 3.7. Under the said clause the engineer shall consult with both the parties jointly or separately to reach an agreement. If the agreement is achieved within 42 days, the engineer shall notify the same, failing which the engineer shall determine the matter or claim fairly and notify his determination within 42 days under Clause 3.7.3. If the engineer rejects the claim or fails to give notice of his decision, he shall be deemed to have rejected the claim and the matter shall be deemed to result in a dispute to be referred to DAAB for its decision.

Caution

Care should be taken by the parties at the time of making changes in the standard form contract. In a case decided by the UK Court it was held that in the modified contract, while Clause 2.5 is widely drafted and applies to "*any*" payment in respect of the contract, the right to liquidated damages under Clause 8.7 was not subject to the mechanism set out in Clauses 2.5 and 3.5. Clause 8.7 set out a self-contained regime for the trigger and payment of delay damages and the obligation to pay contained therein was unqualified. It was clear that there were substantive inconsistencies between Clauses 2.5 and 8.7.4. The judge found that they could all be resolved by construing Clause 8.7 as an independent regime. It was held that the employer was entitled to recover payment of liquidated damages from the contractor under Clause 8.7 without agreement or determination by the engineer of the employer's entitlement under Clauses 2.5 and 3.5.[1]

7.1.2 Ministry of Statistics and Programme Implementation, Government of India (MOS & PI)

The provisions in Clause 9 – Liquidated Damages and Incentives – are self-explanatory and read as follows:

---

[1] J. Murphy & Sons Ltd., v. Beckton Energy Ltd., [2016] EWHC 607 (TCC).

Liquidated Damages

9A. In case of delay in completion of the contract, liquidated damages (L.D.) may be levied at the rate of half per cent (½%) of the contract price per week of delay, subject to a maximum of 10% of the contract price.

9A (i) The owner, if satisfied that the works can be completed by the contractor within a reasonable time after the specified time for completion, may allow further extension of time at its discretion with or without the levy of L.D. In the event of extension granted being with L.D., the owner will be entitled without prejudice to any other right or remedy available in that behalf, to recover from the contractor as agreed damages equivalent to half per cent (½%) of the contract value of the works for each week or part of the week subject to the ceiling defined in sub Clause 9A.

9A (ii) The owner, if not satisfied that the works can be completed by the contractor, and in the event of failure on the part of the contractor to complete work within further extension of time allowed as aforesaid, shall be entitled, without prejudice to any other right, or remedy available in that behalf, to rescind the contract.

9A (iii) The owner, if not satisfied with the progress of the contract and in the event of failure of the contractor to recoup the delays in the mutually agreed time frame, shall be entitled to terminate the contract.

9A (iv) In the event of such termination of the contract as described in clauses 9A (ii) or 9A (iii) or both the owner shall be entitled to recover L.D. up to ten per cent (10%) of the contract value and forfeit the security deposit made by the contractor besides getting the work completed by other means at the risk and cost of the contractor.

9A (v) The ceiling of LD shall be 10% of the project cost in turnkey contracts. Lower limits for LDs should be clearly justified while formulating the contract. Each public sector undertaking/Ministry will take a considered view for adopting any deviations on LDs with necessary legal advice.

9A (vi) Ministries/Departments/Project Enterprises may adopt a suitable percent of the contract price as liquidated damages and allowable time-limit depending upon the nature of turnkey contract.

If the liquidated damages are levied at the rate of half per cent per week, the maximum delay tolerable works out to 20 weeks, that is five months. Thereafter the contract can be terminated and compensation payable will be liquidated damages plus forfeiture of the security deposit made by the contractor besides getting the work completed by other means at the risk and cost of the contractor.

Incentives or Bonus (Optional Clause)

9B For early completion of the contract before the stipulated date of completion, an incentive amount at the rate of half per cent (½%) of the contract price per week of early completion, subject to a maximum of five per cent (5%) of the contract price may be paid to the contractor.

(i) The incentive or bonus (optional clause) would be applicable in time-critical projects.

(ii) The owner (Project Enterprise/Ministry/Department) may determine accurately the quantum of incentive and the period of early completion as the eligibility criteria before the award of contract.

(iii) Each Public Sector Undertaking/Ministry will consider and take a considered view whether the clause regarding incentives is to be included in the contract along with justifications based on legal advice.

If the delay in completion occurs on account of defaults on the part of the employer, the contractor may not be able to claim bonus under the contract provision but can claim compensation for the loss of bonus due to breach of contract by the employer.

The provisions of PWD, CPWD, MES, Railway and other forms adopted by Public Sector Undertakings are similar to the above provisions.

7.1.3 NITI Aayog, Government of India, EPC of Civil Works – Model Form

The Niti Aayog Model Form includes a provision in Clause 10.3.2 which reads as follows:

> 10.3.2 The Contractor shall construct the Project in accordance with the Project Completion Schedule set forth in Schedule-I. In the event that the Contractor fails to achieve any Project Milestone or the Scheduled Completion Date within a period of 30 (thirty) days from the date set forth in Schedule-I, unless such failure has occurred due to Force Majeure or for reasons attributable to the Authority, it shall pay Damages to the Authority in a sum calculated at the rate of 0.05% (zero point zero five per cent) of the Contract Price for delay of each day reckoned from the date specified in Schedule-I and until such Project Milestone is achieved or the Works are completed; provided that if the period for any or all Project Milestones or the Scheduled Completion Date is extended in accordance with the provisions of this Agreement, the dates set forth in Schedule-I shall be deemed to be modified accordingly and the provisions of this Agreement shall apply as if Schedule-I has been amended as above; provided further that in the event the Works are completed within or before the Scheduled Completion Date including any Time Extension, the Damages paid under this Clause 10.3.2 shall be refunded by the Authority to the Contractor, but without any interest thereon. For the avoidance of doubt, it is agreed that recovery of Damages under this Clause 10.3.2 shall be without prejudice to the rights of the Authority under this Agreement including the right of Termination thereof. The Parties further agree that Time Extension hereunder shall only be reckoned for and in respect of the affected Works as specified in Clause 10.4.2.
>
> 10.3.3 The Authority shall notify the Contractor of its decision to impose Damages in pursuance of the provisions of this Clause 10.3. Provided, however, that no deduction on account of Damages shall be effected by the Authority without taking into consideration the representation, if any, made by the Contractor within 20 (twenty) days of such notice. The Parties expressly agree that the total amount of Damages under Clause 10.3.2 shall not exceed 10% (ten percent) of the Contract Price.

The notable features of the above provisions are:

1. The maximum extension with liquidated damages works out to 200 days that is about 6.66 months.
2. No deduction shall be effected by the authority without taking into consideration the representation, if any, made by the contractor within 20 days of such notice.
3. The engineer has, *apparently*, no role to play in deciding the employer's entitlement of damages under clause 10.3.2.
4. However, clause 10.3.2 takes into account extension of time under clause 10.4.2. Under the said clause the authority's engineer is the one to decide and forward in writing to the contractor its determination of time extension no later than 60 days from the date of receipt of the contractor's claim for time extension.
5. The provisions of clause 11.14 "Delays during construction" are also relevant which contemplate issue of notice by the engineer to the contractor if the rate of progress of works is such that completion of the project is not likely to be achieved by the end of the scheduled completion date. The contractor shall, within 15 days of such notice, by a communication inform the authority's engineer in reasonable detail about the steps the

contractor proposes to take to expedite progress and the period within which it shall achieve the project completion date.

6. Collective reading of all the above relevant provisions suggest that it is ultimately the engineer's decision of extension of time which will be deciding factor for determining the recovery of damages by the authority, in the first instance, subject of course to the dispute adjudication clause.

7.2 Purpose of Liquidated Damages

The purpose of liquidated damages is to compensate the injured party for the loss suffered. The provision enables the parties to know in advance what their position will be if a breach occurs and so avoid litigation altogether, but if litigation cannot be avoided, it eliminates what may be the very heavy legal costs of proving the loss actually sustained.[2]

The fundamental provision upon which the rule of damages is based is compensation. Compensation is the value of the performance of the contract, that is what the injured party would have earned had the contract been performed. The injured party is not to make profit out of this provision. It, therefore, follows that although parties may have agreed to a fixed sum payable as damages, if such a sum exceeds the actual loss, the injured party will get nothing more than the actual loss suffered. And this precisely is the meaning of the term "penalty".

It needs to be noted that if a party terminates the contract on account of fundamental breach of contract by another party, the innocent party is entitled to claim damages for the entire contract, that is for the part which is performed and also for the part which is prevented from being performed.[3]

7.3 Penalty and Liquidated Damages Distinguished

The purpose of liquidated damages, makes it clear that there is no distinction between the two ways of naming a sum as compensation, namely penalty and liquidated damages. What is payable is the actual loss not exceeding the sum named. English law, however, recognizes a distinction between these two provisions. The common law doctrine of damages is considered here to show that the Indian law on the subject differs from it and the latter is, in a way, an improved form of the former.

7.3.1 The Common Law Doctrine of Damages

Where parties name a penal sum (penalty) as due and payable on the breach of a contract, the real damages and no more are payable. The sum named is only the maximum of damages. On the other hand, if parties, by consent, assess a fixed measure of damages (liquidated damages), to avoid the difficulty that must often be found in setting a pecuniary value on obligations not referable, on the face of them, to any commercial standard, the entire sum so fixed becomes due and payable.[4] The strong initial presumption behind the liquidated damages is that in a negotiated contract between properly advised parties of comparable bargaining power, the parties themselves are the best judges of what is legitimate in a provision dealing with the consequences of breach. It is in both parties' interests to carry out the quantification exercise themselves by pre-estimating

---

[2] Robophone Facilities Ltd. v. Blank, [1966] 1 WLR 1428 (CA), page 1447.

[3] Maharashtra State Electricity Distribution Company Ltd. v. M/s. Datar Switchgear Limited, 2018(1) Scale 303: 2018(1) JT 361: 2018 AIR (SC) 529: 2018 AIR (SCW) 529: 2018(1) ArbiLR 236: 2018(3) SCC 133.

[4] J. Murphy & Sons Ltd. v. Beckton Energy Ltd., [2016] EWHC 607 (TCC).

their losses at the outset; the actual loss suffered may be more or less but, if the parties have made accurate calculations and negotiated effectively, the liquidated damages payable under the contract should at least be approximately right.

The distinction so made, on the face of it, looks very well. The trouble is that even now the courts have not arrived at clear or certain rules for deciding to which of these two classes a given stipulation of penal or seemingly penal sum belongs. In construing these terms, the court will not of necessity accept the phraseology of the parties but either term may be substituted for the other, at the discretion of the court, after a full consideration of the intention of the parties at the time the contract was made as expressed by the whole instrument. Even the addition of negative words to purposely exclude the other alternative, for example "as liquidated damages and not as penalty" will not be decisive. This unhappy position in interpreting the Common Law doctrine of damages is completely eliminated in the Indian Contract Act of 1872.

7.3.2 Section 74 of the Indian Contract Act

The provision made in the Indian Contract Act (Section 74) reads as follows:

> 74. *Compensation for breach of contract where penalty stipulated for*
>
> When a contract has been broken, if a sum is named in the contract as the amount to be paid in case of such breach, or if the contract contains any other stipulation by way of penalty, the party complaining of the breach is entitled, whether or not actual damage or loss is proved to have been caused thereby, to receive from the party who has broken the contract a reasonable compensation not exceeding the amount so named or, as the case may be, the penalty stipulated for.
>
> *Explanation* – A stipulation for increased interest from the date of default may be stipulation by way of penalty.
>
> *Exception* – When any person enters into any bail-bond, recognizance or other instrument of the same nature, or, under the provisions of any law, or, under the orders of the Central Government or of any State Government, gives any bond for the performance of any public duty or act in which the public is interested, he shall be liable, upon breach of the condition of any such instrument, to pay the whole sum mentioned therein.
>
> *Explanation* – A person who enters into a contract with the Government does not necessarily thereby undertake any public duty, or promise to do an act in which the public is interested.

The language of the section has made its interpretation difficult. The section provides that reasonable compensation is payable. It provides at the same time, that such reasonable compensation would be receivable whether or not actual damage or loss is proved to have been caused by the breach. If it is a case of reasonable compensation then surely such reasonableness would depend on the facts and circumstances of each case. But the section provides that it makes no difference whether or not actual damage or loss is proved. Then again, the last part of the sentence provides that such reasonable compensation is not to exceed the amount named in the contract or the penalty stipulated for in the contract.[5]

With the help of decided cases, the main provisions of the section can be analysed as follows:[6]

i. *The section does not make any distinction between penalty and liquidated damages.* The section is clearly an attempt to eliminate the somewhat elaborate refinements made under English law in distinguishing between stipulations providing for liquidated damages and stipulation in the nature of penalty.

---

[5] Anand Construction Works v. State of Bihar, AIR 1973 Cal. 550.

[6] Mahadeo Prasad v. Siemen Ltd., AIR 1934 Cal. 285.

[T]he Indian Legislature has sought to cut across the web of rules and presumptions under the English Common Law, by enacting a uniform principle applicable to all stipulations naming amounts to be paid in case of breach, and stipulations by way of penalty.[7]

ii. Although the parties may have named a sum payable as damages, the court will award "reasonable" compensation. But that does not mean that compensation can be awarded even though no loss whatsoever has been caused.[8] "For the very concept of award of compensation is bound up with loss of damage that results from a breach of contract."[9] All that Section 74 permits is award of compensation, even where the extent of the actual loss or damage is not proved and it gives discretion to the court to fix the amount.

iii. Even if, in a case the reasonable compensation in the opinion of the court is more than the sum named in the contract, the court is at liberty to grant no more than the sum named. In assessing damages the court has, subject to the limit of the penalty stipulated, jurisdiction to award such compensation as it deems reasonable having regard to all the circumstances.

iv. If the sum is named in a contract payable as damages on the breach of the contract, it is open to the claimant of compensation to prove that the said amount represents a *bona fide* pre-estimate of the damages consequential to the breach complained of. The burden is always on the plaintiff to prove the extent of damages.

v. In a recent case, Hon'ble Mr. Justice R.F. Nariman, after referring to previous leading cases on Section 74 observed:[10]

On a conspectus of the above authorities, the law on compensation for breach of contract under Section 74 can be stated to be as follows:

1. Where a sum is named in a contract as a liquidated amount payable by way of damages, the party complaining of a breach can receive as reasonable compensation such liquidated amount only if it is a genuine pre-estimate of damages fixed by both parties and found to be such by the Court. In other cases, where a sum is named in a contract as a liquidated amount payable by way of damages, only reasonable compensation can be awarded not exceeding the amount so stated. Similarly, in cases where the amount fixed is in the nature of penalty, only reasonable compensation can be awarded not exceeding the penalty so stated. In both cases, the liquidated amount or penalty is the upper limit beyond which the Court cannot grant reasonable compensation.

2. Reasonable compensation will be fixed on well-known principles that are applicable to the law of contract, which are to be found inter alia in Section 73 of the Contract Act.

3. Since Section 74 awards reasonable compensation for damage or loss caused by a breach of contract, damage or loss caused is a sine qua non applicability of the Section.

4. The Section applies whether a person is a plaintiff or a defendant in a suit.

5. The sum spoken of may already be paid or be payable in future.

---

[7] S.A. Bhat v. V.N. Inamadar, AIR 1959, Bom. 452; Fateh Chand v. Balkishan Dass, AIR 1963 SC 1405.

[8] Indian Oil Corpn. v. Lloyds Steel Industries Ltd., AIR 2008 (NOC) 866 (DEL); M/s Haryana Telcom Ltd. v. Union of India, AIR 2006 Del. 339; State of Kerala v. M/s. United Shippers and Dredgers Ltd., AIR 1982 Ker. 281 relied upon.

[9] Maharashtra State Electricity Board v. Sterilite Industries (India) & Another, AIR 2000 Bom. 204, 2000 (3) BomCR 347.

[10] M/s. Kailash Nath Associates v. Delhi Development Authority, 2015(1) Scale 230; 2015(4) SCC 136; AIR 2015 (SCW) 759.

6. The expression "whether or not actual damage or loss is proved to have been caused thereby" means that where it is possible to prove actual damage or loss, such proof is not dispensed with. It is only in cases where damage or loss is difficult or impossible to prove that the liquidated amount named in the contract, if a genuine pre-estimate of damage or loss, can be awarded.

7. Section 74 will apply to cases of forfeiture of earnest money under a contract.

7.3.3 Reasonable Compensation

The words of Section 74 give a wide discretion to the court in the assessment of damages. Although the discretion of the court in the matter of reducing the amount of damages agreed upon is left unqualified by any specific limitation, the expression "reasonable compensation" used in the section necessarily implies that the discretion so vested must be exercised with care, caution and on sound principles. For example, the measure of damages for failure to complete the building in time, where time is of the essence of the contract, is the rental value of the building for the period of delay and not the rent paid by the owner for the house he occupied.[11] In a case, which provided for maximum sum of 10% of the value of the order, the court allowed 1% by exercising the discretion.[12] Where a loaded vehicle carrying rice met with an accident and the food grains were looted by the unknown miscreants causing loss, it was held that the injured party was entitled to compensation at the market price of the food grains at the relevant time and not three times the cost as stipulated.[13]

EXAMPLE FROM CASE LAW

The appellant in a case failed to execute the work of construction of a sewerage pumping station within the stipulated or extended time. The said pumping station certainly was of public utility to maintain and preserve clean environment, absence of which could result in environmental degradation by stagnation of water in low lying areas. Delay also resulted in loss of interest on blocked capital. In these circumstances, loss could be assumed, even without proof, and burden was on the appellant who committed the breach to show that no loss was caused by delay or that the amount stipulated as damages for breach of contract was in the nature of penalty. Even if technically the time was not of essence, it could not be presumed that delay was of no consequence. The Apex Court held:

> Once it is held that even in absence of specific evidence, the respondent could be held to have suffered loss on account of breach of contract, and it is entitled to compensation to the extent of loss suffered, it is for the appellant to show that stipulated damages are by way of penalty. In a given case, when highest limit is stipulated instead of a fixed sum, in absence of evidence of loss, part of it can be held to be reasonable, compensation and the remaining by way of penalty. The party complaining of breach can certainly be allowed reasonable compensation out of the said amount if not the entire amount. If the entire amount stipulated is genuine pre-estimate of loss, the actual loss need not be proved. Burden to prove that no loss was likely to be suffered is on party committing breach, as already observed.[14]

---

[11] Lindberg v. Brandt et al 112, NE.2d 746, Appellate Court of Illinois, 1953.

[12] Macbrite Engineers v. Tamil Nadu sugar Corporation Ltd., 2002(3) Arb. LR 368 (Madras); AIR 2002 Mad. 429.

[13] Pranab Kr. Saha v. F.C.I. and Others, 2002(3) Arb. LR 73 (Gauhati).

[14] Construction & Design Services v. Delhi Development Authority, AIR 2015 SC 1282; 2015(1) Recent Apex Court Judgments 504.

In absence of any evidence by the party committing the breach that no loss was suffered by the party complaining of breach, the court has to proceed on guess work as to the quantum of compensation to be allowed in the given circumstances.

If there is no loss the compensation can also be nil. The Bombay High Court upheld an award denying compensation on this ground after discussing the case law in detail.[15] In another case granting leave to defend a summary suit, it was held:

> Applying the ratio of these precedents to the facts of the present case, it appears prima facie, that the claim of damages to the tune of Rs. 12,00,000/- in an agreement for purchase of a property valued at Rs. 30,00,000/- where the earnest money was only Rs. 3,00,000/-, in a climate where prices of real estate was on the decline, would be unreasonable, and therefore in the nature of a penalty, which would be hit by the embargo contained in S. 74 of the Contract Act. Unconditional leave to defend must therefore be granted.[16]

The above otherwise well-settled position in the law stands disturbed by the Supreme Court of India's decision in the ONGC v. Saw Pipes Ltd. case:[17]

In Fateh Chand v. Balkishan Dass,[18] the five Judge Bench of the Supreme Court of India, laid down the law that S.74 eliminates the distinction of penalty and liquidated damages. However, in the ONGC v. Saw Pipes case an attempt is made to reopen the distinction between penalty and liquidated damages. It was observed:

> If the compensation named in the contract is by way of penalty, consideration would be different and the party is only entitled to reasonable compensation for the loss suffered. But if the compensation named in the contract for such breach is genuine pre-estimate of loss which the parties knew when they made the contract to be likely to result from the breach of it, there is no question of proving such loss or such party is not required to lead evidence to prove actual loss suffered by him. Burden is on the other party to lead evidence for proving that no loss is likely to occur by such breach.

The arbitral tribunal in the said ONGC case, after considering the decisions rendered by the Supreme Court in the cases[19] of Fateh Chand, Maula Bux and Rampur Distillery (supra) concluded that it was for the respondents to establish that they had suffered any loss because of the breach committed by the claimant in the supply of goods under the contract and that as the appellant had failed to prove the loss suffered because of delay in supply of goods as set out in the contract between the parties, it was required to refund the amount deducted by way of liquidated damages from the specified amount payable to the respondent.

The award was set aside. It was observed:

> Section 74 emphasizes that in case of breach of contract, the party complaining of the breach is entitled to receive reasonable compensation whether or not actual loss is proved to have been caused by such breach.

---

[15] O.N.G.C. v. Macqreqor Navire, 2002 (2) Arb. LR 151 (Bombay). Also see: G.M.T.P.C.A. Ltd. v. Deputy Registrar of Co-operative Societies, AIR 1998 Karnataka 354; Electronic Enterprises v. Union of India, AIR 2000 Del.55; A. Murali and Co., Plaintiff v. State Trading Corporation of India Ltd, AIR 2001 Mad. 271.

[16] Roshan Lal, Plaintiff v. Manohar Lal, AIR 2000 Del. 31. Also see Harbans Lal v. Daulat Ram, AIR 2007 (NOC) 2470 (Del.).

[17] Oil and Natural Gas Corpn. Ltd. v. Saw Pipes Ltd., AIR 2003 SC 2629; also see: Jagson International Ltd. v. Oil and Natural Gas Corpn. Ltd., 2004 (1) Arb. LR 663 (Bombay).

[18] Fateh Chand v. Balkishan Dass, AIR 1963 SC 1405.

[19] Fateh Chand v. Balkishan Dass, AIR 1963 SC 1405; Maula Bux v. Union of India, AIR 1970 SC 1955 and Union of India v. Rampur Distillery and Chemical Co. Ltd., (SC) AIR 1973 SC 1098.

Section 74 is to be read along with Section 73 and, therefore,

> in every case of breach of contract, the person aggrieved by the breach is not required to prove actual loss or damage suffered by him before he can claim a decree. The Court is competent to award reasonable compensation in case of breach even if no actual damage is proved to have been suffered in consequences of the breach of a contract.[20]

The finding in the ONGC v. Saw Pipes case above is patently opposed to the well-established principle that if there is no loss, there is no compensation or at the most token compensation but not the sum named, it is respectfully submitted.

The three-judge bench of the Supreme Court has previously held in the Maula Bux case:[21]

> Where the Court is unable to assess the compensation, the sum named by the parties if it be regarded as a genuine preestimate may be taken into consideration as the measure of reasonable compensation, but not if the sum named is in the nature of a penalty. *Where loss in terms of money can be determined, the party claiming compensation must prove the loss suffered by him.*
>
> (Emphasis supplied)

The decision of the Supreme Court in the case of ONGC v. Saw Pipes Ltd. has been much criticized for increasing the scope of public policy and is at variance with the larger bench decisions on the earlier occasions in respect of interpretation of Section 74 and as such does not lay down good law, it is respectfully submitted.

The authors are in respectful agreement with the view expressed in a case by Hon'ble Mr. Justice K.S. Radhakrishnan, of the Supreme Court of India, wherein it was observed:[22]

> 23. I must immediately clarify that it would require a Bench larger than a Five-Judge Bench to alter the legal position from what has been enunciated in Chunilal V. Mehta and Fateh Chand. The decisions of smaller Benches are relevant only for the purpose of analysing the verdict in a particular case on the predication of the elucidation of the law laid down by the Constitution Benches. This would include an oft-quoted decision in Maula Bux v. Union of India, 1969(2) SCC 554, as well as Union of India v. Raman Iron Foundry, 1974(2) SCC 231, and BSNL v. Reliance Communication Ltd., 2011(1) SCC 394, etc.

7.4 Stipulations Made in Construction Contracts

Standard forms of contracts referred to hereinabove, in general (and in cases of public works in particular), stipulate a sum, such as one per cent or half per cent of the contract price per day (or per week), payable as damages by the contractor for the period the work remains unfinished after the stipulated or extended date. Almost all standard forms limit the liability of the contractor to pay liquidated damages from 5% to 10% of the contract sum. These stipulations are open to criticism. The main objections that can be raised against them are considered below:

A fixed sum when named as liquidated damages in a contract, should be a genuine estimate of the likely damages. The loss that the owner of a proposed construction work will suffer, due to delay in completion of work, will obviously depend upon factors like type of work, purpose of construction, etc. Even where two works of the same type are concerned the loss would not be the same but would depend upon other considerations. Contract price and the time limit are not the

---

[20] Maula Bux v. Union of India, AIR 1970 SC 1955; 1969(2) SCC 554.
[21] Maula Bux v. Union of India, AIR 1970 SC 1955; 1969(2) SCC 554.
[22] Maya Devi v. Lalta Prasad a high: 2015(5) SCC 588: 2014(2) SCR 1129: 2015(3) SCC (Civil) 168: 2014(2) Recent Apex Judgments (R.A.J.) 350: 2014 AIR (SC) 1356.

only criteria. Thus, stipulating a fixed percentage of estimated cost for all types of works under different circumstances is not justified. Although it is true that the fixed sum named is only the maximum amount and the actual amount may vary and would be fixed by the competent authority; the amount so fixed would have to be reasonable. If it is not, the contractor may take the matter to the court. In this dispute, as already seen, the court would award reasonable damages.[23] Thus, ultimately, what the owner would get will be reasonable compensation. There is no benefit to the owner of naming a sum higher than the genuine probable loss. If time limit within which the proposed work is to be completed is, in the opinion of the contractor who is willing to undertake the work, inadequate, he would provide for damages likely to be payable. It would not be illogical to consider that the contractors would increase the tendered sums or prices so as to account for damages at the highest rate named in the contract. It may so happen that during the discharge of the contract there would be no delay. Even if there is delay in completion, the contractor would pay reasonable compensation and not the sum named. The owner is thus put to loss for he has to pay the contractor at the enhanced contract price or rates. On the other hand, such a provision, in the opinion of the authors, puts the employer in a disadvantageous position if the actual loss is more than the sum named because the compensation cannot exceed the sum named.

EXAMPLES FROM CASE LAW

1. It was the term of the agreement that the difference in quantity of cement actually issued to the contractor and the theoretical quantity including the authorized variation, if not returned by the contractor, shall be recovered at twice the issue rate without prejudice to the provision of the relevant conditions regarding return of materials governing the contract. So, relying on this clause, it was contended that the department rightly recovered the penal rate for the excessive use of material. On the other hand, in defence it was contended that the difference between the actual consumption and theoretical had been gone into by the arbitrator. He found that the actual loss had not been proved. Hence, relying on Section 74 of the Indian Contract Act he concluded that the respondent, since he could not prove the loss, was not entitled to recover any amount on this account. The court held that the objection to the award was not sustainable.[24]

2. The defendant in a suit had placed orders on 20th February 1985 and 28th February 1985 for supply of five machines with the plaintiff, subject to the terms and conditions that if supply was not effected on or before 15th March 1985, a penalty of 1% on the total value of the supply would be imposed for every three days of delay subject to the maximum of 10%. The machines were supplied between 29th August 1985 and 23rd January 1986. The defendant settled the bills after deducting the penalty amount of 10%. Hence, the suit was filed. On behalf of the appellant/plaintiff it was argued that even if the plaintiff had supplied the machines earlier, the defendant could not have used them as there was no electricity connection in their factory. The trial court did not accept the defence. On appeal it was pointed out that admittedly electricity connection was given to the defendant's factory premises only on 5th June 1986 whereas the machines were supplied in the year 1985 itself and no evidence was led in by the defendant to show the actual loss sustained by them. It was held:

> The plaintiff failed to prove the loss suffered by him in consequence of the breach of the contract committed by the defendant and we are unable to find any principle on

---

[23] Himachal Pradesh State Electricity Board v. M/s Ansal Properties, AIR 2008 (NOC) 2773 (H.P.).

[24] M/s. Jagan Nath Ashok Kumar v. Delhi Development Authority, AIR 1995 Del. 87.

which compensation equal to ten per cent to the agreed price could be awarded to the plaintiff. In the absence therefore of any proof of damage arising from the breach of the contract, we are of opinion that the amount of Rs. 1,000/- (earnest money) which has been forfeited, and the advantage that the plaintiff must have derived from the possession of the remaining sum of Rs. 24,000/- during all this period would be sufficient compensation to him.[25]

7.5 Stipulation by Way of Fixed Sum – If Justified?

The idea of stipulating a fixed sum as damages seems to have its origin in the common law doctrine of damages. As already mentioned, liquidated damages are payable in cases of contracts of uncertain value, where it is difficult to assess the damage and decide the amount of compensation in advance. However, most construction contracts are neither of uncertain value nor is it very difficult to ascertain the probable loss due to delay, barring certain exceptions. It will be wrong to assume that courts would ordinarily award total damages agreed to between the parties. As such the stipulation by way of fixed sum is not justified.

However, there is one reason which justifies stipulating the fixed sum. The injured party can claim damages for the breach of contract under the provisions of Section 73 of the Indian Contract Act. But in that event, for a successful claim, the injured party has to prove that: (i) the loss or damage has been caused, (ii) the loss or damage arose in the normal course of things from such breach, or (iii) which the parties knew when they made the contract to result from the breach of it. Also such compensation is not payable for any remote or indirect loss or damage sustained by reason of such breach. Thus in order to ensure that the contractor does his best to complete the work within stipulated or extended time, it is better to draft the clause so that Section 74 of the Indian Contract Act is applicable, rather than Section 73. The stipulation by way of penalty or liquidated damages is, therefore, justified. But to name a fixed sum arbitrarily and without regard to the probable loss is not justified. A better way to handle this problem of damages is suggested below.

7.6 Alternative Method Suggested

Whenever possible, the liquidated damages to be stated in the form of a specific number of rupees per week, should be the owner's genuine estimate of what he will lose in income by reason of the delay in completion. When it is not possible to ascertain correctly the probable future loss, the contract may incorporate the provision for payment of "reasonable damages".

The stipulation of penalty or liquidated damages suggested above is likely to be criticized as an "uncertain provision". In other words, it may be said that the contractors would not know what exactly they are agreeing to. In the opinion of the authors, the provision would be no more uncertain than a provision which states "one per cent of contract price or such smaller amount as may be fixed". It may be wrong to think that the parties would have to take a recourse to legal action for deciding as to what is a "reasonable sum". It costs both parties if a dispute is to be referred to the court. Therefore, if the reasonable sum is worked out on sound principles suggested below, there should be no difficulty in parties agreeing to it. The authority that is to determine the amount of compensation will have naturally to account for the following:

---

[25] M/s. Macbrite Engineers v. Tamil Nadu Sugar Corporation Ltd., AIR 2002 Mad. 429.

i. Estimated losses to the owner because he could not utilize the property on the time scale he had anticipated

ii. Interest charges which the owner must pay during delay and

iii. Cost and expenses of litigation.

If the amount of compensation is so decided, the court would have no reason to reduce it and the contractor would give as much consideration before he decides to challenge it.

7.7 Liquidated Damages Clause – When Can Be Evoked?

To attract the provisions of Section 74 it is not necessary that the entire contract should come to an end; breach of each term thereof can be visualized in advance and taken care of by providing an adequate clause for liquidated damages so that the parties to the contract can proceed to work out the contract in future and settle the question of damages that have accrued on the basis of the rate that has been put as a pre-estimate at the commencement of the contract. Some construction contracts provide for completion of milestones and liquidated damages for delay in completion of each milestone with the provision that if the project is completed within the agreed time the damages recovered would be refunded. Even where such provision is not there, if the very purpose of putting a genuine pre-estimate is to avoid litigation and introduce certainty in computation of the difficult question of assessment of damages, it seems to be unlikely in the highest degree that the intention would have been that the clause of liquidated damages will not come into operation until the entire contract has been broken.[26]

The petitioner in a case claimed for award of liquidated damages before the arbitral tribunal. The question was whether the period of limitation for raising the counter-claim on account of liquidated damages was to be reckoned from the date of filing of the statement of claims before the arbitrator or three years prior to invocation of the arbitration clause. It was held that the defendant had invoked the arbitration clause on 17th December 2012 and, therefore, the petitioner could only raise claims that stem from a cause of action arising within a period of three years prior to the said date.[27]

EXAMPLES FROM CASE LAW

1. **State of Rajasthan v. Chandra Mohan**

In January 1955, "C" entered into an agreement with the state, undertaking the construction of an aqueduct. According to the terms of the agreement the work was to be completed by "C" on or before 9th January 1956. There were heavy monsoons in Rajasthan in the year 1955, which caused damage and delayed the completion of the work. "C" continued the work and applied for an extension of time. The extension was not formally granted, but he was allowed to continue the work. The work was completed on 8th March 1957. The final bill was also prepared, but it was not paid. "C" claimed a refund of the security deposit and the amount of the final bill. The PWD (Irrigation), though it accepted the amount, claimed to deduct a penalty of Rs. 15,000/- alleged to have been imposed by the Chambal Control Board. After making the deduction, the balance amount was paid to "C". The representation was made by "C" to the department, but having failed, the suit was instituted for the recovery of the balance amount, and interest thereon. After having considered the evidence

---

[26] Mohanlal v. B.G. Deshmukh, AIR 1985 Bom. 188.

[27] Pawan Hans Helicopters Limited v. Ideb Projects Private Limited, 2017(9) Apex Decisions (Delhi) 317.

on record, the learned District Judge held that the penalty of Rs. 15,000/- was unreasonable. He held that 1% of the estimated cost of the suit work which comes to Rs. 3,750/- will be a reasonable compensation. It may be noted that the chief engineer had come in to the witness box. He had clearly and categorically stated that on account of delayed completion of the work, there occasioned, in fact, no loss to the government. Aggrieved by the judgment the state filed an appeal, "C" cross-objected. Dismissing the appeal filed by the state and allowing the claim the Rajasthan High Court held:[28]

1. In view of the statement of the Chief Engineer (Irrigation) himself, it was difficult for the court to hold that the department sustained any loss on account of the breach of the contract in not having completed the work within the stipulated time.

2. The penalty was imposed by the Chambal Control Board on 9th August 1958.

According to the terms of the contract, the Chambal Control Board had no authority to fix any penalty in terms of the contract. It was the Chief Engineer (Irrigation) who alone was authorised to determine the compensation for a breach of the contract and, as a matter of fact, it should have been done soon after 9th November 1956 when the plaintiff was not able to complete it. Clause 2 of the agreement provides that the compensation shall be deducted from time to time as the delay would occur in the progress of the work.

It was finally held that the plaintiff was not at all liable to pay the compensation.

2. A contract for supply of 5,500,000 bricks involving average lead of 11 miles within six months was awarded to a contractor who did not possess the requisite number of trucks to complete the job. The correspondence showed that the matter came to such a pass that it was not possible for the contractor to complete the job without incurring a heavy loss. At the earliest opportunity he stopped the work and virtually compelled the government to annul the contract so that some other contractor might complete the job. At this stage it was practically impossible to assess the actual amount of damages suffered by the government in view of the nature of the work involved in the contract. If a contract of this nature is annulled before completion it is not a question of finding out the difference between the contract rate and the market rate of the materials to be supplied. There could be no ready market for the same. It was held[29] "that the amount mentioned in Clause 2, of the conditions of the contract it was not assessed by way of penalty but was a genuine pre-estimate of the damages in the facts and circumstances of this case".

3. In a case involving a failure to supply sugar cane for which the contract had provided for a Rs. 50/- per tonne compensation, the Madras High Court held that if there is no evidence to show that the amount fixed is unconscionable, the full sum is payable and the party claiming compensation need not prove that he has actually suffered any loss or damage.[30]

4. Where the contractor in a case failed to complete the work in the extended period and the contract was terminated, the award of compensation payable by the contractor was upheld by P. & H. High Court.[31]

---

[28] State of Rajasthan v. Chandra Mohan, AIR 1971 Raj.229; N.C.T. of Delhi v. R.K. Construction 2003(1) Arb. LR 465 (Delhi).

[29] Anand Construction Works v. State of Bihar, AIR 1973 Cal. 550.

[30] T. K. Sundaram v. Co-operative Sugar Ltd. Chittor, AIR 1988 Mad. 167.

[31] Yashpal Dahiya, Contractor v. State of Haryana, AIR 2009 (NOC) 1528(P. & H.).

7.8 Parties Free to Provide for Liquidated Damages to Cover Some Breaches and Not All

There is neither any impediment nor any obstacle for the parties to a contract to make provision of liquidated damages for specific breaches only, leaving other types of breaches to be dealt with as unliquidated damages. There is no principle of law that once the provision of liquidated damages has been made in the contract, in the event of breach by one of the parties, such clause has to be read covering all types of breaches although parties may not have intended and provided for compensation in express terms for all types of breaches. For example, if a clause in a contract provided for compensation to the respondent for failure to supply or delayed supply of the materials, the said clause cannot be deemed to contemplate refusal to supply materials. It is not a question of giving restrictive or wider meaning to a damages clause, but the question is what is intended by the parties by making such a provision and does such a clause cover all situations of breaches. If the answer is in the affirmative, obviously compensation cannot be awarded beyond what is provided therein. On the other hand, if breaches are not covered by the said clause, provision therein with regard to liquidated damages will not be applicable at all.

In construction contracts, the liquidated damages clause is generally attracted in cases of delay in performance and, where there is total failure or termination of the contract by either party prior to full performance, the actual loss resulting from such termination deserves to be compensated and it cannot be limited to the liquidated damages for delayed completion.

7.9 Extension of Time and Penalty for Delay

Invariably construction contracts include provisions for levy of penalty or recovery of liquidated damages on the one hand, and extension of time for completion on the other hand. The PWD forms include such provisions in Clauses 2 and 5 of the B-l & B-2 forms. CPWD, like other departments such as the MES and the Railways also incorporate similar provisions. The FIDIC Form includes such Clauses at Sr. No. 44.1 to 44.3 and 47.1 and 47.2. One improvement noticed in the FIDIC Form is the provision for reduction of the rate of compensation if part of the work has been taken over by the employer, on a pro rata basis. However, the total compensation is not reduced.

The importance of extension of time provisions vis-à-vis the recovery of liquidated damages cannot be ignored. The extension of time clause is of benefit to both the contractor and to the owner. Its benefit to the contractor is that in the event of delay by bad weather or strikes etc. for which he is not responsible, his liability to pay liquidated damages gets reduced to the extent that delay is caused by such factors. This clause is also of substantial benefit to the owner, since in the absence of an applicable cause of this kind, the liquidated damages provisions will cease to have effect where even a part of the delay is due to some act or default on the part of the employer or his agents or any other matter for which he would he held responsible.

Generally the terms of the contract confer power on the engineer to determine the length of the extension of time. Some contracts make his decision final and binding. In the latter capacity his role is quasi-arbitral in character. The two provisions of extension of time and recovery of liquidated damages often become a great source for a number of litigations. In almost every project there will be some reasons for justifying extension of time. The reasons could be the factors "beyond the control of the contractor" or breaches of contract conditions committed by the owner or his engineer/architect. The quantum of extension of time granted is invariably considered inadequate and disputed by the contractor. At times there is considerable delay in granting an extension of time. Extra work, excessive variation in quantities and, in particular, the time of ordering them, add to difficulties in assessing the extension. There can be delays caused by the contractor. The question is how to assess the quantum of liquidated damages recoverable from the contractor. Or, can it be said that the time of completion is left open and if so, under what circumstances? This question assumes importance because if the time of completion is left open, the owner cannot recover penalty or liquidated damages. There is no precise answer to the difficult question inasmuch as all the circumstances may have to be taken into account and the matter decided on a case-by-case basis. Where there is delay in completion of a work and it is not wholly attributable to the contractor, compensation determined without giving the contractor an opportunity of hearing is violative of rules of natural justice.[32]

The following illustrations, however, can serve as guides in helping one to reach a correct answer in a given case.

EXAMPLES FROM CASE LAW

1. Where the work is delayed due to non-availability of materials and delay in supply of drawings, it was held by the Gauhati High Court:[33]

> Under Cl. 2 of the contract, penalty could be imposed in the form of compensation for delay if the work was "not done by the contractor with due diligence" which is to be read with the other relevant provision, Cl. (5), which provided for extension of time. It appears from the evidence that the plaintiff had from time to time brought to the notice of the defendants the causes due to which the execution of the contract was being delayed and on 20–6–66 he also wrote to the defendants for extending the time of completion of the work. This request for extension was not rejected and therefore, the defendants were not justified in imposing the penalty for delayed execution of the work.

2. T contracted to execute a road construction work for F; the work to be commenced 14 days after the date of work order and to be completed within 15 weeks, that is on or before 30th December 1975. The agreement incorporated provision for the recovery of liquidated damages at $500 a week payable by T to F for delay. The agreement further provided for time to be extended for, inter alia, "delays caused by other contractors on the site, delays in installation of services or exceptional circumstances". T applied for extension of time on account of additional work and delays by other contractors. The engineer granted an extension up to 1st June 1976. The delays by other contractors caused the summer job go into winter, resulting in further delays on account of exceptionally bad weather. In the process the completion got delayed by 46 weeks. The question to be decided was whether F's breach relieved T of liability under the liquidated damages clause. The authority of Perini was relied upon. In the said case,[34] the British Columbia Court of Appeal held that where a building contract contains a clause providing that the owner may extend the time for completion upon application by the builder, by reason of extras or delays occasioned by strikes, lockouts, *force majeure* "or other causes beyond the control of the contractor", the concluding words must be construed narrowly and with reference to the preceding specific causes of delay and so as not to include defaults of the owner which would unreasonably result in making him the judge of the extent of his own default. Accordingly, the court held that in view of the default under the implied terms of the contract, the employer having delivered defective machinery for incorporation

---

[32] Mahanadi Coalfields Ltd. v. M/s. Dhansar Engineering Co. Pvt. Ltd., 2016(9) Scale 339: 2016(4) R.C.R.(Civil) 603: 2016(5) Recent Apex Judgments (R.A.J.) 570: 2016(9) JT 385: 2016 AIR (SCW) 4509. 2016(4) J.C.R. 229: 2016 AIR (SC) 4509; M/s Mountain Movers v. State of H.P.& Ors., AIR 2008 (NOC) 2007 (H.P.).

[33] O. & N.G. Commission Nazira v. M/s S.S. Agarwalla & Co., AIR 1984, Gau. 11.

[34] Perini Pacific Ltd. v. Greater Vancouver Sewerage and Drainage District, [1967] SCR 189.

in the plant being erected by the contractor, which caused delay in completion, the contractor was thereby left on his own to decide the date of completion and was released of his liability to pay liquidated damages even though the contractor's own tardiness may have caused some of the delay. The case was ultimately decided on the authority of Perini and it was held that there was no jurisdiction in the engineer to extend time on account of F's breach and that that breach set the completion date open with the effect that T was not liable to pay liquidated damages.[35]

The law seems to be well settled by a series of authorities[36] that where the reason for the contractor's failure to complete on time is wholly or partly due to the fault of the owner, the owner cannot recover liquidated damages unless the contract provides otherwise. If, for example, the owner instructed the main contractor to enter into a subcontract with a company that will be responsible for design of the project and the main contractor's scope of work excluded design, the owner will not be able to recover liquidated damages for delays resulting from issues relating to the design from the main contractor.[37] The provision of liquidated damages is unenforceable if expressed in a manner which is inconsistent with the other clauses to which liquidated damages relate.

EXAMPLE FROM CASE LAW

A contract for construction of 123 dwellings stipulated the date for completion as 6th December 1976. The agreement provided for recovery of "Liquidated and Ascertained Damages" at the rate of £20 per week for each incomplete dwelling. The architect granted various extensions and gave an extended completion date of 4th May 1977. Possession of dwellings was taken over by the owner as and when completed. The last houses were not taken over until 29th November 1977. However, between 4th May and 29th November 1977 the architect issued various instructions to the contractors. The owner retained £26,150 as liquidated damages but did not claim damages for the period 4th May to 29th November 1977. The contractor disputed both the extension of time granted and the owner's right to deduct liquidated damages. An arbitrator issued an interim award on both the points and the matter came before the court. The court had to decide the main contention that since the agreement contained no express provision for sectional completion, were the liquidated and ascertained damages provisions rendered unenforceable? It was held that there was no provision for sectional completion in the Articles of Agreement, and as such the contractual provisions of the contract for liquidated damages, were unenforceable. However, it was observed that the finding of the court "does not in any event prevent" the owner from claiming damages for breach of the contract.[38]

Provisions of standard forms like FIDIC stipulate that the contractor shall pay the relevant sum stated in the Appendix to Tender as liquidated damages. If the sum stipulated therein is nil, it

---

[35] Fembrook Trading Co. Ltd. v. Taggart, Supreme Court of New Zealand, [1979] I NZLR 556.

[36] Wells v. Army & Navy Co-operative Society Ltd., K.B. Div. (1902) 86 LT 764; Miller v. London County Council, K. B. Division (1934) 151 LT 425; Peak Construction (Liverpool) Ltd. v. McKenney Foundations, Court of Appeal (1970) I BLR III; Percy Bilton Ltd. v. Greater London Council, House of Lords (1982) 20 BLR; Forbes Gokak Ltd. v. Central Ware Housing Corporation, 2003 (1) Arb. LR 279 (Delhi).

[37] Sinclair v. Woods of Winchester Ltd., [2006] EWHC 3003 (TCC). Also see, Braes of Doune Wind Farm (Scotland) Ltd. v. Alfred McAlpine Business Services Ltd., [2008] EWHC 426 (TCC).

[38] Bramall & Ogden Ltd. v. Sheffield City Council, QB Div (1985) I Con LR 30.

constitutes an exhaustive agreement as to damages payable or not payable by the contractor in the event of his failure to complete the work on time. The amount nil does not permit a claim for damages in general.[39]

7.9.1 Questions of Importance

Two questions arise for determining, which are of considerable importance: The first is to what extent the decision of the engineer is binding, if at all, on payment of liquidated damages. The answer to the first question is that, subject to scrutiny by the court or arbitrator (if the decision is referable to the arbitrator), the decision is final. However, most modern forms exclude finality to certification or deduction of liquidated damages by the engineer. For example, Clause 20 of FIDIC conditions 1999 includes the dispute in respect of extension of time referable to the Dispute Adjudication Board or arbitration. As far as the contractor is concerned, he can also challenge the deduction of liquidated damages in arbitration or a court.

In most of the cases where the engineer is not specifically empowered to take the liquidated damages into account in his certificate for payment, the absence of any reference to them in the final certificate obviously cannot affect the owner's right to deduct liquidated damages. However, an event which could qualify as an act of God might suspend or reduce the damages.[40]

The second question needs consideration of the provisions in the contract and the case, decided by the Supreme Court of India. The contract stipulations are as follows:

Clause 2 of the General Conditions of Contracts of the CPWD and the state PWD empowers the superintending engineer to decide the rate of compensation and/or penalty for delay in completion of the work by the contractor and further makes his decision final and binding. Clause 5 or 6 of the printed forms provides for suitable extension of time to be granted to the contractor for appropriate reasons shown to the satisfaction of the divisional or executive engineer, who generally is the engineer-in-charge of the work.

Strictly speaking under Clause 2 there is no role to be played by the engineer-in-charge and, conversely, under Clause 5 or 6 by the superintending engineer. The two officers are likely to decide differently on the same factual matrix. The engineer-in-charge, on the one hand, may grant extension of time on the grounds on which the contractor has submitted an application and the superintending engineer, on the other hand, may impose penalty for the same extended period. The resultant situation did not pose any problem in the field prior to the decision of the Supreme Court in Vishwanath Sood v. Union of India.[41]

The Supreme Court in the case declared that the decision of the superintending engineer is excluded from the scope of the arbitration clause.[42] This, with great respect, it is submitted, is an erroneous decision based on certain presumptions which themselves can be proved wrong or will be proved wrong by improper exercise of authority by the engineer-officers. This case has been fully studied and discussed in the author's title *The Law of Arbitration* (second edition) by Sarita Patil and the authors are in respectful agreement with the views expressed therein.[43] It is hoped that in an appropriate case, the Supreme Court will reconsider the above decision and at least clarify or restrict its scope if not holding clearly that it was not correctly decided. Until then the courts would have to find ways and means to obviate the likely injustice, an excellent example of which is to be found in the decision of the Andhra Pradesh High Court in the case dealt with in Chapter 14.[44]

---

[39] Temloc Ltd. v. Errill Properties Ltd., Court of Appeal, (1987) 12 Con LR 10 15A.

[40] Ryde v. Bushell, (1967) EA 817 (Bast Africa).

[41] Vishwanath Sood v. Union of India, AIR 1989 SC 952.

[42] Vishwanath Sood v. Union of India, AIR 1989 SC 952.

[43] B.S. Patil on *The Law of Arbitration*, revised and enlarged second edition 1992 by Sarita Patil, pp. 66–70.

[44] Yelluru Mohan Reddy v. Rashtriya Ispathnigam Ltd., Vishakhapatanam, AIR 1992 A.P. 81.

The Supreme Court of India observed that a provision in an agreement

> that quantification of the liquidated damages shall be final and cannot be challenged by the supplier is clearly in restraint of legal proceedings under S. 28 of the Indian Contract Act. So the provision to this effect has to be held bad.[45]

Reliance was placed on the earlier decision in State of Karnataka v. Shree Rameshwara Rice Mills wherein it was observed:[46]

> Even assuming for argument's sake that the terms of clause 12 afford scope for being construed as empowering the officer of the State to decide upon the question of breach as well as assess the quantum of damages, we do not think that adjudication by the Officer regarding the breach of the contract can be sustained under law because a party to the agreement cannot be an arbiter in his own cause. Interests of justice and equity require that where a party to a contract disputes the committing of any breach of conditions, the adjudication should be by an independent person or body and not by the other party to the contract. The position will, however, be different where there is no dispute or there is consensus between the contracting parties regarding the breach of conditions.

7.10 Liquidated Damages Provision – A Mere Right to Sue

Any stipulation for unliquidated damages in a contract gives rise only to a right to sue on default and no right to the amount. That right to sue will furnish a cause of action for the suit. This position has been well settled.[47] In Union of India v. Raman Iron Foundry,[48] the observations of the Bombay High Court were cited with approval, it is submitted. In the said case, Chagla C.J., as he then was, observed:

> In my opinion it would not be true to say that a person who commits a breach of the contract incurs any pecuniary liability, nor would it be true to say that the other party to the contract who complains of the breach has any amount due to him from the other party .... As already stated, the only right which he has is the right to go to a Court of law and recover damages. Now, damages are the compensation which a Court of law gives to a party for the injury which he has sustained. But, and this is most important to note, he does not get damages or compensation by reason of any existing obligation on the part of the person who has committed the breach. He gets compensation as a result of the fiat of the Court. Therefore, no pecuniary liability arises till the Court has determined that the party complaining of the breach is entitled to damages. Therefore, when damages are assessed, it would not be true to say that what the Court is doing is ascertaining a pecuniary liability which already existed. The Court in the first place must decide that the defendant is liable and then it proceeds to assess what that liability is. But till that determination there is no liability at all upon the defendant.

The Supreme Court fully concurred and further observed:

> The claim is admittedly one for damages under Clause 14, but so far as the law in India is concerned there is no qualitative difference in the nature of the claim whether it be for liquidated damages or for unliquidated damages. ... It therefore makes no difference in the present case that the claim of the appellant is for liquidated damages. It stands on the same

---

[45] Bharat Sanchar Nigam Ltd. v. Motorola India Pvt. Ltd., AIR 2009 SC 357.

[46] State of Karnataka v. Shree Rameshwara Rice Mills, AIR 1987 SC 1359.

[47] State v. M/s M.K. Patel & Co., AIR 1985 GUI 179; Union of India v. Raman Iron Foundry, AIR 1974 SC 1265.

[48] Union of India v. Raman Iron Foundry, AIR 1974 SC 1265.

footing as a claim for unliquidated damages. Now the law is well settled that a claim for unliquidated damages does not give rise to a debt until the liability is adjudicated and damages assessed by a decree or order of a Court or other adjudicatory authority.[49]

EXAMPLE FROM CASE LAW

A contract with the State of Karnataka included a clause which made the purchaser thereunder liable to pay damages to the state as may be assessed by the state in addition to the forfeiture of the security deposit, in case of breach of any of the conditions of the contract. By reference to the words "for any breach of conditions set forth hereinbefore", the Supreme Court observed:[50]

> On a plain reading of the words it is clear that the right of the second party to assess damages would arise only if the breach of conditions is admitted or if no issue is made of it. If it was the intention of the parties that the officer acting on behalf of the State was also entitled to adjudicate upon a dispute regarding the breach of conditions the wording of clause 12 would have been entirely different. It cannot also be argued that a right to adjudicate upon an issue relating to a breach of conditions of the contract would flow from or is inhered in the right conferred to assess the damages arising from a breach of conditions.

It was, therefore, held:

> the powers of the State under an agreement entered into by it with a private person providing for assessment of damages for breach of conditions and recovery of the damages will stand confined only to those cases where the breach of conditions is admitted or it is not disputed.

7.11 Release from Liquidated Damages

It is in the essence of liquidated damages and penalties that they shall run as from a fixed date. It is, therefore, essential to specify the time limit within which the work is to be completed. It is not necessary to state "time shall be of the essence of the contract". Even if so stated it shall cease to be of the essence if provision of liquidated damages is incorporated.

The owner may lose his right to liquidated damages under the following circumstances:

1. The provision for extension of time is made in the contract but extension of time is not granted in spite of just reasons.[51] Most contracts call for a written notice from the contractor communicating to the owner about the happening of an event causing delay within the stated time limit. It must be noted that this is a condition precedent to the consideration of the contractor's request for an extension of time, and should be scrupulously followed by him.

---

[49] AIR 1974 SC 1265 (1272–73).

[50] State of Karnataka v. Rameshwara Rice Mills, AIR 1987 SC 1359 (1316).

[51] Westwood v. The Secretary of State for India, (1963) 7, L.T. 736.23. See paragraph 7.8 in this chapter and examples under it.

2. When any act of the owner or the engineer/architect, not covered by the terms of the contract, renders it impossible for the contractor to complete the work on schedule, he is relieved of his obligation to pay damages.[52]

3. If the architect/engineer gives a certificate of satisfactory completion and certifies the balance due to the contractor, such a certificate is likely to be held as conclusive and the owner cannot deduct liquidated damages.[53] However, such a certificate must have been given by the architect/engineer at his own discretion and without undue influence of one of the parties.

4. The penalty clause for delay in completion applies only to completion by the contractor himself. If according to provisions of the contract, the work is taken out of the hands of the contractor, the clause ceases to apply and the original contractor is freed from its liabilities.[54]

7.11.1 Effect of Payment without Deduction

The right to recover liquidated damages will not be lost, unless special provision is made in the contract by paying the contractor moneys otherwise due to him, or by permitting the completion of the work after the due date for its completion. However, if provision for deduction from money due is made, it may be construed to be mandatory and exclusive, and in such an event failure to make the deduction may disentitle the employer from recovering the damages. Under some contracts, all payments of interim invoices are made on a without prejudice basis. If this is the case, it would take something in the parties' conduct beyond mere approval and payment of interim invoices to show estoppel against the owner.[55]

EXAMPLE FROM CASE LAW

A contract contained a stipulation that in the event of delay the contractor undertook to forfeit and pay to the owner £20 a week, to be paid and retained by the employer for each and every week during which such work shall remain unfinished. Interim payments were released for 14 months after the completion date without retaining any sum for liquidated damages. On these facts it was held that by conduct the employers had disentitled themselves from recovering liquidated damages.[56]

7.11.2 Owner's Obligations

It would be worthwhile to consider what happens when the contractor quits the unfinished job after the time limit has expired. It would not be incorrect to assume that liquidated damages would be recoverable from the time fixed for completion (originally agreed or extended) until the work was abandoned, but thereafter the owner will be entitled only to unliquidated damages.[57] The owner in such a case is, however, under the obligation not to increase the amount of damages by:

---

[52] State of A.P. and Others v. Singam Setty Yellanda, AIR 2003 A. P. 182; 2003(2) Arb. LR 92 (AP); N.C.T. of Delhi v. R.K. Construction, 2003(1) Arb. LR 465 (Delhi); also see paragraph 7.9 in this chapter and examples under it.

[53] See *Hudson's Building and Engineering Contracts*, 10th ed., p. 642.

[54] Re Yeadon Waterworks Co. & Wright (1895) 72 LT 538.

[55] HSM Offshore BV v. Aker Offshore Partner Ltd., [2017] EWHC 2979 (TCC).

[56] Laidlaw v. Hastings Pier Co., (1874). *Hudson's Building and Engineering Contracts*, 10th ed., p. 428, etc.

[57] See *Hudson's Building and Engineering Contracts*, 10th ed., p. 633.

1. Unreasonable delay in taking over the job, and/or
2. Failure to complete it diligently.

7.12 Right to Liquidated Damages Excludes Right to Unliquidated Damages

By providing for compensation in express terms under Section 74 of the Indian Contract Act, the right to claim unliquidated damages under Section 73 is necessarily excluded, and, therefore, in the face of that clause it is not open to the owner to contend that the right is left unaffected. In Chunilal V. Mehta & Sons Ltd. v. C.S. and M. Co. Ltd.,[58] it was held:

> When parties name a sum of money to be paid as liquidated damages they must be deemed to exclude the right to claim an unascertained sum of money as damages. ... Again the right to claim liquidated damages is enforceable under S. 74 of the Contract Act and where such a right is found to exist no question of ascertaining damages really arises. Where the parties have deliberately specified the amount of liquidated damages there can be no presumption that they, at the same time, intended to allow the party who has suffered by the breach to give a go-by to the sum specified and claim instead a sum of money which was not ascertained or ascertainable.

7.13 Effect of Termination/Forfeiture

It now remains to be seen if the above observations of the Supreme Court of India can apply to interpreting normal provisions in standard form contracts which provide for liquidated damages for delay in completion and forfeiture of security deposit and right to recover compensation for the extra expenditure incurred. For example, Clauses 2 and 3 of B-1 and B-2 forms of the CPWD form, or Clauses 63.1 to 63.3 of (1992 amended) FIDIC conditions of contract. The observations in the case above (AIR 1962 SC 1314) were made where the parties had provided for liquidated damages in the event of termination of the contract. As such it is doubtful that the said observations will be attracted to cases and contracts in which liquidated damages are provided for delay in completion and unliquidated damages for terminating or abandoning the contract or committing a fundamental breach by the contractor. It is submitted that the provisions of the agreement will be given effect to unless the termination is based on delay alone, in which event the above observations of the Supreme Court of India can be relied upon. If there is abandonment of the contract or any other breach mentioned in the agreement, there is usually a provision empowering the owner to enter the premises, seize the contractor's tools and plant, and execute the balance work either by himself or through another contractor. Under conditions of the contract in common use, including FIDIC, the owner's damages would probably arise under two main heads, namely (i) damages for delay, and (ii) additional cost, if any, for completion.

The provisions of liquidated damages in most contracts cover only the first of these two conditions. Normally the effect on this clause will be that the owner will be entitled to liquidated damages prior to the date of rescission or determination of the contract and thereafter he will be entitled to recover unliquidated damages only. However, if the contractor disputes the recovery of liquidated damages, then, in effect the entire question is to be adjudicated by a court or arbitrator as the case may be. This is, of course, subject to express stipulations otherwise made in the contract.

---

[58] Chunilal V. Mehta & Sons Ltd. v. C.S. and M. Co. Ltd., AIR 1962 SC 1314.

7.13.1 Forfeiture

A typical construction contract generally provides for payment of earnest money with the offer submitted by a tenderer and performance security deposit on signing of the contract. The provision further empowers the employer to forfeit the earnest money or the security deposit in case there is a breach of an obligation by the bidder/contractor. Such contract stipulations attract the provisions of Section 74 of the Indian Contract Act. A forfeiture clause is not to be construed strictly, but is to receive a fair construction. These are matters of contract between the parties, and should be construed as other contracts conditions.[59] For example, imposing of penalty on a contractor who did not complete work within the stipulated period without giving him an opportunity of hearing is liable to be set aside.[60] Similarly loss must be proved to justify forfeiture of either earnest money or security deposit. If there is a breach of contract of sale by a purchaser but no loss caused to the seller, the seller cannot claim damages from the purchaser for breach of contract. If damage or loss is not suffered, the law does not provide for a windfall.[61]

The forfeiture clauses are invariably linked with breach of some stipulation of bid invitation or term of a contract and as such are dealt with further details and illustrations in Chapter 12.

7.14 Bonus Provisions

Construction contracts sometimes contain provisions designed to give the contractor a financial incentive (bonus) to complete the contract before the date scheduled for completion of the works. Such provisions, when inserted, are to be found in association with the liquidated damages clause. To earn bonus under such a provision, the contractor must comply with the terms of the bonus clause. Invariably such provisions stipulate that in case the completion is delayed for any reason, the bonus will not be payable and if the contractor can prove that there was any hindrance by the owner or his agent, he is entitled to recover damages to the extent he incurred losses under the bonus clause.[62] Claim for payment of bonus *simpliciter* is not tenable but compensation for its loss can be claimed.

If the contract contains provision of a bonus for expedition, and a power to extend time, the intention of the parties as to whether the bonus should be calculated with reference to the new completion date, substituted by the operation of the extension of time clause, has to be ascertained from the provisions of the particular contract. No general rule can be laid down. In one case it was held that the extension of time clause only applied to save the contractors from liquidated damages, and not to give them additional bonuses.[63] In that case the contractors having received bonus for early completion, claimed further bonus in respect of time taken, as they alleged, by extra work.

7.15 Innovative Incentives in Road Maintenance/Improvements Contracts

In England or the USA, contracts dealing with highway improvements or maintenance sometimes use a clause providing for a site occupation or lane rental charge. The contractor includes, in his

---

[59] Doe D. Davis v. Elsam (1828) Moo. & M. 189 cited with approval in Suresh Kumar Wadhwa v. State of M.P., AIR 2017 SC 5435; 2018 All SCR 217.

[60] Mahanadi Coalfields Ltd. v. M/s. Dhansar Engineering Co. Pvt. Ltd., 2016(9) Scale 339: 2016(4) R.C.R.(Civil) 603: 2016(5) Recent Apex Judgments (R.A.J.) 570: 2016(9) JT 385: 2016 AIR (SCW) 4509; 2016 AIR (SC) 4509: 2016(3) Apex Court Judgments (SC) 483: 2016(167) AIC 84: 2016(10) SCC.

[61] M/s. Kailash Nath Associates v. Delhi Development Authority, 2015(1) Scale 230: 2015(1) JT 164: 2015(4) SCC 136: AIR (SCW) 759. Also see M/s. A.S. Motors Pvt. Ltd. v. Union of India, 2013(2) Recent Apex Judgments (R.A.J.) 218: 2013(3) JT 316: 2013 AIR (SCW) 3830: 2013(4) SCR 409: 2013 AIR SC (Civil) 1937.

[62] See *Hudson's Building and Engineering Contracts*, 10th ed., p. 577.

[63] Ware v. Lytterton Harbour Board, 1882 NZLR SC 191.

bid, the total amount he will pay for site occupation based on a daily/hourly rate set out in the contract. This rate can vary based on the time of week or day and peak or off-peak traffic conditions. If the contractor does not close the lane(s) of the road at peak hours or closes them for fewer hours than provided for in the tender, there is a saving of the charge and the clause operates as an incentive to him. If the contractor takes longer to complete the work, the charge works as liquidated damages for the delay that the contractor is responsible for.[64] This system could perhaps be used in contracts for utility companies that may need to close and dig roads, for example, for laying telecom cables.

---

[64] Cho A. Houston Contractor Nabs Hefty Performance Bonus. ENR: Engineering News-Record. 2007; 259(9): 22.

8

Quality of Work, Defects and Maintenance

8.0 Introduction

In a classic form of item rate contracts, the owner and his professional advisers such as architects, structural engineers, etc. assume responsibility for the design of the building or the structure as the case may be. The professional experts prepare detailed drawings and specifications of materials and workmanship, which are to be the basis for executing the work. The duty of the contractor is to carry out the work exactly in accordance with the drawings and specifications supplied to him. In this regard most contracts provide that if any work has been executed with unsound, imperfect or unskilled workmanship, or that any materials or articles provided by the contractor for the execution of the works are unsound or of a quality inferior to that contracted for, the contractor must rectify the work at his own expense. Often the contractor defends such an allegation by a counter-allegation of faulty designs or inadequate specifications provided by the employer. Litigation consuming loss of time and finance results in ascertaining the real cause of failure.

Where an employer desires to pass on to the contractor the responsibility of design and construction, and also to ensure a fixed price, recourse is taken to adopting a turnkey or entire contract on a lump sum basis called Engineering, Procurement and Construction or EPC for short. More recently even the responsibility of financing the project has been passed on to the contractor under Public Private Partnership (PPP) contracts on a Build-Operate-Transfer (BOT) basis. Design-Build-Finance-Operate-Transfer (DBFOT) is one of the latest forms whereunder the contractor does all the operations including maintenance during the concession period by recovering his investment and profit from the users of the facility.

In this chapter it is proposed to discuss the obligations of the parties associated with the execution of a contract work in relation to quality of materials, workmanship, defective work and maintenance. The discussion is with reference to the conditions usually incorporated in the standard form contracts.

8.1 Provisions in Standard Form Contracts

Every tenderer prior to submission of tender must read the provisions of the agreement to understand the duties and liabilities cast on him. Some such provisions relating to the subject matter of this chapter include:

Provisions in Standard Form Contracts	Paragraph No.
FIDIC Form:	8.6
1992 amended edition, Clause 49.2	8.1.1
1999 1st edition, Clauses 7.1–7.8, 9–11	8.1.2
2017 2nd edition, Clauses 7.1–7.8, 9–11	

(Continued)

(Cont.)

Provisions in Standard Form Contracts	Paragraph No.
Ministry of Statistics and Programme Implementation, Government of India (MOS & PI), Clauses 33–36	8.1.4
NITI Aayog Model Form, Clauses 11.1–11.17	8.1.3
PWD B-1 & B-2 Standard Forms of State Government, Clauses 6, 7, 11, 14, 17	8.1.5, 8.7
Military Engineering Services Form, Clauses 44, 46	8.1.5
Railway Engineering Department Form, Clauses 27, 50	8.1.5, 8.7

8.1.1 FIDIC Form 1999 Edition – Clauses: 7.1 to 7.8, 9 to 11

The 1999 edition of the FIDIC Form contains the provisions relevant to the subject matter of this chapter in Clauses 7, 9, 10 and 11 of the General Conditions of Contract. Clauses 7.1 to 7.6 are relevant to the present discussion.

Clause 7.1, under the heading Manner of Execution, stipulates that the contractor shall carry out the work in the manner specified by the contract, in a proper workmanlike and careful manner in accordance with recognized good practice and with properly equipped facilities and non-hazardous materials.

Clause 7.2 deals with "samples" of materials and relevant information to be submitted by the contractor for approval by the engineer.

Clause 7.3 empowers the employer's personnel to have full access at reasonable time to the site for inspection. It further provides that the contractor shall give notice of any work "ready for inspection" (RFI) before it is covered up. The engineer may inspect the work, carry out measurements or testing or may inform the contractor, without unreasonable delay, that he may not require to inspect the work. If the RFI notice is not given, the engineer has a right to direct the contractor to uncover the work for his inspection and reinstate and make good at the contractor's cost.

Clause 7.4 deals with all tests specified in the contract except test after completion. It expressly provides that if as a result of any delay on the part of the engineer to attend the tests (after having received 24 hours' notice to that effect), the contractor suffers delay or incurs any cost, the contractor is entitled to extension of time and payment of additional cost. To ensure this entitlement, the contractor must give notice to that effect to the engineer. Giving notice may be held to be a condition precedent.

Clauses 7.5 and 7.6 make the contractor liable to make good defects pointed out by the engineer and to ensure that the rejected item or work complies with the contract. The contractor's liability continues to remove and replace any plant or materials which are not in accordance with the contract and to remove and re-execute the defective work or any work urgently required for the safety of the works for any cause, notwithstanding any previous test or certification by the engineer.

Clause 9 deals with tests after completion. It provides for the contractor giving notice of 21 days or more indicating the date when he will be ready to carry out the test. If there is an undue delay in carrying out the tests by the contractor and/or failure by the contractor to carry out the tests, after receiving 21 days' notice from the engineer to that effect, the provision empowers the engineer to carry out the tests at the risk and cost of the contractor.

Clause 9.3 provides for retesting in case the works fail to pass the tests on completion.

Clause 9.4 stipulates consequences of any work's failure to pass the tests on completion, payment at reduced rates/cost or other remedies are provided.

Clause 11 deals with defects liability. Apart from making the contractor liable for rectification of the defective works at the cost of the contractor, it provides that if the defective work is attributable to any cause, for which the contractor is not responsible, he shall be notified and the provisions of the variation clause will be attracted. The employer is entitled to an extension of defects liability period on account of any works, section or item of plant that are not put to use for the purpose for which they are intended by reason of a defect or damage, which shall not exceed two years. The wording of this clause may entitle the employer to point out the defects noticed during the extended period, which is not to exceed two years, it is submitted.

Consequences of failure to remedy defects by the contractor are the subject matter of Clause 11.4 which contains provisions empowering the employer to adopt any one of the three alternatives as follows:

(a) Carry out the work himself or by others in a reasonable manner at the cost of the contractor and the contractor shall pay the employer the costs

(b) Require the engineer to agree or determine a reasonable reduction in price or

(c) If the defect or damage deprives the employer of substantially the whole benefit of the works, terminate the contract and recover all sums paid for the works, cost of removal and financing costs including clearing the site and returning the materials and plant to the contractor.

Where, however, the cause of defect is not attributable to the contractor he is entitled to access the site, search for the reasons and will be entitled to payment of additional sum for the expenses incurred by him as specified in Clauses 11.7 and 11.8.

Clause 11.9 pertains to the issuance of a performance certificate by the engineer to the contractor within 28 days after the latest of the expiry dates of the defects notification periods.

Clause 11.10 declares that for unfulfilled obligations, the contract will continue to subsist even after the performance certificate is issued and Clause 11.11 deals with the contractor's liability to clear the site and remove his materials, equipment, rubbish, etc. within 28 days, failing which the employer may sell or otherwise dispose of the remaining items at the cost of the contractor.

8.1.2 FIDIC Form 2017 Edition – Clauses: 7.1 to 7.8, 9 to 11

The 2017 edition of the FIDIC Form retains the provisions of the 1999 edition with variations in some clauses as follows:

Clause 7.3: The provisions by way of addition of sub-clauses (b) (i), (ii) and (iii) expressly stipulate the activities that may be carried out by the employer's personnel at the site and the places from which materials are procured and include the checking of progress of plant, manufacture of materials and making photographic and/or video recordings.

If the contractor has given an RFI notice and the engineer fails to give notice to the contractor that no inspection is intended and the employer's representatives do not attend at the time stated or agreed, the modified condition empowers the contractor to cover the work and/or proceed with packaging for storage or transport of the articles meant to be inspected.

If the contractor fails to give notice under this sub-clause and the engineer requires the work to be uncovered and thereafter reinstated and made good at the contractor's cost, in a sub-clause of the 2017 edition a subtle modification is substitution of the words "risk and cost" in place of "cost". The modification will ensure that the element of uncertainty in the undertaking of uncovering and restoring the work is to be borne by the contractor. The risk may be moral, physical or economic. For example, if the uncovering of the work is likely to damage the other completed part of the work, the contractor will have to reinstate and make good even that part of the work at his cost.

Clause 7.4: The following changes made in the 2017 edition are noteworthy.

1. The contractor's liability to provide everything to carry out the tests efficiently and *properly*. By properly is meant that which is fit, suitable, appropriate, adaptable, correct and reasonably sufficient.

2. In the 1999 edition, the time and place of testing was to be agreed by the contractor with the engineer. In the 2017 edition the provision has been modified. The contractor is required to give notice, allowing reasonable time, stating the time and place of testing. This is a welcome change, it is submitted.

3. The cost of varied or additional tests to be borne by the contractor in the event of results of the tests not being as per the terms of the contract, as per 1999 edition, was to be borne by the contractor "*notwithstanding other provisions of the Contract*". This phrase "*notwithstanding other provisions of the Contract*" is deleted in the 2017 edition. The other provisions of the contract remain unaffected by this change, it is submitted.

4. The notice period of the engineer informing the contractor of his intention to attend the meeting has been increased from not less than 24 hours to not less than 72 hours.

5. In the event the contractor suffers delay or incurs cost, his entitlement to "payment of such cost plus *reasonable* profit" under the 1999 edition has been modified to read "such cost plus profit". Deletion of the word reasonable is justified by inclusion of the definition of "cost plus profit" in Clause 1.1.20. According to the said definition, the applicable profit percentage is to be agreed and mentioned in the contract data and in the absence of such agreed percentage, 5% is applied.

6. The sub-clause ends by addition of the sentence that Clause 7.5 shall apply where the tests fail to yield specified results. It appears to be by way of a caution as Clause 7.5 is clear enough.

Clause 7.5: The modified version is more elaborate, providing for notice to be issued by the engineer pointing out defects, response to the said notice by the contractor and review of the proposal of the contractor by the engineer. If in the review the engineer concludes that defects would not be eliminated by the proposed remedial measures, he would notify his findings to the contractor. The contractor has to submit a fresh proposal. If the contractor fails to submit a proposal or revised proposal and/or rectify the defects, the engineer will give notice with reasons invoking the provisions of sub-clause 7.6 and/or 11.4, as the case may be.

Where the remedial measures to remove the defects have been carried out by the contractor, the engineer is empowered to order carrying out retests under Clause 7.4. If as a result of rejection and retesting the employer incurs additional costs, the contractor has to pay the same to the employer.

Clause 7.6: The only notable difference from the 1999 edition is the express provision, which otherwise also the law will imply, that in case the remedial work is urgently required to be carried out for the safety of the works and it is attributable to any act by the employer or the employer's personnel, or an exceptional event covered by Clause 18.4, the cost shall not be borne by the contractor.

Clause 7.7: In accordance with the provision in the 1999 edition, the ownership in the plant and materials vests in the employer when the contractor becomes entitled to payment of its value; the 2017 edition stipulates "when the contractor is paid the value" thereof. In the period between being entitled to and paid for, the ownership will vest in the contractor and rightly so, it is submitted. When a contract is prematurely terminated the cost and ownership of plant and materials at site is invariably disputed, resulting in the employer not allowing the contractor to shift or take away the plant and materials promptly on termination and mitigate the loss, because the contract declares the plant and materials to be the property of the employer. The change is thus desirable.

The provisions of the other standard forms are more or less similar to the above two forms and are not reproduced or analysed to avoid repetition.

8.1.3 NITI Aayog Government of India MODEL AGREEMENT 2017 EPC (Engineering, Procurement and Construction) of Civil Works

In addition to the provisions of the FIDIC Form discussed above, this form has incorporated some existing and/or new features including the following provisions:

Art. 11.2 Quality Control System

The contractor shall, within 30 (thirty) days of the appointed date, submit to the authority's engineer its Quality Assurance Plan (QAP) which shall include (a) organization, duties and responsibilities, procedures, inspections and documentation; (b) quality control mechanism including sampling and testing of materials, etc. The authority's engineer shall convey its comments to the contractor within a period of 21 (twenty-one) days of receipt of the QAP stating the modifications, if any, required, and the contractor shall incorporate those in the QAP to the extent required for conforming with the provisions of Clause 11.2.

Art. 11.3 Methodology

The contractor shall, at least 15 (fifteen) days prior to the commencement of construction, submit to the authority's engineer for review the methodology proposed to be adopted for executing the works, giving details of equipment to be deployed, traffic management and measures for ensuring safety. The authority's engineer shall complete the review and convey its comments, if any, to the contractor within a period of 10 (ten) days from the date of receipt of the proposed methodology from the contractor.

11.4 Inspection and Technical Audit by the Authority

The authority or any representative authorized by the authority in this behalf may inspect and review the progress and quality of the construction of works and issue appropriate directions to the authority's engineer and the contractor for taking remedial action in the event the Works are not in accordance with the provisions of this agreement.

11.5 External Technical Audit

At any time during construction, the authority may appoint an external technical auditor to conduct an audit of the quality of the works. The findings of the audit, to the extent accepted by the authority, shall be notified to the contractor and the authority's engineer for taking remedial action in accordance with this agreement. The contractor shall provide all assistance as may be required by the auditor in the conduct of its audit hereunder. Notwithstanding anything contained in this Clause 11.5, the external technical audit shall not affect any obligations of the contractor or the authority's engineer under this agreement.

11.7 Monthly Progress Reports (MPRs)

During the construction period, the contractor shall, no later than 10 (ten) days after the close of each month, furnish to the authority and the authority's engineer a monthly report on the progress of works and shall promptly give such other relevant information as may be required by the authority's engineer. The Monthly Progress Reports (MPRs) include useful information such as the plant and machinery deployed at the site each month.

8.1.4 MOS & PI, GOI Form 2005: Clauses: 33 to 36

This form under the heading "Quality control", in four short and simple terms, lays down the rights and liabilities of the employer and the contractor. Clause 33 empowers the nodal officer or his nominee (equivalent to engineer or engineer's representative) to check the work and notify the defects that are found. The clause further stipulates that such power and checking shall not affect the contractor's responsibilities. The contractor, if directed, will be bound to uncover and test any work that the nodal officer or his representative considers may have defects.

Clause 34 provides:

> If the Nodal Officer or his nominee instructs the Contractor to carry out a test not
> specified in the Specification to check whether any work has a Defect and the test shows
> that it does, the Contractor shall pay for the test and any samples. If there is no Defect, the
> test shall be a Compensation Event.

The use of the words "compensation event" in place of cost of tests and sample can be
construed to entitle the contractor to claim loss of time and productivity if caused by the direct
result of such instruction besides the cost of tests and samples. Clause 44 lists the compensation
events and this head figures under sub-clause (d).

Clause 35.1 deals with correction of the defects which have been notified by the nodal officer or
his representative before the end of the defects liability period, which period begins at completion
of the work and is defined in the contract data. It further stipulates that the defects liability period
shall be extended for as long as defects remain to be corrected. This stipulation gives rise to
a doubt as to whether the extended period empowers the nodal officer or his representative to
notify further defects during such extended period. It is submitted that the extension is only
for the purpose of correcting the defects notified within the original period specified in the
contract data.

Clause 35.2 stipulates that the contractor shall rectify the defect every time it is notified, within
the period specified in that respect by the nodal officer or his nominee in the notice.

Clause 36 provides for contingencies in the event the contractor has not corrected any defect
that has been notified. This clause empowers the nodal officer or his nominee to assess the cost of
defect rectification and makes the contractor liable to pay the said amount. In a Canadian case,
a contractor's repeated failure to comply with the specifcation to rectify defects pursuant to the
consultant's instructions was held to amount to a fundamental breach of contract.[1]

8.1.5 Other Standard Form Contracts

Standard forms of contracts of state and central PWDs, Military Engineering Services and the
Railway Engineering Department of the Government of India incorporate provisions similar to
those discussed above in paragraphs 8.1.1 to 8.1.4 and have not been specifically reproduced.
However, the contents of the rest of this chapter are relevant to the said conditions as well.

8.2 Contractor's Implied Obligations

The implied obligations of a contractor, or interpretation of express warranties, as to the quality
of work, generally fall under three heads:

  i. Materials
  ii. Workmanship
  iii. Design

The contractor's responsibility to rectify the defective work at his own cost arises only when it
can be proved that he has not complied with the drawings, specifications and directions of the
engineer. The general principle to be adopted can be stated thus: where a contractor agrees to
produce a certain result, he is responsible if the result is not produced; but, where the owner
assumes to specify the manner of construction, to produce the result, the owner assumes the

---

[1] R.F.M. Electric Ltd. v. University of British Columbia and Martina Enterprises, [1992] B.C.J. No. 1810 – BCSC.

responsibility if the work does not turn out as expected. If the plans supplied by the employer are insufficient to produce the result desired, the contractor is not responsible if he carries out the plans.[2]

EXAMPLES FROM CASE LAW

1. In a case, the contract contained a clause that the waterproofing shall be watertight. The specifications provided how the walls should be built and how the waterproofing should be done. However, when one of the walls cracked and was found not to be watertight, the owner brought in the suit to recover damages. It was held by the court that "where an owner specifies how foundation walls should be built and waterproofed, the contractor does not warrant that they shall be watertight, though the contract represents that they shall be watertight."[3]

2. In Lynch v. Throne (1956)[4] the specifications expressly provided that the walls were to be nine-inch brick walls. The work was done with good quality materials and workmanship and exactly in accordance with the specifications. However, the walls did not keep out the driving rain. The builder was held not liable.

8.2.1 When Contract Is Silent

Where a contractor agrees to build a work for a specified purpose, and the contract is silent as to the manner of construction, the law reads into the contract a stipulation that

  i. The contractor will do his work in a good and workmanlike manner
 ii. The contractor will supply good and proper materials
iii. The completed structure will be reasonably fit for its intended purpose.

EXAMPLES FROM CASE LAW

1. In one case, the contractor agreed to erect a garage with a concrete floor. The floor as built was ruined by frost. The owner refused to accept the work so done. The contractor, thereupon, brought the action to compel the owner to accept the work as built. The trial court granted the contractor judgment on the grounds that the owner did not specify any depth for the foundation, nor did the specifications provide that the concrete floor should not be laid during freezing weather. The Appellate Court reversed the judgment.[5]

2. In Test Valley Borough Council v. The Greater London Council,[6] the court was to lay down a standard of duty to be implied in the agreement. It was held that the contractor undertakes to provide dwellings constructed in good and workmanlike manner, of materials which are of good quality and reasonably fit for human habitation.

---

[2] Kuhs v. Flower City Tissue Mills Co. 104, Misc 243, 171 NYS 688; also see: IRCON International Limited v. Patil Rail Infrastructure Pvt. Ltd., (Delhi), AIR 2018 Del. 98. For facts see paragraph 8.10 below.
[3] Kuhs v. Flower City Tissue Mills Co., 104, Misc. 243, 171 NYS 688.
[4] Lynch v. Throne, [1956] 1 W.L.R. 303.
[5] Minemount Realty Co. v. Ballentine, 162 A 594, 111 N.J. Ed. 2. 398.
[6] Court of Appeal (1979) 13 BLR 63.

3. The owner in one case, got a roof designed by an architect and entered into an agreement with a contractor to construct it. No architect was engaged to supervise the construction. The design was faulty and made insufficient provision for ventilation of the roof space and timbers. The result was a serious attack of rot. The contractor was held liable for failure of his duty to warn the owner of the obvious inherent danger.[7]

4. Builders were proposing to develop an estate of some 40 houses. The plots were laid out. Prospective purchasers purchased the plots but at the time no contract was entered into for the construction of the houses. In the next three months the builders did a good deal of work in laying the foundations. They put in hardcore and a layer of four-inch concrete above it. They started to build houses. After this, contracts were signed. According to the terms of the agreement, the builders had undertaken at their own cost and charge and in a proper and workmanlike manner to erect, build and complete a messuage or dwelling house in accordance with the plan and specification supplied to the purchaser, who agreed to buy the same for the named price. There was nothing in the specifications about the foundation, except that the site concrete was to be four inches thick and laid to required levels. On the plan it was mentioned "four inch site concrete on hardcore". The houses were completed and the purchasers occupied the houses. All appeared well, but two years later there was trouble in one house; three or four years after in the other houses. The floors were cracking and breaking up. The cause was found to be the hardcore, which contained a large quantity of sodium sulphate. The builders were not to be blamed for they had no way of suspecting the hardcore. The builders used all the reasonable care, skill and judgement in their work. It was held:[8]

> The quality of materials is left to be implied; and the necessary implication is that they should be good and suitable for the work. I knew that the builders were not at fault themselves. Nevertheless this is a contract; it was their responsibility to see that good and proper hardcore was put in as it was not put in, they are in breach of their contract.

It was observed that the builder can try to sue their suppliers, if they can prove against the suppliers but they have to take responsibility so far as purchasers are concerned.

8.2.2 Nominated Supplier/Subcontractor

The next question that may be considered is the responsibility of the contractor for the materials supplied for use in work by the supplier nominated by the owner and/or his architect/engineer. The answer to this question *prima facie* appears to be that for defective materials the contractor cannot be held responsible. However, express terms of the contract and any admissible surrounding circumstances have to be taken into account.

EXAMPLE FROM CASE LAW

In a contract, in RIBA form, which contained no express provision as to the liability for quality of materials supplied by a nominated supplier, the contractor was duty bound to accept and

---

[7] Brunswick Construction Ltd. v. Nowlan, Supreme Court of Canada, (1974) 21 BLR 27.
[8] Hancock v. B.W. Brazier (Anerley) Ltd., Court of Appeal [1966] 2 All ER 901.

comply with the instructions of the architect. The architect nominated S as supplier of concrete units and sent to the contractor a quotation from S instructing to accept it, which the contractor did. The quotation contained provisions limiting the liability of S, if the goods supplied proved defective. Under the main contract, the contractor could make reasonable obligations to nominated subcontractors, but no such right in respect of nominated suppliers. The concrete units supplied by S proved defective, causing delay in completion. The contractor exercised his right to determine the contract on the ground that the work got delayed for more than one month "by reason of architect's instructions". The owner claimed damages for wrongful repudiation. The question was whether the contractor was responsible for the implied warranty, in which event it was conceded that the contractor was not entitled to terminate the contract. It was observed, "The employers, through their architect, having directed the contractor to buy from a manufacturer who had substantially limited his liability, it would not be reasonable to suppose that the parties were intending the contractor to accept an unlimited liability" and held that the contractor was not liable.[9]

The above decision is based on the main fact that the contractor did not have any freedom to negotiate the supply terms with the nominated supplier. If it could be shown that the contractor could, and it would be the expectation that he would, or at least it would be his responsibility if he did not, deal with the manufacturer on terms attracting the normal conditions or warranties as to quality or fitness, the contractor would be held liable by the nominated supplier. In such a situation, a contractor is well advised to give his supplier specifications of the main contract and buy on such terms that he may not have to bear the loss if the materials prove to be defective.

EXAMPLES FROM CASE LAW

1. In a contract for supply of work and materials for roofing of the houses, the contract specified use of "Somerset 13" tiles, which were manufactured by only one manufacturer. The contractor obtained supplies from their own suppliers, who in turn obtained them from the manufacturer. Some of the tiles had manufacturing defects which were not patent, but latent. In less than a year the tiles began to disintegrate and the owners claimed the cost of re-roofing of the houses. It was urged on behalf of the contractor, that since he did not choose the tiles and the defect could not have been discovered by any reasonable inspection, he was not liable. It was observed: "The question is whether he warrants the material against latent defects." It was held:

> The loss was not caused by Somerset 13 tiles being unsuitable for the contract purpose; it was caused by the tiles which were supplied being of defective quality. ... There are in my view good reasons for implying such a warranty if it is not excluded by the terms of the contract. If the contractor's employer suffers loss by reason of the emergence of the latent defect, he will generally have no redress if he cannot recover damages from the contractor. But if he can recover the damages, the contractor will generally not have to bear the loss: he will have bought the defective material from a seller who will be liable under – the Sale of Goods Act – because the material was not of merchantable quality.[10]

---

[9] Gloucestershire County Council v. Richardson, House of Lords [1969] 1 A.C. 480 (HL).
[10] Young & Marten Ltd. v. Mc Manus Childs, [1969] 1 A.C. 454.

2. A contract for the construction of a factory building provided for the design, supply and erection of a superstructure including the roof and roof lights to be carried out by the nominated subcontractors who were specialists in "system built" superstructures. The subcontractor completed his work in May 1967 and the main contractor in August 1968. A major leak occurred in the roof of the factory. The cause of the leak was found by an arbitrator to be due 85% to the defective design of the roof lights, 12% to use of inferior material and 3% to bad workmanship. It was held by the trial court that: (a) the main contractor was responsible for so much of the damage as was due to the use of inferior material by, and bad workmanship of, the subcontractor; and (b) the main contractor was not responsible for the loss which was due to the defective design of the roof lights. The appeal from the above decision was dismissed.[11]

3. The main contractor had carried out the construction of buildings with RCC frames. The white ceramic tile external cladding was done by nominated subcontractors. The cladding began to fall. The architect blamed bad workmanship. The main contractor blamed the design. Eventually, the owner decided to remedy the defects by a resin injection process. The architects recommended the sole British licensee of the process be employed by the main contractor to remedy the defects. The main contract was varied accordingly. The licensee was employed by the main contractor and not technically nominated a subcontractor. The remedial works were not successful. The owner alleged breach of implied warranty of fitness for purpose of the resin injection process. It was held that the main contractor was not liable in the contract. It was suggested that the owner should have taken an express warranty from the main contractor or direct warranty from the specialist sole licensees.[12]

8.2.3 Guarantee by Contractor

The principle forming the basis of the above cases is likely to be varied by express terms of the contract. For example, if a contractor is to submit tender in accordance with specifications which include "a guarantee that all material employed in the work will be first-class and without defect and that all work specified" would remain defect-free for a certain period; the contractor will accept the risk involved.

EXAMPLE FROM CASE LAW

A roofing contractor's tender was accepted for providing roofs on buildings under construction. The roofs of the buildings were to be covered by a fluted steel deck and a layer of insulation and roofing felt was to be attached to the steel deck by a compound of fire-resistant material known as "Curadex". The contract contained a stipulation as a part of which the contractor guaranteed "for a period of five years that all work specified above will remain weather-tight and that all material and workmanship employed are first class and without defect". It was held that the guarantee extended to damage to the roof arising out of faulty design and the owner was entitled to recover the cost of repairs to the roof.[13]

---

[11] Norta Wallpapers (Ireland) Ltd. v. John Sisk & Sons (Dublin) Ltd., Irish Supreme Court, (1977) 14 BLR 49.

[12] University of Warwick v. Sir Robert McAlpine, QB (1989) 42 BLR 6.

[13] Steel Co. of Canada Ltd. v. Willand Management Ltd., Supreme Court of Canada (1966) 58 DLR (2d) 595.

8.3 Interpretation of Specifications

The specifications are written instructions which supplement the drawings and set forth the complete technical requirement of the work. The requirements such as arrangement, dimensions and types of constructions which can more readily be expressed graphically are shown on the drawings. The specifications, on the other hand, cover those features that can be described in words. Thus, quality of materials and workmanship are usually the subject matter of the specifications.

Both drawings and specifications are integral parts of the contract documents. As such great care should be exercised to see that there is no conflict between the drawings and specifications. In spite of due care, occasionally, conflict may result between the drawings and specifications. In such cases, it is customary in engineering and architectural practice to apply a rule of interpretation according to which specifications are said to govern. That is, so far as the conflicting part of the work is concerned, the intention of the parties to the contract is to be gathered from the specifications and not from the drawings. This is not, it may be noted, a rule of law and, to avoid a dispute, the contract should include an express provision to this effect.

Occasionally, conflicts between the various parts of the specifications may be the causes of disputes in construction contracts. More often than not such a conflict arises on account of violation of a basic rule of specification writing, namely either specify the methods or specify the results. Consider, for example, the specifications of cement concrete. The specifications usually state the proportion in which cement, sand, coarse aggregate and water shall be mixed, together with the time for mixing and allowable slum. It also specifies an ultimate 28-day compressive strength that the concrete must develop. Such specification, as already mentioned, may relieve the contractor of the responsibility of the final results, if the contractor has faithfully followed the method.

Where construction of a finished product is split into a number of independent items and the contract includes separate specifications for each item, the specifications of each item are to be read in conjunction with trade practice and in relation to all the specifications.

EXAMPLE FROM CASE LAW

A contract called for the construction of concrete spillways for a flood control project. The specifications of the concrete work permitted deviation in line and grade for the finished spillways. However, the specifications for erecting the formwork for laying concrete provided that "forms shall be true to line and grade". The specifications of formwork were interpreted by the government to mean that no deviation in the formwork was permitted. Accordingly, the contractor was required to erect the forms with extreme accuracy. The contention of the contractor was that the specifications of the formwork were not to be read literally. His interpretation was that the specifications for the formwork read with specifications of the finished spillways permitted him to erect the formwork within the tolerances allowed for the structure. The government's insistence and continuous inspection and directions that all forms meet an exact standard were excessive. He, therefore, claimed compensation on that ground. It was held that the contractor's interpretation was reasonable and correct. The court observed:

> The end product contemplated by the contract was a specified mass concrete structure with the wooden forms being but the means to be employed in achieving that end product. It would be unreasonable to assume that the parties intended to contract for a less than perfect end product while intending that the forms be built with mathematical exactness.[14]

---

[14] Kenneth Reed Const. Corp. v. The United States, U.S. Courts of Claims, No. 144–70 (16th March 1973).

8.4 Design Failure

In the final analysis, the word "design" can be considered to be wide enough to include materials, their quality, strength and also workmanship. Thus in a typical building contract wherein the contractor is supplied with a full set of contract documents prepared by the owner's architect/engineer consisting of bill of quantities (BOQ), drawings, specifications, etc., and the contractor faithfully follows the same in executing the work, he is not responsible for defects, if any.

EXAMPLES FROM CASE LAW

1. A contractor successfully tendered for the design, supply and installation of bulk storage and handling plant. The plant was to be built on reclaimed harbour land. The contractor did, in fact, propose the design of the foundation and prepared the first drawings. But by the time the agreement was made the "design" of the foundation had been agreed, after participation of the owner's consulting engineer in the determination of its form. That agreed design extrapolated in a drawing, was expressed to be part of the contract work. These drawings showed ring beam foundations for the storage bins. Subsidence occurred when the bins were erected and filled. It was held that the contractor's liability was to execute the work as per the drawings in a workmanlike manner. The arbitrator held that the contractor did execute the work accordingly.[15]

2. M contracted with B to build a warehouse. Consultant structural engineers carried out the actual design work. The BOQs described the work to be watertight. The basement floors subsequently developed cracks and water penetrated necessitating remedial work resulting in delay and losses. The two issues to be determined were:

    (a) whether the defects were the result of inadequate design and/or faulty workmanship and

    (b) if the cause was design failure, whether M was liable because of performance specifications in the BOQs.

It was held:

> I should require the clearest possible contractual condition before I should feel driven to find a contractor liable for a fault in the design, design being a matter which a structural engineer is alone qualified to carry out and for which he is paid to undertake, and over which the contractor has no control. ... I decline to hold that the specification in the bill of quantities makes the contractor liable for the mistakes of the engineer.[16]

3. A contract was entered into for the supply, and to fit and hang, doors for shops "to suit opening, steel sheet outside area, fit Rivers 4-point non-breakout locking system operated by key from inside, hinge, stops, etc." The contractor fitted the door as per the express terms but upon an existing wooden frame made out of a soft wood. Thieves later broke into the shops by forcing the door away from the frame with a jemmy. It was held that the contractor was in breach of an implied term to provide burglar-proof protection.[17]

---

[15] Cable (1956) Ltd. v. Hutcherson Brothers Pty Ltd., High Court of Australia, (1969) 43 ALJR 321.

[16] John Mowlem & Co. Ltd v. British Insulated Callenders Pension Trust Ltd., Q. B. Div. (1977) 3 Con. LR 63.

[17] Reg. Glass Pty. Ltd. v. Rivers Locking Systems Pty. Ltd., High Court of Australia (1968) 120 CLR 576.

In a package deal or industrialized buildings, the builder is responsible for the design failure, if any. However, he can sue the design engineer/architect to recover loss, if any. The common intention of all parties to the contract being that the engineer should produce a design that would be fit for the purpose for which it was required.[18]

The Model Form introduced by NITI Aayog has done away with the above traditional provision in an item rate contract by introducing the EPC contract. Clause 10.1 puts the entire responsibility of survey, design, preparation of drawings and designs, specifications, etc. on the contractor besides construction. The contractor has to appoint a design director (the "Design Director") who will head the contractor's design unit and shall be responsible for surveys, investigations, collection of data and preparation of preliminary and detailed designs.

8.5 Removal of Improper Work

A contract usually contains many clauses dealing with improper work, defects, maintenance, etc. According to one of the standard clauses, the opinion of the architect or engineer or the internal or external technical auditor as to improper work is made final. Another clause normally provides for making good defects after completion. The arbitration clause usually provides that defects after completion will be subject to review by the arbitrator. The proper and harmonious interpretation of these different clauses will be to make the first condition mentioned above applicable to improper work noticed in the course of construction until the work is "green". All work may be considered green for this purpose until the engineer or architect has examined it during one of his visits. If, after inspecting the work during one of his visits, the architect or engineer has not condemned the work, and if the contractor has obviously treated it as finished, the first clause will no longer be applicable.

It is obvious that, however careful an engineer or architect may be while inspecting the work at his periodical visits, defects in materials and work may possibly be overlooked or covered up in the intervening period. If in fact the work is badly done and defects follow, the engineer or architect is not without power and the owner is not without his protection.

The defence often attempted to be raised on behalf of the contractors who are guilty of executing defective work, is that the owner or his architect/engineer saw the work but no disapproval was expressed. Such a defence is not available, unless some matter was expressly brought to an engineer's attention, he was expressly asked to give his approval and that he gave it. Even in such a case the estoppel would only be for the patent defects and not for the latent defects, which the engineer was unaware of when he gave his approval. The rule is that the architect is not duty bound to the contractor to point out defective work during the progress.[19] The defects and maintenance clause will be applicable in such cases. Under that clause, however, the engineer or architect no longer has the last say in the matter, his decision can be challenged by the contractor and the matter referred to arbitration.

8.6 Defects and Maintenance Clause

As mentioned above, the defects and maintenance clause is inserted in a contract to empower the engineer or architect for protecting the interest of the owner in case defects appear in the work

---

[18] Greaves & Co. (Contractors) Ltd. v. Baynham Meikle & Partners, Court of Appeal (1975) 4 BLR 56; CFW Architects v. Cowlin Construction Ltd., [2006] EWHC 6 (TCC).

[19] East Ham. Corp. v. Bernard Sunley & Sons, [1966] A.C. 406. Also see [1965] 3 W.L.R. 1096.

subsequent to completion. The clause, as seen in the example clauses referred to above, usually provides for the rectification of defects by the contractor at his own cost.

The word "defect" in this context includes any breach of contract affecting the quality of work, whether structural or merely decorative and whether due to faulty material or workmanship, or even design, if the design is the contractor's obligation under the contract.

Obviously the contractor could not be held responsible for rectifying defects for an indefinite period following completion. The clause usually stipulates a period of six months to one year as defects liability period during which it is expected that latent defects, if any, may possibly show themselves. When the guarantee period is named in the contract, the contractor's obligation to rectify the defects extends only to defects which have been discovered during the defects liability period. It is not necessary that the cause or the reason of these defects be discovered during the fixed period.[20]

It is customary for the "defects period" to run from the date of completion. Completion usually means certified completion. The engineer or architect, therefore, must give the completion certificate, without delay. Where completion of "several works" is mentioned in the clause, the expression "several works" refers to the whole of the works collectively, and not to each part severally.

The defects and maintenance clause is not only advantageous to the owner but also to the contractor, inasmuch as it ends the contractor's liability at a definite date unless very special circumstances exist. A contractor should not regard this clause with disfavour because even in the absence of a "defects maintenance clause", the contractor remains liable for rectifying the defects after completion. If he fails to make good the defects brought to his notice, the owner may deduct charges incurred in remedial work from sums due to the contractor under the contract. If full payment has already been made, the owner can sue for damages for defective work.

In the absence of an express stipulation to the contrary the contractor is not liable to make good the "wear and tear" during the maintenance period. His liability will arise only if it can be shown that wear and tear were the direct result of defects for which he is liable.

Further, under the defects and maintenance clause, the contractor is not liable to restore works if they are damaged by fire, storm, etc., during the maintenance period. The precise obligation of the contractor under a defects liability clause depends upon the construction of the clause incorporated in the contract by the parties.

It is made clear that whatever may be the wording of the clause, the contractor should give notice to the engineer with a copy to the owner of his having substantially completed the work, accompanied by a written undertaking to finish with due expedition any outstanding work during the defects liability period. It is then for the engineer to issue a taking-over certificate or a substantial completion certificate or to give instructions as to all works which are required to be completed by the contractor before the issue of such a certificate. The FIDIC Form, as already seen, stipulates a 28-day time limit for the engineer to do so. In other cases, the contractor can maintain the plea that the engineer unduly delayed the issue of the completion certificate to prolong the defects liability period, if he follows this procedure.

According to the FIDIC Form 1992 (Amended Edition) Clause 49.2, the contractor's obligations under the clause are:

a) To complete the work, if any, outstanding on the date stated in the taking-over certificate, and

b) To execute all such work as amendment, reconstruction, and remedying the defects, shrinkages, or other faults. It needs to be noted that the liability extends to defects noticed by the engineer within the defects liability period even though notified within 14 days after the expiry of the defects liability period.

---

[20] Cunliffe v. Hampton Wick Local Board, (1893) 9 TLR (378).

8.6.1 Cost Incurred in Providing Remedial Measures

A question arises: can the owner refuse to pay the contract sum or so much of it as has not been paid, if the contractor's work is defective? The answer to this question is considered below in the light of the usual stipulations made in the contracts and referred to in paragraph 8.1 above.

Usually the contracts include express stipulations, one of which requires the contractor to remove and reconstruct bad, unsound, imperfect or unskilful work. Where, however, the defective work is structurally sound and cannot practically be removed except at an expense out of proportion to the good to be attained, the owner has the option to accept the work and pay for it at lower rates than agreed to in the contract. In other words, in the case of a substantial but defective performance, the contractor can claim and will be entitled to the contract price less the damages for the defective performance. The owner can, upon proof of defective construction, have the contractor's recovery reduced by the amount that would reasonably be required to remedy the defects and make the structure conform to the plans and specifications. Of course, the owner, in making repairs, is not permitted to charge the contractor with the cost of materials, which are more expensive, or have the building placed in a better condition than what was called for in the contract between the two.[21]

Generally, the contractor is responsible for rectification or remedial work:

i. If the use of materials, plant or workmanship is found to be not in accordance with the contract

ii. Where the contractor is responsible for the design of part of the permanent works and there is any fault in such design, or

iii. Where there is neglect or failure on the part of the contractor to comply with any express or implied obligation on his part under the contract.

If the necessity of rectification is due to other causes, the contractor deserves to be paid for such work as he may have to carry out. The FIDIC Form expressly provides this. In other forms it will be an implied obligation of the owner to pay for any remedial work, for which the contractor is not responsible.

8.6.2 Taking Possession Is No Bar to Claim Damages

The owner is not estopped from claiming damages for a breach of a building contract by taking possession and moving into the building. The failure of the contractor to construct the building in accordance with the plans and specifications was a violation of the contract, and the taking of possession of the premises by the owner cannot be considered as a discharge of the contractor's liability.[22] The position may not be the same if at the time of taking possession the owner is aware of the defects.

8.6.3 Contractor May Not Be Liable for Patent Defects

If the defects are obvious, or if circumstances are such that knowledge of imperfections may be imputed to the owner, he may be liable to pay for the defective work. Unless he promptly objects to the shortcomings in the work and makes payment at reduced rates, he may not be able to recover compensation for such patent defects after accepting the work done.

---

[21] Talbot Quevereanx Construction Co. v. Tandy, 260 S.W. 2d 314 St. Louis Court of Appeals, 1953.

[22] Michel v. Efferson, 65 So, 2d, 115, Supreme Court of Louisiana 1962.

EXAMPLE FROM CASE LAW

In a case,[23] the dispute was regarding improper expansion joint in a concrete runway. The contract had a clause which stated that "acceptance shall be final except as regards latent defects". The specifications provided for 6-inch depth whereas the contractor, relying on alleged trade practice, provided the joints equal to only one-sixth of the runway depth. The contention of the government (owner) was that it did not discover the shallow joints because they were covered immediately with a joint sealer, and as such the defect was latent, and it was entitled for reimbursement of its cost to correct the defect. The Armed Services Board of Contract Appeals held that no latent defect existed because the deficiencies were readily discoverable by placing a sharp object into the cuts. In fact, it said, evidence showed that the government's on site inspector did this and "thus knew or should have known of the alleged deficiency". In conclusion it was held that since the deficiency was patent the government's acceptance was final and conclusive.

Some of the propositions stated above stand supported by decisions of the courts under Defects Liability Clauses as seen from the following examples.

EXAMPLES FROM CASE LAW

1. In a case the parties entered into contract for construction of the petitioner's (PHHL) office complex building. A bank guarantee was furnished for securing PHHL for any defect liability for one year after completion. The works were completed on 17th March 2011 and, therefore, the defect liability period had come to an end on 16th March 2012. The arbitrator held that since no evidence of defect in the works executed was established, PHHL ought to have released the said amount and withholding the same was wrongful and that, in any event, the defect liability period could not extend beyond March 2012 and PHHL was duty bound to refund any sum deposited as security deposit on the expiry of the said period. The arbitrator had also awarded interest at the rate of 10% p.a. on the amount with effect from 1st April 2012; that is after expiry of the defect liability period. Held: no interference with the impugned award was warranted.[24]

2. In a case, admittedly a defect liability certificate was issued by the appellant. In terms of the contract, on issuance of defect liability certificate, the appellant was liable to refund the original bank guarantee for discharge, which was not done. Not only was the original bank guarantee not returned, the appellant even wrote to the bank by its letters not to release/ discharge the bank guarantee. On account of non-return of original bank guarantee and letters written by the appellant, the bank continued to charge commission from the respondent. Despite these facts, the arbitral tribunal awarded only 50% of the amount paid by the respondent. It was held that the appellant was liable for the entire amount and, as such, the order of the single judge to the said effect does not call for any interference.[25]

3. In a case, the defect liability period had commenced from 22nd July 2014 and expired in March 2016. Respondent no. 1 by two separate letters dated 6th April 2015 addressed to the respondent no. 2 bank, invoked the bank guarantees in question on the ground that the contractor

---

[23] Appeal of Federal Const. Co. A.S.B.C.A. No. 17599, 73-I BCA 10003 (1973).

[24] Pawan Hans Helicopters Limited v. Ideb Projects Private Limited, 2017(9) AD(Delhi) 317.

[25] National Highways Authority of India v. M/s BSC-RBM-PATI Joint Venture, 2016(2) AD(Delhi) 261: 2016(155) DRJ 50: 2016(4) R.A.J. 261.

was in breach of the terms and conditions contained in the agreement *inter alia* on account of delays and defects. The defence substantially relied on the "experience certificate" dated 18th August 2014, issued by respondent no. 1, which certified that the appellant had completed the work under the contract and that the same was satisfactory. However, in the note appearing in the said certificate it was stated that this experience certificate was issued at the written request of the contractor (appellant) to be submitted for tender and registration purposes only and is not valid for any other purpose. Held: It was not appellant's case that the defects were noticed after a period of 18 months. Thus, demand raised by the respondent was in accordance with the terms of contract and could not be faulted.[26]

4.   Where an employer did not take over possession of premises on completion but was directly handed over to allottees with six months delay and at the time of handing over possession, the contractor had to do the painting, whitewashing for the second time, the additional cost awarded was upheld. Award of extra cost for watch and ward for the extended period was, however, set aside as there was no proof tendered on record such as salary slips, attendance register, etc.[27]

5.   As per Special Conditions of Contract, claims, if any, were to be made within 30 days of expiry of the defect liability period, which was 365 days from the date of completion. The defect liability period ended on 31st March 2004. The claims to be made within 30 days, that is on or before 30th April 2004, were however raised by letter dated 20th October 2004. The claims were barred by time. Held: award made without adherence to the contract terms and therefore award set aside.[28]

6.   After the end of defect liability period the contractor is not liable to maintain the road. Further, the contractor was not given 15 days' notice for repair work as per Clause 23 of the tender condition. An earlier order of the High Court directing authorities to make payment of admitted amount with interest was also not complied with. Held: the writ petition was maintainable and writ of *mandamus* issued to ensure payment with interest.[29]

In order to be entitled to get repair and maintenance cost, the employer has to prove:

  i. Documentary evidence on record to show that the defects/discrepancies existed for which the contractor was responsible

  ii. An inventory/list prepared by the engineer to show what were the defects/discrepancies, if any, left by the contractor

  iii. Value of the work done by the employer/cost incurred which can be recovered from the contractor

  iv. If the contract provision envisages a process of consultation between the employer, engineer and contractor before any liability can be saddled upon the contractor and, if yes, the proof thereof

---

[26] M/s S. Satyanarayana & Co. v. M/s. West Quay Multiport, (Bombay)(DB) 2015(4) BCR 274: 2015(4) ArbiLR 531: 2015(4) BC 682: 2016(1) BankJ 551: 2016(1) R.A.J. 424: 2016(1) D.R.T.C. 195: 2016(2) ALL MR 280: 2015(26) R.C. R.(Civil) 294.

[27] Shiv Kumar Wasal & Co. v. Delhi Development Authority, (Delhi) 2016(154) DRJ 420: 2016(4) R.A.J. 157: 2015(54) R.C.R.(Civil) 717.

[28] Chief Engineer v. Chandagiri Construction Company (Madras), 2009(6) R.A.J. 500: 2009(2) ArbiLR 609: 2009(45) R.C.R.(Civil) 172.

[29] M/s Satyam Engineering & Construction v. State of Jharkhand (Jharkhand), 2015(3) J.C.R. 67: 2015(27) R.C.R. (Civil) 42.

v. The defects liability clause stood attracted as the contract had not been terminated before completion of the work.

Where in a case, neither requisite proof was tendered nor there was determination of the engineer placed on record, the claim was rejected by the arbitral tribunal. The Delhi High Court upheld the award.[30]

8.7 Engineer's/Architect's Decision Final

In standard form contracts listed in the table in paragraph 8.1 above, which contain an arbitration clause and also those which do not include an arbitration clause, provisions are normally included which stipulate that the decision of the architect or engineer as regards the quality of work, as the case may be shall be final and binding on all parties to the contract. The arbitration clause in such contracts, if provided, normally opens with a statement, "Except where otherwise provided in the contract all questions and disputes ... shall be referred to ... arbitration ...". The excepted matters usually include quality of materials, workmanship and removal of improper materials and/or work, etc. In some contracts, financial claims arising out of such decisions are not excluded from the scope of the arbitration clause.

The reasons for such a provision appear to be that: (i) disputes on these matters could be so innumerable that it may be virtually impossible to have the contracts concluded to the satisfaction of all concerned in a reasonable time; and (ii) the disputes may be on certain points about which the opinions of experts may vary.

As to the validity of this provision, the following observations of the M. P. High Court[31] lay down the correct legal position, it is submitted:

The rule is well settled that where parties to building or construction designate a person who is authorized to determine questions relating to its execution, and stipulate that his determination shall be final and conclusive, such parties are bound by his determination of those matters which he is authorized to determine, except in cases of fraud or such gross mistake on his part as would necessarily imply bad faith, or a failure to exercise an honest judgment, i.e. on grounds of collusion or misconduct.

Where the contract provides, as here, that the work shall be done to the satisfaction, approval, or acceptance of an architect or engineer, such an architect or engineer is thereby constituted the sole arbitrator between the parties, and the parties are bound by his decision in the absence of fraud and mistake. More so, where there is a stipulation that his certificate made or approval is a condition of the contractor's right to receive payment, such certificate or approval must be, as it is, conclusive as to all matters within his authority.

The illustration below is a noteworthy exception to the rule that the engineer's decision shall be final and the matter cannot be referred to arbitration.

---

[30] National Highways Authority of India v. Bridge & Roof Co. Ltd., (Delhi) 2017(4) R.A.J. 1.
[31] Dandakaranya Project v. P. C. Corporation, AIR 1975 M.P. 152.

EXAMPLE FROM CASE LAW

Cracks developed in sleepers supplied by the respondent to the petitioner. The respondent could be held liable only if the cracks were developed due to manufacturing defects. After the cracks were noticed, the sleepers were tested again by the parties themselves. An expert retested the sample sleepers and gave his report on the causes for the defects. It was not shown that the cracks were developed due to any manufacturing or transportation defects on the part of the respondent. The arbitrator considered all evidence led by the parties in the form of report of expert and photographs produced by the respondent and came to the conclusion that none of the test reports indicated manufacturing defects in the sleepers. Held:

> [M]erely because the decision of the petitioner is stated to be final and binding on the respondent, it cannot be said that it takes away the right of the respondent to seek remedy there against in the arbitration. The petitioner does not contend that the decision of the petitioner on this aspect falls under the "excepted matters" as far as the arbitration is concerned. Therefore, any decision of the petitioner in this regard would be subject to adjudication through arbitration, which has happened in the present case with the arbitrator concluding against the petitioner.

The petition challenging the award was dismissed.[32]

8.7.1 Engineer/Architect, Must Act Judicially

Where a person is appointed as an authority to adjudicate on certain matters and whose decision becomes final and binding on both the parties and where such a person happens to be an employee or agent of one of the parties, it is all the more necessary for him to act in a judicial capacity and not in an administrative capacity, or departmental capacity.[33]

Though the contractor is bound by his contract, still he has a right to demand that, notwithstanding those preformed views of the engineer,

> "that gentleman shall listen to the argument and determine the matter submitted to him as fairly as he can, as an honest man; and if it be shown in fact that there is any reasonable prospect that he will be so biased as to be likely not to decide fairly upon those matters, then the contractor is allowed to escape from his bargain and to have the matters in disputes tried by one of the ordinary tribunals of the land.[34]

If "the engineer has put himself in such a position that it is not fitting or decorous or proper that he should act as arbitrator in any one or more of those disputes", the contractor has the right to appeal to a court of law to exercise a discretion not to submit that dispute to the engineer as an arbitrator.

8.7.2 Court Can Examine the Matter

If the contractor is not satisfied with the decision of the engineer/architect, he can have the matter examined by the court in an appropriate proceeding. The court, when the matters come before it, has always the power to examine whether the officer whose decision is made final has acted judicially or administratively. If the court finds that the officer has acted judicially, the court may

---

[32] IRCON International Limited v. Patil Rail Infrastructure Pvt. Ltd., (Delhi), AIR 2018 Del. 98.

[33] Hickman & Co. v. Roberts, 1913 AC 229. See Heavy Electrical (India) Ltd. v. Pannalal, AIR 1973 MP 7.

[34] Bristol Corp. v. John Arid & Co., [1913] AC 241; see AIR 1973 MP 7.

uphold his decision. But if the court finds that the officer has acted administratively, in that event the court has jurisdiction to set aside the acts not performed judicially.[35]

8.7.3 Whether the Owner Is Bound by Certificate, Satisfaction or Approval

This aspect requires further consideration. The reason is that there can be two possible views: First, it can be said that the contract together with drawings, BOQs (if enclosed), specifications, etc. precisely define the contractual requirements as to sufficiency of the work. In this view, if the work under a contract has to be completed to the satisfaction of the engineer or architect, it may mean that his duty is merely to see that the requirements of the contract are met. This we may call the objective satisfaction of the engineer/architect. Second, it may be that the engineer's/architect's standard is that to which the parties have submitted themselves and that it constitutes the only provision in the contract about quality, or it may be that his standard (which we may call subjective satisfaction) is an added protection, so that the performance under the contract must satisfy both the contract requirements and the certifier. If the first view is taken as correct, the owner can show, despite the satisfaction of the engineer/architect, that the work is nevertheless not in accordance with the contract, then he is not precluded from suing for breach of contract, or from setting up this allegation as a defence to a claim by the builder for the price of the work.

On the other hand, if the second view is considered correct, not only is the owner precluded from disputing sufficiency of the work but also, in some cases, he may not be permitted to allege that under special provisions of the contract the certificate or approval should not have been given, for example, because of the contractor's failure to obtain written authorization for extras, etc.

These two apparently conflicting views may not, in practice, pose any serious problem. The courts treat the provisions as "quasi-arbitration" and strict rules of arbitration may not be applied. Therefore, where it can be shown that the certifier, while acting with complete impartiality and good faith has taken an incorrect view of the consideration which should govern his decision, and has erred in the principles which he applied in reaching it, invariably the decision of the engineer/architect given in his certificate or approval will be disregarded by the court; very much so in cases where he has given the decision violating the principles of natural justice or by fraud or collusion.

In conclusion, it can be said that if there is some injustice likely to be caused by the provision in the contract making the engineer's or architect's decision final and binding, both the contractor and the owner may have recourse to court proceedings or arbitration, against the decision or certificate given by the engineer or architect.

EXAMPLE FROM CASE LAW

The background leading to the filing of writ applications was briefly stated thus. It was the case of the petitioners that in response to the advertisement, they procured polythene rolls/sheets as per the specifications laid down therein and transported the goods to Bhubaneswar. However, for reasons best known to the authorities, they avoided to take prompt action in putting the material to test in order to ascertain as to whether or not it was as per the specification laid down in the advertisement, with the result that hundreds of trucks loaded with polythene rolls and sheets were stranded in and around the stadium. In the meanwhile, however, the SRC and others continued to receive material supplied by the PRC and OSIC which, according to the petitioners, was not subjected to any test. In substance, it is the case of the petitioners that by this act, though material not in accordance with the specification supplied by PRC and OSIC, who had in turn procured material from certain private parties,

---

[35] South India Rly Co. Ltd v. S. M. Bhashyam Naidu, AIR 1935 Mad. 356; see AIR 1975 MP 7.

were accepted, no action was taken to accept their (petitioners') material in spite of their repeated requests made in that behalf. The inaction on the part of the authorities resulted in huge loss to the petitioners not only because of demurrage and truck hire charges but also by way of loss of profits as they had purchased the material from outside the state on payment for being supplied to the SRC.

It was held:[36]

> In the above facts and circumstances, we are of the view that this is a matter which requires a thorough investigation by an independent agency, which in our opinion, is the Central Bureau of Investigation; more so, when crores of public money is involved.

However, the prayer of the petitioners for acceptance of the polythene tendered by them without subjecting the same to the test of tensile strength was not allowed.

> It is no doubt true that the advertisement in question did not prescribe for such a test. However, this was introduced as the quality checking squad was of the view that the material not having a particular tensile strength will not serve the purpose for which it was intended. Thus, it cannot be said that the action of the authorities in laying down the specification for putting the material to the test of tensile strength, was unreasonable. However, the action of the authorities in accepting the material tendered by the PRC or through the OSIC without subjecting the same to proper test and the test of tensile strength, was unreasonable. ... It will, however, be open to the petitioners if they are so advised, to approach the common law forum for the loss/damage, if any, incurred/suffered by them on account of the inaction of the authorities in not issuing a corrigendum and thereby putting them (the petitioners) in a situation which is not of their making.

8.8 Distinction between Engineer (Certifier) and Arbitrator

It is necessary to keep in mind the distinction between the functions of the arbitrator and the engineer (or certifier) dealt with in the same contract. Essentially, the engineer or certifier is performing an administrative rather than a judicial function. He has been described as a "preventor of disputes" while the arbitrator is a person who comes into the scene after a "dispute" (which could not be prevented) comes into existence. It is the duty of the engineer or certifier to apply the provisions of the contract strictly. Any evidence showing that he has taken extraneous or irrelevant matters into consideration will deprive his certificate of validity.[37] He may become disqualified on the ground of interest in the subject matter and not known to the party and not merely because of being employed by the opposite party; or on the ground of fraud and collusion before or after the contract; or because of undue interference by or improper pressure of the employer; conduct which falls short of the high standard of fairness, discreetness and impartiality expected of him in relation to the issue or refusal of the certificate or decision; unreasonable refusal to give consideration to the matter upon which he is requested to certify;

---

[36] M/s. Narendra v. Special Relief Commissioner, Orissa, AIR 2001 Ori 95.
[37] Panamena Europea Nevigaction v. Leyland, [1947] AC 428.

breach of contract by the employer having the effect of preventing the contractor from obtaining a certificate.[38]

The engineer or certifier does not become an arbitrator merely because the words "adjudge" or "exclusive judge" are used. The position may generally be different if, disputes are to be decided by a single person, and the words "reference", "dispute", "decision" are used.[39]

It is also possible that in some contracts, the same person is described as an "engineer" or "certifier" at one stage and later as an "arbitrator".[40] An engineer or certifier who is to decide certain questions with reference to a contractor would be liable for negligence unless, of course, he is also clothed with judicial functions as an arbitrator.[41]

---

[38] *Hudson Building Contracts*, 10th Ed. 1970, pp. 470–471, 498–499. Also see Chapter 14 for further discussion.

[39] See Y. Parthasarathy v. State of A.P (1988) 1 Andh LT 809.

[40] Eaglesham v. Mc Master, (1920) 2 KB 169; Neale v. Richardson, [1938] 11 All ER 753.

[41] Sutcliffe v. Thackrah, [1974] AC 724; Arenson v. Casson Beckman Rutley, [1977] AC 405.

9

Measurement, Valuation and Payment

9.0 Introduction

In a building or construction contract, the consideration for the performance by the contractor is the employer's promise to pay. It is also provided that payment on account is to be made from time to time by instalments and on the certificate of the engineer or architect, unless the contract is on a BOT basis. In a BOT contract, the consideration is right to collect toll/fee from users for a specified duration called the concession period. Even in PPP contracts if a project is not financially viable by collection of fee from users alone, viability funding is adopted whereby the employer agrees to finance the project partially. Thus, essentially the basis and interval of release of interim payments form an important part of the construction contract. As regards the interval of payments, most forms provide for monthly payments. The amount to be paid depends upon the type of contract adopted by the parties. The form of contract may be lump sum, lump sum with bill of quantities or measurement contract.[1] Occasionally there may be a contract in which the price to be paid has not been agreed upon and is to be decided on the basis of *quantum meruit*, fair and reasonable cost.

This chapter refers to the provisions made in this regard in the frequently used standard form contracts. The disputes sometimes arise in respect of measurement of the work or the certificate of the engineer or architect, or payment to the contractor. Legal principles governing such disputes vis-à-vis provisions in the standard form contract are considered in this chapter and illustrated with court cases.

9.1 Provisions of the Law

The following provisions of the law are dealt with in this chapter.

Provision of the Law	Paragraph No.
Indian Contract Act, Sections 28 and 72	9.4.1, 9.5
Order VIII Rule 6 of the Civil Procedure Code (CPC)	9.7
Provisions in Standard Form Contracts	
FIDIC Form 1992, Clauses 55.1–58.3; 60.1–60.10; 63.1–64.1	9.2.3(v), 9.2.3(vi),
1999 1st edition, Clause 14.2	9.9
2017 2nd edition, Clauses 3.7.3, 12.0–12.4	9.8
	9.1.1, 9.2.3(ii)
Ministry of Statistics and Programme Implementation, Government of India (MOS & PI), Clauses 5–8	9.1.2
NITI Aayog Model Form, Clauses 17.1–17.16	9.1.3

---

[1] See Chapter 2 Types of Contracts.

9.1.1 FIDIC Form

Where the parties have adopted an item rate or percentage rate contract, the BOQ forms an integral part of the contract. The provision in such contracts opens with the stipulation that the quantities in the BOQ are approximate, likely to vary, and payments will be released on the basis of actual quantities as measured at the site. The value of the work shall be based on measured quantities multiplied by unit rates mentioned in the BOQ. Even in a lump sum contract, measurements are necessary for valuation of change of scope or variations ordered.

To avoid disputes in respect of the measurements and quantities, the provision requires the engineer to give reasonable notice (not less than seven days as per the 2017 edition) informing the contractor the date and place when the measurements will be taken and further directing the contractor's representative to remain present for the measurements. The provision also makes it clear that if the contractor's representative fails to attend, the measurements taken would be deemed to have been taken in the presence of the representative.

Where the engineer desires the measurements to be taken of permanent works from records and drawings, the engineer shall prepare the same and give notice to the contractor of not less than seven days, intimating the date and place of measurements, for the contractor's representative to remain present. The provision contemplates possible disputes in the recording of the measurements and stipulates a period of 14 days within which the contractor shall set out the reasons why he considers the measurements inaccurate. The dispute will be resolved under Clause 3.7.3 by the agreement of the parties or determination by the engineer subject to the dispute resolution clause in the agreement that is by reference to Dispute Avoidance/Adjudication Board (DAAB) or arbitration.

One of the common sources of dispute is the mode of measurements. Modern forms eliminate this source by specifying the agreed mode of measurements either in the contract data or schedule attached (see Clause 12.2 of FIDIC Form, 2017 edition) or in the description in the BOQ referring to detailed specifications which, in turn, set out the mode of measurements and the ancillary works deemed to be part of the agreed rate that are not to be separately measured or paid for. Clause 12.3 of FIDIC Form, 2017 edition provides for "Valuation of the Works" more particularly of variations for which rate is not available in the BOQ.

Incidentally, the provision stipulates the agreed variation limit for the BOQ quantities beyond which new rates will be decided and agreed. To qualify for extra payment on account of excessive variation beyond agreed limit Clause 12.3(b) stipulates certain conditions, all of which must be satisfied, it is submitted. These are:

i. Measured quantity changed by more than 10% of the quantity stated in the BOQ
ii. The change in the quantity multiplied by the BOQ rate exceeds 0.01% of the contract sum
iii. The change in the quantity directly changes the cost per unit of the item by more than 1%
iv. The item in question is not specified as fixed price or not subject to adjustment for any change in quantity, and/or
v. The work is instructed under Variations and Adjustment Clause 13 and sub-paragraph (a) or (b) above applies.

The provisions above have been analysed in Chapter 5, paragraph 5.4.3.

9.1.2 Ministry of Statistics and Programme Implementation, Government of India (MOS & PI)

The provisions in respect of release of payments are made in Clauses 5 to 8 as follows.

Clause 5 – Payment of Running Bills

A. Bills should be prepared and submitted by the Contractor. Joint measurements should be taken continuously and need not be connected with billing stage. System of 4 copies of measurements, one each for Contractor, Client and Engineer, and signed by both Contractor and Client can be tried.

B. 75% of bill amount should be paid within 14 days of submission of the bill. Balance amount of the verified bill should be paid within 28 days of the submission of the bill.

C. For delay in payment beyond these periods specified in B) above, interest at a pre-specified rate (suggested rate 12% p. a.) should be paid.

Clause 6 – Payment of Final Bills

A. Contractor should submit final Bill within 60 days of issue of defects liability certificate. Client's engineer should check the bill within 60 days after its receipt and return the bill to Contractor for corrections, if any are needed. 50% of undisputed amount should be paid to the Contractor at the stage of returning the bill.

B. The contractor should re-submit the bill, with corrections within 30 days of its return by the Engineer. The re-submitted bill should be checked and paid within 60 days of its receipt.

C. Interest at a pre-specified rate (say 12%) should be paid if the bill is not paid within the time limit specified above.

Clause 7 – Advance Payment

A. Mobilisation Advance and Construction Equipment Advance should be given at 12% interest or free of interest at the discretion of the owner and against Bank Guarantee for Mobilisation Advance and against hypothecation of Construction Equipment to the Owner for Construction Equipment Advance.

B. Mobilisation Advance should be given up to 10% of Contract price, payable in two equal instalments. The first instalment should be paid after mobilisation has started and next instalments should be paid after satisfactory utilisation of earlier advance(s).

C. Construction Equipment Advance should be paid upto 5% of Contract price, limited to 90% of assessed cost of machinery. For special cases, a higher advance for construction equipment up to 10% of contract price may be considered.

D. Construction Equipment advances should be paid in two or more instalments. First instalment should be paid after Construction Equipment has arrived at the site and next instalments should be paid after satisfactory utilisation of earlier advance(s).

E. Recovery of Mobilisation and Construction Equipment advance should start when 15% of the work is executed and recovery of total advance should be complete by the time 80% of the original Contract price is executed.

Clause 8 – Secured Advance

75% of cost of materials brought to site for incorporation into works only should be paid as Secured Advance. Materials which are of perishable nature should be adequately insured. In case advance is not payable against any particular items, they should be listed in the Contract Document.

The above provisions are commonly adopted by other government departments and Public Sector Undertakings. The provisions are self-explanatory.

Clause 10 – Escalation

This form provides for payment of price adjustment on account of fluctuation of prices of inputs as reflected by Price Indices determined in accordance with the formulae specified in Clause 10 as follows:

- A. (I) All short duration contracts up to 24 months should be awarded on fixed price basis and are not subject to any escalation whatsoever. However, only statutory variation limited to duties and taxes are considered for adjustment in contract price. A (II) For calculating escalation, base prices should be taken as on the date of opening of the Bids.
- B. The Contract document should specify the suitable percentage of input for labor, materials like cement, steel, bitumen, POL and other materials and equipment usage for the purposes of calculating escalation.
- C. Escalation should be calculated, based on
  - Notified fair wages and in the absence of which consumer price index for labour would be applicable,
  - Market rate for cement and steel,
  - Average official retail price of bitumen & POL, and
  - Whole sale price index for other materials.
  - Published Government Documents should be used for calculation of escalation amount.
- D. Escalation Reimbursement should be calculated for to the extent of 85% of the escalation so calculated.

9.1.3 NITI Aayog Model Form

This form, under Clause 17.2.1, also provides for advance payment for mobilization at 10% of the contract price payable in two instalments and recovery thereof as specified. Clause 17.3 spells out in detail the procedure for estimating the payment for the works.

17.3 Procedure for Estimating the Payment for the Works

The form contemplates the lump sum contract with BOQ providing for stagewise payments, as is obvious from the procedure prescribed as follows:

1. Clause 17.3.1: "The Authority shall make interim payments to the Contractor, as certified by the Authority's Engineer on completion of a stage, for a length, number or area as specified, and valued in accordance with the proportion of the Contract Price assigned to each item and its stage and payment procedure in Schedule-G."
2. Clause 17.3.2: "The Contractor shall make its claim for interim payment for the stages completed till the end of the month for which the payment is claimed, valued in accordance with Clause 17.3.1, and supported with necessary particulars and documents in accordance with this Agreement."
3. The claim shall be made with Stage Payment Statement for Works as prescribed in Clause 17.4.

4. Clause 17.5.1 stipulates:

> Within 10 (ten) days of receipt of the Stage Payment Statement from the Contractor pursuant to Clause 17.4, the Authority's Engineer shall broadly determine the amount due to the Contractor and recommend the release of 90% (ninety per cent) of the amount so determined as part payment against the Stage Payment Statement, pending issue of the Interim Payment Certificate by the Authority's Engineer. Within 10 (ten) days of the receipt of recommendation of the Authority's Engineer, the Authority shall make an electronic payment thereof directly to the Contractor's bank account.

Clause 17.3.3 clarifies with an illustration that any Change of Scope or the withdrawal of Works under Clause 8.3 are not to affect the amounts payable for the items or stage payments thereof which are "not affected by such Change of Scope or withdrawal".

5. Clause 17.5.2: "Within 15 (fifteen) days of the receipt of the Stage Payment Statement referred to in Clause 17.4, the Authority's Engineer shall determine and shall deliver to the Authority and the Contractor an IPC certifying the amount due and payable to the Contractor, after adjusting the payments already released to the Contractor against the said statement. For the avoidance of doubt, the Parties agree that the IPC shall specify all the amounts that have been deducted from the Stage Payment Statement and the reasons therefor."

6. Clause17.7.1: "The Authority shall pay to the Contractor any amount due under any payment certificate issued by the Authority's Engineer in accordance with the provisions of this Agreement as specified in Clause 17.7.1."

7. Clause 17.13 Change in law: "Pursuant to Clauses 17.13.1 and 17.13.2, if as a result of Change in Law, the Contractor suffers any additional costs or derives benefits from any reduction in costs in the execution of the Works or in relation to the performance of its other obligations under this Agreement, the Contractor shall, within 15 (fifteen) days from the date it becomes reasonably aware of such addition or reduction in costs, notify the Authority with a copy to the Authority's Engineer of such additional costs/benefit due to Change in Law. The Authority's Engineer has 15 days after receipt of such a notice to determine the amount of addition or reduction to the Contract Price due to the Change in Law."

8. Clause 17.14 entitles the engineer to make any corrections or modifications in any previously issued Interim Payment Certificates.

Other notable relevant provisions include:

Clause 17.5.3: "In cases where there is a difference of opinion as to the value of any stage, the opinion of the Authority's Engineer shall prevail and interim payments shall be made to the Contractor on that basis; provided that the foregoing shall be without prejudice to the Contractor's right to raise a Dispute."

The Authority's Engineer may, for reasons to be recorded, withhold from payment, for the reasons stipulated in Clause 17.5.4, estimated costs of omissions or rectification of defects.

Clause 17.5.5 incorporates the usual provision in all standard form contracts that payment by the Authority shall be deemed to be provisional and shall not be construed as the Authority's acceptance, approval, consent or satisfaction with the work done.

17.8 Price Adjustment for Works

The NITI Aayog Model Form also provides for payment of price adjustment on account of fluctuation of prices of inputs as reflected by Price Indices determined in accordance with the formulae specified in Clause 17.8.4. Clause 17.8.3 further stipulates that:

To the extent that any compensation or reimbursement for increase or decrease in costs is not covered by the provisions of this Agreement, the costs and prices payable under this Agreement shall be deemed to include the contingency of such increase or decrease in costs.

17.16 Bonus for Early Completion

Clause 17.16 contains the following provision for payment of bonus for early completion.

> In the event that the Project Completion Date occurs prior to the Scheduled Completion Date, the Contractor shall be entitled to receive a payment of bonus equivalent to 0.03% (zero point zero three per cent) of the Contract Price for each day by which the Project Completion Date precedes the Scheduled Completion Date, but subject to a maximum of 3% (three per cent) of the Contract Price. Provided, however, that the payment of bonus, if any, shall be made only after the issue of the Completion Certificate. For the avoidance of doubt, the Parties agree that for the purpose of determining the bonus payable hereunder, the Contract Price shall always be deemed to be the amount specified in Clause 17.1.1, and shall exclude any revision thereof for any reason.

The form then includes a provision for the final payment statement as follows.

17.10 Final Payment Statement

17.10.1 Within 60 (sixty) days of receiving the Completion Certificate under Clause 12.4, the Contractor shall submit to the Authority's Engineer six copies of a final payment statement (the "Final Payment Statement"), with supporting documents, in the form prescribed by the Authority's Engineer in respect of:

  (a) the summary of Contractor's Stage Payment Statements for Works as submitted in accordance with Clause 17.4;

  (b) the amounts received from the Authority against each claim; and

  (c) any further sums which the Contractor considers due to it from the Authority.

17.10.2 If the Authority's Engineer disagrees with or cannot verify any part of the Final Payment Statement, the Contractor shall submit such further information as the Authority's Engineer may reasonably require.
      17.10.3 The Authority's Engineer shall deliver to the Authority:

  (i) an IPC for those parts of the Final Payment Statement which are not in dispute, along with a list of disputed items which shall then be settled in accordance with the provisions of Article 24; or

  (ii) a Final Payment Certificate in accordance with Clause 17.15, if there are no disputed items.

17.10.4 The Authority's Engineer does not prescribe the form referred to in Clause 17.10.1 within 15 (fifteen) days of the date of issue of the Completion Certificate, the Contractor shall submit the statement in such form as it deems fit.

The contract includes an express provision for discharge of the authority's obligations to the contractor in Clause 17.11 as follows.

17.11 Discharge

Upon submission of the Final Payment Statement under Clause 17.10, the Contractor shall give to the Authority, with a copy to the Authority's Engineer, a written discharge confirming that the total of the Final Payment Statement represents full and final settlement of all monies due to the Contractor in respect of this Agreement for all the Works arising out of this Agreement, except for any monies due to either Party on account of any Defect. Provided that such discharge shall become effective only after the payment due has been made in accordance with the Final Payment Certificate issued pursuant to Clause 17.12.

The above provisions are further analysed from legal aspects below.

9.2 Interim Payment Certificate (IPC)

Unless the contract provides for a specific form in which an engineer or architect is to issue his certificates, all that is required is that the document must clearly appear to be the expression, in a definite form, of the exercise of the opinion of the engineer or architect, in relation to the matter provided for by the terms of the contract. This point assumes importance particularly when there is a specific time limit laid down for either party to the contract to take some action after receipt of the certificate from an engineer or architect. If the alleged document does not clearly appear to be an exercise of opinion under the relevant terms of the contract, regard being had to "form", "substance" and "intent", either party may insist that the engineer or architect should issue a proper certificate, on receipt of which alone, the time limit will begin to run.

9.2.1 Where a Contract Is Silent on Interim Payments

If a contract makes no mention of interim payments, ordinarily the contractor must complete the work before he is entitled to payment.

> **EXAMPLE FROM CASE LAW**
>
> A contract between the owner and the general contractor for concrete work for a building contained no provision about the time or manner of payment. After two months, when the work had progressed up to the first floor the contractor sent a bill for work done up to that time. The owner refused to pay the bill whereupon the contractor stopped the work. In a suit that followed, the contractor claimed the amount of the bill presented plus damages for breach of contract. The court held that: "Where a contract is made to perform work and no agreement is made as to payment, the work must be substantially performed before payment can be demanded."[2]

9.2.2 Role of an Engineer or Architect in Granting Certificates

Granting or withholding certificates is one of the most important functions of the engineer or architect. Employed by the owner as a skilled servant, the engineer or architect also acts in the capacity of the responsible agent of the owner. Thus, he must not only use his utmost skill to see that the work is done properly, but as an agent must act in the interest of the owner to the fullest extent permitted by the contract.

However, many construction contracts include a stipulation that the decision of the engineer or architect shall be final and binding upon both parties. When the engineer or architect agrees to act under this clause, he undertakes the duty towards both parties of holding the scales even, and

---

[2] Stewart v. Newbury, 220 NY 379 115, N.E. 984, (1917).

deciding between them impartially as to the amount payable by one to the other. Clearly, in this capacity he acts as a quasi-arbitrator. The legal significance of this is that his decisions are held to be final and binding and cannot be set aside except on the ground of fraud or collusion.

9.2.3 Guiding Principles

An attempt is made below to highlight some important principles which need to be noted by the engineer/architect while issuing certificates.

(i) Certificates must be given by the engineer or architect
Unless a contrary provision is made in the contract, the engineer or architect himself must give a certificate. However, he may rely upon the measurements made by another person on his behalf. The certificate will not be binding or conclusive if it is shown that the engineer or architect has acted corruptly or abrogated his duty.[3]

(ii) Independent exercise of judgment
There is no general rule of law prohibiting the influencing of certifiers. Apart from fraud, a duty not to influence can only be imposed by an implication arising from contract. There is, after all, nothing to prevent a party from requiring that work shall be done to his own satisfaction. He might then choose to act on the recommendation of an agent. If an agent is named in the contract, it may be plain that he is to function only as the alter ego of this master and then his master can tell him what to do.[4]

Whether it is the act of the master (owner) or the servant (agent), there may arise a question whether the dissatisfaction must be reasonable, or whether it can be capricious or unreasonable so long as it is conceived in good faith. This question will again depend upon the implication to be drawn from the contract. Most modern contract forms are drafted in such words that it is safe to presume that the dissatisfaction must be reasonable. In any case the tendency of the courts seems to be to require the dissatisfaction to be reasonable. The main question that arises is whether the engineer/architect is, or is not intended to function independently of the owner. If the engineer/architect is satisfied that the contractor has performed certain work, he ought to issue a certificate. Otherwise, his conduct would be capricious and arbitrary.[5]

In public works contracts, the engineer-in-charge is the authorized representative of the owner. The appellate authorities too are representatives of the public bodies. In such cases the engineer functioning "independently" remains so in theory only. However, when forms like the FIDIC are used, the provisions invariably empower the engineer to decide first, and inform the owner of his decision next. FIDIC 2017 edition provides for consultation with both the employer and the contractor first, to help them reach an agreement, and if that fails the engineer has to determine on his own. Undoubtedly, these provisions by implication require the engineer to function independently of the owner. However, it is noticed that when FIDIC conditions are adapted in India, invariably the contract contains Conditions of Particular Applications (COPA) which limit the powers of the engineer to decide certain matters independently of the employer.

An engineer or architect performing this delicate task must attempt to function independently by not imposing his own subjective satisfaction, but by an exercise of judgement to ensure that the requirements of the contract are met. When an engineer is called upon by the role in the contract provision to act as a third party, the satisfaction of a third party operates only within the requirements of the contract; that is to say, it is not permissible for the third party to put on one party to a contract, obligations, such as those related to quality, outside the contract.[6] Where a contract permits the employer to replace a construction manager, the employer cannot use that

---

[3] Hans Haugen et al. v. K. Raupach et al., 43 Wn. 2d 147 (1953)..

[4] Minister Trust Ltd. v. Traps Tractor Ltd., Queens Bench Division [1954] 3 All ER 136.

[5] Hans Haugen et al. v. K. Raupach et al., 43 Wn. 2d 147 (1953).

[6] Commell Laird & Co. Ltd. v. Manganese Bronze & Brass Co. Ltd., 1934 referred to in [1954] 3 All ER 136.

power to appoint itself as the construction manager and deprive the contractor of the protection of the contract separating the duties of the employer and certifier.[7]

(iii) Certificate in respect of quality or quantity, or both

Where the provisions of a contract stipulate merely to certify "that the work has been satisfactorily carried out", the question may arise if the words mean and cover only the quality of work done or also cover the amount and value of materials and labour used. In a case, it was held that the certificate was to confine to actual quality of work done only. In the said case, the surveyor refused to give a certificate unless he was supplied with information about the amount and value of materials and labour. The contractor was held to be entitled to recover the amount due to him without the surveyor's certificate.[8] However, most modern forms of contract in use provide for the certificate to cover the quality as well as the value of the work done.

(iv) Rules of natural justice: if and when required to be followed

An architect or engineer, in a typical building or construction contract, has to discharge a large number of functions, both great and small, which call for the exercise of his skilful professional judgement. As already stated, he must throughout retain his independence in exercising that judgement. As long as he does this, and unless the contract so provides, he need go no further and observe the rules of natural justice, taking due notice of all complaints and giving both parties a hearing. It is the position of independence and skill that gives the parties proper safeguards and not just imposition of rules requiring something in the nature of a hearing. For the rules of natural justice to apply, "there must be something in the nature of the judicial situation".[9] Here again a distinction must be attempted between cases where a professional architect or engineer is retained by the owner for a project, and those where the engineer is in the employment of the owner. In the case of the former, one can assume "independence" but in the latter such an assumption will be open to doubt. As a guiding principle, it may be well to remember that whenever the subject matter of decision is routine in the course of transaction, open to be reviewed either at the time of issue of final certificate or otherwise in arbitration proceedings, following strict rules of natural justice may not be necessary. Interim certificates may fall under these categories. In other cases and particularly when the contract stipulates the decision to be final and binding on one or the other party, it is advisable to follow the rules of natural justice and give hearing in a meeting fixed after due notices to both the parties. In any case the rule must not only be remembered but also followed that "if the engineer or architect hears one side, he must also hear the other". If the contract does not contain an express provision that IPCs are provisional, the engineer/architect should be careful before issuing IPCs to ensure that an overpayment is not made. In Ontario, Canada, for example, the monthly payment certificates for a supplier subcontractor were held to be final and binding. If the subcontractor has frontloaded his costs, and received overpayment under the monthly certificates, then towards the end of the contract there may not be much leverage in witholding the final payment for the rectification of defects or omissions.[10]

(v) Decision on claims

An architect/engineer empowered to issue IPCs, or for that matter final certificates, may have no power to decide liability as to an amount payable to or allowable by the contractor unless the terms of the contract permit him to do so. Often, the question of delay is accompanied with claims for compensation. In such an event the engineer/architect may not be empowered to decide the liability for the amount in variance with the terms of the contract.[11] However, most modern standard form contracts provide for reference of "a dispute of any kind whatsoever" between the

---

[7] Scheldebouw BV v. St. James Homes (Grosvenor Dock) Ltd., [2006] EWHC 89 (TCC).

[8] Panamena Europea Navegacion Compania Limitada v. Frederick Leyland & Co Ltd., [1947] A.C. 428.

[9] Hounslow Borough Council v. Twickenham Garden Developments Ltd., Chancery Division (1970) 7 BLR 81, [1971] Ch. 233.

[10] Greco, P., "Payment Certificates: An Ontario judge decides they are final and binding", *Canadian Consulting Engineer*, Oct/Nov 2007, 48, 6 SciTech Collection, p. 86.

[11] See John Laing Construction Ltd. v. County & District Properties Ltd., Q. B. Div. (1982) 23 BLR I.

owner and the contractor, in the first place, to the engineer. FIDIC form of contract, for example, expressly empowers the engineer to decide the claims by way of an additional sum to be added to the contract price.[12] In such an event the engineer or architect is clothed with the jurisdiction to decide each and every claim, manner, right or issue referred to him for his decision. It is advisable to follow the time schedule specified in the contract.

It has to be kept in view that under the relevant clauses pertaining to settlement of disputes, two types of disputes or differences are contemplated between the parties. They are:

1. Disputes or differences in connection with the contract when the contract work is in progress
2. Disputes or differences after the completion of the contract or its recession when the stage of final bill is reached and the disputes pertaining to the claims arising from such final bill.

In both these disputes and differences between the parties, the procedural gamut and the requirements of clauses would be equally applicable. For example, under FIDIC 1992 amended edition, relevant clauses are 67.1 to 67.4. Clause 67.1 provides for reference of dispute of any kind between the employer and the contractor to the engineer for his decision within 84 days after the day on which he received reference. Upon his failure to give decision or his decision not being acceptable to either party, either the employer or the contractor may give notice to the other party, on or before the 70th day of expiry of the 84-day period or receipt of the decision, of his intention to refer the dispute to arbitration. Clause 67.2 provides for the parties to attempt settlement of such dispute within 56 days after the day the notice to commence arbitration was given. Clause 67.3 provides that if the decision of the engineer has not become final or amicable settlement has not been reached, the dispute shall be settled by arbitration. Clause 67.4 provides that if the decision of the engineer has attained finality because no notice of intention to commence arbitration was given but either party fails to comply with the decision, reference to arbitration can be made without following the procedure under Clauses 67.1 and 67.2.

What happens if the procedure is followed for a dispute which arose when the work is in progress and arbitration is not commenced, but the dispute is revised and claim is updated after the completion of the work? It is advisable to follow the procedure *de novo*.[13] It is pertinent to note that the contractor can raise the claim during currency of the contract for the future period also and the arbitral tribunal will have jurisdiction to decide the future claim as well. It is, therefore, not necessary to wait for completion of the work, if the claim has been raised and the procedure has been followed to commence arbitration proceedings. It was held by the Supreme Court in a case:[14]

> It is not possible to hold that the claim No. 1 in so far as it relates to future period during which the contract work continued is beyond the scope of reference or outside the ambit of arbitration clause. The aim of arbitration is to settle all the disputes between the parties and to avoid further litigation.

(vi) Engineer/architect: when may become *functus officio*
Under the terms of certain standard form contracts, an engineer after having given his final certificate becomes *functus officio* and is thereby precluded from issuing any valid certificate thereafter.[15] Under the settlement of disputes clauses, if the engineer fails to give his decision within the period named in the contract for him to do so, he may not be able to give his valid decision even if the other party has not yet invoked the provisions of the arbitration clause. For

---

[12] See for example, FIDIC Clause 67.1 of the 1992 amended edition.
[13] M/s. Shetty's Construction Co. Pvt. Ltd. v. M/s. Konkan Railway Corpn. Ltd., AIR 2000 SC 122. See Chapter 16.
[14] M/s. Shyama Charan Agarwala and Sons v. Union of India, AIR 2002 SC 2659. Also see Chapter 16.
[15] H. Fairweather Ltd. v. Asden Securities Ltd., Queens Bench Div. (1979) 12 BLR 40.

example, the FIDIC form Clause 67.1 of the 1992 amended edition, gives the engineer 84 days to give notice of his decision after the day on which he received reference and thereafter empowers either party to invoke provisions of the arbitration clause on or before the 70th day, after the day on which the said period of 84 days expired. After the expiry of the limitation of 84 days, the right accrues to either party to refer the matter to arbitration and, as such, any decision given by an engineer after 84 days may not be valid or may not have the effect of prolonging the period of limitation, it is submitted. The matter, however, will be different if both the parties decide to abide by accepting the said decision. Clause 67.2 which provides for amicable settlement and a period of 56 days between the day on which the notice of intention to commence arbitration of such dispute is given and the day on which the arbitration proceedings may commence, will enable the parties to accept such a belated decision which will be the basis of an amicable settlement of their disputes. It should be noted that an owner cannot rely on his failure to appoint or replace an engineer within the 84 days to prevent a contractor from commencing arbitration.[16]

9.2.4 Issue of Interim Certificate Does Not Amount to Acceptance of the Work Done

Almost all standard form contracts stipulate that interim payments are by way of advances. An interim certificate given by the engineer or architect usually indicates a qualified approval by him, of the work done at the time the certificate is given. Such an interim certificate does not preclude the engineer from requiring bad, unsound, imperfect or unskilful work to be removed and taken away and reconstructed. Neither interim certificate nor payment made on its basis prevents the owner from subsequently questioning the quality of work already approved.[17] Interim certificates are subject to readjustment in the final certificate unless the contract can be construed to the contrary. The engineer/architect, in other words, is entitled to take a fresh view as to the state of works each time he issues an interim certificate.[18]

9.2.5 Certificates as Condition Precedent to Payment

The conditions of a contract usually make it abundantly clear that the contractor must produce a certificate from the engineer before he can claim payment from the owner. A certificate is thus a "condition precedent", which must be fulfilled before the payment. Therefore, if a certificate is honestly withheld in good faith the contractor cannot maintain action for the amounts alleged to be due to him.[19] Certificates need to be issued to the contractor in order to be effective.[20]

9.2.6 When Contractor Can Recover Payment in the Absence of Certificates

The certificates are described as "good" and "not good" depending upon whether they are in conformity with the contract or not. It is the duty of the engineer or architect to issue certificates as and when due.

The contractor can recover payment in the absence of a certificate, if he has duly performed his contract but the engineer or architect has wrongfully refused to grant him a certificate to that effect. The rule with reference to obtaining the certificates of an architect, engineer or other third person, when required by the contract, does not apply if it is made to appear that the owner, without sufficient justification thereof, causes the certificate to be withheld. In such a case, the contractor is excused from the necessity of obtaining the certificate. If the engineer/architect is satisfied that there has been a substantial performance of the contract, it then

---

[16] Al-Waddan Hotel Ltd. v. Man Enterprise SAL (Offshore), [2014] EWHC 4796 (TCC).

[17] Newton Abbot Development Co. Ltd. v. Stockman Bros, (1913) 47, T.L.R. 616.

[18] London Borough of Camden v. Thomas McInery & Sons Ltd., Q.B. Div. (1986) 9 Con LR 99.

[19] Dunlop & Ranken Ltd. v. Hendall Steel Structures Ltd., Q.B. Div (1957) 3 All ER 344.

[20] London Borough of Camden v. Thomas McInery & Sons Ltd., Q.B. Div. (1986) 9 Con LR 99.

becomes his duty to issue the certificate, and if he does not do so, his conduct is regarded as arbitrary and capricious. If the architect is in collusion with his principal, or yields to his opposition to the issuance of the certificate when such opposition is not justified, then in such cases the contractor has a legal excuse for not obtaining a certificate as a condition precedent to recovering on this contract.[21]

EXAMPLES FROM CASE LAW

1. In a case, the employer instructed the architect not to issue a certificate until the contractor had submitted his account for extra and he had received it. The architect wrote to the contractor to see the employers "because in fact of their instructions to me I cannot issue a certificate whatever my own private opinion in the matter". On the above facts it was held that the employer improperly influenced the architect and as such the contractor could recover payment in the absence of the certificate.[22]

2. A contract contained a provision empowering the contractor to determine the contract "if the employer interferes with or obstructs the issue of any such certificate" of interim payment. When the work was nearing completion, suspecting overpayments, the employer appointed an independent surveyor to carry out valuations and measurements, which had previously been done. Based on the valuation by the surveyor, the architect issued a certificate for £1,287 as against £5,785. The contractor pointed out several latches and lapses on the part of the surveyor. It was observed:[23]

> I think without attempting an exhaustive enumeration of the acts of the employer which can amount to obstruction or interference, that the clause is designed to meet such conduct of the employer as refusing to allow the architect to go on to the site for the purpose of giving his certificate or directing the architect as to the amount which he should arrive at on some matter within the sphere of his independent duty.

It was held that negligence, errors or omissions by someone who, at the request, or with the consent, of the architect is appointed to assist him in arriving at the correct figure to insert in his certificate does not amount to interference. Interference connotes intermeddling with something, which is not one's business, rather than acting negligently.

3. In a case, the employers instructed their surveyor not to apply day-work rate for part of the works but that he was to use estimated quantities and apply a measured rate. The contract provided that the decision of the surveyor with respect to value was to be final and without appeal. The contractor alleged that the certificate was not honestly made or given in exercise of, or in reliance upon, his own judgement, but was made and given by reason of the interference of and in obedience to, the direction and orders of the employer. It was held that the act of the employer amounted to interference with the surveyor's functions and the final certificate issued by the surveyor was not conclusive and binding on the contractor.[24]

---

[21] Hans Haugen et al. v. K. Raupach et al., 43 Wn. 2d 147 (1953).

[22] Hickman & Co. v. Roberts, House of Lords, [1913] AC 229.

[23] By Lord Tucker in R.B. Burden Ltd. v. Swansea Corporation, House of Lords [1957] 3 All ER 243.

[24] Page v. Lindaff and Dinas Powais Rural District Council, Q. B. Div. (1901) 2 HBC 316.

9.3 Failure to Make Interim Payments Constitutes Breach of Contract

A construction contract usually provides for instalment payments on the basis of certificates to be granted by the engineer or architect. If the owner fails to honour such certificates by making payments to the contractor, his failure amounts to a breach of contract. Upon such default, the contractor can stop the work and may abandon the contract and seek compensation on the basis of fair and reasonable value of a work done.[25] The observations of the United States Supreme Court in this issue in an opinion case are:

> [I]n a building or construction contract like the one in question, calling of the performance of labour and furnishing of materials covering a long period of time and involving a large expenditure, a stipulation for payments on account, to be made from time to time during the progress of the work, must be deemed so material that a substantial failure to pay would justify the contractor in declining to proceed ... As is usually the case with building contracts, it evidently was in the contemplation of the parties that the contractor could not be expected to finance the operation to completion without receiving the stipulation payments on account as the work progressed. In such cases a substantial compliance as to advance payments is a condition precedent to the contractor's obligation to proceed.[26]

The Supreme Court of India also has held that failure to make payment of interim bills and withholding of "substantial amount over a very long period without any reasonable cause" amounts to a breach of contract and may justify stoppage of the work by the contractor and rescinding the contract.[27] This was a contract whereby Durga Datta undertook to transport coal for Rungta and Clause 5 of the contract entitled him to payments by the 10th of the following month. It was observed:

> In commercial transactions, time is ordinarily of the essence and in the agreement, with which we are concerned, the payment of bills, by a particular date was expressly mentioned. The intention, obviously, was that Durga Datt would receive payments for work executed as soon as the amounts became due. Rungta did not pay these amounts, which were also within his own knowledge either by the 10th of the following month or even within a reasonable time after the presentation of the bills. In these circumstances, we are of the opinion that Cl. (5) was breached by Rungta.

It was finally held:

> The case is thus covered by S. 55 of the Indian Contract Act and Durga Datt was entitled to rescind the contract, when the very important condition of the agreement was broken by Rungta. We confirm the finding of the Judicial Commissioner in this part of the case.

From the above cases, the words, "substantial failure to pay" need careful consideration. In the absence of an express provision in the contract empowering the contractor to terminate the contract, on the owner's failure to pay interim bills, the contractor should resort to the last step cautiously. Where the owner is unable to honour an interim certificate due to temporary difficulties, which he is likely to overcome shortly, the contractor's act of terminating the contract may be premature. The contractor must satisfy himself that either the owner is unwilling to pay or unable to pay, before he acts under this provision. The non-payment by the owner may be because the owner disputes the amount certified. It seems that merely because the owner disputes the certificate, he cannot avoid liability to pay if the certificate has been properly issued under the

---

[25] White v. Livingston, 75 N/Y.S.S. 466, 69, App. Div. 361 Affirmed 174 N.Y. 538.

[26] Ceurim Stone Co. v. P.J. Carlin Construction Co., 248U.S.334,63, L Ed. 275, 285.

[27] Mahabir Prashad Rungta v. Durga Datt, AIR 1961 SC 990.

terms of the contract. The certificate must be honoured.[28] The owner has the right under most contracts to invoke the provisions of the arbitration clause. As to which disputes in respect of non-payment can be referred to arbitration, the following observations of the Calcutta High Court are illuminating:[29]

> ... that every kind of non-payment of the price stipulated in a contract containing the arbitration clause cannot be considered to be a repudiation giving rise to a dispute. A non-payment may arise by reason of one's inability to pay, while not disputing liability thereof. A non-payment, on the other hand, may be the result of repudiation or denial of its liability to pay. Thirdly, a non-payment of price may mean failure to fulfill ones' obligation under the contract to pay within the time stipulated. When there is no repudiation or denial of liability a simple non-payment or default in payment may not give rise to a dispute which can be referred to arbitration. On the other hand, when there is denial of liability and by reason thereof payment is not made by a party from whom demand is made by the other party, the same would be a case of repudiation. In our view the third kind of case mentioned by us, that is, failure to pay within the time provided in the contract, depending upon the terms of the particular arbitration clause, could be validly the subject matter of a reference to arbitration.

9.4 Writ Petition if Maintainable to Enforce Payment by State?

Broadly speaking, a writ of *mandamus* cannot be used to enforce an obligation to make payment under a contract. The Supreme Court of India had held that:[30]

> a writ of mandamus may be granted only in a case where there is a statutory duty imposed upon the officer concerned and there is a failure on the part of that officer to discharge that statutory obligation. There is no statutory duty or a public duty cast on the respondents to make payment of the amount to the petitioners in the sense that no payment is enforceable by the issue of a writ of mandamus. It is just an obligation arising out of a contract between the parties in which the proper remedy will be a suit.[31]

This view has not been departed from in the decisions reported which deal with the cases arising under Art. 14 of the Constitution.[32]

The High Court, it is submitted, is not deprived of its jurisdiction to entertain a petition under Art. 226 merely because in considering the petitioner's right to relief, questions of fact may have to be determined. In a petition under Art. 226, the High Court has jurisdiction to try issues both of fact and law. Exercise of the jurisdiction is, it is true, discretionary, but the discretion must be exercised on sound judicial principles. When the petition raises questions of fact of a complex nature, which may for their determination require oral evidence to be taken, and on that account the High Court is of the view that the dispute may not appropriately be tried in a writ petition, the High Court may decline to try a petition.[33] However, "the judicial control over the fast

---

[28] Killby & Gayford Ltd. v. Selincourt Ltd., Court of Appeal (1973) 3 BLR 104.

[29] Nanalal M. Verma & Co. Ltd. v. Alexandra Jute Mills Ltd., AIR 1989 Cal. 6(8).

[30] Lakhrai v. Dy. Custodian, Bombay AIR 1966 SC 334; cited with approval in Namakkal South India Transports v. Kerala State Civil Supplies Corporation Ltd., AIR 1997 Ker. 56.

[31] Namakkal South India Transports v. Kerala State Civil Supplies Corporation Ltd., AIR 1997 Ker. 56.

[32] In Ramana v. I.A. Authority of India, (1974 SC 1628); Dwarkadas Marfatia and Sons v. Bombay Port Trust, 1989 (3) SCC 293: (AIR 1989 SC 1642); and Central Inland Water Transport Corporation Ltd. v. Brojo Nath, AIR 1986 SC 1571: 1986 Lab IC 1312.

[33] Smt. Gunwant Kaur v. Municipal Committee, Bhatinda, (1969) 3 SCC: (AIR 1970 SC 802).

expanding maze of bodies affecting the rights of the people should not be put into watertight compartments".[34] It should remain flexible to meet the requirements of variable circumstances. *Mandamus* is a very wide remedy which must be easily available "to reach injustice wherever it is found". Technicalities should not come in the way of granting that relief under Article 226.[35]

If the action of the state is related to contractual obligations or obligations arising out of tort, the court may not ordinarily examine it unless the action has some public law character attached to it. Broadly speaking, the court will examine actions of state if they pertain to the public law domain, and refrain from examining them if they pertain to the private law field. The difficulty will lie in demarcating the frontier between the public law domain and the private law field. The question must be decided in each case with reference to the particular action, the activity in which the state or the instrumentality of the state is engaged when performing the action, the public law or private law character of the action and a host of other relevant circumstances.[36] The Apex Court of India has held:[37]

> [W]e have no hesitation in saying that the ultimate impact of all actions of the State or a public body being undoubtedly on public interest, the requisite public element for this purpose is present also in contractual matters. We, therefore, find it difficult and unrealistic to exclude the State actions in contractual matters, after the contract has been made, from the purview of judicial review to test its validity on the anvil of Article 14.

Yet another decision referring to interference in contractual matters is one reported in Hindustan Petroleum Corporation Ltd. v. Dolly Das,[38] wherein it is held:

> if the facts pleaded before the Court are of such a nature which do not involve any complicated questions of fact needing elaborate investigation of the same, the High Court could also exercise writ jurisdiction under Article 226 of the Constitution in such matters. There can be no hard and fast rule in such matters.

The position is also well settled that if the contract entered between the parties provides an alternate forum for resolution of disputes arising from the contract, then the parties should approach the forum agreed by them, and the High Court in writ jurisdiction should not permit them to bypass the agreed forum of dispute resolution:[39] Thus, the position is clear that once the state action is attacked as being unreasonable, unfair and against public interest irrespective of the sphere, the state action is amenable to judicial review. Essentially, the only limitation of the High Court is the self-imposed restriction. A few relevant factors in exercising the self-imposed limitation under Art. 226 of the Constitution of India in the matter of payment of contractors' bills are:[40]

1. When there is no disputed question of fact requiring adjudication on detailed evidence.

2. When no alternate form is provided in the resolution of any disputes pertaining to a contract.

3. When claim by one party is not contested by the other and the contest does not require adjudication requiring detailed enquiry into facts.

---

[34] de Smith, Woolf and Jowell's *Judicial Review of Administrative Action*, 5th edition, cited in VST Industries Ltd. v. VST Industries Workers' Union, (2001) 1 SCC 298.

[35] Anadi Mukta Sadguru Shree Muktajee Vandasji Swami Suvarna Jayanti Mahotsav Smarak Trust v. V.R. Rudani, (1989) 2 SCC 691: (AIR 1989 SC 1607), the said issue has been squarely dealt with in paras 17 and 22 (of SCC).

[36] Life Insurance Corporation of India v. Escorts Ltd., (1986) 1 SCC 264: (AIR 1986 SC 1370 at p. 1424).

[37] Kumari Shrilekha Vidyarthi v. State of U.P., (1991) 1 SCC 212: (AIR 1991 SC 537).

[38] (1999) 4 SCC 450.

[39] Life Insurance Corporation v. Smt. Asha Goel, (2001) 2 SCC 160: (AIR 2001 SC 549).

[40] State of Kerala v. T. V. Anil, AIR 2002 Ker. 160. Swaminathan Construction, Petitioner v. The Divisional Engineer, National Highways Thanjavur, AIR 2003 Mad. 202.

9.4.1 Arbitration Clause Bars Court's Jurisdiction

In contracts which contain an arbitration clause, the court has no jurisdiction to go behind a certificate of the engineer or architect or any supervising officer.[41] Under Indian law, an agreement in restraint of trade proceedings is void except in the case of arbitration agreements for which there is an express reservation in Exceptions 1 and 2 of Section 28 of the Indian Contract Act. The said section reads:

> 28: *Agreement in restraint of legal proceedings void*
> Every agreement:
>
> (a) By which any party thereto is restricted absolutely from enforcing his rights under or in respect of any contract, by the usual legal proceedings, in the ordinary tribunal, or which limits the time within which he may thus enforce his rights, is void to that extent; or
>
> (b) Which extinguishes the rights of any party thereto, or discharges any party thereto from any liability, under or in respect of any contract on the expiry of a specified period so as to restrict any party from enforcing his rights, is void to that extent.

> Exception 1: *Saving of contract to refer to arbitration dispute that may arise*
> This section shall not render illegal a contract by which two or more persons agree that any dispute which may arise between them in respect of any subject or class of subjects shall be referred to arbitration and that only the amount awarded in such arbitration shall be recoverable in respect of the dispute so referred.

> Exception 2: *Saving of contract to refer questions that have already arisen*
> Nor shall this section render illegal any contract in writing, by which two or more persons agree to refer to arbitration any question between them which has already arisen, or effect any provisions of any law in force for the time being as to references to arbitration.

9.5 Payment Made under Mistake

If, during the performance of a contract, payment is made under mistake, can it be recovered? The Indian Law on this point is embodied in Section 72 of the Indian Contract Act. The section reads:

> 72. *Liability of person to whom money is paid, or thing delivered, by mistake or under coercion*
> A person to whom money has been paid, or anything delivered, by mistake or coercion, must repay or return it.

In very plain language, the section declares that a person to whom money has been paid by mistake must repay it. However, for proper understanding of this seemingly simple statement, one must have a correct concept of the word "mistake".

9.5.1 Mistake Defined

Mistake may be defined as an erroneous impression, amounting to conviction that some circumstances with regard to the matter in hand are different from what they really are. Further,

---

[41] Oram Builders Ltd. v. M.J. Pemberton and C. Pemberton, [1985] 1 WLUK 408.

a mistake may be either of fact or of law. When a person is induced to do an act by a misconception of facts, there is a mistake of fact. When there is knowledge of the facts and a wrong conclusion is drawn as to their legal effects, it is a mistake of law.

9.5.2 English Law Not Applicable

It is now well settled that the English law relating to this matter is different from the Indian Law as laid down in Section 72. According to principles of English law, money paid under mistake of fact is recoverable, whereas the Supreme Court of India has held that the term "mistake" used in Section 72 of the Contract Act has been used without any qualification or limitation whatever and comprises within its scope a mistake of law as well as a mistake of fact. There is no warrant for ascribing a limited meaning to the word "mistake" as used therein. There is no conflict between the provisions of Section 72, on the one hand, and Sections 21 and 22 of the Contract Act, on the other. Section 21 declares, it may be noted, that a contract is not voidable because it was caused by a mistake as to any law in force in India. Section 22 lays down that a contract is not voidable merely because it was caused by one of the parties to it being under a mistake as to a matter of fact.

The principle is that if one party, under a mistake whether of fact or law, pays to another party money which is not due by contract or otherwise, that money must be repaid. The mistake lies in thinking that the money paid was due when in fact it was not due and that mistake, if established, entitles the party paying money to recover it from the party receiving the same.[42]

It is not to be implied from the observations of their Lordships of the Supreme Court of India, cited above, that a sum paid under a mistake is recoverable no matter what the circumstances may be. The rule laid down was qualified to the effect that the amount paid by mistake is recoverable "subject however to the question of estoppel, waiver, limitation or the like". On these grounds, findings of the Calcutta High Court, in the case in example (1) below are noteworthy, it is submitted.

EXAMPLES FROM CASE LAW

1. In the said case,[43] the contractor paid sales tax believing that sales tax was payable even in respect of construction contracts. Under the terms of the contract if sales tax was payable, it was to be paid by the owner. Accordingly, the owner reimbursed the amount of sales tax paid. Obviously, both parties were labouring under a common mistake of law and fact that sales tax was payable. Immediately after the mistake became known demand was made by the owner for refund of the amount paid to the contractor towards reimbursement of the sales tax paid by him. It was observed by the Calcutta High Court:

> There is no question of estoppel in this case because estoppel must be of the fact and not of law ... this is a case where both parties were labouring under the same mistake and again for this reason there is no case of estoppel. ... There is no question of waiver in this case, because, immediately, after the mistake became known to the plaintiff demand was made by the plaintiff for a refund of the sales tax. There is no question of limitation in this case, because the suit was instituted within three years from the date of payment by the plaintiff to the defendant.

It was held that the owner was entitled to refund of the money paid or advanced by mistake to the contractor.

---

[42] Sales Tax Officer, Banaras v. Kanhaiya Lal Mukund Lal Saraf, AIR 1959 SC 135.
[43] Calcutta Corporation v. Hindustan Construction Co. Ltd., AIR 1972 Cal. 420.

2. The Board of Trustees for the Cochin Port Trust, the plaintiff in a case decided by the Kerala High Court, invited tenders for the purchase of tractors and the defendant, Ashok Leyland Ltd., was one of the parties who submitted its quotation. The subsequent agreement included a provision reading: "Excise duty, Central Sales Tax, Delivery charges, insurance charges and all other duties/taxes as applicable will be paid extra at actuals by the Port Trust." The defendant supplied the tractors and payment was effected to the defendants as per bills with excise duty at 12.5% and 5% special excise duty on the basic excise duty. The plaintiff averred that the defendant had paid the excise duty on a concessional rate and actual amount paid was less by Rs. 11,034.51/- than that paid by the plaintiff. The plaintiff sought the refund of the excess amount wrongfully and improperly levied in the bill. Reliance was placed on Section 72 of the Contract Act. The defence was based on certain statutory notification under which the defendant was entitled to retain an amount representing 25% of the excise duty payable in respect of the motor vehicles cleared in excess of the quantity in a particular year. The trial court dismissed the suit holding that the plaintiff was not entitled to get refund of the excess amount paid. The Kerala High Court allowed the appeal and granted a decree to the plaintiff.[44]

The scope of Section 72 of the Contract Act came up for consideration before the Madras High Court.[45] In that case, a person purchased a motor car at a price controlled under the Madras Civil Motor Cars Control Order 1947, but afterwards the vendee came to know that he paid more than the controlled price upon the false representation of the vendor. The court held that the excess payment was a payment made by mistake, attracting Section 72 of the Contract Act.

The Allahabad High Court, after reviewing the decisions on the question of "mistake" within the meaning of Section 72 of the Contract Act, observed:[46]

> The question as to whether moneys have been paid under a mistake has got to be adjudged with reference to the litigant who claims the refund and not by the yardstick of a prudent and diligent assessee. Therefore, in case the assessee could establish that there was a mistake of law committed by it at the time when the payments were made, the mere fact that it might have been possible for the assessee by the exercise of due care and diligence to have known of the correct position in law would not disentitle it to the benefit of S. 72 of the Contract Act.

A division bench of the Gujarat High Court, in the course of judgment, observed:[47]

> In our opinion, therefore, the person claiming repayment or return of money or thing u/s 72 must establish that justice of the case requires such repayment or return of money or thing .... In our opinion, therefore, in order to succeed in an action for restitution under S. 72 of the Contract Act it is absolutely essential for the person claiming restitution to establish ownership, loss or injury.

---

[44] Board of Trustees of the Cochin Port Trust v. Ashok Leyland Ltd., AIR 1992 Ker. 1 (3–5).

[45] Lakshmanprasad and Sons v. S.V. Kamal Bai, AIR 1960 Mad. 335.

[46] Modern Industries v. State of Uttar Pradesh, Vol. 32 1973 STC 555.

[47] Union of India v. Ahmedabad Manufacturing and Calico Printing Co. Ltd., (1984) (17) BLT 246.

EXAMPLES FROM CASE LAW

1. A bank discovered that there were mistakes in its accounts which resulted in (a) an entry of Rs. 40,000/- being made in the defendant's account for the deposit of a cheque in the sum of Rs. 4,000/- by the defendant, and (b) a debit entry of Rs. 10,000/- being made for a drawing of Rs. 20,000/-. The defendant withdrew these excess amounts that had been wrongly deposited in his account and refused to pay them back. The court did not believe that the defendant's conduct was *bona fide* and awarded Rs. 45,000/- to the bank with 6% per annum simple interest.[48]

2. The plaintiff-respondent had filed a money suit in the Court of the Subordinate Judge, Rourkela against the Union of India and others for refund of Rs. 1,32,87,749/-, which, according to the plaintiff, had been illegally charged to it by the Railway Administration for the transport of coal by rail. A shorter route of 667 km was used but charges were based on a longer route of 1,082 km. The railways relied on the powers of using a "rationalized route" which they claimed had been nationally implemented. They also claimed that a notice had to be given for overcharges under the Indian Railways Act current at the time of the dispute. The Orissa High Court held that there was not an overcharge under the Railways Act; the charge was not in excess of what was prescribed for a route. Rather, there was unjust enrichment by using shorter routes. The Indian Railways Act, it was held, did not supersede Section 72 of the Indian Contract Act. A decree in favour of the plaintiff was upheld on appeal.[49] Deciding the appeal, it was held:

> The claim for differential amount of freight what was payable and what was actually paid would, I feel, be a claim falling under S. 72 of the Contract Act, for it has been held in Union of India v. Steel Stock Holders Syndicate, Poona,[50] that the Indian Railways Act does not supersede the provisions of the Contract Act.

9.6 Overpayments and Underpayments

It is not uncommon to find in a construction contract with government, provision worded as follows:

> The Government reserves the right to carry out post-payment audit and technical examination of the final bill including all supporting vouchers, abstracts etc. The Government further reserves the right to enforce recovery of any overpayment when detected, notwithstanding the fact that the amount of final bill may be included by one of the parties as an item of dispute before an arbitrator appointed under Condition 60 of this Contract and notwithstanding the fact that the amount of the final bill figures in the arbitration award.

---

[48] M/s S.K. Bagra and Co. v. Bank of India, (Rajasthan)(Jaipur Bench) 2017(1) W.L.N. 376.

[49] Union of India and others v. Steel Authority of India Limited, AIR 1997 Ori. 77. Also see Shiv Shakti Brick Kiln Owners Association v. Union of India, (Delhi): 2010(117) DRJ 385: 2010(37) R.C.R.(Civil) 385: 2010(3) ILR (Delhi) 694.

[50] Union of India v. Steel Stock Holders Syndicate, Poona, AIR 1976 SC 879.

If as a result of such audit and technical examination any overpayment is discovered in respect of any work done by the Contractor or alleged to have been done by him under the contract, it shall be recovered by the government from the Contractor by any or all of the methods prescribed above or if any underpayment is discovered the amount shall be duly paid to the Contractor of the Government.[51]

The clause provides for mutual adjustment of payments made, so that no party is put to loss or gains an unfair advantage on account of underpayment or overpayments made under the contract, respectively. The stipulation reserving the right of the government to adjust over-payment against amounts due to the contractor under any other contract with the government is valid and enforceable.

In a case, L and F entered into an oral agreement for one to pay the "build costs" of construction of houses for agreed budgets and to share profits when the houses were sold. F constructed the houses and L paid from time to time on demands raised by F without assessing the precise amounts due or even asking for a breakdown of costs. These payments were made voluntarily but later when the parties' relationship no longer operated smoothly, L tried to recover "overpayments" made by mistake. L had mistakenly believed that the budget and build costs would not be much different from each other. The Court of Appeal held that if a claimant voluntarily pays to the defendant more than he owes, but chooses not to ascertain the precise amount due, he cannot ordinarily recover the overpayment in the absence of fraud or misrepresentation.[52]

EXAMPLES FROM CASE LAW

1. A clause in the contract was: "We agree for any adjustment as may be necessary on account of quality or quantity of supply to be made from our bills or subsequent bills."

A suit was filed for the recovery of an amount due but which was adjusted. The trial court found that though there was such an agreement for adjustment, unless the appellants pleaded either set-off or counter-claim and paid the court fee, they were not entitled to the relief. Consequently, the suit was decreed. On appeal, the High Court found that in the light of the agreement and adjustment from future bills the appellants were entitled to adjust the same from the future supplies since fraud was discovered for the first time in the year 1969. After it was pointed out by the Audit Department that the plaintiff had supplied Grade-II coal but collected the price of Grade-I coal, the appellants were entitled to adjust the same. From the evidence on record, about 12,038 tonnes of coal was supplied but what was the total quantity of the coal supplied between 7th December 1962 and June 1967 had not been brought on record and even the price which prevailed for Grade-II and Grade-I coal during the relevant period was not produced. Consequently, the appellant could not succeed in avoiding the decree. Thus, the appeal was dismissed. The question to be decided by the Supreme Court was whether the High Court was justified in dismissing the appeal and confirming the decree of the trial court on the facts of this case? It was held:[53]

---

[51] Rawla Construction Co. v. Union of India and Another, AIR 1977 Del. 205. Also, see Clause 23 of the General Conditions of Contract used by NTPC Ltd., a Government of India enterprise.
[52] Graham Leslie v. Farrar Construction Ltd., [2016] EWCA (Civ) 1041.
[53] Steel Authority of India Ltd. and others, Appellants v. New Marine Coal Co. (Pvt.) Ltd., AIR 1996 SC 1250.

After the discovery of the fraud, the appellants started adjusting the amounts of over payments from the future bills payable to the plaintiff. Having found this fact, necessarily, the High Court either would have called for a finding from trial Court, after giving opportunity to the parties, and adjudged the rights of the parties or would have remitted the matter to the trial Court to give an opportunity to the appellants to place on record evidence in that behalf. We think that the latter course would be more feasible. Accordingly, we set aside that part of the judgment of the High Court and the decree of the trial Court and remit the suit to the trial Court. The trial Court is directed to give an opportunity to the appellants to adduce evidence of the total supplies made during the period from December 7, 1962 to end of December 1967 and also the prevailing price of Grade-I and Grade-II coal.

2. One of the terms in a government contract read: "When no such amount ... for the purpose of the recovery from the contractor against any claim of the Government is available such a recovery shall be made from the contractor as arrears of land revenue."[54] It was the case of the state government that since the price quoted of certain items exceeded the prescribed price, an overpayment was made. Against those orders passed by the authorities, the petitioners preferred the Special Civil Applications challenging the said notices. Notices were held illegal since the claim was not adjudicated and found due. Against the said decision of the learned single judge, the State of Gujarat filed Letters Patent Appeals. It was observed:

It is not in dispute that the original petitioners had entered into such contract and Clause 20B was in existence at the time when the contract came to be given to the original petitioner. Therefore, when it is an admitted position that Clause 20B was there in the terms and conditions of the contract, the contention that for recovery of the claim of the Government, the same cannot be made as arrears of land revenue cannot be accepted because the original petitioners themselves are party to the said contractual agreement and the petitioners cannot be permitted to back out from the said contract which is admittedly entered into by them.

It was alternatively submitted on behalf of the original petitioners that even if it is assumed that the Government has power to recover as arrears of land revenue, then such powers can be exercised only when the claim is adjudicated by the competent Court and after such order of the competent Court, the amount can be recovered as the arrears of land revenue.

Reliance was also placed upon the judgment of the Apex Court in a case for the recovery of damages for breaches of contract and interpreting Sections 73 and 74 of the Indian Contract Act.[55] which was relied upon by the learned single judge in his judgment. The Apex Court held that

a claim of damages for breach of contract is not a claim for a sum presently due and payable and the purchaser is not entitled in exercise of the right conferred upon it under Clause 18, to recover the amount claimed by appropriating other sums due to the contractor.

---

[54] State of Gujarat v. Pravinchandra C. Khatiwala, AIR 2002 Guj. 374.
[55] Union of India v. Raman Iron Foundry, AIR 1974 SC 1265.

The case before the Gujarat High Court was not that of recovery of damages from the contractor arising out of breach of contract. But the claim of the Government against a contractor was for recovery of excess payment made. The Supreme Court decision was held inapplicable to the case before the Gujarat High Court.

The last plea was that at least the petitioners ought to have been given opportunity to put forward their case to show that the payment made was not an excess payment. In reply it was the case of the state government that umpteen times, notices were given to the original petitioners. It was held:[56]

> So far as the main Special Civil Applications are concerned, we hold that the State Government is entitled to recover the amount as per Clause 20B of the contract agreement as the arrears of land revenue. However, before exercising such powers, the competent authority will afford an opportunity of hearing to the original petitioners and the original petitioners will be heard and thereafter appropriate orders will be passed by the competent authority for recovery of amount as the arrears of land revenue.

9.7 Conditions under Which Legal Set-Off May Be Allowed

The conditions under which a court may allow a legal set-off may be referred to for this purpose. These are mentioned in Order VIII, Rule 6 of the Civil Procedure Code, which reads:

Particulars of set-off to be given in written statement-

(1) Where in a suit for the recovery of money the defendant claims to set off against the plaintiff's demand any ascertained sum of money legally recoverable by him from the plaintiff, not exceeding the pecuniary limits of the jurisdiction of the Court, and both parties fill the same character as they fill in the plaintiff's suit, the defendant may, at the first hearing of the suit, but not afterwards unless permitted by the Court, present a written statement containing the particulars of the debt sought to be set-off.

(2) *Effect of Set off*: The written statement shall have the same effect as a plaint in a cross-suit so as to enable the Court to pronounce a final judgment in respect of the original claim and of the set-off: but this shall not affect the lien, upon the amount decreed, of any pleader in respect of the costs payable to him under the decree.

(3) The rules relating to a written statement by a defendant apply to a written statement in answer to a claim of set-off.

The provisions can be analysed as below:

In order to entitle a defendant to claim a set-off under this rule, the following conditions, must be present:

i. The suit must be one for the recovery of money.

ii. The amount claimed to be set-off must be:

    a. an ascertained sum of money

    b. legally recoverable

---

[56] State of Gujarat v. Pravinchandra C. Khatiwala, AIR 2002 Guj. 374.

c. by the defendant

d. from the plaintiff

e. within the pecuniary limits of the jurisdiction of the court in which the suit is brought; and

iii. Both parties must fill in the defendant's claim to set-off the same character as they fill in the plaintiff's suit.

If a contractor, aggrieved by the fact that a certain amount is deducted from the sum due to him under a contract, files a suit for recovery of the full price, the government can, under the above procedure, claim a set-off for the amount due to it from the contractor under another contract. It may be noted that both parties fill the same character so long as the amounts relate to contracts let out by the government to the contractor. Second, unlike equitable set-off, the cross-demands in a legal set-off need not arise out of the same transaction.

It may, however, be noted that a claim for damages or for breach of contracts is "not a claim for a sum presently due and payable".[57] The amount of such claim, therefore, cannot be recovered by appropriating other sums due to the contractor unless the claim for payment is admitted by the contractor, or in case of dispute, adjudicated upon by a court or other adjudicating authority.

9.8 Final Certificates

The engineer/architect, on completion of the work, grants completion and final certificates to the contractor. These certificates indicate that the work has been completed in accordance with the terms and conditions of the contract and to the satisfaction of the engineer or the architect. Thus the engineer or architect, by way of issuing these certificates, assures the owner that the contractor is entitled to the payment of the balance of his account as finally adjusted. Clause 14.13 of FIDIC Conditions of Contract, 1999/2017 editions provides for issuing such certificate within 28 days after receiving the final statement and written discharge by the contractor.

As already mentioned, the engineer or architect acts in many cases in the capacity of quasi-arbitrator, and as such it is essential that his decision as to the amount finally due must be honest and given without interference by either party.

It may be noted that there is no contractual relationship between the contractor and the engineer or architect. Therefore, if the contractor has agreed to the condition stipulating that the engineer's or architect's decision shall be final, binding and conclusive, no action will lie against the engineer or architect, except on grounds of collusion, fraud, deceit or breach of warranty of authority on his part.[58] Withholding of a final certificate on account of certain parts of the work not having been done in "sound and workmanlike manner" does not amount to a fraud if the contract provides that the work is to be done to the entire satisfaction of the engineer or the architect.[59]

If the engineer's or architect's final certificate is made binding and conclusive on all parties, even an action by the owner against the engineer or architect for negligence in the exercise of these functions will not be maintainable. As an agent of the owner, the engineer or architect would clearly be liable, but not in a case where he acts in the capacity of an arbitrator.[60]

---

[57] Union of India v. Raman Iron Foundry, AIR 1974 SC 1265.

[58] Stevenson v. Watson, (1879) 4. C.P.D 148; Pacific Associates Inc. and Another v. Baxter and Others, [1989] 3 WLR 1150.

[59] Cooper v. Uttoxeter Burial Board (1865) 11 L. J. 565.

[60] Chambers v. Gold Thorpe, Restell v. Nye, (1901) K. B. 624; Sutcliffe v. Thackrah CA(Civ), [1973] 1 WLR 888.

9.9 How Final Are Final Certificates?

The effect of a final certificate issued by an engineer or an architect depends upon the wording used in the contract. Where the contractor accepted the final bill and after two years raised claims, it was held that he was estopped from raising the claims.[61] Most standard form contracts include an arbitration clause for resolving disputes. Such arbitration clauses invariably provide: "The ... arbitrator/s shall have power to open up, review and revise any decision, opinion, instruction, determination, certificate or valuation of the engineer related to the dispute."[62]

In view of the arbitration clause being worded so widely, it incorporates within its fold even the final certificate issued by an engineer/architect. In fact the modern forms, such as the FIDIC form, do not give any finality to the final statement or certificate issued by the engineer. Clause 14.14 stipulates that the contractor can make a claim in respect of the amount due under the Final Payment Certificate (FPC) within 56 days of receiving a copy of the FPC under Clause 20.2. The claims are then processed by the engineer and disputes arising thereunder adjudicated by DAAB/arbitration under Clause 21.1. However, in cases where the final certificate clause and arbitration clause are so worded as to be apparently conflicting, the question of reconciliation between the two provisions may arise.

EXAMPLES FROM CASE LAW

1. A contract for the construction of a school building under RIBA terms provided: "Upon expiration of defects liability period ... the architect shall ... issue a final certificate and such final certificate ... shall be conclusive evidence of sufficiency of the works and materials." An exception was made for latent defects. The arbitration provision gave power to the arbitrator to open up, review or revise any certificate in the manner as if no such certificate had been given. Two years after the issue of the final certificate, stone panels fixed to the exterior walls fell off owing to defective fixing by the contractor. The employer carried out the repairs and sought to recover the cost from the contractor. One of the questions to be decided was whether the arbitrator had the power to reopen the architect's final certificate. It was observed that to read the provision made in the arbitration clause literally would "rob the final certificate almost entirely of its conclusive effect". It was finally held:[63] "the specific provision of conclusiveness of the final certificate prevails over the generality of the words in the final sentence of Clause 27" (namely the arbitration clause).

2. A provision in an agreement stipulated that unless a written request to concur in the appointment of an arbitrator shall have been given by either party before the final certificate has been issued, the said certificate shall be conclusive evidence in any proceedings arising out of the contract to the effect that the works had been properly carried out and completed in accordance with the terms of the contract. The work was completed in April 1967. The architect issued interim certificates in April and July. The employers paid sums on account leaving a balance of £14,861 unpaid on the ground that the floor was faulty. The contractor repaired it and, still not having been paid, started proceedings. The employer did not seek to refer the dispute to arbitration but defended it on the ground that flooring was till then faulty and counter-claimed £13,500. The contractor alleged that he tried to repair it and about 12 months later sought the final certificate. The architect issued the final certificate showing that

[61] Govt. of Gujarat v. R.L. Kalathia and Co., Respondent, AIR 2003 Guj. 185.
[62] See for example, FIDIC Clause 67.3 fourth edition 1987, Reprint 1992.
[63] East Ham Borough Council v. Bernard Sunley & Sons Ltd., House of Lords [1965] 3 All ER 619.

the balance due to the contractor was £2,360. Three days after its issue the employer asked the contractor to concur in the appointment of an arbitrator. The contractor refused, alleging that it was then too late, and issued a second writ claiming the amount of the final certificate. On these facts it was held that the final certificate prevented any further legal action, including the legal proceedings started long before the certificate was issued.[64] As a result of this case the relevant part of the standard form provision was revised to limit the conclusiveness of the final certificate.

3. A contract incorporated a provision that the final certificate shall be conclusive evidence that any necessary effect has been given to all the terms of the contract, which require an adjustment to the contract sum. A completion certificate was issued on 13th May 1986 with an appended list of defects. The contractor failed to rectify the defects which were subsequently rectfied by another agency. Final certificates were issued after adjusting the amount spent in the rectification. However, the deductions shown in respect of the amounts previously certified related to sums certified and not paid. The contractor issued a writ for the sum due. The employer sought a stay of proceedings under the Arbitration Act, raising several cross-claims challenging for the first time, *inter alia*, the validity of the certificate of practical completion issued in May 1986. It was held:

> The final certificate is based on the final measurement and valuation of all the works. It subsumes all interim certificates. If, and to the extent, therefore, that the qualification of the balance due under the final certificate is mistakenly based on sums certified rather than certified and paid, it follows ... that the sum due under an unpaid interim certificate must have the same immunity under Clause 35(3) as the balance due under the final certificate, and the employer ought not be allowed to assert the contrary when the balance in the final certificate has been due to the fault of his agent.

Clause 35(3) was an arbitration clause and, in effect, it was held that the certificate is not reviewable by the arbitrator.[65]

4. In a case, the plaintiff had completed the construction of classrooms. Subsequently the building was found damaged. The advocate commissioner appointed submitted the report that the damage was due to heavy rains and landslide and not due to any fault on the part of the plaintiff. The plaintiff was held entitled to balance amount.[66]

The final bill as the name suggests, is the last claim to be submitted by the contractor to the owner for payment. The claim generally indicates the full amount due to the contractor for the work done under a contract less payment already received by way of interim bills. The contract usually stipulates that no further claims shall be made by the contractor after the final bill except, of course, the refund of the security deposit. By the submission of the final bill, the contractor is deemed to have waived all further claims. A certificate to the effect is generally obtained from the contractor at the time of making the final payment. After scrupulously following this procedure can a contractor succeed in his claim? This question is dealt with below.

---

[64] P. & M. Kaye Ltd. v. Hosier & Dickinson Ltd., House of Lords [1972] 1 All ER 121.
[65] Rush & Tompkins Ltd. v. Deaner, (1989) 13 Con LR 106.
[66] Kunwar Singh Rawat v. State of Uttaranchal & Anr., AIR 2007 (NOC) 1789 (UTR.).

9.10 Claims Made after Payment of Final Bill – If and When Maintainable?

It frequently happens that a contractor, with a view to getting payment of a substantial sum of money under the final bill or certificate, is coerced into signing a no claim certificate. After receipt of the money he initiates arbitration or court proceedings. Generally, by giving a no claim certificate the contractor could not be said to have become disentitled to refer any dispute arising out of the contract to arbitration.[67] The Calcutta High Court has observed:[68]

> It is so well known and notorious a fact that unless a no claim certificate is issued by the contractor the payment of the final bill will not be made. But that does not prevent the contractor from raising his claim before the arbitrator in terms of the arbitration clause for the value of his work or other claims within the scope of the agreement between the parties.

The Supreme Court of India observed:[69]

> that although it may not be strictly in place but we cannot shut our eyes to the ground reality that in the cases where a contractor has made huge investment, he cannot afford not to take from the employer the amount under the bills, for various reasons, which may include discharge of his liability towards the banks, financial institutions and other persons. In such a situation, the public sector undertakings would have an upper hand. They would not ordinarily release the money unless a "No demand Certificate" has been signed.

In another case it was held by the Kerala High Court:[70]

> [W]hen one party says that there was full and final settlement and the opposite party says that it was not voluntary but under compelling circumstances and he had got claims, that is also a matter that could be decided only by the arbitrator.

This view finds support in the decision of the Supreme Court.[71]
The Supreme Court has held:[72]

> The finding of the High Court that prima facie there are triable issues before the Arbitrator so as to invoke the provisions of Section 20 of the Arbitration Act, 1940 cannot be said to be perverse or unreasonable so as to warrant interference in exercise of extraordinary jurisdiction under Article 136 of the Constitution of India.

Remembering the fact that each case is required to be considered on its own merit, the law on this point seems to be well settled in India in the light of several decisions of the Supreme Court of India, some of which are as follows:

1. In National Insurance Company Limited v. Boghara Polyfab (P) Ltd.,[73] the question considered by the Supreme Court of India was whether the contractor gave the "no claim certificates" voluntarily or under any kind of financial duress or fraud/coercion/undue influence. The Supreme Court, observed thus:

[67] See Damodar Valley Corporation v. K.K. Kar, AIR 1974 SC 158.
[68] Jiwani Engineering Works v. Union of India, AIR 1981 Cal. 101 (102). Chairman & M.D., N.T.P.C. Ltd. v. Reshmi Constructions, Builders & Contractors, 2004 (1) R.A.J. 232 (SC).
[69] Chairman & M.D., N.T.P.C. Ltd. v. Reshmi Constructions, Builders & Contractors, 2004 (1) R.A.J. 232 (SC).
[70] Cochin Refineries Ltd. v. S C S Co., EC Kottayam AIR 1989 Ker. 72.
[71] Union of India v. L.K. Ahuja & Co., AIR 1988 SC 1172.
[72] N.T.P.C. Ltd. v. Reshmi Constructions, 2004(1) Arb. LR 156 (SC).
[73] National Insurance Company Limited v. Boghara Polyfab (P) Ltd., (2009) 1 SCC 267.

When we refer to a discharge of contract by an agreement signed by both the parties or by execution of a full and final discharge voucher/receipt by one of the parties, we refer to an agreement or discharge voucher which is validly and voluntarily executed. If the party which has executed the discharge agreement or discharge voucher, alleges that the execution of such discharge agreement or voucher was on account of fraud/coercion/undue influence practiced by the other party and is able to establish the same, then obviously the discharge of the contract by such agreement/voucher is rendered void and cannot be acted upon. Consequently, any dispute raised by such party would be arbitrable.

2. In Chairman & M.D., N.T.P.C. Ltd. v. Reshmi Constructions, Builders & Contractors,[74] the Supreme Court was dealing with a case where submission of the final bill by the respondent contractor was not accepted by the appellant N.T.P.C. Limited. The final bill was prepared by the appellant themselves and forwarded along with the format "no claim certificate". The respondents, though, signed the "no claim certificate" but on the same day sent a letter to the appellant, informing that the signature had been given under coercion and protest, without prejudice to his right. It was held that even when rights and obligations of the parties are worked out, the contract does not come to an end *inter alia* for the purpose of determination of the disputes arising thereunder, and, thus, the arbitration agreement can be invoked.

3. In the case M/s. Ambica Construction v. Union of India[75] the arbitrator held that "no claim certificate" was signed by the contractor under duress and coercion and passed the award in his favour. The High Court set aside the award. When the matter was taken to the Supreme Court, it was held that the contractor had a genuine claim, which was considered in great detail by the arbitrator, and that notwithstanding submission of a no claim certificate by the appellant, he was entitled to claim a reference. The appeal was allowed.

4. In Raj Brothers v. Union of India[76] also the final bill was signed by the contractor "under protest". The contractor subsequently withdrew the protest under duress, but later invoked the arbitration clause. It was held that since withdrawal of protest was made under duress, a triable issue arises and therefore the matter was rightly referred to arbitrator.

5. In Gayatri Project Ltd. v. Sai Krishna Construction,[77] the Supreme Court held that issuance of a full and final discharge/settlement voucher/no-dues certificate, does not preclude arbitration when the said full and final settlement itself is disputed. The question whether a letter dated 6th June 2003 would constitute a "full and final settlement" would have to be determined on proper appreciation of the evidence led by the parties. This issue by itself was subject to arbitration. Hence reference to arbitrator was maintainable, it was held.

6. In Union of India and Another v. M/s. L.K. Ahuja and Co., the contractor executed construction works and accepted payments giving a "no claim declaration". However, he subsequently claimed a certain amount as due under the contract and claimed reference to arbitrator by government within three years, which was denied. The Supreme Court held that it is true that on completion of the work, right to get payment would normally arise and it is also true that on settlement of the final bill, the right to get further payment gets weakened but the claim subsists, and whether it does subsist is a matter which is arbitrable.

7. In Jayesh Engineering Works v. New India Assurance Co. Ltd.[78] the respondents intimated the appellant to receive a cheque in full and final settlement of the works relating to

---

[74] Chairman & M.D., N.T.P.C. Ltd. v. Reshmi Constructions, Builders & Contractors, 2004 (1) R.A.J. 232 (SC).

[75] M/s. Ambica Construction v. Union of India, 2006 (4) Arb. LR 288 (SC).

[76] Raj Brothers v. Union of India, 2009 (1) R.A.J. 146 (SC).

[77] Gayatri Project Ltd. v. Sai Krishna Construction. (2014) 13 SCC 638.

[78] Jayesh Engineering Works v. New India Assurance Co. Ltd.. 2000 (Suppl.) Arb. LR 458 (SC).

Tenders I and II, which he acknowledged by endorsing on the said letter stating that he had received the said amount as full and final settlement and he had no further claim in that regard. Thereafter, he wrote a letter stating that his statement that payment had been accepted by him in full and final settlement is not correct and still there are outstanding dues which need to be paid, otherwise the matter will have to be referred to arbitration in terms of Clause 37 of the agreement. The High Court dismissed the application. The Supreme Court, relying on its earlier decision in Union of India v. L.K. Ahuja and Co., *supra*, held that the view taken by the High Court is not correct. Whether any amount is due to be paid and how far the claim made by the appellant is tenable are matters to be considered by the arbitrator.

8. In Shree Ram Mills Ltd. v. Utility Premises (P) Ltd.,[79] it was held that once it is concluded that there was a live unresolved matter between the parties, the issue of limitation was automatically resolved. Until such time as the settlement talks are going on directly or by way of correspondence no issue arises, and as a result the clock of limitation does not start ticking.

9. In a recent case decided by the Supreme Court of India,[80] the learned single judge had dismissed the application filed by the respondent for setting aside the award. The division bench of the High Court of Calcutta had set aside the judgment and order of the learned single judge on consideration of the plea raised for the first time that appellant issued no claim certificate, thereby forfeiting the right to claim. The respondent had not raised the plea before the chief justice in S. 11 application. Be that as it may, the respondent had not urged the said plea either before the arbitral tribunal or before the learned single judge in the proceedings under Section 34 of the 1996 Act. The Supreme Court, in Mcdermott International Inc. v. Burn Standard Co. Ltd. and Others,[81] has held that the party questioning the jurisdiction of the arbitrator has an obligation to raise the said question before the arbitrator. Therefore, the division bench was not justified while considering the arbitrability of the disputes for the first time, particularly, when the respondent has not urged the issue relating to a "no claims certificate" before the chief justice, arbitral tribunal or before the learned single judge. The appeals were partly allowed.

In conclusion and in the light of the law propounded by the Supreme Court in a number of judgments, the following principles emerge:

(i) Merely because the contractor has issued a "no claim certificate", if there is an acceptable claim, the court cannot reject the same on the ground of issuance of the "no claim certificate".

(ii) It is common knowledge that unless a discharge certificate (usually containing a no claims clause) is given in advance by the contractor, payment of bills is generally delayed, hence such a clause in the contract would not be an absolute bar to a contractor raising claims, which are genuine, at a later date even after submission of such "no claim certificate".

(iii) Even after execution of full and final discharge voucher/receipt by one of the parties, if the said party is able to establish that he is entitled to a further amount for which he holds adequate supporting evidence, he is not barred from claiming such amount merely because of acceptance of the final bill by mentioning "without prejudice" or by issuing a "no claim certificate".[82]

---

[79] Shree Ram Mills Ltd. v. Utility Premises (P) Ltd., (2007) 4 SCC 599. Also see: Parmar Construction Company v. Union of India, (Rajasthan)(Jaipur Bench) 2017(3) DNJ 1321: 2017(4) W.L.C. 780.

[80] Sri Chittaranjan Maity v. Union of India, (SC): 2017(9) JT 423: 2017(12) Scale 216: 2017 DNJ 941: 2017 AIR (SC) 4588: 2017 AIR (SCW). Also see: R.C. Thakkar v. Gujarat Housing Board, AIR 1973 Guj. 34.

[81] Mcdermott International Inc. v. Burn Standard Co. Ltd. and Others, (2006) 11 SCC 181.

[82] R.L. Kalathia and Co. v. State of Gujarat, (SC) 2011 AIR (SC) 754: 2011(2) SCC 400: 2011 AIR (SCW) 703: Recent Apex Judgments (R.A.J.) 571.

The contrary judgments of the Supreme Court considered in illustrations below which were decided on their own facts need also to be noted.

9.11 Accord and Satisfaction – Illustrations

1. The Supreme Court considered the ambit of accord and satisfaction in a case and after considering the entire controversy held.[83]

> Admittedly the full and final satisfaction was acknowledged by a receipt in writing and the amount was received unconditionally. Thus there is accord and satisfaction by final settlement of the claims. The subsequent allegation of coercion is an after-thought and a device to get over the settlement of dispute, acceptance of the payment and receipt voluntarily given. – There is no existing arbitral dispute for reference to the arbitration.

The full facts of this case deserved the finding and the decision, it is submitted.

2. The decision in State of Maharashtra v. M/s Nav Bharat Builders[84] was based on the facts peculiar to the case and does not change the well-established law by the earlier decisions of several High Courts and the Supreme Court of India, it is respectfully submitted.

3. The Supreme Court of India, in a case, summed up the facts thus:

> Admittedly, No-Dues Certificate was submitted by the contractee-Company on 21st September 2012 and on their request Completion Certificate was issued by the appellant-Contractor. The contractee, after a gap of one month, that is, on 24th October 2012, withdrew the No-Dues Certificate on the grounds of coercion and duress and the claim for losses incurred during execution of the Contract at the site was made vide letter dated 12th January 2013, i.e., after a gap of 3–1/2 (three and a half) months whereas the Final Bill was settled on 10th October 2012. When the contractee accepted the final payment in full and final satisfaction of all its claims, there is no point in raising the claim for losses incurred during the execution of the Contract at a belated stage which creates an iota of doubt as to why such claim was not settled at the time of submitting Final Bills that too in the absence of exercising duress or coercion on the contractee by the appellant-Contractor. In our considered view, the plea raised by the contractee-Company is bereft of any details and particulars, and cannot be anything but a bald assertion. In the circumstances, there was full and final settlement of the claim and there was really accord and satisfaction and in our view no arbitrable dispute existed so as to exercise power under Section 11 of the Act. The High Court was not, therefore, justified in exercising power under Section 11 of the Act.[85]

4. In a case,[86] the respondent authorities had raised an objection relating to the arbitrability of the disputed issue before the arbitrator on the ground that the disputed matter was excepted from the scope of the arbitration clause. Despite the objection the arbitrator had

---

[83] M/s P.K. Ramaiah & Co. v. Chairman and M.D. N.T.P.C., 1994 (1) SCAL 1.

[84] C.A. No. 853 of 1994, S.L.P.C. No. 5628 dated 4th February 1994.

[85] M/s ONGC Mangalore Petrochemicals Ltd. v. M/s ANS Constructions Ltd., 2018(2) R.C.R.(Civil) 548: 2018(2) JT 212: 2018(2) Scale 354: 2018 AIR (SC) 796: 2018(1) Law Herald (SC) 358: 2018 AIR (SCW) 796: 2018(1) ArbiLR 597: 2018(2) R.A.J. 157: 2018(3) SCC 373: 2018 All SCR 790.

[86] M/s Harsha Constructions v. Union of India, (SC) 2014(9)SCC 246; AIR 2015 (SC) 270; 2014 (3) Arb LR 482.

rendered his decision on the said "excepted" dispute. On these facts the Supreme Court of India held:

> We, therefore, hold that it was not open to the Arbitrator to decide the issues which were not arbitrable and the award, so far as it relates to disputes regarding non-arbitrable disputes is concerned, is bad in law and is hereby quashed.

5. In another case,[87] the appellants made the full and final payment of the final bill, which the respondent certified by signing the bill without any protest or reservation. The respondent, with the intention of receiving further payments, after two years, raised yet another claim and tried to bring up a dispute. When the claim was denied by the appellants, the respondent requested to appoint an arbitrator. Condition 65 of the general conditions of contract IAFW-2249 states that no further claim shall be made by the contractor after submission of the final bill and these shall be deemed to have been waived and extinguished. Held: the court without considering whether any dispute exists between the parties, could not have appointed an arbitrator.

9.12 Economic Duress under English Law

Under English law, duress,

> whatever form it takes, is a coercion of the will so as to vitiate consent ... in a contractual situation commercial pressure is not enough. There must be present some factor "which could in law be regarded as a coercion of his well so as to vitiate his consent". ... In determining whether there was a coercion of will such that there was no true consent, it is material to inquire whether the person alleged to have been coerced did or did not protest; whether, at the time he was allegedly coerced into making the contract, he did or did not have an alternative course open to him such as an adequate legal remedy; whether he was independently advised; and whether after entering the contract he took steps to avoid it.[88]

There has to be illegitimate pressure that induces a claimant into a settlement in circumstances where there is no other practical alternative open to him.[89] More than the "rough and tumble" of the commercial bargaining, it is also relevant to consider if there was an actual or a threatened breach of contract and whether the pressurizing party acted in good or bad faith.[90] Economic duress only renders an agreement voidable and not void. It is, therefore, important not to ratify the agreement as soon as the economic duress has lifted.

A subcontractor insisted on a settlement of his account before making any further deliveries of cladding materials. The contractor paid a substantial amount of the account as the adjudication of the dispute would have taken long enough to stop the contractor from completing its work by the due date and the contractor was advised against making an application for injunction. The contractor protested about the duress in writing afterwards. He was allowed to rescind the settlement.[91] A memorandum of understanding was not voidable due to economic duress in another case because the party alleging duress had an alternative to terminate the agreement, had not protested about duress in its voluminous correspondence with the other party, and even in its

[87] Union of India v. M/s. Onkar Nath Bhalla & Sons, 2009 (7) SCC350;AIR 2009 SC 3168;, 2009 (6) Scale 602.
[88] Pau On v. Lau Yiu Long, [1980] AC 614, 635.
[89] Universe Tankships Inc. of Monrovia v. International Transport Workers Federation, [1983] AC 366, 400.
[90] Capital Structures Plc. v. Time & Tide Construction Ltd., [2006] EWHC 591 (TCC).
[91] Carillion Construction Ltd. v. Felix (UK) Ltd., [2000] 11 WLUK 152.

internal documents, there was nothing to support the existence of duress.[92] Duress would not assist a party to avoid a bargain when there is the use by the other party of "lawful pressure to achieve a result to which the person exercising pressure believed in good faith it is entitiled, and that is so whether or not, objectively speaking, it has reasonable grounds for that belief."[93] Asplin LJ observed that it was not appropriate to "fetter the lawful use of a monopoly position".[94]

---

[92] DSND Subsea v. Petroleum Geo Services ASA, PGS Offshore Technology AS, 2000 7 WLUK 875.

[93] Times Travel (UK) Ltd. v. Pakistan International Airlines Corporation, [2019] EWCA (Civ) 828, para. 105.

[94] Ibid., para. 117.

10

Breach of Contract

10.0 Introduction

A breach of a contract is failure to perform an obligation arising out of the contract. Where there is failure to perform an obligation, in whole, it is a total breach. When an agreement is broken only in part it is a partial breach. If a party announces, before his performance is due, his definite unwillingness or inability to fulfil the contract, he thereby admits he is guilty of a breach. The breach in such a case is called an anticipatory breach. Occasionally a party may deliberately incapacitate himself or render impossible the performance of his contract duties; or may so interfere to render performance by the other party impossible. Such tactics also constitute a breach of contract.

Every breach of a contractual obligation confers upon an injured party a right of action. However, there are a number of valid excuses for non-performance of contractual obligations. An actionable breach of contract, therefore, occurs when a promisor, without sufficient excuse or justification, fails to perform in accordance with the dictates of his agreement.

Further, there is a distinction between breach of a contract and termination of a contract. When a contract comes to an end it is said to be terminated. Breach of contract may constitute a means of contract termination. However, there are a number of ways other than a breach by which a contract can be terminated. Full and satisfactory performance by both sides is the usual mode. The other modes include:

1. Release under seal
2. Rescission by consent of parties
3. Accord and satisfaction
4. Exercise of option given to a party in a contract to terminate under certain circumstances or events
5. Rescission by a party on account of repudiation or non-performance by the other party
6. Frustration or impossibility of performance

It may be noted that many common forms of breach of important contract conditions are dealt with in their appropriate places in different chapters in this book. The law applicable to some forms of breach of contracts not dealt with elsewhere in the book is considered in this chapter.

The doctrine of frustration, consequences of breach of contract such as waiver, accord and satisfaction, termination, compensation, applicability of the doctrine of specific performance, the doctrine of substantial performance and the arbitration clause are the important topics discussed in this chapter. Common breaches of contracts are dealt with in Chapter 11.

10.1 Provisions of the Law

The following provisions of the law are dealt with in this chapter.

Provision of the Law	Paragraph No.
Indian Contract Act, Sections: 39	10.2
52–54, 67	10.5.1
62	10.7
64 and 65	10.2
Provisions in Standard Form Contracts	
FIDIC Form 1992 Amended Reprint, Clauses 56, 69	10.10.3
1999 1st edition, Clauses 15, 16	
2017 2nd edition, Clauses 15, 16	
Ministry of Statistics and Programme Implementation, Government of India (MOS & PI), Clause 59	10.10.4,
NITI Aayog Government of India, Engineering, Procurement and Construction of Civil Works	10.10.5
Model Agreement 2017, Clauses 19.20, 19.21	
PWD B-1 & B-2 Standard Forms of State Government, Clauses 3, 4	10.10.1
Military Engineering Services Form, Clauses 11A–11C, 52	10.10.2, 10.5.3
Railway Engineering Department Form, Clause 17A	10.5.3

10.2 Refusal to Perform Contract

In a building or construction contract, the earliest possible situation out of which a breach of the contract may arise would be the refusal of a party to go ahead with the contract. The refusal may be either by the owner or by the contractor. Consequences of refusal of a party to perform a promise wholly are dealt with in Section 39 of the Indian Contract Act. It reads:

> 39. *Effect of refusal of party to perform promises wholly*
> When a party to a contract has refused to perform, or disabled himself from performing his promise in its entirety, the promisee may put an end to the contract, unless he has signified by words or conduct, his acquiescence in its continuance.

ILLUSTRATIONS

(a) A, a singer, enters into a contract with B, the manager of a theatre, to sing at his theatre two nights in every week during the next two months, and B engages to pay her 100 rupees for each night's performance. On the sixth night A wilfully absents herself from the theatre. B is at liberty to put an end to the contract.

(b) A, a singer, enters into a contract with B, the manager of a theatre, to sing at his theatre two nights in every week during the next two months, and B engages to pay her at the rate of 100 rupees for each night. On the sixth night A wilfully absents herself. With the assent of B, A sings on the seventh night. B has signified his acquiescence in the continuance of the contract, and cannot now put an end to it, but is entitled to compensation for the damage sustained by him through A's failure to sing on the sixth night.

Section 39 falls under Chapter 4 of the Contract Act under the heading: "OF THE PERFORMANCE OF THE CONTRACT" with the subheading: "Contracts which must be performed". The provision, however, deals with the situation where the contract need not be performed; the topic covered by Sections 62 to 67 of the Act. At the outset it needs to be noted that Section 39 is applicable to executory contracts where the time of performance has not yet arrived. However, that does not necessarily mean that it covers anticipatory breaches alone. In other words it presupposes the contract under which some substantial part of the contract is yet to be performed. The words "in its entirety" indicate that the kind of refusal contemplated by this section is one which affects a vital part of the contract and prevents the promisee from getting in substance what he bargained for.

The law embodied in this section is that where a party to a contract refuses altogether to perform, or is disabled from performing, his part of it, the other party has a right to rescind the contract unless he has signified his acquiescence in the continuance of the contract.

This section covers three sets of circumstances:

1. Renunciation by a party of his liabilities under the contract
2. Impossibility of performance created by his own act and
3. Total or partial failure of performance.

Renunciation means repudiation of liability before the time of performance has arrived. Any intimation, either by words or by conduct, that a party declines to continue with the contract is repudiation. It may occur or impossibility be created either before or at the time of performance. The third situation contemplates total failure during performance and, if failure is partial, to be of the nature which goes to the root of the contract.

A party injured by refusal of the other party to perform the contract must satisfy itself, before rescission of the contract, that the refusal is absolute and that the other party has made its intention not to perform the contract quite plain. Mere failure by the party to perform the promise without evincing an intention to end the contract would not be repudiation covered by Section 39 but covered by Sections 54 and 55 of the Indian Contract Act. Further, it may be noted that the election to rescind, once made, is conclusive. Also, a contract cannot be terminated in part but must be repudiated altogether. These concepts though *prima facie* appearing simple are often difficult to apply in practice. Every intimation of refusal may not amount to renunciation. A party is likely to think wrongly that it is entitled to terminate the contract and commit a repudiatory breach entitling the other party to terminate the contract.

Under English law, it is not necessary to expressly renounce the contract but the intention to do so can be gathered from words or conduct. A contracting party should be treated as "having refused to perform its obligations if it evinces an intention to perform but "only in a manner substantially inconsistent with [its] obligations and not in any other way".[1] A similar result is possible where a party imposes conditions for its performance that are not required by the contract.[2] The court will look at all the circumstances of the case objectively, i.e. from the point of view of a reasonable person in the position of the innocent party to the contract.[3] Interestingly, pursuing an exceptionally weak claim of repudiation with the knowledge that there were no grounds for asserting a repudiatory breach was a main factor leading to an award of indemnity costs to the defendant.[4]

EXAMPLES FROM CASE LAW

1. The petitioner challenged the order denying refund of the amount deducted by the respondents as a condition for accepting the surrender of plot. Held: Since, it was not in dispute that the petitioner had not paid the instalments of premium in accordance with the conditions enshrined in the allotment letter for the plot, even after receipt of the notice issued by the respondent and demand letters, the action taken by the respondent to deduct 10% of the consideration money was not illegal.[5]

---

[1] Ross T. Smyth & Co. Ltd. v. T.D. Bailey, Son & Co., [1940] 3 All ER 60, p. 72.
[2] BV Oliehandel Jorgkind v. Coastal International Ltd., [1983] 2 Lloyd's Rep. 463.
[3] Eminence Property Developments Ltd. v. Heaney, [2011] 2 All ER (Comm) 223, para. 61.
[4] Imperial Chemical Industries Ltd. v. Merit Merrell Technology Ltd., [2017] EWHC 2299 (QB).
[5] G.C. Agarwal v. Haryana Urban Development Authority (S.C.) 2014(16) SCC 327: 2015(3) SCC (Civil) 657.

2.   The case of the plaintiff, a PWD contractor undertaking the works of constructing the roads and bridges for the PWD for the last ten years was as follows: The defendant allotted the work to the plaintiff and a formal contract was signed as per the work order and the contract dated 5th April 1994. As per the said contract, the work was to be commenced on 14th April 1994 and the time limit of completion of work was 450 days including monsoon and thus the date stipulated for completion of the work was 7th July 1995. Immediately after getting the work order, the plaintiff made arrangements to start the work by mobilizing the labour force, obtaining and arranging for supplies of the required construction material needed for the work. However the defendant No. 1 failed to provide to the plaintiff the land required for the work. The land, where the work was to be carried out, was not acquired by the government even by the stipulated date of completion of work, i.e. 7th July 1995. The defendant No. 1 therefore committed a fundamental breach of contract by its failure to provide the site required for work due to which the plaintiff suffered huge losses, including: anticipated profit thereon at least of Rs. 2,82,234.65/- being 15% of the value of the work, advance paid to supplier, pieceworkers, etc. In reply, it was argued that in the present case, the defendants had not terminated the contract and in fact the plaintiff was called for discussion, and that in the agreement there was no clause for automatic discharge by efflux of time and actually there was a clause for extension of the period of contract. It was further argued that the plaintiff, who himself put an end to the contract, was not entitled to claim estimated profit as compensation. Held:

> In fact, by praying to extend the period of the completion work till the L.A. [land acquisition] problem is finalised, the plaintiff has acquiesced in continuance of the contract. Section 39 of the Contract Act cannot come to the rescue of the plaintiff. Hence, the question of demanding compensation at the rate of 15% of the total value of the contract, towards estimated profits, does not arise.[6]

3.   The architect in a case wrongly advised the employer to deduct liquidated damages from interim certificates. The contractor thereupon wrote terminating the contract for the alleged breach and ordered the subcontractors to cease the work. The employer thereupon served notice for termination on the ground that the contractor was in default by wholly suspending execution of the works without reasonable cause. The contractor alleged that the employer was in default by failing to make payment under interim certificates and then issued a writ. It was held: by issuing an invalid termination notice, ceasing the progress of the works and thereafter issuing a writ, the contractor had indicated an intention not to be bound by the contract, and since the owner had no alternative but to accept the repudiation, the contract was brought to an end. The contractor was held to be in breach, even though the employers were wrong in deducting liquidated damages at that stage.[7]

4.   In a case, the Supreme Court of India observed:

> If agreements permit the financier to take possession of the financed vehicles, there is no legal impediment on such possession being taken. Of course, the hirer can avail such statutory remedy as may be available. But mere fact that possession has been taken cannot be a ground to contend that the hirer is prejudiced.

---

[6] C.K. George v. Executive Engineer, (Bombay) 2012(1) BC 666: 2012(3) BCR 556: 2012(4) Mh.LJ 240: 2012(117) AIC 420: 2012(3) ALL MR 24: 2012(12) R.C.R.(Civil) 889.

[7] Lubenham Fidelities & Investments Co. Ltd. v. South Pembrokeshire District Council & Wigley Foz Partnership, Court of Appeal (1986).

This was a case of a failure to make payments under a hire-purchase agreement which entitled the financier to repossession. In response to the stand of learned counsel for the respondent that convenience of the hirer cannot be overlooked and improper seizure cannot be made, it was observed:

> There cannot be any generalisation in such matters. It would depend upon facts of each case. It would not be therefore proper for the High Courts to lay down any guideline which would in essence amount to variation of the agreed terms of the agreement. If any such order has been passed effect of the same shall be considered by the concerned High Court in the light of this judgment and appropriate orders shall be passed.[8]

10.2.1 Disability to Perform

A party can be said to have disabled itself from performing the contract if:

(i) He deliberately puts it out of his power to perform the contract or

(ii) When by his own act or default, circumstances arise which render him unable to perform his side of the contract or some essential part of it.

If there is a dispute, it is easier to prove renunciation than impossibility of performance. In the case of the former a party has to establish that the conduct of the defaulting party has been such as to lead a reasonable man to believe that he did not intend to perform his promise, whereas in the case of the latter, he must show that the contract is in fact impossible to perform due to fault of the other party.

10.2.2 Doctrine of Anticipatory Breach

This section also embodies the doctrine of anticipatory breach. The rule indicated by this doctrine is that on the repudiation of the contract by one party, even before the time for performance has arrived, the other party may, at his option, treat the repudiation as an immediate breach, putting an end to the contract for the future, and at once bring his action for recovery of damages. However, if the aggrieved party prefers to await the time when the contract is to be performed, and then hold the other party responsible for all the consequences of non-performance, he keeps the contract alive for the benefit of the other party as well as his own. In other words, he remains subject to all his own obligations and liabilities under the contract and enables the defaulting party not only to complete the contract, if so advised, notwithstanding his previous refusal to perform it, but also to take benefit of any supervening circumstance which would justify him in declining to perform it.[9]

EXAMPLE FROM CASE LAW

The facts: On 26th March 1993 the respondents awarded the work of "extension of terminal building" at Guwahati airport to the appellant. As per the contract, the date of commencement of work was 10th April 1993 and the period of completion of the work was 21 months, to be

---

[8] Managing Director, Orix Auto Finance (India) Ltd. v. Jagmander Singh, (SC): 2006(2) SCC 598:: 2006(2) Scale 297: 2006(2) SCR 169; 2006(1) Apex Court Judgments (SC) 675.

[9] Ratanlal v. Brijmohan, (1931) 33 Bom. L.R. 703, 122 IC 861 (31) A.B. 386.

completed in different stages. As the appellant (also referred to as the "contractor") did not complete the first phase of the work within the stipulated time, the respondents terminated the contract by order dated 29th August 1994. The termination was challenged by the appellant in a writ petition filed before the Gawahati High Court. By judgment dated 27th September 1994, the High Court set aside the termination and directed the respondents to grant time to the appellant till the end of January 1995 for completion of the first phase, reserving liberty to the appellant to apply for further extension of time. As the work was not completed, the respondents granted an extension up to 31st July 1995 by letter dated 24th August 1995, without levying any liquidated damages. The contractor proceeded with the work even thereafter. However, as the progress was slow, the respondents terminated the contract on 14th March 1996 on the ground of non-completion even after 35 months. The appellant filed a writ petition, challenging the cancellation. The High Court by order dated 25th June 1996, noticed the existence of the arbitration agreement and referred the parties to arbitration. By his award the arbitrator allowed some claims of the contractor and rejected all counter-claims raised by the respondent, including liquidated damages, risk and cost, etc. The Supreme Court of India held: though the claim levying damages on contractor was falling within excepted clause beyond the scope of adjudication by the arbitrator, the question as to who was responsible for delay was arbitrable. Right to levy liquidated damages would arise only when contractor was responsible for delay. The arbitrator holding that the contractor was not responsible for delay and quashing levy of liquidated damages, not improper.[10]

From the above considerations it is clear that refusal of one party to fulfil his part of the contract renders the contract voidable at the option of the other party. When the other party makes use of that option the contract becomes void. As to what are the consequences of rescission of a voidable contract, Sections 64 and 65 of the Contract Act need to be considered.

64. *Consequences of rescission of voidable contract*
When a person at whose option a contract is voidable rescinds it, the other party thereto need not perform any promise therein contained in which he is the promisor. The party rescinding a voidable contract shall, if he has received any benefit thereunder from another party to such a contract, restore such benefit, so far as may be, to the person from who it was received.

EXAMPLE FROM CASE LAW

Work contract for construction of canals was awarded to the plaintiff in November 1988 and March 1989, who carried out the work of excavation, embankment, murrum lining and completed it. By several letters written to the respondent the contractor requested for release of cement, RCC templates and PCC slabs. Proceedings of Meeting convened revealed that due to injudicious act of the respondent's engineer in not supplying cement and cement materials to the plaintiff, the contract work had to be stopped by the plaintiff. Finding no other option, the plaintiff terminated the contract and asked for finalising bill for executed work and termination of contract without penalty by letter dated 9th May 1991. The contract was subsequently rescinded by the chief engineer on 12th October 1994 and penalty was imposed on 24th October 1994. No notice was issued by the chief engineer, nor was the plaintiff given an opportunity of making representations, though

---

[10] J.G. Engineers Pvt. Ltd. v. Union of India, 2011(2) Arb. LR 84: 2011(5) SCC 758: 2011 AIR (SCW) 2849; 2011 AIR (SC) 2477; 2011(4) Recent Apex Judgments (R.A.J.) 52.

required under clause 3(d) of the Conditions of Contract. A suit was filed for a declaration that the order of rescinding contract for work of canal was not binding and seeking an injunction for restraining the respondent from forfeiting or adjusting earnest money and security deposit. Payment was sought for the work executed. The suit was partly decreed. The decree was set aside by the High Court, holding that reasoning of trial court was erroneous and that material evidence was not considered. The Supreme Court of India observed that the learned trial judge had rightly held that the plaintiff had proved that the contract work could not be completed due to the fault of the Department itself and the plaintiff could not be blamed. The Supreme Court India finally set aside the impugned judgment and decrees passed by the High Court restored the common judgment and decrees of the trial court in the original suits, except modification with regard to rate of interest from 18% to 9%.[11]

The provisions made in the above section require a person who has elected to put an end to the contract to restore the benefits received under the contract to the defaulting party. The other party thereafter need not perform his part of the contract. He is, however, liable to pay damages for the defaulting party's breach under Section 75 which reads as follows:

75. *Party rightfully rescinding contract entitled to compensation*
A person who rightly rescinds a contract is entitled to compensation for any damage which he has sustained through the non-fulfillment of the contract.

ILLUSTRATION

A, a singer, enters into a contract with B, the manager of a theatre, to sing at his theatre two nights in every week during the next two months, and B engages to pay her at the rate of 100 rupees for each night. On the sixth night A willfully absents herself. With the assent of B, A sings on the seventh night. B has signified his acquiescence in the continuance of the contract, and cannot now put an end to it, but is entitled to compensation for the damage sustained by him through A's failure to sing on the sixth night.

EXAMPLE FROM CASE LAW

A formal agreement for surface mining of coal was executed between the appellant and the respondent on 26th May 2003. The relevant clauses of the agreement included a provision that the tendered quantity may be reduced or increased by ± 30%, and that no claim shall lie on the company for such variation in quantity whether increase or decrease. The respondents commenced the work of surface miners at Lakhanpur and completed around 70% of the awarded quantity by the end of February 2004. Due to financial problems faced by the respondents, by letter dated 13th February 2004 they requested the appellants to allow them to close the contract by invoking power to reduce the quantity by 30% of awarded quantity, under the general terms and conditions of the NIT; and to issue fresh tender for the remaining work. The appellants did not accede to the said request and

---

[11] Ramachandra Narayan Nayak v. Karnataka Neeravari Nigam Ltd., 2013(15) SCC 140: 2014(6) R.C.R.(Civil) 600: 2013(12) JT 137: 2013(11) Scale 7: 2013 AIR (SCW) 5913.

informed the respondents that the agreement was for performing the contract up to 100% of awarded value and provision of executing extra 30% quantity on the same terms and conditions. The respondents requested the appellants to extend the time frame for completion of the remaining contract up to 15th July 2004 as the contract period was only until 15th April 2004 with the request to treat the contract as closed with the completion of the above awarded quantity. By a letter, the respondent informed that they had stopped the work with effect from 16th July 2004. The appellant took action under penal clauses including liquidated damages, forfeiture and risk and cost, etc. A writ was filed by the respondent contractor. The Writ Court held that, surprisingly after extending the contract period on 5th June 2004, within six days on 11th June 2004 the appellants decided to enhance the contract quantity by 30%. That was not a *bona fide* act and was unacceptable. The court also held that the appellants had not offered any explanation as to in what circumstances the decision to impose penalty was taken by the board of directors. The court further noted that the respondents had executed the contract up to 108.47% at a very low rate, and incurred heavy losses in that regard. Further, a new contract for Lakhanpur OCP was already awarded and there was no loss of production caused to the appellants. On these bases, the division bench allowed the Writ Petition, quashed the order proposing to levy shortfall penalty. The outstanding dues payable to the petitioner were ordered to be released in its favour within the period of 30 days along with simple interest at 8% per annum to be computed from the date of conclusion of contract, i.e. from 16th July 2004. The bank guarantee furnished by the petitioners was directed to be cancelled.

Aggrieved, the appellants filed the appeal before the Supreme Court. The Supreme Court held that a notice increasing the work up to 30%, as permissible under the agreement, could have been given up to 15th July 2004. It was, in fact, given on 11th June 2004. It then had to follow that the respondents had committed breach of their contractual obligation by not completing the balance work out of 130% of work (i.e. 130 − 108.47%). To that extent, it was held, that the respondents became liable to compensate the appellants for the loss caused to them because they had to assign the unfinished work to a third party for completion at a higher rate. It was observed:

> Since we have reversed the findings and conclusion of the High Court and even if this appeal succeeds, the respondents can be granted an opportunity to make a representation to the Appellants – company, who in turn can deal with the same in accordance with law. If the appellants accept the claim of the respondents about the unjustness of penalty or quantum thereof, they would be free to withdraw or modify their claim for recovery of penalty amount, if so advised. In the event, the appellants reject the representation, they will be free to recover the amount as demanded towards penalty along with interest accrued thereon, as may be permissible in law. However, that would not absolve the respondents from the financial liability arising due to difference of rate of contract and the actual cost incurred by the appellants to complete the unfinished work out of 130% of the contract quantity, through a third agency at a higher rate. That can be recovered by the appellants from the respondents along with interest accrued thereon at such rate, as may be permissible in law, even if the representation made by the respondents for recall or modification of the penalty amount is pending consideration.[12]

[12] Mahanadi Coalfields Ltd. v. M/s. Dhansar Engineering Co. Pvt. Ltd., 2016(9) Scale 339: 2016(4) R.C.R.(Civil) 603: 2016(5) Recent Apex Judgments (R.A.J.) 570: 2016(9) JT 385: 2016 AIR (SCW) 4509; 2016 AIR (SC) 4509: 2016(3) Apex Court Judgments (SC) 483.

Section 75 is to be read as supplementary to Sections 39, 55, 64 and 65. Section 55 has been dealt with in Chapter 6 and Section 65 is considered in paragraph 10.8 below. Application of all these principles to some common forms of breach in cases of building and engineering contracts is considered below.

10.3 Breaches by the Contractor

The provisions of the standard form contracts are so numerous that it may not be desirable to attempt to give an exhaustive list of possible breaches of contract by the contractor. For example, failure to submit the programme, failure to appoint qualified engineers, failure to insure, failure to submit labour returns, etc. can be cited as breaches of a less common kind, which may or may not result in damage to the owner. However, it is possible to group the commonest breaches of contract conditions causing substantial damage and often being the subject of litigations, into the following three categories:

a. Abandonment, or total failure to complete the work, either to start with or midway in execution

b. Delay in completion of the works, and

c. Defective work.

The consequences of the first category are considered in this chapter and also in Chapter 11. The second category is considered in Chapters 6, 7 and 11. The third category is dealt with in Chapters 8 and 11.

10.3.1 Forfeiture of Earnest Money – When Successful Tenderer Refuses to Perform Contract

Upon acceptance of a tender, a contract is concluded. A contract is a contract from the time it is made and not from the time that performance is due. If a tenderer wants to back out after his tender is accepted, he commits a breach of the contract. The owner can terminate the contract under the provisions of Section 39. Section 64 of the Contract Act, however, requires that he should restore any benefits received under the contract. Usually a tender is accompanied by earnest money deposit. This clearly is the benefit received by the owner. The contractor would, thus, be entitled to get back the earnest money. However, the owner is entitled under Section 75 to recover damages, which he sustained by reason of the contractor's refusal to perform the contract.

If, by reason of wrongful repudiation of a contract by the contractor, the owner lawfully puts an end to the contract and in the process suffers loss, the contractor must make good that loss. If the loss is more than the earnest money received along with the tender, the owner can recover the difference from the contractor, by taking adequate legal steps, if necessary. If, however, the loss suffered by the owner is less than the amount received by way of the earnest money, the law is that he should, after deducting the sum necessary to make good his loss, return the balance of the earnest money to the contractor. This is true even where the tenderer has agreed to an express stipulation that "the earnest money will be forfeited to the owner in full", if he fails to adhere to the terms and conditions of the contract.

EXAMPLE FROM CASE LAW

The defendant had invited tenders for the removal of zinc dross and zinc ash at Mettur Dam. The plaintiff submitted a tender along with Rs. 5,000/- by way of earnest money in accordance with the tender notice. Later the parties fell apart and the plaintiff called upon the defendant to return the deposit. The defendant, however, contended that the defendant was entitled to forfeit the amount under Clause 9 of the tender notice. Thereupon the plaintiff filed the suit. Clause 9 of the Instructions to Tenderers provided for the earnest money deposit to be forfeited to the Corporation in full as a penalty besides recovery of compensation for other losses, if any. Considering the above facts, it was held:[13]

> The conclusion, that with the acceptance of the tender ... a contract resulted, carries with it the consequence that the plaintiff committed a breach of the contract, because it was not prepared to share the business with another. The defendant would, no doubt be entitled to recover damages which it sustained by reason of the plaintiff's default, but even so, it must return the sum of Rs. 5,000/- which it got from the plaintiff. This is because of Section 64 of the Contract Act.

The learned judge referred to the provisions of Sections 64 and 75 and interpretation of these sections by their Lordships of the Privy Council in a well-known case.[14] In that case the plaintiff had committed the breach, but it was contended on his behalf that, even so, he was entitled to the refund of the sum paid by him. This contention was rejected by the courts below, but was upheld by their Lordships of the Privy Council, on the wording of Section 64 of the Contract Act. Their Lordships pointed out that, though the defendant might have sustained damages, they would have to prove the same and could only set-off those damages against the plaintiffs' claim. It was observed: "On general principles they may set-off such damages as they have sustained, but the Act requires that they give back whatever they received under the contract".

The contention of the defendant in the Madras case under discussion, that even without proving any damage sustained by it, it would be entitled to forfeit the sum of Rs. 5,000/- under Clause 9 of the tender notice was not accepted by the High Court. This decision of the High Court, it is submitted, is in conformity with the law laid down by the Supreme Court of India. The right to retain security money in full without proof of actual damage can only be entertained under Section 74 of the Contract Act. The scope and applicability of Section 74 was considered by the Supreme Court in a couple of cases.[15] The principle laid down by their Lordships of the Supreme Court amply makes it clear that "when it comes to the question of forfeiture of the security money because of the breach the sum forfeited does not, ipso facto, become reasonable compensation if actual loss can be proved".

It may not, however, be necessary for the owner to file a suit to recover the damages to the extent of the earnest money deposit. The owner, if satisfied with the forfeiture of the earnest money, may retain the earnest money. However, if the contractor files a suit for recovery of the earnest money, the owner should set up a plea of equitable set-off in his written statement, if it is not time-barred on the date of the plaintiff's suit. It would be appropriate to explain the juristic principle further. Equitable set-off is allowed where it arises out of the same transaction, which is the basis of the plaintiff's claim and where it would be inequitable to drive the defendant to a separate suit.

Where, however, the loss to the owner is more than the earnest money, the owner must take the initiative in filing a suit if he desires to recover damages. Such a suit must be filed within the period of limitation. The loss sustained by the owner may include a sum of money by which the tender sum of another contractor, whose tender the owner had to accept, exceeds the sum

---

[13] Maheswari Metals and Metal Refinery v. The Madras State Small Industries, AIR 1974 Mad. 39 (42).

[14] Murlidhar Chatterji v. International Film Co., AIR 1943 P.C. 43.

[15] S.A. Bhat v. V.N. Inamdar, AIR 1959, Bon. 452; Fateh Chand v. Balkishan Dass, AIR 1963 SC 1405; and Maula Bux v. Union of India, AIR 1970 SC 1955.

tendered by the defaulting contractor. The fact that the defaulting contractor's liability is not limited to the extent of earnest money/security deposit but extends to make good the full loss is illustrated by the following cases.

EXAMPLES FROM CASE LAW

1. The government advertised bids for sale of surplus materials. The important terms of the invitations were: (1) Each bidder to deposit a specified amount with the government agency; which would stand forfeited by withdrawal of a bid, (2) If the person to whom the contract was awarded refused to go ahead with the purchase, the government could: (a) terminate the contract, (b) sell the material to another party, (c) charge the party who refused to enter into the contract of purchase with any damages the government might sustain. On these facts, it was held that the contractor's liability was not only limited to the security deposit which he would forfeit but also extended to making good the full loss the government sustained by selling the materials to someone else at a lower price.[16]

2. Where in respect of works contract for transportation of coal by road transport the tenderer refused to accept the work order unless a condition for revision of rates was incorporated in the contract, it was held by the Patna High Court that no contract was concluded between the parties and as such earnest money in the form of National Savings Certificates could not be forfeited under Section 74 of the Act.[17]

10.4 Breach by the Owner

Similar to breaches of contract by the contractor, the owner too is likely to commit various defaults amounting to breaches of the standard form contract conditions, either express or implied. Some of the common breaches likely to be committed by the employer are discussed below:

10.4.1 When Owner Cancels a Contract before Work Begins

If the owner, after having accepted a tender, abandons the work or otherwise annuls the contract before the work is started by the contractor, the contractor is entitled to receive damages from the owner, subject of course to the terms and conditions of their agreement. In the absence of a stipulation to the contrary in the agreement, the contractor may, in such a case, be entitled to the actual damages sustained. The actual damages may include, in addition to preliminary expense, which the contractor may have incurred for starting the work, the amount of profit which he would have received had he performed the contract to its completion.

The amount of profit that would be lost by the contractor is "the difference between the contract price and what it would have cost to perform the contract".[18] But in estimating the cost of performing the contract, a contractor may not use the prices submitted by his subcontractors. The advantages and benefits of subcontracts are too uncertain and contingent to be considered in estimating the profit, a general contractor might have made under his contract.[19]

---

[16] United States v. P.J. O'Donnel and Sons, 228 f. 2d, 162 U.S. Court of Appeals, 1956.

[17] Arvind Coal & Construction Co. v. Damodar Valley Corpn, AIR 1991 Pat. 14; also see AIR 1970 SC 1986.

[18] Devlin v. Mayor, 63 NY 9.

[19] Delvin v. Mayor of the City of New York, 63 NY 8 P25.

EXAMPLE FROM CASE LAW

Under the contract entered into between the owner and a contractor, the contractor undertook to carry out repairs to a building for $2,000. The contract provided that, if the owner breached the contract, the contractor was entitled to receive 30% of the contract price as liquidated damages or to sue for the actual damages sustained. The owner cancelled the contract before the work began. The contractor brought in an action to recover $600, i.e. 30% of the contract price. The court dismissed the suit on the grounds that the amount of liquidated damages was disproportionate to the actual damages and further actual losses could readily be determined. The suit was, however, dismissed without prejudice, enabling the contractor to start new action and prove the actual damages sustained.[20]

10.5 Defaults by Owner and/or Contractor

The breaches, which lead to failure to perform an obligation in whole, are considered in paragraphs 10.3 and 10.4 above. These are rare cases. In a great majority of cases there is no total breach but there are defaults in the performance of the contract by either side which result in loss or delay in completion. Building and engineering contracts invariably contain reciprocal promises either express or implied. The legal aspects of breaches of reciprocal promises are considered below, and the most common breaches are considered in Chapter 11. The remedies for breaches are dealt with in Chapter 12.

10.5.1 Reciprocal Promises

Section 52 of the Indian Contract Act, which deals with reciprocal promises, reads as follows:

> 52. *Order of performance of reciprocal promises*
> Where the order in which reciprocal promises are to be performed is expressly fixed by the contract, they shall be performed in that order; and where the order is not expressly fixed by the contract, they shall be performed in that order which the nature of the transaction requires.

The examples below will help in understanding the provisions of the section.

EXAMPLES FROM CASE LAW

1.   In terms of the agreement of sale for immovable property, (a) the purchaser on the basis of whatever initial examination she had taken of the documents, had unconditionally agreed to pay the purchase price in three instalments on or before the agreed dates and (b) if the purchaser was not thereafter satisfied with the title or found the title unacceptable and if the vendors failed to satisfy her about their title when she notified them about her dissatisfaction, the vendors had to refund all payments made within three months. The agreement provided that the time for payment of the balance price was of the essence of the contract and such payment was not dependent upon the purchaser's satisfaction regarding the title. On the other hand, in the terms relating to performance of sale, there is a clear indication

---

[20] Weatherproof Improvement Contracting Corpn. v. Kramer, 172 N.Y.S. 2d 688, Municipal Court of the City of New York, Borough of Bronx, 1958.

that time was not intended to be of the essence for completion of the sale. Clause 3 provided that the execution of the sale deed depended upon the purchaser's satisfaction regarding the title to the lands, and also the nil encumbrances on it. On these facts the Supreme Court of India held:

> We are therefore of the view that the failure of the appellant to pay the balance of Rs. 75,000 on 6th April 1981 and failure to pay the last instalment of Rs. 75,000 on or before 30th May 1981 clearly amounted to breach and time for such payment was the essence of the contract, the respondents were justified in determining the agreement of sale which they did by notice dated 2nd August 1981 ... Therefore rejection of the prayer for specific performance is upheld.[21]

2. The contractor had to carry out certain construction work within two months, the society had to perform various obligations, the substantial obligations being to subdivide their plot, to transfer a portion of their plot to the contractor in lieu of the amount of Rs. 2,84,164/- being lieu already due and payable by the society to the contractor and to that end make an application for subdivision of its plot and to unconditionally make the contractor its nominal member. All these obligations were to be performed upon the execution of the agreement itself. It was held that the first breaches of the agreement were committed by the society, and therefore, the society could not call upon the contractor to perfume its obligations:

> Consequently, if the society committed breach of its reciprocal obligations in subdividing or transferring the plot of land to the contractor, the contractor need not perform his further obligation by way of the reciprocal promises of putting up further construction within the time frame mentioned in the contract.[22]

10.5.2 Order, Which the Nature of the Transaction Requires

The nature of some transactions may demand the fulfilment of an implied condition. Say, for example, whereby a statute property is not transferable without the permission of an authority, an agreement to transfer the property must be deemed subject to the implied condition that the transferor will obtain the sanction of the authority concerned.[23] Under Section 47 there can be no bar to the maintainability of a suit for specific performance of an agreement to direct the defendant (seller) to apply for permission under Section 47 and after he obtains it to execute a sale deed. This result would follow whether or not there was a specific term in the agreement that permission would be obtained under Section 47 and that thereafter a sale deed will be executed.[24] Applying these rules to ascertain the order of performance in a case, it was held:[25]

> The permission of the District Officer/Collector was the essential precondition to the competency of the appellant to transfer. Therefore, unless he obtained that permission there was no question and no need of paying the moneys under the terms of the agreement.

---

[21] Saradamani Kandappan v. S. Rajalakshmi, 2011 AIR (SCW) 4092: 2011 AIR SC (Civil) 1812:; 2011 AIR (SC) 3234: 2011(4) R.C.R.(Civil) 130: 2011(4) Recent Apex Judgments (R.A.J.) 524; 2011(12) SCC 18:; 2011(8) JT 129.

[22] Shanti Builders v. CIBA Industrial Workers' Co-op. Housing Society (Bombay), 2012(4) Mh.LJ 614: 2012(6) BCR 209: 209.

[23] Natthu Lal v. Phool Chand, AIR 1970 SC 546.

[24] Syed Jalal v. Targopal Ram Reddy, AIR 1970 AP 19.

[25] Bishambhar Nath Agarwal v. Kishan Chand, AIR 1998 All. 195.

The Supreme Court has acknowledged the well-settled principle that in cases of contract for sale of immovable property, the grant of relief of specific performance is a rule and its refusal an exception based on valid and cogent grounds. Thus, specific performance of such an agreement will be granted irrespective of the fact that certain permissions were required from the government authorities for completing the building ...However, to deny a claim for specific performance of an agreement to sell an immovable property in existence or to be brought into existence according to the specification agreed to merely because the vendor had to make applications or move the concerned and competent authorities to obtain permission/sanction or consent of such authorities to make the sale agreed to be made an effective and full-fledged one may not be correct. That unless the competent authorities have been moved and the application for consent/permission/sanction have been rejected once and for all and such rejection made finally became irresolutely binding and rendered impossible the performance of the contract resulting in frustration as envisaged under Section 56 of the Contract Act, the relief cannot be refused for the mere pointing out of some obstacles.[26]

When the landlord is under an obligation to refund the security deposit following termination of a lease, the tenant is within its contractual right to hold on to the property until the refund of security deposit and is not liable to the payment of any rent.[27]

No man can complain of another's failure to do something which he has himself made impossible, is civilized well-settled principle of law. It is embodied in Section 53 of the Indian Contract Act.

10.5.3 Section 53 of the Contract Act

53. *Liability of party preventing event on which the contract is to take effect*
When a contract contains reciprocal promises, and one party to the contract prevents the other from performing his promise, the contract becomes voidable at the option of the party so prevented; and he is entitled to compensation from the other party for any loss which he may sustain in consequence of the non-performance of the contract.

ILLUSTRATION

A and B contract that B shall execute certain work for A for a thousand rupees. B is ready and willing to execute the work accordingly, but A prevents him from doing so. The contract is voidable at the option of B; and, if he elects to rescind it, he is entitled to recover from A compensation for any loss which he has incurred by its non-performance.

The owner of a construction work is well-advised to fulfil his part of the contract and to give no ground to the contractor, which may be used as an excuse for his non-performance and further make the owner liable to pay compensation. Few reported decisions will illustrate the circumstances in which a contractor's claim to compensation may be sustainable under this section.

EXAMPLES FROM CASE LAW

1. A contractor undertook to remove waste rock lying at a dump at the owner's mine within two years, provided there were not more than 50,000 tons, the owner agreeing to supply a crusher. The crusher supplied by the owner was so inadequate, crushing only three

---

[26] Nirmala Anand v. Advent Corporation Pvt. Ltd., AIR 2002 SC 2290. Also see, Express Towers Pvt. Ltd. v. Mohan Singh, IA No. 5912/2004 in CS(OS) 1578 of 2002, decided Del. H.C. 22 August 2006.
[27] Mahendra Kaur Arora v. Rent Appellate Tribunal, (Rajasthan), 2013(2) RLW 1525: 2012(60) R.C.R.(Civil) 669.

tons per hour that the work had to be stopped. It was held that the contractor was entitled to recover the damages for the expense to which he had been subjected in preparing for the work, and for the loss of profit he would otherwise have made by supplying crushed stone to a third party.[28]

2. The plaintiff entered into a contract for the construction of a road for Rs. 67,874.38/- but executed work worth Rs. 50,000/- only. The remaining work in the forest land could not be carried out due to the defendant's failure to obtain a no clearance certificate from the Forest Department, as had been promised when the contract was entered into. The plaintiff instituted a suit for recovery of the total amount of the contract, i.e., Rs. 67, 874.38/- and for security, costs, interest, etc. The trial court decreed the suit. An appeal was filed. Held, on failure of defendant to get clearance from the Forest Department, Section 53 of the Act was attracted, but when the plaintiff had done work worth Rs. 50,000/- only, he was not entitled to the total amount of the contract money. Accordingly, the suit was decreed for Rs. 50,000/- and refund of security, with proportionate costs and interest.[29]

3. In a contract for the sale of ore from a mine, the seller tried to justify his failure to supply ore on the ground that the buyer had not performed its obligations, such as appointing a sampling agent, providing a weighbridge, etc. Rejecting the contention, the Calcutta High Court held that the contract itself indicated the order in which reciprocal obligations were to be performed. It was for the seller to raise the material from the mines and inform the buyer. The obligation of the buyer would arise only thereafter.[30]

Where the owner of a project undertakes to provide site of work free from hindrance, good-for-construction drawings, certain materials like cement, steel, etc. to the contractor for use in the work, the effect of the owner's failure to do what was agreed is governed by Section 54 of the Indian Contract Act. The said section reads:

---

54. *Effect of default as to that promise which should be first performed, in contract consisting of reciprocal promises*
When a contract consists of reciprocal promises, such that one of them cannot be performed or that its performance cannot be claimed till the other has been performed, and the promisor of the promise last mentioned fails to perform it, such promisor cannot claim the performance of the reciprocal promise, and must make compensation to the other party to the contract for any loss which such other party may sustain by the non-performance of the contract.

ILLUSTRATION

... (b) A contracts with B to execute a certain builder's work for a fixed price, B supplying the scaffolding and timber necessary for the work. B refuses to furnish any scaffolding or timber, and the work cannot be executed. A need not execute the work, and B is bound to make compensation to A for any loss caused to him by the non-performance of the contract.

---

[28] Kleinert v. Abosso Gold Winning Co., (1913) 58 Sol. J. 45.
[29] U.P. State Electricity Board v. Chattar Singh Negi (Uttarakhand), 2007(2) U.D. 353: 2007(9) R.C.R.(Civil) 214.
[30] Vijaya Minerals Pvt. Ltd., Plaintiff v. Bikash Chandra Deb, AIR 1996 Cal.67 (79).

The owner of a building or construction project is duty-bound under the contract to afford the contractor reasonable facilities for the performance of the contract. For example, the owner should permit the contractor to enter on, or to take possession of, the site of work without which he cannot perform his work.

The liability of the owner to pay compensation will depend upon the agreed terms of the contract. For example, Clause 17A(iii) of the standard form General Conditions of Contract used by the Indian Railways reads:

> In the event of any failure or delay by the Railway to hand over to the Contractor possession of the lands necessary for the execution of the works or to give the necessary notice to commence the works or to provide the necessary drawings or instructions or any other delay caused by the Railway due to any cause whatsoever, then such failure or delay shall in no way effect or vitiate the contract or alter the character thereof or entitle the Contractor to damages of compensation thereof but in any such case, the Railway may grant such extension or extensions of the completion date as may be considered reasonable.[31]

Clause 17(3) thus clearly shows that if there is delay by the Railways, the terms of the contract can be extended. However, the contractor shall not be entitled to damages or compensation.[32] A similar question was considered by a three-judge bench in Ch. Ramalinga Reddy Vs. Superintending Engg. & Anr.[33] It was held in that case that if the contract is extended under the terms of the contract, compensation cannot be awarded by the arbitrator. The aforesaid judgment has been followed in another decision of the Supreme Court in General Manager (Northern Railway) & Anr. v. Sarvesh Chopra.

Clause 17A is in the nature of an exemption clause. It seems unfair that regardless of who is at fault, the contractor is only entitlted to an extension of time and not damages. A similar provision in Clause 11A of MES Standard Form IAFW for lump sum contracts came before the Delhi High Court.[34] Mr Justice Mehta questioned why the general principles in Sections 55 and 73 should not prevail and used Section 23 to prevent such a clause from defeating the very contract itself, "a matter of grave public interest".[35] He also drew attention to the two judgments of the Supreme Court, namely, Ramnath International Construction (P) Ltd. v. Union of India[36] and Asian Techs Ltd. v. Union of India.[37] The former decision upheld an exemption clause but the latter held that Clauses 11A to C prevented the department from granting damages, not the arbitrator.

Section 67 of the Indian Contract Act deals with the effect of neglect of the promisee to afford the promisor reasonable facilities for performance of the contract. The section is reproduced below along with its illustration which is self-explanatory.

---

67. Effect of neglect of promisee to afford promisor reasonable facilities for performance
If any promisee neglects or refuses to afford the promisor reasonable facilities for the performance of his promise, the promisor is excused by such neglect or refusal as to any non-performance caused thereby.

---

[31] Retrieved from www.indianrailways.gov.in.

[32] Union of India v. Chandalavada Gopalkrishna Murthy and others, CA No .s 926-7 of 2002, decided by the Supreme Court on 10th April 2008. Also see: M/s. Sree Kamatchi Amman v. The Divisional Railway, OSA Nos. 109 and 247 of 2005, decided by Mad. H.C. on 18th July 2007. An award granting compensation was set aside.

[33] Ch. Ramalinga Reddy v. Superintending Engg. & Anr., 1999(9)SCC 610.

[34] Simplex Concrete Piles (India) Ltd. v. Union of India, 2010 (1) CT LJ 255 (Delhi). = 2010 (2) ILR Delhi 699.

[35] Ibid., para. 15.

[36] (2002) SCC 453.

[37] (2009) 10 SCC 354.

ILLUSTRATION

A contracts with B to repair B's house. B neglects or refuses to point out to A the places in which his house requires repair. A is excused for the non-performance of the contract, if it is caused by such neglect or refusal.

10.6 Excuses for Non-Performance

Every failure to perform an obligation arising out of a contract may not amount to an actionable breach. There are a number of valid excuses for non-performance including the following ones embodied in Section 62 of the Indian Contract Act.

62. *Effect of novation, rescission and alteration of contract*
If the parties to a contract agree to substitute a new contract for it, or to rescind or alter it, the original contract need not be performed.

ILLUSTRATIONS

(a) A owes money to B under a contract. It is agreed between A, B and C that B shall thenceforth accept C as his debtor instead of A. The old debt of A to B is at an end, a new debt from C to B has been contracted.

(b) A owes B Rs. 10,000. A enters into an agreement with B and gives B a mortgage of his (A's) estate for Rs. 5,000 in place of the debt of Rs. 10,000. This is new contract and extinguishes the old.

(c) A owes B Rs. 1,000 under a contract. B owes C Rs. 1,000. B orders A to credit C with Rs. 1,000 in his books, but C does not assent to the arrangement. B still owes C Rs. 1,000, and no new contract has been entered into.

Section 62 provides for discharge of contractual obligations where the contract is substituted by a new contract or rescinded or altered by all the parties to it. The law is well settled that once the parties reduce the terms of an agreement to writing to form a contract, the said terms are binding on the parties and cannot be unilaterally altered and that too to the detriment of one of the parties by other because the other had not agreed to the alteration.[38]

10.6.1 Novation

The meaning of novation has been thus defined by the House of Lords:[39] "[t]hat, there being a contract in existence, some new contract is substituted for it either between the same parties or between different parties, the considerations mutually being the discharge of the old contract." Thus, where the parties substitute a new contract in place of the old one, they mutually discharge each other from performing obligations arising out of the old contract; which then neither party

---

[38] M/s Scorpion Express Pvt. Ltd. v. Union of India, AIR 2009 Pat. 108; ABL International Ltd. v. Export Credit Guarantee Corpn of India Ltd. (2004) 3 SCC 553; Food Corporation of India Ltd. v. SEIL Ltd. (2008) 3 SCC 440; AIR 2008 SC 1101.
[39] Scarf v. Jardine (1882) 7, App. Cas. 345, 351.

need perform. It should be noted that a party that wants to novate a contract ought not to terminate an existing contract by signing a deed of release; there would then be nothing left to novate.[40]

Novation contemplated in Section 62 of the Act therefore involves an annulment of one debt and the creation of another. In every case of this nature one has to consider not only whether the new debtor has consented to assume liability but also whether the creditor has agreed to accept the liability of the new debtor in substitution of the original debtor. In other words, novation is not consistent with the original debtor remaining liable in any form, since the essential element of novation is that the rights against the original contractor shall be relinquished and the liability of the new contracting party accepted in its place. Under law, therefore, one of the requisites of novation is the agreement of the parties to the new contract. Obviously there must be consent of all parties involved.[41]

The law seems to be well settled that novation of a contract could take place *sub silentio*. For example, a party may be held bound if it accepted the new rates or the periods either expressly or *sub silentio*.[42]

It is clear that where parties to a contract agree to substitute a completely different contract for the first, or to rescind a contract, the performance under the original contract and/or rescinded contract comes to an end. When parties to a contract "alter" a contract, the question that has to be answered is as to whether the original contract is altered in such a manner that performance under it is at an end.[43]

An assignment of a contract might result by transfer either of the rights or of the obligations thereunder. There is a well-recognized distinction between these two classes of assignments. The Supreme Court held that obligations under a contract cannot be assigned except with the consent of the promisee and when such consent is given it is really a novation resulting in substitution of liabilities. But rights under a contract are assignable unless the contract is personal in its nature or the rights are incapable of assignment either under the law or under the agreement between the parties.[44]

Where the contract itself contains a provision for the payment of enhanced rates dependent upon a contingency, for example, liability of the tenant to pay a proportionate increase in the municipal taxes, there is no novation.[45]

Thus, in conclusion it needs to be noted that one of the essential requirements of "novation", as contemplated by Section 62, is that there should be complete substitution of a new contract in place of the old.[46] It is in that situation that the original contract need not be performed. Substitution of a new contract in place of the old contract which would have the effect of rescinding or completely altering the terms of the original contract, has to be by agreement between the parties. A substituted contract should rescind or alter or extinguish the previous contract. But if the terms of the two contracts are inconsistent and they cannot stand together, the subsequent contract cannot be said to be in substitution of the earlier contract. In one case, the rights under the original contract were not given up as it was specifically provided in the subsequent contract that the rights under the old contract should stand extinguished only on payment of the entire amount of R. 9,51,000/-.[47]

---

[40] Wessely v. White, [2018] EWHC 1499 (Ch).

[41] State Bank of India, Appellant v. Mrs. T.R. Seethavarma, AIR 1995 Kerala 31; Appukutm Panicker v. Anantha Chettiar, 1966 KLJ 708: (AIR 1966 Ker. 303); T.M. and Co. v. H.I. Trust, AIR 1969 Calcutta 238.

[42] BSNL v. BPL Mobile Cellular Ltd., (2008) 13 SCC 597; Kanchan Udyog Limited v. United Spirits Limited, 2017(3) R.C.R.(Civil) 433: 2017(4) Recent Apex Judgments (R.A.J.) 126: 2017(7) Scale 69: 2017(5) MLJ 616; 2017(8) SCC 237.

[43] Chrisomar Corporation v. MJR Steels Private Limited, 2017(11) Scale 453: 2017(10) JT 475: 2017 AIR (SC) 5530: 2017 AIR (SCW) 5530. Also see "Alteration" below in paragraph 10.6.3.

[44] Khardah Co. Ltd. v. Raymon and Co., AIR 1962 SC 1810.

[45] Savita Dey v. Nageshwar Majumdar (1995) 6 SCC 274.

[46] Ayodhya Prasad v. Phulesara Bhagwan Das, AIR 2008 Allahabad 169.

[47] Lata Construction v. Dr. Rameshchandra Ramniklal Shah, AIR 2000 SC 380; 1999 AIR SCW 4518.

10.6.2 Rescission

Where parties mutually agree to terminate their contract, no party need perform the obligations arising out of such a cancelled contract. If the supplemental agreement is valid, it will follow that the same puts an end to the original agreement. Nothing will survive.[48]

A contract, however, cannot be cancelled or terminated unilaterally unless the contract empowers one party alone to terminate it. It needs to be noted that the right to one party to terminate the contract does not render the contract void. The Supreme Court of India held:

> Under general law of contracts any clause giving absolute power to one party to cancel the contract does not amount to interfering with the integrity of the contract. The acceptance of the argument regarding invalidity of contract on the ground that it gives absolute power to the parties to terminate the agreement would also amount to interfering with the rights of the parties to freely enter into the contracts. A contract cannot be held to be void only on this ground. Such a broad proposition of law that a term in a contract giving absolute right to the parties to cancel the contract is itself enough to void it cannot be accepted.[49]

A division bench of Bombay High Court held in a case that the provision empowering a party to put an end to the contract after one year by giving 30 days' notice is not unconscionable or opposed to public policy.[50]

It is noteworthy that if a contract is wrongfully terminated, suit against the officer terminating the contract may be maintainable. The officer terminating contract if not signatory to the contract he cannot be termed to be an agent of defendant No. 1, here the Airports Authority of India. He is an official and cause of action arises against him individually not in the capacity of an official of defendant No. 1.[51]

Rescission: Effect on Arbitration Clause

An arbitration clause is a collateral term of a contract as distinguished from its substantive terms; but nonetheless it is an integral part of it. However comprehensive the terms of an arbitration clause may be, the existence of the contract is a necessary condition for its operation, it perishes with the contract. The contract may be *non-est* in the sense that it never came legally into existence or it was void *ab initio* or the contract was validly executed, the parties may put an end to it as if it had never existed and substitute a new contract for it solely governing their rights and liabilities thereunder. In the former case, if the original contract has no legal existence, the arbitration clause also cannot operate, for along with the original contract, it is also void. In the latter case, as the original contract is extinguished by the substituted one, the arbitration clause of the original contract is extinguished by the substituted one, the arbitration clause of the original contract perishes with it. In between the two fall many categories of disputes in connection with a contract, such as the question of repudiation, frustration, breach, etc. In those cases it is the performance of the contract that has come to an end, but the contract is still in existence for certain purposes in respect of disputes arising under it or in connection with it. As the contract subsists for certain purposes, the arbitration clause operates in respect of these purposes.[52]

---

[48] Unikol Bottlers Ltd., M/s v. M/s. Dhillon Kool Drinks, AIR 1995 Del. 25.

[49] H.H.M. Shantidevi P. Gaikwad v. Savjibhai Haribhai Patel, AIR 2001 SC 1462= 2001 AIR SCW 1240.

[50] O.N.G.C. Ltd. v. Streamline Shipping Co. Pvt. Ltd., 2002 (Suppl.) Arb. LR 145 Bombay (DB).

[51] Maheswari Brothers Ltd. v. Airports Authority of India Ltd., AIR 2006 Cal. 227.

[52] Union of India v. Kishorilal Gupta and Bros., AIR 1959 SC 1362.

10.6.3 Alteration

Alteration in the contract, mutually agreed to, frees the parties from performing the old contract. It needs to be noted that the original terms of the contract continue to be part of the contract and are rescinded or superseded except insofar as they are inconsistent with the modifications.[53]

The consideration for the variation lies in the mutual abandonment of existing rights or conferring new benefits by each party on the other or in the assumption of additional obligations or incurring additional liabilities or increased detriment. While assuming "increased detriment" a party may gain a "practical advantage", which can be considered as sufficient consideration. A unilateral declaration by one party alone cannot constitute a variation unless the contract empowers a party to do so, and will amount to breach of contract entitling the other party to damages or right to repudiate the contract, as the case may be. Where the government department agrees to supply certain machineries to the contractor at fixed rates, it cannot enhance the rates unilaterally unless there is a provision to that effect in the contract.[54]

10.6.4 Non-Performance Not Excused because of Abnormal Increase in Prices, and Difficulties, Strikes or the Like

The Contract Act does not enable a party to a contract to ignore the express covenants thereof, and to claim payment of consideration for performance of the contract at rates different from the stipulated rates, on some vague plea of equity. If, on the other hand, a consideration of the terms of the contract in the light of the circumstances existing when it was made, shows that the contractor never agreed to be bound in a fundamentally different situation which has now unexpectedly emerged, the contract ceases to bind at that point not because the court in its discretion thinks it just and reasonable to qualify the terms of the contract, but because on its true construction it does not apply in that situation.[55] For example, where a contractor's lump sum quote was "subject to final re-measurement on issue of finalised construction drawings", the full works were directed to be remeasured, not just those subject to change orders on an add/omit basis.[56] The facts of the case below and its decision only confirm entitlement of price rise to a party in spite of the contract to the contrary if the factual matrix under which the contract was entered into totally changes to upset the calculations, it is submitted.[57]

EXAMPLE FROM CASE LAW

Briefly stated, the facts involved in the case were:[58] The defendant entered into a contract with the plaintiff for the supply and installation of 18 power transformers, etc. on a firm price basis without provision for any escalation on any account. The delivery of equipment was to commence after ten months from the date of the purchase order and was to be completed in the fourteenth month. Admittedly, six out of 18 power transformers were dispatched on various dates after the delivery period was over. The value of the equipment dispatched and delivered was about two-thirds of the value of the entire transformers and the rest of the 12 transformers were not supplied, though four were tested. Extension of time was granted up to 31st March 1975. The defendant terminated the contract by letters dated 3rd September 1975 and 18th September 1975. The plaintiff contended that they were

[53] Juggilal Kamalpat v. NV Internaional Credit-En-Handels Vereeniging "Rotterdam", AIR 1955 Cal. 65.
[54] State of A.P. v. Pioneer Construction Co., AIR 1978 AP 281; also see: Magnum Films v. Golcha Properties Pvt. Ltd., AIR 1984 Del. 162.
[55] Alopi Parshad v. Union of India, AIR 1960 SC 588.
[56] BHC Ltd. v. Galliford Try Infrastructure Ltd. (t/a Morrison Construction), [2018] EWHC 368 (TCC).
[57] Tarapore & Co. v. Cochin Ship Yard Ltd. (1984) 2 SCC 680 (715).
[58] Easun Engineering Co. Ltd. v. Fertilisers and Chemicals, Travancore Ltd., AIR 1991 Mad. 158.

prevented from supplying due to *force majeure* conditions, namely, strikes, power cuts and phenomenal increase in the cost of transformer oil (400%) due to war conditions, etc. It was contended that the termination was unilateral and illegal. The disputes were referred to an umpire by the order of the Madras High Court. The umpire allowed the claims of the plaintiff for the price variation applicable to the six transformers supplied and also for oil transformers so supplied, etc. The defendants contended in the High Court, holding that the contract was a firm price contract with no provision for escalation on any account and, at the same time, by allowing escalation the umpire committed an error apparent on the face of the record. It was held:

> [I]n the present case, it can be safely held that the above mentioned increase in price due to war condition, is an untoward event or change of circumstances which "totally upsets the very foundation upon which parties rested their bargain." Therefore, EASUN (the plaintiff) can be said to be finding itself impossible to supply the transformers which it promised to do. ... In the circumstances, I do not think there is any error apparent on the face of the record, in the award passed by the Umpire.

This case is not an authority on Section 56 and one must view it with caution, it is submitted. However, it could as well be a beginning in paving the way to bring the case of "financial impossibility" within the scope of Section 56, which sooner or later the judiciary has to uphold in the interest of justice, it is respectfully submitted.

Unexpected Difficulties

There is a fairly established principle that a contract is not discharged by frustration merely because the promisor suffers loss or meets with difficulties in the execution of the contract. It is a settled rule that where a person by his contract or agreement charges himself with an obligation possible to be performed, he must perform it, and he will not be excused therefrom because of unforeseen difficulties, unusual unexpected expense, or because it is unprofitable or impracticable.[59]

A party to a contract can always safeguard his interest against unforeseen contingencies by express provisions made in the contract. If he voluntarily undertakes an absolute and unconditional obligation he cannot complain merely because events turned out to his disadvantage. Mere difficulty or the need to pay abnormal prices cannot excuse a party from fulfilling his part of the contract unless the factual matrix on which the contract was based has changed as already stated.[60]

EXAMPLE FROM CASE LAW

The contractors in a case undertook to build 78 houses for a fixed sum of £94,424. Owing to an unexpected shortage of skilled labour and certain materials the contract took 22 months to complete and cost some £115,000. The House of Lords did not accept the contention of the contractors that, under the circumstances, the contract had been frustrated and they were entitled to claim on a *quantum meruit* for the cost actually incurred. It was held by their Lordships, that the mere fact that unforeseen circumstances had delayed

---

[59] Clinchfield Stone Co. v. Stone, 254. S.W. 2d Court of Appeals of Tennessee, 1952.

[60] Mahadeo Prasad v. Calcutta D & C Co., AIR 1961 Cal. 70; Satyabrata v. Mungerneeram Bangur and Co., AIR 1954 SC 49.

the performance of the contract, and rendered it more onerous to the contractors, did not discharge the agreement. The ultimate situation was still within the scope of the contract. "The thing undertaken was not, when performed, different from that contracted for."[61]

10.6.5 Doctrine of Waiver

If, on facts there is a waiver of a provision of an agreement by one of the parties to the contract, Section 63 of the Contract Act will operate in order to give effect to such waiver.[62] The said section reads:

> 63. *Promisee may dispense with or remit performance of promise*
> Every promisee may dispense with or remit, wholly or in part, the performance of the promise made to him, or may extend the time for such performance or may accept instead of it any satisfaction which he thinks fit.

ILLUSTRATIONS

(a) A promises to paint a picture for B. B afterwards forbids him to do so. A is no longer bound to perform the promise.

[…]

(d) A owes B, under a contract, a sum of money, the amount of which has not been ascertained. A, without ascertaining the amount, gives to B, and B, in satisfaction thereof, accepts, the sum of Rs. 2,000. This is a discharge of the whole debt, whatever may be its amount.

Waiver is an intentional relinquishment of a known right or advantage, benefit, claim or privilege which except for such waiver the party would have enjoyed. Waiver can also be a voluntary surrender of a right. The doctrine of waiver which the courts of law will recognize is a rule of judicial policy that a person will not be allowed to take an inconsistent position to gain advantage through the aid of courts. Waiver sometimes partakes of the nature of an election. Waiver is consensual in nature. It implies a meeting of the minds. It is a matter of mutual intention. The doctrine does not depend on misrepresentation. Waiver actually requires two parties, one party waiving and another receiving the benefit of waiver. There can be waiver so intended by one party and so understood by the other. The essential element of waiver is that there must be a voluntary and intentional relinquishment of a right. The voluntary choice is the essence of waiver. Waiver can be by conduct. If a party entitled to a benefit under a contract is denied the same, resulting in violation of a legal right, and does not protest, forgoing its legal right, and accepts compliance in another form and manner, issues will arise with regard to waiver or acquiescence by conduct.[63] Waiver can also be deduced from acquiescence.[64]

[61] Davis Contractors Ltd. v. Fareham U.D.C., [1956] A.C. 696.
[62] All India Power Engineer Federation v. Sasan Power Ltd., 2017(2) Recent Apex Judgments (R.A.J.) 386: 2016(12) Scale 553; 2017(1) SCC 487. Relied upon: Jagad Bandhu Chatterjee v. Nilima Rani, (1969) 3 SCC 445.
[63] P. Dasa Muni Reddy v. P. Appa Rao, 1976() R.C.R.(Rent) 58: (1974) 2 SCC 725.
[64] Kanchan Udyog Limited v. United Spirits Limited, 2017(3) R.C.R.(Civil) 433: 2017(4) Recent Apex Judgments (R.A.J.) 126: 2017(7) Scale 69: 2017(5) MLJ 616: 2017(3) BC 534: 2017(8) SCC 237:; also see: Waman Shriniwas Kini v. Ratilal Bhagwandas & Co., 1959 Supp (2) SCR 217.

English law not to be applied: It is important to remember that when Section 63 of the Contract Act is to be applied, the High Courts in India[65] have cautioned that, being a wide departure from English law, the section alone should be enforced according to its terms and not in accordance with English law.[66]

Public interest and waiver: It is also clear that if any element of public interest is involved and a waiver takes place by one of the parties to an agreement, such waiver will not be given effect to if it is contrary to such public interest.[67] It is thus clear that if there is any element of public interest involved, the court steps in to thwart any waiver which may be contrary to such public interest.[68]

Waiver distinguished from estoppel: The principle of waiver is akin to the principle of estoppel; the difference between the two, however, is that whereas estoppel is not a cause of action – it is a rule of evidence –, waiver is contractual and may constitute a cause of action – it is an agreement between the parties and a party fully knowing its rights has agreed not to assert a right for a consideration. Whenever waiver is pleaded it is for the party pleading the same to show that an agreement waiving the right in consideration of some compromise came into being. Statutory right, however, may also be waived by conduct.

10.7 Doctrine of Frustration

Most legal systems make a provision for the discharge of a contract where, subsequent to its formation, a change of circumstances renders the contract legally or physically impossible of performance. In India, the law of frustration of contract is embodied in Section 56 of the Contract Act. It reads:

56. *Agreement to do impossible act*
An agreement to do an act impossible in itself is void.

Contracts to do acts afterwards becoming impossible or unlawful. – A contract to do an act which, after the contract is made, becomes impossible, or, by reason of some event which the promisor could not prevent, unlawful, becomes void when the act becomes impossible or unlawful.

Compensation for loss through non-performance of act known to be impossible or unlawful. – Where one person has promised to do something, which he knew, or with reasonable diligence, might have known, and which the promisee did not know to be impossible or unlawful, such promisor must make compensation to such promisee for any loss which such promisee sustains through the non-performance of the promise.

ILLUSTRATIONS

(a) A agrees with B to discover treasure by magic. The agreement is void.

(b) A and B contract to marry each other. Before the time fixed for the marriage, A goes mad. The contract becomes void.

---

[65] New Standard Bank, Ltd. v. Probodh Chandra Chakravarty, AIR 1942 Calcutta 87 at 90–91; Anandram Mangturam v. Bholaram Tanumal, AIR 1946 Bombay 1 at 6.

[66] Chrisomar Corporation v. MJR Steels Private Limited, 2017(11) Scale 453: 2017(10) JT 475: 2017 AIR (SC) 5530: 2017 AIR (SCW) 5530.

[67] Lachoo Mal v. Radhey Shyam, 1971 R.C.R.(Rent) 320: (1971) 1 SCC 619, Indira Bai v. Nand Kishore, (1990) 4 SCC 668, Indira Bai v. Nand Kishore, (1990) 4 SCC 668.

[68] All India Power Engineer Federation v. Sasan Power Ltd., (SC) 2017(2) Recent Apex Judgments (R.A.J.) 386: 2016(12) Scale 553: 2017(1) SCC 487.

(c) A contracts to marry B, being already married to C, and being forbidden by the law to which he is subject to practice polygamy. A must make compensation to B for the loss caused to her by the non-performance of his promise.

(d) A contracts to take in cargo for B at a foreign port. A's Government afterwards declares war against the country in which the port is situated. The contract becomes void when war is declared.

(e) A contracts to act at a theatre for six months in consideration of a sum paid in advance by B. On several occasions A is too ill to act. The contract to act on those occasions becomes void.

According to the first paragraph, where parties purport to agree to do something obviously impossible, they must be deemed not to be serious or not to understand what they are doing. Also a promise to do something obviously impossible is no consideration. According to the second paragraph, if due to some event the performance becomes impossible or illegal, the contract becomes void when such an event takes place. It is clear from the third paragraph of the section that a contractor must take care, before entering into an agreement, to see that what he is agreeing to is possible to perform. If he fails to use reasonable diligence to determine whether the performance would be possible and legal he would be liable to the owner, if at a later stage the performance becomes impossible or illegal. Where the owner had obtained a licence to execute certain repair works to his house and engaged the contractor to do the work who, having innocently exceeded the amount of the licence, claimed the excess amount spent, was denied the recovery. It was held that the prohibition on doing unlicensed work was absolute and did not depend on the contractor's state of mind. The owner, thus had the benefit of the work without having to pay for it.[69] Similarly, it may be noted here that performance is not rendered impossible for the purpose of Section 56 of the Indian Contract Act merely on account of a strike of the workmen employed in executing the work under a contract.[70] The doctrine of frustration of contract cannot apply where the event which is alleged to have frustrated the contract arises from the act or election of a party.[71]

10.7.1 Essential Conditions for Applicability of Section 56

In order that the principle of impossibility of performance may apply, the following conditions must be satisfied:

(a) There must be a valid and subsisting contract between the parties

(b) Some part of the contract must yet be unfulfilled

(c) The contract after it is made becomes impossible

(d) That the impossibility is on account of some event which the promisor could not prevent or anticipate, and

(e) That the impossibility is not self-induced by the promisor or due to his negligence.

When the above conditions are satisfied, the contract becomes void when the act becomes impossible.

---

[69] Bostel Brothers Ltd. v. Hurlock, Court of Appeal, [1949] 1 KB 74. This case was distinguished in Westgates (Norwich) v. Frost, [1949] 1 WLUK 238.

[70] Hari Laxman v. Secretary of State, (1927), 52 Bom. 142, 30, Bom. L.R. 49. 108, I.C. 19(28) A.B. 61.

[71] Boothalinga Agencies, Appellant v. V.T.C. Poriaswami Nadar, Respondent, AIR 1969 SC 110.

10.7.2 Meaning of Impossible – Death or Illness of Party

Impossibility contemplated by Section 56 of the Contract Act is both subjective and objective. When impossibility is due to the incapacity of the party who has undertaken the work, it is known as subjective impossibility. Such incapacity may be due to death or serious illness of the party. It becomes a valid excuse for non-performance of contractual obligations in respect of contracts for strictly personal services. For example, death or serious illness of a consulting engineer would relieve him of his contractual obligations of providing professional services. Subjective impossibility may not *ipso facto* become an excuse in the case of a typical building or construction contract. Most contracts, therefore, include a clause which empowers the owner, upon death of the contractor, to terminate the contract without liability to pay damages, if the legal representatives of the contractor are not capable of carrying out and completing the contract.

10.7.3 Bankruptcy of Contractor

On the bankruptcy of a contractor, or liquidation of a company where the contractor is a company, the burden and benefit of his contract pass to the trustee in bankruptcy or to the liquidator, respectively. Most construction contracts, however, provide for such contingencies by including a "forfeiture" clause in the contract which can be enforced in such cases.

10.7.4 Bankruptcy of Owner

Apart from the special provisions of any formal contract, on the bankruptcy of the owner, the contractor is entitled to determine the contract and claim in the bankruptcy proceedings as an ordinary creditor.

10.7.5 Objective Impossibility

The word impossible has not been used, in the sense of physical or literal impossibility alone. The performance of an act may not be literally impossible but it may be impracticable and useless from the point of view of the object and purpose which the parties had in view; and if an untoward event or change of circumstances totally upsets the very foundation upon which the parties rested their bargain, it can very well be said that the promisor finds it impossible to do the act which he promised to do. In a case, their Lordships of the Supreme Court of India observed:[72]

> Section 56 of the Contract Act lays down a rule of positive law and does not leave the matter to be determined according to the intention of the parties. The impossibility contemplated by Section 56 of the Contract Act is not confined to something which is not humanly possible. If the performance of a contract becomes impracticable or useless having regard to the object and purpose the parties had in view then it must be held that the performance has become impossible. But the supervening events should take away the basis of the contract and it should be of such a character that it strikes at the root of the contract.

For an example, reference to a case decided by the Supreme Court may be useful. In the said case, a textile undertaking was nationalized and thereby assets got vested in government. The plea by surety that vesting of assets in government frustrated his contract of guarantee was held as not tenable. The contract of guarantee had no co-relation with the Nationalization Act. It is an independent contract and in all fairness had to be honoured to fulfil the contractual

---

[72] Satyabrata Ghose v. Mugneeram Bangur and Co. and another, AIR 1954 SC 44.

obligation between the surety and the creditor – Recourse to Section 141 was also not available to surety.[73]

There can be no doubt that a man may by an absolute contract bind himself to perform, which performance subsequently becomes impossible, or to pay damages for the non-performance. This interpretation is to be placed upon an unqualified undertaking, where the event which causes the impossibility was or might have been anticipated and guarded against in the contract, or where the impossibility arises from the act or default of the promisor. But where the event is of such a character that it cannot reasonably be supposed to have been in the contemplation of the contracting parties when the contract was made, they will not be held bound by general words which, though large enough to include, were not used with reference to the possibility of the particular contingency which afterwards happened. It is on this principle that the act of God is in some cases said to excuse the breach of a contract. A few examples, wherein the contracts were held to be frustrated and not frustrated will make this point clear.

EXAMPLES FROM CASE LAW

1. The Delhi Development Authority (DDA) proposed a public-private partnership project for the development of an area of 14.3 hectares of prime land for the construction of 750 premium residential flats in a self-contained community to be sold by private real estate development on a free sale basis. In addition to the premium residential flats, the developer would have to construct 3,500 resettlement houses for the economically weaker sections of society. Disputes arose and the contractor submitted a petition to the High Court, and the court summed up the facts thus:

> The petitioner's stand is that it had made the bid for the project and had paid the entire amount of Rs. 450.01 crores on the clear understanding that the project site was residential. This understanding, according to the petitioner, was based on the representation made by the DDA as the detailed facts referred to above would reveal. In fact, the DDA has maintained and continues to maintain its stand that the project site is not within the ridge area and the land use of the same has been clearly shown as residential. ... The stand of the DDA is, however, not accepted either by the DPCC or the Department of Forests, Government of NCT of Delhi. In fact, both the DPCC and the Department of Forests (respondents 2 and 4 herein) along with the Government of NCT of Delhi (respondent No. 3) have taken a unified stand that the land in question falls within the ridge ... and no construction activity can be carried out in the land in question.

On appeal the Supreme Court of India observed:

> On a conspectus of the facts and the law placed before us, we are satisfied that certain circumstances had intervened, making it impracticable for Kenneth Builders to commence the construction activity on the project land. Since arriving at some clarity on the issue had taken a couple of years and that clarity was eventually and unambiguously provided by the report of the CEC, it could certainly be said that the contract between the DDA and Kenneth Builders was impossible of performance within the meaning of that word in Section 56 of the Contract Act. Therefore, we reject the contention of the DDA that the contract between the DDA and Kenneth Builders was not frustrated.

---

[73] I. F. C. I. Ltd. v. Cannanore Spg. and Wvg. Mills Ltd., AIR 2002 SC 1841 = 2002 AIR SCW 1822.

The DDA was directed to refund the deposit made by developer with interest at 6% annum from the date of deposit till its realization.[74]

2. The appellant was a successful bidder in an auction conducted on 24th March 1994 for sale of privilege to vend arrack in three shops for the period 1st April 1994 to 31st March 1995. Her bid was for a sum of Rs. 25,62,000/-. Being declared as auction purchaser, she deposited 30% of the bid amount i.e. Rs. 7,68,600/- on the same date and executed a temporary agreement in terms of Rule 5(10) of the Kerala Abkari Shops (Disposal in Auction) Rules, 1974, which was subject to confirmation by the Board of Revenue. Pursuant to Rule 5(19), this deposit would stand as security for due performance of the conditions of licence. Kalady is the holy birth place of Adi Sankaracharya and adjoining thereto existed a Christian pilgrim centre associated with St. Thomas. The residents of those areas objected to the running of any *abkari* (liquor) shop. A large number of people collected and offered physical resistance to the opening of the *abkari* shops and the law and order enforcing agency could not assure smooth conduct of business. The aforesaid circumstances led the appellant to believe that it was impossible for her to run the arrack shop in the locality in question. The appellant, therefore, by her letter dated 3rd April 1994, addressed to the Board of Revenue, District Collector and Assistant Commissioner of Excise, informed them that because of mass movement it was not possible for her to open and run the shops. Accordingly, she requested them not to confirm the sale in her favour as it was impossible for her to execute the privilege for reasons beyond her control. She also requested that the proposed contract may be treated as rescinded. She further reserved her right to claim refund of the security amount. There is nothing on record to show that after the appellant refused to carry out her obligations, the state government took any step to re-sell or re-dispose the arrack shops in question. Writ was filed. The learned single judge directed to refund the security deposit. On appeal, the division bench set aside the order. The Supreme Court finally held:

> Accordingly, we are of the opinion that in a contract under the Abkari Act and the Rules made thereunder, the licensee undertakes to abide by the terms and conditions of the Act and the Rules made thereunder which are statutory and in such a situation, the licensee cannot invoke the doctrine of fairness or reasonableness. Hence, we negative the contention of the appellant. Appeal dismissed.[75]

A different view was taken by the Andhra Pradesh High Court in the case discussed in Example (3) below.

3. A contract grant of leasehold rights for mining of sand at a river bed was conferred but villagers prevented the petitioner contractor from lifting the sand. The villagers were undeterred even by registration of a criminal case against them and also by grant of injunction order by Civil Court. It was held that performance of contract was thereby

---

[74] Delhi Development Authority v. Kenneth Builders & Developers Ltd., 2016 AIR (SCW) 3026: 2016(6) Scale 14: 2016(4) Recent Apex Judgments (R.A.J.) 447: 2016(7) JT 412: 2016 AIR (SC) 3026; 2016(3) Apex Court Judgments (SC) 283; 2016(13) SCC 561. Also see: Sri Ram Builders v. State of M.P. (SC): 2014 AIR (SCW) 2550: 2014(6) JT 134: 2014(14) SCC 102: 2014(6) Recent Apex Judgments (R.A.J.) 396.

[75] Mary v. State of Kerala, 2014(14) SCC 272: 2013(9) SCR 1126: 2013 AIR (SCW) 6082: 2013(6) Recent Apex Judgments (R.A.J.) 565: 2014 AIR (SC) 1. Also see Gian Chand v. M/s. York Exports Ltd., 2014(6) Scale 35:: 2014(4) Recent Apex Judgments (R.A.J.) 108; 2014 AIR (SC) 3584: 2015(5) SCC 609: 2014(6) JT 616: 2014 AIR (SCW) 4771.

rendered impossible and the doctrine of frustration would apply. The petitioner was held to be entitled to refund of bid amount.[76]

4.   The plaintiff in one case agreed to supply a certain quantity of eucalyptus firewood from plantations which were leased to the plaintiff by the TN Forest Plantation Corporation. The plaintiff did not supply the total quantity of specified quality of firewood. He alleged that he was not aware of the shortage of that type of firewood at the time of entering into the contract. It was held:[77] "As the plaintiff, in the instant case, could have very well ascertained how much firewood was there in the property before entering into the contract with the defendant, he cannot contend that he was not aware of its details." The plea of frustration was not accepted.

5.   An agreement clearly provided that the dispute between the parties should be settled by arbitration in accordance with the Rules of Conciliation and Arbitration of the International Chamber of Commerce. It was urged by the petitioner that the cost of arbitration was prohibitive and hence the agreement stood frustrated. It was held:

> The parties knew what they had bargained for. Therefore, there is no event which has intervened to frustrate the agreement. In these circumstances, it is not a case in which the doctrine of frustration of the arbitration agreement can be invoked.[78]

10.8   Effect of Frustration

It is settled that when there is frustration of contract, the dissolution of the contract occurs automatically and it does not depend on the ground of repudiation or breach or on the choice or election of either party, but it depends on the effect of what had actually happened on the possibility of performing the contract.[79] For example, a contract for construction of a reservoir to be completed within six years was awarded in 1914. The contract contained the usual provision empowering the engineer to grant extension of time. In February 1916, the work was ordered to be stopped under wartime powers by the concerned ministry. Considering the fact that if the power to extend time was exercised it was not impossible to perform, the House of Lords held that a contract, if resumed after the war, would amount to a new contract.[80]

Section 65 of the Indian Contract Act deals with the consequences of void contract rendered void under Sections 32 and 56 of the Act. It reads:

> 65. *Obligation of person who has received advantage under void agreement or contract that becomes void*

---

[76] Alluri Narayana Murthy Raju v. Dist. Collector, Visakhapatnam, AIR 2008 AP. 264.

[77] M/s Gwalior Rayon Silk Mfg Co. Ltd. v. Andavar & Co., AIR 1991 Ker. 134. Also see: C.T. Xavier v. P.V. Joseph, AIR 1995 Ker. 140.

[78] Eacom's Controls (India) Ltd., Petitioner v. Bailey Controls Co., AIR 1998 Del. 365.

[79] Ahmed Khan v. Jahan Begum, AIR 1973 All. 529.

[80] Metropolitan Water Board v. Dick Kerr & Co. Ltd. House of Lords, [1918] AC 119. See Morgan v. Manser, [1948] 1 KB 184.

When an agreement is discovered to be void or when a contract becomes void, any person who has received any advantage under such agreement or contract is bound to restore it or to make compensation for it, to the person from whom he received it.

ILLUSTRATIONS

(a) A pays B 1,000 rupees in consideration of B's promising to marry C, A's daughter. C is dead at the time of the promise. The agreement is void, but B must repay A the 1,000 rupees.

(b) A contracts with B to deliver to him 250 maunds of rice before the 1st of May. A delivers 130 maunds only before that day and none after. B retains the 130 maunds after the 1st of May. He is bound to pay A for them.

(c) A, a singer, contracts with B, the manager of a theatre to sing at his theatre for two nights in every week during the next two months, and B engages to pay her a hundred rupees for each night's performance. On the sixth night A willfully absents herself from the theatre, and B, in consequence, rescinds the contract. B must pay A for the five nights on which she had sung.

(d) A contracts to sing for B at a concert for 1,000 rupees, which are paid in advance. A is too ill to sing. A is not bound to make compensation to B for the loss of the profits which B would have made if A had been able to sing, but must refund to B the 1,000 rupees paid in advance.

The section makes a distinction between an agreement and a contract. According to Section 2 of the Contract Act an agreement which is enforceable by law is a contract and an agreement which is not enforceable by law is said to be void. Therefore, when the earlier part of the section speaks of an agreement being discovered to be void it means that the agreement is not enforceable and is, therefore, not a contract. It means that it was void. It may be that the parties or one of the parties to the agreement may not, when they entered into the agreement, have known that the agreement was in law not enforceable. They might have come to know later that the agreement was not enforceable. The second part of the section refers to a contract becoming void. That refers to a case where an agreement which was originally enforceable and was, therefore, a contract, becomes void due to subsequent happenings. In both these cases any person who has received any advantage under such agreement or contract is bound to restore such advantage, or to make compensation for it to the person from whom he received it. For example, once the sale is declared bad, the transaction of sale fails and, therefore, the seller has no right to retain the sale consideration and has to refund the sale consideration to the buyer.[81]

Where, however, even at the time when the agreement is entered into, both the parties knew that it was not lawful and, therefore, void, there was no contract but only an agreement and it is not a case where it is discovered to be void subsequently. Nor is it a case of the contract becoming void due to subsequent happenings. Therefore, Section 65 of the Contract Act will not apply.[82]

10.8.1 Frustration and Terms of Contract

The question of the effect of express terms in the agreement may have to be decided before concluding that a particular event is a frustrating event. A couple of examples may help in answering this question.

---

[81] Srinivasaiah v. H.R. Channabasappa, 2017(3) Recent Apex Judgments (R.A.J.) 135: 2017(4) JT 513: 2017(5) Scale 306: 2017 AIR (SC) 2141: 2017 AIR (SCW) 2141; 2017 All SCR 2282.

[82] Kuju Collieries Ltd. v. Jharkhand Mines Ltd., AIR 1974 SC 1892.

EXAMPLES FROM CASE LAW

1. By contracts entered into between March and November 1971, the respondents, who were building two blocks of flats in Hong Kong, agreed to the sale of the flats under construction. The construction work began in December 1971. The date stipulated in the contract for completion of the work was 17th May 1973. The agreement, which made time to be of the essence of the contract, also incorporated provisions in certain circumstances for extension of time for not more than one year. In June 1972, part of the hillside above the building site slipped taking with it a 30-storey block of flats. The entire debris landed on the site of work obliterating the building works already completed. Work was stopped and because it could not be recommenced within three months, the building permit of the vendors came to an end. A new permit was not issued until November 1975, by which time the maximum extension contemplated by the parties had elapsed. The agreement contained an express stipulation

> that notwithstanding anything herein contained – should any unforeseen circumstances beyond the vendor's control arise whereby the vendor is unable to sell the undivided share and apartment to the purchaser – the vendor shall be at liberty to rescind the agreement forthwith and to refund the purchaser all installments of the purchase price paid – without interest or compensation.

It was held that the above express stipulation should not be read as applying to the kind of "frustrating event". The contract was frustrated.[83]

2. A contract for the construction of an underground railway to be completed within a fixed period was entered into on the basis that the work was to be carried out on a seven-day, three-shift basis. The work was accordingly started but it was noisy and third parties affected by the work obtained injunctions restraining the activities at the site on evenings and on Sundays. The contracting company commenced arbitration proceedings claiming an addition to the contract price. Two contentions were raised. First, it was contended that an implied term deserves to be read in the contract that if the contractor was restrained by injunction from working as planned, the railway authority would indemnify the contractor against the additional costs. Second, in the alternative, it was contended that the contract had been frustrated by the grant of the injunction. It was held by the High Court of Australia that no term was to be implied but the contract had been frustrated by the grant of injunction. It was held that the performance of the contract with the new stipulation was fundamentally different from the performance in the situation contemplated in the contract.[84]

10.9 Consequences of Breach of Contract

When breach of a contract takes place, the injured party has several alternatives. The injured party may waive the breach or release the other party, or terminate the contract or accept the performance subject to recovery of damages. These alternatives are briefly considered below.

---

[83] Wong Lai Ying v. Chinachem Investment Co. Ltd. P.C., [1979] 13 BLR 81.
[84] Codelfa Construction Proprietary Ltd. v. Steel Rail Authority of New South Wales, High Court of Australia (1982) 149 CLR 337.

10.9.1 Waiver of Breach

Waiver, as already stated[85] is an intentional relinquishment of a known right, claim or legitimate plea. Breach of contract, as already seen, is a failure to perform an obligation under a contract. A breach of the contract by one party may empower the other party to terminate the contract. If the injured party, instead of terminating the contract, continues to treat the contract as a subsisting obligation, by insisting upon further performance by the wrongdoer or by accepting it, notwithstanding the breach, it will be deemed to have waived the breach; in other words, abandoned its right to terminate the contract. Where a contract provided for compensation for a partial breach of the contract, waiver of breach would not preclude the injured party from recovering any damages suffered on account of the partial breach. For example, failure of a contractor to complete the work within the stipulated period amounts to breach of the contract conditions. When the owner waives the breach by granting an extension of time, he thereby may not abandon his right to recover damages for the breach of the contract. He can be well within his rights to accept further performance subject to recovery of damages. However, a few points need to be noted.

In the first place, it is elementary that waiver is a question of fact and it must be properly pleaded and proved. No plea of waiver can be allowed to be raised unless it is pleaded and the factual foundation for it is laid in the pleadings. Second, waiver means abandonment of a right and it may be either express or implied from conduct, but its basic requirement is that it must be "an intentional act with knowledge".[86] There can be no waiver unless the person who is said to have waived is fully informed as to his right and with full knowledge of such right, he intentionally abandons it.[87] The "law protects persons making the choice from stumbling" into waiver as the election is irrevocable.[88] Third, to sustain a defence of waiver of breach there must be something in the nature of consideration or an element of estoppel.[89]

10.9.2 Accord and Satisfaction

An accord and satisfaction emerges from an arrangement under which the party in default promises to render, and the party who holds the right of action, agrees to accept, some performance differing from that which was originally contracted for and which might legally have been enforced. "The accord is the agreement by which the obligation is discharged. The satisfaction is the consideration, which makes the agreement operative."[90] The "accord" being an agreement and "satisfaction" (being) its execution or performance, the arrangement appears parallel to novation. However, the accord (i.e. the new agreement) itself does not ordinarily discharge the rights and duties arising out of the original contract but simply holds these in abeyance pending satisfaction (i.e. performance) of such accord. Parties may discharge any kind of contract by this recognized legal method.

> Formerly it was necessary that the consideration should be executed. ... Later it was conceded that the consideration might be executory. ... The consideration on each side might be an executory promise, the two mutual promises making an agreement enforceable in law, a contract. ... An accord, with mutual promises to perform, is good, though the thing be not performed at the time of action; for the party, has a remedy to compel the performance, that is to say, a cross-action on the contract of accord. ... If, however, it can be shown that what a creditor accepts in satisfaction is merely his debtor's promise and not

---

[85] See paragraph 10.6.5 above.

[86] Per Lord Chelmsford, L. C. in Earl of Darnley v. London, Chatham and Dover Rly. Co., (1867) 2 HL 43 at p. 57.

[87] *Halsbury's Laws of England*, 4th ed., Vol. 16 in para. 1472 at p. 994; Craine v. Colonial Mutual Fire Insurance Co. Ltd., (1920) 28 CLR 305.

[88] H.B. Property Developments Ltd. v. Secretary of State for the Environment, (1999) 78 P. & C.R. 108.

[89] M.P. Sugar Mills v. State of U.P., AIR 1979 SC 621.

[90] British Russian Gazette and Trade Outlook Ltd. v. Associated Newspapers Ltd. and Talbot v. Same, [1931] B. 3155, [1933] 2 KB 616.

the performance of that promise, the original cause of action is discharged from the date when the promise is made.[91]

When, there is full and final settlement, which is acknowledged by a receipt in writing and the amount is received unconditionally, and apart from having received the amount in full and final settlement, the petitioner had received the security amount also, there is accord and satisfaction by final settlement of the claims.[92]

Much depends upon the wording used by the parties in their settlement agreement as is clear from the case in example (1) below.

EXAMPLES FROM CASE LAW

1.   There were three agreements each of which were settled by the parties on acceptable terms. The first two settlement contracts provided that the respondents would pay to the appellant certain sums in settlement of the disputes relating to the two original contracts.

> The relevant part of the settlement of the second contract was: "The contractor expressly agrees to pay ... The contract stands finally determined and no party will have any further claim against the other".
>
>   The relevant part of the settlement in respect of the third contract read: "The firm will pay a sum of Rs ... In order to provide cover for money payable ... the firm undertakes to hypothecate their movable and immovable property ... The firm further undertakes to execute the necessary stamped documents ... The contracts stand finally concluded in terms of the settlement and no party will have further or other claim against the other."

Upon failure of the firm to adhere to the terms of the settlements, the government wrote a letter to the firm demanding the payment of Rs. 1,51,723/- payable under the original three contracts, ignoring the three settlements. The government referred the matter to arbitration in which the firm participated after challenging the jurisdiction and also the correctness of the claims made. The arbitrator made the award in favour of the government for a total sum of Rs. 1,16,446/- in respect of the first and the third contracts and gave liberty to the government to recover the amount due under the second contract in a suit. The award was filed in the Calcutta High Court. On receiving the notice, the firm filed an application for setting aside the award and in the alternative for a declaration that the arbitration clauses in the three contracts ceased to have any effect and stood finally determined by the settlement of the disputes between the parties. The High Court held that the first contract was to be finally determined only on payment in terms of the settlement, and, as such payment was not made, the original contract and its arbitration clause continued to exist. As regards the third contract it was held that by the third settlement, there was accord and satisfaction of the original contract and the substituted agreement discharged the existing cause of action and, therefore, the arbitrator had no

---

[91] British Russian Gazette and Trade Outlook Ltd. v. Associated Newspapers Ltd. and Talbot v. Same, [1931] KB. 3155, [1933] 2 KB 616, cited with approval in The Union of India v. Kishorilal Gupta and Bros, AIR 1959 SC 1362.

[92] A.K. Construction, Banda, M/s. v. U.P. Power Corporation Ltd., AIR 2008 All. 117; also see M/s ONGC Mangalore Petrochemicals Ltd. v. M/s ANS Constructions Ltd., 2018(2) JT 212: 2018(2) Scale 354: 2018 AIR (SC) 796: 2018(1) Law Herald (SC) 358: 2018 AIR (SCW) 796: 2018(1) ArbiLR 597; 2018(3) SCC 373; Cauvery Coffee Traders, Mangalore v. Hornor Resources (Intern.) Co. Ltd., 2012(1) R.C.R.(Civil) 1: 2011(5) Recent Apex Judgments (R.A.J.) 587: 2011(4) ArbiLR 1: 2011(10) SCC 420 2011 AIR (SCW) 6350; Union of India v. M/s. Master Construction Co., 2011(2) ArbiLR 105:: 2011 AIR (SCW) 2669: 2011 AIR SC (Civil) 1312: Recent Apex Judgments (R.A.J.) 430: 2011(12) SCC 349.

jurisdiction to entertain any claim with regard to that contract. As the award was a lump sum award, not severable, the whole award was held bad.

The government, by special leave, filed an appeal against the said order of the High Court in the Supreme Court. Dismissing the appeal, it was held: (as per majority opinion):[93]

> We have, therefore, no doubt that the contract dated 22nd February 1949, was for valid consideration and the common intention of the parties was that it should be in substitution of the earlier ones and the parties thereto should thereafter look to it alone for enforcement of their claims. As the document does not disclose any ambiguity, no scrutiny of the subsequent conduct of the parties is called for to ascertain their intention.

The second question as to whether the arbitration clause in the earlier contracts survived after the settlement contract, was answered in the following words:

> A repudiation by one party alone does not terminate the contract. It takes two to end it and hence it follows that as the contract subsists for the determination of the rights and obligations of the parties, the arbitration clause also survives.

2. On default by the debtor to repay his loan, the creditor bank filed a suit to recover the amount with interest. During pendency of the suit, the parties arrived at a compromise according to which the bank agreed to withdraw the suit, provided the debtor paid the amount within the stipulated time. The debtor requested for extension of time, which was granted. The debtor paid the full amount in instalments and requested the bank to record the full and final settlement of the suit claim. The bank, however, claimed overdue interest for delayed payments. On these facts it was observed by the Madras High Court:[94]

> I may observe that where there has been a true accord under which the creditor agrees to accept a lesser sum, in satisfaction of the debtor and acting upon that accord by paying the lesser sum and the creditor accepts it, then, it is inequitable for the creditor to insist afterwards for the balance or so ... having received the said compromise amount, it is not open for the plaintiff to claim the overdue interest to the extent of Rs. 69,571.20 for which there was no agreement and that under the circumstances, the claim, if any, made by the plaintiff has been directly hit by S.63 of the Indian Contract Act.

10.9.3 Termination of Contract through Breach

It needs to be noted that to justify termination of a contract, by the injured party, the breach of it must be so material as to defeat or render unattainable the very object of the contract. A termination of a contract by the innocent party on account of a breach, which can be characterized as casual, technical or insignificant, might be premature and may render the innocent party liable to pay compensation to the other party. Repudiation by one party does not terminate the contract. It takes two to end it, and hence it follows that the contract subsists for determination of the rights and obligations of the parties.[95] It is indeed very difficult to lay down

---

[93] Union of India vs. Kishorilal, AIR 1959 SC 1362.
[94] Central Bank of India v. V.G. Naidu & Sons (Leather) Pvt. Ltd., AIR 1992 Mad. 139 (147).
[95] Damodar Valley Corporation v. K.K. Kar, AIR 1974 SC 159.

general rules to decide whether a particular default is substantial or insignificant, inasmuch as much will depend upon the nature and terms of the agreements. The following cases may be considered only as illustrations.

EXAMPLES FROM CASE LAW

1. In a contract for installation of energy saving devices, the employer electricity distribution company failed to furnish the list of locations for the same, resulting in breach of contract by the employer leading to a termination of contract by the contractor. Disputes were referred to arbitration. Categorical findings were arrived at by the arbitral tribunal to the effect that as far as the contractor was concerned, it was always ready and willing to perform its contractual obligations, but was prevented by the appellant from such performance. Another specific finding which was returned by the arbitral tribunal was that the appellant had not given the list of locations and, therefore, its submission that respondent No. 2 had adequate lists of locations available but still failed to install the contract objects was not acceptable. The Supreme Court upheld the arbitral tribunal's conclusion that the termination of contract by the respondent was in order and valid.[96]

2. In a case, the Supreme Court of India, with regard to the issue whether the appellant contracting company (Susme) who had undertaken a scheme of slum rehabilitation was entitled to continue with the scheme, held:

> We are clearly of the view that Susme is not entitled to continue with the rehabilitation Scheme on account of the fact that it has been responsible for the delay in completion of the project for an inordinately long time. Susme has not been able to explain the delay. We are dealing with slum dwellers and Susme cannot take the benefit of technical points to defeat the rights of the slum dwellers. The claim of Susme that it had the support of 70% slum dwellers, was contested before Justice Srikrishna and his findings clearly reveal that Susme does not have the support of 70% of the slum dwellers. In writ proceedings, the petitioner must show that both in law and in equity it is entitled to relief. In this case, both equity and law are against Susme. It has dealt with slum dwellers in a highly inequitable manner. The law and the conditions of the letter of intent as well as the conditions imposed in the various letters issued by the SRA clearly required Susme to produce agreements with at least 70% of the slum dwellers. This, Susme has miserably failed to do. We may also add that though Susme may have remained the same entity in name, there have been, at least, three changes in the promoters of Susme and these transfers of shareholdings obviously must have been done for consideration. It is more than obvious that Susme, as a legal entity, was treating the slum dwellers only as a means of making money and, therefore, we are clearly of the view that Susme is not entitled to any relief.[97]

3. Upon the subcontractor's failure to deliver certain materials on the agreed date, the prime contractor granted a 15-day extension. The subcontractor was specifically told by the prime contractor, while granting the extension, that the extended date was essential to permit him (the prime contractor) to complete the job within the time specified in his

[96] Maharashtra State Electricity Distribution Company Ltd. v. M/s. Datar Switchgear Limited, 2018(1) Scale 303: 2018(1) JT 361: 2018 AIR (SC) 529: 2018 AIR (SCW) 529; 2018(3) SCC 133.
[97] Susme Builders Pvt. Ltd. v. Chief Executive Officer, Slum Rehabilitation Authority, 2017(6) Recent Apex Judgments (R.A.J.) 608: 2018(1) Scale 104: 2018(1) JT 93: 2018 AIR (SC) 237: 2018 AIR (SCW) 237: 2018(2) SCC 230.

contract. When the subcontractor failed to deliver materials within the extended date, the prime contractor cancelled the contract. While dismissing the action brought by the subcontractor to recover damages for breach of contract, the court held that the failure of the subcontractor to deliver the parts within the specified time was a material breach of contract, which justified the cancellation of the contract, by the prime contractor.[98]

10.10 Express Provisions Included in the Contract

As mentioned earlier, consequences of a breach of contract may depend upon the express provisions included in the contract by the parties. Even so, the matter may not be easy because the provisions may be scattered in a lengthy document. Further intention of the parties is to be gathered from the document as a whole. As an example, consider an Item Rate Tender and Contract Form B-2 used for public works in various states. Clause 3 of the form, which deals with this aspect, is reproduced here. Provisions in other public works contracts are more or less similar.

10.10.1 Clause 3 of STATE PWD B-2 Form

In any case in which under any clause or clauses of this contract the contractor shall have rendered himself liable to pay compensation amounting to the whole of his security deposit (whether paid in one sum or deducted by installments) or in the case of abandonment of the work owing to serious illness or death of the contractor or any other cause, the Executive Engineer, on behalf of the Governor of ... shall have power to adopt any of the following courses, as he may deem best suited to the interests of the Government:

(a) To rescind the contract (of which rescission notice in writing to the contractor under the hand of the Executive Engineer shall be conclusive evidence) and in that case the security deposit of the contractor shall stand forfeited and be absolutely at the disposal of the Government.

(b) To employ labour paid by the Public Works Department and to supply materials to carry out the work, or any part of the works, debiting the contractor with the cost of the labour and the price of the materials (as to the correctness of which cost and price the certificate of the Executive Engineer shall be final and conclusive) and crediting him with the value of the work done, in all respects in the same manner and at the same rates as if it had been carried out by the contractor under the terms of his contract; and in that case the certificate of the Executive Engineer as to the value of the work done shall be final and conclusive.

(c) To order that the work of the contractor be measured and to take out of his hands such part thereof as shall remain unexecuted and to give it to another contractor to complete, in which case any expenses which may be incurred in excess of the sum which would have been paid to the original contractor, if the whole work had been executed by him (as to the amount of which excess expenses a certificate in writing by the Executive Engineer shall be final and conclusive) shall be borne and paid by the original contractor and shall be deducted from any money due to him by the Government under the contract or otherwise or from his security deposit or the proceeds of sale thereof, or a sufficient part thereof.

---

[98] Vette v. McBride, 118 A, 2D 640, Court of Appeals of Maryland, 1955.

(d) In the event of any of the above courses, being adopted by the Executive Engineer, the contractor shall have no claim to compensation for any loss sustained by him by reason of his having purchased, or procured any materials, or entered into any engagements, or made any advances on account of, or with a view to the execution of the work or the performance of the contract. And in case the contract shall be rescinded under the provision aforesaid, the contractor shall not be entitled to recover or be paid any sum for any work thereto actually performed by him under this contract unless and until the Executive Engineer shall have certified in writing the performance of such work and the amount payable to him in respect thereof, and he shall only be entitled to be paid the amount so certified.

The provisions of the above clause are applicable under the following circumstances:

i. The contractor shall have rendered himself liable for payment of compensation amounting to the whole of his security deposit.

ii. In the case of abandonment of the work by the contractor owing to: serious illness, death, or any other cause.

iii. According to the stipulation in Clause 4 of the same form, Clause 3(b) will be applicable if progress of part of the work is not satisfactory.

iv. According to Clause 26 of the same form, if the contractor "shall assign or sublet, or attempt to assign or sublet, or become insolvent, or commence insolvency proceedings, or make any composition with his creditors, or attempt to do so any bribe, gratuity, gift, loan, perquisite, advantage, pecuniary or otherwise shall either directly or indirectly be given, promised or offered by the contractor or any of his servants or agents to any public officer or person in the employment of the Government in anyway relating to his office or employment or if any such officer or person shall become in any way directly or indirectly interested in the contract".

More often than not, the circumstances contemplated in (i) above, provide the ground for an action to be taken against the contractor. It contemplates the situation when the contractor is liable to pay damages to the extent of 10% of the estimated cost of the work as shown in the tender. There are several clauses which empower the engineer-in-charge to levy penalty for defaults, but the most important among them is Clause 2. That clause provides for compensation for delay in commencing the work, or failure to maintain the progress, or delay in completion. The clause empowers the superintending engineer to decide the amount of penalty to be recovered and further makes his decision final and binding. The superintending engineer, while deciding the penalty, cannot ignore the true interpretation of Clause 27 of the B-2 Form. That clause states that all sums payable by a contractor by way of compensation under any of the conditions shall be considered as a reasonable compensation without reference to the actual loss or damage sustained, and whether any damage has or has not been sustained. This clause is so worded as to stamp "reasonable" any quantum of compensation fixed by the superintending engineer. Properly construed, it means the compensation recoverable should be reasonable under the facts and circumstances of each case. The clause should be so interpreted because it can only provide what the law is on the matter. It is lawful to recover only reasonable compensation and no more. Where there is in fact no loss sustained, the reasonable compensation may be nil or nominal. Further, the compensation should be decided by the superintending engineer and soon after the happening of the case. Before levying penalty, the superintending engineer must satisfy himself that there is no valid ground for granting an extension of time (see Chapter 6). If due care is exercised in deciding reasonable compensation and notice thereof is given to the contractor, steps under Clause 3 can be taken soon after the amount of compensation equals the maximum allowed.

Clause 3 presents three alternatives: (i) to rescind the contract, (ii) to carry out the remaining work departmentally at the risk and cost of the contractor; (iii) to employ another contractor to complete the remaining portion.

The clause gives full power to the executive engineer to decide the course of action, once the decision to apply Clause 3 is taken. He will, therefore, be required to weigh advantages in making a choice of the three alternatives open to him. The choice would depend upon the facts and circumstances of each case, of which the concerned executive engineer is undoubtedly the best judge.

Choice of Alternatives

1. **Termination of Contract:** Clause 3(a) provides an advantageous course of action if the contractor has refused to sign the agreement and pay a security deposit. It may be a good choice even if he has started or partly done the work and then abandoned the contract leaving a substantial portion unfinished. One factor which must weigh in favour of deciding to treat the contract as discharged under Clause 3(a) is the probable loss the government is likely to incur. If the loss is within the limit of the security deposit and can be made good from the amount in the hands of the department, it is in the best interest of the parties to annul the contract.

   In this respect it may be noted that the stipulation made in Clause 3(a) that the security deposit shall stand forfeited is qualified by the provisions made in the Indian Contract Act. Section 64 requires the security deposit to be returned if the contract is terminated. However, compensation can be deducted from it under Section 75 of the Contract Act. Thus, if reasonable compensation is less than the security deposit, the balance may have to be returned to the contractor. Further, if he files a suit for its recovery the actual damage will have to be proved.

2. **Execution of Work Departmentally:** The very fact that it was decided, in the first instance, to let out the work to a contractor rather than execute it departmentally might lead one to the conclusion that it was found to be the best course of action in the interest of the department. It would not, therefore, be illogical to consider that the department would, in the event of abandonment of the contract by the contractor, like to get the remaining work completed by some other contractor. However, where the first contractor has substantially completed the work, it would be difficult to get a new contractor to complete the work within a reasonable time and cost. Under the circumstances, the best course would be to keep the contract subsisting and carry out and complete the work departmentally at the risk and cost of the contractor under the provisions of Clause 3(b).

   Clause 4 rightly recommends the choice of adopting Clause 3(b) when only part of the work is to be taken out of the hands of the contractor. Because a contract cannot be partly terminated, Clauses 3(a) and (c) are not applicable. Also, no other contractor can be employed with advantage because there would possibly be a lack of co-ordination between the two contractors causing hindrance to each other, responsibility of which would naturally be on the department.

3. **To Employ another Contractor:** Having discussed the circumstances under which action under Clause 3(b) can be taken, the choice of adopting Clause 3(c) becomes obvious. The only care that needs to be exercised is that prior to measuring up the work done by the contractor, the contractor should be notified in writing to remain present either personally or through an authorized representative at the appointed time for taking joint measures. This would avoid future disputes, which would be difficult to resolve. If, in spite of the notice, the contractor remains absent he would have to bear the consequences.

Provisions in other standard forms are discussed below.

10.10.2 Clause 54 of MES Standard Form Contract

The said clause empowers the appellant to cancel the contract, only if the contractor "fails to complete the works, work order and items of work, with individual dates for completion, and clear the site on or before the date of completion". The said clause further stipulates:

The Government shall also be at liberty to use the materials, tackle, machinery and other stores on Site of the Contractor as they think proper in completing the work and the Contractor will be allowed the necessary credit. The value of the materials and stores and the amount of credit to be allowed for tackle and machinery belonging to the Contractor and used by the Government in completing the work shall be assessed by the G. E. and the amount so assessed shall be final and binding.

In case the Government completes or decides to complete the works or any part thereof under the provision of this condition, the cost of such completion to be taken into account in determining the excess cost to be charged to the contractor under the condition shall consist of the cost or estimated cost (as certified by G.E.) of materials purchased or required to be purchased and/or the labour provided or required to be provided by the Government as also the cost of the Contractor's materials used with an addition of such percentage to cover superintendence and establishment charges as may be decided by the C.W.E., whose decision shall be final and binding.

For taking action under the said clause, the "failure" must be on the part of the contractor and not by reason of acts of omissions and commissions of the department. The said clause could, thus be invoked only on default on the part of the contractor and not otherwise.[99]

10.10.3 FIDIC Forms 1999 and 2017

These forms include under Clause 15 most of the circumstances listed in PWD Clause 3 above, empowering the employer to terminate the contract. Clause 15.1 stipulates issue of notice to make good the failure and to remedy it in a specified time. In response to the said notice the contractor is expected to give notice to the engineer detailing the measures he intends to take. It needs to be noted that the time stipulated in the notice shall not imply the extension of time.

Clause 15.2.1 provides for notice of intention to terminate to be given by the employer, if the contractor fails to comply with the notice to correct, binding agreement or determination by the engineer under Clause 3.7 or decision of the DAAB. The clause lists the material breaches. If the contractor fails to remedy the shortcomings listed in the notice of intention to terminate within 14 days, the employer may, by giving the second notice, terminate the contract. The balance of the provisions, namely Clauses 15.6 and 15.7, spell out the valuation after termination and payment after termination, respectively.

It is noteworthy that whereas notice under Clause 15.1 is to be issued by the engineer, notices under Clauses 15.2.1 and 15.2.2 are to be issued by the employer, as defined in Clause 1.1.31.

It also includes Clause 16, which empowers the contractor to suspend or terminate the contract. The procedure is similar to Clause 15. Suspension by the contractor is allowed under the grounds mentioned in Clause 16.1 after giving 21 days' notice to the employer followed by notice of intention to terminate or termination in certain specified eventualities in the provision.

The consequences of termination by either party are also included in the respective provisions, which are self-explanatory. FIDIC Form 1992 includes the similar provisions in Clauses 63 and 69, respectively.

10.10.4 MOS & PI, GOI Form

The MOS & PI, GOI Form includes Clauses 59 and 60 dealing with termination by either the employer or the contractor. The said provisions are also self-explanatory and read as follows:

[99] Union of India v. M/s V. Pundarikakshudu, AIR 2003 SC 3209.

59. Termination

59.1 The Employer or the Contractor may terminate the Contract if the other party causes a fundamental breach of the Contract.

59.2 Fundamental breaches of Contract include, but shall not be limited to the following:

(a) the Contractor stops work for 28 days when no stoppage of work is shown on the current Program and the stoppage has not been authorised by the Nodal Officer or his nominee:

(b) the Nodal Officer or his nominee instructs the Contractor to delay the progress of the Works and the instruction is not withdrawn within 28 days.

(c) the Employer or the Contractor becomes bankrupt or goes into liquidation other than for a reconstruction restructure or amalgamation.

(d) a payment certified by the Nodal Officer or his nominee is not paid by the Employer to the Contractor within 50 days of the date of the Nodal Officer or his nominee's certificate:

(e) the Nodal Officer or his nominee gives Notice that failure to correct a particular Defect is a fundamental breach of Contract and the Contractor fails to correct it within a reasonable period of time determined by the Nodal Officer or his nominee.

(f) the Contractor does not maintain a security which is required.

(g) the Contractor has delayed the completion of works by the number days for which the maximum amount of liquidated damages can be paid as defined in the Contract data and

(h) if the Contractor, in the judgement of the Employer has engaged in corrupt or fraudulent practices in competing for or in the executing the Contract.

(i) if the contractor has contravened clause 7.1 and clause 9.00

For the purpose of this paragraph: "corrupt practice" means the offering, giving, receiving or soliciting of anything of value to influence the action of a public official in the procurement process or in contract execution. "Fraudulent practice" means a misrepresentation of facts in order to influence a procurement process or the execution of a contract to the detriment of the Employer, and includes collusive practice. Bidders (prior to or after bid submission) designed to establish bid prices at artificial non-competitive levels and to deprive the Employer of the benefits of free and open competition.

59.3 When either party to the Contract gives notice of a breach of contract to the Nodal Officer or his nominee for a cause other than those listed under Sub Clause 59.2 above, the Nodal Officer or his nominee shall decide whether the breach is fundamental or not.

59.4 Notwithstanding the above, the Employer may terminate the Contract for convenience subject to payment of compensation to the contractor including loss of profit on uncompleted works. Loss of profit shall be calculated on the same basis as adopted for calculation of extra/additional items.

59.5 If the Contract is terminated the Contractor shall stop work immediately, make the Site safe and secure and leave the Site as soon as reasonably possible.

60. Payment upon Termination

60.1 If the Contract is terminated because of a fundamental breach of Contract by the Contractor, the Nodal Officer or his nominee shall issue a certificate for the value of the work done less advance payments received up to the date of the issue of the certificate, less other recoveries due in terms of the contract, less taxes due to be deducted at source as per applicable law and less the percentage to apply to the work not completed as indicated in the Contract Data. Additional Liquidated Damages shall not apply. If the total amount due to the Employer exceeds any payment due to the Contractor, the difference shall be a debt payable to the Employer.

60.2   If the Contract is terminated at the Employer's convenience or because of a fundamental breach of Contract by the Employer, the Nodal Officer or his nominee shall issue a certificate for the value of the work done, the reasonable cost of removal of Equipment repatriation of the Contractor's personnel employed solely on the Works, and the Contractor's costs of protecting and securing the Works and loss of profit on uncompleted works less advance payments received up to the date of the certificate, less other recoveries due in terms of the contract and less taxes due to be deducted at source as per applicable law.

61. Property

61.1   All materials on the Site, Plant, Equipment, Temporary Works and Works for which payment has been made to the contractor by the Employer, are deemed to be the property of the Employer, if the Contract is terminated because of a Contractor's default.

62. Release from Performance

62.1   If the Contract is frustrated by the outbreak of war or by other event entirely outside the control of either the Employer or the Contractor, the Nodal Officer or his nominee shall certify that the Contract has been frustrated. The Contractor shall leave the Site and stop work as quickly as possible after receiving this certificate and shall be paid for all work carried out before receiving it and for any work carried out afterwards to which commitment was made.

10.10.5 NITI Aayog Model Form: Part V Force Majeure and Termination

Part V under the above heading lists out three grounds that may lead to termination of the contract, namely: Clause 19 Force Majeure; Clause 20 Suspension of Contractor's Rights; and Clause 21 Termination. After listing *force majeure* events, Clause 19.5 provides for the mandatory duty to report a *force majeure* event thus "19.5.1 Upon occurrence of a Force Majeure Event, the Affected Party shall by notice report such occurrence to the other Party forthwith. Any notice pursuant hereto shall include full information as called for by the clause". Clauses 19.5.2 and 19.5.3 stipulate:

> The Affected Party shall not be entitled to any relief for or in respect of a Force Majeure Event unless it shall have notified the other Party of the occurrence of the Force Majeure Event as soon as reasonably practicable, and in any event no later than 10 (ten) days after the Affected Party knew, or ought reasonably to have known, of its occurrence, and shall have given particulars of the probable material effect that the Force Majeure Event is likely to have on the performance of its obligations under this Agreement.
>
> 19.5.3 For so long as the Affected Party continues to claim to be affected by such Force Majeure Event, it shall provide the other Party with regular (and not less than weekly) reports containing information as required by Clause 19.5.1, and such other information as the other Party may reasonably request the Affected Party to provide.

19.6 Effect of Force Majeure Event on the Agreement

19.6.1 provides for the risk of a *force majeure* event as follows: "upon the occurrence of any Force Majeure (a) prior to the Appointed Date, both Parties shall bear their respective Force Majeure costs. (b) after the Appointed Date, the costs incurred and attributable to such event and directly relating to this Agreement (the 'Force Majeure costs') shall be allocated and paid as" provided in the Agreement.

19.7 Termination Notice for Force Majeure Event

If a Force Majeure Event subsists for a period of 60 (sixty) days or more within a continuous period of 120 (one hundred and twenty) days, either Party may in its discretion terminate this Agreement by issuing a Termination Notice to the other Party without being liable in any manner whatsoever, save as provided in this Article 19, and upon issue of such Termination Notice, this Agreement shall, notwithstanding anything to the contrary contained herein, stand terminated forthwith; provided that before issuing such Termination Notice, the Party intending to issue the Termination Notice shall inform the other Party of such intention and grant 15 (fifteen) days time to make a representation, and may after the expiry of such 15 (fifteen) days period, whether or not it is in receipt of such representation, in its sole discretion issue the Termination Notice.

19.8 Termination Payment for Force Majeure Event

Clause 19.8.1 provides "in the event of this Agreement being terminated on account of a Non-Political Event, the Termination Payment shall be an amount equal to the sum payable under Clause 21.5.

19.8.2 If Termination is on account of an Indirect Political Event, the Termination Payment shall include" as stated in the Clause including the reasonable cost, as determined by the Authority's Engineer, of the Plant and Materials procured by the Contractor and transferred to the Authority and under Clause 19.8.3, if "Termination is on account of a Political Event, the Authority shall make a Termination Payment to the Contractor in an amount that would be payable under Clause 21.6.2 as if it were an Authority Default".

The second ground that may lead to termination is under Clause 20.1. The relevant parts of the clause read:

20.1 Suspension upon Contractor Default

Upon occurrence of a Contractor Default, the Authority shall be entitled, without prejudice to its other rights and remedies under this Agreement including its rights of Termination hereunder, to (a) suspend carrying out of the Works or any part thereof, and (b) carry out such Works itself or authorise any other person to exercise or perform the same on its behalf during such suspension (the "Suspension"). Suspension hereunder shall be effective forthwith upon issue of notice by the Authority to the Contractor and may extend up to a period not exceeding 90 (ninety) days from the date of issue of such notice.

20.2 Authority to act on behalf of Contractor: During the period of Suspension hereunder, all rights and liabilities vested in the Contractor in accordance with the provisions of this Agreement shall continue to vest in the Contractor and all things done or actions taken, including expenditure incurred by the Authority for discharging the obligations of the Contractor under and in accordance with this Agreement shall be deemed to have been done or taken for and on behalf of the Contractor and the Contractor undertakes to indemnify the Authority for all costs incurred during such period ....

20.3.1 In the event that the Authority shall have rectified or removed the cause of Suspension within a period not exceeding 60 (sixty) days from the date of Suspension, it shall revoke the Suspension forthwith and restore all rights of the Contractor under this Agreement ....

20.4 Termination

20.4.1 At any time during the period of Suspension under this Article 20, the Contractor may by notice require the Authority to revoke the Suspension and issue a Termination Notice. The Authority shall, within 15 (fifteen) days of receipt of such notice, terminate this Agreement under and in accordance with Article 21 as if it is a Contractor Default under Clause 21.1.

20.4.2 Notwithstanding anything to the contrary contained in this Agreement, in the event that Suspension is not revoked within 90 (ninety) days from the date of Suspension here-under, the Agreement shall, upon expiry of the aforesaid period, be deemed to have been terminated by mutual agreement of the Parties and all the provisions of this Agreement shall apply, mutatis mutandis, to such Termination as if a Termination Notice had been issued by the Authority upon occurrence of a Contractor Default.

The third ground of termination is under Clause 21: Clause 21.1 deals with termination for contractor default. The possible defaults are listed in detail. Clause 21.2 deals with termination for employer defaults which are also listed in detail. This form provides a fourth ground for termination in Clause 21.3 Termination for Authority's convenience: "Notwithstanding anything hereinabove, the Authority may terminate this Agreement for its own convenience. The termina-tion shall take effect 30 (thirty) days from the date of notice hereunder and shall be deemed to be termination on account of Authority Default." The rest of the provisions deal with: 21.4 Requirements after Termination; 21.5 Valuation of Unpaid Works; and 21.6 Termination Pay-ment. The provisions are self-explanatory.

10.10.6 Doctrine of Specific Performance Not Applicable

Under certain well-defined circumstances, an injured party to a contract may institute a suit for, and obtain a decree of, specific performance whereby the defaulting party is required to live up to the bargain he has made. A party cannot claim specific performance of a contract after repudiating the same and electing to sue for damages. In a suit for specific performance the plaintiff has to be ready and willing to perform his part of the contract and has to treat the contract as subsisting at all times.[100] Such a remedy is by no means available in all breach situations and certainly not in respect of construction contracts. The main reasons are:

(i) Payment of damages by the defaulting party to the injured party affords adequate remedy
(ii) Superintending performance of a construction contract by a court would be extremely difficult, if not impossible.

10.10.7 The Doctrine of Substantial Performance

While according to the rule exacting strict performance, a person who has failed to fully perform his contract is not entitled to any recovery; construction contracts form an exception to this rule. The doctrine of substantial performance is applicable to such contracts. This doctrine usually allows a builder, upon substantial performance, to recover the contract price, notwithstanding the work may have been defective or incomplete. The remedy of the owner is the recovery of damages on account of incomplete or defective work.

Meaning of Substantial Performance

What amounts to substantial performance depends upon the nature of the contract and is a question of fact and degree in each case and decided cases may be of little use. It is a doctrine judicially evolved to overcome the difficulties of the rule of strict performance of "entire contracts" before being entitled to payment. It can, therefore, mean that on the facts in a case, a court or tribunal can come to the conclusion that there has been sufficient performance of an entire contract notwithstanding defects and omission of non-important parts of the obligations, that is covering cases of both nonfeasance as well as misfeasance but of unimportant matters.

---

[100] Ayissabi v. Gopala Konar, AIR 1989 Ker. 134, follows AIR 1928 PC 208 & (1904) ILR 31.

This rule may not be available to a party which is guilty of a fundamental breach of having abandoned the contract. This rule can also be excluded by an express provision in the contract.[101]

The only guiding principle of some use that can be cited is: "In considering whether there was substantial performance ... it is relevant to take into account both the nature of the defects and the proportion between the cost of rectifying them and the contract price."[102]

EXAMPLE FROM CASE LAW

In an action brought by a contractor to recover a balance due under a contract, the owner interposed the defence that the contractor had not completely performed the contract and therefore there could be no recovery. The court found as a fact that the contractor had substantially performed the contract. The court, therefore, held that the contractor was entitled to recover the balance due under the contract less the cost of providing remedy to the alleged defects in the work. This judgment was affirmed by the appellate court.[103]

In India, the consequences of rescission of a voidable contract are spelled out in Section 64. Under that section, the owner is obliged to pay for the work done by the contractor. The sum payable may, however, be reduced by an amount necessary to remedy the defective performance.

10.10.8 Settlement of Accounts after Termination of Contract; Can a Contractor File Suit for Accounts against the Employer?

A contractor, who is engaged to execute work, is expected to maintain his own accounts. At all events, there is no bar for a contractor to keep an account of the work done. Even where the contract between the employer and independent contractor may provide for payment on the basis of measurements to be recorded by the employer, nothing prevents the contractor from measuring the work done by him and then suing for the value of the work done. The contractor may also demand joint-measurements to determine the quantum of work done. If the employer for some reason does not co-operate or prevents the contractor from taking physical measurements, the contractor can seek appropriate legal remedy which will enable him to take measurements or to secure the information from the measurement book in the custody of the employer. Therefore, either the fact that the measurement book is maintained by the employer, or the fact that the contractor does not possess the exact measurements, will not entitle the contractor to file a suit for rendition of accounts against the employer. The independent contractor is not an agent of the employer. Nor is the employer in the position of a trustee with reference to the independent contractor. Such a right is not created or recognized by any statute. For the reasons that the contractor is expected to keep his own accounts, the claim also cannot be supported in equity by stating that the relationship is such that rendition of accounts is the only relief which will enable the contractor to satisfactorily assert his legal right.[104]

10.10.9 Effect on Arbitration Clause

The case law to answer the question as to what happens to the arbitration clause contained in a contract in which a party pleads full and final settlement or accord and satisfaction was

---

[101] Hoenig v. Isaccs, T.L.R. 1360 [1952] 2 All ER 176.

[102] Bolton v. Mahadeva, [1972] 1 W.L.R. 1009 [1972] 2 All ER 1322. See. *Chitty on Contracts*, 24th ed., p. 613.

[103] Meador v. Robinson, 263, S.W. 2d, 118, Court of Appeals of Kentucky, 1953.

[104] K.C. Skaria v. Govt. of State of Kerala, AIR 2006 SC 811= 2006 AIR SCW 265.

summed up by the Bombay High Court in the Union of India v. M/s. Ajii Mehta & Associates. One dozen cases were considered[105] and it was held:[106]

> Thus the authorities discussed above can be said to lay down the law that in spite of full and final settlement of the claim, the arbitration clause in the contract may subsist where the party invoking it alleges that in fact there was no accord and satisfaction for some reasons such as the final bill was submitted or receipt was given under coercion, mistake or misrepresentation, without prejudice, under protest etc. For then, that itself becomes a dispute arbitrable under the clause. However when there is no such allegation made when invoking the arbitration clause, and it is invoked simpliciter, it will have to be held that the contract itself had come to an end and with it the arbitration clause which is a part and parcel of it. We have come across no decision which has taken a contrary view. On the other hand the decisions discussed above support our conclusion.

The Andhra Pradesh High Court rejected the request for appointment of an arbitrator, where it was found that a full and final settlement was amicably and voluntarily reached by the parties, in such a case the matter ceases to be an arbitrable dispute. However, the division bench of the said High Court held that though the full and final payment was received and "no demand certificate" was issued by the respondent, unless it is clearly shown that the parties have given a clear go by to the earlier agreement, the party will have a right to recourse thereto.[107]

It is submitted with respect that the view taken by the Bombay High Court and the division bench of the Andhra Pradesh High Court above is the correct position in law. Under the Arbitration and Conciliation Act, 1996, an arbitral tribunal is empowered to decide the existence and validity of the arbitration agreement. This aspect has been dealt with in Chapter 16.

10.10.10 Consequences of Common Forms of Breach of Contract by the Owner and the Contractor and Remedies for Breach of Contract

The common breaches of construction contracts are dealt with in Chapter 11, and Chapter 12 covers the remedies available to an innocent party.

---

[105] Kapurchand Sodha v. Himayatalikhan Aza-mjah, AIR 1963 SC 250; Damodar Valley Corporation v. K.K. Kar, AIR 1974 SC 158; Union of India v. L.K. Ahuja & Co., AIR 1988 SC 1172; Bombay High Court in Arbitration Petition No. 123 of 1980; decision in Arbitration Petition No. 81 of 1985 in Award No. 19 of 1985 delivered on 17th March 1986; Union of India v. D. Bose, AIR 1981 Cal. 95; Jiwani Engineering Works (P) Ltd. v. Union of India, AIR 1981 Cal. 101; decision of the Supreme Court in AIR 1974 SC 158; decision of the division bench of the Calcutta High Court in AIR 1981 Cal. 95; Vipinbhai R. Parckh v. General Manager, Western Railway, Bombay, AIR 1984 Guj. 41; Cochin Refineries Ltd. v. C.S. Company, Engineering Contractors, Kottayam, AIR 1989 Kerala 72; and a Division bench decision of the same court reported in (1987) 1 Ker. LT 241.

[106] Union of India v. M/s. Ajii Mehta & Associates, AIR 1990 Bom. 45.

[107] NTPC v. V. Subbarao & CO., 2001(3) Arb. LR 320 (AP) (DB)

11

Common Breaches of Contract

11.0 Introduction

Under the complicated provisions of many building and engineering contracts, possible breaches of contract by the owner and the contractor are numerous. Typical breaches of the common kind, by both the owner/employer and the contractor, are considered in this chapter. This chapter is, in fact, a continuation of Chapter 10 and, as such, the provisions of the law applicable to the discussion in this chapter are considered in the earlier chapters and also in Chapter 12, to which reference may be made, if need be.

11.1 Provisions in Standard Form Contracts

The following provisions of the standard from contracts are dealt with in this chapter.

Provisions in Standard Form Contracts	Paragraph No.
PWD/CPWD, Clauses 2(a), 2(b), 10, 11, 12, 13, 14, 15, 15(A), 17, 18(A), 21, 33, 42	11.3, 11.8.1
Military Engineering Services Form, Clauses 1, 3, 7, 9, 10, 18, 24, 61–68	11.8.1
FIDIC Form 1992 Amended, Clauses 1.1(b)(ii), 1.1(b)(iii), 1.1(f)(vii), 3.1, 4.1, 6.1–6.5, 7.1–7.3, 12.2, 13.1, 14.1–14.4, 39.1, 39.2, 40.1–40.3, 42.1–42.3, 51.1–53.5, 69.1–69.5	11.3, 11.9
FIDIC Form 1999 and 2017 editions: All the conditions relevant to the earlier edition including Clauses 15 and 16	11.4, 11.10.2
Ministry of Statistics and Programme Implementation, Government of India (MOS & PI), Clauses 7, 13, 16, 17–23, 27–32, 33–36, 37–62 NITI Aayog Model Form, Clauses 19, 20, 21, 22	11.3, 11.9

11.2 Breaches by the Owner/Employer

The most common forms of defaults likely to be committed by the owner include:

1. Failure to hand over possession of the site to the contractor.
2. Failure/delay in appointing an architect or an engineer or in filling the vacancy.
3. Delay in supply of working drawings, details, designs and decisions.
4. Delay in supply of materials listed in a schedule incorporated into the contract – that is, the materials agreed to be supplied by the owner to the contractor.
5. Ordering suspension or stoppage of work or interfering with the progress of work in any manner.

6. Failure/delay in making payments of mobilization advance/machinery advances, R.A. bills, extra items, excess quantities, including settlement of final bill.

7. Failure/delay in nominating and approving specialist subcontractors and suppliers.

8. Delay caused by other agencies employed at the site of the work by the owner in addition to the contractor.

9. Wrongful deduction of liquidated damages/penalty.

10. Termination of contract wrongfully and illegally.

Some of the above breaches not dealt with in other chapters are considered in this chapter in detail. The common breaches of contract by the contractor are considered from paragraph 11.10 onwards.

11.2.1 Delay in Appointing an Architect or an Engineer or in Filling the Vacancy

Government authorities and public sector undertakings generally appoint an independent engineer or consulting engineering firm or company to perform the contractual role of the engineer. The contract conditions generally stipulate for the apppointment of the engineer right from the stage of issuing an order to commence the work, approving the work programme submitted by the contractor, issuing working drawings and instructions, supervising the work and certifying payments, etc. It has been observed that in some cases, though the tenders are accepted and the time limit for completion of the work begins to run, appointment of the engineer is not finalized. The delay in appointment of an engineer invariably has a knock-on effect on the delay in supply of drawings and designs. In practice, generally the contractor lodges claims on account of delay in supply of drawings and designs, which default on the part of the employer overlaps with delay in the appointment of an engineer. However, it hardly needs to be stressed that the employer is liable to appoint an engineer in time or within a reasonable time so that the progress of the work is not adversely affected on that count. The same care needs to be taken to fill any potential vacancy so that there is no delay on account of non-availability of the services of the engineer in case of serious illness, death, insolvency or dismissal during the progress of the work.

11.3 Failure to Hand over Possession of Site

Building and construction contracts usually define the "site" to mean

> the lands and other places on, under, in, or through which the works are to be executed or carried out and any other land together with such other places as may be specifically designated in the contract as forming part of the site.

The other places normally designated in the contract include quarries for obtaining material such as rubble, sand, etc., and also land for camp sites, storage of material, plant and equipment.

Though the term "handing over of possession of the site" to the contractor is commonly used in engineering practice, legally the owner is always in possession of the site and the term only means that the contractor is allowed entry upon the site for the purpose of performing his contractual obligations.

When tenders are invited from the contractors for execution of work, the invitation impliedly represents that the owner would be in a position to hand over the whole site to the contractor immediately upon the making of the contract, unless the tender invitation stipulates to the contrary. For example, the CPWD Form No. 6 of Notice Inviting Tenders in Clause 2A provides: "*The site for the work is available/or the site for the works shall be made available in parts as specified below*". The modern forms impose an express duty on the owner to give possession of the site to the contractor. For example, the FIDIC Form 1992 amended edition expressly provides in Clause 42.2 for grant of extension of time and payment of "such costs which shall be added to the contract price" by the engineer after due consultation with the employer and the contractor, in the event that handing over the possession of the site is delayed. Similar provision is made in FIDIC Form 1999 edition in Clause 2.1 read with 3.5. Clause 21 of the Government of India form stipulates that the failure by the employer to hand over the site free from encumbrances will be deemed to have delayed the start and will be a compensation event. The NITI Aayog Model Form includes a similar provision in Clause 8 under the heading: "Right of Way".

Many contracts are silent in respect of availability of the site for the obvious reason that it is a precondition to be fulfilled by the owner so essential for performance of the contract within the stipulated time limit for completion that if a third person were to suggest to both the parties to include this stipulation, they would testily suppress him with a common "Oh! Of course". It, therefore, follows that in a construction contract stipulating time limit for completion, there is an implied obligation on the part of the owner to hand over the site to enable the contractor to commence the work immediately upon making the contract or within a period such as a week or two specified in the agreement for commencement of the work by the contractor. Recently developed forms, such as FIDIC and the NITI Aayog Model Form, expressly stipulate that the appointed date (the date of commencement) will be reckoned only after 80 to 90% of the site is handed over; failing which the contractor is entitled to either terminate the contract and/or claim compensation.

Failure of the owner to hand over the possession of the entire site to the contractor will amount to breach of contract by the owner, in the absence of express stipulation to the contrary. This must particularly be so when a date for completion is specified in the contract.

> If in the contract one finds the time limited within which the builder is to do the work, that means, not only that he is to do it within that time, but it means also that he is to have that time within which to do it.[1]

This breach invariably causes delay in completing of the works and claims of compensation by the contractor on that count.[2] Example (1) below is a typical case of what happens in a majority of cases involving public authorities.

EXAMPLES FROM CASE LAW

1. The Government of India through the Ministry of Shipping, Road Transport and Highway launched a national highways development programme which envisaged 4/6 laning and strengthening of the existing national highways which included a 44 km stretch from km 26.00 to km 70.00 of a section of NH-1A in the State of Punjab. The respondent's bid for undertaking to complete the work at the contract price of Rs. 201 crores was accepted. The unit rate contract agreement was entered into between the parties on 4th October 2005. The date of start of work was 22nd November 2005. The stipulated date of completion was 21st May 2008. According to NHAI, the respondent abandoned the work after 22nd April 2008. The contractor, on the other hand, contended that in terms of the contract, NHAI was to give the respondent 10 km of land at the time of commencement of the project, 12 km after six months of commencement and 22 km after 12 months of commencement. It is stated that on account of various breaches committed by NHAI, the work could not be completed. The work front availability till September 2008, i.e. the date of termination of the contract, was about 25% of the area that should have been handed over to the respondent.

---

[1] Wells v. Army & Navy Co.-op Society (1902) 86 L.T. 764.
[2] D.D.Sharma v. Union of India; 2004(2) Arb. LR 119 (SC).

It was admitted that the availability of the land was delayed on account of land acquisition formalities involved between the NHAI and the state authorities. There were ten claims filed by the respondent for the aggregate value of Rs. 185.45 crores. This was apart from interest and costs of arbitration. While NHAI could have preferred a counter-claim in this arbitration itself, it chose to file an independent statement of claim before another arbitral tribunal. Twelve claims were filed by NHAI. As far as the first arbitration was concerned, the majority of the first arbitral tribunal, by their award, allowed the claims of the respondent contractor to an extent of Rs. 200.64 crores for performance cost, delay damages, termination losses, etc. As far as the second arbitration concerning the claims of NHAI was concerned, all claims were rejected except the claim for refund of advance which was allowed but made subject to the first award concerning the claims of the respondent. The Delhi High Court held:

> This was the right order to be passed since the issue was gone into by both ATs [arbitral tribunals] ... In view of the overwhelming evidence before it on the issue, the conclusion arrived at independently by the majority in both ATs that there was a failure on the part of NHAI to hand over stretches of land to the Respondent as envisaged under the contract cannot be held to be patently illegal or perverse or shocking to the judicial conscience so as to attract any of the grounds of invalidation of an Award as set out under Section 34 of the Act.[3]

2.   Where indisputably there was a delay of six months in handing over of the site due to finalization of the location as well as materials that were agreed to be supplied by the employer not being supplied due to non-availability in the market, it was held that the delay would endure for the benefit of the contractor. Further, no case was made out against the contractor who was allowed extension of time for completion without any reservations or conditions. The contract did not contain any prohibitory clause denying the contractor any enhanced compensation. Award by an arbitrator allowing escalation charges in favour of the contractor was upheld by the High Court.[4]

3.   A contract for demolishing old houses and erecting new ones provided a time limit of six months for completion of the work. It further provided that all brickwork for the new houses was to be simultaneously raised, no part being raised more than five feet higher than the remainder, a provision normally found in specifications to prevent unequal settlement. At the request of the owner, the contractor agreed to a fortnight's delay from the date of signing the agreement. The owner could give possession of part of the site some weeks after the expiry of the extended time of 15 days for commencing the work. It took nearly five months before the contractor got possession of the whole site. The contractor claimed damages for the breach of contract. It was held:

> The contract clearly involves that the building owner shall be in a position to hand over the whole site to the builder immediately upon making of the contract. There was an implied undertaking on the part of the building owner that he would hand over the land for the purpose of allowing the contractor to do that which he has bound himself to do.[5]

4.   In a case the contract expressly stipulated that "No implied obligation of any kind by or on behalf of Her Majesty shall arise from anything in the contract". The contract with the Crown in the said case was one of six contracts for the project as a whole. The work required under two of the later contracts interfered with the contractor's work because they

---

[3] National Highways Authority of India v. Bridge & Roof Co. Ltd., (Delhi) 2017(4) R.A. J. 1.
[4] The Superintending Engineer, T.N.H.B. v. M. Pramasivam, 2003(2) Arb. LR 546 (Madras) (DB).
[5] Freeman & Son v. Hensler C.A., (1900) 64 JP 260.

encroached on the site. The contractor claimed damages for breach of implied terms relating to possession. It was held that the condition that "no implied obligation of any kind shall arise" has no application in the case at bar because it is fundamental to a building contract that work space be provided unimpeded by others.[6]

In a case, where no time limit for completion is specified, the site would be required to be made available to the contractor within a reasonable time. What is a reasonable time is a question of fact and will vary from case to case in view of the facts and circumstances involved in each case. In the case of a new project, the main contractor will normally be entitled to exclusive possession of the entire site in the absence of an express stipulation to the contrary. A few of the common express exceptions to be found in the standard form contracts include permitting:

i. The presence on the site of other contractors employed by the owner
ii. The owner and his representatives, such as architects, engineers to enter the site for supervision
iii. The owner to get the work or any part of it done through other agencies, in case the contractor refuses to comply with any relevant instructions of the engineer/architect.

However, whether, in any given case, the contractor was given possession of adequate site, is a question of fact to be determined in the light of all the circumstances.[7] Common instances of the projects getting delayed on account of the site being not available are indicated below:

1. A building contract contemplating demolition of old structures and constructing new ones in its place; old structures continue to be in use and not vacated.
2. Construction of a bridge, road or remodelling of existing road being contracted for on the presumption that the land acquisition proceedings would be completed before the commencement of the work but the presumption proving to be erroneous.
3. Acquisition of land for quarries, waste weir, canal, setting up of workers' camp and site offices of the contractor, etc. in case of irrigation projects often gets delayed more than anticipated.
4. Delay in removing and relocating obstructions such as trees, telephone or electricity supply poles, towers, water supply pipeline, underground cables and conduits, etc. located in the alignment.
5. Last minute changes in the location or alignment of structures encroaching on the adjoining land not acquired.
6. The quarries indicated in the agreement do not meet the expected requirements of materials in respect of quality or quantity or both and the alternative sites are not made available promptly. Where the agreement contemplates the contractor to procure materials from other sources, "approval" to the new quarries, their opening, extra lead, royalty or compensation, etc. pose special problems.

All the above and similar instances will amount to breach of contract by the owner and he is well advised to take precautions to avoid the same, as these factors invariably result in delayed

---

[6] The Queen in Right of Canada v. Walter Cabbott Construction Ltd., Canadian Federal Court of Appeal (1975) 21 BLR 42.
[7] London Borough of Hounslow v. Twickenham Garden Developments Ltd., Chancery Division, [1970] 7 BLR 81.

completion, claims of compensation put forth by the contractor and sometimes litigation as already stated.

Apart from the provision of the site, the owner/employer may have to deal with any planning or other approvals required. This would depend on where the land is situated and who can apply for the permissions. An individual employer and a contractor specialist in the extension and alteration of buildings in a conservation area in London, UK, entered into a JCT Building contract with Quantities, 2005 edition incorporating Rev. 2(2009). The work included demolition and reconstruction of buildings to create a single dwelling. The local council wrote to the contractor that some of the work would require a conservation area consent; this led to a suspension of the work for over a year. The design had to be changed and planning permission obtained. The contractor claimed an extension of 53.2 weeks which the employer disputed. In the absence of an express contractual term about the planning permission required in the conservation zone, the Court of Appeal observed:

> It is not the law that ... a term is always to be implied that the employer is responsible for obtaining the necessary planning approvals, or ensuring that all such approvals have been obtained, before work is begun. In a building contract there is usually no implied warranty by the employer that the contractor will not suffer delay caused by the fault of a local authority or statutory undertaker ... there is no implied warranty that a third party will refrain from interfering with a contractor's work.[8]

Despite this observation, and considering the circumstances of the case, the Court of Appeal upheld an implied term in the following words:

> The Employer will use all due diligence to obtain in respect of the Works any permission, consent, approval or certificate as is required under, or in accordance with, the provisions of any statute or statutory instrument for the time being in force pertaining to town and country planning.[9]

It was held that it was unnecessary to introduce further qualification or exemption about the local authority being capricious or acting unlawfully or unreasonably. The parties and the judge agreed that the responsibility for obtaining approval lay with the employer; the work would otherwise have been illegal. It should be noted that under English law one can make a planning application without owning an interest in land.

11.3.1 Obstructions by Third Person

In many cases, it has been observed that the government departments claim to have handed over the possession of the site to the contractors but there are obstructions by third persons, invariably persons claiming to be the original owners of the land to whom compensation was either not paid or it was paid but the sum was considered inadequate by them. There have been instances where a work order was given when crops were standing and no work could in effect be commenced until after the crops were harvested. The cases of the contractor being excluded from the site by a third person need to be divided into two categories:

1. Obstruction by a third person for whom the owner is not responsible in law and over whom he has no control.
2. Where it is in the power of the owner to have the situation rectified, if necessary by legal action.

---

[8] Clin v. Walter Lilly & Co. Ltd., [2018] EWCA Civ. 490.
[9] Ibid.

The contractor's claim for compensation is likely to fail in the case falling under the first category, and the employer is likely to be held guilty of breach of either express or implied conditions, in the cases falling under the second category.

EXAMPLES FROM CASE LAW

1. A contract for construction of a school building for the council on its land provided that the contractor should be entitled to enter on the site immediately and that he should complete the work by the agreed date. The only access to the site was from a road. The soil between the road and the site of the proposed work was soft. The agreement had, therefore, incorporated a provision that the contractor lay a temporary sleeper roadway from the road to the site for access, and subsequently provide a permanent pathway. The contractor commenced the work but was forced to suspend it because of a threatened injunction from an adjoining owner, who claimed that the road was his property. Subsequently, after the third party's claims were held to be unfounded, the contractor resumed and completed the works. The contractor claimed damages from the council for the loss suffered due to delay caused by the third party's action. It was held that there was no implied warranty by the council against wrongful interference by third parties with free access to the site.[10]

2. A contract to construct five blocks consisting of over 100 dwellings contained express provision (Clause 21 of JCT 63) that possession of the site should be given to the contractor on 23rd June 1980. The defendant (owner) could not give possession because squatters occupied one corner of the site. The owner took eviction proceedings and it took about 19 days before the site was cleared of squatters so as to enable the plaintiff contractor to occupy the whole of the site. Referring to Clause 21, it was held that "unarguably there was a clear breach of that term by the failure of the defendants, *for whatever reason*, to remove these squatters until an appreciable time after they had promised to give the plaintiffs possession of the site" (emphasis supplied).[11]

11.3.2 Information about Site

In major projects, the basic information in the site information document is generally the result of highly technical efforts on the part of the owner, including for underground subsoil exploration. It is such information that the contractors have neither the time nor the opportunity to obtain. It is doubtful if the contractors could be expected to obtain it by their own efforts as a potential or actual tenderer. But the information is indispensable, to form a judgement as to the extent of the work to be done. When such information is supplied to the tenderers with a proviso that the contractor is to satisfy himself and to obtain his own information on all matters which can in any way influence his tender, the contractor may be able to seek redress on the plea of misrepresentation or fraud as the case may be.[12]

The engineers and officers responsible for the execution of public works often neglect taking prompt steps either to eliminate the hindrances to, or to order suspension of, the contract under appropriate provision (where agreement incorporates such a provision) which entitles the contractor to exercise his option to treat the contract as at an end, or to call the contractor for negotiations and terminate the contract, on mutually agreed terms, if necessary. It is suggested

---

[10] Porter v. Tottenham Urban District Council, Court of Appeal [1915] 1 KB 1041.

[11] The Rapid Building Group Ltd. v. Ealing Family Housing Association Ltd., Court of Appeal [1984] 1 WLUK 630.

[12] Pearson & Son Ltd. v. Dublin Corporation; House of Lords, [1907] AC 351; Howard Marine and Dredging Co. Ltd. v. A. Ogden & Sons (Excavations) Ltd., Court of Appeal [1977] 9 BLR 34; Taylor v. Hamer, [2002] EWCA Civ. 1130.

that prompt action at the right stage would keep the liability of the owner due to breach of this important condition of contract to the minimum.

Unfortunately, certain provisions in the agreement, which stipulate that the delay in handing over the possession of the site to the contractors would entitle him to an extension of time only, and no compensation would be given to him, result in litigation or arbitration. If the delay is reasonable, the condition may be treated as valid, depriving the contractor of claims of compensation. If, however, the delay is abnormal, amounting to fundamental breach, these stipulations will amount to "exemption clauses" and the contractor may succeed in spite of the negative stipulation. These aspects are fully discussed in Chapters 6 and 12.

The best course of action available for the owner is, of course, not to enter into an agreement until after the site is available for execution of the work by the contractor. On the other hand, a contractor is well advised to protect himself by inserting a suitable condition in his tender to take care of such an eventuality.

In practice, occasionally, one comes across a situation where the extent and time of possession of the site of work as discussed above does not figure but the "state of the site" becomes a point of dispute. This aspect has been duly considered under the heading "extras caused by misrepresentation" in paragraph 5.14 of Chapter 5, which may also be referred to in addition to what follows.

11.3.3 Condition of Site

The development of the law over a period of more than 100 years in England is worth nothing. The earlier decisions attempted to ascribe the risk more to the contractor than the owner. As a result, of course, over the years, the contract conditions had to be modified to keep the cost of construction within reasonable expectation. It is clear that if the contractor is asked to bear unexpected risk the cost is bound to be quoted high. One of the earliest decisions by the House of Lords, which deserves taking notice of in this connection, is Alexander Thorn v. London Corporation.[13] In the case, Thorn contracted with the London Corporation to demolish an existing bridge and to construct a new one, in accordance with plans and specifications prepared by the corporation's engineers. The plans and specifications proved defective in important respects, so much so, that the bridge had to be built in a different way. The contractor, in an action claiming reimbursement for his loss of time and labour, could make no recovery. The contractor's contention that there was an implied warranty that the bridge could be built inexpensively in accordance with the plans and specifications was also rejected. It was held that the obligation of the contractor to complete the works covered substantial deviations from what had been planned.

In an even earlier case,[14] the contract was for execution of sewerage works. The contractor intended to use poling boards for the excavation. Neither party had taken boreholes to ascertain the nature of the substrata. However, before signing of the contract the owner had been advised that the contractor's price was such that he was bound to lose money because of the type of soil conditions expected. During the execution of the work the soil turned out to be unsuitable and necessitated extra works. The contractor, after the engineer refused to authorize extra, as a variation, abandoned the works and sued for the value of the work. It was held that the contractor was not entitled to abandon the contract on discovering the nature of the soil or because the engineer refused to certify, and his claim must fail.

The conditions of the contract were bound to change in order that the law as laid down in the above cases gave way to a just approach which must take into account a simple fact that the contractor's price is built up dependent upon the data supplied to him or assumed by both the parties prior to or at the time of signing of their agreement. If the subsoil conditions are materially different from those

---

[13] Alexander Thorn v. London Corporation, (1876) 1 App Cas. 120.
[14] Bottoms v. Lord Mayor of York, Court of Exchequer, (1812) 2 HBC (4th ed.) 208.

described in or implied by the contract, it is unreasonable to expect the contractor to bear the cost of variation. The following examples lay down the guiding principles, it is submitted.

It should be noted that English law may be at variance here with Indian law in terms of risk allocation. This may be a reflection of the sophisticated users of contract forms by commercial entities who are assumed to negotiate more or less on equal terms. For example, the Court of Appeal set aside a decision of the court of first instance that extensive variations had frustrated an original contract such that the contractor ought to be paid on a *quantum meruit* basis.[15] Coulson, J. has recently observed that the Court of Appeal's decision was "a signpost that, in modern times, the courts will uphold the contractual mechanism agreed by the parties wherever possible, and avoid, if they can, relying on extra-contractual concepts such as frustation or time being at large."[16]

EXAMPLES FROM CASE LAW

1. A contract for the construction of a convalescent home expressly provided that

    the quality and quantity of the work included in the contract sum shall be deemed to be that which is set out in the bills of quantities; these bills, unless otherwise stated shall be deemed to have been prepared in accordance with the Standard Method of Measurement.

    The Standard Method, as is usual, included, where practicable, the nature of the soil to be described, with reference to existing trial holes and rock excavation, to be paid separately. The bills required the contractor to satisfy himself as to the local conditions, etc. but did not include a separate item for rock excavation. The possibility of rock excavation, though within the knowledge of the architect, was also not shown on any drawings. On these facts it was held that the contractor was entitled to extra payment for excavation in rock plus a fair profit.[17]

2. A contract for the construction of blocks of dwellings contained sub-structure designs and priced bills of quantities (BOQs) prepared by the contractor in selected foundation conditions. The foundation designs were based on the borehole data supplied to the contractor by the owner. During construction, the nature of the soil (represented to be a mixture of Northamptonshire sand and clay) was found to be different from that assumed. The changed soil strata conditions necessitated redesign of the foundations and execution of additional work in the areas where soil strata were different from that originally assumed. The Court of Appeal took notice of the following basic engineering factors: First, before designing the foundations for any building it is essential to know the nature of the soil conditions. Second, where the contract is for a comprehensive development of the kind in question, the contractor must know the site conditions at the site of each projected block in order to be able to plan his timetable and to estimate his requirement of materials. Third, the above stated matters relate directly to the contract price. Fourth, if the work is interrupted or delayed by unforeseen complications, the contractor is unlikely to be able to complete his contract in time, making the contractor liable to pay liquidated damages. It was held that on the basis of the above stated strong commercial reasons, and on proper interpretation of the contract, there was an implied term or warranty that the ground conditions should be in accord with the hypotheses upon which the contractor had been instructed to design the foundations. The contractors were held entitled to damages for breach of the implied term.[18]

---

[15] McDermott International Inc. v. McAlpine Humberoak Ltd., [1992] 58 BLR 1.

[16] Severfield (UK) Ltd. v. Duro Felguera UK Ltd., [2017] EWHC 3066 (TCC).

[17] C. Bryant & Sons Ltd. v. Birmingham Hospital Saturday Fund, KB (1938) 1 All ER 503

[18] Bacal Construction (Midlands) Ltd. v. Northhampton Development Corporation, Court of Appeal (1976) 8 BLR 88.

It is submitted that the above decisions lay down good law, which will apply to all contracts, except where the agreement makes express provisions to the contrary by excluding the implied term.

11.4 Delay in Supplying Working Drawings, Decisions

In engineering practice, a minimum of three sets of drawings are contemplated as follows:

1. Tender drawings;
2. Working drawings; and
3. Record drawings.

These are briefly discussed below.

11.4.1 Tender Drawings

This set of drawings is primarily prepared for obtaining technical sanction to the project and invitation of tenders. They are meant to give the intending tenderer a fair and reasonable idea regarding the scope and nature of the work, various materials, and workmanship to be incorporated in the work so as to enable him to estimate a realistic cost of the work before submitting his tender. Certain PWDs do take care to stamp or write a bold note on these drawings such as "*Not for Execution*".

11.4.2 Working Drawings

These drawings, compared to the tender drawings, are more detailed and accurate and give the exact location, size, shape and directions of the work and are, in fact, the drawings strictly in accordance with which work is to be executed by the contractor and as such are by far the most important drawings. In a majority of the contracts and especially large civil engineering projects involving millions of rupees, these drawings are prepared by the owner and supplied to the contractor only after an agreement is signed. Occasionally, in some specialized projects, nowadays, the tendency is to ask the contractors to quote the cost of execution on the basis of their own design and drawings. These are called: Engineering, Procurement and Construction (EPC) contracts, under which it is the responsibility of the contractor to prepare the working drawings, get them approved by the owner/engineer or his representative, if the contract so stipulates.

11.4.3 Record Drawings

As the name suggests, these drawings are prepared to show exactly how the work was carried out at the site for the purpose of record and future use. Changes made from time to time during the execution of the work in the working drawings, necessitate preparation of these drawings. They are also called "As Built Drawings".

11.4.4 Normal Provisions in Construction Contracts

Under the definition and interpretation clause, normally incorporated in the construction contracts, "drawings" is usually defined to mean "the drawings referred to in the specifications and any modifications of such drawings as may from to time be furnished or approved by the Engineer-in-charge". The general conditions of contract provide wording to the effect that.

> The contractor shall carry out the whole and every part of the work ... in accordance with the specifications. The contractor shall also conform exactly and fully to the designs,

drawings and instructions in writing, in respect of the work, signed by the Engineer-in-charge. The Engineer-in-charge shall have the power to make any alterations in, omissions from, additions to or substitutions for, the original specifications, drawings, designs and instructions that may appear to him to be necessary during the progress of the work, and the contractor shall carry out the work in accordance with any instructions which may be given to him in writing signed by the Engineer-in-charge, and such alterations, omissions, additions, or substitutions shall not invalidate the contract.

The necessity of these stipulations has already been emphasized in Chapter 5. These provisions when read with other contractual terms mean:

i. That the contractor is to execute the work strictly as per specifications, designs, drawings and instructions issued to him from time to time

ii. That the contractor is to commence the work on or before the date stipulated in the contract, and

iii. That the whole of the work is to be completed on or before the specified date.

In view of these provisions, although the contract may be silent about the time of supply of the designs and drawings, it is an implied term of the agreement that detailed drawings have to be issued at the proper time. There is an implied contract by each party that he will not do anything to prevent the other party from performing a contract or to delay him in performing it. Generally such a term is, by law, imported into every contract.

EXAMPLE FROM CASE LAW

In a contract for construction of 287 dwellings, the completion of the work got delayed. The parties were in dispute regarding the causes for the delay. The contractor contended that the delay was almost entirely due to lack of diligence and care, and also lack of co-operation by the owner's architect. The contractor alleged that the owner was in breach of implied terms of the contractor as follows: (i) the owner would not hinder or prevent the contractor from carrying out his obligations in accordance with the terms of the contract and from executing the works in a regular and orderly manner; (ii) the owner would take all steps reasonably necessary to enable the contractor to discharge his obligations and to execute the works in a regular and orderly manner. When the dispute was referred to arbitration, the arbitrator held that the above two terms ought to be implied. The findings of the arbitrator were confirmed.[19]

It is expected that a full set of working drawings will be kept ready by the owner to be supplied to the contractor within a reasonable time after making the contract, and in any case before the date of commencement of the work by the contractor.

At the time of signing of an agreement the tender drawings alone are available. If there is no major change expected the contractor should ask, and the engineer should clarify, if the contractor can commence the work on the basis of the tender drawings. When the tender drawings clearly carry a note "not for construction", the contractor even prior to moving on to the site of the work should ask for a full set of working drawings. This is advisable because the engineer may require adequate notice for supply of these drawings. The same procedure should be followed whenever asking for any design or other details, instructions and clarifications during the progress of the work. This is for the reason that it is possible, in the absence of

---

[19] London Borough of Merton v. Stanley Hugh Leach Ltd., Chancery Division (1985) 32 BLR 51; also see M/s. Multimetals Ltd. v. K.L. Jolly, AIR 2003 Raj. 8.

express provisions, to be held that details should be supplied on request only. It must be understood that the necessity for a full set of working drawings arises even prior to execution of the work, as these drawings enable the contractor to make advance arrangements for procuring men, materials and machinery.

Failure to supply working drawings before or at the time of commencement of the work would amount to a breach of contract for which the owner will be liable in damages to the contractor.[20] Similarly, a change or variation order, if any, ought to be given sufficiently in advance so as not to hamper the planned progress of the work. Any delay in issuing such instructions would also amount to a breach of contract for which the owner will be liable in damages to the contractor. It is well settled that the courts will imply a duty to do whatever is necessary in order to enable a contract to be carried out. The implementation of a building contract does require close co-operation between the contractor and the engineer/architect. The provision in the agreements imposes on the architect an obligation to furnish the contractor with drawings and details as and when necessary. As such, it must have been in the contemplation of the parties that the engineer/architect would act with reasonable diligence and would use reasonable care and skill in providing the information. This will apply equally to EPC contracts as well, inasmuch as though the designs and drawings are to be prepared by the contractor, the engineer's approval should not be delayed. Clause 1.9 of FIDIC Form 2017 provides for notice by the contractor to the engineer and payment of compensation or extension of time or both to the contractor, if the engineer delays his supply or approval and the contractor incurs cost as a result of such delay.

11.5 Delay in Ordering Variations

It is also safe to presume that under the terms of construction contracts, variations cannot be ordered after practical completion, in the absence of an express provision, unless of course the contractor is willing to carry them out. If it were otherwise, the employer could use the power to vary or change the contract into a project containing two contracts rolled into one.[21]

When the work is not completed within the stipulated time, the fact that changes were ordered in drawings and designs even after the elapse of the stipulated period is likely to be relied upon by the contractor either as a defence justifying the delay or as evidence of breach of contract by the owner.

11.5.1 Earlier Completion, if Planned by Contractor – Duty of the Employer, if Any

It happens, though not frequently, that the contractor submits a programme showing completion earlier than the date stipulated in the contract for completion. The question arises as to whether a term has to be implied in the contract that the owner by himself, or his servants or agents should so perform the contract to enable to contractor to carry out the works in accordance with the programme as submitted by the contractor, i.e. on the earlier date proposed by the contractor. Under the terms of most standard form contracts, the contractor is obliged to complete the work on or before the contractual date of completion. As such the contractor is entitled to submit such a programme showing an earlier completion date and complete the work accordingly; however, the above implied term cannot be read in the contract.[22]

[20] Neodox Ltd. v. Borough of Swinton & Pendlebury, Q.B. Div. (1977) 5 BLR 34; Roberts v. Bury Commissioners, [1870] 2 WLUK 23: (1869–70) L.R. 5 C.P. 310.
[21] Russell v. Sa Da Bandeira, [1862] 1 WLUK 49; *Hudson's B.C.* 10th ed., pp. 326, 552.
[22] Glenlion Construction Ltd. v. The Guinness Trust, Q. B. Div. (1987) 11 Con. L.R. 126.

11.6 Delays in Carrying Out Work or Supplying Materials

Where, under the terms of a contract, the employer undertakes, either by himself or through other contractors, to do work or supply materials in connection with the contract works, and fails to do so at the proper time, it amounts to a breach of contract by the employer. As a result he will make himself liable for damage suffered by the contractor as a result of his failure to do so, unless the contract provides otherwise.

EXAMPLES FROM CASE LAW

1. The contract work in a case consisted of four parts, namely, (i) cross-drainage work, (ii) earthwork for forming the roadway, (iii) protective works and (iv) supply of material like soling stone, metal, etc. The cross-drainage work required cement and steel and, under the contract, it was for the department to issue those materials. The site was delivered on 17th August 1982 and the work had to be completed within 18 months, i.e. by 16th February 1984. The schedule prescribed for progress/executions required the cross-drainage work to be carried out first. But, the first batch of cement was issued only on 9th November 1983 and the first batch of steel was issued only on 26th October 1984. When the period stipulated for completion was 18 months and if the first supply of cement was made after 15 months and first steel supply was made after 26 months, very little is required to conclude that there was inordinate delay and consequential breach on the part of the department in supplying the material. Several letters were written by the contractor. The trial court held that there was breach by the department. The High Court upset the said finding of fact. In appeal the Supreme Court held:[23]

> The delay of 16 months in issuing cement and 26 months in issuing steel is clearly established by oral and documentary evidence. The fact that after the initial delay, steel and cement were progressively supplied will not wipe out the breach on account of the initial delay in supply.

Damages for breach awarded by the trial court were also upheld.

2. The contractor's tender for execution of five items of work for flood control was accepted by the state. The contract was signed on 4th July 1972, for a total sum of Rs. 1,64,300/-. The time fixed for completion of the work was before 3rd September 1972. The contractor, by a letter dated 31st July 1972, requested the executive engineer to release the required quantity of cement. He repeated his demand in another letter dated 16th August 1972. The executive engineer, though insisting upon the contractor to commence the work, did not release the cement. After the exchange of a few letters, the executive engineer cancelled the contract by a letter dated 9th May 1973. The contractor filed a suit. The trial court held that the state committed breach of contract and allowed damages at 10% of the total amount of the contract to be paid by the state to the contractor. The state filed an appeal in the Kerala High Court. The High Court confirmed the judgment of the trial court, and dismissed the appeal.[24]

3. A contract was executed between the parties to the suit on 7th December 1977 for construction of a new shed in the premises of the defendant. As per the averments of the plaint, the duration of contract was up to 6th October 1978. The plaintiff had to execute the

---

[23] K.C. Skaria v. Govt. of State of Kerala, AIR 2006 SC 811 = 2006 AIR SCW 265.

[24] State of Kerala v. K. Bhaskaran, 1984 A.L.R. 289; K.C. Skaria v. Govt. of State of Kerala, AIR 2006 SC 811 = 2006 AIR SCW 265.

construction work on the instructions of the architect and as per the drawings and the designs, so provided by them. The plaintiffs alleged collusiveness between the architect and the defendant since the drawings and designs were not provided to them within time, as a result of which the delay had occurred in execution of the work. It was alleged that the architect changed the designs repeatedly and further that the architect had instructed the plaintiff to dismantle the constructed work and to reconstruct the same. As a result, the plaintiff sought extension of time for completion of the work, which was not granted to him by the architect. He further alleged that his machineries were detained in the factory premises of the defendant on the instructions of the architect. The plaintiff also impugned the deductions made by the architect in his final bill. He also impugned the genuineness of the recovery of Rs. 91,500/- on account of damages for delay in construction of the shed. In the written statement filed by the defendant, he had specifically denied the contention of the plaintiff. It was held:

> From the perusal of the evidence, it is fully established that if there was any delay in execution of the work or completion thereof, it was primarily on account of the attitude of the defendant and for which the plaintiff cannot be blamed.[25]

11.6.1 Exemption Clauses

The exemption clauses providing that no compensation would be payable to the contractor due to delay in supplying materials, apart from extension of time, would be strictly construed against the owner. There have been cases where, in spite of such exemption clauses, substantial damages were required to be paid by the PWDs. These aspects have been fully discussed in Chapters 4 and 6.

11.7 Delays Caused by Other Agencies

Where the contract contemplates getting a part of the work done by other agencies, the owner may be required to make good the losses suffered by the contractor due to delay in getting the work done by other agencies. Such situations are likely to arise due to lack of proper planning and coordination. For example, when civil works are either not started or not progressing well, contracts with other agencies, such as plumbing or electrification works, are finalized, with the result that these specialized agencies are rendered idle at the site of work. There have been, on the other hand, instances of the main contractor's work suffering due to non-finalization of the other agencies or due to slow progress by the other agencies.

EXAMPLE FROM CASE LAW

The contractor, in a case, complained about late supply of certain components to be used in the construction of houses and those supplied were also of inferior quality. The contractor had no contract with the supplier. The contract overran by 25 weeks and the contractor claimed damages. The agreement had provided that the components were to be delivered to the site in accordance with a programme to be agreed between the architect, the contractor and the supplier. It was held that as the owner had undertaken to supply the components

---

[25] M/s. Multimetals Ltd. v. K.L. Jolly, AIR 2003 Raj. 8.

they were liable to the contractor on the basis of an implied term, that the components would be supplied in time to avoid disruption and delay. It was observed that the owner could pass that obligation over to the supplier.[26]

These eventualities, it is believed, can be avoided by proper planning and scheduling of the contract works and by strict control and supervision.

11.8 Stoppage or Suspension of Work

It must be remembered that the contractor is not a servant and the employer is not the master. The distinction between a servant and an independent contractor can be considered. The Supreme Court of India cited with approval, it is submitted, the distinction between a servant and an independent contractor brought out in *Pollock's Law on Torts* as follows;[27]

> A master is one who not only prescribes to the workman the end of his work but directs or at any moment may direct the means also, or, as it has been put, retains 'the power of controlling the work', a servant is a person subject to the command of his master as to the manner in which he shall do his work. ...
>
> An independent contractor is one who undertakes to produce a given result so that in the actual execution of the work he is not under the order or control of the person for whom he does it, and may use his own discretion in things not specified beforehand.

In short, a master can order his servant "what to do, when to do and how to it"; whereas an independent contractor is bound by his contract, but not by his employer's orders. As such, unless the contract expressly provides to the contrary, the owner has no right to suspend the work or to stop the work or to direct the contractor to carry out the work in a particular order, and if he does so he commits a breach of contract.[28] There is a general principle applicable to building and engineering contracts that in the absence of any indication to the contrary, a contractor is entitled to plan and perform the work as he pleases, provided always that he finishes it by the time fixed in the contract.[29]

11.8.1 Provisions in Standard Form Contracts

Certain standard form contracts in use in some state PWDs/MES, etc. in India do provide for suspension of work or power reserved to the owner or his engineer to direct "at what point or points and in what manner the works are to be commenced, and from time to time carried on". However, in practice it is seen that many a time the engineer-in-charge fails to give a written order for suspension or written direction for doing an item of work out of turn. Failure to exercise these powers and allowing the contract to remain in force during the suspension period would amount to a hindrance to the smooth progress of the work and a breach of contract for which the owner will be liable to pay damages to the contractor.

---

[26] Thomas Bates & Son Ltd. v. Thurrock Borough Council, Court of Appeal dated 22nd October 1975, [1976] 1 WLUK 660.

[27] 15th ed., pp. 62 and 63, cited in Shivanandan v. Punjab National Bank, AIR 1955 SC 404.

[28] See *Hudson's B. C.*, 10th ed., pp. 326, 327.

[29] Greater London Council v. Cleveland Bridge & Eng. Co. Ltd., Court of Appeal, (1986) 8 Con. LR.30.

11.9 Failure to Pay as per Agreement

Where the contract expressly provides for interim payments to be made to the contractor by the owner upon certificates of an architect or scrutiny of the bill by the engineer, non-payment by the owner of the sum so certified or billed will amount to a breach of contract. The remedy for this breach of contract will be arbitration or civil suit because writ is not maintainable, particularly if there are counter-claims.[30]

J. M. Hill and Sons Ltd. v. London Borough of Camden,[31] is an example of what is likely to happen, in many cases. There was a delay in the payment of the amount certified by the architect. The contractors shifted some of their equipment and men, but not the subcontractor's men, from the site. The owner apprehended repudiation by the contractor and refused to pay. The question arose: who was right? It was held:

> The very essence of the provisions of the contract about payment on the architect's certificate was to maintain the cashflow of the contractor and when the cashflow is cut off without good reason – it does not lie in the mouth of the employers to say that plaintiffs (contractors) were acting unreasonably or vexatiously.

Whether this breach will permit the contractor to rescind the contract and sue for damages is a difficult question, but if the contractor contends that due to non-payment by the owner his working capital got eroded and progress of the work suffered adversely, it may entitle him to an extension of time and consequential compensation, if he is able to prove the losses incurred on this count. However, there remains the vexed question as to what course of action the contractor should take upon the owner's failure to make payments in terms of the agreement. In the absence of knowledge of a definite course of action, the course adopted is to slow down the rate of progress and wait for the owner to take the initiative in terminating the contract, attempt to establish the termination as unjustified, wrongful and illegal and recover due payment in an arbitration or a civil suit. The most harmful effect of this course of action, especially in public works, is that the engineer-in-charge allows considerable time to pass before taking the final action and, many times, the action taken also does not meet all the requirements of the contract and the law so much so that the termination is ultimately held to be illegal. The public exchequer is made to bear considerable extra expenditure due to price rise in the intervening period, besides losing all the benefits of early completion of the project.

If the contractor, under the circumstances, is to repudiate the contract and take recourse to arbitration or law suit for recovery within a reasonable time of this breach, he is in grave danger of facing the consequences of wrongful repudiation by him. This is so because there has been a tendency to hold that mere non-payment of due instalment does not constitute repudiation by the owner. If the contractor borrows money and continues to carry out the work, as indeed he is duty bound to do, with a view to mitigate the losses, he is afraid that if the owner persists in his breach, he would create further liability by way of loans or overdrafts. Sooner or later, however, he will usually be obliged in his own interest to rescind the contract. He can certainly do so, but is well advised to take the following precautions:

1. In normal times, it is observed, that both the parties bypass the term of the contract requiring the contractor to submit his bill each month or at an agreed interval to the engineer-in-charge. However, when there is a failure to pay, the contractor, in fulfilment of the term of the contract, must prepare and submit the bill. He should continue to do so at the intervals specified in the contract.

---

[30] Namakkal South India Transports, Petitioner v. Kerala State Civil Supplies Corporation Ltd., AIR 1997 Ker. 56.
[31] J.M. Hill and Sons Ltd. v. London Borough of Camden Court of Appeal, (1980) 18 BLR 31.

2. If he gives notice, claiming interest on outstanding sums from the date it is due for payment, he may succeed in getting the interest unless the contract stipulates to the contrary. Besides, the contract stipulation of non-liability to pay interest may not hold good if the breach is during the extended period.[32]

3. If the agreement includes an arbitration clause, the owner's failure to make payment even after notice, amounts to a difference and he can have the matter referred to arbitration.[33]

4. Generally the contractor's contribution for working capital is 12 to 20% of the cost of the work, depending upon the total cost, part of which is received by him by way of mobilization advance. The amount withheld should be "substantial" in relation to the working capital.

5. The contractor must be able to prove that the outstanding payments have not been made over a long period in relation to the period of instalment payment and total time limit stipulated for completion of the work. For example, 15% to 25% of the time limit stipulated for the completion of the work is likely to be considered a long period.

6. The contractor should ensure that he can prove that the amount was withheld without any reasonable cause. In this connection it is to be noted that whether the owner has a *bona fide* and arguable counter-claim entitling him to a set-off against payment due to the contractor is always to be determined on the interpretation of the actual wording of the contract.[34]

    However, if the owner raises a *bona fide* arguable contention that a payment certificate issued by the engineer may have been overvalued, he is entitled to have that issue referred to arbitration or judicial determination and the contractor may not be able to claim summary judgment on the certificate.[35]

7. If small sums are paid from time to time and not the whole of the bill amount, the amount should be accepted under protest, so as not to have waived payment as per the bill/certificate.

8. In general the contractor will have to show that the owner made quite plain the intention not to be bound by the terms of the contract.

If the above conditions are fulfilled, the contractor can repudiate the contract on account of persistent failure of the owner to pay due instalments. The authors feel fortified in the above proposition by the decision of the Supreme Court of India in Mahabir Prasad Rungta v. Durga Datta[36] and also of the Supreme Court of United States.[37]

In the case decided by the Supreme Court of India, the contractor had clearly established that there was inordinate delay in making payment for the work done. The first bill was submitted on 25th August 1984 for Rs. 5,36,800/-. After certain deductions, a sum of Rs. 4,04,628/- towards the said bill was released only on 26th March 1986, i.e. after 19 months. This delay remained unexplained. It was held:

> The trial court has examined the evidence in detail and has recorded clear findings of fact about the delays and the breach committed by the Department. The finding of the High Court without consideration of the evidence cannot be sustained. We therefore restore the finding of the trial court that respondents committed breach of their obligations and the appellant was

---

[32] Hyderabad Municipal Corporation v. M. Krishnaswamy, AIR 1985 SC 607.

[33] M/s Jhabby Mal Jang Bahadur v. Nanak Chand, AIR 1982 Del. 55.

[34] Tubeworkers v. Tilbury Construction, [1985] 1 WLUK 661; 30 BLR 67. Also see, Rush & Tompkins v. Deaner [1989] 1 WLUK 369 and Smallman Construction v. Redpath Dorman Long (No.1), [1989] 11 WLUK 276.

[35] C.M. Pillings & Co. Ltd. v. Kent Investments Ltd., [1985] 1 WLUK 67.

[36] Rungta v. Durga Datta, AIR 1961 SC 990.

[37] Cuerini Stone Co. v. P.J. Carlin Const. Co., 248 US, 63 L Ed 275, see Chapter 9 for details.

justified in refusing to complete the work, and the consequential finding that the respondents could not therefore recover the extra cost in getting the work completed from the appellant.

Some more points incidentally decided in the said case are of vital importance. In the said case the contractor had paid court fees only on Rs. 2 lacs and unsuccessfully sought rendition of accounts and permission to pay the balance fee. Although the evidence showed that the value of the work done was in excess of Rs. 10 lacs the decree was restricted to Rs. 2 lacs on which fee was paid. Reference should be made to the observations of the Supreme Court on the contractor's own obligation to keep measurements or to ask for joint measurements as discussed in detail in pargraph 10.10.8 of Chapter 10.

Under RIBA standard form, the builder has a right to determine the contract if the employer does not pay any sums actually certified by the architect, after receipt of written notice to do so. FIDIC Form Clauses 16.1 and 16.2 of the 1999 and 2017 editions makes a similar provision. In addition, Clause 16.1 of the 2017 edition permits a contractor to suspend the work or slow down the progress besides entitlement of interest under Clause 60.10. A somewhat similar provision is also incorporated in Clause 17.7.2 of the NITI Aayog Model Form. The GOI Form, too, incorporates similar provision for payment of interest in Clause 5.

11.10 Breach of Contract by Contractor

Failure to adhere to a term of the contract by the contractor would be a breach of the said provision by the contractor. Almost every term of a construction contract stipulates directly or indirectly an obligation of the contract to be performed by the contractor. Care should be taken not to breach any such terms. Some major breaches of contract by the contractor are considered below.[38]

The breaches of construction contract by the contractor which cause substantial damage and give rise to litigation include:

1. Abandonment or total failure to complete
2. Delay in completion
3. Defective design, materials and/or workmanship.

Typical breaches of a less common kind are:

1. Failure to submit planned programme
2. Unauthorized subcontracting
3. Failure to insure as required
4. Failure to employ qualified engineers
5. Failure to maintain and submit labour reports
6. Payment of unauthorized wages
7. Failure to take safety precautions
8. Causing damage to property or work of other agencies
9. Misappropriation/extra consumption of Schedule A materials such as cement, steel, etc.
10. Failure to account for or return Schedule A materials. Some of the above breaches are discussed below.

---

[38] Also see Ch. 10.

11.10.1 Abandonment or Total Failure to Perform

The essence of a construction contract is a promise by the contractor to carry out and complete the work in consideration of a promise by the building owner to pay for it. Failure to complete the work in accordance with the terms of the contract without reasonable cause or excuse is therefore a breach of contract, which would entitle the owner to claim damages. In the case of a lump sum or entire contract, if the contractor does not complete the work substantially he will not be entitled to payment even for the work done.[39]

In practice, however, certain difficulties are likely to be faced in establishing the fact of the contractor having abandoned the work. Occasionally, the contractor may remove tools and materials, intimating an intention not to return to the site, and thus place himself in a fundamental breach of the contract.[40] The contractors, knowing fully the contractual and legal provisions and their implications, rarely admit of having stopped the work or abandoned it. Invariably a skeleton of men and machinery would always be at the site to bring the case under "slow progress". As such before the owner acts presuming that the contractor has stopped the work or abandoned it, certain precautions may be taken, including the following:

1. Issue of notice to the contractor pointing out the fact that the work had been stopped or abandoned from a particular date and directing him to start the work or improve the progress and to show cause as to why action as provided for in the agreement should not be taken against the contractor. A reasonable time should be given for improving the progress, if it appears that the work has not been totally stopped.

2. Upon receipt of the reply to the show cause notice, an impartial assessment of the reasons should be undertaken. If indeed there are certain causes or hindrances for stoppage or slow progress, including those discussed under the heading breaches by the owner, the same should be eliminated and the contractor should be given a chance to restart the work or improve his performance.

3. If there is no reply to the show cause notice, provisions of the agreement should be carefully studied or, if need be, legal opinion sought before an action is taken against the contractor. For example, in private works where a standard form contract prepared by the Indian Institute of Architects is used, there is a provision for obtaining a certificate from the architect that the circumstances exist under which the owner is entitled to rescind the contract. Whether any notice prior to termination is contemplated or not, the same should be served upon the contractor and the full notice period should be allowed to elapse.

4. If the contractor has virtually abandoned the work but does not admit it and the progress of work is negligible and the contract provides for recovery of liquidated damages for slow progress, the same provision should be invoked and penalty/liquidated damages imposed and recovered from any money due, and payable to the contractor.

5. Only after the full amount due under the liquidated damages (for example, in many contracts there is a limit of 10% of the contract sum up to which liquidated damages can be recovered) is recovered; or if the full notice period elapses and there is neither resumption of the work by the contractor nor improvement or progress and the intention of the contractor not to do the work is clear, suitable action can be taken.

6. Invariably, construction contracts make express stipulations as to the right of the owner to terminate the contract or to get the work done through another agency or department at the risk and cost of the contractor.[41]

---

[39] Ibmac v. Marshall (Homes), [1968] 1 WLUK 223. (1968) 208 E.G. 851.
[40] McManus v. Kelly, [2010] 8 WLUK 269 – a case where the contract was oral and did not provide for interim or staged payments.
[41] For full discussion see Chapter 10.

11.10.2 Delay in Completion

Where the contract stipulates the dates for commencement of the work and for completion of a certain part or whole of the work, if there is delay in commencement or completion of the work by the contractor, he commits a breach of contract and the owner is entitled to recover compensation for the damage caused due to such delay.

11.10.2.1 *"Regularly and Diligently" – Meaning of*

Certain contracts include provisions to the effect that "the contractor shall proceed with the works regularly and diligently" failing which, action can be taken against him. The FIDIC Form uses the words, "the contractor shall proceed with works with due expedition and without delay". These or similar words are elusive words on which the dictionaries help little. True, the words convey a sense of activity, progress and so on, but such language provides little help on the question of how much activity, progress and so on is expected.[42] The meaning given to "due diligence" in the Greater London Council v. Cleveland Bridge and Engineering Co. Ltd.[43] by the Court of Appeal, is the best one can do, it is submitted. It was observed:

> If there had been a term as due diligence, I consider that it would have been, when spelt out in full, an obligation on the contractors to execute the works with such diligence and expedition as were reasonably required in order to meet the key dates and completion date in the contract.

Most standard form construction contracts contain provisions for levy of penalty or liquidated damages. Where the delay is caused due to the acts of omission and commission of the contractor, he renders himself liable for action under this provision. For detailed discussion see Chapters 6 and 7.

11.10.3 Defective Design, Materials and/or Workmanship

This aspect has to be considered in relation to defects under three heads: design, materials and/or workmanship, noticed (i) during the performance of the contract and (ii) after completion of the work done by the contractor. However, it should be appreciated that the term "design" itself is wide enough to include not merely structural calculations and dimensions, but the choice of materials, specifications and drawings as well.

The cases on suitability of designs are, in a majority of instances though not all, likely to arise where an architect or engineer is not engaged by the owner for designing the structure. "Package deal" contracts, and contracts for sale of houses to be erected or in the course of erection by the private developers, are instances wherein a liability for design on the part of the contractor may arise. The following statement of Lord Denning may be taken as authoritative and of general application.

> It is clear from Lawrence v. Cassell and Miller v. Carman Hill Estates that where a purchaser buys a house from a builder who contracts to build it, there is a threefold implication that the builder will do the work in a good and workmanlike manner, that he will supply good and proper materials and that the house will be reasonably fit for human habitation.[44]

---

[42] See Hounslow Borough Council v. Twickenham Garden Development Ltd., Chancery Division, (1970) 7 BLR 81.
[43] Greater London Council v. Cleveland Bridge and Engineering Co. Ltd., [1986] 5 WLUK 178; 8 Con. L.R. 30.
[44] Hancock v. B.W. Brazier (Anerely) Ltd., [1966] 1 W.L.R. 1317.

In general, the principle to be applied in construction contracts can be summarized thus: that a person contracting to do the work and supply materials implicitly undertakes:

i. To do the work in workmanlike manner, that is, with care and skill
ii. To use materials of good quality, and where specifications of quality are agreed this will mean good of their expressed kind
iii. That both the work and materials will be reasonably fit for the purpose for which they are required, unless there are circumstances to exclude any such obligation, under the contract.

Defective design, materials or workmanship observed during the construction ought to be brought to the notice of the contractor and he should be directed to remove the defects and make good the work. Most standard form contracts provide for the removal of defective work/materials by the owner through another agency or otherwise at the risk and cost of the contractor, if the contractor, after a notice served on him, fails to take the necessary steps to do so within the reasonable time granted to him in the said notice.

The liability of the contractor for defects after completion would be limited to defects pointed out to him within the defects liability period, and further, only in respect of defects arising out of his breach of contract. For further discussion see Chapter 8.

11.10.4 Indemnities and Insurance

The contractor's general obligations, under most modern standard form contracts, include:

i. To insure the works together with materials and plant for incorporation therein to the full replacement cost including profit, plus agreed additional sum to cover additional costs of and incidental to the rectification of loss or damage including professional fees and the cost of demolishing and removing any part of the work or debris, etc.
ii. To insure the contractor's equipment and other things brought on to the site by the contractor, for a sum sufficient to provide for their replacement at the site.
iii. The insurance is to be in the joint names of the contractor and the owner and to remain in force until the liability under the defects maintenance clause is over. The contract conditions generally incorporate excepted risks (for which the contractor is not obliged to insure) such as war, hostilities, and invasion, act of foreign enemies, civil war and sometimes cyclones, storms of unprecedented nature.
iv. To indemnify the owner against all losses and claims in respect of death or injury to any person, or loss or damage to any property resulting from any act or neglect of the owner, his agents, servants or other contractors not employed by the contractor.

Failure by the contractor to insure or indemnify as per the terms of the contract will be a breach of contract but may not entitle the owner to repudiate the contract on that ground. Some contracts include provisions, in the event of the contractor failing to take insurance, empowering the owner to take insurance and deduct the costs thereof from the sum due and payable to the contractor. These provisions sometimes present serious difficulties of interpretation.

EXAMPLE FROM CASE LAW

A contract contained two provisions relevant for the discussion. The first provision required the contractors to indemnify the owner in respect of injury to persons and property (subject to certain exceptions). The second required the contractor to "effect – such insurances – as may be specifically required by the bills of quantities". The BOQs required "insurance of adjoining properties against subsidence or collapse". The contractor took out an insurance policy, which covered the contractor's liability for subsidence but not that of the plaintiff owner. A neighbouring owner brought a claim against the plaintiff in respect of subsidence caused by the building operations. The owner brought an action claiming that the contractor was in breach of his obligation under the second of the two above mentioned conditions. It was held: All the items in the list for "the contractor shall insure", would appear to be cases where the contractors must insure themselves and therefore there was insufficient reason for thinking that the last item "insurance of adjoining properties against subsidence or collapse" gave an obligation to insure not the contractors themselves but also the building owner.[45]

The problem of interpreting complex contract provisions becomes difficult if the owner is to take possession of part of the premises and start using it and the contract stipulates that to the extent possession is taken the obligation of the contractor to insure ceases, as illustrated in the case below:

EXAMPLE FROM CASE LAW

Tenders were invited for carrying out extensive extension to the existing factory (making it three times as large as before) without disrupting the production. The contractors submitted a tender in RIBA form, and BOQs for £687,860 which was accepted and a formal contract signed. The work was started in 1969 and completed to a great extent by January 1970. New machines were installed and stock of reels of paper in the new-reel warehouse was stored, though the contractors had not completed the works in full. On 18th January 1970, there was a big fire and much of the new factory was gutted with the estimated loss at £250,000. The question was who was to bear this loss. The answer depended upon proper interpretation of Clause 20A(1) and 16 of the RIBA form. Clause 20A(1) provided that the contractor should insure against loss and damage by fire all work executed and all unfixed materials and goods and should keep them insured under practical completion of the works or until the employer should authorize in writing the cancellation of such insurance. Clause 16 provided for empowering the owner to take possession of any part or parts of the work, with the consent of the contractors, before practical completion of the work; the architect to issue a certificate stating the estimate of approximate total value of the part so taken over within seven days of the employer taking possession; and entitling the contractor to reduce the value insured under 20(A) by the full value of the relevant part and the said relevant part shall, as from that date when possession was taken, to be at the sole risk of the employer. The great question was, did the employer take possession of the part as claimed by the contractors? The contention of the contractor was "taking possession" is a question of fact which is answered by the fact that the employers were using and occupying the factory at the time of the fire. The employers, on the other hand, contended that "taking possession" could only take place when the relevant part was handed over to them and accepted by them; and that never took place. Simply using or occupying a place was not enough. The architect had issued a certificate of completion to his satisfaction only in respect of the car park. It was observed that the agreement contemplated definite machinery to determine handing over. The practice was for the contractor to inform the architect that a part was ready, the architect was to inspect it and, if ready, accept it and

---

[45] Gold v. Patman & Fotheringham Ltd., Court of Appeal, [1958] 2 All ER 497.

issue a certificate to that effect. It was held the contractors at the time of the fire had not handed over some parts to the employer; it was the responsibility of the contractors to insure them; the risk had not passed to the employers. The contractors, or their insurers, therefore must bear the loss.[46]

It is to be noted that even where a party to the contract agrees to indemnify the other party, such a provision cannot be read to impose on the party indemnifying the other against the other's own negligence unless clear words to that effect are used.[47] In the case of a construction contract, an indemnitee's failure to spot defects perpetrated by its contractor or subcontractor should not defeat the operation of an indemnity clause "even if that clause fails expressly to encompass damage caused by the negligence of the indemnitee."[48]

In the case of building or engineering contracts, where numerous different subcontractors may be engaged, there can be no doubt about the convenience from everybody's point of view, including perhaps insurers, of allowing the main contractor to take out a single policy covering the whole risk, that is to say covering all contractors and subcontractors in respect of loss or damage to the entire contract works. However, a subcontractor who is engaged in contract works may insure the entire contract works as well as his own property.[49]

11.11 Forms of Discharge by Breach

If one of the two parties to a contract breaks the obligation which the contract imposes, a new obligation in every case will arise – a right of action conferred upon the party injured by the breach. In addition to the right of action there are circumstances under which the injured party will also stand discharged from such performance as may still be due from him. In order to have the effect of discharge by breach, the breach must be such as to constitute repudiation by the party in default of his obligations under the contract. A breach of contract does not automatically terminate the contract but renders it voidable at the option of the injured party; he has the option either to treat the contract as still in existence or to regard himself as discharged by reason of repudiation of the contract by the other party.[50] Repudiation by one party alone does not terminate the contract. It takes two to end it.[51] However, the option can be said to be available to the injured party if the injured party can carry out the contract to completion without the co-operation of the party in default and sue for the price. Where performance of the contract cannot be carried out when co-operation is refused, the injured party must accept the repudiation and sue for damages. The construction contracts would invariably fall under the latter category.

Repudiation is an act or omission of a party, which can fairly be regarded as evincing an intention by that party no longer to be bound by the terms of the contract. Repudiation of a construction contract by the owner will give the contractor a right to bring an action for the recovery of money due for the value of the work done together with damages. Similarly repudiation of the contract by the contractor would entitle the owner to bring an action for damages forthwith. In practice, however, quite often the cases are of partial failure of performance, giving rise to the difficult question as to whether such failure by one party amounts to repudiation so as to entitle the other to regard him as discharged.

---

[46] English Industrial Estates Corporation v. George Wimpay & Co. Ltd., Court of Appeal (1972) 7 BLR 122.
[47] Walters v. Whessoe Ltd. and Shell Refining Co. Ltd., Court of Appeal, (1960) 6 BLR 23.
[48] Greenwich Millenium Village Ltd. v. Essex Services Group Plc, [2014] EWCA Civ 960, para. 94.
[49] Petrofina (VK) Ltd. v. Magnagold Ltd., Q.B. Div. (1983) All ER 35.
[50] Heyman v. Darwins, [1942] AC 356.
[51] Damodar Valley Corporation, v. K.K. Kar, AIR 1974 SC 159.

Differing terminology is in use to describe the nature of the failure of performance to discharge the innocent party from the performance of his own obligations under the contract. The breach, it has been said, must be fundamental, "it must go to the root or essence of the contract", or "to the foundation of the whole" or "is such that it deprives the innocent party of substantially the whole benefit which it was intended that he should obtain from the contract".[52]

11.11.1 Fundamental Breach

The expression "fundamental breach of contract" or "breach of a fundamental term" is used in two quite different senses. In one sense these denote a breach by one party, which is sufficiently serious to entitle the other party not merely to claim damages, but to elect to treat himself as discharged from further performance under the contract. In another sense, these terms are used as a supposed principle of law that there are certain breaches of contract which are so totally destructive of the obligations of the party in default, that liability for such a breach cannot be limited or excluded by means of an exemption clause. In this sense it is a rule of construction and is dealt with in Chapter 4.

11.11.2 Anticipatory Breach

The renunciation of a contract by one of the parties before the time for performance has come is known as "anticipatory breach" of contract. It also discharges the innocent party if the party so chooses, and entitles it to sue for damages. If, however, the innocent party refuses to accept the renunciation and continues to insist on exercising his right on the performance of the contract, he loses his right to rely on the anticipatory breach and the contract remains in existence for the benefit, and at the risk of, both the parties. For example, if the circumstances subsequently give rise to frustration of contract, the injured party would lose the right of action under anticipatory breach.[53]

---

[52] For full discussion see Chapter 10.
[53] For full discussion see Chapter 10.

12

Remedies for Breach of Contract

12.0 Introduction

Following on from the discussion of discharge of a contract by breach in Chapter 11, this chapter will focus on the various remedies which are available to a person injured by the breach, whether the contract can be treated as discharged or not. Every party injured by a breach of contract is entitled to damages for the loss he may have incurred. Even if such a party may not be entitled to contractual compensation, he may be entitled to claim at least the value of the work done on a *quantum meruit* basis. In certain circumstances the injured party may obtain a court order for the specific performance of the contract or an injunction to restrain its breach. Standard form contracts include provisions of forfeiture which take effect upon a breach of contract. Given the importance of this topic, this chapter focuses on each of these in the context of building and construction contracts by reference to case law.

12.1 Provisions of the Law

Provision of the Law	Paragraph No.
Indian Contract Act, Sections 23, 73, 74, 148, 170, 171	12.3, 12.8.1, 12.17, 12.19
Provisions in Standard Form Contracts	
PWD/CPWD, Clauses 3, 4, 13, 29	12.8.1, 12.8.2, 12.12, 12.17.4, 12.21.1, 12.21.5
Military Engineering Services Form, Clauses 53, 54, 57; MES standard form IAFW 2249	12.8.1, 12.17.4
Railway Engineering Department Form, Clauses 32, 52, 61, 62	12.17.4
FIDIC 1992 (amended), Clauses 54.1–54.8	12.9, 12.17, 12.20
FIDIC 1999/2017 editions, Clauses 4.2, 15.1–15.5, 16.1–16.4, 20	12.17, 12.9, 12.20
Ministry of Statistics and Programme Implementation, Government of India (MOS & PI), Clauses 44, 52, 59, 60, 61	12.17, 12.20
NITI Aayog Model Form, Clause 21	

12.2 Damages

The Indian Contract Act, besides Section 55 which deals with delay in performance, contains three provisions in Chapter 6 under the heading: "Of the consequences of breach of contract":

1. Section 73 deals with unliquidated damages
2. Section 74 deals with liquidated damages, and

3. Section 75 deals with compensation for any damage which a person who rightfully rescinds the contract is entitled to.

Section 74 dealing with liquidated damages has been considered in Chapter 7. Reference to Section 74 is necessary while discussing forfeiture provisions. Section 75 has been discussed in Chapter 10. Section 73 is considered below:

12.3 Section 73

Section 73 of the Indian Contract Act reads:

Compensation for loss or damage caused by breach of contract

When a contract has been broken, the party who suffers by such breach is entitled to receive, from the party who has broken the contract, compensation for any loss or damage caused to him thereby, which naturally arose in the usual course of things from such breach or which the parties knew, when they made the contract, to be likely to result from the breach of it.

Such compensation is not to be given for any remote and indirect loss or damage sustained by reason of the breach.

Compensation for failure to discharge obligation resembling those created by contract

When an obligation resembling those created by contract has been incurred and has not been discharged, any person injured by the failure to discharge it is entitled to receive the same compensation from the party in default, as if such person had contracted to discharge it and has broken his contract.

Explanation – In estimating the loss or damage arising from a breach of contract, the means which existed of remedying the inconvenience caused by the non-performance of the contract must be taken into account.

ILLUSTRATIONS

(a) A contracts to sell and deliver 50 maunds of saltpetre to B, at a certain price, to be paid on delivery. A breaks his promise. B is entitled to receive from A, by way of compensation, the sum, if any, by which the contract price falls short of the price for which B might have obtained 50 maunds of saltpetre of like quality at the time when the saltpetre ought to have been delivered.

(b) A hires B's ship to go to Bombay, and there takes on board on the first of January, a cargo, which A is to provide, and to bring it to Calcutta, the freight to be paid when earned. B's ship does not go to Bombay, but A has opportunities of procuring suitable conveyance for the cargo upon terms as advantageous as those on which he had chartered the ship. A avails himself of those opportunities, but is put to trouble and expense in doing so. A is entitled to receive compensation from B in respect of such trouble and expense.

(c) A contracts to buy of B, at a stated price, 50 maunds of rice, no time being fixed for delivery. A afterwards informs B that he will not accept the rice if tendered to him. B is entitled to receive from A, by way of compensation, the amount, if any, by which the contract price exceeds that which B can obtain for the rice at the time when A informs B that he will not accept it.

(d) A contracts to buy B's ship for 60,000 rupees, but breaks his promise. A must pay to B, by way of compensation, the excess, if any, of the contract price over the price which B can obtain for the ship at the time of the breach of promise.

(e) A, the owner of a boat, contracts with B to take a cargo of jute to Mirzapur, for sale at that place, starting on a specified day. The boat, owing to some avoidable cause, does not start at the time appointed, whereby the arrival of the cargo at Mirzapur is delayed beyond the time when it would have arrived if the boat had sailed according to the contract. After that date, and before the arrival of the cargo, the price of jute falls. The measure of the compensation payable to B by A is the difference between the price which B could have obtained for the cargo at Mirzapur at the time when it would have arrived if forwarded in due course, and its market price at the time when it actually arrived.

(f) A contracts to repair B's house in a certain manner, and receives payment in advance. A repairs the house, but not according to contract. B is entitled to recover from A the cost of making the repairs conform to the contract.

(g) A contracts to let his ship to B for a year, from the first of January, for a certain price. Freights rise, and, on the first of January, the hire obtainable for the ship is higher than the contract price. A breaks his promise. He must pay to B, by way of compensation, a sum equal to the difference between the contract price and the price for which B could hire a similar ship for a year on and from the first January.

(h) A contracts to supply B with a certain quantity of iron at a fixed price, being a higher price than that for which A could procure and deliver the iron. B wrongfully refuses to receive the iron. B must pay to A, by way of compensation, the difference between the contract price of the iron and the sum for which A could have obtained and delivered it.

(i) A delivers to B, a common carrier, a machine, to be conveyed, without delay to A's mill informing B that his mill is stopped for want of the machine. B unreasonably delays the delivery of the machine, and A, in consequence, loses a profitable contract with the Government. A is entitled to receive from B, by way of compensation, the average amount of profit which would have been made by the working of the mill during the time that delivery of it was delayed, but not the loss sustained through the loss of the Government contract.

(j) A, having contracted with B, to supply B with 1,000 tons of iron at 100 rupees a ton, to be delivered at a stated time, contracts with C for the purchase of 1,000 tons of iron at 80 rupees a ton, telling C that he does so for the purpose of performing his contract with B. C fails to perform his contract with A, who cannot procure other iron and B in consequence rescinds the contract. C must pay to A 20,000 rupees, being the profit which A would have made by the performance of his contract with B.

(k) A contracts with B to make and deliver to B, by a fixed day, for a specified price a certain piece of machinery. A does not deliver the piece of machinery at the time specified, and, in consequence of this, B is obliged to procure another at a higher price than that which he was to have paid to A, and is prevented from performing a contract which B had made with a third person at the time of his contract with A (but which had not been then communicated to A) and is compelled to make compensation for breach of contract. A must pay to B, by way of compensation, the difference between the contract price of the piece of machinery and the sum paid by B for another, but not the sum paid by B to the third person by way of compensation.

(l) A, a builder, contracts to erect and finish a house by the first of January, in order that B may give possession of it at that time to C to whom B has contracted to let it. A is informed of the contract between B and C. A builds the house so badly that, before the first of January, it falls down, and has to be rebuilt by B, who, in consequence loses that rent which he was to have received from C, and is obliged to make compensation to C for the breach of his contract. A must make compensation to B for the cost of rebuilding the house, for the rent lost, and for the compensation made to C.

(m) A sells certain merchandise to B, warranting it to be of a particular quality, and B, placing reliance upon this warranty, sells it to C with a similar warranty. The goods prove to be not according to the warranty, and B becomes liable to pay C a sum of money by way of compensation. B is entitled to be reimbursed this sum by A.

(n) A contracts to pay a sum of money to B, on a specified day. A does not pay the money on that day. B in consequence of not receiving the money on that day, is unable to pay his debts, and is totally ruined. A is not liable to make good to B anything except the principal sum he contracted to pay, together with interest up to the day of payment.

(o) A contracts to deliver 50 maunds of saltpetre to B on the first of January, at a certain price. B afterwards, before the first of January, contracts to sell the salpetre to C at a price higher than the market price of the first of January. A breaks his promise. In estimating the compensation payable by A to B, the market price of the first of January, and not the profit which would have arisen to B from the sale to C, is to be taken into account.

(p) A contracts to sell and deliver 500 bales of cotton to B on a fixed day. A knows nothing of B's mode of conducting his business. A breaks his promise, and B, having no cotton, is obliged to close his mill. A is not responsible to B for the loss caused to B by the closing of the mill.

(q) A contracts to sell and deliver to B, on the first of January, certain cloth which B intends to manufacture into caps of a particular kind, for which there is no demand, except at that season. The cloth is not delivered till after the appointed time, and too late to be used that year in making caps. B is entitled to receive from A, by way of compensation, the difference between the contract price of the cloth and its market price at the time of delivery, but not the profits which he expected to obtain by making caps, nor the expenses which he has been put to in making preparation for the manufacture.

(r) A, a shipowner, contracts with B to convey him from Calcutta to Sydney in A's ship, sailing on the first of January, and B pays to A, by way of deposit, one-half of his passage money. The ship does not sail on the first of January, and B, after being, in consequence detained in Calcutta for some time and thereby put to some expense, proceeds to Sydney in another vessel, and, in consequence, arriving too late in Sydney, loses a sum of money. A is liable to repay to B his deposit, with interest, and the expense to which he is put by his detention in Calcutta, and the excess, if any, of the passage money paid for the second ship over that agreed upon for the first, but not the sum of money which B lost by arriving in Sydney too late.

The section comprises three paragraphs and one explanation besides illustrations. The first two paragraphs and the explanation lay down the law with respect to compensation for loss or damage caused by a breach of contract. Paragraph 3 sets out the law with regard to compensation for failure to discharge obligations resembling those created by a contract.

12.4 Essential Conditions to Be Fulfilled by Injured Party to Be Entitled to Damages

A plain reading of Section 73 makes it clear that for an injured party to be entitled to damages from the party who has breached the contract, all the following conditions must be fulfilled:

1. That there exists a contract or an obligation resembling those created by a contract.
2. That the defaulting party was under an obligation to perform that part which is alleged to have been breached. For example, if it is alleged by the contractor that the owner committed

a breach of contract by not supplying materials agreed to be supplied, such a condition in the agreement has to be proved. If there is no express provision, say for handing over of the site, that it is an implied obligation undertaken by the owner has to be alleged and established from the other relevant conditions of the contract.

3. That there is a breach of contract or any condition of the contract must be established by the party making the allegation.

4. That the breach of contract caused "loss or damage" to the injured party. In law merely because there is a breach of contract such breach is not actionable unless, because of the breach, loss is caused to the aggrieved party. There can be forfeiture of an amount paid under a contract only when the aggrieved party is caused loss. Two relevant judgments in this regard are the those of the Constitution Bench of the Supreme Court in the cases of Fateh Chand v. Balkishan Dass and Kailash Nath Associates v. Delhi Development Authority and Another.[1]

5. That the loss or damage caused by the breach of contract naturally arose in the usual course of things from such breach.

6. That the parties to the contract knew, when they made the contract, such loss or damage "to be likely to result from the breach of it", must be established.

7. That the loss or damage sustained by reason of breach is neither remote nor indirect.

8. That the injured party did take precautions to mitigate the losses to the maximum extent possible.

When all these eight conditions are fulfilled, the injured party is entitled to receive, from the party who has broken the contract, compensation for any loss or damage caused to him thereby. It must, however, be remembered that when a person is claiming damage, he has to specifically plead the manner in which he suffered the loss to justify the damage in the manner it is claimed.[2] Second, if there is no loss proved, damages are not payable.[3] The case in example (1) below is a typical case of breach by employer and the consequences thereof.

EXAMPLES FROM CASE LAW

1. In a case the arbitral tribunal recorded its findings thus:

> I have perused all the relevant provisions of the agreement and rival submissions of the parties and I have no hesitation to say that it is a case of massive delays and defaults on the part of the respondent in regard to its most fundamental obligations. In fact, it is more than obvious that the respondent called tender without doing its proper home work because of which not only there was monumental delay in making available hindrance free site and working drawings and deciding the GAD [General Arrangement Drawing], there also arose necessity of subsequent changes, such as, change from originally proposed cast-in-situ RCC girders to pre-cast girders, shifting of location of abutment 'A2' and consequent reduction in the

---

[1] Fateh Chand v. Balkishan Dass, AIR 1963 SC 1405; Kailash Nath Associates v. Delhi Development Authority and Another, (2015) 4 SCC 136.
[2] State of Orissa v. Pratibha Prakash Bhawan, AIR 2005 Ori. 58.
[3] Thyssen Stahlunion Gmbh v. Steel Authority of India, AIR 2002 Del. 255; Jalpaiguri Zilla Parishad v. Shankar Prasad Halder, AIR 2006 Cal. 1.

length of end span on Sarai Kale Khan side and change from originally proposed girder & slab to only solid slab in the end span on Sarai Kale Khan side. The poor planning by the respondent also gave rise to deviation items which caused further delay .... Based on my finding above, it is held that the respondent committed a serious breach which is of the most fundamental nature and which goes to the very root of the contract and that the respondent is entirely responsible for the delay in completion of the work.

The contractor was awarded compensation on various grounds such as prolongation of work, change in scope, escalation in price, etc. The Delhi High Court upheld the award.[4]

2. In a case, the appellant had received from the respondent a sum of Rs. 10 lacs, and claimed the right to forfeit it as earnest money deposit for the sale of a property. The respondent argued that it was an advance towards the purchase price. The appellant had not pleaded or proved the actual damage caused to him, nor had he made an attempt to show the nature of the damage claimed. There was just a statement in his pleadings that since he did not obtain the sale price in time, he was not able to invest it in his business and that he was forced to take further loans. It was observed:

> These submissions have not been proved or established by cogent evidence and the court below was, therefore, justified in refusing to accept these contentions and in ordering return of the money paid by the respondent. ... We have no doubt in our mind ... that the amounts accepted by the appellant is not an earnest money deposit but only an advance of the sale consideration. That being so, obviously, the court below has not erred in holding that the respondent is entitled to return of the advance amount, concluding rightly that the money does not represent earnest money deposit.[5]

12.5 Distinction between "Compensation" and "Damages"

The word "compensation" would embrace in its purview any actual loss suffered by a party. For example, if trees had to be cut or a certain structure had to be altered or demolished, in that case a question of paying compensation would arise but the question as to what loss a party would suffer in case he is prevented from making any constructions or using the roof would not come within the meaning of the word "compensation". It would come under the wider definition of the word "damages".[6] The distinction between the two terms can be stated thus: While the term "damages" is used in reference to pecuniary recompense awarded in reparation for a loss or injury caused by a wrongful act or omission, the term "compensation" is used in relation to a lawful act which caused the injury in respect of which an indemnity is obtained under the provisions of a particular statute.

Section 73 uses the words "compensation for any loss or damage" which words mean, it is submitted, "damages". According to the *Concise Oxford Dictionary* damages mean: "Sum of money claimed or adjudged in compensation for loss or injury". Similarly *Webster's*

---

[4] Union of India v. M/s. Jia Lall Kishori Lall Pvt. Ltd., 2018 (248) DLT 366.
[5] Soji Peter v. K.B. Vijayan, (Kerala)(DB) 2017 (3) Ker. LJ 816; 2017(4)ILR (Kerala) 475.
[6] Union of India v. Ram Chandra, AIR 1975, All. 221.

Comprehensive Dictionary Encyclopedic Edition gives the meaning "Money recoverable for a wrong or an injury". For all practical purposes the distinction between the words used by the parties namely, "compensation" and "damages" in their correspondence or in pleadings would not matter much, it is submitted.

Where an agreement to build included a provision that the builder could terminate the contract and retain 10% of contracted amount on account of failure of prospective purchaser to pay instalment and the builder refused to refund, it was held that it was prerogative of the buyer to ask for refund and the refusal by the builder amounted to a breach of contract. The builder was directed to refund the amount after deducting 10% along with interest at 10% p.a.[7]

12.6 Kinds of Damages

The function of damages in a contract, as already seen, is to compensate for the legitimate expectation and losses. Sometimes parties incorporate express provisions in their contracts to exclude or restrict damages. Damages for distress or mental agony, vexation or loss of enjoyment are rarely awarded in a case of breach of contract. Similarly, punitive damages sometimes awarded in tort are not awarded for a breach of contract. In England, gain-based restitutionary damages were awarded in a case the House of Lords believed that ordinary damages were inadequate.[8] After this case, there has been academic discussion over whether punitive damages should be introduced in English law for a breach of contract.

Damages, in general, may be classified as:

1. General or ordinary damages – these may be proximate or remote
2. Special damages
3. Nominal damages
4. Exemplary or vindictive damages.

12.6.1 General or Ordinary Damages

Proximate damages are the immediate and direct damages naturally resulting from the act complained of, and such as are usual and might have been expected. Remote damages are damages from an injury not occurring directly from and as a natural result of the wrong complained of. Remote damages are such as arise from the unusual and unexpected result, not reasonably to be anticipated from an accidental or unusual combination of circumstances, a result beyond and over which the negligent party has no control. The terms "remote damages" and "consequential damages" are not synonymous nor to be used interchangeably; all remote damage is consequential, but it is by no means true that all consequential damage is remote.[9] Proximate damages are those contemplated under Section 73 namely, arising naturally in the usual course of things. Where by an agreement the authority agreed to hire plinth for three years and subsequently it was terminated before the three years, the Supreme Court of India upheld the plea of damages at a rate equal to rent payable.[10]

---

[7] Prashant Sanghi & Anr. v. M/s SNS Interbuild Pvt. Ltd., AIR 2008 (NOC) 679 (NCC). Also see Rajendra Singh Wadhwa v. State M.P. & Ors., AIR 2008 (NOC) 682 (M.P.).

[8] Attorney General v. Blake, [2000] UKHL 45.

[9] Eaton v. Railroad Co., 51 N. II. 511, 12 Am. Rep. 147.

[10] Food Corporation of India v. M/s Babulal Agrawal, AIR 2004 SC 2926.

EXAMPLE FROM CASE LAW

The appellant and the respondent entered into a contract for the purpose of setting up a project for conversion of Menthone to Menthol, the appellant to provide technical know-how. The time agreed was five months. The respondent incurred huge expense in setting up the project but the appellant failed to supply the technology not only within the five months agreed but even after three years. Arbitration followed and an award was made in favour of the respondent for payment of Rs. 90 lacs. The trial court refused to set aside the award and the division bench of the Uttarkhand High Court dismissed the appeal.[11]

12.6.2 Special Damages

Special damages, on the other hand, are such as are natural and proximate consequences of the breach, also not in general following as its immediate effects, but which the parties knew, when they made the contract, to be likely to result from the breach of it or the defendant had undertaken an obligation under the contract to pay them. The knowledge must, however, be such and acquired under such circumstances as to amount to evidence of an actual contract to pay the exceptional loss arising from the breach of contract.

Illustrations (j), (k) and (l) given under Section 73 of the Indian Contract Act and mentioned above will help clarify this aspect. Special damages it must be noted, have to be specified in the pleadings and proved by evidence.

12.6.3 Nominal Damages

Usually only a small sum of money is what an injured party would be entitled to receive where he has not, in fact, suffered any loss by reason of the breach or has failed to prove the loss actually incurred. Nominal damages is a technical phrase which means that a party has not incurred anything like real damage, but that it is affirming by claiming nominal damages that there is an infraction of a legal right which, though it gives it no right to any real damages at all, yet gives a right to the verdict or judgment because its legal right has been infringed.[12] It is based on the famous dictum, "Every injury imports a damage though it does not cost the party one farthing".[13]

For example, in a case where three work orders totalling to about Rs. 3 lacs were placed, the contract signed and subsequently the work orders were withdrawn, in a suit for damages, having regard to paucity of evidence in respect of the loss suffered, compensation of Rs. 25,000/- was held to be just and reasonable.[14]

12.6.4 Exemplary or Vindictive Damages

These are sometimes awarded in the law of torts as a sort of punishment and have no place in the law of contract. Canadian courts may award exemplary damages for a breach of contract in extremely rare cases.[15] The court used to require that there was a concurrent case in tort and contract; this requirement was relaxed to award exemplary damages where other remedies would

---

[11] Council of Scientific & Industrial Research v. Goodman Drug House P. Ltd., AIR 2007 58 after AIR 2007 All.

[12] Lord Halsbury, LC in The Mediana, [1900] AC 113.

[13] Holt, C.J., Ashby v. White (1703), 2 Raym. 955.

[14] Vikas Electrical Services v. Karnataka Electricity Board, 2008(2) CTLJ 78 (Karnataka).

[15] Whiten v. Pilot Insurance Company, (2002) 209 D.L.R. (4th) 257.

be inadequate deterrence.[16] American law does not preclude an award of punitive damages for a breach of contract, if an independent tort is also committed in the context of a contract.[17]

12.7 General Principles of Assessment of Damages

The object of awarding damages for breach of contract is to put the injured party into the position in which he would have been had the contract been performed. The Supreme Court of India, laid down this principle in these words:[18]

> The two principles on which damages in such cases are calculated are well settled. The first is that, as far as possible, he who has proved a breach of a bargain to supply what he contracted to get is to be placed, as far as money can do it, in as good a situation as if the contract had been performed; but this principle is qualified by a second, which imposes on a plaintiff the duty of taking all reasonable steps to mitigate the loss consequent on the breach, and debars him from claiming any part of the damage which is due to his neglect to take such steps: These two principles also follow from the law as laid down in S. 73 read with the Explanation thereof.

The claim arising out of termination of a contract should be assessed as on the date of termination of the contract and future conduct of the parties or subsequent events may be wholly irrelevant.[19]

The measure of damages in a contract, is compensation for consequences which flows as a natural and capable consequence of the breach or, in other words, which could be foreseen and are not remote.

EXAMPLES FROM CASE LAW

1.   Where a party committed breach of a contract for supply of imported bales of cotton and the injured party produced documents to establish their claims, including clearance and other charges and resale of the stated goods, the prevailing exchange rate, etc., it was held that the claim for the losses actually incurred deserved to be compensated and was accordingly decreed.[20]

2.   In a case of works contract, the time to complete the works expired and application for extension made by the appellant contractor on 4th September 1967 was rejected by the respondent on 14th February 1968. The appellant filed a suit for the sum due to him for the work done and also claimed Rs. 10,000/- as damages. Damages were claimed for the loss of profit on the ground that if the department had given the decision on the application of extension, on the expiry of the stipulated time limit, the appellant would have been able to get some work elsewhere, would have executed work worth Rs. 1,00,000/- and would have easily earned a profit of Rs. 10,000/-. The Allahabad High Court observed that the statement that the appellant would have got another contract and earned a profit, was not supported by any evidence. Nothing was brought on record to demonstrate that a future contract was even available. There was no material on record to prove the quantum of damages. The claim was held to be remote.[21]

---

[16] Royal Bank of Canada v. W. Got & Associates Electric Ltd., (2000) 178 D.L.R. (4th) 385.

[17] Formosa Plastics Corp. USA v. Presidio Engineers & Contractors Inc., 960 S.W. 2d. 41 (Tex. 1998).

[18] M/s. Murlidhar v. M/s. Harish Chandra, AIR 1962 SC 366; Union of India & Ors. v. Sugauli Sugar Works (P) Ltd., (1976) 3 SCC 32.

[19] M.D. AWHO v. Samangal Services Pvt. Ltd., 2003(3) Arb. LR 361 (SC).

[20] Cotton Corporation of India Ltd. v. Ramkumar Mills Pvt. Ltd., 2008(2) CTLJ 95 (Bombay).

[21] Devender Singh v. State, AIR 1987 All. 306.

The threshold for claiming damages on the basis of a loss of chance in English law is "showing a real or substantial chance of the benefit accruing from the third party had the contract not been broken".[22] In a case involving a termination of a contract for the building and installation of the roof for the Millennium Dome, a question arose as to the contractor's ability to claim for loss of profit caused by the termination, given that the contractor had become insolvent after the termination. Arguably, it would not have been able to earn any profit and its insolvency would have led to the termination of the contract anyway. Voss, J. held that the insolvency should not affect the assessment of damages after the termination of the contract of which loss of profit was a natural consequence. It had to be assumed that the contract would have been performed but for the termination.[23]

12.8 Assessment of Damages in Construction Contracts – Breach of Contract by the Owner

From the point of view of damages, breaches of contract by the owner would be of two kinds:

1. Refusal to perform the contract having the effect of bringing the work to an end or preventing its starting, thereby depriving the contractor of the right to his profit upon work never actually carried out, and
2. The breaches which reduce the contractor's profits or increase the cost of work done by him.

The first kind of breach is dealt with in Chapters 10 and 11. Before considering the effect of the second kind of breach, it is necessary to deal with an exemption clause generally found in public works contracts, which deny payment of compensation to the contractor in the event of the owner not requiring a part of the work or the whole work as specified in the agreement or is in breach of one or the other condition of the contract.

12.8.1 Exclusion of Right to Claim Damages by Exemption Clauses in a Contract

In certain standard form contracts used for public works, it is common to include express provisions denying the contractor the right to compensation even though the employer may commit breach of certain contract conditions. Consider, for example, Clause 13 of the CPWD Standard Form contract. The said clause reads:

> *No compensation for alteration in or restriction of work to be carried out*: If at any time after the commencement of the work the President of India shall for any reason whatsoever not require the whole thereof as specified in the tender to be carried out the Engineer-in-Charge shall give notice in writing of the fact to the contractor who shall have no claim to any payment of compensation whatsoever on account of any profit or advantage which he might have derived from the execution of the work in full, but which he did not derive in consequence of the full amount of the work not having been carried out, neither shall he have any claim for compensation by reason of any alterations having been made in the

---

[22] Dymoke v. Association for Dance Movement Psychotherapy UK Ltd., [2019] EWHC 94 (QB).

[23] Chiemgauer Membran und Zeltbau GmbH (formerly Koch Hightex GmbH) v. New Millennium Experience Co. Ltd. (Formerly Millennium Central Ltd.) No. 2, [2000] 12 WLUK 440.

original specifications, drawings, designs and instructions which shall involve any curtailment of the work as originally contemplated.

The clause contains a proviso entitling the contractor to set-off "the charges on the cartage only
of materials actually and bona fide brought to the site of work" and the cost of the materials at
"their purchase price or local current rates whichever may be less".

The above provision is inserted in the standard form with a view to protect the government
from claims of compensation by the contractor in the event there is breach of contract by the
government. In other words, it is admitted that but for this provision the contractor would be
entitled to recover compensation under Section 73. It means that this is an exemption clause.

Validity of Exclusion Clauses

If fundamental breach is established, the next question is what effect, if any, that has on the
applicability of other terms of the contract. This question has often arisen with regard to
clauses excluding liability, in whole or in part, of the party in breach. The Supreme Court has
observed:

> [W]here the innocent party has elected to treat the breach as a repudiation, bring the
> contract to an end and sue for damages, then the whole contract has ceased to exist,
> including the exclusion clause, and I do not see how that clause can then be used to
> exclude an action for loss which will be suffered by the innocent party after it has ceased to
> exist, such as loss of the profit which would have accrued if the contract had run its full
> term.[24]

The validity of an exemption clause can also be tested on the background of Section 23 of the
Indian Contract Act, which reads:

> *What considerations and objects are lawful, and what not*
> The consideration or object of an agreement is lawful, unless it is forbidden by law or is of
> such a nature that, if permitted, it would defeat the provisions of any law; or is fraudulent; or
> involves or implies injury to the person or property of another; or the Court regards it as
> immoral, or opposed to public policy. In each of these cases, the consideration or object of an
> agreement is said to be unlawful. Every agreement of which the object or consideration is
> unlawful is void.

The provisions of Section 23 show that the consideration or object of an agreement is
unlawful if it is of such a nature that, if permitted, it would defeat the provisions of any law
and the said agreement is void. Sections 55 and 73 of the Contract Act provide for
compensation to be paid by the party committing breach of contract to the injured party. In
a decision of the Delhi High Court an important issue of far-reaching consequences was
raised and answered thus.[25]

The Issue: Can contractual clauses disentitle a person from claiming damages which a person is
otherwise entitled to under law; and whether the rights created by Sections 73 and 55 can or
cannot be contractually waived?

The court was dealing with Clauses 11(A) to (C) of MES standard form IAFW 2249 that
provide for an extension of time and, specifically, that even if there is a breach of contract by
the employer/Union of India, in causing delays by its own faults damages cannot be claimed
by a contractor and such damages cannot be awarded by the arbitrator. It was noted that the
very same clauses were dealt with by the Apex Court in two different cases. In the first case,

---

[24] Union of India & Ors. v. Sugauli Sugar Works (P) Ltd., (1976) 3 SCC 32.
[25] Simplex Concrete Piles (India) Ltd. v. Union of India, 2010 (1) CTLJ 255 (Delhi).

decided in 2007,[26] the provisions were held to be valid and binding on the contractor. In a later decision it was held that the said provisions prevent only the department from granting damages, but the provisions do not prevent the arbitrator from granting damages which are otherwise payable by the employer on account of its breach of contract.[27]

The question was raised as to what the High Court should do in such a situation. Placing reliance on a decision of the full bench of the Patna High Court,[28] and various other High Courts' decisions, including the decision of the Apex Court[29] wherein it was held that it is expected of the High Court to decide the case on merit according to its own interpretation of judgment, it was held that the contract clauses in question would be void, being violative of Section 23 of the Contract Act. Reliance was placed on the Apex Court's decision holding that a clause in a contract cannot prevent the award of damages although the same are otherwise payable in law.[30]

The authors, with respect, state that simple justice demanded that the exclusion clauses be declared void under Section 23 of the Contract Act, in the absence of specific provisions like the Unfair Contract Terms Act, enacted in the United Kingdom and given the imbalance of bargaining positions in India when contractors deal with government departments on standard form contracts that tend to be one-sided. There can be an argument, to oppose the judgment that the contractor in such a case is deemed to have accounted for the contingency and accordingly tendered the cost/rates. It can be countered by the argument that the parties in the contract agree to the terms of the performance of the contract and not as to consequences of its breach. It is unreasonable to expect a party to estimate the nature and gravity of the breach likely to be committed by the other party to the contract in future, and to assess its impact on the contract price and to include it in the tendered sum. In an open competitive bidding it is not likely to be done and it is generally not done. In the absence of a direct authority like the above, justice was aimed at by pressing into service the rules of strict interpretation.

EXAMPLE FROM CASE LAW

In a case decided by the Delhi High Court, a limited legal issue arose for determination as regards the entitlement claimed by the appellant/defendant to not pay interest in view of Clause 9 of the contract which provided: "No interest shall be payable to the contractor in case of delay in payment on account of non-availability of fund in the particular head of account of MCD [Municipal Corporation of Delhi]". It was observed:

> A clause in a contract which disentitles payment of interest, although payment is made with delay, has been the subject matter of a decision by this Court in the case of Union of India v. M/s N.K. Garg & Co. OMP No. 327/2002 decided on 2nd November 2015: 2015 (224) DLT 668.... In the said detailed judgment this Court has held by reference to the provisions of section 23 of the Indian Contract Act, 1872 as also provisions of the Interest Act, 1978 and judgments of the Supreme Court, that a clause in a contract whereby a guilty party refuses to pay interest for delayed payment is illegal and is hit by section 23 of the Indian Contract Act.[31]

[26] Ramnath International Construction (P) Ltd. v. Union of India, (2007) 2 SCC 453 = 2006 SCACTC 862(SC) = 2006(4) Arb. LR 385 (SC).
[27] Asian Techs Limited v. Union of India and others, (2009) 10 SCC 354.
[28] Amar Singh Yadav v. Shanti Devi and Ors., AIR 1987 Pat. 191.
[29] Indian Petrochemicals Corp. Ltd. v. Shramik Sena, AIR 2001 SC 3510.
[30] G. Ramchandra Reddy v. Union of India, (2009) 6 SCC 414.
[31] North Delhi Municipal Corporation v. Prem Chand Gupta, RFA Nos. 623/2017 and 628/2017.

The noteworthy observations in the Delhi High Court judgment, it is submitted are:

> ... (ii) In addition to the example given above which shows that over a period of time actually a person ends up receiving much lesser than what the person actually should have received, there is the additional fact that whatever moneys which may come to the hands of a person after many years and decades will also be along with fall in the value of money i.e. purchasing power of money due to inflation. Also depreciation of the Rupee is a well known fact. (iii) It is further to be also noted that if a person does not receive his moneys on time, then, for the purpose of running of his business enterprise he will necessarily have to borrow moneys from someone and for which borrowing of moneys he will have to pay interest. This rate of interest definitely will be higher than the rate of interest payable on fixed deposit by nationalised banks. Therefore there is not only a double whammy of receiving a small percentage along with fall in money value, but a triple whammy upon the person who is entitled to the payment moneys on time i.e. firstly, a person ends up getting lesser percentage of money every year than what he ought to have got, even that lesser percentage of money is received with a lesser purchasing power and thirdly the person who is entitled to the moneys is forced to pay moneys in the form of interest by borrowing moneys to keep his enterprise going. (iv) The illegal retention of moneys is in fact in my opinion even a quadruple whammy because an aggrieved person suffers the distress of knowing and understanding that the person who is illegally and unlawfully retaining his moneys, uses such moneys either in investment or to grow his business/enterprise, and which otherwise have benefited the aggrieved person if he would have received the moneys due on time.[32]

It was finally held:

> The above referred prohibitory clause came for consideration before this Court in a number of judgments and in all those judgments, this Court took a view that the arbitrator was justified in awarding interest on the amount withheld by the Department. Reference is made to a Division Bench judgment of this Court in FAO (OS) No. 187/2006 titled Union of India v. Pradeep Vinod Construction Company decided on 3rd April 2006. At this stage, ... , the learned counsel appearing on behalf of the petitioner has pointed out that the Railways have taken the above referred judgment of Division Bench of this Court in Special Leave Petition before the Hon'ble Supreme Court and according to him the Special Leave Petition has been admitted for hearing and in that Special Leave Petition, the Hon'ble Supreme Court has stayed the payment of interest. Having regard to the facts that the Hon'ble Supreme Court has stayed the payment of interest in Pradeep Vinod Construction's Case, the award of interest by the learned arbitrators in the present case is upheld subject to the condition that the interest shall be paid to the respondent depending on the final outcome of the above mentioned case pending in the Supreme Court.[33]

---

[32] Ibid., para. 14.

[33] North Delhi Municipal Corporation v. Prem Chand Gupta, AIR 2017 Del. 171. Single judge's decision in the same case, cited by the division bench. Also see: M/s. D. Pal and Company Engineers and Contractors and General Order Suppliers v. South Delhi Municipal Corporation, (Delhi): 2018(1) R.A.J. 157.

12.8.2 Clause 13 of the CPWD Standard Form

According to the relevant rules of construction,[34] for an exemption or exclusion clause, such as Clause 13 of the CPWD Standard Form, reproduced above at the start of the discussion, to become operative, the following conditions have to be satisfied:

1. "If at any time *after the commencement of the work*" (emphasis supplied) are the opening words which will make this clause applicable only if the contractor has already commenced the work and not before. As such the government cannot successfully resist the claims of the contractor if the termination order is issued after making of the contract but before commencement of the work by the contractor. The proviso read with this opening clause supports this contention, it is submitted.

2. Second, the next few words, "the President of India shall for any reason whatsoever not require the whole (work) thereof as specified in the tender to be carried out", suggest that this clause will not cover the cases where the project or work as a whole is abandoned or the contract in its entirety is terminated. The paragraph heading "No compensation for alteration in or restriction of work to be carried out" supports this view inasmuch as the clause deals with "restriction of work" and not its total abandonment. The words "which (profit or advantage) he did not receive in consequence of the full amount of work not having been carried out" suggest that the clause contemplates a situation when the work in part is carried out by the contractor.

3. Third, the words "not require the whole thereof as specified in the tender" will make this clause applicable to cases where "for alteration in" (the words in the paragraph heading) the work as specified in the tender, the work is proposed to be deleted from the scope of the contract. Even part of the work should (i) either be *bona fide* not required at all or (ii) involve major changes resulting in curtailment of the contract work.

The contractor, in the light of the above discussion, will be entitled to receive compensation, in spite of Clause 13, when:

1. The contract is terminated before he has commenced the work, or
2. After he has commenced the work, the whole of the work is abandoned and the contract is terminated, or
3. Part of the work is deleted from the scope of his contract for getting it done through another agency or departmentally without any alterations,[35] or
4. The curtailment in the work is not as a result of any *bona fide* alterations in the original specifications, drawings, designs and instructions.

In other words the clause will be applicable to *bona fide* curtailment of the work due to:

(i) part of the work not at all required to be executed or
(ii) any alterations having been made in the original specifications, drawings, designs and instructions.

Reference to a few other cases decided under the Arbitration Act needs to be made in which the award was set aside in view of the agreement containing prohibitive conditions denying compensation or escalation in cost.

---

[34] See Chapter 4.
[35] Gallagher v. Hirsch, (1899) N.Y. 45 App. Div. 467. N.Y. Sup. 61, 607; Carr v. J.A. Berriman Pty. Ltd., (1953) 27 A.L. J. 273 (Australia). See *Hudson's B. C.*, 10th ed., pp. 532, 533.

EXAMPLES FROM CASE LAW

1. Associated Engineering Co. v. Government of Andhra Pradesh[36] – In this case, award in respect of four claims, not payable under the contract, was set aside holding that a conscious disregard of the law or the provisions of the contract from which the arbitrator had derived his authority vitiates the award.
2. Steel Authority of India Ltd. v. J.C. Budharaja[37] – In this case, Clause 32 specifically stipulated that no claim whatsoever for not giving the possession of the entire site on award of contract will be tenable and the contractor would have to arrange his working programme accordingly. The award granting compensation was set aside, holding that it was not open for the arbitrator to ignore the said conditions. The distinguishing factor in the said case was after the claim was raised a supplementary agreement was signed by the parties for the same work at increased rates. In another case where such supplementary agreement was signed with the knowledge that there was a claim for damages but the agreement only extended the time, the Kerala High Court upheld the decision of the trial court that by accepting the final bill there was no estoppel and the contractor's claim for compensation due to delay on the part of the employer in performing its obligations was justified.[38]
3. State of Orissa v. Sudhakar Das[39] – The agreement did not contain an escalation clause and yet the arbitrator awarded escalation. It was held that the award was not sustainable and suffered from a patent error. It is not clear if the escalation awarded was by way of compensation for the delay or not.
4. National Building Construction Corporation Ltd. v. Décor India Pvt. Ltd.[40] – In this case, the tender was submitted with a letter stating that the rates were valid for 30 days and thereafter the right was reserved to revise them according to the market conditions. The said condition did not form part of the agreement subsequently signed by the parties. The agreement did not contain an escalation clause, rather Clauses 18 and 32 clearly stipulated that no escalation should be given. The award allowing escalation was set aside relying on the authority of State of Orissa v. Sudhakar Das.
5. D.S.A. Engineers v. Housing and Urban Dev. Corporation[41] – The agreement contained provisions for extension of time if the site was not made available in accordance with the agreed programme or for *force majeure* or suspension of work by the employer, but clearly disentitled the contractor from making any clams of compensation whatsoever on the said counts. The award allowing compensation towards extra overheads was set aside. The contention that the prohibition would not be valid for extension beyond 20 to 25% of the time allowed was rejected by holding that the clause cannot be said to be qualified by the words "reasonable period".

The authors, with respect, defer from the above view. Interpreting or construing a contract quite often involves inserting or reading words not used by the drafters.[42] One

---

[36] Associated Engineering Co. v. Government of Andhra Pradesh, JT 1991 (3) SC 123 = 1991(2) Arb. LR 180 (SC); also see The New India Civil Erectors (p) Ltd. v. Oil and Natural Gas Corporation, JT 1997(2) SC 633 = 1997(1) Arb. LR 292 (SC); V.G. George v. Indian Rare Earths Ltd., JT 1999(2) SC 629 = 1999(2) Arb. LR 47(SC); Rajasthan State Mines and Minerals Ltd. v. Eastern Engineering Enterprises and another, AIR 1999 SC 3627 = 1999(3) Arb. LR 350 (SC). M/s Crompton Greaves Ltd. v. Dyna., AIR 2007 (NOC) 2121(Mad.).

[37] Steel Authority of India Ltd. v. J.C. Budharaja, (1999) 8 SCC 122 = 1999(3) Arb. LR 335 (SC).

[38] State of Kerala v. Mohammed Kunju, 2009(1)CTLJ 207 (Kerala) (DB).

[39] State of Orissa v. Sudhakar Das (2000) 3 SCC 27 = 2000(1) Arb. LR 444 (SC).

[40] National Building Construction Corporation Ltd. v. Décor India Pvt. Ltd., 2004 (2) Arb. LR 1 (Delhi).

[41] D.S.A. Engineers v. Housing and Urban Dev. Corporation, 2004 (2) Arb. LR 33 (Delhi).

[42] See Doctrine of Presumed Intent in Chapter 4.

must not overlook the fact, while interpreting the construction contracts, that the cost/rates tendered have a definite relationship with the time limit stipulated for completion of the work. The cost/rates cannot be said to be valid for an indefinite period and, as such, when exemption clauses such as those in the case under study do not stipulate a limit on the extension of time, it is necessary to interpret that the parties agreed for a reasonable extension of the originally stipulated period beyond which the exemption clause would not be applicable. Whether 20% or 25% can be considered as a reasonable extension of time may be debatable, but to hold that the clause is not qualified at all is erroneous, it is submitted. Courts have always held that what is "reasonable" is a question of fact to be decided in each case.

6. Osnar Paints & Contracts Pvt. Ltd. v. NBCC[43] – Three contracts for painting 665 houses contained the provision under the heading "Rates to be firm" which stipulated that the painting contractor would have no right to claim any increase in the agreed contract rates because of increase in the cost of living, or increase in price of oil or any other increase whatsoever. The delay in completion of the work was alleged to be on the part of the contractor. The award allowing compensation was set aside.

7. State of Kerala v. Mathai[44] – In this case, decided under the 1940 Act, an award allowing enhanced rates on extra items of work by holding that the supplementary agreement containing prohibitive condition was signed under compulsion of the department, was set aside. It was held that once the agreement had been entered into, the arbitrator was not called upon to adjudicate as to the circumstances under which the parties had come to the agreement; that was a matter for an appropriate court to consider.

It is to be noted that under the 1996 Act, the arbitrator is empowered to decide the validity of the agreement and this decision will be of doubtful utility to cases to be decided under the 1996 Act, it is respectfully submitted.

8. Union of India v. Mohanlal Harbanslal Bhyana & Co.[45] – An award granting revision of rates for the work done when the work could not be completed within the stipulated period was set aside on the ground that the agreement Clause 10(CC) provided for escalation and the arbitrator could not ignore the said clause and award escalation over and above the escalation paid and extension of time granted. Reliance was placed on the Division Bench decision in Delhi Development Authority v. K.C. Goyal & Co. and Delhi Development Authority v. U. Kashyap.[46]

The authors, with respect, submit that the division bench decision in Metro Electric & Co. v. Delhi Development Authority[47] had laid down good law by holding that Clause 10 (C) incorporated in the agreement contemplates completion of the work within the agreed time schedule and if the work prolongs beyond the completion date so stipulated, the clause is not attracted to prohibit payment of compensation for breach of contract, if the actual loss is in excess of the escalation payable under Clause 10(C). Clause 10(CC) also stipulates payment of escalation "only for the work done during the stipulated period of the contract including such period for which the contract is validly extended". Of course, as stated above, if the extended period vis-à-vis the originally stipulated period is "reasonable" the claim may not be tenable.

---

[43] Osnar Paints & Contracts Pvt. Ltd. v. NBCC, 2002 (3) Arb. LR 653 (Delhi).

[44] State of Kerala v. Mathai, 2003(3) Arb. LR 29 (Kerala) (DB).

[45] Union of India v. Mohanlal Harbanslal Bhyana & Co., 2002 (Suppl) Arb. L.R. 474 (Delhi).

[46] Delhi Development Authority v. K.C. Goyal & Co., 2001 II AD (Delhi) 116; and Delhi Development Authority v. U. Kashyap 1999 (1) Arb. LR 88 (Delhi).

[47] Metro Electric & Co. v. Delhi Development Authority, AIR 1980 Del. 266.

12.8.3 "No Damage" Clauses, an American Expression – Not Applicable under Indian Law

The above decisions are based on express contract stipulations in age-old contract forms and the courts have interpreted them strictly. The contract terms are being revised and the courts too are accepting the reality of the situation resulting in injustice to an injured party. The law in respect of exclusion of right or no damage clauses has been analysed and laid down in the decisions of the Apex Court in the case of General Manager, Northern Railway v. Sarvesh Chopra.[48] In paragraphs 14 and 15, their Lordships expressed their view as under:

> 14. In Hudson's Building and Engineering Contracts (11th Edn. Pp. 1098–1099) there is reference to, "no damage" clauses, an American expression, used for describing a clause which classically grants extension of time for completion for variously defined delays including some for which as breaches of contract on his part, the owner would prima facie be contractually responsible, but then proceeds to provide that the extension of time so granted is to be the only right to remedy of the compensation are not to be recoverable therefore. These damages clauses appear to have been primarily designed to protect the owner from late start or coordination claims due to other contractor delay, which would otherwise arise. Such clauses originated in the federal Government contracts but are now adopted by private owners and expanded to cover wider categories of breaches of contract by the owner institution which it would be difficult to regard as other than oppressive and unreasonable. American jurisprudence developed so as to avoid the effect of such clause and permitted the contract to claim in four situations namely (i) where the delay is of a different kind from that contemplated by the clause, including extreme delay, (ii) where the delay amounts to abandonment (iii) where the delay is a result of positive acts of interference by the owner, and (iv) bad faith. The first of the said four exceptions have received considerable support from judicial pronouncement in England and common-wealth. Not dissimilar principles have enabled some commonwealth Court to avoid the effect of "no damage" clauses.

The Apex Court of India further observed:

> 15. In our country question of delay in performance of the contract is governed by Sections 55 and 56 of the Indian Contract Act, 1872. If there is an abnormal rise in prices of material and labour, it may frustrate the contract and then the innocent party need not perform the contract. So also, if time is the essence of the contract, failure of the employer to perform a mutual obligation would enable the contract to avoid the contract as the contract become voidable at his option where "time is of essence" of an obligation, Chitty on Contract (Twenty-Eighth Edition, 1999, at p. 1106, para. 22.015) states: "a failure to perform by stipulated time will entitle the innocent party to (a) terminate performance of the contract and thereby put an end to all the primary obligations of both parties remaining unperformed; and (b) claim damages from the contract-breaker on the basis that he has committed a fundamental breach of the contract (a breach going to the root of the contract) depriving the innocent party of the benefit of the contract ('damages for loss of the whole transaction')." If, instead of avoiding the contract, the contractor accepts the belated performance of reciprocal obligation on the part of the employer, the innocent party i.e. the contractor, cannot claim compensation for any loss occasioned by the non-performance of the reciprocal promise by the employer at the time agreed, "unless at the time of such acceptance, he gives notice to the promisor of his intention to do so". *Thus it appears that under the Indian law in spite of there being a contract between the parties whereunder the contractor had undertaken not to make any claim for delay in performance of the contract occasioned by an act of the employer, still*

---

[48] General Manager, Northern Railway v. Sarvesh Chopra, (2002) 4 SCC 45: (AIR 2002 SC 1272).

a claim would be entertainable in one of the following situations: (i) if the contractor repudiates the contract exercising his right to do so under Section 55 of the Contract Act, (ii) the employer gives an extension of time either by entering into supplemental agreement or by making it clear that escalation of rates or compensation for delay would be permissible, (iii) if the contractor makes it clear that the escalation of rates or compensation for delay shall have to be made by the employer and the employer accepts performance by the contractor in spite of delay and such notice by the contractor putting the employer on terms.

(Emphasis supplied)

In paragraph 16 their Lordships have further held:

Thus, it may be open to prefer a claim touching an apparently excepted matter subject to a clear case having been made out for excepting or excluding the claim form within the four corner of expected matters.[49]

Relying on the above decision, the division bench of the M.P. High Court held:[50] "The owner allowed the claimant to carry on with the work ... As performance has been accepted we are disposed to think ... the third para of Clause 2.1.22[51] would not be a impediment for entertaining his claim." The award granting relief was upheld.

The division bench of the Orissa High Court held in a case that endorsement made by the contractor while seeking extension of time for completion of work that he will not claim any compensation on account of escalation would not debar him from claiming increased labour cost on account of statutory increase in minimum wages.[52]

12.9 Delay in Completion/Termination of Contract by the Owner

Experience shows that the majority of the cases which lead to arbitration and/or court proceedings arise out of the facts which develop somewhat as under:

   (i) Due to defaults committed by the owner amounting to breaches of contract, the completion of the work gets delayed, or is likely to be delayed.

  (ii) The breach, if of a serious nature, the contractor treats himself discharged from the further performance of the work or repudiates the contract himself.

 (iii) The owner after granting an extension of time or without granting it wrongfully and illegally terminates the contract.

 (iv) The contractor completes the work or pending completion of the work commences legal proceedings for recovery of damages caused by the delay in completion of the work.

In all the above cases, if it is established that the owner has committed a breach of contract which goes to the root of the contract and, as a result, the contract is terminated by either party, or if the contractor completes the work and stakes his claim for compensation, the owner will be liable to pay compensation to the contractor.[53] Some standard form contracts will require the

---

[49] General Manager, Northern Railway v. Sarvesh Chopra, (2002) 4 SCC 45: (AIR 2002 SC 1272).

[50] Mintoolal Brijmohandas, M/s. v. State of Madhya Pradesh, AIR 2005 M. P. 205.

[51] If the materials are not supplied in time, the Contractor will not be allowed any claim for any loss which may be caused to him, but only extension of time will be given at the discretion of the Executive Engineer and Superintending Engineer if applied for by the Contractor before the expiry of the contract.

[52] Suryamani Nayak v. Orissa State Housing Board, AIR 2005 Ori. 26.

[53] D.D. Sharma v. Union of India, 2004(2) Arb. LR 119 (SC).

contractor to give prompt notice to the employer of any delay events that may give rise to a claim for money. The contract should be checked for any such provisions and their instructions followed carefully. This would avoid a failure of a claim simply on the basis that a prerequisite to the claim was not observed.[54] Examples of pre-notification include Clause 20.1 of the FIDIC Red Book 1999 and Clause 20.2.4 of the FIDIC Red Book 2017. Even if there is no contractual obligation to do so, it is good practice to keep the employer in the loop about the potential consequences of any ongoing delays and disruptions. This would automatically ensure that the contractor's job of proving the claim will be easier in case of any future litigation. Records may be kept not only in terms of measurements, photographs, documents including invoices and correspondence, site books minuting meetings, and videos.

12.10 The Assessment of Damages Payable to the Contractor

In all such cases, the main question will be as to what amount of compensation the contractor will be entitled to, due to the breach of contract committed by the owner. Damages are likely to be as follows:

1. Loss of overheads and profit:

    (i) During the stipulated period for completion of the work and
    (ii) During the extended period.

2. Direct losses on account of idle labour and/or machinery and equipment reducing productivity

    (i) during the stipulated period, and
    (ii) during the extended period.

3. Increased cost of materials, and labour

    (i) during the stipulated period
    (ii) during the extended period, or in the alternative, the claim for enhancement in the agreed rates for the work done beyond the stipulated period.

As an alternative to all the above claims the contractor may claim payment for the work done on the basis of "*quantum meruit*".

These claims are further explained below.

12.10.1 Loss of Overheads and Profit

For proper appreciation of this claim, a little understanding of the contractor's price build-up is necessary. The cost of any item of work consists of the cost of materials, labour, and machinery, which is called direct cost. After a detailed analysis of each item rate, the prime cost of the whole project is found by multiplying the quantity of work under each item by the rate so worked out. To this prime cost or direct cost is added generally a single percentage for both the overheads and profit. "Overheads" are indirect expenses required to be incurred by the contractor for the execution of the work and are of such nature that they cannot be specifically attributed to any particular item of work or contract. There are literally dozens of items of expenses such as office rent, power bills, salaries of office and supervisory staff, stationery, postage, telephone bills, interest on borrowed capital, travel expenses, insurance, etc. In a similar manner, the costs of small tools and plant such as iron baskets, ladders, rope etc. are also included under this heading.

---

[54] For an example of such a failure see Glen Water Ltd. v. Northern Ireland Water Ltd., [2017] NIQB 20.

Every firm maintains or is expected to maintain the accounts of expenses under this heading. At the end of each financial year, it is, therefore, possible to find out the percentage these costs bear on the total turnover of the firm. For example, if the overheads of a firm in a year amounted to Rs. 9,50,000/- and the total turnover of the firm in that year was Rs. 100,00,000/- the firm worked with overheads amounting to 9.5% of the total turnover.

This percentage figure will vary from firm to firm depending upon the size of the firm and efficiency of its site organization and management.

Loss of Profit

The profit is the only remuneration that a contractor receives for execution of a works contract. It has been said that "the profit is the cost of being in business today and remaining in business tomorrow". The margin of profit a contractor may expect depends upon many factors, such as the nature and volume of the work, works in hand, etc. The percentage figure may vary from 15% on small jobs to 5% on big projects.

The Cost Committee Report published by the Government of India, the Ministry of Irrigation and Power (Central Water and Power Commission) in 1950, after investigation and survey of the various major irrigation projects in India, has laid down and recommended the percentage of overheads and profit to be considered in the estimates to be prepared by the various departments of the Government as below:

> **Overheads**: An allowance of 10% would be adequate for the contractor's actual expense on supervisory establishments, field office and share of head office charges, travelling expenses, insurance of damage to plant and injury to labour.
>
> **Profit**: We believe that in normal circumstances an allowance of 10 per cent of the prime cost as contractor's profit is reasonable.

In view of the above, compensation is often awarded on the basis of 20% allowance for overheads and profit. In exceptional cases, wherein the contractor's tendered rates are higher or lower than the prevailing market rates, the percentage to be adopted would be accordingly higher or lower, depending upon the proof tendered. In an English case, the work was undertaken by a contractor on the basis of a pre-contract letter which set up with a cap of £500,000 that "all direct costs and directly incurred losses shall be underwritten and reimbursed". The parties had intended to use JCT conditions for the contract which, in the event, was not concluded. The contractor demanded payment under the pre-contract letter including overheads and profit. This was resisted on the basis that overheads and profit are not "direct". It was held that overheads and profit were included in the phrase "direct costs and directly incurred losses".[55]

In this respect, reference needs to be made to the decision of the Bombay High Court wherein it was held that when estimated loss in profit was claimed by the contractor *vide* claim no. 1, it was not open for him to again claim losses suffered on account of idle labour and machinery separately under claim no. 2.[56] This proposition may be applicable to the facts of the case before the court and does not amount to general application as a rule of law, it is respectfully submitted. There are, as already stated, four distinct heads of expenses in addition to expected profit, and if the loss under idling labour or machinery is not awarded the profit to the said extent would be reduced.

In a landmark decision, the law has been settled by the Supreme Court of India in Mcdermott International Inc. v. Burn Standard Co. Ltd.[57] It was held:

---

[55] Robertson Group (Construction) Ltd. v. Amey-Miller (Edinburg) Joint Venture, 2006 SCLR 772.

[56] Municipal Corporation of Greater Bombay, Mumbai v. Bharat Construction, a registered firm, Mumbai, (Bombay) (DB): 2017 (4) AIR Bom. R-9; 2017 (5) Arb. LR 472.

[57] Mcdermott International Inc. v. Burn Standard Co. Ltd., AIR 2006 SCW 3276; 2006 (11) SCC 181.

107. A claim for overhead costs resulting in decrease in profit or additional management costs is a claim for damages.

108. An invoice is drawn only in respect of a claim made in terms of the contract. For raising a claim based on breach of contract, no invoice is required to be drawn ... .

110. While claiming damages, the amount therefor was not required to be quantified. Quantification of a claim is merely a matter of proof.

[...]

METHOD FOR COMPUTATION OF DAMAGES

113. The arbitrator quantified the claim by taking recourse to the Emden formula. The learned arbitrator also referred to other formulae, but, as noticed hereinbefore, opined that the Emden Formula is a widely accepted one.

It was finally held:

116. We do not intend to delve deep into the matter as it is an accepted position that different formulas can be applied in different circumstances and the question as to whether damages should be computed by taking recourse to one or the other formula, having regard to the facts and circumstances of a particular case, would eminently fall within the domain of the Arbitrator ... .

119. Sections 55 and 73 of the Contract Act do not lay down the mode and manner as to how and in what manner the computation of damages or compensation has to be made. There is nothing in Indian law to show that any of the formulae adopted in other countries is prohibited in law or the same would be inconsistent with the law prevailing in India.

It was also held that the claim for damages raised by the claimant could not be said to be consequential damages.

Different Formulae

Different formulae have evolved to compute compensation payable to a contractor and are in use. Some commonly used formulae are discussed below.

1. **Hudson formula**: In the Hudson formula, the head office overhead percentage is taken from the contract. Although the Hudson formula has received judicial support in many cases up to the late 1980s,[58] it has been criticized, principally because it adopts the head office overhead percentage from the contract as the factor for calculating the costs, and this may bear little or no relation to the actual head office costs of the contractor. This widely used formula is considered after considering the other formulae with slight variations, namely Emden formula and Eichleay formula.

2. **Emden formula**: In *Emden's Building Contracts and Practice*, the Emden formula is stated in the following terms:

$$\left(\frac{\text{O.H/Profit percentage}}{100}\right) \times \left(\frac{\text{Contract sum} \times \text{Period of delay in weeks}}{\text{Contract period in weeks}}\right)$$

---

[58] For example, see the discussion in Walter Lilly & Company Ltd. v. Giles Patrick Cyril Mackay, DMW Developments Ltd., [2012] EWHC 1773 (TCC).

This formula has been widely applied and has received judicial support in a number of cases decided by the American courts.[59] It is occasionally used in India to cross-check a contractor's claim.[60]

Using the Emden formula, the head office overhead percentage is arrived at by dividing the total overhead cost and profit of the contractor's organization as a whole by the total turnover. This formula has the advantage of using the contractor's actual head office and profit percentage rather than those contained in the contract.

3. **Eichleay formula**: The Eichleay formula also evolved in America and derives its name from a case heard by the Armed Services Board of Contract Appeals, Eichleay Corp.

$$\left(\frac{Total\ Contract\ Billing}{Total\ Company\ Billing\ over\ contract\ period}\right) \times Home\ Office\ O.H.\ for\ contract\ period = Project\ Specific\ H.O.O.H.$$

$$\left(\frac{Project\ Specific\ H.O.O.H.}{Contract\ Performance\ Period}\right) = Daily\ H.O.O.H.\ for\ the\ contract$$

The claim will then be equal to the daily H.O.O.H. rate multiplied by the number of days of delay. This formula is used where it is not possible to prove loss of opportunity and the claim is based on actual cost. It can be seen from the formula that the total head office overheads (H.O.O.H) during the contract period are first determined by comparing the value of work carried out in the contract period for the project with the value of work carried out by the contractor as a whole for the contract period. A share of head office overheads for the contractor is allocated in the same ratio and expressed as a lump sum to the particular contract. The amount of head office overheads allocated to the particular contract is then expressed as a weekly amount by dividing it by the contract period. The period of delay is then multiplied by the weekly amount to give the total sum claimed. The Eichleay formula is regarded by the Federal Circuit Courts of America as the exclusive means for compensating a contractor for overhead expenses. It per se does not compensate for the loss of expected profit. It has been criticized on the basis that it does not even reliably measure the contractor's actual loss over time; it does not include any variable costs at all although in reality a contractor's home office will have some.[61] It requires waiting until the project has been completed before it can be applied. It has also been criticized on the ground that it applies regardless of when the delay takes place; a delay may cause less disruption in the monsoon in India when a contractor is less active anyway than in the summer.

12.10.2 "The Hudson Formula" Critically Examined and Explained

For the assessment of the contractor's losses due to delay, under this head, the formula as stated by Hudson in his book *Building and Engineering Contracts*, 10th edition on page 599 is as follows:

$$\left(\frac{H.O/Profit\ Percentage}{100}\right) \times \left(\frac{Contract\ sum \times Period\ of\ delay\ in\ weeks}{contract\ period\ in\ weeks}\right)$$

[59] Nicon Inc. v. United States, decided on 10th June 2003 (USCA Fed. Cir.); Gladwynne Construction Company v. Balmimore, decided on 25th September 2002; and Charles G. William Construction Inc. v. White, 271 F.3d 1055. Norwest Holst Construction Ltd. v. Cooperative Wholesale Society Ltd., decided on 17th February 1998; Beechwood Development Company (Scotland) Ltd. v. Mitchell, decided on 21st February 2001; and Harvey Shopfitters Ltd. v. Adi Ltd., [2003] EWCA Civ. 1757, decided on 6th March 2003.

[60] National Highways Authority of India v. PNC-BEL(JV), OMP (Comm) 41/2019.

[61] "Home Office Overhead as Damage for Construction Delays", 17 *Georgia Law Review* 761, 7904 (1983).

The above formula is based on the simple rule of three. If, for example, overheads and profit combined contribute Rs. 20/- in the contractor's tendered rate of Rs. 100/-, the contract sum is Rs. 10,00,000/- and the time limit allowed is 40 weeks, the pro-rata amount per week the contractor expects towards overheads and profit from this contract would be:

$$\frac{20}{100} \times \frac{10,00,000}{40} = \text{Rs. } 5000/-$$

This amount represents the loss per week for each week's delay at the site of the work.

Let us consider, that there is a delay of ten weeks in completion. The compensation under this head, according to this formula works out to Rs. 5,000 × 10 weeks = Rs. 50,000. Let us analyse the contractor's financial position considering different situations.

On completion of the work, the contractor had received, considering only this aspect, the losses on the other counts apart, towards overheads and profit at 20% of Rs. 10,00,000/- = Rs. 2,00,000/- plus Rs. 50,000/- extra by way of compensation, totalling to Rs. 2,50,000/-.

Now, it is to be understood that the overheads of an organization are somewhat in the nature of "standing charges" and remain more or less the same irrespective of the actual turnover. The profit-earning capacity also, therefore, must be presumed to remain the same. In other words, in the above example, the contractor is actually capable of getting Rs. 5,000/- per week (say Rs. 2,500/- by way of actual expenses under "overheads" and Rs. 2,500/- by way of profit-earning capacity) and his organization remains at the site for a period of 50 weeks. If he has to be put in the same position, as far as payment of money could do it, as he would have been if the owner had not committed breaches of contract, the amount of compensation as per the principles laid down by the Supreme Court of India[62] would be:

$$\text{Rs. } 5,000 \times \ 50 \text{ weeks} = \text{Rs. } 2,50,000 \text{ Less received in final bill } 20\% \times 10,00,000$$
$$= \text{Rs. } 2,00,000 \text{ Compensation payable} = \text{Rs. } 50,000$$

which is the same amount worked out by the Hudson formula.

Thus the formula holds good for compensating the contractor for the shortfall in progress during the stipulated period and also during the extended period. Similarly, it compensates both the profit actually lost and loss in profitability. The formula will not yield good results if the delay is not proportionate to the balance cost of work on the date of expiry of the original stipulated period as explained below.

The Hudson formula assumes the delay is in proportion to the cost. In a bulk of the cases, it is seen that the actual turnover during the stipulated period is never proportional to the total time taken for actual completion vis-à-vis the time limit stipulated for completion. For example, in the above illustrative example, the actual turnover in the first ten months may be only Rs. 2 lacs or less, or maybe say Rs. 10 lacs, the full contract sum due to extra items and excess quantities with the work incomplete. Also, the balance work sometimes lingers for a period too far in excess of the originally stipulated period or is done more expeditiously than the originally contemplated average speed. The final bill value and the contract sum may also differ considerably. The application of the above formula becomes a little difficult in the light of the defence arguments and evidence available if the relief is to be modified to suit the varying factors.

In practice, while drafting pleadings and arguing cases or giving awards some difficulties are likely to arise. For example, the formula assumes that the profit expected by the contractor in

---

[62] See paragraph 12.7. M/s. Murlidhar v. M/s. Harish Chandra, AIR 1962 SC 366; Union of India & Ors. v. Sugauli Sugar Works (P) Ltd., (1976) 3 SCC 32.

his prices was in fact capable of being earned by him elsewhere had the contractor been free to leave the delayed contract at the proper time. It requires two factors to be established:

(i) The rate tendered by the contractor contained the profit percentage and the percentage was realistic; and

(ii) There was no change in the market, so that the work of at least the same general level of profitability would have been available to him at the end of the contract period.

There is no doubt that satisfactory evidence on these matters is necessary, even if one succeeds in persuading that the loss is not remote and imaginary. More often than not it is difficult to establish the second one at least, of the above factors. If an arbitrator or court were to reject the claim for profitability, there is a chance that the formula would be adopted eliminating the profit margin altogether whereby the claim is likely to be upheld up to 50%, in which event the loss in the profit margin during the stipulated period, which can be easily established as the direct loss, is also likely to be denied to the contractor.

English courts have observed that the Hudson formula is problematic in terms of the causation link beween the owner's wrongful act/omission and the contractor's loss.[63] English courts expect that there should be "no material causative factor for which the defender is not liable".[64] The use of a formula is accepted as a legitimate mode of ascertaining the contractor's entitlement to loss of overheads and profit so long as, on a balance of probabilities, the contractor can show that the loss being allowed was incurred as a result of the delay or disruption for which the defendant is responsible. The controversy and confusion tend to be associated with "total" or "global" claims. Sometimes it may involve disproportionately expensive proof, given that several factors may cause delays and losses. Ramsey, J. recognized the times when a global claim may have to be advanced in the following words: "The essence of a global claim is that, whilst the breaches and the relief claimed are specified, the question of causation linking the breaches and the relief claimed is based substantially on inference, usually derived from factual and expert evidence."[65] This also shows why sometimes an expert engineer's presence on a tribunal is helpful. A global approach to a claim may be justified in some circumstances but should not be chosen simply as a short-cut to save costs or preparation time for the lawyers.

The author has tried, tested and developed the following form based on the universally accepted principles of ascertaining compensation payable, which permits partial reliefs to be granted on the basis of actual proof tendered and the facts and circumstances of the case, with ease and without the possibility of any error creeping into the judgment. It also incidentally incorporates the advantages of all the three accepted formulae mentioned above.

12.10.3 Suggested Basis: "Patil Form"

The principle used is the one generally accepted by the judiciary, including the Supreme Court of India,[66] namely, ascertainment of that sum of money which, when paid by the owner to the contractor, would place the contractor in the position in which he would have been if the project would have been completed within the stipulated period, as far as overheads and profit elements are concerned.

---

[63] John Holland Construction & Engineering Pty Ltd. v. Kvaerner RJ Brown Pty Ltd., (1996) 82 BLR 81, at 85E. Also see Bernhard's Rugby Landscapes Ltd. v. Stockley Park Consortium Ltd., (1997) 82 BLR 39.

[64] Walter Lilly & Co. Ltd. v. Giles Patrick Cyril Mackay, DMW Developments Limited, [2012] EWHC 1773 (TCC), para. 37.

[65] London Underground v. Citylink Telecommunications, [2007] All ER (D) 318 (Jul.).

[66] See Paragraph 12.7: *Anson's Law of Contract*, 26th ed., p. 494; M/s. Murlidhar v. M/s. Harish Chandra, AIR 1962 SC 366; Union of India & Ors. v. Sugauli Sugar Works (P) Ltd., (1976) 3 SCC 32.

ILLUSTRATIVE EXAMPLE

(A) Compensation for the stipulated period

(1) Contract sum	Rs. 10,00,000/-
(2) Prime cost	Rs. 8,00,000/-
(3) Overheads & profit	Rs. 2,00,000/-
(4) Overheads & profit percentage in the contract sum	20%
(5) Time limit	10 months
(6) Value of the work done within the stipulated period	Rs. 3,50,000/-
(7) Pro-rata overheads & profit deemed to be received: 20% of (6) Rs. 3,50,000/-	Rs. 70,000/-
(8) Net loss suffered as on the date stipulated for completion **(3) – (7), i.e. 2,00,000 – 70,000**	**Rs. 1,30,000/-**

(B) Compensation for the extended period

(9) Rate of overheads per month (3) ÷ (5)	Rs. 20,000/-
(10) Delay in months, say	8 months
(11) Amount pro-rata due for the extended period (9) × (10)	Rs. 1,60,000/-
(12) Value of the work done in the extended period (including extra work)	Rs. 8,00,000/-
(13) Pro-rata overheads & profit deemed to be received: 20% of (12)	Rs. 1,60,000/-
(14) Net loss suffered during the extended period	
(11) – (13)	**Rs. NIL**
(15) Grand total of loss incurred in the total period: **(8) + (14)**	**Rs. 1,30,000/-**

Check the above evaluation by the Hudson formula. Compensation for 8 months' delay will work out to Rs. 1,60,000/- as against Rs. 1,30,000/- under the Patil Form. The difference is due to the execution of the extra work valued at Rs. 1,50,000/- during the extended period. If due to the curtailment of the scope of work, the final value of the work is reduced to say, Rs. 8,50,000/-, the loss suffered in the extended period in the above example would be:

(12) Value of the work executed in the extended period: Rs. 8,50,000 – Rs. 3,50,000	Rs. 5,00,000/-
(13) Pro-rata overheads & profit received: 20% of Rs. 5,00,000/-	Rs. 1,00,000/-
(14) Loss suffered during the extended period (11) – (13): (Rs. 1,60,000 – Rs. 1,00,000)	Rs. 60,000/-
(15) Grand total of loss suffered during the completion period: (= Rs. 1,30,000 + Rs. 60,000)	Rs. 1,90,000/-

This is against Rs. 1,60,000/- evaluated by the Hudson formula. The Patil Form has the advantage of no separate claim being required to be raised for curtailment of the scope of the work. It is possible to establish each of the elements reflected in the above analysis by tendering appropriate proof, and the method is flexible enough to permit correct evaluation wherever differing elements enter into the analysis or the proof falls short and a part of the claim is to be rejected on that ground or on any other count. This method would help eliminate the stock argument that assessment of loss is done on the basis of ad hoc formula without due consideration of the evidence on record or without there being any evidence introduced at all.

This form has a further advantage inasmuch as in a case where the owner wrongfully and illegally terminates the contract immediately after the stipulated date for completion of the work, the contractor will be entitled to compensation worked out by the above method up to the date of completion originally stipulated in the contract. For use of the Hudson formula in such a case, pro-rata delay has to be assumed.

Loss of Profit

This is a common head of claims in construction contracts involving allegations of breach on the part of the owner of a project. The Supreme Court of India has laid down the law in the following words:[67]

> It was not disputed before us that where in a works contract, the party entrusting the work commits breach of the contract, the contractor would be entitled to claim damages for loss of profit which he expected to earn by undertaking the works contract. What must be the measure of profit and what proof should be tendered to sustain the claim are different matters. But the claim under this head is certainly admissible ... What would be the measure of profit would depend upon the facts and circumstances of each case. But that there shall be a reasonable expectation of profit is implicit in a works contract and its loss has to be compensated by way of damages if the other party to the contract is guilty of breach of contract cannot be gainsaid.

EXAMPLES FROM CASE LAW

1. The facts as stated and the findings thereon by the Supreme Court of India in the Dwaraka Das judgment are as follows:

> The claim of the petitioner for payment of Rs. 20,000/- as damages on account of breach of contract committed by the respondent-State was disallowed by the High Court as the appellant was found to have not placed the material on record to show that he had actually suffered any loss on account of the breach of contract.

The High Court had disallowed the claim for the contractor had not shown that he lost money paid to labourers or for materials on account of the breaches of contract. No evidence was tendered for the loss of profit claimed; instead, the contractor relied on the assessment made by the income tax officer. The Supreme Court held:

> The appellant had never claimed Rs. 20,000/- on account of alleged actual loss suffered by him. He had preferred his claim on the ground that had he carried out the contract he would have earned profit of 10% on Rs. 2 lacs which was the value of the contract. This Court in A.T. Brij Pal Singh and others v. State of Gujarat, [1984]

---

[67] M/s. A.T. Brij Paul Singh & Bros. v. State of Gujarat, AIR 1984 SC 1703 = (1984) 4 SCC 59.

(4) SCC 59 while interpreting the provisions of Section 73 of the Contract Act, has held that damages can be claimed by a contractor where the Government is proved to have committed breach by improperly rescinding the contract and for estimating the amount of damages court should make a broad evaluation instead of going into minute details.

It was specifically held that where in the works contract, the party entrusting the work committed breach of contract, the contractor is entitled to claim the damages for loss of profit which he expected to earn by undertaking the works contract. Claim of expected profits is legally admissible on proof of the breach of contract by the erring party. The Supreme Court, in the A.T. Brij Pal Singh case, observed:[68]

> What would be the measure of profit would depend upon facts and circumstances of each case. But that there shall be a reasonable expectation of profit is implicit in a works contract and its loss has to be compensated by way of damages if the other party to the contract is guilty of breach of contract cannot be gainsaid. In this case we have the additional reason for rejecting the contention that for the same type of work, the work site being in the vicinity of each other and for identical type of work between the same parties, a Division Bench of the same High Court has accepted 15 per cent of the value of the balance of the works contract would not be an unreasonable measure of damages for loss of profit. ... Now if it is well-established that the respondent was guilty of breach of contract inasmuch as the rescission of contract by the respondent is held to be unjustified, and the plaintiff-contractor had executed a part of the works contract, the contractor would be entitled to damages by way of loss of profit. Adopting the measure accepted by the High Court in the facts and circumstances of the case between the same parties and for the same type of work at 15 per cent of the value of the remaining parts of the work contract, the damages for loss of profit can be measured.[69]

The appellate court refused to interfere with the finding of facts by the trial court even if the quantum of damages awarded was based on guesswork. In two cases,[70] 15% of the contract price was awarded to the contractor.

Having considered these cases, the Supreme Court in the Dwaraka Das case held:

> In the instant case however the trial court had granted only 10% of the contract price which we feel was reasonable and permissible, particularly when the High Court had concurred with the finding of the trial court regarding breach of contract ... It follows therefore as and when the breach of contract is held to have been proved being contrary to law and terms of the agreement, the erring party is legally bound to compensate the other party to the agreement. The appellate court was, therefore, not justified in disallowing the claim of the appellant for Rs. 20,000/- on account of damages as expected profit out of the contract which was found to have been illegally rescinded.

2. In A.T. Brij Pal Singh and others v. State of Gujarat,[71] a contract for providing a cement concrete surface for a 22 mile stretch of the then existing road, a tender 7%

---

[68] [1984] 4 SCC 59.

[69] Also see, Mohd. Salamatullah and others v. Government of Andhra Pradesh, AIR 1977 Supreme Court 1481.

[70] Mohd. Salamatullah and others v. Government of Andhra Pradesh, AIR 1977 Supreme Court 1481; A.T. Brij Pal Singh and others v. State of Gujarat, 1984(4) SCC 59.

[71] A.T. Brij Pal Singh and others v. State of Gujarat, 1984(4) SCC 59.

lower than the estimated cost was accepted for a sum of Rs. 16,59,900/-. The time limit for completion of the work was 14 months from the date fixed by the written order to commence the work. The contractor commenced the work, completed sub-grade work over five miles and concrete was laid for 2 1/2 miles. Certain disputes arose between the parties as a result of which the state rescinded the contract, imputing that as time was of the essence of the contract and as the contractor failed to execute the work within the stipulated time, he was guilty of committing a breach of the contract. The contractor, after accepting payment of the final bill under protest, filed a suit for recovery of damages on various counts including cost of alleged extra work done. The trial court held that although a witness on behalf of the contractor had orally given the minutest details and measurements of the work executed by the contractor, no documentary evidence was produced to substantiate the claim and, therefore, the plaintiff contractor failed to prove the damages as claimed. On appeal, the High Court reversed the findings of the lower court on the question of justification of rescission of the contract by the state, examined the principal contention whether the contractor was entitled to recover damages for expected profits and rejected the claim for want of proof. The Supreme Court of India held that the state was guilty of breach of the works contract, part of which was already performed and as such the contractor would be entitled to damages. The Supreme Court of India observed:

> In our opinion, while estimating the loss of profit that can be claimed for the breach of contract by the other side, it would be unnecessary to go into the minutest details of the works contract. A broad evaluation would be sufficient. The law does not need absolute certainty of data upon which lost profits are to be estimated. All that is required is such reasonable certainty that damages may not be based wholly upon speculation and it is sufficient if there is a certain standard or fixed method by which profits sought to be recovered may be estimated with a fair degree of accuracy.

The contractor was allowed 15% compensation for loss of profit of the balance work.[72]

3. In a case of wrongful rejection of the lowest tender and award of contract to another party was held illegal and the contract was not terminated since the agency had already commenced the work, the party to whom the contract was awarded was directed to pay a part of profit, namely Rs. one crore (which would be less than 10 per cent of the profit likely to be earned), to the wronged lowest tenderer by way of compensation.[73]

Awards of 10% of the contract sum for loss of profit on account of illegal termination of contract have also been upheld by Madras, Andhra Pradesh, Bombay and Kerala High Courts.[74]

12.10.4 Loss Due to Idle Labour/Machinery and/or Reduced Productivity

When a breach of the contract by the owner adversely affects the planned progress of the contractor's work, it is inevitable that the contractor's labour and machinery will be either rendered idle or will give reduced productivity. In the process, the contractor will suffer loss

---

[72] M/s. A.T. Brij Paul Singh & Bros. v. State of Gujarat, AIR 1984 SC 1703 = (1984) 4 SCC 59.

[73] Subhash P. and M. Ltd. v. W.B. Power Devpt. Corpn. Ltd., (SC): AIR 2006 SC 116.

[74] Superintending Engineer, T.N.U.D.P., Madras v. A.V. Rangaraju, AIR 1994 Mad. 217.
   Government of Andhra Pradesh v. V. Satyam Rao (A.P.)(D.B.), AIR 1996 AP 288; also see ONGC Ltd. v. Comex Services SA, 2003 (3) Arb. LR 197 (Bom) (DB). State v. K. Bhaskaran (Kerala)(DB), AIR 1985 Ker. 49.

which he will be entitled to claim by way of damages. This right to claim damages will be independent of the fact whether there is or is not any overall delay. This damage is usually very hard to assess. Particularly reliable evidence in respect of idle labour wages is rarely available, though it is possible to establish the loss due to machinery and equipment. While assessing the loss of productivity of plant, "standing" charges alone are to be considered and "operating" charges component from the hire charges should be considered as possible to be mitigated. It is not necessary that the equipment should have been procured by the contractor on hire because even in respect of equipment owned by the contractor his loss will be the "standing charges" of the machinery and equipment deployed. It is not unusual in the absence of any more precise method, to claim this type of loss as an arbitrary percentage on total labour or plant expenditure during the period of dislocation. In the contractor's price build-up, the normal component for machinery, in the case of building works will be about 5%, and in the case of big engineering projects such as dams, canals, highways, etc. requiring use of heavy machinery and equipment, the component may be of the order of 10–15%. The labour component in building works is generally 20–25%. In plumbing contracts, it may be 12–15%. In cases where heavy machinery is deployed, the labour cost percentage may be 10–12%. On the basis of evidence tendered in arbitral cases, the combined percentage for overheads, expected profit and deployment of machinery and equipment works out to 40% at average rates of 15% for overheads, 15% for machinery deployment and 10% for profit.

12.10.5 Loss Due to Increased Cost of Materials and Labour

(A) Loss during the Stipulated Period

If, due to breach of contract by the owner, the contractor's planned programme is adversely affected it is possible that he may have to pay more for materials and labour due to delay and price fluctuation. He is entitled to receive this increased cost by way of damages. This claim, however, can be conceived in theory but difficult to establish in practice, first, for want of proper proof of the contractor's planned programme for procurement of materials, the increase in cost, etc., and second, because most present day works contracts provide for a price escalation clause in the contract. It is true that the contractor under these facts and circumstances will be bound by his tendered rates subject to price adjustment according to "escalation" provision, for the full period stipulated for completion of the work and the claim may not be tenable.

(B) Loss during the Extended Period

The contention that the contractor will be bound by the agreed rates until completion of the project, when the contract contains a provision for extension of time and the extension of time is granted to him, needs careful consideration. The fact that the contractor has undertaken to complete the work at the agreed rates (i) within the stipulated period, and (ii) with a provision for extension of time, lends support to the above contention that his agreed rates will be binding until completion of the work. In fact, on proper construction of the terms of contracts, this contention appears sound but within a reasonable limit. In practice, one comes across, and quite frequently too, situations where this contention is attempted to be extended beyond its elastic limit and abused to the point of shocking the conscience of a reasonable man. The authors have, in their practice, to deal with cases where completion of the project got delayed by over 100% of the originally stipulated time limit, which in some cases itself was 3–6 years. The delay was caused by the breaches of contract committed by the owner. The work to be executed during the extended period amounted to several crores of rupees. With prices of building materials and labour costs skyrocketing and the value of the rupee going down the cavernous depths each year, if the above principle were to be applied it would have resulted in total financial disaster to the concerned contractors. With the use of standard form contracts of adhesion containing exemption clauses and other provisions drafted to protect the owner under any eventuality and the judicial pronouncements attempting to uphold the sanctity of the contract, to get relief of a thorough revision of rates by way of damages becomes a difficult task but not an impossible one. For, one must begin with a basic supposition that in a civilized nation with a rule of law prevailing, there is no law in existence which upholds the "wrong" and punishes the "right". The concepts of "right" and "wrong" keep changing with time and the judiciary takes note of it and the law becomes to that extent flexible. In the realm of works contracts, this process takes place under the name of "proper construction of contracts". Applying the well-established principles of construction of contracts, it is possible to lay down the general propositions, some of which have also found recognition, though impliedly, in certain cases decided by the courts. An attempt is made below in paragraphs 12.11 and 12.12 to formulate the general principles.

12.11 Facts and Circumstances Justifying Payment at the Originally Agreed Rates during the Extended Period

The facts and circumstances under which it can be said that the contractor's claim for enhancement in the original rates will not be tenable under the terms of the contract can be summarized as follows:

1. If the parties have not agreed at the time of making of the contract as to the extent or duration of the extended period, the law will presume that the extension of time contemplated by the parties in their contract was "reasonable time". What is a reasonable time will be a question of fact in each case. Barring exceptional provisions or circumstances, the extended time may be presumed to be 10–25% of the originally agreed time, a small percentage to be adopted in cases where the original time exceeds one year and the high percentage to be adopted where the original time for completion as stipulated in the agreement is less than one year.

2. The project gets completed within the reasonable extended time, and such extension is not necessitated on account of breach of contract committed by the owner.

3. Where the extension of time is caused by an increased quantum of work or extra items and the contract provides for new rates to be agreed between the parties for such excess quantities (beyond reasonable variation of the originally agreed quantity) or extra items.

4. The extension is necessitated due to factors beyond the control of both the parties to the contract and none of which will amount to breach of contract committed by either party.

5. Where the time had been extended and agreement expressly prohibits payment of compensation.[75]

12.11.1 Escalation and Delay in Completion

Certain difficulties encountered in giving effect to contract provisions pertaining to extension of time and payment of escalation need to be considered here. The question commonly asked and possible answers to them are as follows:

---

[75] Sidhardha Constructions Pvt. Ltd. v. Union of India & Ors., AIR 2009 (NOC) 1427 (A.P.).

Question 1

a. The escalation clause in a PWD contract stipulates that no escalation will be paid, if the printed clause is deleted or time limit specified for completion of the work is less than 12 months. If extension without imposing penalty/liquidated damages is granted in a contract whereby the actual completion period exceeds 12 months, will the escalation be payable, and if yes, for which period?

b. If the original period is more than 12 months, whether escalation will be payable for the first 12 months or not?

Suggested answer:

a. The question of payment of escalation under the escalation clause will not and cannot arise if the printed clause is deleted and the deletion is prior to signing of the agreement and bears the signatures of both the parties. If the escalation clause is not applicable due to time limit specified being less than 12 months and if extension is granted, strictly speaking, no escalation is or will be payable under the escalation clause. But, if the delay is due to breaches of contract committed by the department, the contractor will be entitled to the loss of increased cost of executing the work, overheads, profit, etc. for the extended period only. In negotiations, it may be to the advantage of the department to settle such a claim on the basis of the escalation formulae in the agreement, though the same may not be directly applicable.

b. This part of the question can be answered in the light of the language of the escalation clause. If the clause expressly provides no escalation will be payable for the first 12 months, the provision is valid and the intending tenderer is expected to make provision for possible increase in the prices and labour wages during the first 12 months. If the wording indicates that the escalation clause will not be applicable for contracts with a time limit of less than 12 months, and if the time stipulated is in excess of 12 months, escalation will be payable for the full completion period, from day one as per the provisions made.

Question 2

Where an agreement excludes the payment of escalation on the cost of materials supplied by the owner at fixed rates but the owner fails to supply the materials and directs the contractor to buy them from the open market, what is the right course to adopt?

Suggested answer:

The provision in the agreement will have no application and the owner will be bound to reimburse extra expenditure incurred, if any, in full.

Question 3

An escalation clause provides for payment of escalation on the basis of actual consumption of certain materials such as steel, bitumen, etc. during the construction. The agreement also provides for payment of 90% of secured advance. Is escalation to be paid on the basis of the prevailing rate of steel on the date of consumption, or on the date the steel was procured?

Suggested answer:

The answer to this question also cannot be free from doubt and will depend upon the full text of the relevant provisions for a harmonious interpretation. However, *prima facie*, it appears that the agreement provisions will have to be given full effect. Escalation will be payable on the basis of the rate prevailing during the period of actual consumption.

12.12   Claim for Revision of Rates – When Tenable?

In all cases where the extended time exceeds the "reasonable" extension originally presumed to have been contemplated by the parties, and where the delay and extension of time are not caused by the acts of omission and commission of the contractor which would amount to a breach on the part of the contractor, the claim for revision of rates would be *prima facie* tenable[76] and requires to be thoroughly investigated and impartially decided. Where in a case the rates were tendered in the year 2002 and the tender was accepted in April 2004, and despite recommendation by the executive engineer the time was not extended and the contract terminated, the court directed the authorities to reconsider the matter and consider the petitioner's request for an extension of time and pass such other order as to the dispute, keeping in view the financial consequences of paying the new agency at the current rates. In the said case, though fresh tenders were invited no offer was received within the extended time also.[77] Reference to a few decided cases may be helpful to explain the applicable principles.

EXAMPLES FROM CASE LAW

1.   The Delhi High Court upheld the award of 25% escalation of costs to the contractor in the case where an extension of time was necessary due to the delay caused by the respondent authority alone and the contractor had demanded the escalation by various letters. The arbitral tribunal relied on the CPWD cost index showing an increase of 55% and other contracts awarded at 45% increase in rates by the respondent authority during that period.[78]

2.   In another case decided by the division bench of the Orissa High Court, it was held:

> "we are of the considered view that since ... the original tender was floated in the year 2002–03 and the work was awarded in the year 2004, directing the petitioner to complete the work at the old tendered rate would not be feasible since a substantial period of delay is ascribable to the authorities.[79]

The writ was allowed, the termination of the contract and fresh tender invitation were quashed and declared illegal. The petitioner had showed willingness to undertake and complete the balance work subject to waiver of penalty imposed and payment at the lowest rates quoted by any of the bidders in response to fresh tenders received for the balance work.

---

[76] State of Kerala v. T.E. Mohammed Kunju, AIR 2008 (NOC)2008 (Ker.).

[77] B.K. Enterprises v. State of Bihar, 2008(2) CTLJ 388(Patna).

[78] M/s Sudhir Bros. v. Delhi Development Authority & Ors., AIR 2009 (NOC) 1433 (Del.); also see: Union of India v. M/s. Jia Lall Kishori Lall Pvt. Ltd., 2018 (248)DLT 366.

[79] Govardhan Kumar Varjani v. Mahanadi Coalfields Ltd., 2008(2)CTLJ 392 (Orissa) (DB).

3. Performance of a contract got delayed due to inordinate delay on the part of the department to supply design to the contractor. The contractor was persuaded to carry on with the work with an extension of time having been granted and labour escalation paid. On completion of the work, the contractor was asked to refund the labour escalation paid. Held: the state would be estopped and it would be iniquitous to allow the state to unjust enrichment.[80]

4. In a case, the contractor claimed compensation for delay in completion of a project. The board, after constituting a committee and considering the recommendations of the said committee, in its meeting resolved to pay a particular sum to the contractor subject to adjustments of amounts in relation to quantities indicated in the Committee Report. Subsequently, the decision was taken not to honour the recommendations of the committee. The High Court held the earlier resolution should be given effect to. The Supreme Court confirmed the High Court's decision, holding that non-confirmation of minutes of meeting of the committee in the subsequent meeting does not have effect on a decision taken at the earlier meeting.[81]

A positive and just approach by the owner and architects/engineers, as suggested above and supported by the decisions in the cases under illustrations, would not only reduce disputes and delays and consequent losses due to delayed completion, but the contractor could be persuaded to accept enhancement which would be less than that which he would get in a judicial proceedings. In other words the owner might get the benefit of "bargaining power".

It quite frequently happens that during the progress of the work a contractor is persuaded to accept extension of time, giving up his right to claim compensation, and under coercive circumstances he may agree to do so. Subsequently if he intends to raise claims by contending that the supplemental agreement or undertakings were not binding on account of coercion, not only the plea should be raised but note that the party alleging coercion must lead evidence in support of its plea. The Supreme Court of India in a case observed:

> For coming to such conclusion, material had to be placed, evidence had to be led. Mere assertion by the plaintiff without any material to support the said stand should not have been accepted by the trial court and the High Court.[82]

It can be said that the contractor's right to claim and get a revision of the rates, if there is a breach of contract on the part of the owner, is now getting recognition by the courts of law in India and can be said to be well established.

EXAMPLES FROM CASE LAW

1. **Clause in the contract which obliged the plaintiff to continue to work at the rates initially offered by him – validity**
   In a recent decision the Supreme Court of India held:[83]

---

[80] M/s Ram Barai Singh & Co. v. State of Bihar & Ors., AIR 2009 (NOC) 1526 (PAT.).

[81] Kerala State Electricity Board v. Hindustan Construction Co. Ltd., AIR 2007 SC 425.

[82] State of Kerala v. M.A. Mathai, AIR 2007 SC 1537.

[83] M/s Aries & Aries v. T. Nadu Electricity Board, AIR 2017 SC 1897 = AIR 2017 SCW F1897.

The clause in the contract which obliged the plaintiff to continue to offer the rates initially offered by him would, naturally, be for the duration of the contract and cannot work to his peril for the period of delay for which the Department was admittedly responsible. Such a construction of the clause in the contract would not be reasonable. We, therefore, reverse the aforesaid finding of the High Court and hold that the plaintiff would be entitled to the 50% of the escalation charges as decreed by the learned trial Court.

(The delay by the owner was admittedly 50%).

2. Hon. Mr. Justice Sultan Singh in his judgment as early as in 1979, in the case M/s Alkaram, Petitioner v. Delhi Development Authority and another, in no uncertain words upheld the following principle:[84]

> If a contractor is required to do a work within certain time and the department fails to vacate the premises for carrying out the required work and after expiry of contracted period there has been a sharp increase in prices of the materials, the department cannot compel the contractor to carry out the work at the same rates at which he had agreed to do the same within the stipulated period ... Therefore, the contractor's claim for enhanced rate could be entertained by the arbitrator when the same had been referred to him by the parties and the same could not be deemed to be misconduct on the part of the arbitrator.

The above decision, was upheld by the division bench of the Delhi High Court.[85]

3. In Hyderabad Municipal Corporation v. M. Krishnaswami,[86] the Supreme Court of India observed:

> Drainage works for CSIR Laboratory at Uppal was entrusted to the respondent-plaintiff and under the terms of the contract the work was to be completed by the plaintiff within a period of one year, i.e. from 26th March 1951 to 25th March 1952. Admittedly at the instance of the Executive Engineer P.W.D. due to financial difficulties less budget having been provided for in the year 1951–52 the plaintiff was requested to spread over the work for two years more, that is to say to complete the same in three years but the respondent-plaintiff was agreeable to spread over the work for two years more as suggested on condition that extra payment will have to be made to him in view of increased rates of either material or wages.
> The Government did not intimate to the respondent-plaintiff that no extra payment on account of increased rates would be paid to him or that he will have to complete the work on the basis of original rates. In fact no reply was sent by the Government and studied silence was maintained by the Government in regard to the respondent-plaintiff, demand for extra payment, in spite of several reminders in that behalf, till the plaintiff actually completed the work during the spread over period and only when after completion of work the plaintiff-respondent submitted his final bill claiming 20 per cent extra over and above the rates originally agreed upon

---

[84] M/s Alkaram v. Delhi Development Authority, AIR 1980 NOC 47 (Delhi).

[85] Delhi Development Authority v. M/s. Alkaram, AIR 1982 Delhi 36. For presence of escalation clause and claim for revision of rates see Paragraph 6.15 in Chapter 6. Also see Metro Electric & Co. v. D. D. A., AIR 1980 Del. 266.

[86] Hyderabad Municipal Corporation v. M. Krishnaswami, AIR 1985 SC 607.

between the parties, the Government stated that he was not entitled to increased rates. After considering the correspondence exchanged between the parties and the other materials on record the High Court has taken the view that the government was liable to make extra payment for the work done as there was no dispute that the rates of material, etc. had increased during the extended period of two years and plaintiff was entitled to such extra payment. After considering the relevant material on record we are of the view that *both in equity and in law* the plaintiff contractor is entitled to receive extra payment and the High Court was right in deciding the question in respondent-plaintiff's favour.

(emphasis supplied)

4. In a contract, the value of the work was over Rs. 24 crores. The nature of the work required the contractors to import pile driving equipment and the technical know-how, with payment to be made in Japanese Yen at the estimated investment of Rs. 2 crores. The required pile driving equipment and the technical know-how were not available as provided for in the agreement. Ultimately, equipment more or less of the same specifications could be procured from Holland. In the meantime there were variations in the rate of foreign exchange and customs duty also went up. The plaintiff contractor claimed Rs. 61.27 lacs from the Cochin Shipyard Ltd., being the increase in the cost of equipment and technical know-how fees. The defendant declined to pay and the matter was referred to an arbitrator. The arbitrator upheld the claim of the contractors. The award was challenged. The Supreme Court observed:[87]

The rates payable to the contractor were related to the investment of Rs. 2 crores under this head by the contractor. Once the rates become irrelevant on account of circumstances beyond the control of the contractor, it was open to the contractor to make a claim for compensation.

The principle upheld by the Supreme Court, it is submitted, is summed up in the words: "When an agreement is predicated upon an agreed fact situation, if the latter ceases to exist the agreement to that effect becomes irrelevant or otiose".

5. In a building contract, the appellant contractor contended that the work could not be completed as stipulated because the site was not handed over to him. The contractor claimed compensation and extension of time. The respondent – Union of India – denied the liability. The disputes were not referred to the arbitrator by the chief engineer and the appellant took the matter to the engineer-in-chief. His request was not acceded to. The relationship between the parties became strained. The respondent asserted that the appellant had abandoned the work and committed breach of contract. Thereafter, the appellant called upon the engineer-in-chief to appoint an arbitrator. The officer who had terminated the contract was accordingly appointed as the sole arbitrator. Litigation followed because the appellant refused to accept the appointment so made. Ultimately the Supreme Court of India appointed a former judge of the Supreme Court as arbitrator. The arbitrator allowed certain claims including escalation in cost at 20% by way of compensation due to increase in cost of materials, labour, and transportation during the extended period. The award was challenged by the Union of India. The award of the arbitrator allowing compensation for losses suffered due to increased prices of materials and cost of labour, and transport, with

---

[87] Tarapore & Co. Cochin Shipyard Ltd., (1984) 2 SCC 680 (715).

interest was upheld by the Supreme Court of India. The Supreme Court upheld two basic propositions, it is submitted, namely:[88]

    a. Escalation is a normal incident arising out of an increase in time in the inflationary age, while performing any contract.

    b. Once it is found that the arbitrator has the jurisdiction to find that there was a delay in the execution of the contract, due to the default of the owner, the owner is liable for the consequences of the delay, namely, increase in prices.

6. In the State of Karnataka v. R.N. Shetty & Co. the defendant contractors contended that they were entitled to compensation in respect of tendered items after the expiry of the original contract period, the delay being caused by the appellant state. The state contended that there was little or no delay on the part of the state and no damages could be claimed. The dispute was referred to arbitrators. The arbitrators found that the claim of the contractors was justified and awarded damages calculated on the basis of the then prevailing rates of labour, machinery, spares, and petrol, oil and lubricant for the extended period. The Karnataka High Court rejected the contention of the state that the finding of the arbitrators was contrary to law.[89]

7. Where there was increase in the quantum of the work in excess of variation limit of plus/minus 25%, the claim for enhancement of rates by the arbitrator was upheld by the Supreme Court.[90] In the said case, it was observed that the power of the employer to vary the terms relating to the quantum of work cannot be unlimited. Even under the general law of contracts, once the contract is entered into, any clause giving absolute power to one party to override or modify the terms of the contract at his sweet will or to cancel the contract – even if the opposite party is not in breach – will amount to interfering with the integrity of the contract. There is thus good reason as to why, in modern works contracts, a limitation up to 20% (now 25%) has been put on this power of alteration, both plus and minus. Reference was made to a three-judge bench decision of the Supreme Court in S. Harcharan Singh v. Union of India.[91] In the said case it was held that the arbitrator could award higher rates on the analogy of Clause 12A of CPWD contracts for excess variations beyond 20% while construing the provisions of Clause 12.

8. Claim for loss in connection with the price escalation under Clause 32 of the original agreement and claim for revision of rates were allowed by the arbitrator in a case where the original contract time for completion was fixed but, due to fault of the employer, the work could not be completed by that time. It was contended that the appellant contractor was forced to execute the Supplemental Agreement in a form as dictated by the respondent as otherwise the respondents were not clearing the pending bills of the appellant. After hearing the parties and considering the submissions of the respondents, the arbitrator had, by a reasoned award, sanctioned the claims. In the award the arbitrator had held that the delay was not due to the fault of the appellant and that the Supplemental Agreement was executed by the appellant without prejudice to the claims which had already been made. The arbitrator held that the Supplemental Agreement did not debar the appellant from making or maintaining his claims.

---

[88] P.M. Paul v. Union of India, AIR 1989, SC 1034; also see G.S. Kalra, Petitioner v. New Delhi Municipal Committee and another, AIR 1999 Del. 355; T.P. George, Appellant v. State of Kerala and another, AIR 2000 SC 816.

[89] State of Karnataka v. R.N. Shetty & Co., AIR 1991 Kant 96.

[90] National Fertilizers v. Puran Chand Nangia, AIR 2001 SC 53 = 2000 AIR SCW 3860.

[91] S. Harcharan Singh v. Union of India, (1990) 4 SCC 647: (AIR 1991 SC 945).

The High Court held that the arbitrator had misconducted himself by awarding relief contrary to the Supplemental Agreement. On appeal, the Supreme Court held:[92]

> It is to be seen that the question, whether the Supplemental Agreement dated 20th October 1983 debarred the appellant from pursuing his claims, was before the arbitrator. Such a question having been referred to the arbitrator the view of the arbitrator would be binding if it is one which is possible. The arbitrator has taken note of the appellant's letters dated 6th October 1983 and 24th November 1983 and come to a conclusion that the Supplemental Agreement had been got executed and that the same was executed without prejudice to the claims which had already been made. This is a possible view. In this view of the matter the impugned judgment cannot be sustained and is set aside in respect of claims under Items 12(i) and (k).

9. The admitted position in a case was that though the rates in respect of certain items came to be revised by the Municipal Corporation in March 1996 in pursuance of a proposal put up by the Municipal Commissioner to the Standing Committee and a supplementary agreement was entered into, the rates of the 18 items which formed the subject matter of Claim No. 2 had not been revised. In respect of the aforesaid 18 items, the supplementary agreement of 13th March 1996 continued the same rates which prevailed in the year 1989. The question which arose before the arbitrator was whether, as a result of the supplementary agreement, the contractor must be bound to complete the work at the same rates as had been prescribed in 1989 or whether a revision of the rates of those 18 items was called for. The question as to whether the supplementary agreement that was entered into between the parties on 13th March 1996 would exclude the respondent from the benefit of claiming a revision of those rates which were not revised at that stage was a dispute which lay within the jurisdiction of the arbitrator. In holding that the respondent was entitled to a revision, it was held the arbitrator did not act outside his jurisdiction.[93]

10. The appellant company made a demand for payment at an enhanced rate of 421/2% over the basic rates stipulated under the original contract. This claim was made on several grounds, including "substantial deviation" in the nature of work of which the detailed work drawings were supplied to the appellant company after the date of the contract; "great increase in the price of materials and labour on account of undue prolongation of the period of work", etc. The Supreme Court held:[94]

> The appellant Company had undertaken under the terms of the contract to do specific construction work at "basic rates". The Engineer-in-charge was by the terms of cl. 12 of the agreement competent to give instructions for work not covered by the terms of the contract, and it was provided that remuneration shall be paid at the rate fixed by the Engineer-in-charge for such additional work, and in case of dispute the decision of the Superintending Engineer shall be final.

The trial court had decreed the suit accordingly. It was common ground that the claim made by the appellant company was not covered by the arbitration agreement and on that account it was not referred to the arbitrator. Held: The claim in suit related to the revision

---

[92] T.P. George v. State of Kerala, AIR 2001 SC 816 = 2001 AIR SCW 616.
[93] Municipal Corporation of Greater Mumbai v. Jyoti Construction Co., 2003 (3) Arb. LR 489 (Bombay).
[94] Gannon Dunkerley and Co. Ltd. v. Union of India, AIR 1970 SC 1433.

of rates due to the complex nature of the work and due to an increase in the quantity of work and also grant of contracts to other competing parties at substantially higher rates and other related matters. The decree was upheld.

11.   In a case, the claim for refund of excess hire charges of machinery and payment towards losses suffered as a result of poor performance of department machinery and also a direction for the future claim was allowed by the arbitrator and his decision was upheld by the High Court. The Supreme Court, on appeal against the said decision held: "The Government was in terms of the contracts bound to compensate the Contractor for the excess higher charges paid as a result of the poor performance of the machinery supplied by the Government."[95]

12.   The decision of the Supreme Court in Continental Construction Co. Ltd. v. the State of M.P.[96] raises a doubt but cannot be said to upset the principles emerging from the above discussion. The main point in the latter case was that where an agreement expressly provided that no compensation would be payable in spite of a rise in the prices of materials and labour wages, a non-speaking award allowing relief on the same grounds was invalid. The award was set aside. In that case the contractor could not complete the work because of the alleged gross delay on the part of the state government. The trial court and the High Court held that the contractor was not entitled to claim the extra cost in view of the terms of the contract and that the arbitrator misconducted himself in awarding the relief. This decision was upheld by the Supreme Court. After study of the said decision, in the opinion of the authors, instead of a non-speaking award had there been a reasoned award holding that in the light of the facts and circumstances of a case the exemption clauses would not have applied, the court might not have upset the award, the view expressed by the arbitrator being one possible view.

If the delay is attributable to both the parties, and bifurcation of percentage of delay attributable to each party is not possible, recovery of liquidated damages is illegal and the contractor may be entitled to partial relief.[97]

12.12.1 The Quantum of Increase

The increase in the cost of materials, labour, and plant and equipment to which the contractor would be entitled in the event of a breach of contract by the owner resulting in delayed completion of the work, is a question of fact and has to be properly pleaded and proved. Many a contract stipulate that the contractor shall furnish periodical reports of labour employment as well as the machinery and equipment deployed at the site. Such a record, if maintained, can be useful as evidence if need arises. Vouchers showing the rates at which materials, etc. were in fact procured need to be exhibited and proved.

Where the enhancement sought is on the basis of "mutual agreement" in respect of totally new rates for the extended period, the contractor may not be entitled to damages for

---

[95] Associated Engineering Co. v. Government of Andhra Pradesh, AIR 1992 SC 232 = 1991 AIR SCW 2960.
[96] Continental Construction Co. Ltd. v. the State of M.P., AIR 1988 SC 1166.
[97] M/s Mecon Ltd. v. M/s Pioneer Fabricators (P) Ltd., AIR 2008 (NOC) 870 (DEL).

overheads and profit for the extended period as the same would merge in the new agreed rates. Where, however, in a litigation he is being awarded revised rates by way of damages on the basis of increase in the cost of "materials and labour", there is no duplication, if the compensation is awarded separately under the head loss of "overheads and profit". Even in a non-speaking award of an arbitrator, it is advisable to mention that the quantum awarded under revision of rates is over and above the damages for continuing loss on account of overheads, and profit, separately awarded, if both these claims are referred to arbitration and deserve to be awarded.

Where the parties have kept the award of damages for breach of contract outside the purview of an arbitrator, the suit for compensation instituted within three years from the date of rescission is not barred and is also maintainable.[98]

12.13 Alternative Remedy to Damages – *Quantum Meruit*

"*Quantum meruit*" is a right to be paid a reasonable remuneration for work done or goods supplied. The Supreme Court of India has observed:[99] "that a claim for *quantum meruit* is a claim for damages for breach of contract." In cases where the work is partly carried out and the contract is repudiated by the owner, the contractor would get an option either to sue for damages for breach of contract or bring an action in *quantum meruit* for the work done by him. If the contractor's tendered rates are highly profitable, his best option would be to claim damages for breach of contract. If not, and if a substantial part of the work is carried out, his claim for a reasonable price for the work done will be more advantageous to him.

It is to be noted that "*quantum meruit*" arises in a number of situations where for one reason or another no contract exists, either originally or subsequently. For example, a contract voided by mistake, by operation of the doctrine of frustration or similar cases where the law has set aside an apparent contract or declared it unenforceable. In such an eventuality, an arbitration clause in the agreement would also be unenforceable and, therefore, an arbitrator appointed under it will have no power to grant relief by way of *quantum meruit* which a court of law will have under Section 65 of the Indian Contract Act.

Under Australian law, a builder can claim on a *quantum meruit* basis after a repudiation of the contract, and in such a claim the contract price, albeit relevant particularly to the reasonableness of the claim, is not always the best evidence of the value of the benefit conferred on the owner.[100] Under English law, the court will determine whether or not in the facts and circumstances of each case, "the law should, as a matter of justice, impose upon the defendant an obligation to make payment of an amount which the claimant deserved to be paid (*quantum meruit*)".[101] The court will not impose

> an obligation to make payment if the claimant took the risk that he or she would only be reimbursed for his expenditure if there was a concluded contract; or if the court concludes that, in all the circumstances the risk should fall on the claimant. ... The court may well regard it as just to impose such an obligation if the defendant who has received the benefit has behaved unconscionably in declining to pay for it.[102]

As to what is a reasonable price for the work done, it will be a question of fact to be proved in each case. It is suggested that prevailing market rates, assessed on the basis of schedule of rates prepared by

---

[98] H.P. Housing Board v. M/s Rajeev Bros., AIR 2008 (NOC) 1739 (H.P.).

[99] Madras State v. G. Dunkerley & Co., AIR 1958 SC 560 (577).

[100] Sopov v. Kane Constructions Pty Ltd., (No. 2) [2009] VSCA 141 (CA(Vic)).

[101] William Lacey (Hounslow) Ltd. v. Davis, [1957] 1 WLR 932.

[102] Christopher Clarke J. in MSM Consulting Ltd. v. United Republic Tanzania, [2009] EWHC 121 (QB), cited in Moorgate Capital (Corporate Finance) Ltd. v. HIG European Capital Partners LLP, [2019] EWHC 1421.

the PWDs or on the basis of rates tendered by the contractors in competition for similar works, in the vicinity, at the relevant time, may be useful in deciding the "reasonable price".

12.14 Breach of Contract by the Contractor

As already stated, breaches of contract by a contractor entitling the owner to seek compensation may be many. Some important breaches are considered below with a view to point out remedies available against each breach. For illustrations, reference may be made to Chapter 11.

12.14.1 Abandonment or Total Failure to Complete

The basic principles in ascertaining the damages due to breach of contract already outlined in paragraph 12.7 are applicable to a breach of contract by the contractor. In cases, therefore, where the work has been left incomplete, whether by abandonment, termination or otherwise, the direct measure of damage will be the difference between the reasonable cost to the owner of completing the work, together with any sums paid by or due from him under the contract, and the sums which would have been payable by the owner to the contractor if the contract had been properly carried out. In case final cost to the owner to complete the work does not exceed the contract sum, only nominal damages would be recoverable.[103] The reasonable cost of completion usually means the cost of completion of the work substantially, as it was originally intended and in a reasonable manner.

EXAMPLE FROM CASE LAW

In a case involving construction of a tunnel for the city, the contract provided that in case of default by the contractor, the city should be entitled to complete the work at the contractor's expense. On the contractor's default the city constructed a tunnel which was essentially different in plan and cost of construction from that contemplated by the contract. It was held that the city was not entitled to recover damages from the contractor.[104]

12.14.2 Defective Work

The measure of damage for defective work will be the reasonable cost to the owner of repairing the defects. Such damages are recoverable under Section 73 as naturally arising in the usual course of things.

EXAMPLE FROM CASE LAW

In a case, a builder substantially departed from the specification in relation to the foundation of a house, which was consequently unstable. It was held that the measure of damage was not the difference between the value of the building as built and the value if built in accordance with the contract; but the cost, in excess of any amount of the contract price (unpaid), of reasonable and necessary work required to be undertaken to conform to the contract.[105]

---

[103] T.N.W.S. & D.B. v. Satyanarayana Brothers Pvt. Ltd., 2002(1) Arb. LR 444 (Madras) (DB).

[104] Milwaukee City v. Shailer, (1898) 84 Fed. Rep. 106 *Hudson's B. C.*, 10th ed., p. 587.

[105] Bellgrove v. Eldridge, (1954) ALR 929 Vol. 90 CLR 613 (Australia).

The Rule of "Difference in Value"

If there is a breach of the contract on the part of the contractor, the owner is entitled to receive, generally, the market price of completing or correcting the contractor's performance. However, if the cost of completion or correction is grossly and unfairly out of proportion to the outcome to be achieved, the measure of damages will be the "difference in value". In short, this rule means that the measure of damages would be the difference between the value of the defective structure and that of the structure as it should have been completed. In case of substantial performance, this difference in value may work out to be either nominal, or nothing.

The House of Lords had to consider the question of proper compensation payable when a contractor built a swimming pool to the depth of six feet nine inches, despite the contract specifying seven feet six inches. The trial judge only awarded £2,500 for loss of amenity, but the Court of Appeal awarded £21,560 towards the cost of reinstatement of the pool to the contracted depth. The House of Lords decided that the cost of reinstatement was not the appropriate measure of damages in that case and decided to award nominal damages on the basis of difference in value.[106] The House of Lords also had to consider whether the plaintiff's intention (as to the rebuilding of the pool or not) was relevant to the question of damages. Ordinarily, a court is not concerned with the plaintiff's intention to spend the compensation but it was held that the intention was relevant to the question of reasonableness. In this case, Mr. Forsyth, who ordered the pool, had died. If the executors were going to put the property on the market, should they recover the cost of reinstatement? If there are two potential measures of damages, high and low, the court has to decide if in the circumstances of the case the high cost measure is reasonable. The House of Lords view was that the late Mr. Forsyth could not be allowed to create a loss, which did not exist, "in order to punish the defendants for their breach of contract". This goes back to the basic principle of English law that damages are mainly compensatory and not punitive, albeit that there is an exception for gain-based damages.[107] The House of Lords upheld the trial judge's award of £2,500. This approach was followed in several cases, for example, in a case where there was discolouration of concrete roof tiles that were supposed to look like slate and had been purchased as a premium product; the case was not one of a building defect but as a diminution in value of a supplied product.[108]

EXAMPLE FROM CASE LAW

In a case, the contractor had failed to use the prescribed brand of pipe for the plumbing work. It was found by the court that the omission was neither fraudulent nor wilful and that the pipe as furnished was substantially the same in quality, weight, service, ability, appearance and market price. Most of the plumbing was encased within the walls. The replacement of pipes would have required demolition at a great expense, of substantial parts of the completed structure. The court held:[109] "In the circumstances of this case ... we think the measure of the allowance is not the cost of replacement which would be great, but the difference in value, which would be either nominal or nothing."

When, however, a contractor substantially defaults in the contract performance so as to render the finished structure partially unusable or unsafe, the measure of damage would be "the market price of correcting or completing the performance" and not the difference in value.[110]

---

[106] Ruxley Electronics and Construction Ltd. v. Forsyth and Laddingford Enclosures Ltd. v. Forsyth, [1995] 3 WLR 118.

[107] See paragraph 12.6 above.

[108] Peebles v. Rembrand Builders Merchants Ltd., [2017] 4 WLUK 352.

[109] Jacobs v. Kent, 230, N.Y. 239, 129, N.E. 889 (N.Y. 1921).

[110] Bellizi v. Huntley Estates, 3. N.Y. 2d, 1 12 143 NE 2d. 802 (N.Y. 1957).

EXAMPLE FROM CASE LAW

In a case, the foundation of a house built on a concrete raft failed. The only remedy was to rebuild it, using the "pier and beam system" which would cost more than the original house. It was held that the disparity between the original cost of the house and the new foundations was not the reason for departing from the normal rule, but that from the cost of repair should be deducted a sum to take account of the fact that the new foundations would eliminate all danger of sinking, whereas some sinking on an even plane was expected with the original house. Against the claim of £2,200 for the cost of the new type of foundation and £500 for depreciation in value, damages were restricted to £2,000.[111]

12.14.3 Delay in Completion

The measure of damage to the owner in the event of delay in completion by the contractor will be largely governed by the nature of the contract work. If the work involves a commercial building, such as a factory or shop, it is obvious that the delay in completion will affect the profits that the owner is likely to earn from use of the building. In the case of a block of flats or a boarding house, the owner's profit from his rents is likely to be affected. In the case of a dwelling house, the claim for loss of profits from letting will be recoverable only if the fact that the owner intends to let out the house was made known to the contractor at the time of agreement. Otherwise, the reasonable cost of living accommodation or living elsewhere and storing furniture, etc., if in fact expenses of this kind were incurred, would be the measure of damages.[112]

In cases of public works, such as dams, bridges, roads, canals, etc., the actual loss due to delay in completion is difficult to assess and, therefore, such contracts provide for liquidated damages to be recovered for delay. This aspect is considered separately in Chapter 7.

12.15 Damages Difficult to Estimate – No Ground to Award Nominal Damages

The fact that damages are difficult to estimate and cannot be assessed with certainty or precision does not relieve the wrongdoer of the necessity of paying damages for his breach of duty, and is no ground for awarding only nominal damages. A distinction must be drawn, however, between cases where the difficulties are due to uncertainty as to the causation of damage, where questions of remoteness arise, and cases where they are due to the fact that assessment of damages cannot be made with any mathematical accuracy. Lack of relevant evidence may make it impossible to assess damages at all, as where the extent of the loss is dependent upon too many contingencies, and in such cases where the liability is established, nominal damages only may be awarded. Where it is established, however, that damage has been incurred for which a defendant should be held liable, the plaintiff may be accorded the benefit of every reasonable presumption as to the loss suffered. Thus the court, or a jury doing the best that can be done with insufficient material evidence, may have to form conclusions on matters on which there is no evidence, and to make allowance for contingencies even to the extent of making a pure guess; this is of common occurrence in a claim made, for example, in respect of pain and suffering, loss of expectation of life, and the loss of a chance of winning a prize.[113]

---

[111] Cooke v. Rowe, (1950) N.Z.L.R. 410 (New Zealand) also see: *Hudson's*, 10th ed., p. 586.
[112] Lindberg v. Brandt et al., 1 12, N.E. 2d, 746, Appellate Court of Illinois 1953.
[113] Gambhirmul v. Indian Bank Ltd., AIR 1963, Cal. 163.

12.16 Specific Performance – Injunction to Restrain Breach

In building and engineering contracts, payment of compensation by the defaulting party to the injured party affords adequate remedy, and also superintending the performance of such contracts by the court is difficult. For these reasons the remedy of specific performance is not available. For the same reasons, the court would be reluctant to issue an injunction to restrain a party from committing a breach of contract.

12.17 Lien/Vesting Clauses

Many old standard form building contracts contain provisions vesting in the owner, property in unfixed materials and also the contractor's plant, machinery and equipment during the currency of the contract. FIDIC 1999/2017 editions, however, provide that the employer, after termination of the contract, shall give notice to the contractor and release the contractor's equipment and temporary works at or near the site. The Government of India MoS & PI Standard Form 2001 in Clause 61.1 provides as follows:

> All materials on the Site, Plant, Equipment, Temporary Works and Works for which payment has been made to the contractor by the Employer, are deemed to be the property of the Employer, if the Contract is terminated because of a Contractor's default.

The validity and effect of such provisions can be studied along with the law relating to lien and express provisions made in the Indian Contract Act.

12.17.1 Lien Defined

Black's Dictionary gives the meaning of lien as follows:[114]

> A claim, encumbrance, or charge on property for payment of some debt, obligation or duty. Qualified right of property which a creditor has in or over specific property of his debtor, as security for the debt or charge or for performance – Right or claim against some interest in property created by law as an incident of contract.

Where the law itself, without the stipulation of the parties, raises a lien, as an implication or legal consequence from the relation of the parties or the circumstances of their dealings, it is lien by operation of the law; for example, a mechanic's lien.

12.17.2 Mechanic's Lien

According to the *Black's Dictionary*,[115] a mechanic's lien is a claim created by statutes for the purpose of securing priority of payment of the price or value of work performed and materials furnished in erecting, improving, or repairing a building or other structure, and as such attaches to the land as well as buildings and improvements erected thereon. Such lien covers materialmen, tradesmen, suppliers, and the like, who furnish services, labour or materials on the construction or improvement of property.

---

[114] 6th ed. p. 922.
[115] 6th ed. pp. 922 and 1420.

12.17.3 Stop Notice Statute

Stop notice statute is an alternative to the mechanic's lien remedy that allows contractors, suppliers and workers to make and enforce a claim against the construction lender, and in some instances, the owner, for a portion of the undisturbed construction loan proceeds.

12.17.4 General and Particular Lien

The lien can be (i) particular or (ii) general. The former is dealt with in Section 170 and the latter in Section 171 of the Indian Contract Act. To understand the same, Section 148 needs to be perused: The said section reads:

148: *"Bailment" "bailor", and "bailee" defined*
A "bailment" is the delivery of goods by one person to another for some purpose, upon a contract that they shall, when the purpose is accomplished, be returned or otherwise disposed of according to the directions of the person delivering them. The person delivering the goods is called the "bailor". The person to whom they are delivered is called the "bailee".

Explanation – If a person already in possession of the goods of another contracts to hold them as a bailee, he thereby becomes the bailee, and the owner becomes the bailor of such goods although they may not have been delivered by way of bailment.

170: *Bailee's particular lien*
Where the bailee has, in accordance with purpose of the bailment, rendered any service involving the exercise of labour or skill in respect of the goods bailed, he has, in the absence of a contract to the contrary, a right to retain such goods until he receives due remuneration for the services he has rendered in respect of them.

ILLUSTRATIONS

(a) A delivers a rough diamond to B, a jeweller, to be cut and polished, which is accordingly done. B is entitled to retain the stone till he is paid for the services he has rendered.

(b) A gives cloth to B, a tailor, to make into a coat. B promises A to deliver the coat as soon as it is finished, and to give A three months' credit for the price. B is not entitled to retain the coat until he is paid.

171: *General lien of bankers, factors, wharfingers, attorneys and policy-brokers*
Bankers, factors, wharfingers, attorneys of a High Court and policy-brokers may, in the absence of a contract to the contrary, retain as a security for a general balance of account, any goods bailed to them; but no other persons have a right to retain, as a security for such balance, goods bailed to them, unless there is an express contract to that effect.

It is clear from the above provisions that the lien dealt with in Section 170 relates to services done in respect of the very goods retained. The general lien, the subject matter of Section 171, is for a general balance of account and operates in the absence of the contract to the contrary, which may either be express or implied.[116] The following important points need to be noted:

---

[116] Lalchand v. Pyare, 1971 MPLJ 672; 1971 Jab L.J. 601.

1. The essential requisite of lien is possession obtained previously. The word "retain" implies this meaning. The possession must have been lawfully acquired and the lien subsists only as long as possession lasts.

2. The lien consists only in the right to retain and does not extend to a right to sale.[117]

3. It is subject to a contract to the contrary or on giving possession, or by waiver of the lien, as, for example, by making a contract inconsistent with the existence of the lien.

4. Apart from the persons mentioned in Section 171, no other person has the right to retain as security goods bailed to them, unless there is an express contract to that effect.

In view of the above, the right of lien available both to the owner or the contractor will depend upon the express terms of the contract. In the public works contracts most standard forms provide:[118]

> The Government shall also be at liberty to hold and retain in their hands materials, tackle, machinery and stores of all kinds on site, as they may think proper and may at any time sell any of the said materials, tackle, machinery, and stores, and apply the proceeds of sale in or towards the satisfaction of any loss which may arise from the cancellation of the contract.

On the other hand, there is no provision in the law of contract or in the standard form contracts, either express or implied, under which a contractor can have a mechanic's lien on the structure or work done. Section 170 deals with bailees' particular lien and Section 171 does not include a contractor among the persons having general lien. On proper construction of the contract law and the terms, it can be said that no lien can arise from a building contract, whether by operation of law or under the terms of the contract. The contract provisions such as the above, however, may confer contractual rights over materials or plant having some or all of the characteristics of a lien. An example will help clarify this point.

EXAMPLE FROM CASE LAW

A main contractor subcontracted the roofing work of erection of a school; the subcontract provided that "the sub-contractor shall be deemed to have knowledge of all the provisions of the main contract". The main contract contained a provision which read:

> Any unfixed materials and goods delivered to, and placed on, or into the works, shall not be removed except for use upon the works, unless consent in writing be given, and when the value of the goods has been included in any interim certificate under which the contractor has received payment, such material and goods shall become the employer's property.

The subcontractor delivered 16 tons of slates to the site for which the main contractor received payment in an interim certificate. The main contractor did not pay the subcontractor and subsequently became insolvent. The subcontractor was refused permission to take away the slates. The subcontractor successfully sued for the return of the slates, and damages or the value of the slates. It was held:[119] The clause in the main contract presupposes there is privity of contract between the owner and the subcontractor which there was not, or the main contractor

---

[117] Mulliner v. Florence, (1878), 3 Q.B. D. 484.

[118] See Clause 4 of PWD/CPWD, Clauses 53 and 54 of MES Dept. form, Clause 62 of the Railway Dept. form.

[119] Dawber Williamson Roofing Ltd. v. Humberside County Council, Q.B.D. (1979) 14 BLR 70.

had good title to the materials and goods. The clause in the main contract had force only if the title had passed to the main contractor. The slates were on the site but that did not mean that the possession had passed to the main contractor; they were on the site as was the usual practice and custom of the builders. The owner was not entitled to retain the property.

12.18 Forfeiture

To forfeit is to incur the loss through some fault, omission, error or offence. Forfeiture is the act of forfeiting. If there is no provision authorizing forfeiture, security cannot be forfeited.[120] Most construction contracts include express provisions empowering the owner, in certain specified events, to determine the contract and to take possession of the works, materials, tools and plant of the contractor and complete the works himself. Such provisions include the right of the owner to forfeit earnest money or security deposit without prejudice to any other remedy or remedies under the contract. It is very common these days that the security deposit for proper performance of the contract is given by the contractor to the owner in the form of a bank guarantee. When an event entitling the owner to forfeit occurs, a demand is made on the bank which gave the bank guarantee. Invariably the question arises if the demand is legal and justified, when the contractor challenges the demand. What is the duty of the bank in such cases? All these aspects are considered in the remainder of this chapter, starting with the provisions of the Indian Contract Act applicable to the forfeiture.

12.19 Applicability of Section 74 to Security Deposit and Other Penal Stipulation

Section 74 is reproduced in paragraph 7.3 of Chapter 7 and applies to those provisions in the contract which give the injured party the right to recover liquidated damages. Can it be said that it also applies to cases covering the right to forfeit what has already been received by the party aggrieved? In some old cases decided by the High Courts in India[121] it was held that Section 74 applies where a sum is named as penalty to be paid in future in case of breach, and not to cases where a sum is already paid, and by a covenant in the contract it is liable to forfeiture. This view, it is respectfully submitted, was rightly negated by the Supreme Court of India as early as in 1963 in these words:[122]

> There is however no warrant for the assumption made by some of the High Courts in India that Section 74 applies only to cases where the aggrieved party is seeking to receive some amount on breach of contract and not to cases whereupon breach of contract an amount received under the contract is sought to be forfeited. In our judgment the expression "the contract contains any other stipulation by way of penalty" comprehensively applies to every covenant involving a penalty whether it is for payment on breach of contract of money or delivery of property in future, or for forfeiture of right to money or other property already delivered.

[120] Sushil Kumar Karan v. State of Bihar, AIR 2001 Pat. 221.
[121] Abdul Gani & Co. v. Trustees of the Port of Bombay, AIR 1952 Bom. 310(ILR 1952) Born. 747; Natesa Aiyar v. Apparu, Padayachi ILR 38 Mad. 178 AIR 1915 Mad. 896 (FB), M/s. Shyam Bin Works Pvt. Ltd. v. U.P. Forest Corpn., AIR 1990 All. 205.
[122] Fateh Chand v. Balkishan Dass, AIR 1963 SC 1405.

This was confirmed subsequently in 1971 wherein it was held by the Supreme Court:[123]

> Duty not to enforce the penalty clause but only to award reasonable compensation is statutorily imposed upon courts by Section 74 of the Indian Contract Act. In all cases, therefore, where there is a stipulation in the nature of penalty for forfeiture of an amount deposited pursuant to the terms of a contract which expressly provides for forfeiture the court has jurisdiction to award such sum only as it considers reasonable, but not exceeding the amount specified in the contract as liable to forfeiture. The same principles, in our judgement would apply in the case in which there is a stipulation in the contract by way of a penalty, and the damages awarded to the party complaining of the breach will not in any case exceed the loss suffered by the complainant party.

12.19.1 Breach of Contract When Disputed, without Adjudication of Question of Breach of Contract – Recovery of Amount Not Proper

A complaining party is not entitled to presume a breach and proceed with the recovery particularly when it is disputed by the other party. An attempt on the part of the complaining party to do away with the adjudication of the question of breach remains illegal and without adjudication of the question of breach, the intended recovery remains wholly unauthorized.[124]

12.20 Provisions Made in Construction Contracts – Forfeiture of Earnest Money/ Security Deposit

A typical standard form contract, in use by various state PWDs and public undertakings, includes the following provisions:

1. Payment of earnest money along with the tenders which shall be forfeited if the contractor fails to:

   a. Abide by the stipulation to keep the offer open for the period mentioned in the tender notice

   b. Sign and complete the contract documents as required by the engineer and furnish the security deposit as specified in the tender notice.The provisions further stipulate that upon signing of the agreement, the earnest money shall stand converted into security deposit.

2. Payment of security deposit in cash, or government securities endorsed to the engineer-in-charge, and permit the government at the time of making any payment to the contractor to retain such a sum as will make good the full security deposit. In the alternative, the contractor is permitted to furnish bank guarantees on such terms as may be specified in lieu of the security deposit. This security deposit shall stand forfeited in the following cases:

   a. Rescission of the contract by the engineer under Clause 3 of the General Conditions of Contract

   b. Assigning or subletting the contract without the written approval of the engineer in-charge.

FIDIC 1999/2017 edition contains Clause 4.2 which provides for furnishing of performance security by the contractor, within 28 days after receipt of letter of acceptance and to keep it valid and enforceable until completion of the work and rectification of any defects. It further provides that the employer shall not make a claim under the performance security except for

---

[123] N.C. Sanyal v. Calcutta Stock Exchange Assn., AIR 1971 SC 422 (428).
[124] State of Rajasthan v. Nathu Lal, AIR 2006 Raj. 19.

amounts to which the employer is entitled under the contract in certain eventualities listed therein including termination of the contract by the employer.

The Government of India Form 2001, in Clause 52 provides for security deposit in two parts: (i) performance security, 5% of the contract amount, to be submitted at the time of award of the work within 28 days after receipt of letter of acceptance and (ii) retention money to be recovered from running bills. It is to remain valid until a date 14 days from the date of expiry of the taking over certificate.

12.20.1 Distinction between Earnest Money and Security Deposit

"Earnest money" is the term commonly used in a contract of sale where it is considered as "part of the purchase price when the transaction goes forward. It is forfeited when the transaction falls through by reason of the fault or failure of the vendee".[125] Further, it is not merely a part payment, but is then also an earnest to bind the bargain so entered into, and creates by the fear of its forfeiture a motive in the payer to perform the rest of the contract.

The phrases "give something in earnest" or "in part payment", are often treated as meaning the same thing, although the language clearly intimates that the earnest money is "something to bind the bargain, or, the contract", whereas there can be no part payment till after the bargain or contract has been bound, or closed.[126]

The authors further state that there are two distinct alternatives, namely, a buyer may give the seller money or a present as a token or evidence of the bargain quite apart from the price, i.e., earnest, or he may give him part of the agreed price to be set off against the money to be finally paid, i.e., part payment and that if the buyer fails to carry out the contract and it is rescinded, cannot recover the earnest, but he may recover the part payment. But this does not affect the seller's right to recover damages for breach of contract unless it was by way of deposit or guarantee, in which case it is forfeited. It is further stated that an earnest does not lose its character because the same thing might also avail as a part payment.

A deposit, by its very nature, is something which has to be refunded unless some term of a contract enables forfeiture.

EXAMPLE FROM CASE LAW

The two petitioners had submitted offers and paid earnest money. The highest bidder's offer was accepted on opening of the bids. He had backed out. Thereafter conditional acceptance was communicated to the two petitioners being the second and third highest bidders, subject to the condition of outcome of writ petitions filed. The second highest bidder protested and had withdrawn his offer before acceptance and demanded back earnest money. The third highest bidder sought further information. The tender was cancelled and the earnest money of tender-petitioners forfeited. It was held that the second highest bidder having revoked his offer before acceptance, and in respect of the third highest bidder, though the tender was valid when accepted, the acceptance was conditional and not absolute and unqualified as required by Section 7 of the Indian Contract Act, 1872. There being no default on the part of petitioners, the earnest money was ordered to be refunded.[127]

A "deposit" in case of sale of land or other immovable property, on the other hand, is not recoverable by the buyer, for a deposit is a guarantee that the buyer shall perform his contract and

---

[125] Chiranjit Singh v. Harswarup, AIR 1926 PC 1.
[126] Benjamin, *Sale of Goods*, 8th edition, at p. 219, cited with approval in Shree Hanuman Cotton Mills & Ors. v. Tata Air-Craft Ltd., AIR 1970 SC 1986.
[127] Martin Lottery Agencies Ltd. v. State of Goa, 2009(1) CTLJ 242 (Bombay) (DB).

is forfeited on his failure to do so.[128] If a contract distinguishes between the deposit and instalments of price and the buyer is in default, the deposit is forfeited. Part of the price may be payable as a deposit. Part payment towards price is different from a deposit or earnest,[129]

> The deposit serves two purposes – if the purchase is carried out it goes against the purchase money but its primary purpose is this, it is a guarantee that the purchaser means business; and if there is a case in which a deposit is rightly and properly forfeited it is, I think, when a man enters into a contract to buy real property without taking the trouble to consider whether he can pay for it or not.[130]

The Supreme Court of India after a review of a number of decisions, including those cited above, observed that the following principles emerge regarding "earnest"[131]

(1) It must be given at the moment at which the contract is concluded.

(2) It represents a guarantee that the contract will be fulfilled or, in other words, "earnest" is given to bind the contract.

(3) It is part of the purchase price when the transaction is carried out.

(4) It is forfeited when the transaction falls through by reason of the default or failure of the purchaser.

(5) Unless there is anything to the contrary in the terms of the contract, on default committed by the buyer, the seller is entitled to forfeit the earnest.

From the above discussion it follows that the deposit is a guarantee for the performance of the contract for the sale of immovable property. The moment the contract is performed it becomes a part of the price or purchase money. But in a construction contract, it is the contractor who is entitled to receive money consideration for the work done or services rendered. As such earnest money cannot be said to be "a part of consideration" payable to the owner. It is, therefore essentially a security deposit showing an earnestness, that it is an ardent desire, on the part of the contractor to go forward with the contract. In that sense it is a security deposit for performance of the contract and indeed the terms of the contract stipulate that it will be converted into security deposit on signing of the contract by the contractor. By this analysis when it comes to forfeiture of it, it is submitted, there can be no difference between the two: whether it is earnest money or security deposit.

The term in an invitation of tender, the right to forfeit the earnest money for failure of the contractor to keep the offer open, is clearly without consideration and not enforceable. The forfeiture of earnest money can be considered only after the tender is accepted and the contractor refuses to sign the agreement or perform it, for in that case, the acceptance of tender converts it into a legally binding agreement and the contractor becomes liable for breach of it. Even in such cases the owner may not be able to automatically forfeit the earnest money. When there is no concluded contract, the party paying security deposit is entitled to refund of security deposit.[132]

---

[128] Ashok Sharma & Asso. P. Ltd. v. Sr. Divisional Commercial Manager, 2009(1) CTLJ 245 (Bombay) (DB).

[129] Halsbury, in *Laws of England*, Vol. 34 III edition, in paragraph 189 at p. 118.

[130] Soper v. Arnold, (1889) 14 AC 429(435).

[131] H.C. Mills v. Tata Aircraft, AIR 1970 SC 1986(1991–1994).

[132] Rajasthan State Electricity Board and others, v. M/s. Dayal Wood Works, AIR 1998 AP 381. Also see E. Bhagwan Das and others v. Dilip Kumar and another, AIR 1998 AP 374; State of Tripura v. M/s Bhowmik & Company, AIR 2004 Gau. 19 relying on Firm Vijay Nipani Tobacco House v. Sarwan Kumar, AIR 1974 Pat.117; Sardar Surjeet Singh, v. State of U. P. and others, AIR 1995 All.146; AIR 1984 Kant 122, Followed. A. Murali and Co. v. State Trading Corporation of India Ltd., AIR 2001 Mad. 271; P.K. Abdulla v. State of Kerala, AIR 2002 Ker. 108. Also see Baldev Steel Ltd. v. Empire Dyeing and Manufacturing Co. Ltd., AIR 2001 Del. 391. Also see M/s Rose Valley Real Estate and Construction Ltd. v. United Commercial Bank, AIR 2008 Gau. 38 in which the contract was held to be frustrated.

EXAMPLE FROM CASE LAW

According to a tender notice, the tenders were to be opened at a specified date and time by opening the tender box. After opening of the box, the tender forms were to be scrutinized as to whether they were valid tenders and to identify the successful bidder. That process, admittedly, was not gone through and even according to the respondent the sealed covers were not opened on the given date and were kept back in the box. The opening of tenders by removing the seals and scrutinizing the tender forms was postponed because of the interim orders of the court. Before the actual process started, the petitioner had asked for return of the earnest money deposit (EMD). It was observed:

> Though the request of the petitioner did not specifically refer to withdrawal of the tender, still no one will ask for return of the E.M.D. unless there is an intention not to participate in the tender. The respondent had also understood this because in the reply the respondent had clearly stated that the petitioner should participate in the opening of the tenders to which date it was postponed and expressed the inability to return the E.M.D. referring to Clause (12) of the tender notification.[133]

It was held that clause comes into operation only after the tenders are opened and the highest bidder is declared and such a declaration can come only after scrutinizing the tenders. In the present case, that did not take place. The petition was allowed.[134]

Where a promise to keep an offer open until a stipulated period was not shown to be supported by consideration, the condition that tender could not be withdrawn before it was accepted was held invalid.[135] This well-settled position in law *prima facie* appears to have been changed by the decision of the Supreme Court of India which needs to be considered at length.[136] In the said case, in the impugned judgment, the High Court had held that the offer was withdrawn before it was accepted and thus no completed contract had come into existence. The High Court held that in law it is always open to a party to withdraw its offer before its acceptance. The Supreme Court observed: "To this proposition there can be no quarrel. We, therefore, did not permit ... to cite authorities for the proposition that an offer can be withdrawn before it is accepted." The Supreme Court observed:

> The Indian Contract Act merely provides that a person can withdraw his offer before its acceptance. But withdrawal of an offer, before it is accepted, is a completely different aspect from forfeiture of earnest/security money which has been given for a particular purpose. A person may have a right to withdraw his offer but if he has made his offer on a condition that some earnest money will be forfeited for not entering into contract or if some act is not performed, then even though he may have a right to withdraw his offer, he has no right to claim that the earnest/security be returned to him. Forfeiture of such earnest/security, in no way, affects any statutory right under the Indian Contract Act. Such earnest/security is given and taken to ensure that a contract comes into existence. It would be an anomalous situation that a person who, by his own conduct, precludes the coming

---

[133] Aditya Mass Communication Pvt. Ltd. v. A.P. State Road Transport Corporation, Hyderabad, AIR 1998 AP 125. Also see: Omjee Finance Ltd. v. Amravati Municipal Corpn & Anr., AIR 2008 (NOC) 680 (Bom.).

[134] Ibid.

[135] M/s. Krishnaveni Constructions, Petitioner v. The Executive Engineer, Panchayat Raj, Darsi and others, AIR 1995 AP 362.

[136] National Highways Authority of India, v. M/s. Ganga Enterprises and another, AIR 2003 SC 3823; Ranjit Kumar Saha v. State of Tripura, AIR 2006 Gau. 70.

into existence of the contract is then given advantage or benefit of his own wrong by not allowing forfeiture. It must be remembered that, particularly in Government contracts, such a term is always included in order to ensure that only a genuine party makes a bid. If such a term was not there even a person who does not have the capacity or a person who has no intention of entering into the contract will make a bid. The whole purpose of such a clause i.e. to see that only genuine bids are received would be lost if forfeiture was not permitted.[137]

It was further observed:

> There is another reason why the impugned judgment cannot be sustained. It is settled law that a contract of guarantee is a complete and separate contract by itself. The law regarding enforcement of an "on demand Bank guarantee" is very clear. If the enforcement is in terms of the guarantee, then Courts must not interfere with the enforcement of Bank guarantee. ... The Bank guarantee stipulated that if the bid was withdrawn within 120 days or if the performance security was not given or if an agreement was not signed, the guarantee could be enforced. The Bank guarantee was enforced before the bid was withdrawn within 120 days. Therefore, it could not be said that the invocation of the Bank guarantee was against the terms of the Bank guarantee.

The Apex Court reaffirmed the above position, in a subsequent decision.[138]

The authors, with great respect, state that the view expressed by the High Court in the above case is the correct view. The Supreme Court, having accepted the proposition of the right of a party to withdraw its offer before its acceptance, could not have denied the enforcement of the very said right unless there was a separate and distinct consideration for holding the offer open for acceptance for a definite period of time. It is a possible view that the existence or non-existence of an underlying contract becomes irrelevant when the invocation of a bank guarantee is in terms of the bank guarantee and as such the writ court may not exercise its discretion. However, applying the principle of avoiding multiplicity of proceedings, the authors with respect reiterate that the view expressed by the High Court and judgment pronounced is the correct position.

12.20.2 To Justify Forfeiture the Terms of Contract Should Be Sufficiently Explicit

To justify forfeiture of advance deposit, being a part of price as "earnest", the terms of contract should be sufficiently explicit and made known to the party making the deposit. In a case, the validity period of an offer was asked to be extended and the tenderer did so, subject to a condition that the rebate was reduced as indicated in the conditional extension, and the acceptor ignored the said condition and treated the extension unconditional and accepted the tender as originally submitted. It was held that there was no underlying contract between plaintiff and defendant; once the plaintiff's offer lapsed it became entitled to return of earnest money amount. The defendant's invocation of the bank guarantee, in full knowledge that it was not entitled to forfeit the amount, was fraudulent. The plaintiff was granted an injunction against the forfeiture by the defendant. The plaintiff was also entitled to injunction restraining the bank from making payment under the bank guarantee.[139]

Where the tenderer contended that the work order was not given despite acceptance of his tender but admitted his signature on an alleged work order issued and also disclosed his incapacity to carry out the work, the forfeiture of earnest money was held justified in terms of the contract.[140] In the absence of a clause in the instructions to bidders, forfeiture may not be valid as held in Examples (1) and (2) below.

---

[137] Ibid.

[138] State of Maharashtra v. A. P. Paper Mills Ltd., AIR 2006 SC 1788 = 2006 AIR SCW 2037.

[139] D. S. Constructions Ltd., M/s. v. Rites Ltd, AIR 2006 Del. 98.

[140] Western Coalfields Ltd. V. Pradeep Kumar Soni & Anr., AIR 2009(NOC) 2009 (M.P.).

EXAMPLES FROM CASE LAW

1. Where the bank guarantee was conditional and contingent upon three conditions, namely, bidder withdrawing his bid after bid opening, refusing to execute the agreement and failing to provide a security deposit, and none of the three conditions were even alleged but the enforcement was sought on the ground that the bidder failed to perform the agreement with due punctuality; it was held by the Patna High Court that the action amounted to arbitrary, unjust, illegal and unlawful enrichment violative of Art. 14 of the Constitution. The bank was directed to return the amount.[141]

2. A bank guarantee furnished by way of bid security was invoked though the petitioner bidder had neither withdrawn its bid during the period of bid validity nor was he a successful bidder who failed to sign agreement so as to invoke bank guarantee which was condition precedent for forfeiture of bid security. The forfeiture was sought on the alleged ground of disqualification of the tenderer for indulging in fraudulent practice. It was held that in the absence of a clause in the instructions to bidders entitling the respondents to forfeit bid security on the said ground, the action of the respondents in invoking the bank guarantee was not legal.[142]

3. In a case concerning an auction sale, terms of the auction were neither advertised in writing nor supplied at the time of auction. What was stated was that the terms were orally intimated on the date of the auction. One such term intimated orally was that on non-deposit of three-quarters of the bid money, the one-quarter advance deposit would be forfeited as earnest. The applicant in the High Court had sought recovery of the one-quarter advance deposit that was forfeited. It was held by the Madhya Pradesh High Court:[143] "Without proof of any loss to the auctioning authority the forfeiture of 1/4th deposit, therefore, cannot be supported by provisions of S.74 of the Contract Act."

4. A society deposited fixed earnest money and formally applied for allotment of land. On acceptance of the offer, formal allotment was made by the authority. The authority, however, before delivery of possession, communicated enhancement of premium of land. The allottee society failed to make payment within the time extended by the High Court. It was held that forfeiture of earnest money by the authority was not illegal as communication by the authority as to enhancement was in continuation of the earlier offer and the society had accepted the offer contained in the said communication.[144]

5. Where the defendant in a case was found guilty of (a) fraud played by preponing date of lodging of goods to obtain benefit of early payment as per conditions of the letter of credit, (b) non-supply of goods of approved quality in specified time, and (c) short supply of goods, it was held by the Gujarat High Court that the plaintiff was entitled to injunction against encashment of the letter of credit.[145]

6. The petitioner, in a case, was the successful bidder and his bid was accepted subject to the right of the authorities to correct any inadvertent or typographical error. The belated contention of the petitioner that the letter of acceptance was a counter-offer was rejected

---

[141] Hindusthan Steel Works Constructions v. State of Bihar, AIR 2009 (NOC) 1527 (PAT.).

[142] M/s Sanicons v. Government of A.P., AIR 2006 A.P. 282.

[143] Bhanwarlas v. Babulal, AIR 1992 M.P. 6 (12).

[144] Delhi Development Authority v. Grihsthapana Co-operative Group Housing Society Ltd., AIR 1995 SC 1312. G. Ram v. Delhi Development Authority, AIR 2003 Del. 120. Also see: Narendrakumar Nakhat, Petitioner v. M/s. Nandi Hasbi Textile Mills Ltd. and others, Respondents, AIR 1997 Kant. 185. Also see M/s New Media Broadcasting (Pvt.) Ltd. v. Union of India, 2008 (NOC) 967 (DEL.).

[145] Adani Exports Ltd. v. Marketing Service Incorporated, AIR 2005 Guj. 257.

and further opportunity given to him to sign the contract. Eventually the contract was rescinded and earnest money was forfeited. The petitioner was disallowed from participating in the new tender. The Delhi High Court held the petitioner made himself liable for the penal action and dismissed the writ petition.[146]

7. The petitioner, in a case, sought an order restraining the respondents from encashing bank guarantees, when the respondents, without giving any notice, arbitrarily invoked bank guarantees. The claims of petitioners had not been finally adjudicated and there was an apparent attempt on the part of the respondent to frustrate findings recorded by the Disputes Review Board (DRB). Held: Action of the respondents in insisting upon encashment of bank guarantees was bound to cause irretrievable injustice to the petitioner and the petition was allowed.[147]

8. Where the owner had not made any claim for breach of contract against the contractor during the validity of the bank guarantee in question, the bank guarantee has outlived its purpose. The owner's threat to encash the guarantee (after the contractor had filed a Section 9 application) on the ground that the contractor did not agree to renewing the said guarantees was held beyond the terms of the contract and the injunction order in favour of the contractor against encashment of the bank guarantee was held legal. The order directing the contractor to furnish a fresh bank guarantee to secure ad hoc payments, it was held, amounts to creating a new contract, not permissible in law, as it is neither in ambit of the main contract nor was it agreed upon between parties even at a later stage.[148]

12.20.3 Writ for Refund of Earnest Money May Be Maintainable

Order of forfeiture, passed by an authority other than the one authorized to do so, is not valid.[149] In case the amount is wrongly retained by public authority, writ for refund may be maintainable. However, in a case the Supreme Court held: "It is settled law that disputes relating to contracts cannot be agitated under Article 226 of the Constitution of India. It has been so held in the cases …".[150] The dispute in this case was regarding the terms of offer. They were thus contractual disputes in respect of which a writ court was not the proper forum.[151]

Wrongful forfeiture of earnest money is generally when there exists no contract and, as such, the authors with respect state that the earlier view allowing writ is the correct view, when there is no contract between the parties. This view, expressed in the fifth edition of this title, found support in the decision of the Bombay High Court Division Bench by which a writ petition seeking relief in the nature of declaration that the order of forfeiture of the earnest money passed by the respondent is null and void. It was held:[152]

> As regards the earnest money, it is undisputed fact that the terms and conditions regarding the tender did not contain any term to the effect that the offerer would not be entitled for refund of the earnest money in case of withdrawal of the offer and certainly not in case of

---

[146] S.P. Singla Constructions Pvt. Ltd. v. Union of India, 2009(1) CTLJ 258 Delhi (DB).

[147] Hindustan Construction Co. Ltd. v. Satluj Jal Vidyut Nigam Ltd., AIR 2006 Delhi 169

[148] Satluj Jal Vidyut Nigam Ltd. v. Jai Prakash Hyundai Corsortium, AIR 2006 Delhi 239.

[149] State of Orissa and others, Appellants v. Ganeswar Jena, AIR 1994 Ori 94.

[150] Reference was made to: Kerala State Electricity Board v. Kurien E. Kalathil, reported in (2000) 6 SCC 293; State of U. P. v. Bridge and Roof Co., (India) Ltd., reported in (1996) 6 SCC 22; and B.D.A. v. Ajai Pal Singh, reported in (1989) 2 SCC 116.

[151] National Highways Authority of India v. M/s. Ganga Enterprises, AIR 2003 SC 3823; also see: Prasant Kumar Khuntia v. Union of India, 2009(1) CTLJ504 (Orissa) (DB).

[152] V.J. Agarwal v. C.O. Mira-Bhyander Municipal Council, AIR 2006 Bom. 254.

withdrawal of the offer beyond the period of validity of the bid. Once it is not in dispute that the period for the bid had expired on 16–5–1994 and thereafter but before the date of opening of the tenders, the offer was withdrawn by the petitioner, ... the same was liable to be refunded to the petitioner ... Taking into consideration the prevailing rate of interest, in our considered opinion, the flat rate of 5% per annum would be the appropriate rate of interest on the said amount.

12.21 Forfeiture of Security Deposit – Loss Must Be Proved

In view of the decisions of the Supreme Court of India, discussed above in paragraph 12.19, the legal position is very well settled that forfeiture of security deposit as well as other stipulations like recovery of costs of excess cement/steel consumption at penal rate would attract the provisions of Section 74 and loss has to be proved.[153] Similarly, if the provision is unreasonable it may be invalid. For example, the conditions in a laundry receipt restricting the launderer's liability for loss to 20 times the laundering charges or half of the value of un-returned clothes, whichever was less, was held unreasonable, arbitrary and opposed to public policy and hence void.[154]

In a number of cases, the Supreme Court of India has made it plain that a provision naming a fixed sum to be paid or amount fixed to be forfeited only gives to the aggrieved party a right to sue[155] and it, therefore, follows that if upon forfeiture of Security Deposit, the contractor is to file a suit for recovery of it, the aggrieved party will have to prove that (i) the contractor committed breach of contract and, further, that (ii) as a consequence of such breach the aggrieved party, in fact, suffered the loss equal to or more than the amount forfeited.

Section 74 of the Indian Contract Act deals with the measure of damages in two classes of cases: (i) where the contract names a sum to be paid in case of breach; and (ii) where the contract contains any other stipulation by way of penalty. Jurisdiction of the court to award compensation in case of breach of contract is unqualified except as to the maximum stipulated; but compensation has to be reasonable, and that imposes upon the court a duty to award compensation according to settled principles. The section undoubtedly stipulates that the aggrieved party is entitled to receive compensation from the party who has broken the contract, whether or not actual damages or loss is proved to have been caused by the breach. Thereby it merely dispenses with proof of "actual loss or damages". It does not justify the award of compensation when in consequence of the breach no legal injury at all has resulted. This is so because compensation for breach of contract can be awarded to make good loss or damage which naturally arose in the usual course of things, or which the parties knew when they made the contract, to be likely to result from the breach. This position was well settled by the Constitution Bench Decision of the Supreme Court of India.[156] Therefore, the party to a contract taking security deposit from the other party to ensure due performance is not entitled to forfeit the deposit on the ground of default when no loss is caused to him on consequence of such default.[157]

Where a bond has been invoked upon a breach of the contract under a clause allowing the performance bond to become "liable to be forfeited", the Court of Appeal held that the words

---

[153] M/s Kamil & Bros. v. Central Dairy Farm, AIR 2008 All. 33.

[154] R.S. Deboo v. Dr. M.V. Hindlekar and another, AIR 1995 Bom. 68.

[155] See Chapter 7, paragraph 7.10.

[156] Fateh Chand v. Balkishan Dass, AIR 1963 SC 1405. Also see: Kailash Nath Associates v. Delhi Development Authority and another, 2015(2) R.C.R.(Civil) 206; 2015(110) ALR 464.

[157] Besides Fateh Chand, also see Maula Bux v. Union of India, AIR 1970 SC 1955; Union of India v. Rampur Distillery and Chemicals Limited, AIR 1973 SC 1089, AIR 1989 M.P. 7 (18); HPMC v. M/s Khemka Containers(P) Ltd., AIR 2008 (NOC) 399 (HP).

meant "liable to be encashed". The right to call on a bond at an early stage of a breach would act as an obvious incentive for the performance of a contract. Subsequently, however, there must be an accounting between the parties to the contract. Thus, the bank must pay when called upon to do so but in the accounting the loss has to be proved. If it is less than the bond, the overpayment ought to be returned.[158]

EXAMPLES FROM CASE LAW

1. Clause 42 of the agreement in a case stipulated that the contractor was to see that only the required quantities of materials are issued. It was further the term of the agreement that the difference in quantity of cement actually issued to the contractor and the theoretical quantity including the authorized wastage, if not returned by the contractor, shall be recovered at twice the issue rate without prejudice to the provision of the relevant conditions regarding return of . materials governing the contract. The difference between the actual consumption and theoretical as pointed out under Clause 42 considered by the arbitrator. He found that the actual loss had not been proved, meaning thereby that the arbitrator, on the basis of the record provided before him, was not satisfied with the calculations arrived at by the department and, therefore, disallowed the double recovery. The arbitrator has given reasons for not accepting the double recovery as the DDA failed to prove the loss. The Delhi High Court upheld the award.[159]

2. The two-judge bench judgment of the Supreme Court of India held:[160]

> [W]hen parties have expressly agreed that recovery from the contractor for breach of the contract is pre-estimated genuine liquidated damages and is not by way of penalty duly agreed by the parties, there was no justifiable reason for the arbitral tribunal to arrive at a conclusion that still the purchaser should prove loss suffered by it because of delay in supply of goods.

This decision, it is stated with utmost respect, being contrary to the decision of the Constitution Bench in Fateh Chand's case, has not laid down good law. It relies on the age-old distinction in the English law between the penalty and liquidated damages, which distinction was clearly eliminated by the express provisions of Section 74 which treat both the stipulations at par insofar as giving effect to them. Until the position changes by a subsequent decision, this case can be distinguished on the basis of peculiar facts which were stated as follows.

The agreement between the parties specifically provided that without prejudice to any other right or remedy if the contractor fails to deliver the stores within the stipulated time, the appellant will be entitled to recover from the contractor, as agreed, liquidated damages equivalent to 1% of the contract price of the whole unit per week for such delay. Such recovery of liquidated damage could be at the most up to 10% of the contract price of whole unit of stores. Not only this, it was also agreed that:

---

[158] Cargill International SA and Another v. Bangladesh Sugar and Food Industries Corporation, [1998] 1 WLR 461.
[159] M/s. Jagan Nath Ashok Kumar v. Delhi Development Authority, AIR 1995 Del. 87.
[160] Oil and Natural Gas Corporation Ltd. v. SAW Pipes Ltd., AIR 2003 SC 2629.

a. Liquidated damages for delay in supplies will be recovered by paying authority from the bill for payment of cost of material submitted by the contractor;

b. Liquidated damages were not by way of penalty and it was agreed to be genuine, pre-estimate of damages duly agreed by the parties;

c. This pre-estimate of liquidated damages was not assailed as unreasonable assessment of damages by the parties.

d. "Further, at the time when respondent sought extension of time for supply of goods, time was extended by letter dated 4–12–1996 with a specific demand that the clause for liquidated damages would be invoked and appellant would recover the same for such delay. Despite this specific letter written by the appellant, respondent had supplied the goods which would indicate that even at that stage, respondent was agreeable to pay liquidated damages."[161]

3. Placing reliance on the above decision, it was observed and held by the Orissa High Court in a case:[162]

> So far as first item of claim is concerned, it appears from the award that the claimant-respondent had claimed Rs. 49,721/- and the appellant in the final bill had proposed for recovery of money for the materials issued and not refunded at penal rate.

The arbitrator held that there was no pleading regarding theft or misuse or pilferage of the material issued to the claimant and consequently there was no proof of any loss. He also found that recovery of money at double rate was hit by Section 74 of the Contract Act and, therefore, allowed single recovery for an amount of Rs. 47,345/-. Held:

> The claimant-respondent having accepted the terms and conditions of the agreement before executing the work, has no right to say that the aforesaid two clauses are hit by any provision of law. ... Therefore, so far as claim No. 1 is concerned, in my view, the Arbitrator has committed an illegality in not allowing recovery as per the terms of the agreement and in this respect the appellant shall be entitled to recover a further sum of Rs. 47,345/-.

12.21.1 Forfeiture of Security Deposit in Addition to Recovery of Excess Cost – Illegal

Outdated PWD standard form contracts include provisions empowering the authority to forfeit security deposit and also recover the excess expenditure incurred in getting the balance work executed by another agency. This would result in conferring double benefit to the state. The amount deposited by way of security for guaranteeing the due performance of the contract could not be regarded as earnest money as already stated. Under Section 74 only a reasonable amount can be forfeited if a contract is not performed.[163] The law is well settled that the party to

---

[161] Ibid.

[162] Bharat Sanchar Nigam Ltd. v. Trinath Singh, AIR 2005 Ori. 70.

[163] State of U.P. v. Chandra Gupta & Co., AIR 1977 All. 28; also see: Maula Bux v. Union of India, AIR 1970 SC 1955; Union of India v. Rampur Distillery and Chemicals Limited, AIR 1973 SC 1908; M/s. Macbrite Engineers v. Tamil

a contract taking security deposit from the other party to ensure due performance of the contract is not entitled to forfeit the security deposit on the ground of default, when the loss caused to him is made good in consequence of such default or where no loss is proved.[164]

The following observations of the Supreme Court of India are worth noting in this connection, it is submitted:[165]

> We find that there is no term in the agreement between the parties enabling the appellant to forfeit the security deposit. The only right given to the appellant was to deduct out of the security deposit the amount of loss incurred by the appellant which was caused to them by reason of non-completion of the work by the respondent in time and to recover the extra cost of the work which had to be completed by the appellant departmentally on account of default of the respondent, subject to certain limits. The appellant has neither proved the amount of damage incurred by it nor has it proved the extra amount of cost incurred by it for getting the incomplete work done departmentally. Thus, the argument that the appellant was entitled to forfeit the security deposit or any part of the same, must fail.

Where the breach by the contractor stands proved and the balance work is executed through another agency and excess cost is incurred, the employer will be entitled to receive the compensation to the extent loss is proved.[166]

12.21.2 Events Justifying Forfeiture

In modern standard form contracts the commonest contingencies upon which the right to forfeit is made dependent include breach of the contract by the contractor, by:

- i. Failure to sign the agreement and commence work
- ii. Making himself liable to pay the full amount of penalty/liquidated damages
- iii. Committing breach of any condition of the contract such as not providing remedy for defective work
- iv. Declaring bankruptcy or winding up of the company.[167]

Invariably the question of deciding whether the event upon which a power of forfeiture may be exercised has occurred or not, is left for determination by the engineer or architect and special provisions of each contract need to be referred to in order to ascertain the event. The drafters of such clauses expect the engineer or architect to be a third person, independent of the two parties to the contract, which may be true in private works but not in the case of public works. Thus, in the case of public works, not infrequently, the engineer becomes the judge of his own cause, which permits a large number of grave injustices and anomalies. Invariably the matters land in arbitration or court. It is to be noted that an arbitrator or the court will strictly construe the forfeiture clauses in order to determine whether the operative event has occurred or not.

In a case where the first highest bidder failed to deposit the balance amount in an auction sale and the second successful bidder who had given Rs. 10 lacs as earnest money stepped into the shoes of the highest bidder, the division bench of the Calcutta High Court held that once the second successful bidder accepted the offer and deposited money the contract stood

---

Nadu Sugar Corporation Ltd., AIR 2002 Mad. 429; State of A. P. and others v. Singam Setty Yellananda, AIR 2003 AP 182.

[164] State of Tamil Nadu v. T.R. Surendranath, AIR 2008(NOC) 974 (Mad.).

[165] State of Rajasthan v. Bootamal Sachdeva, AIR 1989 SC 1811.

[166] Karnataka Electricity Board v. M.S. Angadi, AIR 2008 Kar. 55.

[167] These aspects of breaches have been explained in Chapters 10 and 11.

concluded. It was held that having paid only 10 lacs and bid for property of Rs. 3.5 crores, invocation of forfeiture clause was not oppressive or onerous.[168]

In a case decided by the P. & H. High Court, the facts were peculiar and are briefly stated as follows. The highest bidder did not accept the terms and conditions and filed a writ petition for the return of his bank guarantee. The second highest bidder also sought return of his guarantee, which was not returned. In the meantime the second highest bidder was given the option to match the offer of the highest bidder, which was declined by him. Thereafter his second highest offer was accepted. The second highest bidder declined to sign the agreement whereupon the earnest money was forfeited and encashment of bank guarantee was sought. It needs to be noted that the conditions of the NIT were clear that the offer was to remain valid for acceptance for five months and to be irrevocable during the validity period and earnest money would be adjusted towards the total amount to be received. The plea that the petitioner not being the highest bidder was entitled to refund of the earnest money was rejected. The writ petition was rejected and it was held that the authorities were entitled to encashment of bank guarantee and it would not amount to unjust enrichment.[169]

12.21.3 Waiver and Estoppel

The owner may find himself precluded from enforcing the forfeiture, although the circumstance may have occurred justifying the exercise of such a right, either because he has waived his right or he is estopped from exercising it. The right must be exercised within a reasonable time after the breach unless the breach is a continuing one.

12.21.4 Forfeiture of Any Money Due to the Contractor

Besides forfeiture of earnest money and security deposit which are common provisions in India in the standard form contracts for public works considered above, forfeiture clauses, if any, empowering the owner to forfeit any money due to the contractor at the time of forfeiture, generally, operate as penalties and as such are subject to actual proof of damage. Section 74 of the Indian Contract Act applies to such provisions of the agreement. One such provision is considered in paragraph 12.21.5 below.

12.21.5 Right to Recover Sum from Other Contracts

Standard form contracts for building and engineering contracts in use by the government departments, generally include a provision such as the one embodied in Clause 29 of the General Conditions of Contract used by the CPWD which reads as follows:

> CLAUSE 29 (1) Whenever any claim against the contractor for the payment of a sum of money arises out of or under the contract, the Government shall be entitled to recover such sum by appropriating in part or whole, the security deposit of the contractor, and to sell any Government promissory notes, etc. forming the whole or part of such security. In the event of the security being insufficient or if no Security has been taken from the contractor, then the balance or the total sum recoverable as the case may be, shall be deducted from any *sum then due or which at any time thereafter may become due to the contractor under this or any other contract* with the Government. Should this sum be not sufficient to cover the full amount recoverable, the contractor shall pay to the Government on demand the balance remaining due.
>
> (Emphasis added)

---

[168] Pawanputra Commotrade Pvt. Ltd. v. Official Liquidator, High Court Calcutta & Anr., AIR 2008 (NOC) 398 (Cal.).

[169] INFOTECH 2000 India Limited v. State of Punjab, AIR 2007 P. & H. 58.

A similarly worded Clause 18 of the General Conditions of Contracts contained in the standard form of contract No. D.G.S. & D. 68 was considered by the Supreme Court of India on two occasions. In the first case, the Union of India v. Raman Iron Foundry, it was held:[170]

> A claim for damages for breach of contract, is, therefore, not a claim for a sum presently due and payable, and the purchaser is not entitled, in exercise of right conferred upon it under Clause 18 to recover the amount of such claim by appropriating other sums due to the contractor.

However, in the second case the above decision was reconsidered by the Supreme Court. It was held:[171]

> We are clearly of the view that an injunction order restraining the respondents from withholding the amount due under other pending bills to the contractor virtually amounts to a direction to pay the amount to the contractor-appellant. Such an order was clearly beyond the purview of clause (b) of Section 41 of the Arbitration Act. The Union of India has no objection to the grant of an injunction restraining from recovering or appropriating the amount lying with it in respect of other claims of the contractor towards its claim for damages. But certainly Clause 18 of the standard contract confers ample power upon the Union of India to withhold the amount and no injunction order could be passed restraining the Union of India from withholding the amount.

There is no doubt, that the law as laid down in the later decision of the Supreme Court will be applicable to such cases. As a result thereof it is permissible for the competent court to grant an injunction restraining the government from recovering or appropriating any damages claimed by it from other pending bills of the contractor but no such order can be issued restraining the government from "withholding" the payments due to the contractor under other pending bills. Clause 18 as well as Clause 29 referred to above empower the government to recover the amount or appropriate any sum due and payable under other contract and not to withhold it. An injunction will restrain the government from both "recovering and appropriating". In spite of such an injunction, if the amount is "withheld" by the government the contractor may contend that there is a breach of term of payment of the contract in which the amount is wrongfully and illegally withheld.

12.22 Wrongful Forfeiture

In conclusion of the discussion in respect of forfeiture, it can be said that the exercise of power to forfeit may be invalidated on any one or more of the following grounds:

i. None of the events upon which it is conditioned has occurred

ii. Although such an event has happened, it was caused by the act of the party seeking to exercise it, or by his agents

iii. The notice required to be given in terms of the contract was not given

iv. There has been delay or other conduct recognizing the continued existence of the contract after knowledge of the breach, if the breach is not a continuing one.

---

[170] Union of India v. Raman Iron Foundry, AIR 1974 SC 1265.
[171] H.M. Kamaluddin Ansari and Co. v. Union of India, AIR 1984 SC 29; (1983) 4 SCC, 430.

The exercise of the power to forfeit under the above circumstances may amount to wrongful forfeiture. It has to be noted that a wrongful forfeiture cannot be justified by reference to a subsequent event which would have justified it.

EXAMPLE FROM CASE LAW

A tender for the road work submitted by the petitioner was accepted on 25th June 2007. Time allowed for completion of the work was 19 months. The site was partly handed over. It was full of obstructions. On 4th November 2007, cognizance was taken by the authorities and steps were to be taken with district authorities for removal of the obstructions. On 11th January 2008, the contract was terminated and earnest money forfeited. The writ petition by the tenderer/petitioner was allowed. It was held that the termination was per se arbitrary and violative of Art. 14 of the Constitution. The contract was restored and the notice for re-tender was rescinded.[172]

If a contractor accepts a wrongful forfeiture and termination of a contract and rescinds it, the owner is liable for the possible heavy damages attendant upon wrongful and illegal termination. It is to be noted that if he discovers his mistake, he cannot restore the contract and the status quo ante without the consent of the contractor.

12.22.1 Silence for Long Time Amounts to Refusal to Pay Back Security

Conduct by the owner maintaining silence about refund of security deposit may amount to refusal to refund and give rise to a dispute.

EXAMPLE FROM CASE LAW

By an agreement dated 24th January 1979, the appellant had undertaken the work of construction of the spillway of head regulator of Talsara Irrigation Project. The stipulated date for completion of the work as per the agreement was 23rd July 1980. The appellant abandoned the work on 31st March 1980 and prayed for closure of the agreement. After waiting for three years the appellant made a reference in 1983 to arbitration tribunal. In respect of the refund of security deposit claim, the tribunal came to the conclusion that the claim was premature as no order had been passed on prayer of the appellant by the state. The Orissa High Court held:[173]

> Three years after notice of abandonment, reference was made to Tribunal. Silence for such a long period amounts to refusal to pay back the security. Even if no decision was taken on receipt of notice of arbitration, respondent could have brought all facts why the security had not been refunded. In such a circumstance, terms of agreement by finding that it is premature is an error of law apparent on the face of the award. No further material is necessary to be considered to come to this conclusion.

Award to that extent was set aside.

---

[172] Satav Infrastructure Pvt. Ltd. v. Union of India, 2009(1) CTLJ 481 (Patna).
[173] M/s Vibgyer Structural Construction Pvt. Ltd. v. State, AIR 1991 Ori.323.

12.23 Enforcement of Bank Guarantee

In most construction contracts, especially for big projects, the practice of accepting bank guarantee in place of cash security is followed. The system has an advantage of the working capital of the contractor not getting locked up and as a result, thereby, the owner also gets a benefit of lower rates than the rates which the contractor would have tendered if he were to keep cash security deposit. More often than not when a demand is made for the encashment of a bank guarantee, the court is approached with a view to getting an injunction restraining the bank from making payment under the bank guarantee. The question in such cases is also similar to the question of forfeiture of security deposit. Legal aspects of encashment of bank guarantees need to be considered here because it involves, besides the two parties to a contract, the bank as a third party.

12.23.1 Liability Under Bank Guarantee – Whether Absolute

A bank guarantee is the common mode of securing payment of money in commercial dealings as the beneficiary, under the guarantee, is entitled to realize the whole of the amount under that guarantee in terms thereof irrespective of any pending dispute between the person on whose behalf the guarantee was given to the beneficiary. In contracts awarded to private individuals by the government, which involve huge expenditure, as for example, construction contracts, bank guarantees are usually required to be furnished in favour of the government to secure payments made to the contractor as "Advance" from time to time during the course of the contract and also to secure performance of the work entrusted under the contract. Such guarantees are encashable in terms thereof on the lapse of the contractor either in the performance of the work or in paying back to the government advance money received. When the guarantee is invoked, the beneficiary recovers the amount from the bank. It is for this reason that the courts are reluctant in granting an injunction against the invocation of bank guarantee, except in the case of fraud, which should be established, or where irretrievable injury is likely to be caused to the guarantor. This is the principle laid down by the Supreme Court in various decisions.[174]

What is important, therefore, is that the bank guarantee should be in unequivocal terms, unconditional and recite that the amount would be paid without demur or objection and irrespective of any dispute that might have cropped up or might have been pending between the

---

[174] In U.P. Co-operative Federation Ltd. v. Singh Consultants and Engineers Pvt. Ltd., (1988) 1 SCC 174, the law laid down in Bolivinter Oil SA v. Chase Manhattan Bank, (1984) 1 All ER 351 was approved and it was held that an unconditional bank guarantee could be invoked in terms thereof by the person in whose favour the bank guarantee was given and the courts would not grant any injunction restraining the invocation except in the case of fraud or irretrievable injury. Also see: United Engineers (Malaysia) Essar Projects Ltd. v. NHAI, 2008(1)CTLJ 208(Delhi) (SN); Petroleum India International v. Bank of Baroda, 2008(1) CTLJ 184 (Bombay). State Bank of India v. Nihar Fathima & Ors., AIR 2009(NOC) 110 (Mad.). 12.22.1 BG. Svenska Handelsbanken v. Indian Charge Chrome, (1994) 1 SCC 502/(1993 AIR SCW 4002: AIR 1994 SC 626); Larsen and Toubro Ltd. v. Maharashtra State Electricity Board, (1995) 6 SCC 68: (1995 AIR SCW 4134: AIR 1996 SC 334); Hindustan Steel Works Construction Ltd. v. G.S. Atwal and Co. (Engineers) (P) Ltd., (1995) 6 SCC 76: (1995 AIR SCW 3821: AIR 1996 SC 131); National Thermal Power Corporation Ltd. v. Flowmore (P) Ltd., (1995) 4 SCC 515: (1995 AIR SCW 430: AIR 1996 SC 445); State of Maharashtra v. National Construction Co., (1996) 1 SCC 735: (1996 AIR SCW 895: AIR 1996 SC 2367); Hindustan Steel Works Construction Ltd. v. Tarapore and Co., (1996) 5 SCC 34: (1996 AIR SCW 2861: AIR 1996 SC 2268) as also in U.P. State Sugar Corporation v. Sumac International Ltd., (1997) 1 SCC 568: (1997 AIR SCW 694: AIR 1997 SC 1644: 1997 All LJ 638); Dwarikesh Sugar Industries Ltd., Appellant v. Prem Heavy Engineering Works (P) Ltd., and another, AIR 1997 SC 2477; Fenner (India) Ltd., Appellant v. Punjab and Sind Bank, AIR 1997 SC 3450; Larsen and Toubro Limited, Appellant v. Maharashtra State Electricity Board and others, AIR 1996 SC 334; National Thermal Power Corporation Ltd., Appellant v. M/s. Flowmore Private Ltd. and another, AIR 1996 SC 445; Federal Bank Ltd., Appellant v. V. M. Jog Engineering Ltd. and others, 2001 (Suppl.) Arb LR 572 (SC) the same principle has been laid down and reiterated. Also see: Bank of Baroda v. Ruby Sales Corpn. (Agency), AIR 2006 Guj. 251.

beneficiary under the bank guarantee or the person on whose behalf the guarantee was furnished. The terms of the bank guarantee are, therefore, extremely material. Since the bank guarantee represents an independent contract between the bank and the beneficiary, both the parties would be bound by the terms thereof. The invocation, therefore, will have to be in accordance with the terms of the bank guarantee; or else the invocation itself would be bad.

Where a bank guarantee made amount payable

> on his first demand without whatsoever right of objection on our part and without his first claim to the contractor, in the amount not exceeding Rs. 10,00,000/- (Rupees Ten lakhs only) in the event that the obligations expressed in the said clause of the above mentioned contract have not been fulfilled by the contractor giving the right of claim to the employer for recovery of the whole or part of the Advance Mobilization Loan from the contractor under the contract,

the Supreme Court held:[175]

> The Bank Guarantee thus could be invoked only in the circumstances referred to in Clause 9 whenever the amount would become payable only if the obligations are not fulfilled or there is misappropriation. That being so, the Bank Guarantee could not be said to be unconditional or unequivocal in terms so that the defendants could be said to have had an unfettered right to invoke that Guarantee and demand immediate payment thereof from the Bank.

No distinction can be drawn between the guarantee for due performance of works contract or towards security deposit. The obligation of bank remains the same and has to be discharged in a manner provided in the bank guarantee.[176] The general rule of law is that liability arising out of the unilateral contracts of commercial credits such as letters of credits, performance bonds or bank guarantees, is absolute. An irrevocable letter of credit has a definite implication. It is a mechanism of great importance in international trade. Any interference with that mechanism is bound to have serious repercussions on the international trade of a country. Except under very exceptional circumstances the courts do not interfere with that mechanism.[177] The letter of credit is independent of and unqualified by the contract, sale or any other underlying transaction. The autonomy of an irrevocable letter of credit is entitled to protection.[178] A bank guarantee is very much like a letter of credit. The courts will do the utmost to enforce it according to its terms.[179] Where in a case the parties entered into four contracts and one wrap-up agreement and four bank guarantees were submitted; three against advances paid and one for performance. The agreement stipulated: "In case of any material breach of any or all contracts, BSES shall have the right to embark upon the retentions and encashment of Bank guarantees of all the contracts." The beneficiary had recovered advance payment in full from running bills and yet the three bank guarantees against the same were also encashed. The Supreme Court did not accept the plea that this fact was an "egregious fraud" or in any case created a situation of "special equities". The plea of "irretrievable injustice" was rejected on the ground that arbitral proceedings regarding the same issues were in progress.[180]

---

[175] Hindustan Construction Co. Ltd. v. State of Bihar, AIR 1999 SC 3710.

[176] Hindustan Steel Works Construction Ltd. Appellant v. Tarapore and Co. and another, AIR 1996 SC 2268.

[177] Tarapore & Co. Madras v. Tractoroexport, Mascow, AIR 1970 SC 891; Syndicate Bank v. Vijay Kumar, AIR 1992 SC 1066, reversing the decision of the Delhi High Court and authorities listed in an opinion case National Aluminium Co, Ltd. v. M/s. R.S. Builders (India) Ltd., AIR 1991, Orissa 314.

[178] United Commercial Bank v. Bank of India, AIR 1981 SC 1426, relied on: Jagadish Constructions Ltd. v. M.P. Rural Road Development Authority, AIR 2007 M.P. 266.

[179] Elur v. Matsar, (1966) 2 LILR d95.

[180] BSES Ltd. (now Reliance Energy Ltd.), M/s. v. M/s. Fenner India Ltd., AIR 2006 SC1148 = 2006 AIR SCW 721.

"On Demand and Without Demur" – Interpretation

The forms in use for bank guarantees in construction contracts usually stipulate that the bank shall pay the sum guaranteed "on demand and without demur". A typical illustration and true legal position in such cases is to be found in the case of NPCC Ltd. v. M/s G. Rajan,[181] it is submitted.

The brief facts of the case were that the NPCC Ltd., entrusted a construction contract to a firm. The contract provided for giving of mobilization advance by the NPCC Ltd. to the firm and the recovery thereof guaranteed by a bank guarantee. The firm allegedly having failed to adhere to the agreed work schedule, the NPCC Ltd. rescinded the contract and asked the firm to deposit unrecovered advance within seven days failing which the demand on the bank for encashment of bank guarantees would be made. On receipt of the letter, the firm filed a suit under Section 20 of the Arbitration Act, 1940 for reference of disputes to an arbitrator and during pendency of the said suit prayed for an order of injunction under Section 41 of the Arbitration Act, 1940 restraining the NPCC Ltd. from taking any steps to enforce the bank guarantee. The trial court granted the injunction order taking the view that if the NPCC Ltd. is not allowed to encash the bank guarantee until final adjudication of the matter, no harm or inconvenience would be caused to it. On the other hand, the enforcement of the bank·guarantee at that stage would result in irreparable loss and injury to the firm, particularly if the dispute was ultimately decided in its favour and that there would also be a multiplicity of proceedings if the prayer for injunction was not allowed. The NPCC Ltd. filed a revision petition against the said order of injunction passed by the trial court. It was held:

> The decision of the Supreme Court in MSEB v. Official Liquidator (AIR 1982 SC 1497) (Supra) makes it clear that the Bank guarantee executed by the Bank in favour of the petitioner stands independent of the other connected transactions, namely, the contract between the petitioner and the opposite party and the document under which the opposite party had given security to the Bank for executing the letter of guarantee. It also stands independent of any claim or counter-claim arising out of the contract between the petitioner and the opposite party. That being so, there is prima facie no reason according to us, why the Bank guarantee cannot be enforced by the petitioner subject, of course, to the fulfillment of the conditions as laid down therein.[182]

A number of authorities, about 48, were examined during the course of judgment. There are a number of decisions to support the proposition that the liability under a bank guarantee is absolute and so the law is well established.[183]

The Supreme Court has also settled the law that the bank is not concerned with the recovery of part of the amount or whole of the amount and outstanding amount under running bills by holding: "The right to recover the amount under the running bills has no relevance to the liability of the Bank under the guarantee."[184] With utmost respect, it is submitted that the law as declared is based on the letter and not the spirit and the casualty is the justice. Whereas the party encashing the bank guarantee receives the cash without adjudication, the party paying has to wait for a decade or so to get the amount back even if the party succeeds in arbitration. The well-

---

[181] N.P.C.C. Ltd. v. M/s G. Rajan, AIR 1985 Cal. 23.

[182] Ibid.

[183] Daewoo Motors India Ltd., Appellant v. Union of India and others, AIR 2003 SC 1786; National Highway Authority of India, Appellant v. M/s. Ganga Enterprises and another, AIR 2003 3823; AIR 1982 SC 1497; AIR 1981 SC 1426; AIR 1979 Cal. 44; AIR 1979 Cal. 370; [1973] 2 Lloyds Rep. 437; (1973) 117S.J 483 (C.A.); Bache & Co. v. Banque Vernes, (1970) 74 Cal. W. N. 991; (1966) 2 Lloyds Rep. 495 (CA) Elian v. Master. Also see: Centax (India) Ltd. v.Vinmar Impex Ltd., AIR 1986 SC 1924 (upholds AIR 1986 Cal. 356); also see Allied Resins & Chemicals Ltd. v Mineral & Metal Trading Corpn. of India Ltd., AIR 1986 Cal. 346; M/s Triveni Engineering Work Ltd. v. M/s. Belganga S.S.K. Ltd., AIR 1992 Del. 164 etc.

[184] General Electric Technical Services Company Inc. v. M/s. Punj Sons (P) Ltd., AIR 1991 SC 1994.

known doctrine of balance of convenience gets wholly ignored in such a case, it is respectfully submitted. In England, for example, the accounting after the encashment of a bank guarantee, would take place reasonably quickly and, in case of a disagreement over it, the dispute would be resolved relatively quickly in courts. English courts recognize that a performance bond is not the same as a letter of credit in its importance to day-to-day trade transactions. There is, therefore, less need for a doctrine of strict compliance.[185]

Indeed in the above case a number of foreign judgments wherein the rule and its exceptions have been elegantly summarized were cited.[186] It was also submitted that the Singapore Court has gone so far as to say that the unconscionable calling of a bank guarantee was an exception independent of fraud. However, all these submissions were rejected and it was held: "Whatever may be the law, as to the encashment of bank guarantees in other jurisdictions, when the law in India is clear, settled and without any deviation whatsoever, there is no occasion to rely upon foreign case law." It is noteworthy that the foreign judgments were not declared to be opposed to public policy of India, a ground on which a court can refuse to apply the foreign judgment.

The authors are reminded of the questions raised by the Apex Court in an earlier case; the relevant passages read:[187]

> The law exists to serve the needs of the society which is governed by it. If the law is to play its allotted role of serving the needs of the society, it must reflect the ideas and ideologies of that society. It must keep time with the heartbeats of the society and with the needs and aspirations of the people. As the society changes, the law cannot remain immutable. … This task must, therefore, of necessity fall upon the courts because the courts can by the process of judicial interpretation adapt the law to suit the needs of the society.

In the same judgment, after tracing the development of the law in several countries, it was observed:

> Should then our courts not advance with the times? Should they still continue to cling to outmoded concepts and outworn ideologies? Should we not adjust our thinking caps to match the fashion of the day? Should all jurisprudential development pass us by, leaving us floundering in the sloughs of nineteenth-century theories?

It is heartening to notice that the courts in India do not hesitate to grant stay on encashment of bank guarantees in appropriate cases by devising ways and means to fit a case under recognized exception and render justice. As every rule has a few exceptions, the rule of law in respect of the bank guarantee too has a couple of exceptions.

12.23.2 Exceptions to the General Rule

"General propositions do not solve concrete cases", Justice Holmes has said.[188] While the law generally stated is that liability arising out of the unilateral contracts of commercial credits, such as letters of credit, bank guarantees and performance bonds is absolute, the intention of the parties as gathered from a reasonable construction of the language of the particular contract must ultimately govern the decision of the court as to the arising of the liability thereunder. The terms of a particular document may even constitute an exception to the general rule.[189]

---

[185] IE Contractors Ltd. v. Lloyds Bank Plc, and IE Contractors Ltd. v. Rafidain Bank, [1990] 2 Lloyd's Rep. 496.

[186] (2003) EWHC 762 (TCC): (2003) 1 All ER (Comm.) 914. See also Elian and Rabbath (Trading as Elian and Rabbath) v. Matsas, etc., [1966] 2 Lloyd's Rep. 495; Ibid., at paragraph 46, per Judge Thomton QC.

[187] Central Inland Water Transport Corporation Ltd. v. Brojo Nath Ganguly, AIR 1986 SC 1571.

[188] Lochner v. New York, 198 US 45, 76 (1905) Holmes, J. in a dissenting opinion.

[189] Harprashad & Co. v. Sudarshan Steel Mills, AIR 1980 Del. 175; also see AIR 1982 Del. 357.

(1) *Suppression of Material Facts by Beneficiary*

The duty of the appellant in making the demand on the bank is like the duty of the plaintiff to disclose the cause of action in the plaint. Just as a plaint is liable to be rejected for non-disclosure of the cause of action, a demand by the beneficiary of the bank guarantee is liable to be rejected by the bank if it does not state the facts showing that the conditions of the bank guarantee have been fulfilled. Enforcement of bank guarantee can be prevented in case equity is in favour of a person who has given the guarantee and also when the invocation letter is not in terms with the agreement. The bank guarantee and the letter of invocation must be read together to find out whether invocation has been made in accordance with the terms of the bank guarantee or not. If not, the bank is not liable to pay. Unless the loss or damage is quantified as required by the bank guarantee – there would be no debt payable by the bank as held in Union of India v. Raman Iron Factory AIR 1974 SC 1265. If the damage is not assessed, it remains a claim for unliquidated damage.[190]

EXAMPLE FROM CASE LAW

In a case decided by the Calcutta High Court the petitioner had instructed his bank to issue a bank guarantee against mobilization advance to be received from the beneficiary. The bank negligently issued a guarantee in a form used for a performance guarantee without striking out unnecessary and inapplicable clauses contained therein. The petitioner was not a party to this document, and, therefore, could not detect the mistake on the part of the bank. The guarantee was accepted by the respondent. It was held that the respondent was not entitled to invoke mobilization advance guarantee for the recovery of loss and damage.[191]

(2) *A Wrongful and Fraudulent Enforcement of Bank Guarantee*

If the guarantee is enforced by fraud, misrepresentation, deliberate suppression of material facts, or the like, that will give rise to special equity in favour of the contractor who will then have the right to stop its enforcement by obtaining an order from the court. The law can be said to be well settled that courts will not interfere with the enforcement of conditional or unconditional bank guarantees or letters of credit on mere allegation of fraud or special equity without any *prima facie* proof. For obtaining an order from the court, a very strong *prima facie* case must be made out. In England, even if fraud is shown, the court may consider the balance of convenience in order not to compromise the principle that the bank is autonomous from the contractual parties. The balance of convenience will be in favour of an injunction, for example, if the bank was also aware of the fraud.[192]

EXAMPLES FROM CASE LAW

1. In a case, the petitioner was engaged by the respondent for completion of civil work for which bank guarantees were furnished by the petitioner to the respondents. Disputes arising

---

[190] Banerjee & Banerjee v. Hindustan Steel Works, AIR 1986 Cal. 374.

[191] M/s. G.S. Atwal & Co. Engineers Pvt. Ltd. v. H.S.W.C. Ltd., AIR 1989 Cal. 1 84; also see M/s. Escort Limited v. M/s. Modern Insulators Ltd., AIR 1988 Del. 345. M/s Synthetic Foams Ltd. v. Simplex C.C.P. (India) Pvt. Ltd., AIR 1988 Del. 207.

[192] Tetronics (International) Ltd. v. HSBC Bank Plc, [2018] EWHC 201 (TCC); [2018] 4 WLUK 130.

between the parties were referred to the DRB. The respondents, without giving any notice, arbitrarily invoked the bank guarantees. There was an apparent attempt on the part of the respondents to frustrate findings recorded by DRB. It was held: the action of the respondents in insisting upon encashment of bank guarantees was bound to cause irretrievable injustice to the petitioner and the respondents were therefore liable to be restrained from encashing bank guarantees.[193]

2. The employer in a case, never laid any claim for breach of contract against the contractor during validity of the bank guarantee in question. Beneficiary did not invoke the said guarantees until the contractor moved court under Section 9 of the Arbitration and Conciliation Act, 1996 for injunction order against encashment of bank guarantee in question pending a final decision of the arbitral tribunal. Held: there was no justification for beneficiary to extend threat for invocation of bank guarantee in question on ground that contractor did not agree to renewing said guarantees beyond terms of contract. Injunction order granted by the single judge in favour of the contractor against encashment of bank guarantee was upheld by the division bench of the Delhi High Court. At the same time the order of the single judge directing the contractor to furnish fresh bank guarantee to secure ad hoc payments was held as amounting to creating a new contract, not permissible in law as it is neither in the ambit of the main contract nor was it agreed upon between parties even at a later stage.[194]

3. Where it was not a case of mere breach of contract *simpliciter* but a highly and grossly arbitrary action wrongly and virtually verging on fraud committed by the state in invoking a bank guarantee and the bank honouring such an invocation, the writ petition would be maintainable and the petitioner in such a situation cannot be relegated to onerous remedy of civil court.[195]

4. Seven bank guarantees were given by the petitioner in favour of the respondents, pursuant to the express term of a contract entered into between the petitioner and the respondent for construction work. Out of the seven guarantees, two were in lieu of security deposit and five were for securing mobilization advance made by the respondent to the petitioner. The two guarantees for security deposit could be enforced by the respondent by making a written demand stating that the petitioner had committed a breach of any of the terms of the contract and the extent or quantum of loss or damages suffered. The decision of the respondent was final and could not be questioned by the bank as regards the quantum of damage. The remaining five guarantees could be enforced if the petitioner failed to utilize the mobilization advance for the purpose of the contract or the respondent failed to fully recover the said sum with interest in accordance with the stipulation in the contract. The respondent had already recovered certain amounts towards the mobilization advance and security deposit. The respondent suppressed the facts of recoveries already made. The respondent also failed to discharge its duty as the sole judge, under the terms of the guarantee, to quantify the damages and to mention the extent of recoveries made by it,

---

[193] Hindustan Construction Co. Ltd. v. Satluj Jal Vidyut Nigam Ltd., AIR 2006 Delhi 169.
[194] Satluj Jal Vidyut Nigam Ltd. v. Jai Prakash Hyundai Corsortium, AIR 2006 Delhi 239.
[195] Hindustan Steel Works Construction v. State of Bihar & Ors, AIR 2009 (NOC) 380 (Pat.).

although it was within its special knowledge. The demand letters must be in accordance with the bank guarantees. By suppressing the material fact of recoveries already made by the respondent to the extent of Rs. 3.5 lacs, the respondent attempted to recover the entire sum of Rs. 11.5 lacs under the seven guarantees. The High Court held:[196]

> The suppression of this material fact in the demand letters has given rise to a special equity in favour of the petitioner to stop payment by the bank on the basis of these demand letters. Although in the petition, there is no allegation of fraud, ... this willful false representation by the beneficiary ... is a factor, which must be treated on the same footing as "fraud" giving rise to special equity and must be treated as an exception to the general rule that the Court should not interfere in these matters. ... The respondent ... must wait until the disputes are resolved by the arbitrator.

5. In a case, the tender submitted by the appellant was accompanied by a bank guarantee for Rs. 4,00,000/-. After the tenders were opened, the appellant discovered that they had quoted the price for only one unit by mistake while the NIT specified supply of two units. As soon as the error was discovered, the appellant wrote to the respondent explaining the same and revising the bid. The respondents contended that the price could not be increased after the tenders were opened and sought to enforce the bank guarantee. The application for injunction to restrain the respondents from encashing the guarantee was dismissed. The appellant approached the Bombay High Court. The respondents had not taken any steps to their detriment because of the tender submitted by the appellants and, therefore, the theory of promissory estoppel was held to be not applicable to the case. The High Court observed that there was no contract between the parties to keep the tender open and, therefore, the appellants could revoke their bid before it was accepted, as was in fact done. The guarantee could be invoked only if the contract were to be awarded to the appellants and they failed to pay the amount or to perform their part, which stage was held not to have arisen. The bank was, therefore, restrained from making payment under the guarantees.[197]

6. The Karnataka High Court has laid down the principle that in every case of guarantee given, the bank is not bound to pay the money to the beneficiary notwithstanding any dispute or litigation subsisting between them. If the terms mentioned in the contract do not amount to an unconditional or irrevocable contract and if it does not contain any undertaking to pay on demand, the person concerned has no right to insist on the bank making the payment as mentioned in the guarantee. Where the guarantee amounts to a contract of indemnity, as was the case before the High Court, the beneficiary must show the loss or damage caused to it. It can be enforced only after the amount of loss caused to the beneficiary is determined.[198]

7. The Delhi High Court has held that normally the court would refrain from granting an injunction to restrain the bank from performing the contractual obligation except on the ground of fraud. It was further held that misrepresentation or suppression of material facts or violation of the guarantee can be treated as species of the same genus as fraud. In such

---

[196] M/s. Banerjee & Banerjee v. Hindustan Steel Works Construction Ltd., AIR 1986 Cal. 374; (382); M/s. G.S. Atwal & Co. (Engineers) Pvt. Ltd. v. N.P.C.C. Ltd., AIR 1988 Del. 243; AIR 1986 Mad. 161.
[197] Kirloskar Pneumatics Co. Ltd. v. N.T.P. Corn. Ltd., AIR 1987 Bom. 308.
[198] Kudremukh Iron Ore Co. Ltd. v. Konila Rubber Co. Pvt. Ltd., AIR 1987 Kant. 139.

a case there exists special equity in favour of the plaintiff for grant of injunction. In that case the defendant had not disclosed that: (a) the contract was cancelled without any fault or mistake on the part of the claimant; (b) the fact of withholding a sum of money; (c) some amount of advance had already been recovered; and (d) that the plaintiff had been all along ready and willing to perform his part of the contract. It was held that the above facts gave rise to special equity in favour of the plaintiff.[199]

It is important that the employer/owner makes a full and frank disclosure when defending an application for injunction against a bank guarantee.

12.23.3 Suspension of Main Contract by Statute – Bank Guarantee Not Suspended?

In few matters the question arises as to whether the liability of a guarantor is automatically suspended when the liability of the principal debtor is suspended under some statutory provision. The law in this respect seems to be settled by a couple of Supreme Court decisions in which it was held:[200] The contract of guarantee is an independent contract between the guarantor and the guarantee, and hence, the remedy against the guarantor is not suspended merely because the liability of the principal debtor is suspended.

EXAMPLE FROM CASE LAW

The plaintiff, in a case before the Bombay High Court, was engaged as a subcontractor for carrying out the work of erection, testing and commissioning of L.P. Pipe Work and tube-oil unloading system at the Riband Super Thermal Power Station for a sum of over Rs. 1 crore. The bank undertook to pay on demand "any or all moneys payable by the contractor to the extent of Rs. 10,72,806/- at any time up to 30/06/1989 without any demur, reservation, contest, recourse of protest and/or without any reference to the Contractor." It further stated: "Any such demand made by NPIL (Defendant No. 2) on the Bank shall be conclusive and binding notwithstanding any difference between NPIL and the contractor or any dispute pending before any Court, tribunal, arbitrator." The Government of West Bengal declared the plaintiff SCIL (India) Ltd. as an unemployment relief undertaking under the West Bengal Act XIII of 1972, with effect from 11th May 1990. Under the notification issued, all contracts between the plaintiff and the third parties and all the rights, obligation and liabilities arising therefrom were suspended except in respect of certain contracts specified in the notification. The main contractor stated that huge losses were suffered by reason of the default committed by the plaintiff. The bank guarantee was, therefore, sought to be encashed. The main contention of the plaintiff was that the subcontract stood suspended and the contract of bank guarantee being ancillary to the subcontract was liable to be treated as suspended. It was held:[201] The contract of bank guarantee in question is a contract between the 1st defendant bank and the second defendant beneficiary and the said contract is independent of the subcontract dated 31st March 1987.

---

[199] M/s Synthetic Foams Ltd. v. Simplex C.C.P. (India) Pvt. Ltd., AIR 1988 Del. 207.

[200] M.S.B.B., Bombay v. Official Liquidator, H.C. Emakulam, AIR 1982 SC 1497; and State Bank of India v. M/s. Saksaria Sugar Mills Ltd., AIR 1986 SC 868.

[201] SCIL India Ltd. v. Indian Bank, AIR 1992 Bom. 131.

12.23.4 No Lawful Invocation after Interim Order of Injunction – An Illustration

In a case, a contract was entered into between the plaintiffs and defendant No. 1 for supply of goods. The plaintiffs had duly discharged their contractual obligations and delivered the goods to the buyer. Defendant No. 1 had consumed the said goods and to date no complaint had been received by the plaintiffs regarding the quality thereof. Thereafter, despite repeated requests defendant No. 1 failed to cancel the respective guarantees given by the respective plaintiffs. The plaintiffs were apprehending invocation of the said guarantees. An order of temporary injunction was therefore passed by the court restraining defendant No. 3, from making any payment on the basis of the said bank guarantee. Applications were made by defendant No. 2 asking for vacating interim order of injunction. Held: At that point of time order of injunction passed by court was perfectly justified and there was no reason to recall the same. By subsequent event the plaintiffs were able to make out a very strong *prima facie* case that there must be some element of fraud in connection with the invocation. It was not clear why even after order of injunction passed by court restraining defendant No. 3 from making payment, bank guarantee was invoked. No case was therefore established for vacating interim injunction.[202]

12.23.5 Bank Guarantee Not Enforceable after Its Validity Period?

The law in this respect has been well settled by a series of decisions of the Apex Court that the bank guarantee is not valid beyond the period of its validity prior to the amended Section 28 of the Contract came into effect on 8th January 1997.[203] Whether a guarantee is enforceable or not depends upon the terms under which the guarantor bound himself. But the cardinal rule is that the guarantor must not be made liable beyond the terms of his engagement. The Apex Court upheld the contention of the bank that the guarantee was no longer enforceable after its validity period.[204]

It is well settled by the decisions of the Apex Court and the High Courts that a bank guarantee is a separate and independent contract between the bank and the beneficiary enforceable on its own terms independently of disputes between the parties to the main contract in pursuance whereof the bank guarantee is furnished. The only parties to the contract of bank guarantee are the bank and the beneficiary. The party at whose instance the bank guarantee is furnished is not a party to the contract of bank guarantee. Similarly, the bank is not a party to the main contract, is unconcerned with it. The arbitration clause contained in the main contract cannot bind the bank as the bank is not a party to the main contract.[205]

In view of the above, even if Section 128 of the Contract Act is attracted, it clearly provides:

> 128. *Surety's liability*
> The liability of the surety is co-extensive with that of the principal debtor, unless it is otherwise provided by the contract.

The words "unless it is otherwise provided by the contract" need to be considered. The contract, if it means the contract of guarantee, the provisions are clear. If it means the principal contract it also invariably refers to the bank guarantee in the prescribed form. In other words, the limited liability is accepted by the party in whose favour the guarantee is given. In terms thereof the bank

---

[202] Maheswari Brothers Ltd. v. Airports Authority of India Ltd., AIR 2006 Cal. 227.

[203] Continental Construction Ltd. v. Food Corporation of India Ltd., AIR 2003 Delhi 32; 2002 (Suppl.) Arb. LR 192 (Del.).

[204] State of Maharashtra v. M.N. Kaul, Dr., AIR 1967 SC 1634, relying on the decision of AIR 1963 SC 746 (752) and AIR 1935 PC 21 (24).

[205] Suresh v. U.O.I., 1990 Mah. LJ 1243; Hindustan Paper Corporation Ltd. v. Keneilhouse Angami, (1990) 68 Com Cas 361.

guarantee becomes invalid after the mutually agreed period. The following observations of the Apex Court are noteworthy, it is respectfully submitted. It was observed:[206]

> The terms of the bank guarantee are explicit. The bank guarantee was to remain in force and effect only up to the date stated therein or any extended date and it was to be enforceable for a period of 6 months thereafter. The bank guarantee stated that unless a suit or action to enforce the claim under it was filed within 6 months from the date of its expiry, all the State's rights under the bank guarantee would stand forfeited and the bank would be released and discharged from all liabilities there under. There was, therefore, in our view, no way in law in which it could be declared that the bank guarantee would remain effective and enforceable for so long as the work under the contract was successfully commissioned or the amounts due under the bank guarantee were paid or an injunction could issue to the bank to so renew the period of the bank guarantee. It is correct, as argued by learned counsel for the State that bank guarantee has to be read in conjunction with the terms of the contract in pursuance of which it is issued, *but that is not to say that the rights and obligations of the contractor at whose instance the bank guarantee is issued become the rights and obligation of the bank.* (emphasis supplied). ... When the contractor declines to extend the terms of the bank guarantee, the proper course for the State is to terminate the contract on the ground of breach of the terms thereof, make a claim for damages and recover on the bank guarantee, if necessary by filing a suit.

Can it be said that the above observations of the Apex Court are not attracted in view of the amended Section 28 of the Act? Apparently it appears so. The relevant part reads:

> *Agreements in restraint of legal proceedings, void.* – Every agreement – (a) ... (b) which extinguishes the rights of any party thereto, or discharges any party thereto, from any liability, under or in respect of any contract on the expiry of the specified period so as to restrict any party from enforcing his rights, is void to that extent.

The Bombay High Court, in a case, upheld the limited period of validity of the bank guarantee. It, however, held as void and *non est* the stipulation in a bank guarantee that unless a demand or claim is made before a specified period, the right of the party claiming under the guarantee is extinguished and the bank would stand discharged from its liability. This was so in view of Section 28 of the Act as amended.[207]

In the 6th edition of this title, the author had shown by citing authorities that by application of the doctrine of waiver the contract stipulation could be held valid and binding. However, the said discussion has been deleted in the present edition since the Indian Parliament had caused an amendment to Section 28 of the Indian Contract Act, 1872 (Contract Act) by inserting Exception 3 by Act 4 of 2013 to be effective from 18th January 2013. The said exception saves a guarantee agreement of the bank or a financial institution by not rendering illegal the contract in writing stipulating a term in the contract for extinguishment of right beyond the period specified in the guarantee or contract, but limiting the said period to "not less than one year" as against the six months that normally used to be stipulated.

Once the bank guarantee in question is invoked in good faith within the prescribed time as per the terms and conditions of the guarantee, the question of settlement of any dispute or conflict, insofar as any claim between the parties to the main contract, is not relevant. The bank has no concern with the said dispute.[208]

---

[206] Makharia Brothers, Appellant v. State of Nagaland and others, AIR 1999 SC 3466.
[207] Union of India v. Bhagwati Cottons Ltd., 2008(2) CTLJ 180 (Bombay).
[208] Mula Sahakari Karkhana Ltd. v. State Bank of India, AIR 2005 Bom. 385.

12.23.6 *Performance Guarantee of One Contract Cannot Be Encashed for Alleged Breach of Other Contract*

The power of the courts to restrain the bank under appropriate circumstances, however, remains valid. For example, when a bank guarantee has been given pursuant to a particular contract, it is not open to the beneficiary to encash the bank guarantee against the alleged breach of another contract.[209]

---

[209] Jivanlal Joitaram Patel v. National Highways Authority of India, AIR 2008 Guj. 181.

13

Subcontracts

13.0 Introduction

A subcontract is a contract between a contractor and his subcontractor. Thus, essentially the relationship between the contractor and his subcontractor is no different from the relationship between the owner/employer and the contractor. Therefore, the points governing the relationship between the owner and the contractor, discussed in earlier chapters are equally applicable to the relationship between two or more contractors.

The use of nominated subcontracts is extensive in modern forms evolved for building and engineering contracts. Generally the selection of the subcontractor rests with the contractor. He may have an agreement with him prior to tendering for the main work or he may enter into an agreement after securing the main contract. Where, however, the contract stipulates nominated subcontractor(s), the freedom to choose the subcontractor is not available to the main contractor. The owner, prior to inviting tenders, gets the quotation from the subcontractor of his choice and nominates him as a subcontractor in the main contract. The main contractor has to submit his tender based on the offer of the subcontractor for a part of the work for which he is nominated. This practice occasionally creates problems and the terms of the main contract need to be carefully drafted to avoid difficulties likely to arise.

13.1 Provisions in Standard Form Contracts

The following provisions of the Standard Form Contracts are dealt with in this chapter.

Provisions in Standard Form Contracts	Paragraph No.
FIDIC 1992 Amended, Clauses 59.1–59.5	13.6.1, 13.6.2
FIDIC 1999/2017 edition, Clauses 5.1, 5.2	13.2, 13.6.1, 13.6.2
Ministry of Statistics and Programme Implementation, Government of India (MOS & PI), Clause 8	13.2.1
NITI Aayog Model Form, Clauses 3.2, 22.1	13.2.1

13.2 Written Permission to Sublet or Assign When Necessary

Unlike a personal contract, a construction contract is considered to be assignable. It can therefore be said that except where the contract stipulates actual performance by the contractor himself he is free to sublet. However, the owner would naturally like the contractor to perform a substantial part of the actual work himself. Thus, to prohibit the contractor from subletting the whole of the work to one or more subcontractors and assuming only the overall control and general super-intendence, the conditions of the contract usually prohibit subletting of the work without the written permission of the owner or his engineer/architect and that, too, is limited to a stipulated percentage.[1] If the contractor who has agreed to this condition in fact sublets the work without obtaining necessary permission, he commits a breach of a term, which may entitle the owner/ employer to cancel the contract or claim damages. The International Chamber of Commerce (ICC) published revised Infrastructure Conditions of Contract in 2018. Clause 3 of the Design and Construction version of the ICC Conditions forbids subcontracting without consent and Clause 7 provides that the contractor can raise objections to the nominated subcontractors for well-defined reasons.

Invariably the construction contracts also contain clauses against assignment. There is a difference between the assignment and subcontract. By "assignment" is meant: "The act of transferring to another all or part of one's property, interest or rights."[2] At common law, a party cannot assign the burden of a contract without the consent of the other party. Where the contract contained an express term against assignment without written consent of the contractor, the House of Lords held that the clause was not against public policy.[3] "Subcontract" on the other hand means:

> A contract subordinate to another contract, made or intended to be made between the contracting parties, on one part, or some of them, and a third party (i.e. subcontractor). ... The term "subcontractor" means one who has contracted with the original contractor for the performance of all or a part of the work or services which such contractor himself has contracted to perform.[4]

Prohibition to sublet is obviously incorporated to let the owner retain control of who does the work. Closely associated with this right of the owner, is the right at the end to balance claims for money due against the counter-claims, for example, for bad workmanship. The contractor may say that such a counter-claim may be made against the assignee. The owner may not like to deal with unknown people for recovery of such counter-claims. Where, therefore, the contract contains provision to the effect, "The contractor shall not assign the contract or any part thereof or any benefit or interest therein or thereunder without the written consent of the owner", and if the contractor purports to assign his right to payment under the contract to someone, the assignment will be ineffective.[5]

13.2.1 Prior Approval When Necessary – An Illustration

The relevant contract provisions generally restrict the power to subcontract the entire work, and a part can be sublet with prior approval of the authority named in the contract. For example, NITI Aayog Model Form Clause 3.2.1 stipulates that "the Contractor shall not sub-contract Works comprising more than 70% (seventy per cent) of the Contract Price".

The MOS & PI, Government of India, form in Clause 8 provides: "8.1 The Contractor may subcontract with the approval of the Nodal Officer or his nominee but may not assign the Contract without the approval of the Employer in writing. Subcontracting does not alter the Contractor's obligations." A similarly worded clause was included in a contract as Clause 7.1. The contract in the said case was terminated and the contractor blacklisted allegedly for breach of the above clause. The single judge upheld the action and an appeal against the said judgment was filed before the division bench of the Madras High Court. It was observed:[6]

---

[1] See, for example: FIDIC 2017 edition, Clause 5.1. Similar provisions are made in other standard forms.
[2] *Black's Law Dictionary*, 6th ed., pp. 119.
[3] Linden Gardens Trust Ltd. v. Lenesta Sludge Disposals Ltd. [1993] AC 85.
[4] *Black's Law Dictionary*, 6th ed., pp. 324–325.
[5] Helstan Securities Ltd. v. Hertfordshire County Council, Q.B.D. [1978] 3 All ER 262.
[6] M/s. P.T. Sumber Mitra Jaya, Appellant v. The National Highways Authority of India, New Delhi, AIR 1998 Mad. 221.

A cursory reading of Clause 7.1 would suggest that the concept of "subcontracting" was not abhorred in this contract agreement. ... The meaning is, therefore, clear:

1. In the first place, there is no user of the words "prior approval" either in respect of the subcontract or in respect of the assignment of the contract.
2. Then, there is a definite difference in the concept of subcontracting of the portion of the work and assigning the original contract itself. While for subcontracting only the approval of the Engineer is required, for assignment of the contract itself, the approval of NHAI that too in writing, is necessary. There is undoubtedly a dichotomy in the concepts of "subcontracting" and "assignment of the contract". It is, therefore, obvious that for such subcontracting, there would be no need of the approval of NHAI though unfortunately, the parties have understood otherwise.

We have already pointed out that the learned single Judge also deduced that firstly there was an agreement between the appellant and M/s. CECON and that the said agreement was without the prior approval of NHAI. Even presuming that there was any such agreement, in our opinion, there was no necessity of having a prior approval of NHAI as per the express language of clause 7.1. That could at the most be an agreement for subcontracting which would be undoubtedly an agreement different from the "assignment" of the contract itself. The approval of the Engineer was necessary for subcontracting the work while the approval of the employer i.e. NHAI in writing was necessary for the assignment of the contract. ... Learned single Judge has also not addressed the question from these angles which, in our opinion, were the necessary angles emanating from the language of clause 7.1. The finding of the learned single Judge as also the finding arrived at by NHAI on the issue of subcontracting thus are erroneous.

13.2.2 When a Third Party Can Sue on Contract

It is interesting to see the effect of the employer assigning the right under his contract with the contractor to a third party. It remained the law, although much criticized, that a third party could not sue for damages on a contract to which he was not a party. In English law, a statutory exception was created to the doctrine of privity under the Contracts (Rights of Third Parties) Act 1999. This Act allows third parties to enforce a benefit conferred on them by a contract if the contract expressly provides that a third party may enforce its term(s) or where a contract purports to confer a benefit on the third party. In both cases, the third parties have to be identifiable within the parameters of the Act. In the commercial context, many times contracts expressly exclude a right to third parties by providing something along the lines: "Unless expressly stated otherwise in this Agreement, nothing in this Agreement confers or is intended to confer any rights on any third party pursuant to the Contracts (Rights of Third Parties) Act 1999."[7]

The general position was that if a plaintiff contracted with a defendant to make a payment or confer some other benefit on a third party who was not a party to the contract, the plaintiff could not recover substantial damages from the defendant for breach of that obligation on the part of the defendant.[8] The plaintiff could *prima facie* recover only for his own loss. However, to this general principle there were a couple of exceptions even before the Contracts (Rights of Third Parties) Act 1999. First, where the plaintiff made the contract as agent or trustee for the third party and was enforcing the rights of a beneficiary, in the fiduciary relationship;[9] and second, the rule in Dunlop v. Lambert,[10] which provided "a remedy where no other would be available to

---

[7] Hurley Palmer Flatt Ltd. v. Barclays Bank Plc, [2014] EWHC 3042 (TCC).

[8] Woodar Investment Development Ltd. v. Wimpey Construction UK Ltd., [1980] I WLR 277.

[9] M/s. Continental and Eastern Agencies, Plaintiff v. M/s. Coal India Limited and others, AIR 2003 Del. 387.

[10] Dunlop v. Lambert, (1839) 6 Cl. & F. 600.

a person sustaining loss which under a rational legal system ought to be compensated by the person who caused it".[11] This rule was applied in a case by the House of Lords in 1994.[12] The Court of Appeal applied the above rule in a case decided on 28th June 1994. The facts of the case are given in the example below.

EXAMPLE FROM CASE LAW

D.B. Council wished to build a new recreational centre and to avoid financial constraints for the requisite borrowing, M.G. Ltd. entered into building contract for the construction of the centre with W for the benefit of the Council. Collaterally to the building contracts, a covenant agreement recorded that M.G. Ltd. as employer would pay W and the council would reimburse M.G. Ltd. who in turn assigned to the council all rights it had against W. The council claimed that there were serious defects in the centre due to bad workmanship or other breaches by W and expenses of £2 million would be incurred to remedy the defects. The council obtained a deed of assignment from M. G. Ltd. of all rights and causes of actions, which M. G. Ltd. might have. The trial judge decided that M. G. Ltd., having no proprietary interest in the centre and no obligation to the council for the quality of W's workmanship, had suffered no damage or loss from the defects, and could transfer no claim for substantial damages to the council. The council, it was held, was precluded by the privity of contract rule from claiming the damages it suffered. The Court of Appeal held that the council, as assignee, had a valid claim against W for more than nominal damages for breach of contract and the damages should be assessed on the normal basis as if the council had been the employer under the building contract.[13]

13.3 Agreement to Subcontract

At the outset it may be noted that all essentials of a valid contract discussed in Chapter 2 must be fulfilled before there exists a contractual relationship between the parties to a subcontract. For example, in one case the tender notice of the general contract provided that if the contractor intended to sublet part of the work, he should submit the name of the proposed subcontractor in his tender. The general contractor wrongly specified the name of a subcontractor whose bid was higher than the one who was intended to be named. It was held that there was no contract between the general contractor and the subcontractor whose name was specified in the tender. In the said case, the school board inviting tenders also had authority and did consent to the substitution of the name of the subcontractor to whom the general contractor intended to award the subcontract.[14]

As already stated, the main contractor can enter into an agreement with the subcontractor, prior to tendering, stipulating that on the basis of his success in getting the main contract he would, in turn, get the work done by the subcontractor. Such a contract is binding and breach thereof will entitle the injured party to compensation. If a subcontractor wants to protect its position, it would have to qualify its bid. Also, the main contractor needs to ensure that a binding subcontract only comes into effect if the main tender is successful. Care needs to be taken in

---

[11] St Martins Property Corporation Ltd. v. Sir Robert McAlpine Ltd., [1994] 1 AC 85, pp. 114–15.
[12] Ibid.
[13] Ibid.
[14] Hotel China & Glassware Co. v. Board of Public Instruction of Alachua County, 130 So. 2d 78 (Fla. 1st DCA 1961).

negotiations of subcontracts as a binding subcontract may come into effect upon oral negotiations.

EXAMPLES FROM CASE LAW

1.   In a case, an oral agreement was entered into by two contractors, according to the terms of which one was not to tender for the general contract but to supply the steel necessary for the work and erect the same. The second contractor was to submit the tender for the general contract and, if successful, to sublet the steel work to the other contractor. The contractor who submitted the tender for the general contract was awarded the same. He, however, sublet the steel work to a contractor other than the one to whom he had agreed to sublet it. The contemplated original subcontractor brought an action to recover damages for breach of contract and obtained a recovery. Although this judgment was reversed on appeal and a new trial ordered because the evidence did not justify the large judgment obtained; the right of the subcontractor to receive compensation for the breach of the alleged contract was not affected.[15]

2.   The general contractor about to submit a tender entered into a written agreement with another contractor according to the terms of which it was provided that in the event of the general contractor getting the contract, the other contractor would perform certain parts of the work at agreed specified prices. The general contractor was awarded the contract. However, when called upon to perform the part of the work called for by the subcontract, the subcontractor refused to do so. The general contractor then subcontracted that part of the work to another subcontractor at a price in excess of the original subcontract. The general contractor brought the action to recover the excess cost from the original subcontractor and he obtained recovery.[16]

3.   In Canada, the building industry practice involves the subcontractor submitting its price to the main contractor a short time before the deadline for the submission of the main tenders. The main contractor used such a price in his bid but after its bid was accepted by the government, the subcontractor realized that he had seriously mistaken the amount it bid by telephone. The subcontractor was held bound by the telephoned amount and would be liable for the difference between the price he bid and the actual price the replacement subcontractor charged for the same work.[17]

4.   The employer's project manager negotiated with a piling subcontractor before the appointment of the management contractor. The project manager then gave a letter of intent to the subcontractor to commence the piling work. The finalization of the subcontract was made conditional on the finalization of the management contract. The subcontractor completed the work substantially before the management contract was finalized. The employer, however, became insolvent and the management contractor refused to pay the subcontractor whose work was all carried out under the letter of intent, without a final subcontract. Hale, J. held that the piling subcontract came into effect between the management contractor and the subcontractor as soon as the management contract was concluded.[18]

---

[15] Fitcher Steel Corpn. v. P.T. Cox Const. Co., 42 N.Y.S. 2d. 225, 226. App. Div. 347.

[16] North Weston Eng. Co. v. Ellerman *et al.*, N.W. 2d, 879.

[17] Northern Constr. Co. v. Gloge Heating (1986), 67 A.R. 150 (CA).

[18] Stent Foundations Limited v. Carillion Construction (Contracts) Ltd., (2000) 78 Com LR 199.

13.4 Provisions of General Contract – How Far Applicable to Subcontract

It is in the best interest of the general contractor that all the material terms and conditions, to which his own contract with the owner is subject, are included in the subcontract. Otherwise the provisions of the general contract may not apply to the subcontract.

EXAMPLES FROM CASE LAW

1.   A contract contained the provision, which empowered the owner to cancel the contract upon the engineer certifying that the progress of work would be delayed. The contractor while subletting part of the work did not stipulate that the subcontract should be dependent upon continued existence of the general contract. The general contract was terminated for slow progress before the subcontractor had completed his work. There was no criticism of the progress of the subcontractor's work. In a suit that followed it was held by the court that the subcontractor was prevented from performing his contract by the termination of the general contract; and because the general contractor did not protect himself against such a contingency, the subcontractor could recover the benefit of his contract, less the reasonable cost of completing his work.[19]

2.   A subcontract contained a provision according to which the subcontractor was to perform the work at such time or times and in such manner and in such quantities as might be required by the general contractor to meet its time schedule. The subcontractor brought an action against damages for the loss caused by delays. It was claimed by him that the general contractor failed to comply with the provisions and specifications of the general contract. Had he followed those provision the delay would not have been caused. The court, while denying the subcontractor a recovery, held that the provisions of the general contract had no application to the subcontractor who was not a party to the contract that contained the provision in question. The court further observed that the reference made by the subcontract to the plans and specifications of the general contract was only for the purpose of defining the work. The provision as to how the work was to proceed prevailed.[20]

3.   A contract for construction of a tunnel sewer for a city through the mountain, was sublet to subcontractors who were miners. The subcontract provided for the tunnelling to be done according to the dimensions and specifications as set forth in the contract between the city and the main contractor. It was held that these words would undoubtedly require the tunnel to be dug out as per the dimensions shown, and any other provision affecting the physical characteristics of such tunnel no doubt would be applicable to and binding upon the subcontractors. It was further held: The reference to specifications as expressed in this context does not import into the subcontract, sweeping and extraordinary powers exercised by the city engineers.[21]

4.   A subcontract did not include a term in the general contract giving the owner a right to require the dismissal of the subcontractor. The Court of Appeal refused to imply a term into the subcontract that such a demand for dismissal of the subcontractor would empower the contractor to terminate the subcontract without compensation. If the parties wished to express such a term, they could have done so. The clause that the subcontractor's work would be executed "in accordance with the terms" of the main contract did not go so far as to include all the terms into the subcontract. All that the subcontractor had to do was to

---

[19] Soley v. Jones, 208, Mass. 561, 95 N.E. 94 (Mass 1911).

[20] Arrow Steel Metal Works v. Bryant & Detwiler, 61 N.W. 2 d., 125, Supreme Court of Michigan, 1953.

[21] Smith and Montgomery v. Johnson Bros. Co. Ltd., Ontario High Court (1954) 1 DLR 392.

provide the work of the quality stipulated in the main contract within the specified time frame. It was held by the court that the subcontractor, if so dismissed, may sue the general contractor for breach of contract or on a *quantum meruit*.[22]

5.   In a case, the subcontractor undertook to carry out work in accordance with certain specifications in the main contract. One of these provided for disputes to be settled by arbitration. The subcontractor sued for his work, and the contractor pleaded that the matter was covered by the arbitration clause. It was held that the arbitration clause was incorporated only to the extent of matters of dispute between the main contractor and his employer, and was not incorporated as regards matters between the contractor and the subcontractor.[23] Courts are generally reluctant to allow arbitration clauses being incorporated into subcontracts unless there is clear evidence that the parties expressly agreed to it.

6.   The appellant sought from the respondent reimbursement of income tax paid by the appellant to the subcontractor, contending that the reimbursement of tax liability of the subcontractor by the appellant was necessary and reasonable extra cost arising due to change of law and that the respondent was liable to compensate the appellant to that extent. The respondent contended that they had not taken over the liability to pay any taxes which may be due to be paid by subcontractors of the appellant. Dispute referred to arbitration. The umpire allowed the claim. The award was set aside by the trial court and the appeal was also dismissed. The Apex Court of India held: the award of the umpire being a well reasoned award and one within his jurisdiction giving a meaningful interpretation to all the clauses of the contract, no interference was called for by the court. The appeal was allowed.[24]

7.   The main contract included a complex, staged dispute resolution clause eventually culminating in a provision for a reference to arbitration. The court refused to incorporate that long process into the subcontract which had its own dispute resolution clause providing for an amicable settlement. The incorporating clause stated that the main contract would apply to the "Subcontract Works" except where amended by the subcontract. The "back-to-back" provision was intended to apply to the execution of the work but not to "subcontract rights and obligations" of the kind of a dispute resolution clause.[25]

Where a general contractor decides to split up his contract into separate items and subcontract these items, care must be taken by him to provide for all the items included in the general contract.

EXAMPLE FROM CASE LAW

The general contractor for a road job subcontracted certain definite items of work to a subcontractor. The basis of payment to the subcontractor was agreed unit prices. The subcontractor refused to do certain work which was not included in the subcontract and for which the unit price was also not fixed. The general contractor did that work and brought an action against the subcontractor to recover the cost of performing that work. The

---

[22] Chandler Bros. Ltd. v. Boswell, [1936] CA 179.

[23] Goodwins Jardine & Co. Ltd. v. Brand & Son, [1905] 7 WLUK 75. Also see Haskins (Shutters) Ltd. v. D. & J. Ogilvie (Builders) Ltd., 1978 S.L.T. (Sh.Ct.) 64).

[24] Sumitomo Heavy Industries Ltd. v. Oil & Natural Gas Corporation Ltd., (SC): AIR 2010 SC 3400; 2010(11) SCC 296.

[25] Cegelec Projects v. Pirelli Construction, [1997] No. 1997 ORB 646.

contractor relied upon the provision in the subcontract that the subcontractor is bound by the terms, plans and specifications of the general contract. Reversing the judgment of the trial court which had allowed the general contractor a recovery, the court held:

> A mere reference to the general contract and the provision in the subcontract that the subcontractor is bound by the terms, plans and specifications of the general contract cannot enlarge the terms or conditions of the subcontract which by its terms is limited to certain definite items at a specified unit price per unit of quantity.[26]

13.5 Relationship between Owner and Subcontractor

Usually there is no contractual relationship between the owner/employer and the subcontractor; in which case neither party has any right of action against the other. However, a general contract may provide that payments of subcontractors shall be made direct to them, and not necessarily through the general contractor. The employer reserves this right in case the main contractor defaults on this count and the progress is likely to be adversely affected. It may be noted that such a provision is valid and not affected even by the bankruptcy of the general contractor. A subcontractor is entitled to receive the payment certified by the engineer or architect, before the trustee in bankruptcy is entitled to collect the amount due to the bankrupt.[27]

The owner can make direct payment to the subcontractor, due under the general contract, only when the general contract contains an express stipulation to that effect. Even contrary to the express stipulation in the original contract, if the owner subsequently agrees to consult and pay to the subcontractor directly, he would be estopped from denying it.

EXAMPLE FROM CASE LAW

In a case the petitioner, the principal contractor, secured a contract to set up a super Thermal Power Station with a right to award subcontract. The respondent No. 2 obtained a subcontract from the principal contractor. The respondent No. 2 associated respondent No. 1 for carrying out the subcontract. A tripartite agreement was executed between the parties. The agreement *inter alia* provided that both the respondents shall be "contractor" for the purpose of the agreement and that respondent No. 2 shall be the sole representative of the subcontractors in dealing with the principal contract during execution; however, some differences on the submission of bills arose between the respondents 1 and 2. Respondent 1 protested that the bills submitted were not properly and correctly prepared and payment received from the petitioner was not shared by respondent 2 with respondent 1. The petitioner, with a view to sort out the differences, convened and held a meeting with both the respondents, and after deliberations it was agreed between the parties that in view of the differences between respondents 1 and 2 by way of departure to the prevalent procedure of submission of claims by and payments to respondent 2 alone, the petitioner

---

[26] Inland Eng. and Const. Co. v. Maryland Casualty Co. *et al.*, 290 p. 367, 76 Utals 435.
[27] Wilkinson Ex p. Fowler, Re, [1905] 2 KB 713 followed in Glow Heating Ltd. v. Eastern Health Board, [1992] 1 WLUK 455. Note that this case was not followed in McLaughlin & Harvey Plc (In Liquidation), Re. [1996] N.I. 618.

required the claims in category III to be submitted both by respondents 1 and 2 along with the explanations and statements to enable the petitioner to examine, discuss and finalize the same and this arrangement was to continue until differences between the respondents were settled between them. However, in spite of the departure in the procedure relating to payment of bills adopted in terms of the relevant clause in the tripartite agreement, the petitioner made final payment of the bills to the respondent No. 2. It was held[28] that considering the changes in policy, procedure and practice, thus brought about by the petitioner themselves and agreed to by all concerned, the petitioner could not settle accounts with respondent No. 2 or make any payment in consequence without further specific reference to and approval by respondent No. 1.

Another situation in which, despite the general rule of lack of privity, the subcontractor's liability to the owner can arise, is if during the course of negotiations warranties are given or representations made by the supplier or subcontractor on the faith of which the owner instructs the main contractor to place his order with the supplier of the subcontractor.

EXAMPLE FROM CASE LAW

In Shanklin Pier Co. Ltd. v. Detel Products,[29] the director of a paint firm had interviewed the managing director of the pier's owners and later their architects and made representations as to the suitability of his company's products for the protection of the pier from corrosion. On the basis of the said representation the owner directed the main contractors to order the firm's paint. On the paint proving defective, the pier owners sued the paint manufacturers, alleging that the representations were warranties as to the suitability of the paint given in consideration of the pier owners causing their contractors to order the paint. It was held that the contention was correct and the paint manufacturers were liable.

This case can be said to have been decided on rather strong facts and between parties that were at arm's length. It was not followed when the main contractor and the subcontractor were not at arm's length. A Japanese company, Fuji, was persuaded to try CCC's product by company CCE, which was, in fact, a subsidiary of CCC. The parties could have made the contract directly between CCC and Fuji but instead the contract was between CCE and Fuji. This case was held to be a

> far cry from that prevailing in *Shanklin Pier*. It is not appropriate for this court to supplement the contractual arrangements which experienced and well-advised commercial parties choose to take. Fuji and CCC chose not to enter into any direct contract. In my judgment, no form of collateral warranty between those two companies can be read into or derived from the pre-contract documents.[30]

In practice, statements made in brochures or at meetings with architects might, depending on the circumstances, be held to amount to mere representations.

---

[28] M/s. Rolls Royce Industrial Power India Ltd., Petitioner v. M/s. Urmila and Co. Pvt. Ltd.; AIR 1998 Del. 411.
[29] Shanklin Pier Co. Ltd. v. Detel Products, [1951] 2 QB 854.
[30] Fuji Seal Europe Ltd. v. Catalytic Combustion Corporation, [2005] EWHC 1659 (TCC), para. 158.

13.5.1 No Privity of Contract

It needs to be emphasized again that no privity of contract between the owner and the subcontractor can arise out of a subcontract between the main contractor and the subcontractor. This will be so even in the cases of nominated subcontractors.[31]

EXAMPLE FROM CASE LAW

The Food Corporation (FCI) of India entered into a contract with M/s India Machinery Co., the main contractor, who in turn gave the work to a subcontractor who was the petitioner in the High Court. There was delay in execution of the work and the FCI cancelled the contract and called for fresh tenders. Notice was issued to the main contractor by the FCI. No notice was issued to the petitioner. The petitioner filed a suit against the FCI contending that the contract should not be cancelled and a fresh contract should not be entered into with others. He claimed an injunction. The trial court rejected the stay application and that order was confirmed in appeal and revision.[32]

13.6 Nominated Subcontractors

The system of nomination of subcontractor(s) springs from the owner's desire or need to control the quality of specialist work at the competitive price. An ordinary main contractor, who may not be able to carry out specialist work, will generally find it difficult in tendering a competitive price for the specialist work. Thus, it may seem that the practice enables the owner to have control over the price. For such items of work, the main contract generally includes provisional and prime cost (PC) sums.

13.6.1 Provisional and PC Sums

"Provisional sums" provided for in the bill of quantities mean sums provided for work or for costs which cannot be entirely foreseen, defined or detailed at the time of preparation of contract documents. Certain fixtures, specialist works, or finishing items or the like, which are not finally decided by the owner at that stage but which may eventually be required to be incorporated in the work, are covered by such provisional sums.

"Prime cost sums", on the other hand, mean sums provided for work or services to be executed by a nominated subcontractor or for materials or goods to be obtained from a nominated supplier.

In nearly all the modern standard forms, provisions are made for release of money to the contractor from the provisional/PC sums, depending upon the expenses incurred on the work actually carried out. Additions or deductions to the contract sum is thus to be regulated by such provisions. For example, according to FIDIC conditions,[33] a right is reserved enabling the engineer to issue instructions to carry out the work or for the supply of goods, materials, plant and services by: (1) either the main contractor or (2) a nominated subcontractor. The payment is to be made in accordance with Clause 52 in the case of the former and Clause 59.4 in the case of the latter. Thus it will be seen that in the FIDIC Form, the term "Provisional sum" is used to represent PC as well. Clause 59.4 provides, besides the actual cost incurred, the main contractor will be entitled to a percentage rate of the actual cost incurred to cover all the other charges and

---

[31] Concrete Constructions Co. Ltd. v. Keidon & Co. Ltd., 1995 (4) S. A. 313 (South Africa).

[32] Raj Kumar v. Food Corporation of India, AIR 1990 Raj. 64 (67).

[33] FIDIC 1992 amended ed. = Clauses 13.3 and 13.4 of FIDIC 2017 ed.

profit. The contractor is expected to quote this percentage rate in the Appendix at the time of tendering.

13.6.2 Responsibility of Main Contractor for Nominated Subcontractor's Work

It is absolutely necessary, in the case where the subcontractor is nominated, that terms be incorporated in the main contract in sufficiently precise words which will help determine the exact extent of the main contractor's obligations to the owner for the work of nominated subcontractors. As already stated, in such cases, the subcontractors are contracting directly with the owner and, as such, the owner and his architect/engineer rely on the subcontractor's skill and judgement. If any undertakings or warranties are given by the subcontractor the same need to be imported into the contractual relationship between the main contractor and the subcontractor. The possible difficulties likely to arise due to nominated subcontractor's work and the liability of the main contractor are attempted to be eliminated in the FIDIC Standard Form. The said 1992 amended form (Clause 59.1) incorporates the usual provision: "All specialists, merchants, tradesmen and others executing any work or supplying any goods – be deemed to be the subcontractors to the contractor – and are referred to – as nominated subcontractors." However, it provides in Clause 59.2 that the contractor shall not be under any obligation to employ a nominated subcontractor if the subcontractor declines to save harmless and indemnify the main contractor from and against: (i) all claims, proceedings, damages, costs, charges and expenses whatsoever arising out of or in connection with any failure to perform as agreed by the subcontractor; including any matter of design or specifications, included in the subcontract; and also (ii) any negligence by the nominated subcontractor, his agents, workmen, servants, or misuse of any temporary works, etc.

In the absence of proper care in drafting provisions of the main contract, difficulties are bound to arise and the case below is an excellent illustration of such difficulty.

EXAMPLE FROM CASE LAW

J contracted with W for the erection of a multi-storied building. The contract was in JCT 63 form. L was nominated subcontractor for the piling work. L purported to complete the piling work by 20th June 1966 and left the site. J carried out further construction work thereafter. It was discovered that many piles were defective, either due to use of poor materials or as a result of bad workmanship. Remedial works were carried out by L. In the process the main work was delayed over 21 weeks. J claimed an extension of time under Clause 23(g) which provision entitled J to an extension of time on the ground of "[d]elay on the part of nominated subcontractors or nominated suppliers". The owner contended that there is delay only if, by the subcontract date, the subcontractor fails to achieve such completion of his work that he cannot hand over to the contractor. In the case it was contended further that so long as, by the subcontract date, the subcontractor achieves such apparent completion that the contractor is able to take over, notwithstanding that the work so apparently completed may in reality be defective; this may involve a breach of contract but does not involve delay. It was held that J was not entitled to an extension of time.[34] It was observed:

> If such an interpretation were imposed by the words used, it would have to be accepted whatever (short of completely frustrating the contract) the consequences might be – the parties must abide by what they have agreed to and it is not for the courts to make sensible bargain for them.

---

[34] The Lord Mayor, Aldermen and Citizens of the City of Westminster v. J. Jarvis & Sons Ltd. and Another, House of Lords, (1970) 7 BLR 64.

13.6.3 Consequences of Repudiation by Nominated Subcontractor

It cannot be very uncommon that for one reason or another the contract between the main contractor and the nominated subcontractor is terminated before the subcontract work is completed. The question arises: what is to happen in that event? There can be three possibilities:

1. The owner may make a new nomination, and bear the loss to the extent the price in the second subcontract exceeds the price in the first subcontract.
2. The owner may delete the subcontract work and give an instruction to that effect by way of variation.
3. It becomes the right and duty of the main contractor to do the prime cost work itself at the price fixed in the subcontract. This is so on the basis that the main contractor must be held to have undertaken that the nominated subcontractor will complete the work and breach by the subcontractor will be the breach by the main contractor.

To avoid trouble, the agreement must incorporate express provision for such an event. In the absence of an express provision, it is likely to be held that it is implicit in the contract that the owner comes under duty to renominate.[35] However, if the subcontract is wrongfully terminated by the main contractor, the main contractor will be in breach of his contract with the owner and will be liable to pay damages including any loss to the owner.

13.6.4 Delay Caused by Nominated Subcontractor

Where the nominated subcontractor withdraws, there is likely to be delay divisible into two parts. The first part comprises delay arising directly from withdrawing, and the second, due to failure of the owner to renominate a replacement with reasonable promptness. The owner is clearly responsible for the second part. The responsibility of the first part will depend upon the terms of the contract. In one case it was held that the main contractor will be responsible for the first part of the delay so long as an instruction was given for nomination of a replacement within a reasonable time after receiving the main contractor's application.[36] It is advisable that the main contractor should object to renomination if the second subcontractor is not willing to agree to the completion schedule as per the original contract.[37]

13.6.5 Defective Work by Nominated Subcontractor

Under the terms of most standard form contracts, the main contractor is responsible to the owner for defective work, if any, by the nominated subcontractor but will have an action against the subcontractor. In certain circumstances, the owner may also have a direct action against the nominated subcontractor, either in contract, or under collateral warranty, if executed, and in an exceptional case in tort if the nominated subcontractor is guilty of negligence. The implied warranty that a subcontractor's work is reasonably fit for its purpose may not apply if the client is given a suitably clear warning by the subcontractor and still wishes to proceed.[38] This is likely to be the case where there is innovation in design and the risk of failure may shift to the employer.

---

[35] North-west Metropolitan Regional Hospital Board v. T.A. Bickerton & Sons Ltd., House of Lords, [1970] 1 WLR 607; [1970] 1 All ER 1039.

[36] Percy Bilton Ltd. v. Greater London Council, House of Lords, (1982) 20 BLR 1.

[37] See Fairclough Building Ltd. v. Rhuddlan Borough Council, C.A. (1985) 3 Con L.R. 38 [1985] 1 WLUK 172.

[38] Baylis Farms Ltd. v. R.B. Dymott Builders Ltd., [2010] EWHC 3886 (QB).

EXAMPLE FROM CASE LAW

EMI were the main contractors and BICC were the nominated subcontractor for erection of a television mast 1250 ft high. The mast was erected and handed over to IBA (the owner) in November 1966. It collapsed in March 1969. IBA sued both EMI and BICC and EMI sued BICC. The trial court held the cause of the collapse as negligence in design. It was held that EMI had impliedly undertaken that the mast was reasonably fit for the purpose for which it was erected. BICC were held to be in breach of their subcontract with EMI. BICC were held to be liable in tort to IBA as they had been negligent in assuring them that the design was satisfactory. The assurances, which were given two years after the letting of the contracts, did not amount to a direct contract between BICC and IBA.[39]

13.7 Liability of Subcontractor under Indemnity Clause

As a point of law, a prime contractor is liable for the acts of his subcontractor. However, when the contract between the prime contractor and the subcontractor includes a stipulation under the terms of which "the subcontractor agrees to protect, defend, indemnify and hold harmless (prime contractor) from any damages, claims, liabilities, or expenses whatsoever arising out of or connected with the performance of the contract", the subcontractor will be liable, if he were to be the party primarily responsible for an injury.

EXAMPLE FROM CASE LAW

In a case, the dispute involved liability for a pedestrian who allegedly was injured on a public sidewalk that was torn up and made unsafe by the subcontractor. The pedestrian filed a claim against the owner, the prime contractor and the subcontractor. The subcontract included an indemnity clause similar to the one referred to above. The prime contractor filed a claim against the subcontractor under the indemnity provision contained in their contract. It was held that the subcontractor was responsible for damages although the prime contractor might be guilty of negligence as a point of law. The prime contractor had the right under the indemnity clause, only because he was not guilty of the primary negligent act that caused the injury out of which the liability arose. It was also observed that in construing the indemnity provision, the word negligence does not have to appear.[40]

13.8 Rights of Materials Suppliers

Where a general contractor promises to pay for materials and labour furnished to a subcontractor with a view to assure the performance of the subcontractor, such a promise is enforceable.[41]

---

[39] Independent Broadcasting Authority v. EMI Electronics and B.I.C.C. Construction Ltd., House of Lords (1980) 14 BLR 1.

[40] Walter L. Couse & Co. v. Hardy Corp., 274, So. 2d., 322 (1973).

[41] Curtis Mfg. & Asbestos Co. v. W. B. Bates Construction Co. Inc. 94 A 2d., 550, Supreme Court of New Hampshire (1953)

EXAMPLE FROM CASE LAW

In a case, the materialman stopped delivering materials to the subcontractor because of the subcontractor's failure to pay for materials already delivered. The general contractor thereupon promised to pay for all materials already delivered and to be delivered in the future. This promise of the general contractor, it was held, was not invalid.[42] A plea that the materialman was already under obligation to supply the agreed materials, under his contract with the subcontractor, would not invalidate the general contractor's promise.[43]

It may appear so, but the general contractor's promise in the above example was not without consideration. The materialman's promise to deliver the material in spite of the breach of the subcontract (by the failure of the subcontractor to pay for materials delivered) constituted the consideration to support the general contractor's promise.

---

[42] Walter L. Couse & Co. v. Hardy Corp. 274, So. 2d., 322 (1973)
[43] Naples v. Mc Cann, 95 A.2d. 158, Supreme Court of New Jersey, Appellate Division (1953).

14

Contract between Owner and Architect/Engineer

14.0 Introduction

Most construction works fall under one of two categories, namely, public works or private works. Generally public works are executed by the engineering and architectural staff in the employment of the public bodies. However, it is not uncommon that the public bodies, though having their own engineering and architectural staff, seek professional advice or help from independent architects or engineers. In the case of private works, barring works of minor nature such as for repairs or maintenance or construction of small buildings, for which the owner relies upon the contractor to guide him as to what should be done, large building and engineering contracts are usually entered into between the owner and a contractor after designs and specifications are prepared by an architect or engineer for and on behalf of the owner. Even in the cases of "package deals" or "industrialized buildings", the owner may avail of the services of a professional architect or engineer, though this is usually limited to consultative advice.

The advice sought by an owner from his architect or engineer, in the first instance, may be limited to preparation of sketch designs, preliminary drawings and estimates of the works to be carried out with a view to ascertain the practicability and suitability of the works, and to obtain necessary permission from the statutory bodies like the corporation, etc. Only when the owner is satisfied that he wishes the project to proceed and all statutory permissions have been obtained, is he likely to require contract drawings and specifications with bills of quantities to be prepared and tenders invited or other work entrusted to a contractor. After the work is entrusted to a contractor, working drawings including RCC designs, etc. are supplied to him. Generally the work of structural design is entrusted to a structural engineer either directly by the owner, or by the architect. Usually the architect or engineer will also be required by his employer to supervise the work and to certify the payments.

The role that an engineer or architect plays in a construction contract may be fourfold. He is a skilled servant if in the employment of the owner, or as an agent of the owner to administer or manage the project, or if the contract between the owner and the contractor contains stipulation which makes him play the role of quasi-arbitrator or finally as an arbitrator.

The relationship that exists between the owner and the engineer or architect stems out of the contract of the agency. The general law relating to the agency regulates the responsibilities of an engineer or architect in his function as an agent of the owner.

14.1 Provisions of the Law

The following provisions of the law are dealt with in this chapter.

Provisions of the Law	Paragraph No.
Indian Contract Act, Sections 182, 185–187	14.2
Indian Contract Act, Sections 188–210	14.4
Provisions in Standard Form Contracts	
FIDIC, 1992 amended edition, Clauses 2.1–2.6, 67.1 1999/2017 edition, Clauses 3.1–3.5	14.4

14.2 Contract to be in Writing

A contract of agency between the owner and the engineer or architect may be created expressly in writing or implied by words, conduct or necessity. The following provisions of the Indian Contract Act need to be noted.

182. *Appointment and authority of agents*
   An "agent" is a person employed to do any act for another or to represent another in dealings with third persons. The person for whom such an act is done, or who is so represented, is called the "principal."

185. *Consideration not necessary*
   No consideration is necessary to create an agency.

186. *Agent's authority may be expressed or implied*
   The authority of an agent may be expressed or implied.

187. *Definitions of express and implied authority*
   An authority is said to be express when it is given by words spoken or written. An authority is said to be implied when it is to be inferred from the circumstances of the case; and things spoken or written, or the ordinary course of dealing, may be accounted circumstances of the case.

ILLUSTRATION

A owns a shop in Serampore, living himself in Calcutta, and visits the shop occasionally. The shop is managed by B, and he is in the habit of ordering goods from C in the name of A for the purpose of the shop and of paying for them out of A's funds without A's knowledge. B has an implied authority from A to order goods from C in the name of A for the purpose of the shop.

It is thus clear that the law does not, in general, require the agency contract to be in writing. However, it is in the best interest of the engineer or architect to insist upon a written contract, for a number of reasons. The first important reason for insisting on a written contract is that in the case of the owner's death, it would be difficult for the engineer or architect to prove the understanding arrived at between him and the deceased owner. Second in many instances the statute law requires that the contract, to be binding, must be in writing and by a deed. For example, in most instances contracts with corporations must be made under the seal of the corporation if they are to be binding. Third, an oral contract may not leave both parties with the same understanding of its terms and conditions.

   Where an English architect, the defendant, "effectively went on a frolic of his own producing a wonky industrial design rather than the sleek modern design the Claimants were expecting", it was important to have a brief expressed not just in words but with a drawing and/or a mock-up or a 3-D drawing and a detailed written description of the design.

To avoid misunderstandings at the very least, a written brief is essential and changes to that brief must be recorded in writing whether by drawings, sketches and/or minutes of meetings. If that is not done, the absence of such written records must be explained to the clients in writing and they must make an informed decision not to receive a written brief and written records of any changes or developments of that brief. ... The same approach should also be adopted to any changes or variations to the written brief. This approach is not to help manage a client's expectations as to what is being designed and what can be built. This approach is an essential part of the architect's services paid for by the client. This approach, I consider, is all the more important where there is an element of unfamiliarity in the relationship between the client and the architect to ensure that the client knows what is being designed and the architect knows what the client expects to be built.[1]

EXAMPLE FROM CASE LAW

In a case, a young architect did not enter into a written agreement with the owner. Before the work was completed the owner discharged the architect because of difficulty which arose between the two. In his action to recover the fee for services rendered, the architect testified that the owner had expressly agreed to pay him well for his service. The owner, on the other hand, testified that the architect had solicited the job, saying that it would be a great help to him in starting out in his business and had offered to provide his services free. Though the case was decided in favour of the architect, there was the risk of losing a case under comparable circumstances.[2]

It needs to be noted that for an agency contract no consideration is necessary as per Section 185 of the Indian Contract Act. Even without contributing any share capital a person can validly become a partner of a firm. Contribution of capital by a partner is not the *sine qua non* for the validity of partnership, as per Section 18 of the Partnership Act every partner is the agent of the firm. Therefore, the act of a partner would be treated as "act of a firm" which is defined in Section 2(a) of the Partnership Act. As per Section 25 of the Partnership Act, all partners of the firm are jointly and severally liable for the act done by the other partner as partner of the firm.[3]

14.3 Conditions of Appointment

The relation between the donor of the power and the donee of the power is one of principal and agent. The relation of agency arises whenever one person, called the agent, has authority to act on behalf of another, called the principal, and consents so to act. The relationship has its genesis in a contract. It is possible for co-principals jointly to appoint an agent to act for them and in such case become jointly liable to him and may jointly sue him.[4]

The terms of the appointment govern the rights and obligations of the parties. The appointment of an architect or engineer for a particular construction project or for a period of time is often made by exchange of letters with or without incorporation of the terms of engagement or of scales of fees in the standard forms prepared by the national Associations of Architects such as

---

[1] Freeborn v. De Almeida Marcal (T/a Dan Marcal Architects), [2019] EWHC 454 (TCC), paras 58 and 60.

[2] Chapel v. Clark, 117 Mich. 617, 76, NW 62 (Mich. 1898).

[3] M/s. Shivraj Reddy and Brothers v. S. Raghu Raj Reddy, (A.P.): 2002(3) R.C.R. (Civil) 270. Also see: Southern Roadways Ltd., Madurai v. S.M. Krishnan, (SC), AIR 1990 SC 673.

[4] *Halsbury's Laws of England*, Vol. I, 4th ed., para. 726, Syed Abdul Khader v. Rami Reddy, (SC): AIR 1979 SC 553.

the Indian Institute of Architects (IIA), the Royal Institute of British Architects (RIBA) or the American Institute of Architects (AIA). Care should be taken, where terms of the standard form are incorporated by reference, to make it clear whether they apply only to the amount of remuneration or whether the conditions of engagement as a whole, including the arbitration clause are intended. A mere request to act as arbitrator or engineer in relation to a project, without specifying at the outset the services required of the architect or engineer, may lead to doubt or dispute as to what are the respective rights and duties of the parties. It is, therefore, desirable that conditions of appointment should be formulated in greater detail and incorporated in the agreement. These should include:

1. What other consultants, designers and specialists are to be retained and if by the owner directly or by the architect
2. Whether an independent quantity surveyor and/or clerk of works are to be employed and if so on what terms and for what services
3. The use and ownership of drawings and designs
4. Special services like preparing perspective drawings, scale models, etc.
5. The authority of the architect/engineer to negotiate or contract with contractors and subcontractors
6. The architect's or engineer's power in letting out contract and to order variations and deviations, issue payment certificates and in general bind the owner once the contract has been made
7. The duration of the appointment and the terms on which the services, if continued beyond the agreed period, would be rendered
8. The position in the event of death, retirement, or incapacity of the architect or engineer
9. Scale of professional charges, stages of payments, and payment on abandonment of the works by the owner, drastic modification in approved drawings, etc.
10. Termination of agreement by either party
11. Provisions for settlement of disputes by reference to arbitration or otherwise.

In this connection it may be noted that in the absence of a clear agreement as to the purpose of the appointment it will be presumed that when an architect was first employed to prepare plans and estimates and advice on the project or get necessary sanctions, it would not necessarily follow that the owner required his services in letting the contract, if there was to be one, or in supervision of the work. In fact, until the stage of accepting a tender and signing the contract has been reached, the architect is likely to be employed almost on a day-to-day or assignment-to-assignment basis. It can be said that up to this stage the project is of exploratory or tentative character. For example, in the event of the lowest tender received being higher than the budget provision of the owner, the project is likely to be abandoned or drastically modified.

An appointment in explicit terms for a particular project cannot be determined until the purpose for which the appointment was made has been achieved. In a premature termination without cause other than the owner's mere violation, the architect would be entitled to a reasonable remuneration (in the absence of agreed fees or remuneration as per the terms of agreement for the services rendered) and also to damages for the loss of remuneration, which he had been prevented from earning until the work was finished.[5]

---

[5] Thomas v. Hammersmith B.C., [1938] All E. R. 203.

14.4 The Scope and Extent of Engineer's and Architect's Authority

The engineer or architect, under his contract with the employer, acts in the capacity of the agent of the employer. As a result of this relationship, questions frequently arise as to the owner's liability for the acts of the engineer or architect towards the contractor or the orders given by the engineer or architect to the contractor. The best way to avoid such disputes is to make provisions in the contract to meet those eventualities which often give rise to disputes. For example, Clause 3.1 in FIDIC 1999/2017 edition expressly stipulates that the engineer shall have no authority to amend the contract or to relieve either party of any duties, obligations or responsibilities under the contract. Where FIDIC conditions are adopted for major projects in India, the contract generally includes Conditions of Particular Applications which provide that the power to the engineer to grant extension of time or sanction extra item rates in excess of the stipulated limit will be exercised only after approval by the employer.

The knowledge of provisions of the Indian Contract Act that may be applicable to these aspects includes the following provisions, the contents of which are self-explanatory:

---

188. *Extent of agent's authority*
    An agent having an authority to do an act has authority to do every lawful thing which is necessary in order to do such act. An agent having an authority to carry on a business has authority to do every lawful thing necessarily for the purpose, or usually done in the course, of conducting such business.

ILLUSTRATIONS

(a) A is employed by B, residing in London to recover at Bombay a debt due to B. A may adopt any legal process necessary for the purpose of recovering the debt, and may give a valid discharge for the same.

(b) A constitutes B, his agent, to carry on his business of a ship-builder. B may purchase timber and other materials, and hire workmen for the purposes of carrying on the business.

189. *Agent's authority in an emergency*
    An agent has authority, in an emergency; to do all such acts for the purpose of protecting his principal from loss as would be done by a person of ordinary, prudence, in his own case, under similar circumstances.

ILLUSTRATIONS

(a) An agent for sale may have goods repaired if it be necessary.

(b) A consigns provisions to B at Calcutta, with directions to send them immediately to C at Cuttack. B may sell the provisions at Calcutta, if they will not bear the journey to Cuttack without spoiling.

14.4.1 The Engineer/Architect to Act within the Scope of His Authority Expressly Given to Him

An engineer or architect is well advised to limit his acts and directions within the scope of his authority expressly given to him by the owner. If he executes his authority without the knowledge of the contractor that he is acting in excess of his authority and thereby causes the contractor to provide materials or labour or to make additions or alterations, he is liable to an action for breach of warranty, if the owner objects to payment thereof.[6]

In brief, it may be concluded that the engineer or architect should not issue to the contractors instructions that are a material departure from the contract documents without the owner's written approval. Similarly, the engineer or architect should not appoint a substitute or delegate his authority without the express or implied consent of the owner. He may appoint assistants but he must always retain complete control of and personal responsibility for their work. In this respect the following self-explanatory provisions of the Indian Contract Act need to be noted:

---

190. *When agent cannot delegate*
     An agent cannot lawfully employ another to perform acts which he has expressly or impliedly undertaken to perform personally, unless by the ordinary custom of trade a sub-agent may, or, from the nature of the agency, a sub-agent must, be employed.

191. *"Sub-agent" defined*
     A "sub-agent" is a person employed by, and acting under the control of, the original agent in the business of the agency.

192. *Representation of principal by sub-agent properly appointed*
     Where a sub-agent is properly appointed, the principal is, so far as regards third persons, represented by the sub-agent, and is bound by and responsible for his acts, as if he were an agent originally appointed by the principal.
     Agent's responsibility for sub-agent
     The agent is responsible to the principal for the acts of the sub-agent.
     Sub-agent's responsibility
     The sub-agent is responsible for his acts to the agent, but not to the principal, except in case of fraud or willful wrong.

193. *Agent's responsibility for sub-agent appointed without authority*
     Where an agent, without having the authority to do so, has appointed a person to act as a sub-agent, the agent stands towards such person in the relation of a principal to an agent and is responsible for his acts both to the principal and to third persons, the principal is not represented by or responsible for the acts of the person so employed, nor is that person responsible to the principal.

194. *Relation between principal and person duly appointed by agent to act in business of agency*
     Where an agent, holding an express or implied authority to name another person to act for the principal in the business of the agency has named another person accordingly, such person is not a sub-agent, but an agent of the principal for such part of the business of the agency as is entrusted to him.

ILLUSTRATIONS

(a) A directs B, his solicitor, to sell his estate by auction, and to employ an auctioneer for the purpose. B names C, the auctioneer, to conduct the sale. C is not a sub-agent, but is A's agent for the conduct of the sale.

(b) A authorizes B, a merchant in Calcutta, to recover the money's due to A from C & Co. B instructs D, a solicitor, to take legal proceedings against C & Co. for the recovery of the money. D is not a sub-agent but is a solicitor for A.

---

[6] Steel v. Young (1907) S.C. 360

195. *Agent's duty in naming such person*

In selecting such an agent for his principal, the agent is bound to exercise the same amount of discretion as a man of ordinary prudence would exercise in his own case; and if he does this he is not responsible to the principal for the act or negligence of the agent so selected.

ILLUSTRATIONS

(a) A instructs B, a merchant to buy a ship for him. B employs a ship's surveyor of good reputation to choose a ship for A. The surveyor makes the choice negligently and the ship turns out to be unseaworthy and is lost. B is not, but the surveyor is, responsible to A.

(b) A consigns goods to B, a merchant, for sale. B, in due course, employs an auctioneer in good credit to sell the goods to A, and allows the auctioneer to receive the proceeds of the sale. The auctioneer afterwards becomes insolvent without having accounted for the proceeds. B is not responsible to A for the proceeds.

14.4.2 Implied or Ostensible Authority

An implied or ostensible authority of the architect arising out of his appointment as such and the extent to which, without express authority, he can bind his employer in relation to the contractor and third parties needs to be considered. It must be remembered that the architect or engineer, unless there is a contractual provision giving his opinion, decision or certificate finality, has no authority whatever to waive strict compliance with the contract or to bind the employer. Furthermore, even provisions purporting to confer finality can be interpreted as merely giving an additional protection to the owner and not as binding him.[7]

In a conventional building and engineering contract, where it is customary to describe the contract work in detail and with precision, the correct view will be, it is submitted, that the contractor must do the described contract work, and for this purpose the decision of the architect as to what is sufficient performance of the described contract work is to be binding on both parties.

An architect or engineer, in private practice has no implied authority to make a contract with the contractor binding on his employer, or to vary or depart from a concluded contract. For this reason there is a provision usually found in building or engineering contracts under which the owner, or more usually his architect, is given power to order variations. In the absence of such a provision the owner may not be liable to pay for any additional or varied work which has been done under the architect's instructions unless he had knowledge of the architect's instruction and did not countermand it. However, an order to vary, issued by an architect after compliance of the formalities required under the contract, will bind the owner.

The architect or engineer has no implied authority to employ engineering or other consultants, so as either to render the employer liable for their fees or entitle the architect to additional payments in respect of the fees. A complex work might often necessitate such consultants to be brought in early at the design stage, but as the design is the architect's duty, such consultants, it can be said, in effect carry out part of the engineer's or architect's duty. For example, in a case involving the reinforced concrete structure, while holding the architect liable for a defective design provided by a specialist subcontractor, it was observed by Sir Walker Carter Q.C. in Moresk v. Hicks[8] that an architect

---

[7] Billyack v. Leyland, [1968] 1 W.L.R. 471.

[8] Moresk Cleaners Ltd. v. Hicks, [1966] 2 Lloyd's Rep. 338. Also see (a) Sealand of the Pacific v. Robert C. McHaffie Ltd., 51 DLR (3d) 702 regarding the experimental use of vermiculite concrete for an underwater aquarium; and (b) District of Surrey v. Church 76 DLR (3d) 72, where the architects received a report from an engineering firm that was not an expert in soil conditions that deep soil tests should be conducted, but this was not informed to the employer as the architect believed that the employer would not approve of the additional expense.

had three courses open to him if he was not able to design the whole of the work himself. One was to refuse the job; one was to ask the building owner to employ a structural engineer on his part of the work; and one was, while retaining responsibility for the design, himself to seek the advice and assistance of a structural engineer, paying for the service out of his own pocket, but with the satisfaction of knowing that if the advice given was wrong the engineer would owe him the same duty as he owed to the employer.

Each case should be examined on its own facts. Usually, there is no negligence, if a structural and civil engineer acts in accordance with "the practice and views of a reasonable body of other structural and civil engineers."[9] The standard is that of "the reasonable average".[10] The court should judge the conduct by "what was known at the time and not with the wisdom of hindsight".[11]

EXAMPLE FROM CASE LAW

A supermarket site comprising a layer of very soft clay (with a risk of substantial settlement) was under development for CGL by a developer (X). It was treated by vibro replacement stone columns before casting a concrete floor slab. It failed to stabilize the ground and the slab settled at a differential rate by up to 110 mm giving rise to an uneven floor. The main contractor was employed by (X) under a JCT 1980 With Quantities Building Contract. Engineer JAA, the defendant, was in charge of preparing the performance specification which did provide for vibro compacting of the soil for improving ground stability. CGL argued that JAA had committed a breach of its duty to exercise reasonable skill and care by proposing the vibro replacement for that site. JAA's defence was that it was a reasonable proposal based on the advice of a specialist geotechnical company. It was held that there was no breach by JAA of its duty to exercise reasonable skill, care and diligence as engineers. The expert evidence supported their contention that it was not inevitable that the vibro replacement would not work. It was observed:

> In this case JAA do not say, as in Moresk v Hicks that they can avoid liability because they had the right to delegate the question of the suitability of and the detailed design of the vibro replacement to other parties. Rather they say that vibro replacement is an area where competent civil and structural engineers reasonably seek the assistance of specialists ... I consider that JAA are broadly correct in those submissions. The evidence shows that competent civil and structural engineers often seek advice from specialist contractors in the fields of piling and ground treatment.[12]

Having chosen a world renowned expert firm, JAA had acted reasonably in acting on their advice. CGL's claim against them failed. Thus, a court will look at all the facts and circumstances to determine if the architect or engineer was negligent and liable to the client.

---

[9] Bolam v. Friern HMC, [1957] 1 WLR 582 at pp. 587–588.

[10] Per Bingham, LJ in his dissenting judgment in Eckersley v. Binnie & Partners, (1988) 18 Con. L.R. 1 at pp. 79–80.

[11] Cooperative Group Ltd. v. John Allen Associates Ltd., [2010] EWHC 2300 (TCC), para. 150.

[12] Ibid., paras 247–248.

14.4.3 Ratification by Owner/Employer

Where acts done by the engineer/architect exceed the power and authority expressly or impliedly given to him and are without the owner's knowledge, the owner has an option to ratify or disown such acts. He may exercise his option expressly or impliedly. The following self-explanatory provisions of the Indian Contract Act are worth noting.

196. *Ratification*

Where acts are done by one person on behalf of another, but without his knowledge or authority, he may elect to ratify or to disown such acts. If he ratifies them, the same effects will follow as if they had been performed by his authority.

197. *Ratification may be expressed or implied*

Ratification may be expressed or may be implied in the conduct of the person on whose behalf the acts are done.

ILLUSTRATIONS

(a) A, without authority, buys goods for B. Afterwards B sells them to C on his own account; B's conduct implies a ratification of the purchase made for him by A.

(b) A, without B's authority, lends B's money to C. Afterwards B accepts interest on the money from C. B's conduct implies a ratification of the loan.

198. *Knowledge requisite for valid ratification*

No valid ratification can be made by a person whose knowledge of the facts of the case is materially defective.

199. *Effect of ratifying unauthorized act forming part of a transaction*

A person ratifying any unauthorized act done on his behalf ratifies the whole of the transaction of which such act formed a part.

200. *Ratification of unauthorized act cannot injure third person*

An act done by one person on behalf of another without such person's authority, which, if done with authority, would have the effect of subjecting a third person to damages, or of terminating any right of interest of a third person, cannot, by ratification, be made to have such effect.

ILLUSTRATIONS

(a) A, not being authorized thereto by B, demands, on behalf of B, the delivery of a chattel, the property of B, from C who is in possession of it. This demand cannot be ratified by B, so as to make C liable for damages for his refusal to deliver.

(b) A holds a lease from B, terminable on three months' notice. C, an unauthorized person, gives notice of termination to A, the notice cannot be ratified by B, so as to be binding on A.

14.4.4 Revocation of Authority

The authority of an engineer or architect may be determined in any one of the following ways:

1. By the completed performance of his agreement
2. By the expiration of the period, if any, for which the authority was given
3. By the performance becoming unlawful or impossible
4. By either party giving notice to the other without prejudice to any claim for damages that either party may have, one against the other, for breach of the contract of employment.

The following provisions in the Indian Contract Act, which may have application, are self-explanatory.

---

201. *Termination of agency*
   An agency is terminated by the principal revoking his authority; or by the agent renouncing the business of the agency; or by the business of the agency being completed; or by either the principal or agent dying or becoming of unsound mind; or by the principal being adjudicated an insolvent under the provisions of any Act for the time being in force for the relief of insolvent debtors.

202. *Termination of an agency where agent has an interest in subject matter*
   Where the agent has himself an interest in the property which forms the subject matter of the agency, the agency cannot, in the absence of an express contract, be terminated to the prejudice of such interest.

ILLUSTRATIONS

(a) A gives authority to B to sell A's land and to pay himself, out of the proceeds, the debts due to him from A. A cannot revoke this authority, nor can it be terminated by his insanity or death.

(b) A consigns 1,000 bales of cotton to B, who has made advances to him on such cotton, and desires B to sell the cotton, and to repay himself, out of the price, the amount of his own advances. A cannot revoke this authority, nor is it terminated by his insanity or death.

203. *When principal may revoke agent's authority*
   The principal may, save as it otherwise provided by the last preceding section, revoke the authority given to his agent at any time before the authority has been exercised so as to bind the principal.

204. *Revocation where authority has been partly exercised*
   The principal cannot revoke the authority given to his agent after the authority has been partly exercised, so far as regards such acts and obligations as arise from acts already done in the agency.

ILLUSTRATIONS

(a) A authorizes B to buy 1,000 bales of cotton on account of A and to pay for it out of A's money remaining in B's hands. B buys 1,000 bales of cotton in his own name, so as to make himself personally liable for the price. A cannot revoke B's authority so far as regards payment for the cotton.

(b) A authorizes B to buy 1,000 bales of cotton on account of A, and pay for it out of A's moneys remaining in B's hands. B buys 1,000 bales of cotton in A's name, and so as not to render himself personally liable for the price. A can revoke B's authority to pay for the cotton.

205. *Compensation for revocation by principal or renunciation by agent*
Where there is an express or implied contract that the agency should be continued for any period of time, the principal must make compensation to the agent, or the agent to the principal, as the case may be, for any previous revocation or renunciation of the agency without sufficient cause.

206. *Notice of revocation or renunciation*
Reasonable notice must be given of such revocation or renunciation; otherwise the damage thereby resulting to the principal or the agent, as the case may be, must be made good to the one by the other.

Reasonable Notice Must Be Given of Such Revocation or Renunciation – Interpretation

Whether a reasonable notice for determination of the service of the agency is necessary at all, when under the arrangement between the parties, period of agency has been agreed to either expressly or by necessary implication. The doubt arises because of use of the word "such" in Section 206. One version can be that Sections 205 and 206 of the Indian Contract Act are so inter-linked that Section 206 can only be invoked in a case where there is an express or an implied contract that the agency should be continued for any period of time and if for any reason such agency is revoked by the principal or renounced by the agents without sufficient cause, then one must compensate the other. The other version is even in the case when the agency is not for the fixed period, compensation is payable. The division bench of the Madras High Court in a case held:[13]

[W]e are unable to agree with the conclusion that the word "such" in Section 206 of the Act, should only be read for purposes of interpreting Section 205 of the Contract Act. On the other hand, we are of the view that the word "such" has been used so as to be referable to all the other relevant Sections which precede and succeed Section 205 under the sub-head "Revocation of authority".

Particular reference was made to Section 207 of the Act. The said section reads:

207. *Revocation and renunciation may be expressed or implied*
Revocation and renunciation may be expressed or may be implied in the conduct of the principal or agent respectively.

ILLUSTRATION

A empowers B to let A's house. Afterwards A lets it himself. This is an implied revocation of B's authority.

Section 207 does not touch upon the rights or obligations of either the principal or the agent. It only deals with a general term whereunder revocation or renunciation may be by necessary implication also. In fact, Section 205 also refers to an implied term of a contract of an agency. It, therefore, follows that such an implied contract may also be determined by a conduct which is implied as between the parties. The division bench of the Madras High Court in the above case held:[14]

---

[13] J. K. Sayani v. Bright Brothers (Madras)(DB), AIR 1980 Mad. 162.
[14] J. K. Sayani v. Bright Brothers (Madras)(DB), AIR 1980 Mad. 162.

Section 206 in our view has to be understood as being general in scope and not being one which has an impinge [*sic.*]only and only upon Section 205 for the only reason the word "such" is used in Section 206. As revocation and renunciation of the contract of agency may take place in myriad ways, the rights and obligations that flow from such termination can only be after the party who intends to snap such jural relationship, puts the other aggrieved party reasonable notice. This could also be justified on the principles of natural justice.

---

208. *When termination of agent's authority takes effect as to agent, and as to third persons*
The termination of the authority of an agent does not, so far as regards the agent, take effect before it becomes known to him, or, so far as regards third persons, before it becomes known to them.

ILLUSTRATIONS

(a) A directs B to sell goods for him, and agrees to give B 5 per cent commission on the price fetched by the goods. A afterwards, by letter, revokes B's authority. B, after the letter is sent, but before he receives it, sells the goods for 100 rupees. The sale is binding on A and B is entitled to five rupees as his commission.

(b) A, at Madras, by letter, directs B to sell for him some cotton lying in a warehouse in Bombay, and afterwards, by letter, revokes his authority to sell, and directs B to send the cotton to Madras. B, after receiving the second letter, enters into a contract with C, who knows of the first letter, but not of the second, for the sale to him of the cotton. C pays B the money, with which B absconds. C's payment is good as against A.

(c) A directs B, his agent, to pay certain money to C. A dies, and D takes out probate to his will. B, after A's death, but before hearing of it, pays the money to C. The payment is good as against D, the executor.

209. *Agent's duty on termination of agency by principal's death or insanity*
When an agency is terminated by the principal dying or becoming of unsound mind, the agent is bound to take, on behalf of the representatives of his late principal, all reasonable steps for the protection and preservation of the interests entrusted to him.

210. *Termination of sub-agent's authority*
The termination of the authority of an agent causes the termination (subject to the rules herein contained regarding the termination of an agent's authority) of the authority of all sub-agents appointed by him.

14.5 Remuneration of the Engineer and Architect

The right of the engineer or architect to receive compensation for the service rendered arises either under an express or implied contract. As already stated there can be an agency contract without consideration and, as such, an owner may be able to prove to the court that he did not at any time undertake to pay the engineer or architect anything for his services. The engineer or architect can claim compensation only if he can show that there has been some bargain for a consideration either expressed or implied.

14.5.1 Implied Promise of Remuneration

In the absence of an express contract of any kind, however, the presumption is that there was no intention on the part of a practising engineer or architect to give his services free. In such a case all that an engineer or architect need prove is the fact of his employment. The law will imply an obligation on the part of the owner to pay the reasonable value of his services; unless, as already stated, the owner can show that in the particular case it was not intended that the engineer or architect should receive remuneration.

14.5.2 Measure of Amount of Remuneration

If a contract between the owner and the engineer or architect contains an express stipulation as to the amount of remuneration payable and the conditions under which it is to be paid, such a stipulation is conclusive on the parties.

> **EXAMPLE FROM CASE LAW**
>
> A surveyor, at the request of the owner, submitted an estimate of the cost of a survey of the land, which according to him would not be less than $80 and more than $100. The owner then directed the surveyor to proceed with the work. On completion of the work, the surveyor submitted a bill for $587.55. The owner refused to pay the amount of the bill and sent the surveyor $100. The surveyor, thereupon, sued the owner for the recovery of the balance, which he claimed to be due to him. It was held that the surveyor's compensation was limited to $100 and he was denied any recovery in excess of $100 paid to him.[15]

14.6 When Remuneration May Not Be Payable

The engineer or architect may not be able to recover remuneration for reasons including the following.

14.6.1 Absence of Required Licence

In states where statutes require engineers and architect to be licensed, no recovery can be had for the services rendered in the absence of the required licence. In one case involving an architectural partnership, some of the certificates issued by the firm were signed by an unlicensed member of the firm. In a suit by the firm to recover the balance due for the services rendered, it was held by the court that the firm could not recover for services based on the work of an unlicensed architect.[16]

14.6.2 When Plans Are Not Used

Where services are rendered by an architect or work done on approval, or are in the nature of a proposal, the architect may have no claim unless the design is actually approved or used and unless the contract stipulates special provision. It is up to the engineer or architect to make certain that he will be entitled to a compensation for his services rendered. The following

---

[15] Seybolt v. Baber, 97 a 2d. 907, Court of Appeal of Maryland 1953.
[16] Palmer v. Brown, 237 P. 2d 306.

examples will help emphasize the care and diligence with which an engineer or architect should enter into a contract with the owner.

EXAMPLES FROM CASE LAW

1. An engineer undertook to prepare a report including cost estimates and recommendations for a central air-conditioning system in a hotel, in consideration of $4,000. The contract guaranteed an additional fee for plans and specifications only if they were used. The contract further provided that no payment would be made if plans and specifications were not used. The hotel company considered the central air-conditioning plan too expensive, and instead installed room air-conditioners. The engineer was paid $4,000 only, as plans were not used. The court, on the ground that the plans were not used, dismissed the engineer's suit for recovery of payment for his plans.[17]

2. A school board employed an architect to prepare plans and specifications for remodelling and repairing certain school buildings. The architect's compensation was agreed to be 8 per cent on "all work approved and let". The school board reserved the right to discontinue any or all the work at any time, and in that case agreed that the architect's compensation would be determined by the schedule established by the American Institute of Architects. The architect prepared plans and specifications, and tenders were also received for the work. However, the board did not let any contract for the work and on this ground refused to pay the architect any compensation. In an action by the architect to recover compensation, the court allowed him the recovery according to the schedule of the American Institute of Architects upon the lowest tender received for the work.[18]

3. From 2005 onwards a firm undertook work for a hotel developer to assist with a project. No binding agreement was concluded about the architects' remuneration in relation to their preliminary work. In June 2006, the developer engaged another firm but soon after agreed with the architects that a sum of €150,000 would be paid to them as a reward for their services in helping with the negotiations and architectural preliminary planning. Upon refusal to pay this sum, the respondent commenced proceedings. The developer argued no future work was necessary from the architects and that the agreement for payment could not have referred to past work. The judge held that the developer's argument did not make any sense in light of the statement that further performances were not necessary. The judge held that the amount was payable even though no further services were performed by the firm. The Court of Appeal confirmed this.[19]

14.6.3 When Construction Cost Is Limited

If an engineer or architect undertakes to prepare plans and specifications for a building to cost not more than a restricted amount, he may not recover compensation if the drawings and specifications prepared by him are for a building which will cost substantially in excess of that amount. For example, in one case where the cost of construction was 27% more than the limit fixed by the architect's contract, the architect was denied compensation for his services.[20]

---

[17] De Laureal and Moses v. Pensacola Hotel Co., 148, F. Supp 684, U. S. District Court 1959.

[18] Liewelly v. Board of Education of Cicero Stickney High School, Tp. Dist. 154, N. E. 889, 324 el. 254.

[19] D&K Drost Consult GmbH, She Architekten v. Foremost Leisure (Holdings) Limited, [2015] EWCA Civ. 73.

[20] Stevens v. Fanning, 59, III, App. 2d, 285, 207 N.E. 2d. 136 (III 1965).

However, the doctrine of substantial performance is applicable in such cases and the engineer or architect can recover if there is a slight variance between the restricted cost and the actual cost. An architect who prepared plans and specifications for a $75,000 building as against the restricted cost of $70,000, was allowed recovery of his fee from the owner.[21]

14.6.4 Competition Work

An architect or engineer who has submitted plans and designs in response to a general invitation in open or limited competition, cannot claim any remuneration unless:

a. remuneration is offered to all competitors in the terms of the invitation or,
b. he is able to prove that his plans, etc. or any portion of these, have been used by the owner either for the work in question or for any other purpose.[22]

14.6.5 Incomplete Services

The architect's right to remuneration, when his contract of employment with the owner has been partly performed and there is a refusal or failure on the part of the owner to complete the employment, would be dependent upon the express provisions made in the agreement to cover such an eventuality. In the absence of any specific provision he will be entitled to all sums due at the time of breach and for profit lost by reason of the breach or upon a *quantum meruit*.[23]

14.7 Manner of Payment of Remuneration

If a contract for engineering and architectural services contains no stipulation as to the time and manner of the payment of fees, the engineer or architect will, in general, be entitled to payment by instalments as the work proceeds.[24] Where, however, an engineer or architect undertakes to supervise the execution of an entire contract in consideration of a commission on the total cost, he may not be able to claim any compensation until the work is completed.[25]

14.8 Ownership of Plans and Copyright

An architect ordinarily has no right to the ownership of a plan furnished and accepted by and paid for by another. Plans forming the essential part of the building contract, unless proved to be the property of the architect, are deemed to be the property of the employer.[26] However, the engineer or architect has a lien on them until he receives the final payment of his fees.[27] It may be noted that, unless the engineer or architect assigned the copyright in writing or he were under a contract of full time service, the copyright remains his property and not that of the employer.[28]

---

[21] Vaky v. Phillips, 1943 W. 601 (Tex 1919) L.P.
[22] Landless v. Wilson, (1880) 8R (Ct. or Sess 289) [1880] 12 WLUK 65.
[23] Section 205 of the Indian Contract Act.
[24] Appleby v. Myres, (1867), L. r. 2 C.P. 651.
[25] Johnson v. Gaudy, (1855), 26 L. T. (O.S.) 72.
[26] Berlingshot v. Lincoln County, 257, N. W. 373, 128 Neb. 28.
[27] Hughes v. Lenny, (1839) 5M & W 183.
[28] Meikle and others v. Maufe and others, [1944] 3 All ER 144.

14.9 Position of the Engineer and Architect

The role that an engineer or architect plays in a construction contract, as already stated, may be fourfold. He is a skilled servant if in the employment of the owner. He acts as an agent of the owner. The contract between the owner and the contractor generally contains stipulations, which make him play the role of quasi-arbitrator or sometimes as an arbitrator.

14.9.1 The Engineer or Architect as a Quasi-Arbitrator

If the contract between the owner and the contractor contains stipulation to the effect that the decisions of the engineer or architect in respect of claims and disputes relating to the execution or progress of the work or interpretation of the contract documents shall be final and conclusive, the engineer or architect has then the general powers and duties of the arbitrator and to distinguish him from "arbitrator", his position is generally referred to in the strict sense as "quasi-judicial" or "quasi-arbitral" and he functions as a quasi-judge or quasi-arbitrator.

Construction contracts generally contain provisions to the effect that the decisions of the architect/engineer shall be final in respect of certain matters as quality of the materials, workmanship, designs, etc. The matters in respect of which the decisions are made final are called "excepted matters", because the scope of arbitration clauses contained in such contracts excludes these matters from the ambit of the arbitration clause. The purpose of these provisions is to avoid the possible delay in reaching final decisions on matters of day-to-day working and which could be numerous.

14.9.2 Decisions of the Engineer or Architect Are Final in Absence of Fraud

When parties to a construction contract provide that the certificates and decisions of an engineer or architect shall be final and conclusive, such estimates and decisions have the effect of the award of an arbitrator. In the absence of fraud or such gross mistakes as imply bad faith or failure to exercise an honest judgment, they are conclusive and binding upon the parties. Similar contract provisions have been approved and enforced by the United States Courts[29] and also by the M. P. High Court.[30]

14.10 Liabilities of Engineers and Architects

Like any other professional, an engineer or architect, while undertaking work in his field, represents that he possesses the requisite knowledge and skill. If he does not possess the skill he is liable for lack of it and if he fails to use it he is liable for negligence. An architect is one who possesses, with due regard to aesthetic as well as practical consideration, adequate skill and knowledge to enable him (i) to originate, (ii) to design and plan, (iii) to arrange for and supervise the erection of such buildings or other works calling for skill in design and planning as he might in the course of his business reasonably be asked to carry out, or in respect of which he offers his services as a specialist.

---

[29] United States v. Moorman, (1950) 338, U. S 457, 461, 462, 705 C.A 288 94L. Ed. 2561 United States v. Wunderlich 1951, 342, U.S. 98, 72, S.Ct 154 96 L Ed 113.

[30] Dandakaranya Project v. P.C. Corpn., AIR 1975 MP 152; Heavy Electricals (India) Ltd. Bhopal v. Pannalal Devchand Malviya (Madhya Pradesh) AIR 1973 MP 7.

14.10.1 Degree of Care and Skill Required

In the preparation of plans and specifications, the architect must possess and exercise the care and skill of those ordinarily skilled in the business. If he does so, he is not liable for faults in construction resulting from defects in the plans, as his undertaking does not imply or guarantee a perfect plan or a satisfactory result, it being considered enough that the architect himself is not the cause of any failure, and there is no implied promise that miscalculation may not occur. Where, however, the architect does not possess and exercise such care and skill he will not only be liable in damages for defects in his plans but he cannot recover compensation for them.

14.10.2 Duty in Respect of Design

Under the traditional arrangement, architects and engineers are responsible to the owner for the design and suitability of the works for their intended purpose, whereas the contractor is only responsible for bringing the works to completion according to the design. However, it is not uncommon in present-day practice for the contractor to undertake the responsibility of the design as well.

An architect offered help to her friends, a wealthy couple, to landscape their garden for which she did not charge any fee. She did intend to charge a fee for any future design work carried out after the hard landscaping was carried out by a contractor she recommended. She also acted as a project manager. Although there was no contract, the court held that she owed a duty of care under common law as a person of special skill on which the friends had relied. She knew about their budgetary requirements after the initial quote was obtained for the hard landscaping, and she ought to have advised on costs reasonably accurately.[31]

14.10.3 Examination and Subsoil Investigations

To produce a successful construction design, in nearly all cases, a sufficient examination of the site with a view to determining the area available for the proposed works, and the nature of the subsoil for deciding upon the correct design for foundations or methods of underground working, is a necessary preliminary. In a case where an architect is retained to advise, and instructed to build on old foundations, he must employ reasonable tests of the soundness and stability of the underlying structures. If the architect employs others to carry out these investigations he will nonetheless remain liable if he adopts incorrect information furnished to him by others save as discussed in paragraph 14.4.2 above. Further, it goes without saying that in normal circumstances an architect or engineer will not be automatically relieved from liability from his plans and designs by obtaining the owner's approval of them, if the defect of design complained is one of construction or of a technical nature.

In a case where by reason of the known facts relating to soil conditions, etc. there is only one really foolproof type of scheme, and another considerably more economical but involving an element of risk, the architect or engineer is duty-bound to acquaint the owner of the position and leave the decision to the owner. In such a case, if the owner were to approve the less safe course, the architect would not be liable.

14.10.4 Cost of Construction-Excess Over-Estimate

An architect or engineer is not liable for damage if the cost of constructing a building exceeds the expectations and financial liability of his employer, unless the contract of employment contains an upset cost of constructing the structure. However, it is essential that his plans and specifications comply in all particulars with his contract of employment.[32] In practice, at an early stage of the

---

[31] Burgess v. Lejonvarn, [2017] EWCA Civ. 254.
[32] Medical Arts. Bldg. v. Ervin, 257, P. 2d. 969, Supreme Court of Colorado (1953).

employment, the owner will usually indicate or impose limitation on the cost of the proposed project. In such a case there is an express condition of the employment that the project, as designed, should be capable of being completed within the stipulated or reasonable cost. Even if the owner does not mention it, it is suggested that an architect must design works capable of being carried out at a reasonable cost having regard to their scope and function. If he fails to do so, such an implied condition of employment is likely to deprive him of the remuneration for the work done.

An architect has a great role to play in making an estimate. He is expected to neither under-estimate nor over-estimate value of the works. He is bound by his conduct to the owner. He can be sued for his negligence. For his misconduct, fees payable to him may be forfeited. He may incur other liabilities not only under the contract but under the statute.[33]

EXAMPLES FROM CASE LAW

1. In a case, the architect estimated that a school building he had designed would cost $1,10,000. He knew that the estimate was for the purpose of preparing a bye-law to raise the necessary fund. The lowest tender was for $1,57,800. He then eliminated 40% of the cubic content of the school, and said the remainder could be carried out within the limit. The lowest tender for this modified design was $1,32,900. Under these facts and circumstance, the Court of Appeal of British Columbia held that he had been negligent, and he was properly dismissed and liable to pay damages.[34]

2. An architect was instructed to prepare designs for a building not to exceed £4,000 in cost. He prepared plans and the tenders were invited; the lowest tender was £6,000. It was held that the architect was not entitled to recover his remuneration for the work done.[35]

14.10.5 Delay in Supplying Plans

If a contract of employment does not contain a definite stipulation as to when plans and details are to be furnished, the law implies that they shall be furnished within a reasonable time. In one case, the contract between the owner and the engineer provided that the engineer should render his services at such times as the owner might direct. It was held that the engineer was entitled to a reasonable time after he received the direction to perform his services. "Where both contracting parties contribute to the delay, neither can recover damages unless there is a clear proof as to the appointments of the delay and the expenses attributable to each party."[36]

It is submitted that, unless an act or requirement of the owner or some circumstances quite outside the architect's or engineer's control make it impossible, an architect or engineer must, as a matter of business efficacy, impliedly undertake to the owner that he will give instructions, plans, certificates, etc. so as to comply with the express or implied requirement of the building contract, which he himself will normally have recommended to the owner. In short, it can be said that an architect is duty-bound to supply plans, instructions, etc. under his contract with the owner, at the times for supplying the same under the building contract between the owner and the contractor so as to avoid the owner's liability to the contractor for claims on this count.

---

[33] M.D. AWHO v. Sumangal Services Pvt. Ltd., 2003(3) Arb. LR 361 (SC).

[34] Savage v. Board of School Trustees, [1951] 3 DLR 39.

[35] Flannagan v. Mate, (1876) 2 Vict. L.R. (Law) 157.

[36] Coath and Goss Inc. v. United States, 101.Ct. 702, 714, 715.

14.10.6 Supervision

An architect or engineer, who has undertaken to supervise the construction in progress must properly supervise the works and inspect them sufficiently frequently to ensure that the materials and workmanship conform to the contractual requirements. In a case, it was held that the architect's duties

> were to give reasonable supervision, and that meant such supervision that would enable him to certify that the work of the contractor had been executed according to contract, and that having failed to give such supervision he was liable in damages to his employer on account of work which he had passed, but which in fact did not conform to the contract.[37]

The normal practice of architects is to visit a site for which they are responsible, periodically, say about once a week. For day-to-day supervision, the architect relies upon the clerk of works, usually paid and employed by the owner. Thus, other subordinate persons usually supplement his personal supervision. From the legal point of view, it is submitted that the architect or engineer will usually be fully responsible to the employer for his own employees' mistakes or errors. Normally he will not be responsible for errors by persons paid and employed by the owner, such as clerks of works, resident engineers separately engaged, unless the matter was of a kind that the architect or engineer should have seen or dealt with, or he failed to give proper instructions to the subordinate.

In a case of improperly executed work, if defects remain in the completed structure, the owner can recover damages not only from the contractor but also from the architect or engineer, if the architect or engineer has failed to exercise due care during the progress of the work. The contractor is not liable if he relied on the architect's or engineer's instructions.

Negligence – Meaning Of

By negligence is meant neglect of some care and skill, which one is bound by law to exercise towards somebody. Thus there cannot be a liability for negligence unless there is a breach of some duty. The degree of care, which a man is required to use, will be the care that an ordinary prudent man is bound to exercise. However, if a person professes to have special skill or who has voluntarily undertaken a higher degree of duty is bound to exercise more care than an ordinary prudent man. Under a contract between an owner and an engineer or architect, the latter is duty-bound to exercise such diligence and skill as are exercised in the ordinary and proper course of similar employment for which he may or may not receive payment. The engineer or architect is liable for the consequences of ignorance or non-observance of the customs and practices of his profession, and of failure to observe and comply with statutes and bye-laws that affect his work.

EXAMPLES FROM CASE LAW

1. Under a building contract the architect's duty was to supervise the work. The clerk of works, for corrupt purposes of his own, permitted the contractor to deviate from the design and specifications, by laying the ground floor without the necessary precautions against damp and assisted in concealing the deviation. The architect, relying upon the clerk of the works, failed to detect the deviation and to have it rectified. As a result of this deviation dry-rot set in. The defence put forth by the architect was that the owner had appointed an unfit improper clerk of works and that damage was due to the negligence and fraud of the owner's servant, or alternatively that the owner was bound to employ a fit and proper clerk of works, but failed to do so. It was held that the architect was liable in damages to the employer for negligence.[38]

---

[37] Jameson v. Simon, (1899) 1 F (Ct. of Sess.) 1211.
[38] Leicester Board of Guardians v. Trollope, (1911) 75 J. P. 197.

2. The owner had employed an architect to prepare plans and specifications and, later on, also to supervise the work of construction of a home on a hillside plot of California. The owner, on the advice of his architect, accepted the house as constructed. Shortly after taking possession the owner noticed that floors were sloping and there were cracks in the plaster. It was discovered that foundations had settled at least 5 cm. During the construction the owner had noticed the workmen placing foundation forms on loose earth fill. He had brought this to the notice of the architect who had promised to look into the matter and set things right. In fact, after a couple of days, he did inform the owner that things had been fixed up. The owner refused to make the final payment to the contractors. In an action by the contractors to foreclose their lien, the owner counter-claimed against the contractors for negligence. The owner also filed a suit for negligence against the architect. The action against the contractor was tried first. During the trial the architect admitted that although the plans and specifications called for front foundations to be a minimum of 15 cm below the original natural ground level, he had told the men to go ahead when foundations were in fact 45 cm above the natural ground level. It was held by the court that the depth of the front foundation trench was approved by the architect and as such constituted the waiver on the part of the owner (the architect being his agent) of the contractor's failure to follow the plans and specifications in that respect. However, the owner was awarded damages for that portion of the said defects for which the contractors were responsible and neither the owner nor his architect waived.

The action involving the architect was tried before a jury. The jury, guided by the expert's estimates of cost of repairs (which varied from $1,846.50 to $17,354), fixed the damages to the amount of the judgment against the architect in the sum of $4,972.55. The judgment was entered and was appealed by the architect. The main contention of the architect was that, as established by the evidence, the contractors and architects were joint tort-feasors and as such the satisfaction of the judgment against the contractor released the architect as a matter of law. He also urged that the first adjudication against the contractor prevented the owner from any further recovery against the architect, as the subject matter of both the actions was the same. It was held by the appellate court that the architect and contractors were independent wrongdoers and not joint tort-feasors as there was neither unity of purpose nor concert of action between them. The judgment against the architect and that against the contractors did not result in the owner getting double compensation.[39]

14.10.7 Issue of Certificates

Many building and engineering contracts frequently include express terms that the work, in addition to the compliance with the contract requirements, is to be done to the satisfaction of the engineer or architect, and further require the contractor to obtain and produce certificates to that effect. The certificates are of various kinds and generally include:

1. Interim or progress certificates upon which periodical payments or advances to the contractor on account are made

2. Final certificates, which frequently are certificates both of satisfaction with the work and of the sum due to the contractor upon the final discharge of his obligation

3. Various other certificates, e.g. the contractor is not duly proceeding with the work upon which a special contractual right of determination by the owner may depend; or certificate of practical or substantial completion, maintenance during the defects liability, upon which

---

[39] Alexander v. Hammarberg, 103 Cal 2d, 872, 230 P. 2d 399 Cal 1951.

the provisions as to payment of retention money, release of security deposit, maintenance of works or liquidated damages for delay may depend

4. Certificate to the effect that the contractor has committed default justifying termination of his contract, etc.

Where, under the terms of a contract, finality is given to the certificate or decision of the architect or engineer, it may be that the sense of the contract requires that the architect or the engineer act judicially rather than as the agent of one party. Where he is acting in this capacity to determine or value some issue between two or more parties, he is not liable to either party for want of skill, ignorance of law, or negligence. Many architects and engineers misunderstand and exaggerate the effect of this somewhat peculiar status, and appear to consider that it confers upon them an element of discretion, which enables them to give effect to their own opinions or notions of fairness as between the parties, without too particular a regard to the letter of the contract. In some cases, on the other hand, many fail to issue certificates, which are by way of condition precedent to invoke an arbitration clause. In this connection it is pointed out that the IIA standard form includes a provision that if the contractor is either dissatisfied with the decision of the architect or in the case of withholding of a certificate by the architect, he can invoke provisions of the arbitration clause. Similar provisions are also found in other modern forms.

Quite apart from the provisions in the contract, which entitle the arbitrator or court to open up, to review any decision or certificate issued by the architect, such review will be quite objective and without regard to any discretion. It is simply not the case that provisions making use of a certifier as a piece of convenient administrative machinery mean that it is thereby desired to give him any discretion. Even provisions making express reference to his satisfaction or approval are not drafted in the expectation or desire that he should apply any standard other than that required by the contract provisions. It must be remembered that the duty of the certifier is to apply the provisions of the contract exactly and strictly to a situation upon which he is required to give his decision or certificate, and any evidence showing that he has taken any extraneous or irrelevant matter into consideration will deprive his certificate of validity and, therefore, of finality.

14.10.8 Liability to Third Parties

Are architects or engineers and contractors liable to third parties for injuries received as a result of defective design and construction? They are, according to the verdict of a New York Court.

EXAMPLE FROM CASE LAW

An action was brought against the architects and builders of an apartment house to recover damages sustained by an infant who fell from the porch of the apartment. The action was based on the grounds that the porch was so improperly designed and constructed that it was a danger to its users. The defence put forth by the builders and the architects was that the injured party being a stranger to the contract had no cause of action. The trial court sustained this contention and dismissed the action. However, on appeal the judgment was reversed. The appellate court held that the situation was similar in principle to the law that where a contractor builds a defective scaffold, he is liable for damages resulting therefrom, even though the injured party was not a party to the contract.[40]

Where there is no fault in plan or design prepared by an architect, he may still be liable to third parties for negligence in his supervision functions. If he has undertaken supervision works, and if the circumstances are such that he knows, or should know, in the exercise of

---

[40] Inman v. Binghamton Housing Authority, 152 N.Y.S. 2d 79.

reasonable care that dangerous and unsafe conditions exist on the site, he has the right and corresponding duty to stop the work until the unsafe condition has been remedied. If he breaches such a duty he would be liable to persons who could foreseeably be injured by the breach. The architect may be liable in spite of the fact that he has neither the right nor the duty to direct or control the methods or means by which the independent contractor converts the blueprints into reality.

To protect himself and the owner, an architect is well advised to have an indemnity clause included in the contract whereby the contractor should indemnify and hold harmless both the owner and the engineer or architect to any litigation, which may take place. The protection under the clause would be available if any fault occurs during the execution of the contract and is not brought about by design errors or by the architect's failure to give important instructions.

14.10.9 Architect Liable for Structural Engineer's Fault

Can an architect avoid liability for faulty design because he had delegated that part of the work to an independent consulting engineer? To answer this question the principle of vicarious liability needs to be studied. Vicarious liability means liability which is incurred for, or instead of, another. Normally and naturally the person who is liable for a wrong is he who does the wrong. But there are circumstances when liability may be attached to a person for the wrongs committed by others. The most common example is the liability of the master for the wrong committed by his servant. Thus an engineer or architect is liable for the acts of his assistants and subordinates if the owner does not directly employ them.

Another relevant instance of vicarious liability is the case of an owner and his independent contractor. The general rule, as to an independent contractor, is that for the acts and omissions of an independent contractor (except those actually authorized), the owner is not liable to third persons. But to this general rule also there are certain exceptions. The following cases illustrate them.[41]

EXAMPLES FROM CASE LAW

1.   The construction company was used to fabricate and assemble steel girders, an essential structural component at the site of a bridge construction. Each girder was about 220 feet in length, and weighed a massive 55 tonnes. To effect easy movement and launching of this heavy girder, the company had adopted a system by which the girder was placed on two trollies which ran on land rails placed at the approaches of the bridge. The plaintiff wanted to contact the executive engineer for some work in connection with the contract which his master had taken at Gadchiroli. Upon learning that the executive engineer was supervising the work of the bridge, the plaintiff came up at the bridge site on the day. The company engineers were watching the trolley of the launching girders when suddenly the movement of the front trolley was obstructed. Suddenly the girder slipped away from its platform, fell down and hit the plaintiff. The plaintiff had to be taken to various hospitals at Chanda, Nagpur and rushed by air to Bombay where his leg was amputated and an artificial limb was fitted. The plaintiff sued the contractor to recover Rs. 60,000/- as damages, which suit

---

[41] Also see paragraph 14.4.2 above.

having been dismissed, the plaintiff appealed. The defence supported the judgment on the basis that the defendant contractor owed no duty to the plaintiff who was a rank trespasser at the site. Allowing the appeal, it was held:

> It is true that the plaintiff did not go to the site upon the invitation of the occupier viz. defendant company. The plaintiff went there for his own business which he had to transact with the Executive Engineer ... As part of the site was used by the P.W. D. as an office it would be in the nature of things impossible that the P.W.D. office would be used only by the Executive Engineer. The entire paraphernalia of the office of the Executive Engineer viz., his Deputy Engineer, Time keepers. Head Clerks etc., would always go with him. So would the contractors, who verily have to visit the Engineers and his officers on business such as getting day to day instructions, ensure that the measurements are recorded and the bills are passed periodically. The defendant company who is the occupier of the site had impliedly given a permission to all such persons to enter upon the site whosoever had any work with the Executive Engineer. This implied permission given by the contractor would confer on such person the status of a 'visitor' being a person who at common law will be treated either as an invitee or a licensee, and in no event as a trespasser. ... It is common ground that the site on bank of the river Wainganga belongs to the State and the defendant company was working there as an independent contractor and not as servant of the State. It is settled law that unlike vicarious liability arising out of malfeasance or misfeasance of a servant, the employer is not liable for the tortuous act of its independent contractor.

Relying on Wheat v. E. Lacon and Co. Ltd.,[42] it was held that the defendant company was in occupation and control of the premises where the activity was being conducted. The company was held liable to pay compensation.[43]

2. A masonry wall was designed without complying with a City Building Ordinance. A pedestrian was hurt when it collapsed upon him due to high wind. He sued the architect, the owner and the contractor. In his contract with the owner, the architect had assumed overall responsibility for designing the building of which the wall in question was a part. A consulting engineer retained by the architect, in fact, had designed the wall. The trial court held the architect alone responsible for the damages to the pedestrian for injuries sustained when the wall failed. The judgment was affirmed in appeal. The appellate court considered three basic questions. The first question naturally related to proof of negligence on the part of designers of the building. This question was answered in the affirmative on the principle that if the design fails to comply with the City Code, such a violation amounts to a statutory negligence. The second question related to establishing a casual connection between the injuries received by the plaintiff and the alleged design defect. Although, it was contended on behalf of the architect that the hurricane-force wind gusts on the critical day were of such unprecedented velocity as to amount to an "act of God", it was held that a proper wall would have stood up to the unusual winds. There was evidential support for the decision on this point. Regarding the third question, which is the subject of the present discussion, the architect put forth a plea that, as a professional man, the standard of care applicable to him was to be measured by the practices of the reasonably prudent architect in the same locality. Since it was the custom of the architects in the area to refer engineering work to consulting engineers, it was argued, his only obligation was to select a reliable firm. By doing that, he said, he had fulfilled his duty of exercising due care. It was held[44] that the

---

[42] Wheat v. E. Lacon and Co. Ltd., 1966 AC 552 at 589.

[43] Nuruddin v. M/s. Khare and Tarkunde, (Bombay) (DB) (at Nagpur) AIR 1985 Bom. 401.

[44] Johnson v. Salem Title Co., 425, p. 2d 519 (1967).

architect's reliance on local customs among his fellow professionals was misplaced. Selecting a competent engineering firm only indicated that he himself was not negligent but it begged the question of vicarious liability. It was the architect who contracted with the owner to provide all drawings and specifications necessary for construction of the building. The engineers were strangers to the owner. Any engineering work incorporated in the plans became a part of the architect's design. Since he assumed the benefits and burdens of designing the building, he assumed the responsibility of meeting the building code design provisions, including the structural engineering requirements. His duty to meet the minimum safety standards of the building code was, therefore, non-delegable.

The court recognized the general rule governing the owner–independent contractor relationship mentioned earlier but noted that there are significant exceptions to it. Even where an independent contractor is employed the owner remains liable when the work involved is "inherently dangerous" or when, as a matter of law, the owner's duty cannot be delegated. The plaintiff's contention in this case, that the owner (here the architect) was under a non-delegable duty and as such liable for the acts of the independent contractor (here the consulting engineer), was upheld as already mentioned. It may also be noted that if the plan or design is defective and not the construction, the architect or engineer will be liable, irrespective of whether he was retained to supervise the construction or not.

15

Limitation

15.0 Introduction

In this chapter it is proposed to discuss the principle of limitation of actions, whereby the right to a legal remedy, such as a claim for money due under a contract or for damages for its breach, is extinguished by the passage of time.

Limitation is regarded as adjective law, as distinguished from substantive law. It is almost always, if not invariably, statutory. The statutes of limitations are known as statutes of repose or statutes of peace. All statutes of limitation have for their object the prevention of the rearing up of all claims after a long time, when evidences have been lost, and to prevent persons from being harassed, at a distant period of time. The object is to extinguish stale demands.

This chapter will focus on what kinds of demands are likely to be raised in building and engineering contracts and what is a period of limitation in relation to each of them as per the Indian Limitation Act, 1963. Each country specifies its own limitation period. For example, in England the limitation for bringing a claim for a breach of contract is six years. In Finland, it is ten years unless the claim is for a debt, in which case it is three years. The general or common causes of action in building and engineering contracts are given in paragraph 15.2. The relevant provisions of the Indian Limitation Act are reproduced in paragraph 15.3. Illustrative cases are dealt with from paragraph 15.4 onwards. Various issues discussed in this chapter arise in other countries too, so that some of the examples are drawn from non-Indian cases.

15.1 Provisions of the Law

The following provisions of the law are dealt with in this chapter.

Provision of the Law	Paragraph No.
Indian Limitation Act, Articles 13–18, 24, 55, 70, 113	15.3
J & K Limitation Act, Articles 53–56	15.4.1
Provisions in Standard Form Contracts	
FIDIC, 1999 edition, Clause 14	15.4.6

15.2 Causes of Action in Building and Engineering Contracts

The rights in relation to building and engineering contracts for the enforcement of which suits may be commenced include:

1. For the price of materials sold and supplied

2. For the price of work done (labour wages only)
3. For the price of work done and materials supplied, which in turn may include:

    i. Claim under contract for completed work

    ii. Claim for enhanced rates for work done in view of altered circumstances

    iii. Claim for extra items not included in the final bill

    iv. Deductions made in Running Account (RA) bills – adjustments in the final bill

    v. Payment of final bill

    vi. Refund of earnest money/security deposit/retention money including forfeiture

    vii. Compensation for breach of contract

    viii. Any other cause arising out of or in relation to the contract

    ix. To refer disputes to arbitration.[1]

15.3 Relevant Provisions of the Indian Limitation Act 1963

The relevant provisions applicable to the above general causes of actions mentioned in paragraph 15.2 above are reproduced below:

Art. No.	Description of suit	Period of limitation	Time from which period begins to run
13	For the balance of money advanced in payment of goods to be delivered	Three years	When the goods ought to be delivered
14	For the price of goods sold and delivered where no fixed period of credit is agreed upon	Three years	The date of the delivery of the goods
15	For the price of goods sold and delivered to be paid for after the expiry of a fixed period of credit	Three years	When the period of credit expires
16	For the price of goods sold and bill delivered to be paid for by a bill of exchange, no such bill being given	Three years	When the period of the proposed bill elapses
17	For wages in the case of any other person	Three years	When the wages accrue due
18	For the price of work done by the plaintiff for the defendant at his request, where no time has been fixed for payment	Three years	When the work is done
24	For money payable by the defendant to the plaintiff for money received by the defendant, for the plaintiff's use	Three years	When the money is received
55	For compensation for breach of any contract, express or implied, not herein specially provided for	Three years	When the contract is broken or (where there are successive breaches) when the breach in respect of suit is instituted occurs or (where the breach is continuing) when it ceases
70	To recover movable property deposited or pawned from a depository or pawnee	Three years	The date of refusal after demand
113	Any suit for which no period of limitation is provided elsewhere in this Schedule	Three years	When the right to sue accrues.

---

[1] Arbitration may be commenced without filing a suit for the same reliefs.

15.4 Date on Which Cause of Action Accrued – Illustrations

Limitation period begins to run from the date on which the cause of action accrued. It is not always easy to establish the date on which the "cause of action accrued" in a given case. The following examples and subsequent paragraphs will help lay down the guiding principles.

EXAMPLES FROM CASE LAW

1. A block of flats built in 1962 with defective foundations (which were either not noticed or inspected carelessly by the defendant council through its inspector), were taken by the plaintiff on long lease. In February 1970 structural movement began, resulting in cracks. The plaintiff brought an action against the council in 1972. It was held: The cause of action can only arise when the state of the building is such that there is a present or imminent danger to the health or safety of the persons occupying it. An action could lie after the danger became obvious.[2]

2. A boiler flue chimney, designed negligently by the defendant, a firm of consulting engineers, was built between June and July 1969. In due course cracks occurred in the internal lining of the chimney. Action was brought in October 1978. The trial court found that the cracks occurred not later than April 1970. It was not proved that the plaintiffs could with reasonable care have discovered the cracks before October 1972. The action was time barred if time started to run out when the cracks occurred, but not if it only started to run out when the cracks could with reasonable care have been discovered. Referring to the decision in Example (1) above, it was observed:

> His Lordship did not say, nor – did he imply, that the date of discoverability was the date when the cause of action accrued. The date which he regarded as material – was, of course, related to the particular duty resting on the defendant as the local authority, which was different from the duty resting on the builders or the architects.

It was held that the limitation period began when the cracks occurred and that the action was statute barred.[3] This case was disapproved by the Privy Council in a New Zealand case.[4] It continues to be applied in England.[5] The Court of Appeal has clarified that the case was not overruled (expressly or by implication) by Murphy v. Brentwood DC[6] so that under English law, an owner's cause of action for negligent design starts the limitation period when the physical damage to the building first occurs.[7]

15.4.1 Recovery of Cost of Materials to Be Used in Works – Cause of Action

Where, in a construction contract, a bill of quantities includes certain items for supply and delivery of materials to the site, and separate items for execution of the work with the use of such materials, the limitation for recovery of the cost of materials supplied would begin not when the materials were supplied, but when the contract was completed or when the contract came to an end.

---

[2] Anns v. London Borough of Merton, House of Lords, (1978) 5 BLR 1, [1978] AC 728.
[3] Pirelli General Cable Works Ltd. v. Oscar Faber and Partners, House of Lords (1982) 21 BLR 99.
[4] Invercargill City Council v. Hamlin, [1996] AC 624.
[5] Oxford Architects Partnership v. Cheltenham Ladies College, [2006] EWHC 3156 (TCC).
[6] [1991] 1 AC 398 – this was a case where the defects were discovered before physical injury occurred but under English law of tort pure economic loss does not give rise to a claim.
[7] Abbott v. Will Gannon & Smith Ltd., [2005] EWCA Civ. 198.

EXAMPLES FROM CASE LAW

A contract provided for the supply and consolidation of shingle for a roadwork. It provided separate rate for the supply of shingle and its consolidation, adding,

The contractor agrees to execute the work ... and to complete the work as under:

(a) Supply within one month; and

(b) Consolidation within one month from the date of supply of roller, and to complete ... as per the time schedule given, failing which the contract will be cancelled and work got done departmentally or through some other agency at the cost of the original contractor.

It was held that

the intention of the parties was that the price of shingle shall not be claimed as of right on the date when it was supplied but rather when the contract was completed or when the contract came to an end.

It was further held:

Accordingly the plaintiff could file a suit for the recovery of the price of shingle supplied by him or, for the portion of such price remaining unpaid, within three years from the date the contract was completed: and, if the contract proved abortive, from the date it was cancelled to the knowledge of the plaintiff.[8]

15.4.2 Recovery of Cost of Work Done

Art. 18 of the Limitation Act deals with the subject matter of a suit for the price of work done, where no time has been fixed for payment. It declares the cause of action "when the work is done". It further qualifies that the plaintiff, at the request of the defendant, must do the work. The period of limitation is three years from the date of doing the work.

In most standard form contracts, the time for payment of the final bill is fixed by the agreement and to such cases Art. 18 will not apply. In such cases, Art. 113, which stipulates a three-year period of limitation from the date "when the right to sue accrues", will apply. The right to sue would accrue when the bill was due for payment. The question as to when a bill is due for payment, in construction contracts, is peculiar and not always easy to answer.

EXAMPLES FROM CASE LAW

1. To file a suit against government for recovery of money, limitation is three years, but suit could be filed within a period of two months after expiry of three years provided notice under Section 80 CPC was given to the government. The Apex Court of India held:[9]

---

[8] Ghulam Hassan v. State, AIR 1976 J. & K. 93.
[9] Disha Constructions v. State of Goa, (SC): AIR 2012 SCW 2252; 2011(6) RAJ 449.

In our view, proper interpretation of Section 15(2) of the Act would be that in computing the period of limitation, the period of notice, provided notice is given within the limitation period, would be mandatorily excluded. That would mean a suit, for which period of limitation is three years, would be within limitation even if it is filed within two months after three years, provided notice has been given within the limitation period. In such a case, the period of notice cannot be counted concurrently with the period of limitation. If it is done, then period of notice is not excluded. Any other interpretation would be contrary to the express mandate of Section 15(2) of the Act.

2. The original plaintiff and the State of Gujarat entered into the contract for doing earthwork and lining for the canal. The time to complete the work was extended up to 30th June 1988. The final bill was submitted by the original plaintiff contractor on 31st March 1989 and the payment under the final bill was made on 31st March 1989, which the plaintiff had accepted under protest. Thereafter, the plaintiff served the statutory notice under Section 80 of the CPC on 28th March 1992, which was received by the defendant on 30th March 1992 and thereafter, after excluding period of vacation, the original plaintiff instituted Special Civil Suit No. 196 of 1995 before the learned trial court on the first day of reopening, i.e. 15th June 1992. The learned trial court, by impugned judgment and decree, dismissed the suit also on the ground that the suit was barred by law of limitation, more particularly Arts 55 and 113 of the Limitation Act. Feeling aggrieved and dissatisfied with the impugned judgment passed by the learned trial court, the original plaintiff had preferred an appeal. Held: Therefore, considering Section 15(2) of the Limitation Act, it cannot be said that the suit was barred by Arts 55 and 113 of the Limitation Act. Reliance was placed on the decision of the Hon'ble Supreme Court in the case of Disha Constructions and others.[10]

15.4.3 Deductions Made in RA Bills – Limitation Begins When?

The standard form contracts invariably provide for release of payment once every month or at a lesser interval than one month. If payment is not correctly made for the work done, or if wrongful deduction is made, it would appear that the legal proceedings for recovery of the full price of the work done or deductions made must be commenced within a period of three years. If the provision were to be so read there would be a multiplicity of proceedings with each month some new cause of action arising. Giving due consideration to the contract stipulation of treating all RA bill payments as advances and to be adjusted in the account to be settled once and for all in the final bill, it is possible to construe that the fresh cause of action would accrue at the time of settlement of the final bill in respect of all disputes in the payment of RA bills, including disputed recoveries made, and the period of limitation for them also will begin to run from the date the final bill was due for payment. Thus, failure to pay the final bill constitutes a new cause of action and the starting point of limitation for payment will arise from the date of default in the payment of the final bill. In other words, even if the cause of action to enforce payment of intermediate or RA bills is barred by lapse of time, the payment of the same can, nevertheless, be enforced as a part of the payment of the final bill.[11]

---

[10] Virvijay Construction Company v. State of Gujarat, (Gujarat)(DB): AIR 2016 Guj. 167.

[11] M.L. Dalmia & Co. v. Union of India, AIR 1963 Cal. 277.

EXAMPLES FROM CASE LAW

In respect of three cases heard together, the contract work was completed in August 1974 for the first, in 1975 for the second and in 1967 for the third, but the reference to arbitration was made by the parties after notice was served by the contractor in 1979. On the above facts, it was held that the final measurements had not been effected for many years and when the public officers took measurement which was not to the satisfaction of the contractor, disputes arose. There can be no doubt that dispute could arise in these cases when the final measurement was done and the plaintiff's claims were denied *in toto*. In fact, in each of these cases, as rightly stated for the contractor, the dispute arose in relation to the work done, which was not being reflected in the final measurement. That being the position, in none of these cases could the claim be barred by limitation.[12]

15.4.4 Recovery of Cost of Extra Work

Extra work ordered to be carried out by the contractor under the provisions of the conditions of contract can be said to be the work done at the request of the owner. If the order for the extra work specifies the time limit for its completion, Art. 18 will not be attracted. But normally an order for extra work does not stipulate a time limit and, as such, Art. 18 will be attracted. But here again, in view of numerous requests for executing extra work on several different occasions, it can be said that a fresh cause of action will arise for the recovery of the cost of the work done at the time of the settlement of the final bill. Extra work is indeed supplemental work. When such is the position, the supplementary works must be taken as a part of the main work and the arbitration clause would ordinarily apply to disputes relating to such extra work being supplementary to the main work.[13]

15.4.5 Claim for Enhanced Rate – Which Article of Limitation Act Applies?

Art. 55 of the 1963 Act will cover situations where the enhanced rates are sought by way of compensation for the extra expenses incurred on account of delay in completion due to breaches of contract conditions committed by the employer. Where the suit filed is not a suit for compensation for breach of contract, express or implied, but it is a suit for enhanced rate because of change of circumstances, and in respect of work not covered by the contract, the claim is not covered by any specific article under the First Schedule, and must fall within the terms of Art. 113, and it provides a period of only three years.[14] Similarly, in a suit for payment of escalation cost for labour, Art. 113 of the 1963 Act applies.[15]

15.4.6 Final Bill – Limitation Begins When?

Art. 18 of the Limitation Act, 1963 provides for filing of a suit for recovery of money for work done by the plaintiff, within three years from the date when the work is done, in a situation where no time has been fixed for payment. Every standard form contract provides for submission of the final bill by the contractor within the period stipulated in the relevant condition for that purpose. The said clause further stipulates the period when the money due under the final bill will be

---

[12] Executive Engineer, R.E. Div. Dhenkanal v. J.C. Bhudaraj, AIR 1981 Ori. 172.

[13] Executive Engineer, R.E. Div. Dhenkanal v. J.C. Bhudaraj, AIR 1981 Ori. 172.

[14] Gannon Dunkerley & Co. v. Union of India, AIR 1970 SC 1433.

[15] State of Bihar v. Thawardas Pherumal, AIR 1964 Pat. 326.

payable to the contractor. For example, FIDIC 1999 edition includes Clause 14 which provides for submission of application for final payment certificate within 56 days after receiving the performance certificate along with the draft final statement and, thereafter, if any disputes arise the same are to be resolved by DRB or by amicable settlement and, after resolution of disputes, the final statement is to be submitted. The engineer is duty-bound to issue the final payment certificate within 28 days after receiving the final statement. The final payment is to be made within 56 days after the employer receives the final certificate. As such, the limitation should begin from the date the money was due and payable but was not so paid. For a suit for claim of the final bill, Art. 113 applies, which stipulates limitation of three years from the date on which the right to sue accrues.[16] Where the petitioner invoked arbitration after ten years of completion of the work and the court superseded the arbitration clause on account of limitation, it was held by the Supreme Court that the subordinate courts had rightly exercised the power.[17]

In practice, it is noticed that some cases of settlement of final bills linger for years and with a view to render substantial justice and not to deny it on technicalities, the court is required to take cognizance of various provisions of the agreement and also of the defaults committed by both the sides. For example, if the completion/performance certificate is not issued, it can be said that an occasion to submit the final bill by the contractor did not arise. Similarly, if an acknowledgement is made, the period of limitation will stand extended from the date of acknowledgement. An acknowledgement should be of a present subsisting liability. An acknowledgement made with reference to a liability cannot extend limitation for a time-barred liability or a claim that was not made at the time of acknowledgement or some other liability relating to other transactions. Any admission of jural relationship in regard to the ascertained sum due or a pending claim cannot be an acknowledgement for a new additional claim for damages. These and other points are covered by the following examples.

EXAMPLES FROM CASE LAW

1. In a case, the Supreme Court of India summed the facts thus:[18]

> De hors the correspondences that had been exchanged by and between the parties after the date of final payment i.e. 13th January 1981, the aforesaid date of final payment would have been crucial for determination of the period of limitation for filing the instant suit. However, in the case, from the correspondences that had been exchanged after the date of final payment it clearly appears that the plaintiff after receipt of the payment on 13th January 1981, reiterated its claim for additional payment on different counts including escalation and for extra works done. The defendant instead of rejecting the said claim entertained the same and kept the matter pending. Finally on 6th November 1981 ... the said claims were rejected. If the claims raised by the plaintiff were entertained and rejected finally on 6th November 1981, it would be reasonable to assume that the cause of action for the suit in respect of the said rejected claims arose on 6th November 1981 and the suit could have been filed at any point of time prior to the expiry of three years from the said date i.e. 6th November 1981 in view of Article 113 of the Limitation Act, 1963. The suit having been filed on 6th November 1984, the same, therefore, will have to be considered to be within the period of limitation. The High Court, therefore, was

---

[16] State of Bihar v. Rama Bhushan, AIR 1964 Pat. 326.

[17] Panchu Gopal Bose v. Board of Trustees for Port of Calcutta, AIR 1994 SC 1615, 1993(3)ALT6(SC), JT 1993(3) SC 537, 1993(2) SCALE696, (1993) 4SCC338; also see: Steel Authority of India Limited v. Respondent: J.C. Budharaja, AIR 1999 SC 3275, JT 1999(6) SC429, 1999(5) SCALE351.

[18] M/s Aries & Aries v. T. Nadu Electricity Board, AIR 2017 SC 1897=AIR 2017 SCW F1897.

not justified in holding the contrary. This will require the Court to consider the additional plea urged on behalf of the plaintiff, namely, that the High Court was not justified in reversing the decree passed by the learned trial Court so far as 50% of the escalation charges is concerned.

It was further held:

> The clause in the contract which obliged the plaintiff to continue to offer the rates initially offered by him would, naturally, be for the duration of the contract and cannot work to his peril for the period of delay for which the Department was admittedly responsible. Such a construction of the clause in the contract would not be reasonable. We, therefore, reverse the aforesaid finding of the High Court and hold that the plaintiff would be entitled to the 50% of the escalation charges as decreed by the learned trial Court. (The delay by the owner was admittedly 50%.)

2. The work in a case was completed on 15th June 1975 as per the contractual agreement between the parties. Disputes arose between the contractor-appellant and the respondent regarding various claims. The contractor signed the final bill under protest on 14th April 1977. The respondent wrote a letter on 28th October 1978 acknowledging the pending claims of the contractor. The contractor issued notice invoking arbitration on 4th June 1980. The arbitrator awarded certain amount in favour of the contractor. The contractor filed a suit for making the award rule of the court. Objections to the award filed by the respondent were overruled by the court and the court directed that the award of the arbitrator be made the rule of the court and a decree be drawn in terms of the award. The respondent filed a miscellaneous appeal, challenging the decision of the civil judge refusing to set aside the award. The contractor also filed a Miscellaneous Appeal and Civil Revision claiming future interest from the date of decree. The High Court allowed the appeal filed by the respondent and dismissed the appeals filed by the contractor as barred by limitation. In appeal before the Apex Court the issue to be decided was: whether the claim made before the arbitrator or any part thereof was barred by limitation? Held: By virtue of Section 18(1) of the Limitation Act, where, before the expiration of the prescribed period for a suit or application in respect of any right, an acknowledgement of liability in respect of such right has been made in writing signed by the party against whom such right is claimed, a fresh period of limitation shall be computed from the time when the acknowledgement was so signed. In the present case, in view of the acknowledgement in writing on 28th October 1978 and payment of certain amount on 4th March 1980, the limitation stood extended by three years from 4th March 1980. As such, whatever claims made before the arbitrator within three years from 28th October 1978 and 4th June 1980 were within the limitation. The impugned judgment was set aside.[19]

3. In a case decided by the Rajasthan High Court, the contractor had accepted payment under the final bill on 10th February 1956 after protesting on 23rd April 1955 and the suit for recovery was filed on 13th April 1959. On these facts it was held that time began to run from 10th February 1956 and the suit was barred by limitation.[20]

---

[19] J.C. Budhraja v. Chairman, Orissa Mining Corporation Ltd. and Anr., AIR 2008 SC 1363, 2008 (1) ARBLR 238 (SC), 2008 (3) MhLj 33, 2008(1) SCALE 597, (2008) 2SC C444.
[20] Bhawani Shankar v. State, AIR 1970 Raj. 268.

Preconditions for Submission of Final Bill Must Be Fulfilled

Provisions of many a contract stipulate that the final bill shall be submitted by the contractor within the period stated in the contract, such as one month of the date fixed for completion of works, otherwise the engineer-in-charge's certificate of measurement and of the total amount payable for the works accordingly shall be final and binding on all parties. Another clause provides that a final certificate of completion of works shall be furnished to the contractor; however, the government will be entitled to withhold the certificate until the contractor removes certain materials and scaffolding and does some clearing-up operation. In default of the contractor's removing the materials and doing the clearing-up operation, the government will be entitled to do it itself and the contractor is liable to pay the expenses incurred by the government. It follows that in cases where the contractor fails to remove materials and scaffolding, etc. and to carry out the clearing-up operation, the penalty is that the contractor will not be entitled to a certificate of completion, and without such certificate the contractor is not entitled to submit his bill for payment. In such cases, it is the duty of the government, after doing the work departmentally, to give the contractor a certificate of completion. Until such certificate is given, the contractor cannot submit his bill and get payment and the cause of action for payment, in terms of the contract, does not arise and no claim for payment can be made by the contractor, even after the completion of the work. If no such certificate is given, no claim could be made on the basis of the final bill and the question of time running against the contractor's claim for the final bill would not arise on that basis.

EXAMPLES FROM CASE LAW

The final bill appeared to have been passed by the government on 18th February 1949, but no intimation was given to the contractor of the passing of the final bill until 28th April 1954, when the executive engineer informed the contractor, by a letter of the same date, the government's decision in answer to the contractor's claim submitted on 9th May 1952. It was held that in the instant case, the government did not intimate its decision till 28th April 1954 and time cannot, therefore, run before that date. Reference of the dispute to arbitration having been made shortly thereafter, well within three years from that date, the amount claimed in the final bill is well within time and is not barred by limitation.[21]

15.4.7 Suit for Compensation for Breach of Contract

A breach of contract occurs where a party repudiates or fails to perform one or more of the obligations imposed upon him by the contract. If one of the two parties to a contract breaks the obligations which the contract imposes, a new obligation will in every case arise, giving rise to a right of action conferred upon the party injured by the breach. The period of limitation for claim of compensation for breach of contract is governed by Art. 55.[22] The provisions of Art. 55 are self-explanatory. The period of limitation for filing a suit under Art. 55 shall begin with effect from the date breach of contract was committed by the party in breach.[23]

---

[21] M.L. Dalmia & Co. v. Union of India, AIR 1963 Cal. 277.

[22] Dhapai v. Dalla, AIR 1970 All. 206.

[23] Sundaram Finance Limited (M/s.) v. Noorjahan Beevi (SC): 2016 (6) JT 224=2016(4)RAJ 227. H.P. Financial Corporation v. Smt. Pawna, 2015 (5) SCC 617.

EXAMPLES FROM CASE LAW

1. In the case State v. P.K. Jain,[24] involving two contracts for supply of bricks to the state government by a fixed date, the government had directed the contractor to suspend the supplies. It was held by the Patna High Court that there was a breach of the contract and the suit for compensation was governed by the provision of Art. 115 and not Art. 120 (corresponding to Art. 55 and 113 of the 1963 Act, respectively). It was based on whether there was a breach of contract, and not whether there was an end of the contract. It was further held that if specific goods are contracted to be delivered at or by a fixed date, it will not be a continuing contract and there can be no question of a continuing breach.
2. In a case, tenders to supply bricks for construction of roads and buildings were called for on behalf of the defendant and the plaintiff's tenders were accepted. Two separate agreements were executed in connection with two road projects, but no written agreement was entered in respect of the building projects. The contract in respect of the buildings was found to be void for want of compliance with Art. 299 of the Constitution. Bricks, however, had been delivered and the defendants had made part payments. The contracts, including the void contract, were a case of running accounts with credits and debits being adjusted from time to time. On these facts the Patna High Court held that the cause of action could arise only after giving of refusal to the amount claimed by the plaintiff and, as such, the period of limitation of three years would run from the date when the plaintiff's claim had been repudiated by the defendant.[25]

Counterclaim

A counterclaim is required to be treated as a separate suit and therefore the period of limitation would be three years from the date of accrual of the cause of action.[26]

Suit against the Government

In the case of a suit against the government in respect of a breach of contract, which is required to comply with the formalities provided under Art. 299 of the Constitution, the time does not begin to run until the date on which the contract is signed as required by the provisions of the law, though the provisional contract and the breach thereof might have taken place prior thereto. It was so held in an opinion case by the Calcutta High Court. The facts were as follows: The contractor entered into three separate contracts with the government for the manufacture and supply of bricks. As is usual in case of government contracts, there was in each case what may be characterized as a provisional agreement of an informal character entered into by one of the officers of the government, which was subsequently embodied in a formal document in compliance with the formalities provided by Section 175(3), Government of India Act. Held:

> It is elementary that the cause of action in a suit for damages for breach of contract arises at the date of the breach. In ordinary cases, therefore, the date of breach and the date of the cause of action for damages for breach of contract synchronise. But it may not synchronise as in the instant case when cause of action to institute a suit becomes complete after the contract is signed in compliance with the provisions of Section 175 (3), Government of India Act, and this date is long after the date of breach. In such cases could it

---

[24] State v. P.K. Jain, AIR 1981 Pat. 280.
[25] Hindustan Construction Co. v. State of Bihar, AIR 1963 Pat. 254.
[26] Thomas Mathew v. The Construction Engineer, K.L.D.C. Ltd., 2016(9) Scale 379.

possibly have been the intention of the Legislature that limitation would begin to run and complete its full course even before the accrual of the cause of action? To my mind the answer must clearly be in the negative.[27]

15.4.8 Claim for Refund of Security Deposit

The building and engineering contracts and especially the standard form contracts expressly provide that earnest money, security deposit/retention money shall be refunded after the expiry of a particular period. In such cases, the contractor's right to claim the refund accrues immediately on lapse of the period fixed, or happening of the event stipulated, and will give rise to a cause of action. The suit, if necessary, has, therefore, to be filed within the period prescribed by the article of the Limitation Act applicable.

The Bombay High Court has held in a case that Art. 113 is applicable and not Art. 24 or 70.[28] The Andhra Pradesh High Court, on the other hand, has held in a case that Art. 24 applied.[29] The Patna High Court has held in a case that Art. 55 applied.[30] It will thus be seen that the terms of the contract and the facts and circumstances of a case will be the guiding factors in deciding as to which article of the Limitation Act is applicable to the given case. For example, in the case decided by the Bombay High Court, the facts involved are similar to the point under discussion. It was observed:

> In the present case, according to the terms of the contract, the amount was to become due on completion of the contract. It was to become due by reason of subsequent events. The right to refund did not arise immediately on receipt by the defendant. Accordingly, Article 62 will have no application to the facts of the present case.[31] We have also stated hereinabove that Article 145 would not be applicable. The only article that would appear to us to be applicable is Article 120, which is the residuary article.[32]

15.4.9 Limitation For Demand For Arbitration

The limitation for demand for arbitration begins to run from the date on which a difference arises to which arbitration agreement applies.[33] Many contracts specify that if the arbitral proceedings are not demanded within a period of a few weeks from the date of decision given, the decision will attain finality. The question arises in such cases, if the demand should be made within the limitation and the arbitral proceedings must commence. It is submitted that if the right to demand arbitration is exercised within the period stipulated, the claim will revive at the time of the final bill, even if the claim stands barred by the elapse of the original period of limitation. Where the agreement stipulated that demand for arbitration shall be made within 90 days of intimation of the final bill by the government but thereafter amounts were withheld and payments were made, it was held that limitation for demand for arbitration would run from the date of the last payment.[34] Amended Section 28 of the Contract Act will render invalid the provisions such as 90 days or a shorter time than the three-year period allowed by the Limitation Act, it is submitted.[35]

---

[27] India Trades Corporation v. Union of India, AIR 1957 Cal. 153.

[28] Shankar v. State, AIR 1970 Bom. 8.

[29] Mohd. Dahil Khan v. State of A.P., AIR 1963 AP 216.

[30] Union of India v. M/s Gangadhar Mimraj, AIR 1962 Pat. 372.

[31] Reliance was placed on A. Venkata Subba Rao v. State of Andhra Pradesh, (1965) 2 SCR 577; (AIR. 1965 SC 1773).

[32] Shankar v. State, AIR 1970 Bom. 8.

[33] Oriental Bldg. & Furnishing Co. Ltd. v. Union of India, AIR 1981 Del. 293.

[34] M/s Ram Nath Mehra & Sons v. Union of India, AIR 1982 Del. 164.

[35] See paragraph 15.5 below for full details. B.L. Kashyap and Sons Ltd. v. Airport Authority of India, (Delhi) 2016 (234) DLT 14. M/s. Tarapore & Co. v. M/s. United India Insurance Co., Ltd., (Madras): 2017(5)LW 535=2018(1) Mad WN (Civil)113.

The Rajasthan High Court in a case held:[36]

> Once it is held that provision in Clause 24.2 of the agreement to the effect that upon failure of the contractor/employer to give notice to commence arbitration proceedings within 28 days after decision of the Project Manager is directory and not mandatory, it may perhaps be not necessary to go into any other question. Even then, the matter can be examined from another angle whether at all such condition would be valid and enforceable in view of the amended provision of Section 28, amended by the Indian Contract (Amendment) Act, 1996 (Act 1 of 1997) with effect from 8th January 1997.

Held:

> In view of above discussion and analysis of facts and law, such condition of Clause 24 or any of its sub-clauses and any such construction thereof, which debars the reference of dispute for arbitration, would be void in view of Section 28(b) of the Contract Act.

In a case decided by the Supreme Court of India it was held:[37] under Art. 137 of the Limitation Act, the postulated period of limitation is three years. In the instant case, the period of limitation would be, three years prior to the date of invocation of arbitration. After the appellant issued the notice dated 1st February 2002, it invoked the arbitral clause on 8th May 2002, and therefore, the period of limitation in terms of Art. 137, would bar all claims prior to 9th May 1999.

It needs to be noted that Section 28(b) as amended with effect from 8th January 1997 has prospective effect. It will not apply to agreements entered into prior to the amendment.[38]

15.4.10 Applicability of the Limitation Act for Filing Applications under the Arbitration and Conciliation Act, 1996

The Arbitration and Conciliation Act, 1996 under Section 43 provides as follows:

43. *Limitations*

(1) The Limitation Act, 1963 (36 of 1963), shall apply to arbitrations as it applies to proceedings in court.

(2) For the purposes of this section and the Limitation Act, 1963 (36 of 1963), an arbitration shall be deemed to have commenced on the date referred in Section 21.

(3) Where an arbitration agreement to submit future disputes to arbitration provides that any claim to which the agreement applies shall be barred unless some step to commence arbitral proceedings is taken within a time fixed by the agreement, and a dispute arises to which the agreement applies, the court, if it is of opinion that in the circumstances of the case undue hardship would otherwise be caused, and notwithstanding that the time so fixed has expired, may on such terms, if any, as the justice of the case may require, extend the time for such period as it thinks proper.

(4) Where the court orders that an arbitral award be set aside, the period between the commencement of the arbitration and the date of the order of the Court shall be excluded in computing the time prescribed by the Limitation Act, 1963 (36 of 1963), for the commencement of the proceedings (including arbitration) with respect to the dispute so submitted.

---

[36] M/s JIL-Aquafil (JV) v. Rajasthan Urban Infrastructure Development Project, (Rajasthan)(Jaipur Bench) 2016(4) WLC 474.

[37] Grasim Industries Ltd. v. State of Kerala, (SC) 2017(6) Scale 443 = 2017(4) RAJ 698.

[38] Union of India v. M/s Indusind Bank Ltd., (SC): 2016(5) RAJ 439 = 2016(9) SCC720.

Section 21 of the Act of 1996 declares that "Unless otherwise agreed by the parties, the arbitral proceedings in respect of a particular dispute commence on the date on which a request for that dispute to be referred to arbitration is received by the respondent."

For the purpose of filing an application for making the award rule of the court under the Arbitration Act, 1940, it is not necessary that the application should contain any or all other papers apart from the signed award. Under Art. 119 of the Limitation Act, 1963, two clauses (a) and (b) provided for 30 days limitation under the Arbitration Act, 1940 for filing in the court of an award and for setting aside an award or getting it remitted for reconsideration, from the date of service of the notice of the making of award and of filing of the award, respectively.

EXAMPLES FROM CASE LAW

An award was filed on 29th October 1991. Notice was issued to the respondent on 30th October 1991. Notice was sought to be served on the respondent on 6th November 1991 who refused to accept it. Substituted service was ordered on 24th December 1991. It was held that the issuance of fresh notice by substitute service would not take away the effect of the deemed notice effected on 6th November 1991. The 30 days limitation would start from 6th November 1991. The second order, dated 24th December 1991, did not make the issuance of the notice by the court on 30th October 1991 ineffective.[39]

15.4.11 Limitation for Filing Application under Section 34 of the Act of 1996

Under the 1996 Act, Section 34(3) provides that an application for setting aside an award may not be made after three months have elapsed from the date on which the party making that application had received the arbitral award or, if request had been made under Section 33, from the date on which that request had been disposed of by the arbitral tribunal. The proviso empowers the court, if satisfied that the applicant was prevented by sufficient cause from making the application within the said period, it may entertain the application within a further period of 30 days, but not thereafter. An application for condonation of delay under Section 34 filed with affidavit evidence is sufficient for consideration by the court.[40] An application, if filed in the wrong court, in good faith and within limitation, may attract Section 14(2) of the Limitation Act. Section 14(2) permits exclusion of time during which the applicant has been prosecuting with due diligence in the court without jurisdiction.[41]

EXAMPLES FROM CASE LAW

1. An application for setting aside award was filed after more than four months. The application was rejected. Dismissing the revision petition, it was held that, for sufficient reasons, application could have been entertained within 30 days after the three months limitation but not thereafter. Section 5 of the Limitation Act is not applicable.[42]

2. The counsel for the petitioner was present in the court on 5th July 1991 and he was informed of the filing of the award. Though the court was conscious of the fact a notice was

---

[39] Gurubax Singh v. Punjab Mandi Board, 2004(1) Arb. LR 73 (SC).
[40] Transparent Packers v. The Arbitrator-cum-Managing Director, 2002 (2) Arb. LR 637 (SC).
[41] HMP Engineers Ltd. v. Rallis India Ltd., 2003(3) Arb. LR 452 (Bombay); Anas Abdul Khader v. Abdul Nasar, 2000 (Suppl.) Arb. LR 382 (Kerala).
[42] Union of India v. Som Dutt Gargi, 2004 (1) Arb. LR 278 (P. & H.).

served on 13th August 1991. It was held that when the notice was specifically ordered by the Court, the date of service of notice should be reckoned as the date for the purpose of computation of period of limitation.[43]

15.5 Section 28 of the Indian Contract (Amendment) Act

Amended Section 28 in the Amendment Act No. 1 of 1997 published in the Gazette of India, Extraordinary, Part II, Section 2 dated 8th January 1997, reads thus:

> 28. *Agreements in restraint of legal proceedings void*
>   Every agreement,
>
>   (a) by which any party thereto is restricted absolutely from enforcing his rights under or in respect of any contract, by the usual legal proceedings in the ordinary tribunals, or which limits the time within which he may thus enforce his rights; or
>
>   (b) which extinguishes the rights of any party thereto, or discharges any party thereto from any liability, under or in respect of any contract on the expiry of a specified period so as to restrict any party from enforcing his rights, is void to that extent.

In order to appreciate the need for amendment, it is necessary to refer to the unamended Section 28. The relevant part of the unamended Section 28 of the Indian Contract Act, 1872, read:

> 28. *Agreements in restraint of legal proceedings void*
>   Every agreement, by which any party thereto is restricted absolutely from *enforcing his rights* under or in respect of any contract, by the usual legal proceedings in the ordinary tribunals, or which limits the time within which he may thus *enforce his rights*, is void to that extent.
>
>   (Emphasis supplied)

With reference to the condition of the policy providing a shorter period than the period prescribed by law, namely, Section 28 of the Indian Contract Act, before its amendment, the Supreme Court had held thus:[44]

> From the case law referred to above the legal position that emerges is that an agreement which in effect seeks to curtail the period of limitation and prescribes a shorter period than that prescribed by law would be void as offending Section 28 of the Contract Act. That is because such an agreement would seek to restrict the party from enforcing his right in Court after the period prescribed under the agreement expires even though the period by law for the enforcement of his right has yet not expired. But there could be agreements which do not seek to curtail the time for enforcement of the right but which provide for the forfeiture or waiver of the right itself if no action is commenced within the period stipulated by the agreement. Such a clause in the agreement would not fall within the mischief of Section 28 of the Contract Act.

---

[43] Chairman, Visakhapatnam Port Trust v. Gurucharan Singh, 2004 (1) Arb. LR 319 (AP)(DB).
[44] National Insurance Company Ltd. v. Sujir Ganesh Nayak and Co., AIR 1997 SC 2049.

It is clear that by the Indian Contract (Amendment) Act, 1997, the original Section 28 has been replaced by a new paragraph in which such extinction of right unless exercised within a specified period of time, if not beyond the period of limitation, is also rendered void. As observed earlier, in the absence of any specific reference in the amended Act, it is prospective in nature and the same cannot affect the contract made earlier.[45]

Section 28 of the Act was further amended in 2013 by inserting Exception (3) to make an exception of bank guarantee agreements from the amended Section 28, which exception is self-explanatory and reads as follows:

Exception 3: *Saving of a guarantee agreement of a bank or a financial institution*
This section shall not render illegal a contract in writing by which any bank or financial institution stipulate any term in a guarantee or any agreement making a provision for guarantee for extinguishment of the rights or discharge of any party thereto from any liability under or in respect of such guarantee or agreement on the expiry of a specified period which is not less than one year from the date of occurring of or non-occurring of a specified event for extinguishment or discharge of such party from the said liability.

Explanation

(i) in Exception 3, the expression "bank" means

    (a) a "banking company" as defined in clause (c) of section 5 of the Banking Regulation Act, 1949;

    (b) "a corresponding new bank" as defined in clause (da) of section 5 of the Banking Regulation Act, 1949;

    (c) "State Bank of India" constituted under section 3 of the State Bank of India Act, 1955;

    (d) "a subsidiary bank" as defined in clause (k) of section 2 of the State Bank of India (Subsidiary Bank) Act, 1959

    (e) a "Regional Rural Bank" established under section 3 of Regional Rural Banks Act, 1976;

    (f) "a Co-operative Bank" as defined in clause (CCI) of section 5 of the Banking Regulation Act, 1949;

    (g) "a multi-State co-operative bank" as defined in clause (cciiia) of section 5 of the Banking Regulation Act, 1949; and

(ii) In Exception 3, the expression "a financial institution" means any public financial institution within the meaning of section 4 A of the Companies Act, 1956.

---

[45] Continental Construction Ltd. v. Food Corpn. of India, 2002 (Suppl) Arb. LR 192 (Delhi); The Oriental Insurance Company Limited v. Karur Vysya Bank Limited, AIR 2001 Mad. 489. Union of India v. M/s Indusind Bank Ltd., (SC): 2016(5) RAJ 439 = 2016(9) SC C720.

16

Claims, Disputes and Their Resolution by DRE/ DRB/DAAB/Conciliation and/or Arbitration

16.0 Introduction – "Disputes Arising Out of Contract"

A dispute essentially involves a disagreement. A contract, on the other hand, means an agreement. "Disagreements arising out of agreements" does sound like a contradiction in terms but is common and especially in construction contracts. The standard form contracts, therefore, incorporate elaborate provisions laying down a procedure for the resolution of the disputes. Generally a dispute results when one party to the contract denies a claim made by the other party to the contract. With the enormous number of contracts and equally large number of disputes, the traditional means of dispute resolution through the intervention of a regular court of law is thought to be expensive and time consuming. Alternative Dispute Resolution (ADR) methods are evolving and growing in popularity. The cheapest and the best mode of dispute resolution method under ADR is amicable settlement by direct negotiations between the disputant parties. Given human nature as it is, more often than not direct negotiations fail to yield settlements and therefore the need for a third party involvement (such as a conciliator) in the process of settlement sometimes helps. If that also fails, a judicial method, such as arbitration or litigation in a court of law, needs to be followed.

The modern standard form contracts, encourage parties to a contract to follow step by step the above said methods till the dispute is resolved. This chapter mainly focuses on staged dispute resolution clauses in standard form contracts frequently used in the construction industry. The law of arbitration is a vast subject and because a separate treatise by the authors is available as an accompanying volume to this book, the exhaustive coverage of the said law, as was attempted in the fourth edition of this book, is dropped and instead stress is laid on the pre-arbitration provisions in the construction contracts. If and when the need arises for the reader to resort to arbitration, the book on the law of arbitration by the same authors or any other standard treatise may be referred.

16.1 Provisions in Standard Form Contracts

Provisions in Standard Form Contracts	Paragraph No.
FIDIC Form 1999/2017 edition, Clauses 20, 21	16.2, 16.4, 16.6
NITI Aayog Model Form, Clause 24	16.3, 16.6.2
Ministry of Statistics and Programme Implementation, Government of India (MOS & PI)	16.4
Military Engineering Services Form, Clause 70	16.7
Clauses for Domestic Bidding Contracts 2015 edition, Clauses 11, 11A	16.6.2
CPWD Standard Form, Clause 25	16.6.3, 16.7
State PWD Maharashtra,* Clause 30	16.6.3

*Similar provisions are found in other states' PWD forms.

16.2 FIDIC Form 1999/2017 Edition: Clauses 20 and 21

The FIDIC Form is commonly adapted in a number of countries, including India, for major projects let out on international bidding competitions. Incidentally it also incorporates exhaustive provisions for claims, disputes and resolution of disputes.

16.2.1 Clause 20.1 – Contractor's Claims

Clause 20.1 contains the provisions, which were earlier contained in Clause 53 of the 1992 amended edition. The said provisions require either party to give notice to the engineer, as soon as possible and not later than 28 days after the party becomes aware of the event or circumstance that entitles, in the opinion of the party to be entitled to:

1. Any extension of time for completion of the works by the contractor; and/or
2. Any additional payment, under any clause of the conditions of the contract or in connection with the contract.

Upon failure on the part of the claiming party to give such notice within such period of 28 days, neither the time for completion will be extended nor shall the claiming party be entitled to additional payments and the other party shall be discharged from all liability of the claim. Clause 20.2.2 provides the procedure to be followed for a notice submitted by the claimant party beyond the period of 28 days. In such an event, the engineer has to give notice of late submission within 14 days. On the claimant party's reply to the said notice either denying delay or justifying delay, the engineer will review, under Clause 20.2.5, this aspect at the time of determination of the detailed claim submitted under Clause 20.2.4.

In cases where such notice is given within the stipulated time of 28 days, the claiming party has to take follow-up actions as stated below:

1. To submit any other notices required by the contract with supporting particulars for the claim relevant to such event or circumstance.
2. To keep such contemporary records as may be necessary to substantiate any claim and/or such other records which the engineer may direct him to maintain without prejudice to the right of the employer to dispute liability. Further, the contractor shall permit the engineer to inspect the records or, if directed, to submit copies thereof to the engineer.
3. To submit a fully detailed claim within 84 days (or such other period proposed by the claiming party and approved by the engineer) of when the claimant became aware or should have become aware of the event or the circumstance giving rise to the claim. Clause 20.2.6 stipulates that if the circumstances are of continuing effect, this claim is to be considered as an interim claim and should be followed up with further interim claims at monthly intervals and the final claim within 28 days after the end of the effect.

After all formalities are completed, including submission of additional particulars etc., under Clause 3.7.3, within a period of 42 days (or such other time limit proposed by the engineer and agreed by the parties), the engineer is required to give his decision either approving or disapproving of the claim, with detailed comments. He may seek further particulars but nonetheless must give his response within the time of 42 days after receiving the claim. Each payment certificate to be issued to the contractor shall include the amount of the claim upheld by the engineer. The determination of the engineer as to the extent of extension of time or additional payment to be made to the contractor is to be in accordance with the provisions of Clause 3.7.1, which *inter alia* requires the engineer to consult both the employer and the contractor in an endeavour to reach agreement. In the absence of such an agreement the engineer has to decide the matter in accordance with the contract and the facts and circumstances of the case. Each party is expected to give effect to the decision of the engineer unless and until it is revised under Clause 20.

Clause 3.7.2 stipulates that if the engineer fails to give notice of his determination within the time limit, it may be presumed that the engineer rejected the claim, the matter shall be deemed to be a dispute to be referred to the DAAB (Dispute Avoidance/Adjudication Board) constituted under Clause 21.1 read with Contract Data. The reference shall be under Clause 21.4. The DAAB shall decide the matter under Clause 21.4.3 within 84 days or such other period proposed by the DAAB and agreed by both parties. If the decision requires payment of an amount by one party to the other party, the DAAB is empowered to direct the payee to produce appropriate security in respect of such amount.

The provisions make it clear that the DAAB proceedings shall not be deemed to be arbitration, nor shall the DAAB act as arbitrator(s).

If the DAAB fails to give its decision within the stipulated time, or either party is dissatisfied with the decision of the DAAB, it shall within 28 days of receiving the decision give Notice of Dissatisfaction (NOD) to the other party with a copy to the engineer and the DAAB under Clause 21.4.4. The claims in respect of which a party is dissatisfied, as indicated in the NOD, will then be in the first instance attempted to be amicably settled under Clause 21.5, and if not so settled the matter shall be referred to arbitration which may commence on or after 28 days after the date on which NOD was given.

Clause 21.6 provides for arbitration to be under the Rules of Arbitration of International Chamber of Commerce (ICC). Clause 21.7 deals with failure to comply with the DAAB's decision, whether final or final and binding, and declares the failure qualifying for reference to arbitration directly. Clause 21.8 provides for the contingency where the disputes arise and there is no DAAB set up for whatever reason, the disputes may be referred to the ICC directly without need to obtain the DAAB's decision or amicable settlement.

16.2.2 The Stipulation as to Time Limit in Clause 20 – If Valid?

The main reason for stipulating a short time limit for raising claims in construction contracts is to enable the opposite party to verify the factual position at the relevant time and take corrective steps. Also for the party raising any claim it is easy to establish the claim by evidence of the relevant time.

It is interesting to compare the provision of Clause 20 of 2017 edition with the provision of Clause 53.4 of the earlier 1992 amended edition of the said form. The said provision under the paragraph heading "Failure to comply" restricted the contractor's entitlement in respect of the claim so as not to "exceed such amount as the engineer or any arbitrator or arbitrators appointed pursuant to Sub-Clause 67.3 assessing the claim considers to be verified by contemporary records" irrespective of the fact whether such records were brought to the notice of the engineer or not. It is submitted that the earlier version was not only simple and clear to understand but also made the provision legally valid. The present version in the 1999/2017 edition, after making provision similar to the earlier version, in Clause 21.2.1 (second paragraph) extinguishes the rights and the liabilities of the parties to the contract, if the action is not taken within the stipulated period of limitation which period is less than the period stipulated by the law of limitation. Section 28 of the Indian Contract Act as amended is discussed in Chapter 15. For ready reference, the relevant part of the amended Section 28 of the Indian Contract Act, is reproduced as follows:-

> 28. *Agreements in restraint of legal proceedings void*
>   Every agreement,
>
>   (a) by which any party thereto is restricted absolutely from enforcing his rights under
>   or in respect of any contract, by the usual legal proceedings in the ordinary
>   tribunals, or which limits the time within which he may thus enforce his rights; or

(b) which extinguishes the rights of any party thereto, or discharges any party thereto from any liability, under or in respect of any contract on the expiry of a specified period so as to restrict any party from enforcing his rights, is void to that extent.

It is respectfully submitted that in view of amended Section 28 of the Indian Contract Act, the condition in Clause 20.1 in paragraph no. 2 of the FIDIC Form 1999/2017 is void and of no effect in India.

16.2.3 Reference of Dispute to Engineer in the First Instance

Clause 67.1 of the FIDIC 1992 (Amended) Form or Clause 20.2 read with Clause 3.7 of the 2017 edition, stipulates that if a dispute of any kind arises between the employer and the contractor in connection with or arising out of the contract or execution of the work at any time during the currency of the contract or after termination, the matter in dispute shall, in the first place, be referred in writing to the engineer. The engineer is given a certain number of days to decide the dispute and notify his decision to the parties. If any party is not satisfied with the said decision, or the engineer fails to give his decision within the time allowed, either party may, on or before the period stipulated after the date on which the party received the notice of such decision or 70th day after the day on which the stipulated period expired, give notice of his intention to commence arbitration. In some other variants of similar provisions in contracts, such decision of the engineer or lack of it, empowers the dissatisfied party to approach Dispute Review Expert (DRE) or Dispute Review Board (DRB) or DAAB as the case may be, within the stipulated time limit, failing which the decision attains finality.

It needs to be noted that these or similar provisions are mandatory and need to be strictly complied with to invoke further provisions of dispute resolution including arbitration.[1]

The amended Section 28 of the Indian Contract Act will not be attracted to avail remedy like arbitration because the right to challenge the decision may not be extinguished but the contractual remedy of dispute resolution may not be available. Arbitration being a voluntary forum, the parties are entitled to limit its scope to particular kinds of disputes and, if the said contractual remedy is not available, the affected party can always invoke the compulsory forum of a court of law having jurisdiction to try the matter within the period of limitation prescribed by the Limitation Act.

This provision of referring the dispute to the engineer in the first place under the dispute resolution clause makes sense inasmuch as the clause contemplates disputes between the employer and the contractor, with the engineer being the third person. In practice it is found to be an exercise in futility because mostly the disputes involve the decisions given by the engineer or his representative called Team Leader, which some contracts expressly stipulate as the employer's decision. A dispute comes into existence only when the contractor does not accept the decision of the engineer or his representative. The 1999/2017 updated edition retains the role of the engineer to resolve the dispute, in the first instance, albeit for attempting to secure amicable settlement or perhaps power to review his decision in capacity as a quasi-arbitrator to enable the parties to implement it until altered by DRB/DAAB or arbitral tribunal.

16.2.4 Amicable Settlement

Amicable settlement of a dispute by the parties by direct negotiations, as already stated, is the best mode available. However, direct negotiations invariably fail for various reasons. One of the main reasons for failure of direct negotiations, one will be surprised to know, is lack of full and complete understanding by a party of its rights and liabilities vis-à-vis the facts and circumstances

---

[1] This aspect is further discussed in detail in paragraph 16.13 below.

of the issue involved. Negotiations are generally conducted in the traditional manner or method known as "positional bargaining". It is only when the parties are required to present their respective cases before third persons that the full facts and evidence relevant to the issues are gathered and an attempt is made to ascertain the law on the point. After the third person gives its decision, the party who lost its case invariably challenges it. However, by that time enough insight is available to the parties to make them see the reason and try to settle the matter amicably. At that stage there is more possibility of the parties reaching the settlement by following "principled negotiations". Clause 20.5 of FIDIC 1999 edition, suggesting amicable settlement at this stage and not to start with, can be appreciated in the light of this discussion.

In India, in public works contracts, this provision also is virtually rendered redundant because of the lack of initiative on the part of the public authorities to take decisions, for various reasons, including reasons such as audit objection or insinuation of *mala fide*, etc. It is unfortunate that huge amounts of public funds, which otherwise could be saved by timely amicable settlement of disputes, are spent due to this attitude of the public authorities. The Arbitration and Conciliation Act, 1996 (ACA) has given statutory recognition to conciliation, an effective mode of resolution of disputes, which can come in handy for public authorities to achieve savings of public funds if the suggestions made above are adopted.

16.3 NITI Aayog Model Form

The dispute resolution clause incorporated in the form is Clause 24. The provision is self-explanatory and reads as follows:

24.1. *Dispute resolution*

24.1.1 Any dispute, difference or controversy of whatever nature howsoever arising under or out of or in relation to this Agreement (including its interpretation) between the Parties, and so notified in writing by either Party to the other Party (the "Dispute") shall, in the first instance, be attempted to be resolved amicably in accordance with the conciliation procedure set forth in Clause 24.2.

24.1.2 The Parties agree to use their best efforts for resolving all Disputes arising under or in respect of this Agreement promptly, equitably and in good faith, and further agree to provide each other with reasonable access during normal business hours to all non-privileged records, information and data pertaining to any Dispute.

24.2. *Conciliation*

In the event of any Dispute between the Parties, either Party may call upon an officer of the Authority, not below the rank of Secretary to the Government or Chief Engineer, as the case may be, or such other person as the Parties may mutually agree upon (the "Conciliator") to conciliate and assist the Parties in arriving at an amicable settlement thereof. Failing conciliation by the Conciliator or without the intervention of the Conciliator, either Party may require such Dispute to be referred to the Secretary or Chief Engineer of the Authority and the Chairman of the Board of Directors of the Contractor for amicable settlement, and upon such reference, the said persons shall meet no later than 7 (seven) business days from the date of reference to discuss and attempt to amicably resolve the Dispute. If such meeting does not take place within the 7 (seven) business day period or the Dispute is not amicably settled within 15 (fifteen) business days of the meeting or the Dispute is not resolved as evidenced by the signing of written terms of settlement within 30 (thirty) business days of the notice in writing referred to in Clause 24.1.1 or such longer period as may be mutually agreed by the Parties, either Party may refer the Dispute to arbitration in accordance with the provisions of Clause 24.3.

24.3. *Arbitration*

24.3.1 Any Dispute which is not resolved amicably by conciliation, as provided in Clause 24.2, shall be finally decided by reference to arbitration by an arbitral tribunal constituted in accordance with Clause 24.3.2. Such arbitration shall be held in accordance with the Rules of Arbitration of the International Centre for Alternative Dispute Resolution, New Delhi or Construction Industry Arbitration by the Parties, and shall be subject to the provisions of the Arbitration and Conciliation Act, 1996. The place of such arbitration shall be the capital of the State, and the language of arbitration proceedings shall be English.

The Arbitration shall be in Institutional mode & would not be ad hoc, in any case, and the online mode of dispute resolution may also be resorted to as per the latest notification of Ministry of Law & Justice, Government of India.

24.3.2 There shall be an arbitral tribunal comprising three arbitrators, of whom each Party shall select one, and the third arbitrator shall be appointed by the two arbitrators so selected and in the event of disagreement between the two arbitrators, the appointment shall be made in accordance with the Rules.

24.3.3 Wherever possible or required, On-line dispute resolution mechanism as prescribed by Ministry of Law by their notification is to be resorted to, in place of the traditional in-situ arbitration procedures. The decision on which system of procedures is to be followed (viz. traditional or on-line) would rest with the decision to be made by Third Arbitrator (appointed by the two nominated arbitrators).

24.3.4 The arbitral tribunal shall make a reasoned award (the "Award"). Any Award made in any arbitration held pursuant to this Article 24 shall be final and binding on the Parties as from the date it is made, and the Contractor and the Authority agree and undertake to carry out such Award without delay.

24.3.5 The Contractor and the Authority agree that an Award may be enforced against the Contractor and/or the Authority, as the case may be, and their respective assets wherever situated.

24.3.6 This Agreement and the rights and obligations of the Parties shall remain in full force and effect, pending the Award in any arbitration proceedings hereunder.

24.3.7 In the event the Party against whom the Award has been granted challenges the Award for any reason in a court of law, it shall make an interim payment to the other Party for an amount equal to 75% (seventy five per cent) of the Award, pending final settlement of the Dispute. The aforesaid amount shall be paid forthwith upon furnishing an irrevocable Bank Guarantee for a sum equal to the aforesaid amount. Upon final settlement of the Dispute, the aforesaid interim payment shall be adjusted and any balance amount due to be paid or returned, as the case may be, shall be paid or returned with interest calculated at the rate of 10% (ten per cent) per annum from the date of interim payment to the date of final settlement of such balance.

Adjudication by Statutory Tribunal

In some states of India, for example Orissa, Madhya Pradesh, Gujarat, for resolution of disputes statutory tribunals or other forums with powers to adjudicate upon disputes between the contractor and the authority/state are constituted. The clause provides that all disputes arising after such constitution shall, instead of reference to arbitration under Clause 24.3, be adjudicated upon by such tribunal or other forum in accordance with applicable laws and all references to dispute resolution procedure shall be construed accordingly. This is a valid provision.[2] In a case where the contract contained an arbitration clause under the ACA, 1996, it was held:

---

[2] Madhya Pradesh Rural Road Development Authority v. M/s. L.G. Chaudhary Engineers and Contractors, (SC), 2018 (5) Scale 103; = 2018 (3) RAJ 184.

The scheme of Sections 8, 9 and 22 of the State Act shows that in the absence of an agreement stipulating the applicability of the Central Act, the State Act applies to works contracts. Since in the present cases, an arbitration agreement exists and stipulates applicability of the Central Act, the State Act will not apply.[3]

16.4 Government of India Form

The Government of India Form Clause 11 – Disputed Items and Arbitration, stipulates "Disputes between the Employer and the Contractor shall first be submitted to Conciliation. The procedure outlined in the Arbitration and Conciliation Act, 1996 shall be followed."[4] It further provides that if conciliation proceedings have become infructuous or have been terminated, the party which initiated the conciliation shall refer the disputes for arbitration. The reference to arbitration should be made preferably within 28 days of the termination of conciliation proceedings. The arbitration shall be conducted in accordance with the Indian ACA, 1996. It suggests for contracts costing up to Rs. 10 crores, a sole arbitrator should be appointed. Under the Act he must be appointed by mutual consent or by the appointing authority, in the absence of agreement between the parties. For contracts costing over Rs. 10 crores, a committee of arbitrators should be appointed comprising of one arbitrator to be nominated by the contractor, one to be nominated by the owner and the third arbitrator, who will act as a chairman but not as umpire, to be chosen jointly by the two nominees. The decision of the majority of the arbitrators shall be final and binding on both parties.

16.4.1 Appointment of DRE/DRB/DAB/DAAB

Most standard form contracts in use for major projects incorporate provisions for avoiding time-consuming and expensive disputes in the first place, and to resolve them amicably and quickly in the second place. To this end: an ongoing dispute resolution or adjudication process is embodied in the conditions of the contract right at the time of signing of the contract. The parties agree to jointly appoint an expert in the field of contract works as DRE, or three mutually agreed members forming a DRB, who are appointed by the parties within the time specified to do so after signing of the contract, and preferably soon after commencement of the work under the contract. The contract also provides for fees and expenses payable to the members (which are equally shared), and also filling of the vacancy due to failure by any party to appoint its nominee member. The contract generally names an independent body or organization as appointing entity who is authorized to appoint members for and on behalf of the erring party.

The duties of the DRE or DRB include periodical visits (generally one in three months at least) to the site and inspection of the project and holding joint meetings with the representatives of the parties to identify problem areas, if any, and suggest measures to remedy the situation before it turns into a dispute. Also, any party can refer a decision of the engineer not acceptable to it to the DRE or DRB. The DRE or DRB will conduct arbitration-like proceedings and, after hearing the parties, will give recommendations to the parties within a stipulated time limit generally of not more than eight weeks. The parties are bound to follow the recommendations given by the DRE or DRB pending resolution of the dispute, if any, by the mode suggested in the contract. The conditions of the contract do not make the decision of the DRE or DRB final and binding on the parties, if any party dissatisfied with the decision serves notice on the other party, within 28 days after receiving the recommendation/decision of the DRE or DRB, of its intention to refer the

---

[3] State of Bihar v. M/s. Brahmaputra Infrastructure Limited, AIR 2018 SC 2640 = 2018 (5) Scale 529.
[4] For further discussion see paragraph 16.3 below.

matter to arbitration. If no such notice is given the decision attains finality. Again in such a case a party may not have lost its right to challenge the decision in a court of law, it is respectfully submitted, though the remedy by arbitration may not be available to the party under the terms of the contract and if the court is satisfied that the matter deserves its consideration. The Government of India Form, for example, includes Clause 11A as follows:

Clause 11A – Dispute Resolution Board

If a dispute of any kind whatsoever arises between the Employer and the Contractor in connection with, or arising out of the Contract or the execution of the Works, whether during the execution of the Works or after their completion and whether before or after the repudiation or other termination of the Contract, including any disagreement by either party with any action, inaction, opinion, instruction, determination, certificate or valuation of the Engineer, the matter in dispute shall, in the first place, be referred to the Dispute Review Board.

The Board shall be established by signature of the Dispute Review Board Agreement ("the Board Agreement") which shall occur at the same time as the signature of the Contract Agreement.

[...]

Any dispute on which the Board has not issued a Recommendation within 42 days of its final hearing on the dispute, or regarding which the Recommendation(s) are not accepted, may be referred in writing by either party to arbitration in accordance with this Clause, by written notice to the other party with copies to the Engineer and the Board. Such notice shall state that it is being made pursuant to this Clause and shall establish the entitlement of the party giving it to commence arbitration provided that no such arbitration may be commenced until such notice is given. Such reference shall be made within 14 days of receipt of the Board's recommendation(s), or within 14 days of the day on which said period of 42 days expired, as the case may be, failing which reference any Recommendation(s) previously rejected or not accepted shall be deemed accepted despite such previous rejection or non-acceptance and shall be final and binding upon the parties.

All Recommendations, which have become final and binding, shall be implemented by the parties forthwith; such implementation shall include any relevant action of the Engineer.

Whether or not accepted or deemed accepted, all of the Recommendations shall be admissible in any subsequent dispute resolution procedure, including any arbitration or any litigation having any relation to the dispute or disputes to which the Recommendation(s) relate.

Unless the Contract has already been repudiated or terminated, the Contractor shall, in every case, continue to proceed with the Works with all due diligence and the Contractor and the Employer shall give effect forthwith to every decision of the Engineer unless and until the same shall be revised, as hereinafter provided, in an arbitral award.

It seems obvious that Clause 11A is an alternative to Clause 11.

FIDIC 1999 edition incorporates the above provisions in Clauses 20.2 to 20.4 as already stated above.

16.5 Conciliation

The dictionary meaning of the word conciliation is "The act of bringing into agreement". Conciliation can be defined as a voluntary, non-binding, private, dispute resolution process in which a neutral person helps the parties try to reach a negotiated agreement. The word is sometimes used interchangeably with the word "mediation". A distinction between the two words is sometimes drawn in terms of the process adopted. This distinction, however, is not real and the ACA permits the conciliator, at any stage of the proceedings, to make proposals for a settlement of disputes. The ACA expressly confers confidentiality to all matters relating to conciliation proceedings.

Conciliation introduces a powerful new structure and dynamic into any negotiation which facilitates communication, helps overcome deadlocks and emotional blockages, restores the negotiation process, separates the people from the problem and helps separate needs of the parties from wants and reassess their case. This process gets the right people and right information to the table, thereby increasing the options for resolution of disputes. Above all, it keeps the ownership and settlement of the problem with the parties and thereby restores and safeguards relationships.

Conciliation, in essence, is a process of settlement of dispute by negotiations. It has already got worldwide recognition. The question arises: when negotiations have failed, how and why does conciliation succeed? There are many reasons. The main ones are:

1. Quite often parties will think they don't want to settle; yet once in a conciliation environment and with a skilled conciliator they often realize common ground.

2. Skill of the conciliator and improved techniques of negotiations make the parties change their negotiation style from the traditional "positional" to a "principled" one. The parties are made to ascertain their respective BATNA (Best Alternative to Negotiated Agreement) and WATNA (Worst Alternative to Negotiated Agreement). This makes the parties see clearly the advantage of immediate resolution of their dispute.

3. In addition, a conciliator is expected to generate a good working relationship.

4. Last but not least, the process invariably succeeds because of the four principles which a conciliator is required to follow and make the parties follow:

    i. Separate the people from the problem; be soft on the people and hard on the problem.

    ii. Stress and focus on the interests and not on positions.

    iii. Seek or invent options for mutual gains by exploring the mutual needs distinguished from wants.

    iv. Insist on objective criteria and not on subjective or positional criteria.

Advantages of conciliation include: savings in costs, speed of disposal, control over the process and the outcome, minimization of deterioration of relationship, creative and forward looking solution, confidentiality, satisfaction and quick compliance.

Not all kinds of disputes and matters can be settled by conciliation. Cases not suitable for conciliation include: matrimonial matters, testamentary matters, insolvency matters, guardianship of lunatic or minor, criminal matters, questions relating to charities and charitable trusts, matters within the purview of MRTP Act, dissolution or winding up of a company, etc. If a claim is of an 'all or nothing' nature, it is unlikely to be settled in a conciliation.

Risks of conciliation, according to some, include: early disclosure indicating to the other party the weakness of a case; a party can use conciliation as a delaying tactic in which event there is no real cost saving; settlement if reached is under pressure. Each of these criticisms has a good counter-argument in favour of the system.

16.5.1 Commencement of Conciliation

The parties to a dispute can commence conciliation at any stage by sending a proposal in writing to the other party. The other party is at liberty to reject the invitation to conciliate or accept it. If the other party rejects the invitation there will be no conciliation. Similarly if the party sending the invitation does not receive a reply within 30 days from the date on which he sends the invitation, or within such other time specified in the invitation, he may elect to treat this as rejection and, if he so elects, he shall inform in writing to the other party. However, if the other party communicates acceptance in writing, not only the conciliation agreement comes into existence but the proceedings also commence.[5] The number of

---

[5] Section 62 of the ACA 1996.

conciliators can be one, two or three. When the parties agree upon one conciliator, all parties should agree upon his name. When the number agreed upon is two, one is to be nominated by each party. In the case of three, the third name is agreed upon and he is to act as presiding conciliator. The parties may agree upon an institution or authority or individual who may help the parties by recommending suitable names to the parties.[6]

The conciliators so appointed may direct the parties to submit written statements supplemented by any documents and other evidence. The statement to be submitted should be short, maybe five or six pages, in the form of a note a junior officer submits to his senior explaining the case. Further particulars can be asked for during the process. The formal rules of procedure and evidence are not applicable to the proceedings.[7] The conciliator(s), after conducting the first joint meeting of all parties and framing the issues, can meet each party in privacy and ascertain their views, which the party may not disclose in front of the other party. All disclosures before the conciliators are confidential and not to be disclosed to the other party without the permission of the party making the disclosure and that too when the conciliators feel that such disclosure may result in settlement agreement.[8]

The conciliators are expected to help the parties to reach settlement by providing assistance in an independent and impartial manner. The parties may, on their own or at the instance of the conciliators, suggest drafts of settlement agreements for consideration by the other parties at any stage of the proceedings. Such proposals need not be in writing.[9]

In the case of public authorities, if the conciliator is an eminent person or jurist, such as a retired judge of the Supreme Court or any High Court, and he were to suggest settlement agreement in writing, acceptance of such proposal should not create any difficulty in future for having accepted the settlement.

The conciliator may formulate or help the parties to formulate the terms, draw up and sign the settlement agreement. The conciliator shall authenticate the settlement agreement so signed and a copy thereof shall be furnished to each of the parties. The settlement agreement so signed shall be final and binding upon the parties and have the same status and effect as if it is an arbitral award.[10]

In case the conciliation proceedings fail, the parties should not worry about disclosures made in the proceedings. Section 75 of the ACA declares that the conciliator and the parties shall keep confidential all matters relating to the conciliation proceedings, including the settlement agreement, except where the disclosure is necessary for the purpose of implementation and enforcement of the settlement agreement.

The success rate of conciliation depends mainly upon the skill of the conciliator and, as such, the conciliators are advised to undergo a few days training, which is now available in many countries including India. In England, if the contract stipulates that conciliation or mediation should be a pre-condition of beginning a litigation or an arbitration, courts will try to enforce such an agreement. Although it may appear to be an "agreement to agree", if the clause provides sufficient details, there is an obligation on the parties to attempt mediation in good faith before commencing litigation or arbitration. Cost sanctions may be used against a successful party that fails to satisfy such a pre-condition.

16.6 Arbitration

When direct negotiations, third party intervention, or conciliation fails to bring about settlement of disputes, the parties have two forums available, public or private. The former is a court of law and the latter is arbitration. Section 28 of the Indian Contract Act, as already seen declares:

---

[6] Sections 63 and 64 of the ACA, 1996.
[7] Sections 65 and 66 of the ACA, 1996.
[8] Sections 69 and 70 of the ACA, 1996.
[9] Section 67(4) of the ACA, 1996.
[10] Sections 73 and 74 of the ACA, 1996.

Every agreement, (a) by which any party thereto is restricted absolutely from enforcing his rights under or in respect of any contract, by the usual legal proceedings in the ordinary tribunals, or which limits the time within which he may thus enforce his rights is void to that extent.

No one can exclude himself from the protection of courts by a contract. That is why the court of law is called a compulsory forum. However, the said section has exceptions, the first two exceptions are relevant to the discussion and read as follows:

Exception 1. – This section shall not render illegal a contract by which two or more persons agree that any dispute which may arise between them in respect of any subject or class of subjects shall be referred to arbitration, and that only the amount awarded in such arbitration shall be recoverable in respect of the dispute so referred.

Exception 2. – Nor shall this section render illegal any contract in writing by which two or more persons agree to refer to arbitration any question between them which has already arisen, or affect any provision of any law in force for the time being as to references to arbitration.

Exception 1 excludes the contracts for referring future disputes and Exception 2 excludes the reference to arbitration of the past or present disputes from the operation of Section 28.

Arbitration is thus an optional forum subject to an agreement between the parties. The agreement has to be in writing. Section 7 of the Arbitration and Conciliation (Amendment) Act, 2015 reads:

7. Arbitration agreement

(1) In this Part, "arbitration agreement" means an agreement by the parties to submit to arbitration all or certain disputes which have arisen or which may arise between them in respect of a defined legal relationship, whether contractual or not.

(2) An arbitration agreement may be in the form of an arbitration clause in a contract or in the form of a separate agreement.

(3) An arbitration agreement shall be in writing

(4) An arbitration agreement is in writing if it is contained in

(a) a document signed by the parties;

(b) an exchange of letters, telex, telegrams or other means of telecommunication including communication through electronic means which provide a record of the agreement; or

(c) an exchange of statements of claim and defence in which the existence of the agreement is alleged by one party and not denied by the other.

(5) The reference in a contract to a document containing an arbitration clause constitutes an arbitration agreement if the contract is in writing and the reference is such as to make that arbitration clause part of the contract.

From the above provision it is clear that to constitute "an arbitration agreement" it is not necessary that there should be a formal agreement or that the terms should all be contained in a formal document. If the original agreement does not contain an arbitration clause, the parties can agree for reference to arbitration after the disputes have arisen. However, if there is no such agreement, there cannot be reference to arbitration. Where there was no recital in contract to refer any dispute or difference present or future to arbitration, an impugned order directing appointment of arbitrator was set aside.[11]

---

[11] State of Orissa and another etc. v. Sri Damodar Das, 1996 IAD (SC) 589, AIR 1996 SC 942, 1996 (1) ARB LR 221 (SC), JT 1995 (9) SC 419, 1996 (1) SCALE 68, (1996) 2 SCC 216, [1995] Supp 6 SCR 800.

The construction contracts and standard forms in general contain an arbitration clause. For example, FIDIC Form 1999/2017 edition contains the provision for reference to arbitration as already mentioned hereinabove.[12] The 1999/2017 editions incorporate some welcome additions providing for reference to arbitration any decision of DRB/DAAB which had attained finality but has not been honoured by the other party and direct reference to arbitration where for any reason there is no DRB/DAAB in place.

16.6.1 Subsection (5) Reference in a Contract to a Document – Interpretation

Though general reference to an earlier contract is not sufficient for incorporation of an arbitration clause in the later contract, a general reference to a standard form would be enough for incorporation of the arbitration clause. In a case the Apex Court of India observed:

> We are of the opinion that a general reference to a consensual standard form is sufficient
> for incorporation of an arbitration clause. In other words, general reference to a standard
> form of contract of one party will be enough for incorporation of arbitration clause.

In the said case, the purchase order was issued by the appellant in which it was categorically mentioned that the supply would be as per the terms mentioned therein and in the attached standard terms and conditions. The respondent, by his letter dated 15th December 2012, confirmed its acceptance of the terms and conditions mentioned in the purchase order except delivery period. The dispute arose after the delivery of the goods. No doubt, there is nothing forthcoming from the pleadings or the submissions made by the parties that the standard form attached to the purchase order is of a trade association or a professional body. However, the respondent was aware of the standard terms and conditions which were attached to the purchase order. Held: "The purchase order is a single contract and general reference to the standard form even if it is not by a trade association or a professional body is sufficient for incorporation of the arbitration clause."[13]

16.6.2 FIDIC Form with COPA and Other Standard Form Contract Clauses

In India, when FIDIC conditions are adopted for public works, they are supplemented by COPA (Conditions of Particular Application) which alter or modify the standard conditions, and it has been the practice in India to replace arbitration by the ICC with a domestic forum of three arbitrators one each to be appointed by the employer and the contractor and the third presiding arbitrator to be appointed by the two arbitrators. The Standard Contract Clauses for Domestic Bidding Contracts prepared by the Government of India also incorporates a similar provision. Certain other public works departments such as, railways, military engineering services and state and central public works departments include provisions for reference of disputes to sole arbitrator or a panel consisting of its employees. Some forms expressly provide for decisions of its employee to be final and binding, giving rise to questions as to whether such a provision amounts to an arbitration agreement or not. Some forms, which expressly provide for arbitration, also exclude certain matters from the scope of arbitration clause, called "excepted matters". These aspects are considered below.

16.6.3 Decision of Engineer to Be Final and Binding on Parties to the Contract – If Amounts to Arbitration Agreement?

Certain state PWDs in India use standard forms containing a provision reading somewhat as follows:

---

[12] See paragraph 16.3 above. Clause 24.3 of FIDIC Form.

[13] M/s. Inox Wind Ltd. v. M/s Thermocables Ltd., (SC); 2017 (6) RAJ 628 = 2018 (1) Scale 77.

Direction of Works (or alternatively Decision of the Engineer) to be final

Except where otherwise specified in the contract the decision of the Superintending Engineer for the time being shall be final, conclusive and binding on all parties to the contract upon all questions relating to the meaning of the specifications, design, drawing and instructions herein before mentioned. The decision of such Engineer as to the quality of workmanship, or materials used on the work, or as to any other question, claim, right, matter or things whatsoever, in any way arising out of or relating to the contract, designs, drawings, specifications, estimates, instructions, orders, or these conditions, or otherwise concerning the works, or the execution or failure to execute the same, whether arising during the progress of the work, or after the completion or abandonment of the contract by the contractor, shall also be final, conclusive and binding on the contractor.

The question arises, whether the above clause amounts to an arbitration clause. The plea that it amounts to an arbitration clause found favour with the trial court as well as the appellate court, in a case, but was rejected by the High Court in revision on the ground that it merely conferred power on the superintending engineer to take decisions on his own and that it did not authorize the parties to refer any matter to his arbitration. In this connection the High Court particularly adverted to the marginal note to the said clause, which was to the following effect: "Direction of work". Disposing of the appeal, a three-judge bench of the Supreme Court of India held:

After perusing the contents of the clause and hearing learned counsel for the parties we find ourselves in complete agreement with the view taken by the High Court. Admittedly the clause does not contain express arbitration agreement. Nor can such an agreement be spelled out from its terms by implication, there being no mention in any dispute, much less of a reference thereof. On the other hand, the purpose of the clause clearly appears to be to vest the Superintending Engineer with supervision of the execution of the work and administrative control over it from time to time.[14]

The relevant clause in another contract read:

15. Whenever any doubt, difference or dispute shall hereafter arise touching the construction of these presents or anything herein contained or any matter or things connected with ... the matter in differences shall be decided by the lessor whose decision shall be final.[15]

It was held by the Supreme Court that the clause spells out an arbitration agreement.

From these decisions the test to be applied to decide if a provision amounts to an arbitration agreement or not is that if the provision mentions *dispute*, its *reference* and gives *finality to the decision* of the person to whom reference is to be made, it amounts to an arbitration clause.[16]

In the case of State of Orissa v. Damodar Das[17] the Supreme Court observed that agreement to refer disputes or differences to arbitration must be expressly or impliedly spelled out from the clause. The relevant Clause 25 of the agreement reads as follows:

25. Decision of Public Health Engineer to be final – Except where otherwise specified in this contract, the decision of the Public Health Engineer for the time being shall be final, conclusive and binding on all parties to the contract upon all questions relating to the meaning of the specifications; drawings and instructions herein before mentioned and as to the quality of workmanship or materials used on the work, or as to any other question, claim, right matter or thing whatsoever in any way arising out of, or relating to, the

---

[14] State of Orissa and another etc. v. Sri Damodar Das, AIR 1996 SC 943; U.P. v. Tipper Chand, AIR 1980 SC 1522.

[15] Rukmanibai v. Collector, Jabalpur, AIR 1981 SC 479.

[16] Also see: Chabbel Dass & Co. v. State of U.P., IR 1077 ALL. 143.

[17] State of Orissa v. Damodar Das, (1996) 2 SCC 216 (217): (AIR 1996 SC 942).

contract, drawings, specifications, estimates, instructions, orders or these conditions, or otherwise concerning the works or the execution or failure to execute the same, whether arising during the progress of the work or after the completion or the sooner determination thereof of the contract.

It was held that Clause 25 of the agreement does not contain an arbitration agreement nor does it envisage any difference or dispute that may arise or had arisen between the parties in execution of the works for reference to an arbitrator.

Clause 30 of Maharashtra State PWD Form, reading as follows, has also been subject matter of controversy:

> The decision of the Superintending Engineer of the ... Circle for the time being shall be final, conclusive and binding on all parties to the contract upon all questions relating to ... or as to question, claim, right, matter, or thing whatsoever, in any way *arising out of or relating to the contract*, designs, drawings, specifications, estimates, instructions, orders or those conditions, or otherwise concerning the works or the execution, or failure to execute the same, whether arising during the progress of the work, or after the completion or abandonment thereof *in case of dispute arising between the contractor and Gulbarga University.*
>
> (Emphasis supplied)

A three-judge bench of the Supreme Court of India held:[18]

> A plain reading of the aforementioned clause would show that in case of dispute between the contractor and Gulbarga University, the decision of the Superintending Engineer of the Gulbarga Circle, Gulbarga, shall be final and binding to all parties to the contract, such dispute may embrace within its fold all questions relating to the matters specified therein as also any other question, claim, right, matter or thing whatsoever in any way arising out of or relating to the contract. ... Such disputes may be referred to for decision of the Superintending Engineer; whether arising during the progress of the work after the completion thereof.

Held: "There cannot, thus, be any doubt whatsoever that Clause 30 aforementioned fulfills all the criteria of a valid Arbitration agreement. It is further not in dispute that the Superintending Engineer, Gulbarga Circle, Gulbarga is an independent person." Reference was made to the decision in Bihar State Mineral Development Corporation and Anr. v. Encon Builders (1) (P) Ltd., which decision spelt out the essential elements of the arbitration agreement, as follows:[19]

> (i) There must be a present or a future difference in connection with some contemplated affair; (ii) There must be the intention of the parties to settle such difference by a private tribunal; (iii) The parties must agree in writing to be bound by the decision of such tribunal; and (iv) The parties must be ad idem.

In a later decision the Supreme Court of India, after review of several cases decided earlier, held[20]:

> In view of the above discussion, we hold that the High Court had rightly held that Clause 30 of B-I Agreement is not an Arbitration Agreement and the trial Court was not right in appointing the Chief Engineer as an Arbitrator.

---

[18] Vishnu v. State of Maharashtra, AIR 2013 SCW 5811 = 2014 (2) RAJ 122. Mallikarjun v. Gulbarga University, 2003 (3) Arb. LR 579 (SC); Lachmanna B. Horamani, Petitioner v. State of Karnataka and others, AIR 1998 Kant. 405.
[19] Bihar State Mineral Development Corporation and Anr. v. Encon Builders, (1) (P) Ltd., 2003(7) SCC 418 = 2003 (3) Arb. LR 133(SC).
[20] Vishnu v. State of Maharashtra, AIR 2013 SCW 5811 = 2014 (2) RAJ 122.

It was observed that the principles laid down in the aforementioned decision are not in question. The clauses in another contract, which did not satisfy the above essential elements will not amount to arbitration clauses.[21]

16.7 Excepted Matters

If the parties to an arbitration agreement expressly exclude certain matters as beyond the scope of the arbitration clause, obviously the law will give effect to this intention of the parties.[22] This well-established position in law still poses some questions, including the following:

a. Can matters in respect of which the decisions of the engineer/architect are made final be referred to arbitration?

b. If not, is the contractor bound to accept the said decisions as final without an opportunity to challenge them at some stage?

c. Can the contractor challenge such decisions in a court of law?

d. Should the court decide the question on merits?

e. What if such decision of the court renders the entire arbitration or other interrelated issues invalid?

f. Should the court refer the matter to arbitration to avoid consequences contemplated in question (e) above?

A study of court decisions in the following examples will help answer the above questions authoritatively.

EXAMPLE FROM CASE LAW

1. As per the arbitration agreement (contained in Clause 25 of the contract) all questions and disputes relating to the contract, execution or failure to execute the work, whether arising during the progress of the work or after the completion or abandonment thereof, "except where otherwise provided in the contract", had to be referred to and settled by arbitration. The High Court held that claims 1, 3 and 11 of the contractor were not arbitrable as they related to excepted matters in regard to which the decisions of the superintending engineer or the engineer-in-charge had been made final and binding under clauses (2) and (3) of the agreement. The Supreme Court deciding the appeal held:

> Thus what is made final and conclusive by clauses (2) and (3) of the agreement, is not the decision of any authority on the issue whether the contractor was responsible for the delay or the department was responsible for the delay or on the question whether termination/rescission is valid or illegal. What is made final, is the decisions

---

[21] Bharat Bhushan Bansal v. U.P. Small Industries Corporation Ltd., Kanpur 1999 (2) SCC 166 = 1999 (l) Arb. LR 326 (SC). M/s. Veer Construction Technocrats (P) Ltd., Petitioner v. M/s. Saraswati Enclave C.G.H. Society Ltd., AIR 1996 Del. 12. M/s. Garg Builders and Engineers, Petitioner v. U.P. Rajkiya Nirman Nigam Ltd., and others, AIR 1995 Del. 111; State of M.P. v. K.K. Shukla and Co., 2002 (Suppl.) 538.

[22] M/s. P. C. Corpn. Ltd., v. The Chief Administrator, Dandakaranya Project, AIR 1991 SC 957. See Example (1) below; Vishwanath Sood v. Union of India, AIR 1989 SC 952; also see: DDA v. Sudhir Brothers, 1995 (57) DLT 474 = 1995 (2) Arb. LR 306 (Delhi); Wee Aar Constructive Builders v. Delhi Jal Board, 2004 (2), Arb. LR 418 (Delhi).

on consequential issues relating to quantification, if there is no dispute as to who committed breach. That is, if the contractor admits that he is in breach, or if the Arbitrator finds that the contractor is in breach by being responsible for the delay, the decision of the Superintending Engineer will be final in regard to two issues. The first is the percentage (whether it should be 1% or less) of the value of the work that is to be levied as liquidated damages per day. The second is the determination of the actual excess cost in getting the work completed through an alternative agency. The decision as to who is responsible for the delay in execution and who committed breach is not made subject to any decision of the respondents or its officers, nor excepted from arbitration under any provision of the contract.

Placing reliance on Bharat Sanchar Nigam Ltd. v. Motorola India Ltd.,[23] it was held:[24]

> In view of the above, the question whether appellant was responsible or respondents were responsible for the delay in execution of the work, was arbitrable. The arbitrator has examined the said issue and has recorded a categorical finding that the respondents were responsible for the delay in execution of the work and the contractor was not responsible. The arbitrator also found that the respondents were in breach and the termination of contract was illegal. Therefore, the respondents were not entitled to levy liquidated damages nor entitled to claim from the contractor the extra cost (including any escalation in regard to such extra cost) in getting the work completed through an alternative agency.

2.   A firm of contractors entered into a contract for the collection and supply of hard granite chips for the construction of a highway at four reaches. The firm experienced difficulty in supplying the materials of the agreed quality. The firm, therefore, negotiated with the superintending engineer for substitution of hard stone chips for hard granite chips. The superintending engineer permitted, under items of their contract, to deviate from agreed terms. He held that the firm would be entitled to payment for the substituted item according to the Schedule of Rates of Dandakaranya Project which rate was including delivery to the site and stockpiling. It was almost half the rate quoted by the firm for the agreed quality of chips. The firm unequivocally accepted payments with respect to two reaches, without any reservation whatsoever. In respect of the third reach the firm claimed that payment of lead should be made according to the terms of the contract. This was obviously contrary to the decision of the superintending engineer and, as such, the claim was rejected. The firm raised a dispute with reference to lead and claimed that dispute be referred to arbitration. The two material clauses in the conditions of the contract read as follows:

> Clause 13-A reserved the right to the Engineer-in-charge to make any alterations in, omissions from, additions to, or substitution for, the original specifications, drawings, designs and instructions, that may appear to him to be necessary or advisable during the course of the supply of the materials and the contract shall be bound to supply the materials accordance with any instructions which might be given to him in writing signed by the Engineer-in-charge ... And if the altered, additional and substituted materials include any class of materials for which no rate is specific in this contract, then such class of material shall be supplied at the rates entered in the schedule of rates of the D.N.K. Project on which the estimated cost shown on page 2 of the tender was based.

---

[23] Bharat Sanchar Nigam Ltd. v. Motorola India Ltd., 2008 (5) R.A.J. 408: (2009 (2) SCC 337).
[24] J.G. Engineers Pvt. Ltd. v. Union of India, AIR 2011 SC 2477 = 2011 (5) SCC 758 = 2011 (2) Arb LR 84.

It concluded with the sentence: "In the event of dispute the decision of the superintending engineer shall be final."

Clause 14 opened with the sentence "Except where otherwise provided in the contract all questions and disputes ... shall be referred to sole arbitration ... ".It was held by the M.P. High Court that all matters covered by Clause 13-A were outside the arbitration Clause 14. The Supreme Court of India confirmed the decision.[25]

3. Clause 2 of the CPWD Form stipulates for recovery of penalty/liquidated damages and authorizes the superintending engineer to decide the amount, which decision is made final and binding on the contractor. Clause 25 dealing with settlement of disputes by arbitration, opens with the words "Except where otherwise provided in the agreement all questions and disputes ... shall be referred to the sole arbitration of ... ". The Supreme Court of India, in a case, held that Clause 25 clearly excludes the matters like those mentioned in Clause 2 and the compensation decided by the superintending engineer will not be capable of being called in question before the arbitrator.[26]

4. The appellant allotted the contract for construction of houses, the work to commence on 9th October 1984 and to be completed up to 7th February 1985. There was 11 months delay in completion. The appellant issued the letter of levy of compensation after 29 months. In the arbitration that followed, the recovery was challenged by the contractor and was not opposed by the appellant. The contractor succeeded in some claims and obtained an award dated 15th November 1993. The award was challenged on the ground that the levy of compensation fell within the excepted matters. On behalf of the contractor it was contended that the superintending engineer having failed to exhaust all the requirements and options provided in the clause, the matter did not fall within the excepted matters. The contention found favour with the trial court. Dismissing the appeal, the division bench of the Delhi High Court observed:

> We are also not impressed by the contention that levy of compensation of Rs. 57,000/- and odd on respondent in terms of Clause 2 of the agreement fell within "excepted matters" under Clause 25, even if it was accepted that it excludes such matter from arbitration on which Superintending Engineer's decision was final. Because the compensation leviable for late completion of work was dependent on number of factors which provided options to the Engineer concerned in the matter. He had the discretion to explore and take recourse to any one of such requirements/ options and then pass final orders for levy of compensation. It would be only then that his decision on this would become final and binding perhaps bringing it within the "excepted matters" under Clause 25. This is apart from the view taken by the Ld. Single Judge that the levy of compensation without a show cause notice to respondent was not sustainable.

It was held:[27]

> Viewed thus, the Appellant's plea that Respondent's claim was not arbitrable was untenable and represented an attempt to defeat his claim which has remained

---

[25] M/s. P.C. Corpn. Ltd. v. The Chief Administrator, Dandakaranya Project, AIR 1991 SC 957.

[26] Vishwanath Sood v. Union of India, AIR 1989 SC 952.

[27] D.D.A. v. Foundengers P. Ltd., 2003(2) Arb. LR 499 (Delhi); also see: Bhartiy Construction Co. v. Delhi Development Authority, 1997 (2) Arb. LR 369 (Delhi).

pending determination for years. For all this, we find no merit in this appeal which is dismissed.

5.   An award was challenged on the ground that the disputes referred to were not arbitrable. The contract provision in question was Clause 11(C) of the General Conditions of MES Standard Form Contract. The said clause stipulates that no claim in respect of compensation or otherwise howsoever arising, as a result of extension granted under Clauses (A) and (B) shall be admitted. Clause (A) lists certain causes of delay, which justify extension of time if the contractor has given notice of the happening of any event, covered by the causes listed within 30 days. Clause (B) lists some other causes attributable to the department. The extension granted by the authority named in the clauses is made final and binding on the parties. Clause 70 which is an arbitration clause, covers all disputes except those for which the decision of the person named in the contract is expressed to be final and binding. The arbitrator gave the finding that the extension of time was admittedly granted but the delay of five months out of six months could not be attributed to the reasons in Clause (A) or (B) but was independent thereof. Placing reliance on the judgment of the Supreme Court in M/s Shyama Charan Agarwala and Sons v. Union of India, Appeal No. 1249 of 1996 decided on 8th January 1997, dealing with the same contract provisions, the Bombay High Court upheld the award.[28] It was held that even if the view taken by the arbitrator was a possible view, the court would not have interfered.

6.   A contract was entered into between the parties on 27th April 1985. The contract was subject to the general conditions of the contract of railways read with special conditions. Disputes arose between the parties and the respondent moved a petition under Section 20 of the Arbitration Act, 1940 praying for the arbitration agreement being filed in the court and six claims set out in the petition being referred to the arbitrator for settlement. The learned single judge of the High Court of Delhi (original side) directed two claims to be referred, but as to claims numbers 3 to 6 formed an opinion that the claims being "excepted matters" within the meaning of Clause 63 of general conditions of contract were not liable to be referred to arbitration. An intra-court appeal preferred by the respondent was allowed and the four claims were directed to be referred by the division bench to the arbitrator on forming an opinion that they were not covered by "excepted matters". The appellants filed the petition seeking special leave to appeal against the decision of division bench. The Supreme Court of India, after referring to several clauses in the contract observed:[29]

> In our opinion those claims which are covered by several clauses of the Special Conditions of the contract can be categorized into two. One category is of such claims which are just not leviable or entertainable. Clauses 9.2, 11.3 and 21.5 of Special Conditions are illustrative of such claims. Each of these clauses provides for such claims being not capable of being raised or adjudged by employing such phraseology as "shall not be payable", "no claim whatsoever will be entertained by the Railway", or "no claim [will/shall] entertained". These are "no claim", "no damage", or "no liability" clauses. The other category of claims is where the dispute or difference has to be determined by an authority of Railways as provided in the

---

[28] Union of India v. Moti Enterprises and another, 2003 (2) ARB LR 229 Bom, 2003 (2) Bom CR 737, (2003) 1 BOM LR 639, 2003 (1) MhLj 930. Also see: K.R. Ravindranathan v. State of Kerala, (1998) 9 SCC 410 = 1998 (Suppl.) Arb. LR 220 (SC); Yelluru Mohan Reddy v. Rashtriya Ispathnigam, AIR 1992 AP 81.

[29] General Manager, Northern Railways v. Sarvesh Chopra, 2002 (1) Arb. LR 506 (SC).

relevant clause. In such other category fall such claims as were read out by the learned counsel for the respondent by way of illustration from several clauses of the contract such as General Conditions Clause 18 and Special Conditions Clauses 2.4.2.(b) and 12.1.2. The first category is an "excepted matter" because the claim as per term and conditions of the contract is simply not entertainable; the second category of claims falls within "excepted matters" because the claim is liable to be adjudicated upon by an authority of the Railways whose decision the parties have, under the contract, agreed to treat as final and binding and hence not arbitrable. The expression "and decision thereon shall be final and binding on the contractor" as occurring in Clause 63 refers to the second category "excepted matters".

It was observed:

Thus, it appears that under the Indian law, in spite of there being a contract between the parties where under the contractor has undertaken not to make any claim for delay in performance of the contract occasioned by an act of the employer, still a claim would be entertainable in one of the following situations: (i) if the contractor repudiates the contract exercising his right to do so under Section 55 of the Contract Act, (ii) the employer gives an extension of time either by entering into supplemental agreement or by making it clear that escalation of rates or compensation for delay would be permissible, (iii) if the contractor makes it clear that escalation of rates or compensation for delay shall have to be made by the employer and the employer accepts performance by the contractor in spite of delay and such notice by the contractor putting the employer on terms. ... Thus, it may be open to prefer a claim touching an apparently excepted matter subject to a clear case having been made out for excepting or excluding the claim from within the four corners of "excepted matters".

The authors with utmost respect submit that the observations were not necessary to decide the case before the Supreme Court. In construction contracts the time is not of the essence and as such the third paragraph of Section 55 is not attracted.

At this stage an attempt can be made to answer the questions raised earlier.

a. Can matters in respect of which the decisions of the engineer/architect are made final be referred to arbitration? Answer is NO.

b. If not, is the contractor bound to accept the said decisions as final without an opportunity to challenge them at some stage? Answer is NO.

c. Can the contractor challenge such decisions in a court of law? Answer is YES.

d. Should the court decide the question on merits? Answer is both YES and NO. In the first instance the court will ascertain if the decision of the engineer is administrative or judicial. If the court finds the decision administrative, without conducting a hearing and giving the parties a chance to make submissions it will be set aside and decision on merits will be given. If the court finds that the decision of the engineer is judicial and the view taken by the engineer is a plausible view, the decision may be upheld.

e. What if such decision of the court renders the entire arbitration or other interrelated issues invalid? and

f. Should the court refer the matter to arbitration to avoid consequences contemplated in question (e) above? The answer to questions (e) and (f) is that the court cannot refer the disputes in respect of which the decision of the engineer is made final but can open up and review the same. The award will be on the issues other than the issues in respect of excepted matters and, as such, question (e) may not arise.

16.8 After No Claim Certificate Is Given – Can Claims Be Referred to Arbitration?

It is quite common in public works contracts, in India, that the authorities insist on the contractor signing a "no claim certificate" before release of the amount due for payment under the final bill, and also release of security deposit/contract performance guarantee. The wording of such certificate is generally as follows:

> This is to certify that we have received all payment in full and final settlement of the supplies made and services rendered and/or all work performed by us in respect of the above referred LOA/Contract and we have no other claims whatsoever final or otherwise outstanding against … (Authority) … We further confirm that we shall have no claim/demands in future in respect of this contract of whatsoever nature, final or otherwise.

If the outstanding amount is huge, such certificates are virtually signed under financial coercion. If the contractor thereafter attempts to raise the issue and prefer claims or gives notice for reference to arbitration, the question arises as to whether the arbitration clause survives termination of the contract by full and final performance or can the claims given up by signing a no claim certificate be revived and referred to arbitration? The answer to this question will depend upon the conditions of the contract, particularly in respect of the final bill and arbitration clause.

In a case the Supreme Court of India observed:[30]

> What is of some concern is the routine insistence by some Government Departments, Statutory Corporations and Government Companies for issue of undated "no due certificate" or a "full and final settlements vouchers" acknowledging receipt of a sum which is smaller than the claim in full and final settlement of all claims, as a condition precedent for releasing even the admitted dues. Such a procedure requiring the Claimant to issue an undated receipt (acknowledging receipt of a sum smaller than his claim) in full and final settlement, as a condition for releasing an admitted lesser amount, is unfair, irregular and illegal and requires to be deprecated.

It was further held that a claim for arbitration cannot be rejected merely or solely on the ground that a settlement agreement of discharge voucher had been executed by the claimant, if its validity is disputed by the claimant. If an application under Section 11 is filed, the issue will have to be decided by the Chief Justice or his nominee or by the arbitral tribunal as directed by the order under Section 11.

The earlier cases were distinguished as falling under one of two categories, namely:[31]

1. Cases in which the court after considering the facts, found that there was a full and final settlement resulting in accord and satisfaction and there was no substance in the allegation of coercion/undue influence

2. Cases in which the court found some substance in the contention of the claimant that "no due/claim certificate" or "full and final settlement Discharge Vouchers" were insisted and taken (either in a printed format or otherwise) as a condition precedent for release of the admitted dues

3. It was also observed that the cases were decided under the 1940 Act whereas "the perspective of the new Act is different from the old Act. The issue is not covered by the perspective of SBP & Co."

---

[30] National Insurance Co. Ltd. v. M/s Boghara Polyfab Pvt. Ltd., AIR 2009 SC 170.
[31] National Insurance Co. Ltd. v. M/s Boghara Polyfab Pvt. Ltd., AIR 2009 SC 170 (182).

The Delhi High Court held that the issue of fraud, undue influence and coercion has to be adjudicated by the civil court and not by arbitration.[32] The authors respectfully submit that this is not a correct view. The basis of arbitration is party autonomy and if any dispute as to fraud, undue influence or coercion comes within the scope of the arbitration agreement, there is no reason why the disputes should not be arbitrable. The facts of a given case can also be equally important. For example, consider the cases in the examples below:

EXAMPLES FROM CASE LAW

1. In a case the Supreme Court summarized the facts thus:[33]

    Admittedly, No-Dues Certificate was submitted by the Respondent on 21st September 2012 and on their request Completion Certificate was issued by the appellant. The Respondent, after a gap of one month, that is, on 24th October 2012, withdrew the No-Dues Certificate on the grounds of coercion and duress and the claim for losses incurred during execution of the Contract site was made vide letter dated 12th January 2013, i.e., after a gap of 3-1/2 (three and a half) months whereas the Final Bill was settled on 10th October 2012. When the Respondent accepted the final payment in full and final satisfaction of all its claims, there is no point in raising the claim for losses incurred during the execution of the Contract at a belated stage which creates an iota of doubt as to why such claim was not settled at the time of submitting Final Bills that too in the absence of exercising duress or coercion on the Respondent by the appellant.

Held:

    In our considered view, the plea raised by the Respondent is bereft of any details and particulars, and cannot be anything but a bald assertion. In the circumstances, there was full and final settlement of the claim and there was really accord and satisfaction and in our view no arbitrable dispute existed so as to exercise power under Section 11 of the Act. The High Court was not, therefore, justified in exercising power under Section 11 of the Act. In view of the foregoing discussion, we set aside the judgment and order dated 12th January 2015 passed by the High Court.

2. A party, having signed the no claim certificate, wrote on the same day a letter stating:

    We had submitted a pre-final bill in November itself but the authorities denied the bill and insisted on final bill. But when the alleged final bill was prepared the authorities insisted that "No Demand Certificate" should be executed by us in favour of the Corporation. They served us with a printed specimen of the document and insisted that it should be typed in our own letterhead and submitted to the N.T. P.C. We refused to submit such a document. But, the authorities of N.T.P. C. threatened that unless and until we execute the said document in favour of the Corporation, the N.T.P.C. would not effect payment of our bill. We have to comply with the instructions of authorities of N.T.P.C. Ltd. out of our helplessness in order to receive payment. Hence this letter. The certificates, undertakings, etc. as aforesaid have been executed without prejudice to our rights and claims whatsoever on account of the alleged final bill. We are signing the alleged final bill under coercion, undue influence and under protest only without prejudice to our rights and claims whatsoever. There is no accord and satisfaction between the contracting parties.

---

[32] Also see: M/s S.K. Sharma v. Union of India & Ors., AIR 2009 (NOC) 2057 (DEL.).

[33] M/s ONGC Mangalore Petrochemicals Ltd. v. M/s ANS Constructions Ltd.; AIR 2018 SC 796; 2018 (2) Scale 354.

The letter concluded with the request to pass the final bill with claims. Thereafter the final bill and security deposit was accepted without any protest. On these facts the trial court declined to appoint an arbitrator, under Section 20 of the 1940 Act. The High Court of Kerala allowed the appeal. Against the said decision, the appeal was filed before the Supreme Court. Before considering the decision of the Supreme Court it is necessary to reproduce the relevant part of the clause dealing with the final bill payment. The relevant part read:

> After payment of the amount of the final bills payable as aforesaid has been made, the contractor may if he so desires, reconsider his position in respect of the disputed portion of the final bill and if he fails to do so within 90 days, his disputed claim shall be dealt with as provided in contract.

The Supreme Court after discussing the reasons held:[34] "[W]e are of the opinion that there is no infirmity in the impugned judgment. This appeal is, therefore, dismissed."

3. The work was completed on 15th February 1980. A final bill was raised thereafter. There were claims for extra works in the final bill. The respondent was not agreeing to the final bill as raised. The arbitration clauses between the parties enjoined upon them to first sit across the table and endeavour to settle the disputes. Parties negotiated. A settlement was arrived at. The settlement was reduced in writing and it was specifically recorded that with the agreement arrived at, all claims of the appellant are fully and finally settled and nothing is pending against the contract in question. The settlement was arrived at on 11th April 1980. A "no claim certificate" issued on 12th April 1980 followed it. Payments were released pursuant to the settlement on 21st April 1980 and even at the time of receiving payment, the appellant did not allege any coercion. On these facts it was held by the Delhi High Court that there was thus, a complete accord and satisfaction of the disputes. There was no existing arbitrable dispute, which could be referred to arbitration.[35]

4. The applicant in a case was a Co-operative Group Housing Society. After the negotiations, the petitioner engaged the respondent No. 1 for the electrical installation in flats constructed by the society and Clause 44 was one of the terms of the contract entered into between the parties. It stipulated that all disputes and differences should be referred to the employer who shall state his decision in writing, which decision shall be final and binding and without appeal. When the respondent submitted a final bill for Rs. 13,38,189/-, the President of the Society reduced it to Rs. 4,11,674/- and sent a cheque of the net balance towards full and final settlement. The crucial question before the arbitrator was whether the letter dated 27th June 1998 whereby the cheque for the amount of Rs. 4,11,674/- was forwarded to respondent No. 1 by referring it as net balance amount towards full and final settlement, was tantamount to the decision of the President as contemplated under Clause 44 or not, and whether stipulation in the said clause that if the contractor be dissatisfied with the decision of the employer on any matter, question or dispute of any kind except the matters listed above, then and in any such case the contractor may within 28 days after receiving notice of such decision, give a written notice to the employer, is mandatory or directory in nature. The learned arbitrator after hearing both the parties came to the conclusion that the letter dated 27th June 1998 cannot be termed as the final decision of the President as contemplated under Clause 44 and since no letter was given by the

[34] N.T.P.C. Ltd. v. Reshmi Constructions, 2004(1) Arb. LR 156; also see Bharat Coking Coal Ltd. v. Annapurna Construction, 2003 (3) Arb. LR 119 SC.
[35] Jain Refractory Erectors v. Cement Corpn. of India Ltd., 2003 (3) Arb. LR 256 (Delhi) (DB).

respondent accepting the amount towards the full and final settlement, mere acceptance of the cheque for the said amount cannot be said to be acceptance of the bill towards full and final settlement. The learned arbitrator reasoned that "since reduction in the payment of final bill was because of deduction, this itself is a dispute to be looked into on the merit of controversy". The other relevant observation of the learned arbitrator was with regard to the plea of the respondent that since the claimant did not dispute the decision of the President within 28 days, he is debarred from approaching the arbitrator. The arbitrator held that whether this clause is directory or mandatory can also be gone into by the arbitrator. The Delhi High Court held:

> The contention of the learned counsel for the petitioner that with the settlement of disputes through the letter dated 27th June 1998, the contract itself came to the end and therefore arbitration clause ceased to exist has neither any substance nor any legal force as the Section 16 was specifically brought on the statute book to do away with such objections in order to remove obstacles or obstructions created by the parties in the suit in smooth, efficient and expeditious running of the arbitration proceedings.

The petition was dismissed.[36]

16.9 Arbitration Clause in Original Contract Applies to Extra/Additional Work

Most standard form construction contracts include express provisions empowering the engineer or the employer to order extra or additional work and further make the contractor liable to execute the same. The arbitration clause in the main contract will obviously govern the disputes arising out of the extra or additional work.

EXAMPLE FROM CASE LAW

Original contract was entered into for the construction of seven towers. The work of three more towers and finally four more towers was subsequently awarded. Dispute and differences had arisen in construction of the latter seven towers. The contractor, invoking Clause 32 of the agreement, filed an application under Section 20 of the Arbitration Act, 1940 calling upon the owner (petitioner) to produce the agreement and to have the dispute referred to arbitration. In response, the petitioner prayed to dismiss the suit. Two grounds were mainly pressed for consideration in support thereof. The first contention urged was that the petitioner being a co-operative society registered under the Delhi Co-operative Societies Act, 1972, the dispute is arbitrable under Section 60 of that Act and Section 93 thereof puts an embargo on the power of the Civil Court to decide the dispute and that, therefore, the suit was not maintainable. The learned single judge and the division bench of the High Court rejected the contention. The Supreme Court upheld the said judgment by holding:

> By operation of the non-obstante clause, the bar of suit attracts only if the dispute falls within the parameters of clauses (a) to (d) thereof and the bar of jurisdiction of

---

[36] Rajanigandha Co. Op. G.H.S. Ltd. v. Chand Constn. Co., 2002 (1) Arb. LR 160 (Delhi).

the Court under Section 93 gets attracted in respect of the specified subjects in sub-section (1) of Section 93 thereof. Therefore, the plea of the bar of Sections 60 and 93 is devoid of substance.

As regards the other contention raised, the Supreme Court observed:

> A serious contention raised … is that contract for arbitration is a pre-condition to avail arbitration. Since the agreement for the 7 towers does not contain such arbitration clause, the application under Section 20 is not maintainable and, there-fore, the suit deserves to be dismissed. We find no force in the contention. … It is seen that the above quoted terms of the agreement and Cl.32 of the contract read together clearly indicate that the award of the work during the course of execution is a part of the agreement originally entered into. Therefore, Cl.32 of the agreement containing settlement of disputes by arbitration is an arbitration agreement within the meaning of Section 2(a) of the Act. Therefore, application under Section 20 would be maintainable. We do not find any substance in the special leave petition.[37]

16.10 Two Contracts – One Award – if Valid?

Where there are separate contracts and separate work orders, even if each contract contains an arbitration clause, there must be a separate reference under each contract and separate award under each contract.[38] Similarly there cannot be more than one award in one reference unless otherwise agreed by the parties.[39]

16.11 Arbitration Clause Survives Termination of Contract

As to when, on termination of an agreement, an arbitration clause would also cease to be operative, has been well explained by a three-judge bench of the Supreme Court of India in Union of India v. Kishorilal Gupta and Bros.[40] In that case the discussion of the law on the subject led to the emergence of the principles which were reiterated in Damodar Valley Corporation v. K.K. Kar,[41] as below:

> As the contract is an outcome of the agreement between the parties it is equally open to the parties thereto to agree to bring it to an end or to treat it as if it never existed. It may also be open to the parties to terminate the previous contract and substitute in its place a new contract or alter the original contract in such a way that it cannot subsist. In all these cases since the entire contract is put to an end, the arbitration clause, which is a part of it, also perishes along with it. However, when a question of breach of contract, inter alia, is raised after the termination of the contract, it is the performance of the contract

[37] Supreme Co-operative Group Housing Society, Petitioner v. M/s. H.S. Nag and Associates (P) Ltd., AIR 1996 SC 2443.
[38] Bodh Raj Daulat Ram v. Food Corporation of India, 2003 (2) Arb. LR 677 (P. & H.); I.G.H. Ariff and others v. Bengal Silk and another, AIR 1949 Cal. 350.
[39] Prabartak Commercial Corp. v. Ramsahaimuull More Ltd., AIR 1963 Cal. 137.
[40] Union of India v. Kishorilal Gupta and Bros., (1960) 1 SCR 493: (AIR 1959 SC 1362).
[41] Damodar Valley Corporation v. K.K. Kar, (1974) 2 SCR 240: (AIR 1974 SC 158).

that comes to an end on termination of the contract, but the same remains in existence for certain purposes in respect of disputes arising under it or in connection with it.[42]

16.12 Transfer of Appointed Arbitrator – Continuation of Proceedings before Him – If and When Valid?

It is quite common in building and construction contracts, in use by the public authorities in India for execution of public works, to include an arbitration clause which names the arbitrator by designation such as "the Superintending Engineer of the named Circle" or "the Chief Engineer for the time being". When disputes arise and are referred to the person so designated, can he cease to be arbitrator, if he is transferred and some other person comes to hold the designated post? This question, in the absence of clear provisions to the contrary in the arbitration clause, was answered by the author in the negative, under the provisions of the 1940 Act.[43] Under the said Act, the authority of the appointed arbitrator could not be revoked except with the leave of the court. The same view was held by the Orissa High Court.[44] A few decisions holding to the contrary were distinguished on the facts of the cases.[45] The Supreme Court held in a subsequent case that it cannot be said that continuance of the proceedings and rendering of award by the arbitrator after his transfer was in disregard of any provision of law, much less a mandatory one but, at the highest, in breach of agreement. Therefore, by their conduct by participating in the arbitration proceedings without any protest, the parties would be deemed to have waived their right to challenge validity of the proceedings and the awards. Consequently, the objections taken to this effect did not merit any consideration and the High Court was not justified in allowing the same and setting aside the award.[46]

16.13 Condition Precedent to Invocation of Arbitration Clause – Mandatory

Standard form contracts invariably provide for certain steps to be taken by the parties prior to invocation of the arbitration clause. For example, the engineer has to decide the claims in the first instance. If any party is dissatisfied with the said decision, it has to appeal to the DAAB and if the DAAB fails to give decision within the time allowed or the decision given by the DAAB is not acceptable to any party, it must give a NOD and thereafter the parties must try amicable settlement. Arbitration can commence only on or after a specified number of days after the day on which the NOD was given, even if no attempt at amicable settlement has been made.

It must be remembered that the steps preceding the coming into operation of the arbitration clause are essential. However, the steps preceding the coming into operation of the arbitration clause, though essential, are capable of being waived and if one party has by its own conduct or the conduct of its officials disabled such preceding steps being taken, it will be deemed that the procedural prerequisites were waived. The party at fault cannot be permitted to set up the bar of

[42] Union of India v. Kishorilal Gupta and Bros., (1960) 1 SCR 493: (AIR 1959 SC 1362); M/s. Indian Drugs and Pharmaceuticals Ltd. v. M/s. Indo Swiss Synthetics Gem Manufacturing Co. Ltd., AIR 1996 SC 543.

[43] See: *B.S. Patil on the Law of Arbitration*, 2nd ed., by Sarita Patil, pp. 122–123.

[44] Union of India v. Ch. Radhanath Nanda, AIR 1961 Ori. 143.

[45] Krishan Lal v. Haryana S.A.M. Board, AIR 1986 P. & H. 376; M/s. Neelakantan & Bros. Construction v. Superintending Engineer, National Highways, Salem and others, (1988) 4 SCC 462 1989 (1) Arb. LR 34.

[46] Inder Sain Mittal v. Housing Board, Haryana, 2002 (1) Arb. LR 431; also see: N. Chellappan v. Secretary, Kerala State Electricity Board and another, 1975 (2) SCR 811; Chowdhri Murtaza Hossin v. Mussumat Bibi Bachunnissa, [3 I.A. 209]; Prasun Roy v. Calcutta Metropolitan Development Authority and another, (1987) 4 SCC 217 = 1987 (2) Arb. LR 196; State of Punjab v. Hardyal, (1985) 2 SCC 629 = 1985 Arb. LR 297.

non-performance of prerequisite obligations so as to exclude the applicability and operation of the arbitration clause.[47]

EXAMPLE FROM CASE LAW

1. The main objection raised by the 1st respondent was that the applicant had filed the arbitration request without complying with the procedure laid down in Clauses 24 and 25 of the agreement and, therefore, this arbitration request was not maintainable, being premature. It was found that the applicant did not refer the dispute for decision by the engineer and prevented the adjudicator from adjudicating the disputes as per the terms of the agreement. The 1st respondent had been asserting that the applicant should follow the preceding steps before enforcing the clause regarding arbitration incorporated in the agreement. On these facts it was held:

> [I]t is clear that the applicant either refused to comply or prevented compliance of the procedure laid down or the preceding steps to enforce the arbitration clause in the agreement. The applicant, who has been responsible for preventing or frustrating the operation of the earlier or preceding steps for enforcing the arbitration clause in the agreement, cannot seek arbitration by contending that the 1st respondent has not complied with the request for appointment of arbitrator as provided in the agreement.[48]

The arbitration request was dismissed. However, the order, it was stated, "will not preclude the applicant from enforcing the arbitration clause in the agreement after due compliance of the prerequisites for enforcing the arbitration clause provided in the agreement."[49]

2. In a case, the contractor had referred 16 claims during the currency of the contract for decision by the authority. After elapse of 90 days, the contractor requested for demand in writing for reference of disputes to arbitration. The authority failed to refer the disputes to arbitration. The contractor did not treat the claims raised earlier as final and submitted final claims with updated rates and sums. There was no demand made in writing in respect of final claims for reference to arbitration. He rushed to the court on 24th August 1995 and filed an application for appointing an arbitrator for the claims finally submitted. The High Court directed the authority to comply with the procedure of appointment of arbitrators and declined to appoint independent arbitrators as claimed by the contractor. The contractor filed an appeal in the Supreme Court of India. Held:

> [T]he High Court was perfectly justified in relegating both the parties to the procedure of arbitration as laid down under the contracts binding on them. The final Orders passed by the learned Single Judge ... and as confirmed by the Division Bench remain well sustained on record.[50]

16.14 The Law of Arbitration

The law of arbitration is too vast a subject to be covered within the compass of one chapter. The authors have given word by word commentary in their book on *The Law of Arbitration* of which

---

[47] M.K. Shah Engineers & Contractors v. State of M.P. (1999) 2 SCC 594 = 1999 (1) Arb. LR 646.
[48] Nirman Sindia v. Indal Electromelts Ltd. Ciimbatore, 2000 (1) Arb. LR 144 (Kerala).
[49] Ibid.
[50] M/s. Shetty's Construction Co. Pvt. Ltd. v. M/s. Konkan Railway Corpn. Ltd., AIR 2000 SC 122.

seventh edition is under preparation. It is advisable to refer to any book on the law of arbitration. What follows is a bird's eye view of the provisions of the ACA as amended by the (Amendment) Act 2015.

16.14.1 Arbitration Agreement

Since arbitration is a voluntary forum of dispute resolution, it cannot be invoked if there is no agreement. However, the parties can enter into an arbitration agreement even after the disputes have arisen. Section 7 of the ACA, defines arbitration agreement. The arbitration agreement must necessarily be in writing. Section 7(4) further clarifies as to what can be construed to be a written agreement. Section 40 declares that an arbitration agreement shall not be discharged by death of a party thereto. Section 10 gives the parties to an arbitration agreement freedom to determine the number of arbitrators, provided that such number shall not be an even number. If the agreement is silent the arbitral tribunal shall consist of a sole arbitrator. Section 11 deals with the procedure for appointment of arbitrators. The ACA gives freedom to the parties to agree on a procedure for appointing the arbitrators. If there is no agreement and reference is to three arbitrators, each party shall appoint one arbitrator and the two appointed arbitrators shall appoint the third arbitrator who shall act as the presiding arbitrator. If in spite of 30 days' notice a party fails to appoint its nominee or the two arbitrators fail to appoint the third arbitrator, the appointment shall be made by the Chief Justice of the High Court or any person or institution designated by him in the case of domestic arbitration and by the Chief Justice of the Supreme Court of India or any person or institution designated by him, in the case of international arbitration.

16.14.2 Date of Commencement of Arbitration Proceedings

Section 21 of the ACA declares: "the arbitral proceedings in respect of a particular dispute commence on the date on which a request for that dispute to be referred to arbitration is received by the respondent." The provision is self-explanatory. Section 3 may be referred to for receipt of written communication.

The date of commencement of arbitral proceedings is important if the parties have in their agreement specified any time limit for completion of the arbitral proceedings and in any case for the purpose of limitation. The (Amendment) Act, 2015 in Section 29A stipulates the time limit for completion of the arbitral proceedings as 12 months subject to further extension by the parties for the period not exceeding 6 months. Subsection 4 of Section 29A provides for further extension of the mandate by the court.

16.14.3 Challenge to Arbitrators

Any party may challenge an arbitrator on the ground stated in Section 12. Section 13 prescribes the procedure for challenge, which challenge the arbitral tribunal itself is empowered to decide, if the challenged arbitrator declines to withdraw from his office or the other party does not agree to the challenge. If the challenge is unsuccessful, the party challenging the arbitrator is entitled to apply for setting aside the award in accordance with Section 34. If the award is set aside on the grounds of challenge, the court is to decide if the arbitrator challenged is entitled to any fees.

Section 14 covers the situation when an arbitrator is unable to perform or fails to perform and if the parties cannot sort out the controversy or the arbitrator does not withdraw from his office, the power is conferred on the court to decide termination of his mandate. Section 15, accordingly, deals with termination of mandate and substitution of arbitrator. Section 20 gives freedom to parties to decide the place of arbitration.

16.14.4 Jurisdiction of Arbitral Tribunal

The arbitral tribunal is empowered, under Section 16 of the ACA to rule on its jurisdiction, including on objections in respect of existence and validity of arbitration agreement. The parties to arbitration should raise the plea of lack of or excess of jurisdiction promptly and in any case within the time limit specified in Section 16 in that respect. The decision of the arbitral tribunal rejecting the plea entitles the aggrieved party to challenge the award under Section 34. Where, however, the arbitral tribunal accepts the plea, an aggrieved party can file an appeal under Section 37 to the court.

16.14.5 Interim Measures Ordered by Arbitral Tribunal

The ACA, under Section 17, empowers an arbitral tribunal to order interim measures, at the request of any party, for protection of the subject matter of the dispute as may be deemed necessary. The lacuna in the ACA, inasmuch as there was no provision which could enforce the interim measure so ordered, has been eliminated by the Amendment Act, 2015. Section 17(2) makes the order of the tribunal enforceable as if it was an order of the court. The parties, in the alternative can resort to the court under Section 9, which empowers the court to order interim measures.

16.14.6 Conduct of Arbitration Proceedings

Chapter 5 contains Sections 18 to 27 regarding conduct of the arbitral proceedings. The said provisions are self-explanatory and include the duty to give equal treatment of parties, non-binding of provisions of the Code of Civil Procedure (CPC) or Evidence Act to arbitral procedure, place and language of arbitration, exchange of pleadings by the parties, hearings and written proceedings, power to hold *ex parte* hearing, appointment of an expert, court assistance in taking evidence, etc.

16.14.7 Making and Publication of Arbitral Award

Chapter 6 contains Sections 28 to 33. Section 28 makes it mandatory to decide the disputes according to the substantive law in force in India. In the case of international arbitration, the law as agreed by the parties, and failing such agreement the proper law as considered applicable by the tribunal as also under the terms of the contract, will be applied. Without express authorization, the tribunal shall not decide as *amiable compositeur.*

Section 29 deals with decision making by the tribunal. The decision of the majority of members of the tribunal, unless otherwise agreed by the parties, will be final and binding on the parties. Section 30 empowers the tribunal to encourage settlement of dispute and if the parties settle the dispute amicably, the tribunal may, if requested by the parties, record settlement in the form of an arbitral award on agreed terms.

Section 31 makes elaborate provisions regarding the form and contents of the award. Unless the parties have agreed otherwise or the award is on consent terms agreed under Section 30, the award shall state the reasons upon which it is based. An unreasoned award is likely to be set aside under Section 34 as being opposed to the public policy of India.

Section 32 declares that the arbitral proceedings will terminate with publication of the final award. The provisions also empower the tribunal to terminate the proceedings under the circumstances listed in the said provisions, including the claimant withdrawing the claims with the respondent not objecting to such withdrawal, or the parties agreeing to terminate the proceedings, or continuation of the proceedings has become impossible or unnecessary.

Section 33 empowers the tribunal to correct any computation, clerical or typographical or other errors of a similar nature or to give interpretation of an award or to give an additional award. The errors can be corrected by the tribunal on its own. The interpretation of the award or additional award can be given if requested by any party in respect of claims presented but omitted from the award.

16.14.8 Recourse against Arbitral Award and Its Enforcement

Section 34 deals with application for setting aside an award. The application can only be made within three months of the party getting the award. The court is empowered to grant further extension of 30 days if satisfied that the applicant was prevented by sufficient cause from making the application. No further extension is possible to be given under the law of limitation as well.[51]

The section lists the grounds on which the award can be set aside. The grounds include incapacity of a party, invalidity of arbitration agreement, lack of proper notice, lack of or excess of jurisdiction, composition of the tribunal or the procedure not in terms with the agreement or Part I of the ACA, subject matter of dispute not capable of being settled by arbitration. The last ground, "award is in conflict with the public policy of India", has virtually thrown open the door for any and every objection that can possibly be raised.[52] However, the recent decisions of the Supreme Court have restricted the scope of challenge to arbitral award and duty of the court while deciding the matter.[53]

The provisions of Section 34 have been amended by the Act of 2015 to the effect that an application under Section 34 should be disposed of expeditiously within a period of one year from the date of service of notice. If Section 34(5) is considered as mandatory, keeping in view the fact that if the time limit of one year is not adhered to under Section 34(6), no consequence thereof is provided. Whereas under Section 29A(4) if an award is made beyond the stipulated or extended period contained in the section, the consequence of the mandate of the arbitrator being terminated is expressly provided. As such the provisions of Sections 34(5) and (6) are directory. It was so held by the Supreme Court of India and qualified as follows:

> However, we may add that it shall be the endeavour of every Court in which a Section 34 application is filed, to stick to the time limit of one year from the date of service of notice to the opposite party by the applicant, or by the Court, as the case may be. In case the Court issues notice after the period mentioned in Section 34(3) has elapsed, every Court shall endeavour to dispose of the Section 34 application within a period of one year from the date of filing of the said application, similar to what has been provided in Section 14 of the Commercial Courts, Commercial Division and Commercial Appellate Division of High Courts Act, 2015.[54]

Sections 35 and 36 deal with the enforcement of the award that has attained finality. The provisions declare that an award shall be enforced under CPC, in the same manner as if it were a decree of the court.

Section 37 provides for appeals against certain orders including the order setting aside an award. No second appeal is permissible but the power of the Supreme Court to allow a Special Leave Petition is not taken away.

16.14.9 Deposits for Fees and Expenses – The Accounts to Be Given in the End

Chapter 10 of the ACA includes Sections 38 to 43. Section 38 pertains to deposits for fees and expenses and the accounts to be given in the end. Section 39 deals with lien on arbitral award and deposits and costs. Section 40 declares, as already stated, that the arbitration agreement shall not

---

[51] Union of India v. Popular Construction, (2001) 8 SCC 470. Also see Chapter 15 on limitation.

[52] Oil & Natural Gas Corporation Ltd. v. Saw Pipes Ltd., 2003 (4) SCALE 92, 2003 (2) Arb. LR 5 (SC); AIR 2003 SC 2629.

[53] M.P. Power Generation Co. Ltd. v. Ansaldo Energia SPA, 2018 (4) JT 371; 2018 (5) Scale 731. Also see Delhi Development Authority v. M/s R.S. Sharma & Co., New Delhi 2008 (13) SCC 80; Associate Builders v. Delhi Development Authority, 2015 (3) SCC 49; 2014 (13) Scale 226.

[54] State of Bihar v. Bihar Rajya Bhumi Vikas Bank Samiti, (SC): AIR 2018 SC 3862; 2018 (9) Scale 291.

be discharged by death of a party thereto. Section 41 incorporates provisions in case of insolvency of a party.

Section 42 deals with the jurisdiction of the courts. Section 2 defines the court to mean the Principal Civil Court of Original Jurisdiction in a district and includes High Court in exercise of its ordinary original civil jurisdiction but no lower courts. For appeal against the order refusing to grant interim measure or of setting aside an award the court is the appellate court authorized to hear appeals from original decrees. However, by the Commercial Courts, Commercial Division and Commercial Appellate Division of High Courts Act, Section 10, the jurisdiction in respect of arbitration matters has been conferred, as specified in the said provisions, in arbitration of commercial dispute of a specified value on commercial courts wherever constituted.

Section 43 deals with the limitation and declares that the provisions of the Limitation Act, 1963 shall apply to arbitration proceedings.

16.14.10 Part II of the ACA Deals with Enforcement of Certain Foreign Awards

Part II consists of two chapters. The first chapter incorporates New York Convention Awards. The second deals with Geneva Convention Awards. The provisions are similar to the earlier The Arbitration (Protocol and Convention) Act, 1937 and The Foreign Awards (Recognition and Enforcement) Act, 45 of 1961.

16.14.11 Part III of the ACA

Part III of the ACA deals with conciliation and its provisions are considered in paragraph 16.5 above.

16.14.12 Part IV of the ACA

Part IV includes supplementary provisions in Sections 82 to 86. Section 82 empowers the High Courts to make rules consistent with the provisions of the ACA. Similarly Sections 83 and 84 enable the central government to remove difficulties in implementing the ACA and to make rules for carrying out the provisions of the ACA. Section 85 repeals the three earlier enactments, namely the Acts of 1937, 1940 and 1961. Section 86 repeals the Ordinance of 1996. The Act of 1996 is to apply to all arbitral proceedings commenced on or after the Act came into force. The ACA was made effective from 22nd August 1996 by the notification of the same date published in The Gazette of India. The first Ordinance was made effective from 25th January 1996. The ACA being in continuation of the Ordinance is deemed to have been effective from 25th January 1996 when the first Ordinance came into force.[55]

16.15 Investment Treaty Arbitration

Where an international bank has given a loan for a construction project in India, or an international company is a party to a construction contract and the employer is either a government or a governmental agency or company for whose actions or omissions the government bears responsibility at international law, there is a possibility of another mode of dispute resolution running in parallel to litigation or commercial arbitration. Regardless of what the parties' contract provides and of the fact that the foreign company or bank has not signed an investment treaty with the Government of India, they may qualify as "investors" and their claims or loans may qualify as "investments" in respect of which an international treaty arbitration may be commenced.

---

[55] Feurst Day Lawson Ltd. v. Jindal Exports Ltd., AIR 2001 SC 2293; 2001 (2) Arb. LR 1 (SC).

International investment treaties may be bilateral or multilateral. An investor can commence one or more investor-state arbitration proceedings in respect of the same claim under different treaties. Some treaties, such as the India-ASEAN Free Trade Agreement, exclude construction and turnkey projects from the definition of assets. Any resulting award can be enforced under the New York Convention, 1958. This type of arbitration gives foreign investors in India's construction sector a powerful remedy over and above that negotiated in the complex contracts for large infrastructure projects. The Government of India will always be a defendant in the case of such an arbitration even if the contracting party is a state government or an autonomous public sector undertaking in India. A breach of contract in such a case may also become a breach of international law. This is a complex subject and will be dealt with in the seventh edition of the authors' book on arbitration. In the meantime, any public sector entity entering into a contract with a significant foreign investment element is well advised to consider the impact of such treaties.

Index